Biochemistry

T.A. BROWN

Biochemistry

T.A. BROWN

University of Manchester, UK

Scion

A CIP catalogue record for this book is available from the British Library.

ISBN 978 1 907904 28 8

Scion Publishing Limited

The Old Hayloft, Vantage Business Park, Bloxham Road, Banbury OX16 9UX, UK
www.scionpublishing.com

Important Note from the Publisher

The information contained within this book was obtained by Scion Publishing Ltd from sources believed by us to be reliable. However, while every effort has been made to ensure its accuracy, no responsibility for loss or injury whatsoever occasioned to any person acting or refraining from action as a result of information contained herein can be accepted by the authors or publishers.

Readers are reminded that medicine is a constantly evolving science and while the authors and publishers have ensured that all dosages, applications and practices are based on current indications, there may be specific practices which differ between communities. You should always follow the guidelines laid down by the manufacturers of specific products and the relevant authorities in the country in which you are practising.

Although every effort has been made to ensure that all owners of copyright material have been acknowledged in this publication, we would be pleased to acknowledge in subsequent reprints or editions any omissions brought to our attention.

Registered names, trademarks, etc. used in this book, even when not marked as such, are not to be considered unprotected by law.

Cover design by Andrew Magee Design Ltd; www.amdesigner.co.uk
Illustrations by Matthew McClements at Blink Studio Ltd; www.blink.biz
Typeset by Phoenix Photosetting, Chatham, UK
Printed in the UK by Charlsworth Press, Wakefield, UK

Contents

Detailed contents

Preface

Biochemistry is an essential component of any degree course in the biological sciences. This has always been the case for courses focusing on the molecular or cellular aspects of biology, but recent developments in structural biology, metabolomics and related biochemical methods are making biochemistry increasingly important in subjects such as zoology, plant science, ecology and environmental biology. The need for students in all areas of biology to gain a good grounding in biochemistry has never been greater. This presents a challenge for many biology students, and for their teachers, because there has always been a tendency for students who are fascinated by biology to express less interest in the physical sciences such as chemistry. Biochemistry – whose very name signals its chemical background – can therefore be perceived as a difficult subject by many biology students. This is a pity because biochemistry, when approached in a sympathetic manner, is one of the most satisfying subjects to study, the insights that it gives into the molecular basis to life enhancing one's appreciation of all aspects of biology.

Biochemistry is well served by textbooks, with a number of excellent volumes available. The majority of these books are large, comprehensive texts, which provide detailed support for biochemistry courses at all undergraduate levels, and to some extent graduate level also. These texts are excellent for students majoring in biochemistry, but are less helpful, and in some respects rather daunting, for those students who are taking a single introductory biochemistry course as part of their studies. The intention in writing my version of *Biochemistry* has been to cater to the needs of this latter group of students. *Biochemistry* therefore covers only the basics of the subject. It makes no attempt to support advanced study in biochemistry, but instead aims to help those students who need to acquire a solid background in biochemistry as part of their broader degree course. In particular, it does not assume an extensive starting knowledge of the principles of chemistry, but introduces those principles as and when they become relevant to the gradual development of the reader's understanding of biochemistry. Additionally, I have tried throughout the book to emphasize the underpinning that biochemistry provides to other areas of biology, by giving specific examples in the text and in the accompanying boxes.

In writing this book I was fortunate enough to receive a number of perceptive and evaluative reviews of the draft chapters, which helped greatly in shaping the final text. In particular, I am very grateful to David Hames for reviewing all aspects of the book and indicating areas to which I had given insufficient emphasis or where I had made simple concepts appear overly complex. The splendid artwork is the creation of Matthew McClements. The team at Scion Publishing – Jonathan Ray, Simon Watkins and Clare Boomer – provided excellent support throughout the writing process, and were remarkably patient when I missed deadlines by weeks and sometimes months. I must also thank my wife, Keri, who displayed equally remarkable patience when I spent evenings, weekends and sometimes entire vacations in pursuit of the elusive deadlines.

T.A. Brown
Manchester

Abbreviations

ACP	acyl carrier protein
ALA	δ-aminolevulinate
ATP	adenosine 5′-triphosphate
CAP	catabolite activator protein
CD	circular dichroism
cDNA	complementary DNA
CF	cystic fibrosis
CFTR	cystic fibrosis transmembrane regulator
CIEP	crossover immunoelectrophoresis
CPSF	cleavage and polyadenylation specifying factor
CRE	cAMP response element
CREB	cAMP response element binding
CstF	cleavage stimulation factor
CTD	C-terminal domain
ddNTPs	dideoxynuceotides
DNA	deoxyribonucleic acid
DNase I	deoxyribonuclease I
EGFR	epidermal growth factor receptor
ELISA	enzyme-linked immunosorbent assay
HAT	histone acetyltransferase
HDAC	histone deacetylase
HIV/AIDS	human immunodeficiency virus and acquired immunodeficiency syndrome
HPLC	high performance liquid chromatography
HRP	horseradish peroxidase
ICATs	isotope-coded affinity tags
IPTG	isopropyl-β-D-thiogalactoside
IRES	internal ribose entry site
ITAF	IRES trans-acting factor
JAKs	Janus kinases
MALDI–TOF	matrix-assisted laser desorption ionization time of flight
MAP	mitogen activated protein
miRNA	microRNA
mRNA	messenger RNA
NHEJ	nonhomologous end-joining
NMR	nuclear magnetic resonance
ORF	open reading frame
PADP	polyadenylate-binding protein
PCNA	proliferating cell nuclear antigen
PCR	polymerase chain reaction
PNPase	polynucleotide phosphorylase
PRPP	phosphoribosyl pyrophosphate
PSE	proximal sequence element
qPCR	quantitative PCR
RISC	RNA-induced silencing complex
RNA	ribonucleic acid
RRF	ribosome recycling factor
rRNA	ribosomal RNA
scRNA	small cytoplasmic RNA
SDS	sodium dodecyl sulfate
SDS–PAGE	sodium dodecyl sulfate polyacrylamide gel electrophoresis
siRNA	small interfering RNA
snoRNA	small nucleolar RNA
snRNA	small nuclear RNA
snRNP	small nuclear ribonucleoprotein
SRP	signal recognition particle
SSB	single strand binding protein
STATs	signal transducers and activators of transcription
TAF	TBP-associated factor
TBP	TATA-binding protein
TFIID	transcription factor IID
TIM	translocator inner membrane
TOM	translocator outer membrane
tRNA	transfer RNA
UBF	upstream binding factor
UCE	upstream control element
UTR	untranslated region

How to use this book

To be of value, a textbook needs to be as user-friendly as possible. *Biochemistry* therefore includes a number of devices intended to help the reader and to make the book a more effective teaching aid.

Organization of the book

Biochemistry is divided into four parts:

Part 1 – Cells, Organisms and Biomolecules begins by providing the biological context to biochemistry, the aim of Chapter 2 being to explain the cellular basis to life and to describe how a eukaryotic cell is divided into subcompartments, each with its own particular biochemical activities. Chapters 3–6 then cover the structural features and main functions of the four main types of biomolecules: proteins, nucleic acids, lipids and carbohydrates. I believe that for an introductory course in biochemistry, it is important to establish these structural and functional features before attempting to explain how these compounds participate in metabolic reactions. The student who is unfamiliar with chemistry is then able to focus on and understand issues such as bonding, ionization and polarity, before being asked to deal with the second suite of chemical principles that underlie biological catalysis and energy generation.

Part 2 – Energy Generation and Metabolism moves the discussion on from structure–function to the central issues in biochemistry, concerning the generation of energy and the nature of the metabolic pathways responsible for the synthesis and breakdown of biomolecules. This part of the book begins with a description of the role and mode of action of enzymes, including the thermodynamic basis to biochemical reactions, the effects of substrate concentration on reaction rate, and the impacts of different types of inhibitors. Three chapters are then devoted to energy generation, Chapter 8 on glycolysis, Chapter 9 on the TCA cycle and electron transport chain, and Chapter 10 on photosynthesis. The following chapters then cover the main metabolic pathways, divided into those involving carbohydrates (Chapter 11), lipids (Chapter 12) and nitrogen-containing compounds (Chapter 13). In each of these last three chapters, I describe how a compound is made before describing how it is broken down: for example, Chapter 12 begins with the synthesis of fatty acids and triacylglycerols and then continues with the breakdown of these compounds. This is largely a personal preference, but one also held by many students, for whom the logical sequence is synthesis followed by breakdown rather than vice versa. Within Part 2 as a whole, I spent a good deal of time trying to find the ideal division of material between these chapters, and I appreciate that in some cases the order is different to that in other textbooks. For example, I felt it was important in Chapters 8 and 9 to maintain a clear focus on the movement of substrates through glycolysis, the TCA cycle and electron transport chain, and therefore I have defined the related topic of gluconeogenesis as an aspect of carbohydrate metabolism, and cover it in Chapter 11. Similarly, the formation of ketone bodies seemed to fit better when considered alongside the breakdown of ketogenic amino acids rather than lipids.

Part 3 – Storage of Biological Information and Synthesis of Proteins keeps to the conventional order of events with the material split into DNA replication and repair (Chapter 14), RNA synthesis (Chapter 15), protein synthesis (Chapter 16) and the control of gene expression (Chapter 17). In these chapters I have tended to focus less on the molecular genetic aspects of DNA replication and gene expression, and more on the role of these processes in synthesizing DNA, RNA and proteins. For this reason, topics such as RNA and protein processing, protein targeting and the turnover of RNA and proteins are given greater emphasis than would be appropriate if this were a genetics rather than biochemistry text. Whether the topics in Part 3 are taught as biochemistry or genetics, or a bit of both, is one of the trickiest decisions for course directors.

Part 4 – Studying Biomolecules attempts to give an overview of the more important of the many methods that are used in biochemistry. In this part of the book the main difficulty is, of course, deciding which methods to include. Bearing in mind that several techniques are described in earlier chapters, I decided to focus in Chapter 18 on the immunological, proteomic and structural methods for studying proteins, together with an overview of 'omics techniques for lipids and carbohydrates, and in Chapter 19 on DNA methods, with emphasis on PCR, sequencing and cloning. These chapters are not, therefore, intended to be a comprehensive survey of the biochemist's toolkit, but instead cover those biochemical techniques (e.g. immunoassay, PCR) that are commonly used in all areas of biological research, as well as those other techniques (e.g. 'omics, protein structure analysis, next generation DNA sequencing) that biological researchers might not perform themselves, but the results of which they might frequently use, or at least need to be aware of, as part of their own projects.

Organization of the individual chapters

As well as attempting to make the text as user-friendly as possible, the organization of the individual chapters is intended to help students structure their learning.

Study goals

Each chapter starts with a set of study goals. These have two roles. First, they provide a synopsis of what the chapter contains, and so can be used by the reader as a quick check during review to ensure that all of the key points from a chapter have been recalled. Secondly, the study goals are intended to give an indication of the level and type of knowledge that the student should gain from reading the chapter, whether this is being able to describe a pathway, distinguish between two or more related processes, or understand why something is the way it is. The intention is that the student knows exactly what they should get out of each chapter, and hence is in no doubt about whether they have dealt satisfactorily with the material.

Principles of Chemistry, Research Highlights, and other boxed material

The main text in each chapter is supported and extended by additional information contained in boxes. There are three types of boxes, given different color coding in the Contents and in the boxes themselves.

Principles of Chemistry (color coded in orange) are a series of boxes that describe key aspects of the chemical basis to biochemistry. It is not possible, nor desirable, to remove all of the chemistry from the main text, but I believe there is an advantage in dealing with several of these topics as boxes, because this avoids disrupting the flow of information in the text with side avenues exploring underlying chemical concepts. Also, having the key chemical principles presented as discrete and self-contained boxes helps the student focus on what they need to know about a particular chemical topic, without having to tease that information out of the text.

Box 1.3   **Atoms, isotopes and molecular masses**
Box 3.2   **The ionization of water and the pH scale**
Box 3.3   **Types of chemical bond**
Box 3.4   **The unusual characteristics of the peptide bond**
Box 4.1   **Base stacking**
Box 4.3   **Sugar pucker**
Box 7.2   **Oxidation and reduction reactions**

Box 8.1 **Units of energy**
Box 9.3 **Redox potential**
Box 10.2 **Atomic orbitals**
Box 15.4 **Transesterification**

Research Highlights (red) are designed to illustrate some of the strategies that are used in biochemical research as well as some of the broader applications of biochemistry in medical research and biotechnology. Each Research Highlight is based on one or a few research papers, the objective being to illustrate the way in which real research is conducted and to show how information on biochemical topics is obtained.

Box 2.3 **Bacteria communicate with one another in a biofilm**
Box 3.6 **Predicting the secondary structure of polypeptide from its amino acid sequence**
Box 3.7 **Using collagen structure to identify extinct animals**
Box 3.9 **Studying protein folding**
Box 4.2 **The discovery of the double helix**
Box 5.7 **The biochemistry of cystic fibrosis**
Box 6.3 **Some humans have recently evolved the ability to digest milk**
Box 7.5 **Exploiting thermostable enzymes in biofuel production**
Box 9.1 **Identification of the mitochondrial pyruvate carrier protein**
Box 9.6 **The rotation of the F_0F_1 ATPase**
Box 10.6 **Increasing the photosynthetic capability of crop plants**
Box 11.6 **Did Pythagoras ban the broad bean because of the pentose phosphate pathway?**
Box 12.5 **The biochemistry of *Lorenzo's Oil***
Box 13.3 **Genetically modified crops that are resistant to a herbicide that disrupts aromatic amino acid synthesis**
Box 14.4 **The interaction between Tus proteins and the replisome**
Box 14.5 **Telomerase and cancer**
Box 16.4 **Antibiotics that target the bacterial ribosome**
Box 17.1 **Transcriptomics – studying changes in gene expression patterns**
Box 19.3 **Did Neanderthals and modern humans meet and interbreed?**
Box 19.5 **Synthesis of recombinant factor VIII protein**

Generic boxes (green) contain discrete packages of information that I have taken out of the main text, either for emphasis or, as with the Principles of Chemistry, to avoid disrupting the flow of the text. Some boxes provide more detailed descriptions or extensions of topics that are also covered in the text, others address issues that are not central to the information content of a chapter but which a student might query, and some are simply included because they are interesting (and still, hopefully, educational). Chapter 9, for example, has four generic boxes supporting the text on the TCA cycle and electron transport chain. Two of these, on succinyl CoA synthetases and the location of the electron transport chain, fall into the first of the three categories listed above, extending and elaborating on topics covered in the main text. A third box, on why the enzyme that synthesizes ATP is called an ATPase, is in the second category, addressing a point that is not essential to one's understanding of energy generation, but which an astute student will query. Finally, the box on skunk cabbage comes into the interesting category, albeit making the important point that there are variations on the standard electron transport chain. In passing, I would say that I included this box after searching the literature for an interesting example of a variant electron transport chain, and was rather pleased with myself when I discovered skunk cabbage. Later, of course, I found it is a standard example included in all the textbooks.

Further reading

Each chapter has a Further reading list which contains books, review articles and some primary research papers that cover the topics described in that chapter. In those cases where the title does not make clear the relevance of an article, I have appended a one-line summary stating its particular value. The Further reading lists are not all-inclusive and I encourage readers to spend some time searching the internet (or even a library) for other books and articles. Browsing is an excellent way to discover interests that you never realized you had!

Self-assessment questions

Each chapter also has three sets of self-assessment questions.

Multiple-choice questions take the usual format, with only one answer being correct for each question. Answers can be found on the website: www.scionpublishing.com/biochemistry.

Short answer questions require 100–500 word answers, or occasionally ask for an annotated diagram or a table. The questions cover the entire content of each chapter in a straightforward manner, and do not require additional reading, so they can be marked simply by checking each answer against the relevant part of the text. A student can use the short answer questions to work systematically through a chapter, or can select individual ones in order to evaluate their ability to answer questions on specific topics. The short answer questions could also be used in closed-book examinations.

Self-study questions vary in nature and in difficulty. They may require calculation, additional reading and/or informal research. Some are reasonably straightforward and merely require a literature survey, the intention with these questions being that the students take their learning a few stages on from where *Biochemistry* leaves off. In some cases, questions of this type also point forwards to later parts of the book, so the student, through their own reading, becomes primed in a particular concept before we deal with it together. Some questions require that the students evaluate a statement or a hypothesis, which could be done by reading around the subject but which, hopefully, will engender a certain amount of thought and critical awareness. A few of the questions are very difficult, to the extent that there is no solid answer. These are designed to stimulate debate and speculation, which stretches the knowledge of each student and forces them to think carefully about their statements. Many of the self-study questions would be suitable for group discussions in problem-based learning or tutorial sessions. There are no answers at the back of book! To provide answers would defeat the purpose – the intention is that the students discover the answers for themselves.

Glossary

I am very much in favor of glossaries as learning aids and I have provided an extensive one for *Biochemistry*. Every term that is highlighted in bold in the text is defined in the glossary, along with a few other entries that the reader might come across when referring to books or articles in the reading lists. Note that, in order not to confuse the reader, the definitions given in the glossary usually only reflect the usage of that term in the main text. The definitions are not intended to be comprehensive and all-embracing descriptions such as one might find in a genuine dictionary of biochemistry.

CHAPTER 1

Biochemistry in the modern world

STUDY GOALS

After reading this chapter you will:

- understand what we mean by the term 'biochemistry'

- be aware of the central position that biochemistry occupies within the biological sciences

- appreciate that to understand biochemistry it is also necessary to understand the basic principles of chemistry

- know that four types of large molecule – proteins, nucleic acids, lipids and polysaccharides – are particularly important in biochemistry

- be aware that metabolism plays a vital role in all living organisms

- recognize that metabolism includes catabolic processes, which break molecules down to generate energy, and anabolic ones, which build up larger molecules from small ones

- know that biological information is stored in DNA and made available to the cell via the process called gene expression

- realize that biochemistry is an experimental science and that understanding the methods used in research projects is a key part of becoming a biochemist

Imagine you mixed together a few kilograms of oxygen, carbon, hydrogen, nitrogen, calcium and phosphorus, with smaller amounts of some 53 other elements from aluminum to zirconium, using the recipe in *Table 1.1*. What would you have? A rather bizarre mix of solid, liquid and gaseous chemical elements, which is odd, because the average adult human is made up of the same elements in the same proportions. But outside of the world of cult movies, no amount of heating, electrifying or irradiating will turn the mixture into a living person. **Biochemistry** tells us why.

1.1   What is biochemistry?

When I studied biochemistry at university it was quite common for a hyphen to be inserted in the word, so it was written as bio-chemistry. The longer term '**biological chemistry**' is still used today. The implication is that biochemistry is simply a combination of the two subjects, chemistry applied to biology, or the 'chemistry of life'. This is a reasonable way to define biochemistry in half a dozen words, but modern biochemistry is, in reality, much more than just the study of the chemicals present in living organisms. Let us explore exactly what we mean by 'biochemistry'.

Table 1.1. **The elemental composition of the average adult human**

Element	Amount in a 70 kg human
Oxygen	43 kg (61%)
Carbon	16 kg (23%)
Hydrogen	7 kg (10%)
Nitrogen	1.8 kg (2.5%)
Calcium	1.0 kg (1.4%)
Phosphorus	780 g (1.1%)
Potassium	140 g (0.20%)
Sulfur	140 g (0.20%)
Sodium	100 g (0.14%)
Chlorine	95 g (0.14%)
Magnesium	19 g (0.03%)
Iron	4.2 g
Fluorine	2.6 g
Zinc	2.3 g
Silicon	1.0 g
Rubidium	0.68 g
Strontium	0.32 g
Bromine	0.26 g
Lead	0.12 g

Trace amounts (less than 100 mg each) of copper, aluminum, cadmium, cerium, iodine, tin, titanium, boron, nickel, selenium, chromium, manganese, arsenic, lithium, cesium, mercury, germanium, molybdenum, cobalt, antimony, silver, niobium, zirconium, lanthanum, gallium, tellurium, yttrium, bismuth, thallium, indium, gold, scandium, tantalum, vanadium, thorium, uranium, samarium, beryllium and tungsten

Those elements shown in red are known to play a role in human biochemistry. Most of the other elements are absorbed from the environment but have no known function within the body.
Data from **Emsley J** (1998) *The Elements, 3rd edn*. Clarendon Press, Oxford.

1.1.1 Biochemistry is a central part of biology

Biochemistry is a part of **biology**, or of the **life sciences** as the subject is now commonly called. Within the life sciences, biochemistry occupies a central position. This is because biochemistry is concerned with the synthesis and structure of the molecules that make up living organisms, and with the way in which chemical reactions provide organisms with the energy they need to survive. Biochemistry therefore explains how the mixture of atoms described in *Table 1.1* can be combined together to make a living, functioning human being. In doing so, biochemistry underpins our understanding of all aspects of biology, and biologists who do not look on themselves as biochemists still have to understand the subject, and often have to use biochemistry in their own studies.

For some areas of the life sciences, it is very easy to see why a knowledge of biochemistry is important. Biologists who study the structures and properties of living cells, for example, cannot proceed very far in their research without considering the molecules contained in those cells. These molecules make up the structure of the cell and are responsible for each cell's particular properties (*Fig. 1.1*). So there is a great deal of overlap between biochemistry and cell biology. The same is true for genetics, which focuses on the genetic information contained in genes. Genes are made of DNA, and understanding how genes work means studying the structure of DNA and the way in which DNA interacts with other molecules so that the information it contains can be used by the cell. These are exactly the same questions that biochemists are interested in, and a large part of genetics could be described as 'the biochemistry of DNA'.

Figure 1.1 Representation of part of a cell of the bacterium *Escherichia coli*.
The cell wall, which is made up mainly of carbohydrate and protein, is shown in green, as is the cell membrane and a flagellum, the latter extending from the cell wall. The flagellum is made of protein, and rotates like a propeller, enabling the bacterium to swim at speeds up to 100 μm per second. Within the cell, the long yellow strands are parts of the bacterium's DNA molecule, which in places is wrapped around barrel-shaped proteins, also shown in yellow. The orange structures are enzymes that are making RNA copies of the genes in the DNA molecule. These copies, called messenger RNA, are shown in white. They move to the purple ribosomes (made of RNA and protein) where they direct the synthesis of new proteins. These proteins include enzymes, in blue, that catalyze the biochemical reactions occurring within the bacterium.
Illustration by David S. Goodsell, the Scripps Research Institute, and reproduced here with permission.

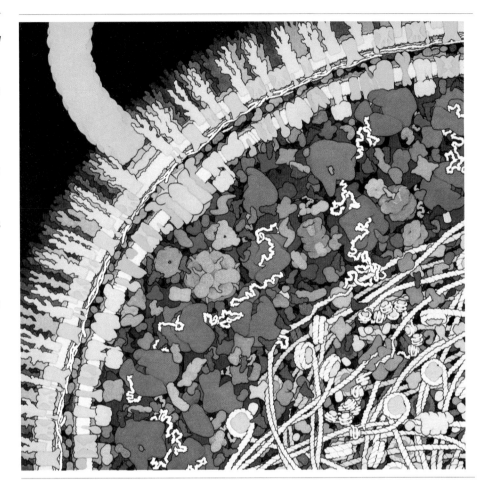

Modern biochemistry also underpins areas of biology that we associate with organisms rather than cells. In ecology, for example, ecosystems are often described in terms of **food webs**, with energy being generated by photosynthesis and then transferred up a food chain through the herbivores to the top carnivores (*Fig. 1.2*). The generation of energy is a central topic in biochemistry, and this aspect of ecosystem ecology is, in reality, biochemistry applied to a community of different species rather than to individual organisms. Similarly, we do not immediately think of biochemistry when evolution is being discussed. But the evolutionary relationships between species are now studied not only by comparing the morphology of those species and the structures of their bones. Today, the relationships are more likely to be probed by comparing the structures of the molecules contained in the organisms (*Fig. 1.3*). Evolutionary biologists must therefore learn biochemical techniques in order to work out those molecular structures, and must understand biochemistry to make sure that the comparisons that they make are based on sound principles.

Because biochemistry is so central to the life sciences, we must begin this book with a brief survey of the key principles of biology, to provide the context for our study of molecules and their biochemical reactions. We do this in *Chapter 2*, where we will look at the variety of life on the planet, examine the structures of cells, and consider how the vast diversity of life arose.

1.1.2 Chemistry is also important in biochemistry

Although biochemistry is part of the life sciences, it depends very much on the principles and analytical methods of **chemistry**. Indeed, biochemistry began when chemists first

Figure 1.2 Movement of energy through the food web of the African savannah.
Energy from the sun is captured by photosynthesis that takes place in the primary producers. Herbivores obtain their energy by eating primary producers, and herbivores are in turn eaten by carnivores. The energy from sunlight is therefore transferred, step by step, up the food chain.

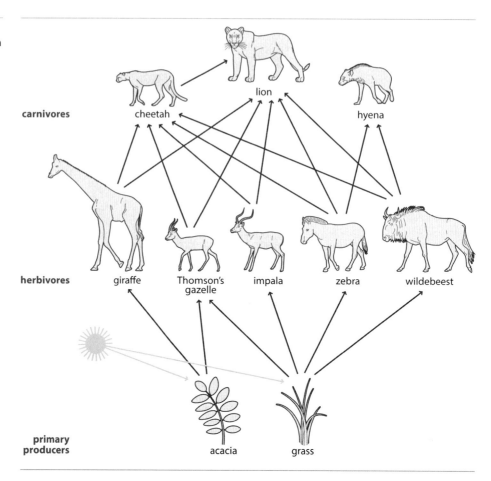

Figure 1.3 Evolutionary relationships among the mammals.
Trees like this used to be constructed from morphological information, but today are more likely to be deduced by comparing the structures of proteins or DNA molecules from the species being studied.

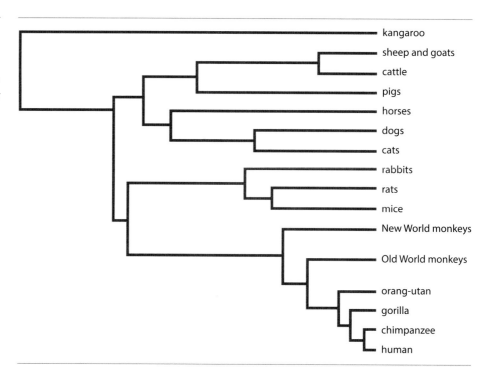

Box 1.1 **The origins of biochemistry**

The German chemist Carl Neuberg is looked on as the father of biochemistry. Neuberg introduced the term 'biochemistry' in 1903, and promoted biochemistry as a distinct subject by setting up and editing the *Biochemische Zeitschrift*, the first journal devoted to biochemistry, now called *FEBS Journal*. However, the origins of biochemistry date back much further, to the mid-1700s when scientists first began to study chemicals and chemical processes in living organisms. This work gradually overturned the long-standing notion that living entities contain a 'vital principle' which cannot be described in chemical or physical terms. By 1900 it had been established that living organisms are subject to the same chemical and physical laws as inanimate matter, enabling all areas of biology, not just biochemistry, to develop into the rigorous scientific disciplines with which we are familiar today.

The key steps in the development of biochemistry prior to 1900 were as follows:

1770s   Carl Wilhelm Scheele isolated citric acid from lemons, malic acid from apples, and lactic acid from milk. These carbohydrates were among the first organic compounds to be identified.

1780s   Antoine Lavoisier and Pierre Laplace showed that the amount of heat and carbon dioxide generated during respiration is identical to that generated during combustion. Lavoisier also proposed that, during photosynthesis, plants take up carbon dioxide and release oxygen. These experiments indicated that energy generation in living organisms is subject to the same chemical laws as energy generation in chemical reactions.

1811–1823   Michel Eugène Chevreul studied the chemistry of animal fats. His work was the first application of chemical and physical analysis to any type of biomolecule.

1820s   William Prout distinguished different types of food as saccharinous, albuminous and oleaginous, which are roughly equivalent to carbohydrates, proteins and fats.

1827   Hans Fischer synthesized porphyrins and showed that these compounds bind oxygen in red blood cells.

1833   Anselm Payen and Jean-François Persoz isolated and studied the first enzyme, 'diastase', which we now call amylase and which converts starch to sugar. We will examine their work in more detail in *Section 7.1*.

1850s   Claude Bernard showed that glycogen is synthesized from glucose in the liver. This was one of the first demonstrations that animals are able to synthesize biomolecules as well as break them down.

1877   Moritz Traube suggested that enzymes are a type of protein.

1880–1900   Emil Fischer identified the structures of many important biomolecules, including the 16 different isomers of glucose, and the purines that are components of DNA and RNA. Later, he showed how amino acids are linked together to form a polypeptide.

1895–1900   The first hormones were discovered. Adrenaline (also called epinephrine) was identified by Napoleon Cybulski, Jokichi Takamine and others.

---

became interested in studying the chemical reactions that occur in living organisms. These chemists discovered, back in the nineteenth century, that biochemistry presents its own unique challenges. Foremost among these is the complexity of the mixture of molecules that are present in a living cell. Chemists were, and still are, more used to studying reactions that occur in relatively simple solutions whose chemical makeup is precisely known. Cells, and extracts prepared from cells, contain many different types of compound, and understanding which parts of the mix are responsible for particular biochemical reactions is a real problem. Solving this problem required the development of new methods and scientific approaches that gave biochemistry a unique flavor. Chemistry, on the other hand, continued to be organized around the traditional split of inorganic, organic and physical. Biochemistry fits into neither of these categories, and so became a subject in its own right.

Although biochemistry grew up as a separate subject, it is not possible to study biochemistry or to be a biochemist without understanding the basic principles of chemistry. For some young biochemists, especially those who are drawn to the subject because of an interest in biology, learning chemistry can be a daunting prospect. The very first lecture that I attended at university was in a Physical Chemistry unit and was on the Schrödinger wave equation. The class began with the lecturer writing the equation (or one of them, there are several versions) on the blackboard. To say I was daunted would be an understatement; at the time I did not even understand what half the symbols meant. It would also be true to say that since completing that unit in Year 1 Physical Chemistry, I have never needed to refer to Schrödinger's equation ever again. In this book we will not follow my undergraduate experience and plunge straight into the deep end of chemistry. Instead we will deal only with those aspects of chemistry that are important in biochemistry, and we will deal with those topics as and when they become relevant. Most of these chemical topics will be presented in bite-sized units as a series of orange 'Principles of Chemistry' boxes that you will meet at the appropriate places as we progress through the book.

See p. xx for a list of all the Principles of Chemistry boxes.

Box 1.2 Schrödinger and biology

Although Erwin Schrödinger is most noted for his work on quantum theory, he was one of several physicists from the early part of the twentieth century who also took a keen interest in biology. One of these, Max Delbrück, changed disciplines mid-career, moving from theoretical physics to genetics and carrying out pioneering work with **bacteriophages** (viruses that infect bacteria) that led to the discovery that genes are made of DNA.

Schrödinger remained a physicist, but in 1944 he wrote a short book called *What is Life?*, in which he speculated about inheritance and the structure of genes. Reading the book today, many of Schrödinger's ideas appear far-fetched. He concludes that genes are crystalline structures, and in places he comes close to reviving the 'vital principle' of pre-twentieth century

biology, by suggesting that living organisms might utilize unknown laws of physics. Despite its errors, in one respect *What is Life?* was an important landmark in the development of twentieth century biochemistry. It had already been established that genes contain information that specifies the developmental plan of an organism and the biochemical reactions that it can carry out. Schrödinger argued that this information must be encoded within the structures of the organism's genes. Again, his specific ideas about how this coding system worked were incorrect, but the notion that there must be some kind of **genetic code** that an organism uses to read the information in its genes was an important insight that set the agenda for research into genes over the next 20 years. In *Section 16.1* we will explore how information is encoded in a gene and how this information is utilized.

1.1.3 Biochemistry involves the study of very large biomolecules

When the first biochemists began to examine the complex mixtures of molecules in living cells, they quickly realized that some of these molecules are very large indeed. The size of a molecule is expressed as its **molecular mass**, measured in **Daltons (Da)**, with 1 Da equal to one-twelfth the mass of a carbon atom. Most of the compounds known in nature, and most of the artificial ones synthesized by chemists, have molecular masses substantially less than 1000. Water, for example, has a molecular mass of 18.02 Da, ethanol is 46.07 Da, and phenol is 94.11 Da. Even a complex organic compound such as quinine, which is used to treat malaria, has a molecular mass of only 324 Da. In contrast, there are many molecules in living cells whose molecular masses are measured in thousands of Daltons, called **kiloDaltons (kDa)**. Relatively small examples of these **macromolecules** include **thrombin**, which is involved in blood clotting and whose molecular mass is approximately 37 400 Da or 37.4 kDa, and **alpha-amylase**, which is secreted in saliva and begins the breakdown of dietary **starch** into **sugar**, which has a mass of 55.4 kDa. Starch itself is variable in size, ranging from 190 kDa to 227 000 kDa depending on which type of plant it comes from.

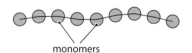

monomers

Figure 1.4 A linear polymer.

Most of these large biomolecules are **polymers**, compounds made up of long chains of identical or very similar chemical units called **monomers** (*Fig. 1.4*). In starch, the monomeric unit is a **glucose** molecule, and the polymer is built up by linking glucose monomers together in branched chains. The greater the number of glucose units, the larger the molecular mass of the starch molecule. One of the smaller starch molecules, with a molecular mass of just 190 kDa, would contain about 1050 glucose units, whereas the larger molecules will have over a million.

Starch is an example of a **polysaccharide**, a polymer made of glucose or similar sugar molecules. Polysaccharides have two main roles in living cells. First, polysaccharides such as starch (in plants) and **glycogen** (in animals) act as stores of energy. This is because the sugar units that they contain can be released from the polymers and broken down further in order to generate chemical energy. The second role of polysaccharides is structural. **Cellulose**, which gives plant cells their rigidity, is a type of polysaccharide (*Fig. 1.5*), as is **chitin**, which forms part of the exoskeleton of insects and of animals such as crabs and lobsters.

As well as polysaccharides, there are three other classes of large biomolecule that are important in biochemistry. The first of these is **proteins**, which are unbranched polymers of **amino acids**. Proteins play an immense range of roles in living organisms and most **enzymes**, which catalyze biochemical reactions, are proteins. Alpha-amylase, which catalyzes the chemical reaction responsible for the release of glucose units from starch, is an example of an enzyme. Another is thrombin, which catalyzes the reaction that converts fibrinogen (itself a protein) into insoluble polymers of fibrin, which bind together as part of the blood clotting process (*Fig. 1.6*).

The third type of large biomolecule is **nucleic acid**, of which there are two types, **deoxyribonucleic acid** or **DNA** and **ribonucleic acid** or **RNA**. DNA is present in chromosomes and contains biological information. In other words, genes are made of DNA. RNA is involved in the way in which the information contained in DNA is read by the cell. Finally, there are the **lipids**, a diverse group of large biomolecules which, like polysaccharides, have structural roles and act as energy stores, but which also have a number of other functions including regulatory ones – several **hormones** are lipids.

Figure 1.5 The structural role of cellulose in a plant cell wall. Cellulose molecules line up and attach to one another, producing microfibrils which in turn form a network that gives the plant cell wall its rigidity.

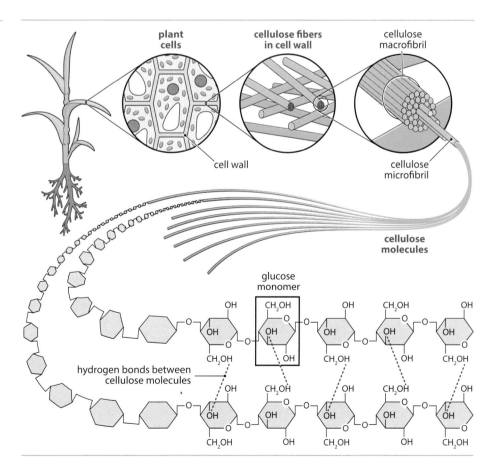

Figure 1.6 The roles of thrombin, fibrinogen and fibrin in the formation of a blood clot.
(A) Thrombin catalyzes a biochemical reaction that modifies the structure of fibrinogen, converting it to fibrin. Fibrin molecules then attach to one another to form polymers. (B) The fibrin polymers form a network over a rupture in a blood vessel, entrapping blood cells. This structure forms the blood clot.

A. conversion of fibrinogen to a fibrin polymer

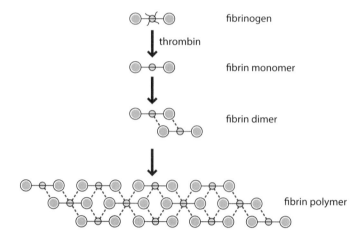

B. a blood clot

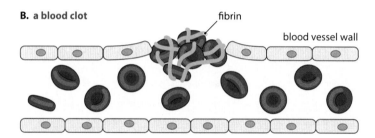

Understanding the structures and functions of these four types of biomolecule will be the objective of *Chapters 3–6* dealing, in turn, with proteins, nucleic acids, lipids and **carbohydrates**, the last of these being the type of biochemical that includes polysaccharides.

Box 1.3 **Atoms, isotopes and molecular masses**

PRINCIPLES OF CHEMISTRY

An atom consists of a nucleus containing positively charged protons and neutral neutrons, surrounded by a cloud of negatively charged electrons. The chemical identity of the element is determined by the number of protons, which is called the **atomic number**. This number is the same for all atoms of that particular element. For example, every hydrogen atom has just a single proton and an atomic number of 1, and every carbon atom has six protons and an atomic number of 6.

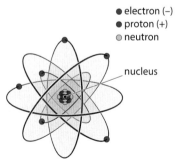

● electron (–)
● proton (+)
○ neutron

nucleus

Although the number of protons is invariant, different atoms of the same element can have different numbers of neutrons. These

different versions of an element are called **isotopes**. Carbon, for example, has three naturally occurring isotopes, each containing six protons but with six, seven or eight neutrons. The total number of protons and neutrons in a nucleus is called the **mass number**, so the three isotopes of carbon have mass numbers of 12, 13 and 14, and are called carbon-12, carbon-13 and carbon-14, or ^{12}C, ^{13}C and ^{14}C. These are the isotopes of carbon that are found in nature. Carbon-12 makes up 98.93% of all the carbon atoms in existence, and carbon-13 contributes most of the remaining 1.07%. Carbon-14 is present in only trace amounts, about one out of every trillion carbon atoms. There are also 12 isotopes from ^{8}C to ^{22}C that do not exist in measurable amounts in the environment but which can be created under laboratory conditions. The majority of elements, but not all, have naturally occurring isotopes, the largest numbers being nine isotopes for xenon and ten for tin.

Carbon-12 is considered to have a molecular mass of exactly 12 Da. The values for other atoms are calculated according to their masses relative to carbon-12. The molecular mass of a compound is worked out simply by adding together the masses of its constituent atoms.

1.1.4 Biochemistry is also the study of metabolism

Living organisms, and the cells they comprise, are dynamic structures. This means that they require energy to power their various activities, and must also be able to synthesize new biomolecules as and when these are needed. These are the processes that constitute 'life'. The fundamental principle of biochemistry is that these 'life processes' are chemical reactions. There are a great many of them, and they are linked together in complicated pathways (*Fig. 1.7*), but by studying the reactions individually it is possible to build up an understanding of the molecular basis of life. This will be our goal in *Part II*.

Any chemical reaction can occur spontaneously, but its rate might be very slow. When chemical reactions are carried out in a test tube, a **catalyst** is often added to

Figure 1.7 The metabolic pathways of a typical animal cell.
Each dot represents a different biochemical compound. The lines indicate the steps in the network, each of these steps resulting in conversion of one compound into another. Copyright 2014 from *Essential Cell Biology*, 4th edition by Alberts *et al*. Reproduced by permission of Garland Science/Taylor & Francis LLC.

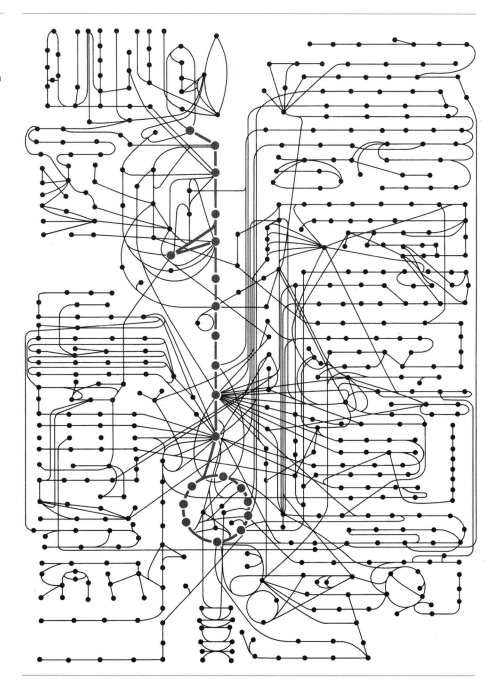

speed up the reaction. An example is the use of vanadium oxide in the industrial production of sulfuric acid, from sulfur dioxide and oxygen, by the Contact Process (*Fig. 1.8*). The reaction rate is increased because the two gaseous reactants (sulfur dioxide and oxygen) become absorbed on the surface of the catalyst, bringing the molecules close together and promoting their combination to form sulfur trioxide, which reacts with water to give sulfuric acid. Biological reactions also make use of catalysts, but these are not metals. They are called enzymes and the vast majority are protein molecules, though a few made of RNA are also known. We will discover how enzymes act as catalysts in *Chapter 7*.

Figure 1.8 The role of a catalyst. In the Contact Process, sulfur dioxide and oxygen are passed through layers of vanadium oxide particles. The two gases become absorbed to the surface of the particles, promoting their combination to form sulfur trioxide. The vanadium oxide catalyzes the reaction, but is not itself used up during the process.

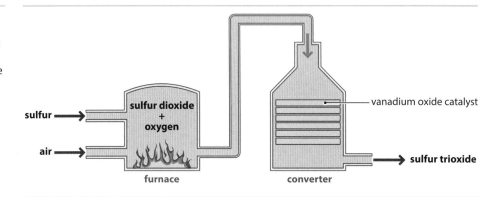

Metabolism is the word used to describe the chemical reactions that occur in living organisms. These reactions are traditionally divided into two broad groups:

- **Catabolism** – this is the part of metabolism that is devoted to the breakdown of compounds in order to generate energy.

- **Anabolism** – this refers to those biochemical reactions that build up larger molecules from smaller ones.

We will study the central energy-generating processes of the cell, called **glycolysis**, the **TCA cycle** and the **electron transport chain**, in *Chapters 8* and *9*. A special type of energy generation, from sunlight by photosynthesis, will be the focus of *Chapter 10*. In *Chapters 11–13* we will examine the metabolic pathways that result in the synthesis and breakdown of carbohydrates, lipids, and nitrogen-containing biochemicals, the latter including the monomeric components of proteins and nucleic acids.

During these chapters, we will also ask how the various metabolic reactions occurring in living cells are regulated. Biochemical reactions do not occur randomly. Individual reactions are carefully controlled so that those that work together in a metabolic pathway operate in a coordinated fashion, to ensure that the substrates for the pathway are converted efficiently into the final end product. The rate of product synthesis can be controlled, and possibly switched off entirely, or the pathway may be modified so that different substrates are used, depending on what is available. Some of the signals that exert control over metabolic pathways originate within the cell in which the pathway is occurring, and others come from outside the cell. The signals might be very specific, altering the rate of a single reaction in a lengthy pathway, or they might be more general, affecting several different pathways at the same time. We will look at these various events at the appropriate places in *Chapters 8–13*.

1.1.5 The storage and utilization of biological information is an important part of biochemistry

One set of biochemical reactions, although strictly speaking a part of anabolism, are regarded as having such special features that they are usually considered separately from the metabolism of the cell. These reactions are the ones responsible for the synthesis of DNA, RNA and protein. Here biochemistry and genetics overlap, because the same reactions are responsible for the replication and utilization of the **biological information** that is contained in genes. This is the information that an organism needs in order to develop, reproduce and carry out all of its metabolic reactions. These are the central topics of modern genetics, and biochemistry is used to study them, as we will see in *Part III* of this book.

First we will examine how DNA molecules replicate, so that precise copies of every gene are made every time a cell divides or an organism reproduces. That will be the focus of *Chapter 14*. The next question we will address is how the biological information contained in the genes is made available to the cell. This process is called **gene expression**, and for all genes it begins with synthesis of an RNA molecule (*Fig. 1.9*). In *Chapter 15* we will discover that, because DNA and RNA are very similar types of molecule, this step in gene expression, called **transcription**, is quite straightforward in chemical terms. We will also learn that the RNA molecules that are made by transcription fall into different groups based on their function. Among these groups is **messenger RNA (mRNA)**, which directs synthesis of proteins by a process referred to as **translation**. Protein synthesis will be the subject of *Chapter 16*.

Not all of the genes in a cell are active all the time. Many are silent for long periods, only being converted into RNA and protein on those particular occasions when their products are needed. All organisms are therefore able to regulate expression of their genes, so that those whose RNA or protein products are not required at a particular time are switched off. We will study the great variety of processes by which gene expression can be controlled in *Chapter 17*.

1.1.6 Biochemistry is an experimental science

In *Parts I–III* we will learn the facts of biochemistry. In *Part IV* we will examine how those facts were discovered. Biochemistry is and always has been an experimental science, and one of the attractions of the subject for new students is the possibility of one day carrying out their own biochemical research projects. Throughout *Parts I–III* of this book we will refer to some of the key experiments that have built up our understanding of biochemistry. In *Part IV* we will look more specifically at the methods and research strategies that are used in the most active areas of modern biochemistry.

The first of these topics is the analysis of large biomolecules, in particular proteins. This is an important area of research because of the role of proteins as enzymes. Working out the detailed structure of an enzyme is often the best way of understanding how that enzyme catalyzes its specific biochemical reaction, and how the enzyme activity, and hence the biochemical reaction, is regulated in response

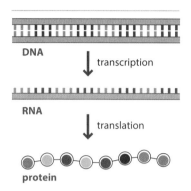

Figure 1.9 Gene expression. The DNA copy of a gene is transcribed into an RNA molecule. For protein-coding genes, the RNA is then translated into the protein.

Box 1.4 'Omes are collections of biomolecules

The proteome – the collection of proteins in a cell or tissue – is just one of several sets of biomolecules that biochemists study. These collections are loosely referred to as 'omes, specific examples being:

- The **genome**, which is the entire complement of DNA molecules in a cell, containing all the organism's genes.
- The **transcriptome**, which is the collection of RNA molecules in a cell or tissue; the name 'transcriptome' derives from the fact that RNA molecules are copies, or **transcripts**, of genes.
- The **lipidome** is the total lipid content of a cell or tissue.
- The **glycome** is the carbohydrate content.

Finally, there is the **metabolome**, which has a more complex composition. The metabolome is the complete collection of metabolites present in a cell under a particular set of conditions. These metabolites are the substrates, products and intermediates of all of the catabolic and anabolic reactions that are occurring in the cell. The metabolome therefore reflects the cell's biochemical activities. These activities are specified by the proteome, and are dependent to at least some extent on the compositions of the lipidome, glycome and transcriptome. The biochemistry of a cell can therefore be looked on as resulting from the interplay between its various 'omes.

to chemical signals from within and outside of the cell. Over the years, biochemists have developed sophisticated methods for studying protein structure, such as **nuclear magnetic resonance (NMR)** and **X-ray crystallography**. We will look at these techniques in *Chapter 18*.

In *Chapter 18* will we also examine the various methods used to characterize the **proteome**, the collection of proteins present in a cell or tissue. The composition of the proteome defines the biochemical capability of a cell and so **proteomics**, the methods used to identify the individual components of a proteome, is an important aspect of biochemical research. To complement our study of proteomics, we will end *Chapter 18* with a brief look at the equivalent techniques used to catalog the lipid and carbohydrate contents of cells.

The second set of biochemical research methods that we will examine are those used to study DNA and RNA molecules. Central to these is **DNA sequencing**, the technique used to determine the structures of genes and the organization of genes in DNA molecules. Sequencing is used to understand the nature of the genetic information possessed by an organism and the way in which it is expressed. A second important technique is **DNA cloning**, which is used to transfer genes from one species to another, and which enables important pharmaceutical proteins such as human insulin to be synthesized by genetically engineered microorganisms. These and other methods used to study DNA and RNA are described in *Chapter 19*.

Further reading

Coley NG (2001) History of biochemistry. *Encyclopedia of Life Sciences*. Wiley Online Library DOI: 10.1038/npg.els.0003077.

Dronamraju KR (1999) Erwin Schrödinger and the origins of molecular biology. *Genetics* **153**, 1071–6.

Hui D (2012) Food web: concept and applications. *Nature Education Knowledge* **3(12)**: 6.

Hunter GK (2000) *Vital Forces: the discovery of the molecular basis of life*. Academic Press, London. *An account of the history and development of biochemistry*.

Patti GJ, Yanes O and Sluzdak G (2012) Metabolomics: the apogee of the omics trilogy. *Nature Reviews Molecular Cell Biology* **13**, 263–9.

Springer MS, Stanhope MJ, Madsen O and de Jong WW (2004) Molecules consolidate the placental family tree. *Trends in Ecology and Evolution* **6**, 430–8. *Explains how comparisons of molecular structures are used to construct evolutionary trees*.

CHAPTER 2
Cells and organisms

STUDY GOALS

After reading this chapter you will:

- understand that although life is diverse, it is built on a limited number of themes, and that similar biochemical processes occur in all species

- be able to describe the key distinguishing features of prokaryotes and eukaryotes

- appreciate that the prokaryotes include two different types of organism, called bacteria and archaea

- know that prokaryotic cells lack an extensive internal architecture

- be able to recognize the main structural components of a eukaryotic cell and describe the functions of those components

- begin to understand the importance of membranes in cell structure

- recognize that viruses are obligate parasites that can only reproduce by infecting a host cell

- be aware of the current theories regarding the origins of life on the planet

- understand that the common themes displayed by the morphologies and biochemistries of all species indicate that those species evolved from a single common origin

The planet teems with life. Latest estimates suggest that there are about 8.7 million species of animals, plants and fungi, and at least 10 million species of bacteria. Both figures are approximate because it is thought that the vast majority of species are unknown to science. Over 1200 different types of beetle were once counted on just 19 trees in Panama, of which almost a thousand were new species. There are at least 10^{11} individual animals alive on the planet and, according to one study, 10^{30} bacteria. The numbers are so astronomical as to be almost incomprehensible.

There is a serious side to these numerical Olympics. How can a biochemist hope to understand the chemical reactions occurring in living organisms when there is such a huge number of different species to study? Fortunately, the vast diversity of life is built on a limited number of themes. To begin to place biochemistry in its biological context we must examine these themes, drawing out the similarities, rather than differences, between the myriad organisms on the planet.

2.1   Cells – the building blocks of life

All living organisms are made of cells. Some organisms are **unicellular**, comprising just a single cell, but many others are **multicellular**. The latter includes virtually all

---

Box 2.1 Units of measurement

The standard SI (from Système International d'Unités) unit of length is the meter (m). Smaller units are as follows:

$1\ cm = 10^{-2}\ m$ (100 cm in 1 m)
$1\ mm = 10^{-3}\ m$ (1000 mm in 1 m)
$1\ \mu m = 10^{-6}\ m$ (1 000 000 µm in 1 m)
$1\ nm = 10^{-9}\ m$ (1 000 000 000 nm in 1 m)

Biochemists also sometimes use two non-SI units of length:

$1\ micron = 1\ \mu m$
$1\ Ångstrom\ (Å) = 10^{-10}\ m$ (10 Å in 1 nm)

Other units of measurements use the same SI prefixes:

- For volume, the standard is the liter (l), and smaller volumes are ml, µl, etc.
- For weight, the standard is the gram (g), and smaller weights are mg, µg, etc.

---

macroscopic organisms, those visible with the naked eye. This is because cells are very small structures and most unicellular organisms are too small to be seen without magnification. The common *Paramecium*, for example, which is found in freshwater environments such as ponds and riverbeds, is quite a large single-celled organism, but is still only 120 µm in length. One µm (micrometer or micron) is one-thousandth of a millimeter, so 120 µm is just 0.12 mm.

Cells are therefore the basic building blocks of all organisms and it is with these that we must begin our examination of the themes that underlie the diversity of life.

2.1.1 There are two different types of cell structure

The multitude of organisms on the planet can be divided into two groups based on the structures of their cells. These groups are called the **prokaryotes** and the **eukaryotes**. The distinction is evident when we look at electron micrographs of the two cell types (*Fig. 2.1*). A prokaryotic cell has very few visible internal features, other than a lightly staining central region, called the **nucleoid**, which contains the cell's DNA. In contrast, a typical eukaryotic cell is larger and more complex, with a membrane-bound **nucleus** containing most of the DNA, and with other membranous **organelles** such as **mitochondria** and **Golgi apparatus**.

Most prokaryotes are unicellular though in some species individual cells associate together to form larger structures. An example is the chains of cells formed by *Anabaena* (*Fig. 2.2*). Eukaryotes can be unicellular or multicellular. Eukaryotes therefore make up all the macroscopic forms of life such as plants, animals and fungi.

Until 1977 it was thought that all prokaryotes were similar to one another. The group includes a great diversity of organisms but the differences were thought to be

---

Figure 2.1 Transmission electron micrographs of (A) a prokaryotic and (B) a eukaryotic cell.

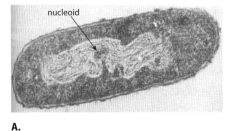

A.

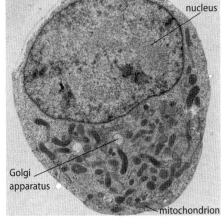

B.

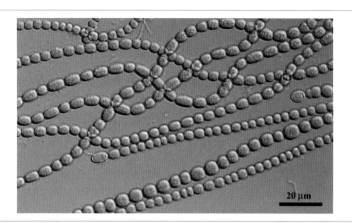

Figure 2.2 *Anabaena*, **a prokaryote whose cells form chains.**
Reproduced from the Culture Collection of Autotrophic Organisms (CCALA) with permission.

variations on the same basic theme. This assumption has been overturned and we now recognize that there are two distinct groups of prokaryotes, the **bacteria** and the **archaea**. The bacteria comprise most of the prokaryotes that are familiar to us as disease-causing organisms, such as the pathogens *Mycobacterium tuberculosis*, which causes tuberculosis in humans, and *Vibrio cholerae*, responsible for cholera. There are also many harmless species of bacteria living in the soil, on the surfaces of vegetation, in the air, indeed virtually everywhere.

The archaea also live in a wide range of habitats. These include the sediments at the bottom of lakes and other bodies of water, hot springs where the temperature can be 60°C or higher, brine pools and high-salt lakes such as the Dead Sea, and acidic streams emerging from old mine workings. Many of these environments are hostile to most other forms of life.

Both the bacteria and the archaea are types of prokaryote and their cells look very similar. The distinction lies in their biochemistries, as we will see below. This means that all the multitude of living organisms on the planet can be divided into just three groups, the bacteria, archaea and eukaryotes. Taxonomically, these are called the three **domains** of life. Each group contains diverse species with different appearances, habitats and their own special features, but within each group there is also a considerable degree of unity, especially at the biochemical level.

2.1.2 Prokaryotes

cocci

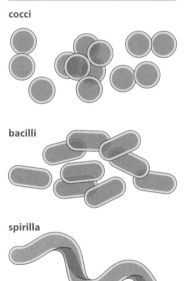

bacilli

spirilla

Figure 2.3 **Three types of prokaryotic cell shape.**

Most prokaryotic cells are either spherical (referred to as **cocci**; singular **coccus**), rod-shaped (**bacilli**; singular **bacillus**), or coiled (**spirilla**; singular **spirillum**) (*Fig. 2.3*). The dimensions are rarely greater than 10 μm. *Escherichia coli*, for example, is a rod with a diameter of approximately 0.5 μm and a length of 2.0 μm. This is a species of bacterium that lives inside the intestines of warm-blooded animals including humans, sometimes causing food poisoning. A harmless strain of *E. coli* is used in the research laboratory as a 'typical' bacterium, what biologists call a **model organism**.

Box 2.2 **Species names**

Species names for all organisms, not just bacteria, use a **binomial nomenclature**. The first part of the name is the **genus** to which the species belongs (e.g. *Mycobacterium*, *Vibrio*, *Homo*), and the second part of the name identifies the species (e.g. *tuberculosis*, *cholerae*, *sapiens*). Italics are used for both the genus and species components of the binomial, and the first letter of the genus name is always given a capital. Examples are *Mycobacterium tuberculosis*, *Vibrio cholerae* and *Homo sapiens*, which can be abbreviated to *M. tuberculosis*, *V. cholerae* and *H. sapiens* for simplicity after the genus name has previously been shown in full.

The biochemistry of *E. coli* has therefore been extensively studied, the assumption being that its biochemical pathways will not be significantly different from those in most other bacterial species. Generally, this is a sound assumption, so long as we bear in mind that some other bacteria have important special features, such as the ability to photosynthesize or to make **antibiotics**, which *E. coli* cannot do and which must therefore be studied in different species.

Prokaryotic cells lack an extensive internal architecture

A bacterial cell is surrounded by a **cell** or **plasma membrane**, which is made up of lipids and proteins and acts as a barrier between the internal parts of the cell and the external environment (*Fig. 2.4*). The cell membrane might display small infoldings called **mesosomes**, which have been assigned various roles but which most microbiologists now believe do not actually exist. Instead these mesosomes appear to be artefacts that arise as a result of the chemical treatments used when bacteria are prepared for electron microscopy. What is not in doubt is that most bacteria also have a **cell wall** on the external surface of the cell membrane, made up largely of a modified type of polysaccharide called **peptidoglycan**. In some species there is a second cell membrane on the outer surface of the cell wall. The entire structure comprising plasma membrane, outer membrane and cell wall is called the **cell envelope**. The envelope provides the cell with rigidity and gives the bacterium its characteristic shape. It also acts as a barrier to the movement of larger molecules into and out of the cell. It is not an impassable barrier, however, because nutrients such as sugars can be transported into the cell when they are needed. This means that the cell envelope plays an important role in controlling the relationship between the bacterium and the environment in which it is living.

> Membrane structure is described in *Section 5.2*.

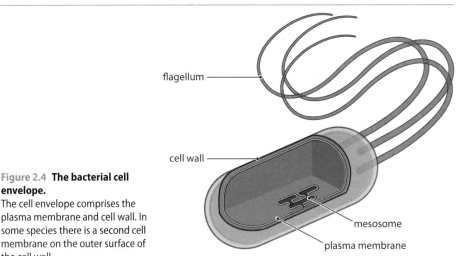

Figure 2.4 The bacterial cell envelope.
The cell envelope comprises the plasma membrane and cell wall. In some species there is a second cell membrane on the outer surface of the cell wall.

flagellum

cell wall

mesosome

plasma membrane

Some species of bacteria have additional structures attached to their surfaces. The most obvious of these is one or more **flagella** (singular **flagellum**). These are long filaments, often much longer than the length of the cell, made of a protein called flagellin. They are a means of propulsion for the bacterium, enabling it to swim through a liquid or semi-liquid medium. Attached to the base of each flagellum is a rotary motor, embedded in the inner cell membrane (*Fig. 2.5*). When the flagellum rotates, it adopts a helical conformation which acts just like a propeller, moving the cell forward. A flagellum can rotate at up to 1000 rpm, allowing the bacterium to swim at speeds of 100 μm per second.

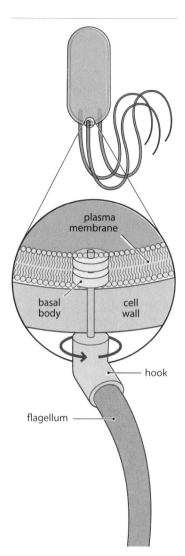

Figure 2.5 **A bacterial flagellum.** The structure at the base of the flagellum is a rotary motor, embedded in the plasma membrane.

Pili and **fimbriae** are other types of filament that are attached to the surfaces of some bacterial species (*Fig. 2.6*). These are made of proteins called pilins and are shorter than flagella. Pili enable bacteria to attach to one another during the process called **conjugation**, when one cell passes some of its DNA to a second cell. At one time it was thought that the DNA threads through the inside of a pilus, which acts like a hollow tube, but this has never been proven. Fimbriae enable bacteria to attach to various surfaces, including other bacteria. Sometimes this leads to formation of a **biofilm**, a collection of bacteria adhered to one another and to a solid surface, usually embedded in a slimy matrix. Biofilms are common in nature and cause a problem in hospitals because they can form inside medical implements such as catheters, increasing the risk of infections being transferred between patients.

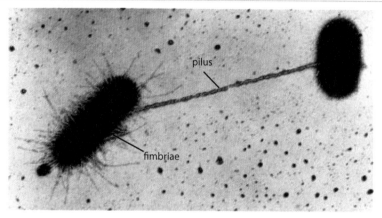

Figure 2.6 **Pili and fimbriae.** Reproduced from http://theultimatebacteria.blogspot.co.uk.

The most distinctive feature of a bacterium is the absence of any visible internal structure when the cell is observed with the electron microscope. The inside of the cell appears simply as a dark granular region surrounding a lighter central area (see *Fig. 2.1*). The dark part is the **cytoplasm**, within which many of the cell's biochemical reactions occur. The lighter region is the nucleoid. This is where the DNA, containing the genes, is located. DNA molecules are very long and thin (the DNA molecule is 1.6 mm in length in *E. coli*) and, to fit it into the nucleoid, the DNA molecule is coiled up tightly and attached to proteins that hold the coils in place.

Archaea have distinctive biochemical features

It was once thought that all archaea were **extremophiles**, living in inhospitable environments such as hot springs and acidic streams. Now we know that they are much more common, being present in many non-extreme environments, including the human gut. This means that more and more research is being carried out with archaea to find out what makes them different from bacteria.

In terms of cell anatomy, there are no significant differences between archaea and bacteria, which is just as we expect because both types of organism are classed as prokaryotes. Some archaea are spherical, some rod-shaped, and others have various diverse forms. Most archaea are smaller than the average bacterium, the largest being less than 1.0 µm in diameter, but otherwise the structural features that we have just described for bacteria also hold true for archaea.

The difference between the two groups of prokaryotes lies with their biochemical, rather than structural, features. These include biochemical differences in both the

Box 2.3 Bacteria communicate with one another in a biofilm

Within a biofilm, bacteria communicate with one another in order to maintain the structure and, when the population density becomes too high, to coordinate the departure of bacteria from the biofilm. For example, a key stage in the causation of cholera by *Vibrio cholerae* bacteria is formation of a biofilm in the patient's intestine. The biofilm protects the bacteria against the patient's immune system, and also to some extent against the action of antibiotics administered in attempts to control the disease. The antibiotic molecules are simply unable to get through the slimy extracellular matrix to reach the bacteria. When the population density in the biofilm reaches a certain size, cells are released into the intestine and excreted, to go on to infect other people.

How do the bacteria know when the population density in the biofilm has reached the critical point when dispersion should occur? The answer is that the bacteria communicate with one another by a process called **quorum sensing**. The cells release signaling chemicals called **autoinducers**. By sensing the concentration of autoinducers in the environment, each bacterium is able to assess the population density in its vicinity. Different species use different types of chemical as autoinducers. With some species these molecules are small proteins called **oligopeptides**, for others it is a class of compounds called *N*-acyl homoserine lactones.

A bacterium can detect the presence of the autoinducer because it has **receptor proteins** that bind the autoinducer. Oligopeptide inducers can cross the cell membrane, and their receptor proteins are located inside the cell. The lactones cannot get inside the cell and so bind to receptors that are on the cell surface, attached to the outer face of the cell membrane. When the population density is low, relatively little autoinducer is present in the environment, so most of the receptors in or on the surface of any individual bacterium are unoccupied. Conversely, if the population density and hence autoinducer concentration are both high, then the receptors become occupied. Binding of the autoinducer to the receptors influences the biochemical reactions occurring within the cell. In a biofilm, the resulting biochemical changes lead to departure of some bacteria from the structure.

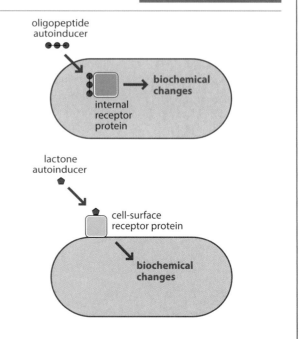

Quorum sensing illustrates the key properties of signaling processes that occur not just between different bacteria, but also between cells in multicellular organisms such as animals. Many of the cells in an animal have receptors of different types, which enable the cells to respond to the presence of various kinds of signaling compound. These compounds include **hormones** such as insulin and glucagon, which control the release of energy from the body's stores of fat and carbohydrate, and **cytokines**, which regulate activities such as cell division. In later chapters, we will encounter several examples of the way in which cell–cell signaling controls biochemical activities in multicellular organisms.

---

cell membrane and the cell wall. Bacterial and eukaryotic membranes have very similar biochemical compositions, but the lipids present in archaea membranes have their own distinctive structures. These structures provide the archaea lipids with a greater degree of heat stability, and might be one of the factors that enable some species to live in extreme environments such as hot springs. The cell walls of archaea are also distinctive because they do not contain peptidoglycan, which is the main constituent in the cell walls in bacteria. In most archaea the equivalent biomolecule is **pseudomurein**, which is also a modified polysaccharide but which has a slightly different chemical structure. Finally, the flagella of archaea are composed of different proteins and are assembled in a different manner.

Those are the main differences between the biochemistries of the bacteria and archaea. Other distinctions become apparent when the processes of DNA replication and gene expression are examined. DNA replication in archaea occurs by a series of events that is more similar to those in eukaryotes rather than bacteria, and the same is true for some steps of the gene expression pathway. So despite their similar

---

Box 2.4 The microbiome

The archaea present in the human gut are members of a wider community of bacteria, archaea and fungi called the human **microbiome**. These are the microorganisms that live on or within the human body. They comprise over 1000 species and contribute up to 3% of the total body weight. Most of the species are harmless, with pathogens making a significant contribution to the microbiome only when an individual has a specific infection. The microbiome does, however, include a number of species that are opportunistic pathogens, ones that are harmless in a healthy individual, but can cause infections during illness or when the immune system is not operating at full effectiveness.

For many years, the microbiome was looked on as unimportant, but increasing evidence is suggesting that at least some of the species carry out useful activities for their hosts. In the digestive tract, it appears that bacteria break down some types of carbohydrate into metabolites that can be further digested by the intestinal cells. Without the bacterial activity, the human host could not use these carbohydrates as nutrients. At present, the human microbiome is being intensively studied, with the aim of cataloging the species that are present, charting how these vary in different people and in different parts of the world, and understanding how the microbiome influences human health.

---

cell structures, which led microbiologists initially to look on archaea and bacteria as closely related to one another, their underlying biochemical and genetic features make it clear that the two groups of prokaryotes are in fact very different types of organism.

2.1.3   Eukaryotes

It is more difficult to define the features of a typical eukaryotic cell, because the individual cells in most multicellular organisms have specialized functions that are reflected by special structural anatomies. The 10^{13} cells present in an adult human, for example, are made up of over 400 specialized types. We have to be cautious even when we consider basic features such as size, but it is generally true that most eukaryotic cells are larger than prokaryotic ones, the majority having diameters ranging from 10 to 100 μm. Note that an increase in the diameter of a sphere from 10 to 100 μm results in a 100-fold increase in volume, so the largest eukaryotic cells are substantially bigger than the average bacterium.

Eukaryotic cells have complex architectures

All eukaryotic cells are surrounded by a membrane, and those of plants, fungi and algae also have a cell wall. The plant cell wall is a complex structure made up of various polysaccharides including cellulose, hemicellulose and pectin, possibly with an extra internal layer containing lignin, which is a tough cross-linked polymer unrelated to the polysaccharides. It is the stiffness of their cell walls that give plants their characteristic rigid structure. Fungal cell walls are equally complex structures made of different types of polysaccharide, one of which is chitin, which is also present in insect exoskeletons. Most algal cell walls are also made of polysaccharide, but those of the diatoms are rather different. These unicellular organisms have cell walls that are made of silica, and so are mineral based rather than biomolecular.

When a eukaryotic cell is viewed with the electron microscope, the most obvious difference compared with a prokaryote is the presence of an internal architecture made up of various structures collectively called **organelles** (see *Fig. 2.1*). Most organelles are surrounded by a membrane, some by two membranes, one inside the other, and many can be recovered intact if the cell is broken open carefully. There are many different types of organelle, each with its own function inside the cell (*Fig. 2.7*). Some of these functions are very specialized and the organelles that carry them out are present only in certain types of cell. Other organelles have functions that are needed by all or most cells. The most important of these organelles are the **nucleus**, which contains most of the cell's DNA, the energy-generating **mitochondria** (singular **mitochondrion**), the **Golgi apparatus** and **endoplasmic reticulum**, and in plants, the **chloroplasts**, which

Figure 2.7 **An animal cell showing the important organelles.**

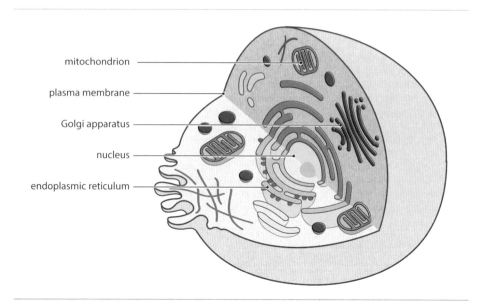

mitochondrion

plasma membrane

Golgi apparatus

nucleus

endoplasmic reticulum

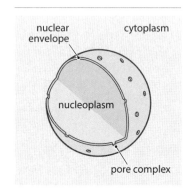

nuclear envelope

cytoplasm

nucleoplasm

pore complex

Figure 2.8 **A cell nucleus.**

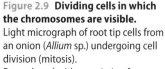

See *Section 4.2* for details about how a DNA molecule is packaged into a chromosome.

carry out photosynthesis. Over the next few pages we will look at the structures and activities of these important types of organelle.

The cell nucleus contains DNA

The nucleus is the largest organelle in most eukaryotic cells, possibly taking up one-tenth of the cell volume. It is surrounded by the **nuclear envelope**, which is a **double membrane**, two membranes one inside the other. The envelope is dotted with **pore complexes**, small channels connecting the cytoplasm outside of the nucleus with the **nucleoplasm** within it (*Fig. 2.8*).

The nucleus contains most of the cell's DNA. As in prokaryotes, each of these DNA molecules is much longer than the diameter of the cell, and so has to be packaged by binding to proteins in order to fit inside the nucleus. In eukaryotes, these packaging proteins are called **histones**, and the complex between them and a single DNA molecule makes up a **chromosome**. Unless the cell is dividing, the chromosomes appear simply as an amorphous mass, and individual ones cannot be distinguished, although they are known to occupy their own specific regions within the nucleus, called **chromosome territories**. When the cell divides, the chromosomes become more compact and can now be seen, even with the light microscope (*Fig. 2.9*).

Figure 2.9 **Dividing cells in which the chromosomes are visible.**
Light micrograph of root tip cells from an onion (*Allium* sp.) undergoing cell division (mitosis).
Reproduced with permission from Science Photo Library (Steve Gschmeissner).

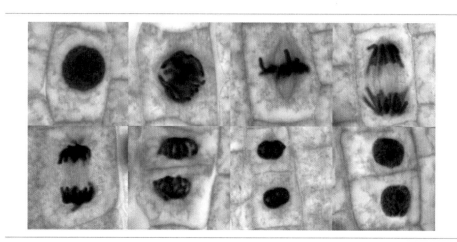

The number of different chromosomes, and hence the number of different DNA molecules, is characteristic of each species, but unrelated to the biological features of the organism (*Table 2.1*). Some unicellular eukaryotes have multiple chromosomes, such as the yeast *Saccharomyces cerevisiae* which has sixteen, while some multicellular organisms have relatively few. The ant *Myrmecia pilosula* has just one chromosome, and the Indian muntjac deer has only four. Humans have 24. Each of these DNA molecules is different – each one carries a different set of genes. Together they make up the organism's **genome**.

Table 2.1. Chromosome numbers for various eukaryotes

Species	Type of organism	Number of chromosomes
Saccharomyces cerevisiae	Yeast	16
Myrmecia pilosula	Ant	1
Agrodiaetus shahrami	Butterfly	135
Gallus gallus	Chicken	40
Arabidopsis thaliana	Plant	5
Muntiacus muntjak	Deer	4
Pan troglodytes	Chimpanzee	25
Homo sapiens	Human	24

As well as being the storeroom for the cell's DNA, the nucleus is also the organelle within which the genes in the DNA molecules are copied into RNA, during the first stage in gene expression. This copying process is called **transcription**, and for certain genes it occurs in distinct regions of the nucleus called the **nucleoli** (singular **nucleolus**). In electron micrographs these appear as dark areas (*Fig. 2.10*) which, when observed at higher magnification, are found to have a fibrillar structure. Careful observation can also sometimes reveal additional small structures inside the nucleus, given names such as Cajal bodies, gems and speckles. Most of these small bodies are involved in transcription or in making structural alterations to RNA molecules before the latter are exported out of the nucleus into the cytoplasm, where protein synthesis, the second stage of gene expression, occurs.

Figure 2.10 Transmission electron micrograph of an animal cell nucleus showing a nucleolus. Nucleus of an acinar cell from the pancreas of the bat, *Myotis lucifugus*. Reproduced with permission from Science Photo Library (Don Fawcett)

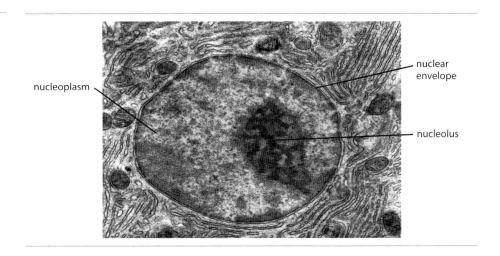

nucleoplasm

nuclear envelope

nucleolus

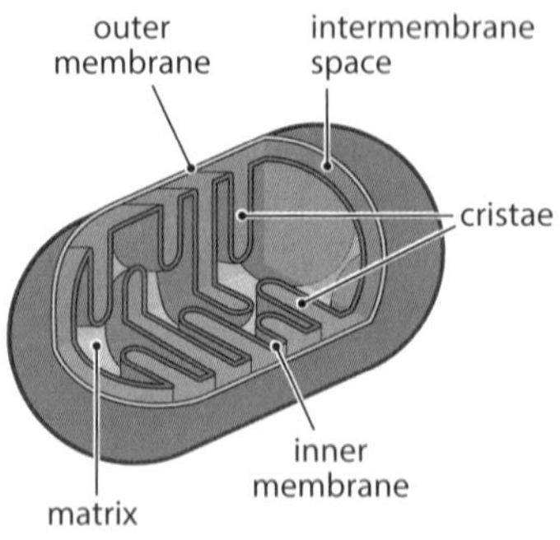

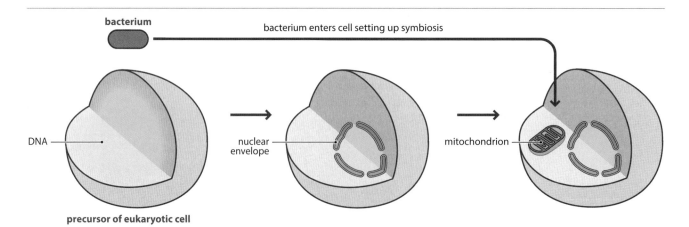

Figure 2.11 **A mitochondrion.**

The biochemical events responsible for energy generation in mitochondria are described in *Chapter 9*.

See *Section 8.1.1* for information on ATP and other energy carrier molecules.

Mitochondria are the energy-generating powerhouses of the cell

After the nucleus, the mitochondria are the most conspicuous organelles when electron micrographs of eukaryotic cells are examined (see *Fig. 2.1*). They are rod-shaped structures, about 1.0 μm in diameter and 2.0 μm in length. Their numbers vary in different cells. Some unicellular eukaryotes have just a single mitochondrion, but most animal cells have 500–2000, making up about 20% of the total cell volume.

Each mitochondrion has two membranes (*Fig. 2.11*). The **outer mitochondrial membrane** forms the external surface of the organelle and gives the structure its rod shape. The **inner mitochondrial membrane** is infolded to form **cristae**, which look like thin fingers in cross-section but in reality are plate-like structures that fill up a large part of the mitochondrion. The area between the two membranes is called the **intermembrane space** and the central region enclosed by the inner membrane is called the **mitochondrial matrix**.

The main function of the mitochondria is to generate energy for the cell. The biochemical reactions that take place in the mitochondria use up oxygen and produce carbon dioxide, and so are referred to as **cellular respiration**. They also result in synthesis of **adenosine 5'-triphosphate** (**ATP**), which is the main energy store for the cell. The synthesis of ATP locks up energy which can subsequently be released when this compound is broken down again. Energy that is generated in the mitochondria can therefore be stored as ATP and transported to other parts of the cell, where the energy is released and used to drive biochemical reactions.

The reactions that result in ATP synthesis involve proteins that are embedded in the inner mitochondrial membrane, with the ATP initially released into the mitochondrial matrix. The infolding of the inner membrane into cristae increases the surface area of the inner membrane, increasing the capacity for ATP synthesis. This means that those cells that have the greatest requirement for energy generation not only have the largest number of mitochondria, but also have mitochondria with the most densely packed cristae. In human liver cells, for example, the surface area of the inner mitochondrial membrane is about five times that of the outer membrane.

There is one final aspect of mitochondria that we should consider. Each mitochondrion contains multiple copies of a small DNA molecule which carries a set of genes that are not present in the chromosomes in the cell nucleus. The number of

Figure 2.12 **The endosymbiont theory for the origin of mitochondria.**
The theory suggests that at an early stage of evolution, the precursor of the eukaryotic cell developed a nucleus in which its DNA was contained. The cell then engulfed a free-living bacterium, setting up a symbiosis. After many cell divisions, the progeny of the bacterium evolved into what we now know as mitochondria.

genes is quite small, just 37 in the human mitochondrial DNA compared with 45 500 in the nucleus, but they specify some important proteins involved in ATP generation. The presence of DNA in mitochondria has led to the suggestion that these organelles are the relics of free-living bacteria that formed a symbiotic association with the precursor of the eukaryotic cell, during the very earliest stages of evolution (*Fig. 2.12*). Support for this **endosymbiont theory** has come from the discovery of organisms which appear to exhibit stages of endosymbiosis that are less advanced than those seen with mitochondria. An example is *Pelomyxa*, a type of amoeba that lacks mitochondria but instead contains symbiotic bacteria, though it is by no means certain that these bacteria provide the amoeba with energy.

Photosynthesis takes place within chloroplasts

The most distinctive biochemical feature of plants is their ability to **photosynthesize**. This is the conversion of sunlight into chemical energy that is stored in carbohydrates such as starch. In plants, photosynthesis takes place in special organelles called **chloroplasts** (*Fig. 2.13*).

We will study photosynthesis in *Chapter 10*.

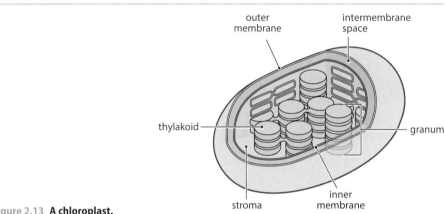

Figure 2.13 **A chloroplast.**

Chloroplasts are larger than mitochondria, about 2.5 μm in diameter and 5.0 μm in length, and less numerous, with there rarely being more than 100 in a single cell. Like mitochondria, they have inner and outer membranes and an inner space called the **stroma**. Unlike mitochondria, within the stroma there is a third membrane system, which forms interconnected structures called **thylakoids**, which are stacked on top of one another like piles of plates to form structures called **grana**. The thylakoids are the actual sites of photosynthesis. They contain pigments such as **chlorophyll** which absorb sunlight, along with the various proteins that convert the absorbed energy into ATP.

Chloroplasts are similar to mitochondria in another way. They too contain their own DNA molecules, usually carrying 200 or so genes, and are thought once to have been free-living bacteria. An early stage in the endosymbiosis that resulted in chloroplasts might be displayed by the protozoan *Cyanophora paradoxa*, whose photosynthetic structures, called cyanelles, are different from chloroplasts and instead resemble bacteria. This alerts us to the fact that plants are not the only organisms able to photosynthesize. Several types of prokaryote are able to carry out photosynthesis. These include the **cyanobacteria**, which contain a blue light-absorbing pigment called phycocyanin, present on a folded membrane structure similar to the thylakoids of plant cells. The cyanelles of *Cyanophora* could well be ingested cyanobacteria. Algae, which are eukaryotes, photosynthesize in a similar manner to plants. They contain chloroplasts and chlorophyll, sometimes with additional pigments which make the cells look red.

The Golgi apparatus and endoplasmic reticulum are involved in protein processing and secretion

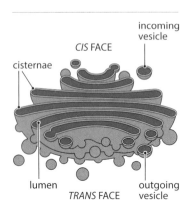

Figure 2.14 **The Golgi apparatus.**

Protein secretion is described in *Section 16.4.*

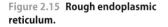

The Golgi apparatus is visible in electron micrographs of eukaryotic cells as stacks of membranous plates called **cisternae** (*Fig. 2.14*). An animal cell typically has about 50 stacks, each comprising 5–10 cisternae. Often in electron micrographs a stack is surrounded by small membrane-bound spheres called **vesicles**. In living cells, these vesicles continually fuse with the cisterna at one side of the stack, called the *cis* face, and bud off the cisterna at the opposite side, the *trans* face. The vesicles that fuse with the stack carry newly synthesized proteins which are transferred from cisterna to cisterna, gradually passing through the stack. During their transport through a stack, the proteins are modified by attachment of new biochemical groups. Many proteins have short chains of sugars added to them, a process called **glycosylation**. The resulting protein–sugar hybrids are called **glycoproteins**. These molecules, along with many other proteins that are processed in the Golgi stacks, are secreted by the cell. When they get to the *trans* face of the stack they enter vesicles that move to the periphery of the cell and fuse with the plasma membrane, so their contents are transferred to the space outside of the cell.

Some secreted proteins stay in the space outside the cell, forming the **extracellular matrix**. This is a fibrous network that provides structure to tissues and carries signals between cells. Other proteins travel away from the cell in which they are made. An example is pepsin, which is made in the **chief cells** of the stomach lining and is secreted into the stomach itself, where it breaks down proteins from the food eaten by the animal. As well as these secreted proteins, the Golgi apparatus also processes proteins that become inserted into the plasma membrane rather than passing through it. Some of these cell membrane proteins help to transport small molecules into and out of the cell, and others act as cell surface receptors, enabling the cell to respond to external signals.

Where do the original protein-containing vesicles, the ones that fuse with the *cis* face of a Golgi stack, come from? The answer is the **rough endoplasmic reticulum**, which is where these proteins are synthesized (*Fig. 2.15*). The rough endoplasmic reticulum is an extensive network of membranous sheets that pervades the entire cell. The 'roughness' comes from the presence of small structures called **ribosomes** on the outer surfaces of the sheets. The ribosomes make the proteins, which are placed inside the endoplasmic reticulum. The vesicles then bud off from the endoplasmic reticulum, carrying the proteins to the Golgi apparatus. There is also **smooth endoplasmic reticulum**, which looks more like tubes than plates and which, as the name implies, does not have any attached ribosomes. The smooth endoplasmic reticulum is not involved in protein synthesis, but instead has a variety of roles including synthesis and storage of some types of lipid.

Figure 2.15 **Rough endoplasmic reticulum.**

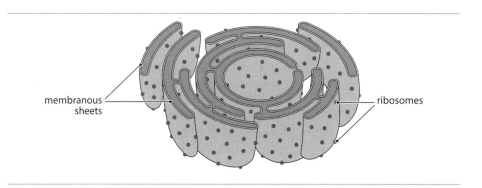

2.1.4 What about viruses?

Earlier in this chapter we stated that all living organisms are made of cells. This is true if we take the view that viruses are not alive. Most biologists would agree with this view, partly because viruses are not made of cells, but mainly because of the nature of the virus life cycle. Viruses are obligate parasites of the most extreme kind. They are able to reproduce only by infecting a host cell, and they make use of many of the host's biomolecules in order to complete their replication cycle. For this reason, most viruses are very specific for a particular host species, because they cannot make use of biomolecules from any other type of organism. Viruses are important in biology because they are responsible for many diseases, such as HIV/AIDS (human immunodeficiency virus infection and acquired immunodeficiency syndrome), some types of cancer, and the common cold. Therefore we must become familiar with their structure and other key features.

Viruses are constructed primarily from two components, protein and nucleic acid. The protein forms a coat or **capsid** within which the nucleic acid, carrying the virus genes, is contained. The nucleic acid is often DNA, as in cellular organisms, but in some viruses it is RNA.

There are three common capsid structures (*Fig. 2.16*):

- **Icosahedral**, in which the individual protein subunits (**protomers**) are arranged into a three-dimensional geometric structure that surrounds the nucleic acid. Despite the name, many viruses of this type have so many protomers in their capsids that they appear spherical. Examples of icosahedral viruses are the human herpes and polio viruses.

- **Filamentous**, in which the protomers are arranged in a helix, producing a rod-shaped structure. Tobacco mosaic virus is an example.

- **Head-and-tail**, which is found only in **bacteriophages**, viruses that infect bacteria. This structure comprises an icosahedral head, containing the nucleic acid, and a filamentous tail which facilitates entry of the nucleic acid into the bacterium. There may also be other components, such as the 'legs' possessed by the *E. coli* bacteriophage T4.

With some eukaryotic viruses, the capsid is surrounded by a membrane, derived from the host when the new virus particle leaves the cell, and possibly modified by insertion of virus-specific proteins.

Viruses are much smaller than cells, with most icosahedral ones being 25–250 nm in diameter. There are different types of virus life cycle, but the basic strategy is the same for all (*Fig. 2.17*). The virus attaches to the outer surface of the cell, and either

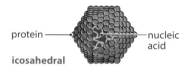

protein — nucleic acid

icosahedral

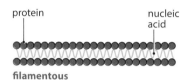

protein — nucleic acid

filamentous

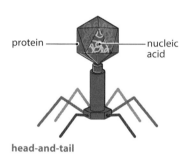

protein — nucleic acid

head-and-tail

Figure 2.16 Three common types of virus capsid structure.

Box 2.5 **Unusual types of infectious particles**

Viruses are not the only type of non-cellular infectious particle. Biologists are aware of four types of **subviral particles**:

- **Satellite RNA viruses** and **virusoids** are short RNA molecules. They do not carry genes for making any capsid proteins, and instead move from cell to cell within the capsids of helper viruses. A satellite RNA virus shares the capsid with the genome of the helper virus whereas a virusoid RNA molecule becomes encapsidated on its own. They are found mainly in plants.

- **Viroids** are small RNA molecules that contain no genes. They do not become encapsidated, spreading between host cells as naked DNA. Again, they are mainly found in plants.

- **Prions** are infectious, disease-causing particles that are made solely of protein. They are responsible for scrapie in sheep and goats, and Creutzfeldt–Jakob disease in humans. A prion protein exists in two forms. The normal version is found in the brains of mammals although its function is unknown. The infectious version has a slightly different structure that forms fibrillar aggregates that are seen in infected tissues. Once inside a cell, infectious molecules are able to convert the normal proteins into the infectious form. Transfer of one or more of the infectious proteins to a new animal therefore results in accumulation of additional infectious proteins in the brain, resulting in disease transmission.

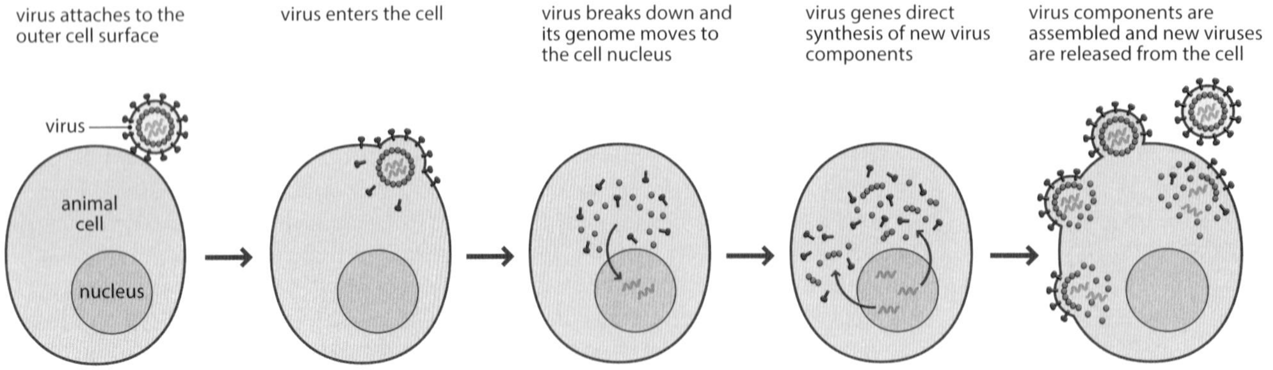

virus attaches to the outer cell surface

virus enters the cell

virus breaks down and its genome moves to the cell nucleus

virus genes direct synthesis of new virus components

virus components are assembled and new viruses are released from the cell

virus

animal cell

nucleus

Figure 2.17 A typical virus life cycle.
In this example, the infecting virus enters the cell. Other viruses inject their DNA or RNA across the cell membrane.

enters the cell itself or transfers its DNA or RNA genome across the cell membrane. The virus genes might become active immediately, subverting the host cell's DNA replication and protein synthesis processes to make copies of the virus genome and capsid proteins from which new viruses are assembled. These new viruses are released from the cell, possibly causing the cell to burst, and go on to infect other cells. Alternatively the virus genome might remain in the cell for some time, possibly years, either as an independent DNA or RNA molecule or inserted into one of the host's chromosomes. The cell might be forced to make new viruses continually, or the virus might be entirely quiescent until suddenly restarting its infection cycle and making a new set of viruses.

2.2   Evolution and the unity of life

We have seen that all living organisms, despite their vast diversity, can be divided into two groups based on the structures of their cells. But when we look more closely at prokaryotes and eukaryotes, particularly when we examine their biochemistries, we see that both types of organism are built according to the same general plan. At the molecular, as at the cellular level, there are distinctive features that distinguish prokaryotes and eukaryotes, but their basic designs are the same. All living organisms use DNA to store genetic information, and that information is organized into genes and read by the cell in very similar ways. The processes by which energy is generated are very similar as are the pathways for synthesis of proteins, nucleic acids, lipids and carbohydrates. Even the distinctive features of the archaea are merely modifications of the same plan.

The underlying similarities shared by all living organisms tell biologists two things. First, all of life descended from a single origin. Secondly, since that origin, evolutionary processes have resulted in a diversification of species. To complete our survey of the biological context to biochemistry, we will examine the origin of life and its subsequent evolution.

2.2.1   Life originated four billion years ago

According to cosmological theory, the universe as we know it is thought to be about 13.7 billion years old. At first, the universe was little more than a featureless, expanding cloud of gas, but after about 4 billion years galaxies began to form. Within those galaxies, condensation of gases formed stars and planets. Our own sun and solar system originated in this way about 4.6 billion years ago.

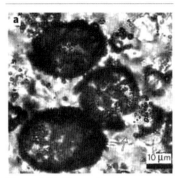

Figure 2.18 The earliest known fossils.
These microfossils, which are thought to be the remains of some of the earliest prokaryotic cells, were found in 3.4 billion year old rocks in Western Australia.
Photo reproduced with permission from Macmillan Publishers Ltd: Wacey *et al.* (2011) Microfossils of sulphur-metabolizing cells in 3.4-billion-year-old rocks of Western Australia. *Nature Geoscience*, **4**: 698.

To begin with, the Earth was covered in water, the first land masses not appearing for another billion years. During this period the first cells evolved in the planetary ocean. We believe this to be the case because tiny microfossils of structures resembling bacteria have been discovered in 3.4 billion year old rocks from Australia (*Fig. 2.18*). The very early fossil record is difficult to interpret because it is easy to mistake a natural microscopic feature in a rock for the fossilized remains of a cell, but these and other discoveries have convinced scientists that living cells evolved during the first billion years of the Earth's history. How could this happen?

The answer lies with the chemical composition of the early Earth's atmosphere. This was very different from today's atmosphere, mainly because the oxygen content was much lower, and indeed remained low until the first photosynthetic organisms appeared many millions of years later. Methane and ammonia were probably the most abundant gases in the early atmosphere, and the ocean would also have contained these gases in dissolved form. Experiments that recreate these early chemical conditions have shown that electrical discharges, lightning for example, in a mixture of methane, ammonia, hydrogen and water vapor can result in spontaneous synthesis of several types of amino acid, the monomeric units from which proteins are constructed (*Fig. 2.19*). Other products including hydrogen cyanide and formaldehyde are also formed, which can initiate further chemical reactions that give rise to more types of amino acids, and to purines and pyrimidines, which form parts of the nucleotides from which nucleic acids are made. It is possible that sugars were also made in this way. Sugars are particularly important because they are the substrates for the energy-generating pathways that evolved in the first cells.

Many of the compounds that make up the polymeric biomolecules found in living organisms could therefore have been created by chemical reactions occurring in the atmosphere and the ocean of the early Earth. Assembly of the polymeric molecules might have been possible in the ocean, or might have occurred between monomers that had become absorbed onto sediments at the ocean bottom. The latter theory is attractive because the concentrations of the building blocks in the ocean itself might have been quite low, so some way of bringing the molecules together would have

Figure 2.19 An experiment recreating the chemical conditions of the early atmosphere.
In this experiment, carried out by Stanley Miller and Harold Urey in 1952, a mixture of methane, ammonia, hydrogen and water vapor (the last of these from the boiling water) was subjected to electrical discharges simulating lightning. The products were collected by passing the gaseous mixture through the condenser. Analysis of the resulting solution revealed the presence of two amino acids, glycine and alanine. Re-analysis of the solution in 2007, using more sensitive techniques, detected over 20 amino acids.

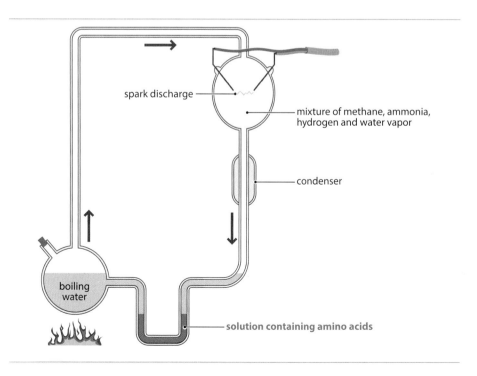

been needed. Absorption onto solid particles would have achieved this (*Fig. 2.20*). Alternatively, polymerization might have been promoted by the repeated condensation and drying of droplets of water in clouds.

Figure 2.20 One possible way in which polymerization of biomolecules might have occurred in the ocean of the early Earth.

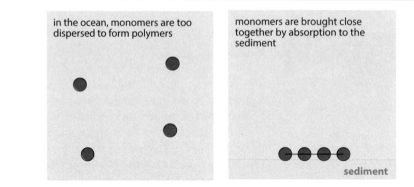

in the ocean, monomers are too dispersed to form polymers

monomers are brought close together by absorption to the sediment

sediment

Synthesis of polymeric biomolecules is, of course, just the first step in the evolution of cellular life. The biomolecules must be assembled, in appropriate proportions, within lipid-bound structures, and these structures must acquire the ability to generate energy and to reproduce. Current thinking is that the first cellular biochemical systems comprised self-replicating RNA molecules enclosed in lipid vesicles. These RNA molecules might have evolved the ability to specify protein synthesis, which is the main function of RNA in today's world (*Fig. 2.21*). Gradually these early cells could have built up a collection of useful proteins, possibly including ones able to break down sugars with the release of energy. At some point the information for making those proteins was transferred from the self-replicating RNA molecules to DNA polymers that formed the primordial genomes.

Figure 2.21 An early type of cellular biochemical system?
Self-replicating RNA molecules that have evolved the ability to polymerize amino acids into proteins have become enclosed in a lipid vesicle.

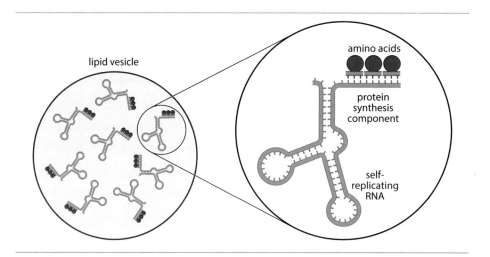

lipid vesicle

amino acids

protein synthesis component

self-replicating RNA

2.2.2 Three and a half billion years of evolution

In the early nineteenth century, biologists became increasingly aware of the similarities between the body plans of different animals. It was realized that structures as seemingly dissimilar as a human arm, a whale's flipper and the wing of a bird are constructed from the same set of bones (*Fig. 2.22*). This led some far-sighted scientists to suggest that related species have evolved from a common ancestor, and gradually the notion

MILLION
YEARS

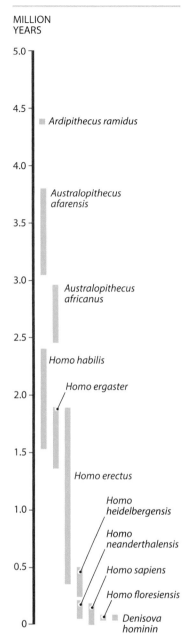

Figure 2.23 **A timeline for human evolution.**

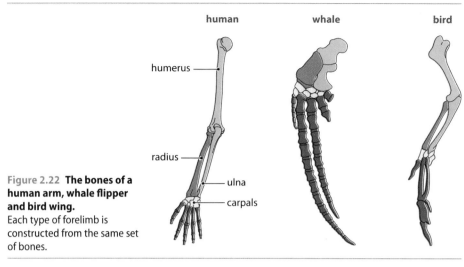

Figure 2.22 **The bones of a human arm, whale flipper and bird wing.**
Each type of forelimb is constructed from the same set of bones.

of a 'tree of life' became popular. All species alive today, and all those known from the fossil record, were therefore linked together in one great evolutionary scheme.

If we follow the evolutionary record forward in time from the early microfossils of 3.4 billion years ago, we see a gap of some 2 billion years before the first eukaryotic cells appear. These are structures resembling single-celled algae, which probably included several types that had evolved the ability to photosynthesize, increasing the oxygen content of the atmosphere and leading to conditions not too different from those we know today. Multicellular algae first appear in the fossil record about 900 million years ago, and multicellular animals 640 million years ago, although there are enigmatic burrows suggesting that animals lived earlier than this. The Cambrian Revolution, when invertebrate life proliferated into many novel forms, occurred 530 million years ago, and ended with the disappearance of many of those novel forms a few million years later. Since then, evolution has continued apace and with increasing diversification, punctuated by mass extinctions of differing scales. The first terrestrial insects, animals and plants were established by 350 million years ago, the dinosaurs had been and gone by the end of the Cretaceous period, 65 million years ago, and by 50 million years ago mammals had become the dominant type of animal on the planet.

What about our own species? Bishop Samuel Wilberforce once famously asked Thomas Huxley, one of Charles Darwin's supporters, if his descent from a monkey was on his mother or father's side. The answer is both, humans and chimpanzees being descended from a common ancestor that lived around 6.0 million years ago. Since this split, the evolutionary lineage leading to *Homo sapiens* has run through a series of species, gradually developing the attributes that we look on as distinctively human (*Fig. 2.23*). The ability to walk upright, as opposed to the knuckle walking locomotion of chimpanzees, was first possessed by *Ardipithecus ramidus*, who lived in east Africa 4.4 million years ago. The first stone tools were manufactured about 2.5 million years ago by *Homo habilis*, the earliest member of our own genus. *Homo sapiens* first appeared in Africa some 195 000 years ago and gradually spread across the globe. We do not know when speech evolved, but humans started sculpting stone for artistic rather than utilitarian purposes about 50 000 years ago, and at the same time might have been making music. And about 130 years ago, humans first began to study biochemistry.

Box 2.6 Mass extinction events

Many of the novel invertebrate species that evolved during the Cambrian period disappeared as a result of a series of mass extinctions, dated to 517, 502 and 485 million years ago. The fossil evidence for these periods is too sparse for geologists to be able to identify the reasons for the extinctions, but oxygen depletion in the oceans could have been a factor – at this time virtually all species lived in the sea.

Since the Cambrian period, there have been five major mass extinctions, each resulting in a dramatic change in the dominant groups of species in the oceans and on the land:

- The **Ordovician–Silurian event**, 443 million years ago, resulted in extinction of 65% of the species alive at that time. It occurred during a period of global cooling, which resulted in extensive glaciation, reduction in sea levels, and changes in the ocean chemistry.
- The **Late Devonian mass extinctions** occurred during a lengthy period 375–355 million years ago, and resulted in loss of 70% of species. The cause is not known, but there may have been a series of events including an asteroid impact.
- The **Permian mass extinction**, 250 million years ago, is the largest known event, with only 4% of species surviving. Again, multiple factors are likely to have been responsible, one possibility being increased volcanic activity that released poisonous gases into the atmosphere.
- The **Triassic–Jurassic event**, 200 million years ago, is also ascribed to multiple events, possibly a combination of asteroid impact and volcanic activity. About 75% of species disappeared.
- The **Cretaceous–Tertiary mass extinction** is the most famous event, as it resulted in extinction of the dinosaurs. It occurred 66 million years ago, and is widely ascribed to an asteroid impact that resulted in the Chicxulub crater in the Gulf of Mexico. Again, approximately 75% of species are thought to have died out.

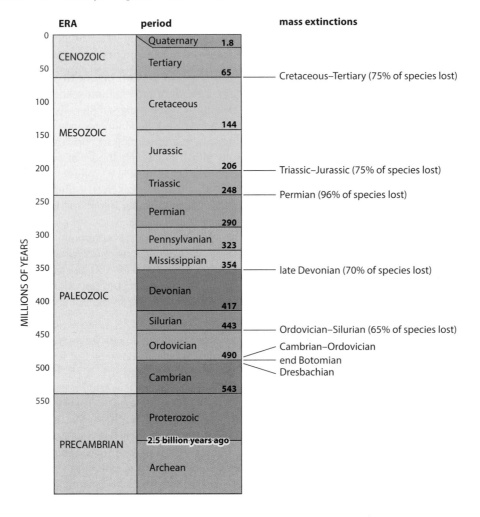

Although catastrophic, mass extinctions are looked on as important acceleration points for the evolution of life. By removing so many species, a mass extinction enables those species that survive to diversify in order to occupy the ecological niches that have become vacant. The best example is the extinction of the dinosaurs during the Cretaceous–Tertiary event, which facilitated the subsequent evolution of mammals. During the Cretaceous period, the dinosaurs were the dominant terrestrial species, and mammals were much less common. However, many of the smaller mammals survived the mass extinction, and were then able to take over the terrestrial niches previously occupied by the dinosaurs.

Further reading

Aguzzi A and Lakkaraju AKK (2016) Cell biology and prions and prionoids: a status report. *Trends in Cell Biology* **26**, 40–51.

Benton ML (2015) *When Life Nearly Died: the greatest mass extinction of all time.* Thames and Hudson, London.

Butler PJG and Klug A (1978) The assembly of a virus. *Scientific American,* **239**(5), 52–9. *Describes the basic features of virus structure.*

Cela-Conde CJ and Ayala FJ (2007) *Human Evolution: trails from the past.* Oxford University Press, Oxford.

Diener TO (1984) Viroids. *Trends in Biochemical Sciences* **9**, 133–6.

Donlan RM (2002) Biofilms: microbial life on surfaces. *Emerging Infectious Diseases* **8**, 881–90.

Macnab RM (1999) The bacterial flagellum: reversible rotary propellor and type III export apparatus. *Journal of Bacteriology* **181**, 7149–53.

Paleos CM (2015) A decisive step towards the origin of life. *Trends in Biochemical Sciences* **40**, 487–8. *Summarizes recent research into this topic.*

Sifri CD (2008) Quorum sensing: bacteria talk sense. *Clinical Infectious Diseases* **47**, 1070–6.

Soto C (2011) Prion hypothesis: the end of the controversy. *Trends in Biochemical Sciences* **36**, 151–8.

Ursell LK, Metcalf JL, Parfrey LW and Knight R (2012) Defining the human microbiome. *Nutrition Reviews* **70**, S38–44.

Ward BB (2002) How many species of prokaryotes are there? *Proceedings of the National Academy of Sciences USA* **99**, 10234–6.

Woese CR and Fox GE (1977) Phylogenetic structure of the prokaryotic domain: the primary kingdoms. *Proceedings of the National Academy of Sciences USA* **74**, 5088–90. *One of the first descriptions of the distinction between bacteria and archaea.*

Zimorski V, Ku C, Martin WF and Gould SB (2014) Endosymbiotic theory for organelle origins. *Current Opinion in Microbiology* **22**, 39–48.

Self-assessment questions

Multiple choice questions

Only one answer is correct for each question. Answers can be found on the website: www.scionpublishing.com/biochemistry

1. Approximately how many species of animals, plants and fungi are there believed to be on the planet?
   (a) 8.7 million
   (b) 100 million
   (c) 8.7 billion
   (d) 100 billion

2. What is the approximate length of a cell of the single-celled organism *Paramecium*?
   (a) 1.2 μm
   (b) 12 μm
   (c) 120 μm
   (d) 120 mm

3. Which one of the following statements about the nucleoid is **incorrect**?
   (a) A nucleoid contains DNA
   (b) A nucleoid is a lightly staining region in a prokaryotic cell
   (c) A nucleoid is surrounded by a membrane
   (d) Nucleoids are visible when cells are observed with the electron microscope

4. Which of these is an example of a prokaryote that forms chains of cells?
   (a) *Anabaena*
   (b) *E. coli*
   (c) *Mycobacterium*
   (d) *Vibrio*

5. Which one of the following statements regarding the archaea is **correct**?
    (a) Archaea are a type of eukaryote
    (b) Many of the environments in which they live are hostile to other forms of life
    (c) Their cells look very different to those of bacteria
    (d) They include the species *Mycobacterium tuberculosis* and *Vibrio chlolerae*

6. Which of the following describes the typical cell shapes for prokaryotes?
    (a) Bacillus, coccus and flagellum
    (b) Bacillus, coccus and spirillum
    (c) Bacillus, coli and spirillum
    (d) Icosahedral and filamentous

7. Peptidoglycan, present in bacterial cell walls, is a type of which of these structures/compounds?
    (a) DNA
    (b) Lipid
    (c) Membrane
    (d) Polysaccharide

8. Which one of the following statements is **incorrect** regarding biofilms?
    (a) Biofilms can form inside medical implements such as catheters
    (b) The bacteria in a biofilm are usually embedded in a slimy matrix
    (c) Only bacteria with flagella are able to form a biofilm
    (d) Within a biofilm, bacteria communicate with one another by quorum sensing

9. Which one of the following statements is **correct** regarding human cells?
    (a) There are 10^{10} cells present in an adult human made up of over 400 specialized types
    (b) There are 10^{10} cells present in an adult human made up of over 1000 specialized types
    (c) There are 10^{13} cells present in an adult human made up of over 400 specialized types
    (d) There are 10^{13} cells present in an adult human made up of over 1000 specialized types

10. Pore complexes are a feature of which type of eukaryotic organelle?
    (a) Chloroplasts
    (b) Golgi apparatus
    (c) Mitochondria
    (d) Nuclei

11. Within the nucleus, which specific regions do chromosomes occupy?
    (a) Cajal bodies
    (b) Nucleoli
    (c) Speckles
    (d) Territories

12. The inner mitochondrial membrane is infolded to form plate-like structures called what?
    (a) Cristae
    (b) Stroma

(c) Thylakoids
(d) The mitochondrial matrix

13. The proteins responsible for ATP synthesis are located in which part of a mitochondrion?
    (a) The outer mitochondrial membrane
    (b) The intermembrane space
    (c) The inner mitochondrial membrane
    (d) The mitochondrial matrix

14. What are the stacks of thylakoid in a chloroplast called?
    (a) Chlorophyll
    (b) Cisternae
    (c) Grana
    (d) Stroma

15. What are the stacks of membranous plates that make up the Golgi apparatus called?
    (a) Cisternae
    (b) Grana
    (c) Lumen
    (d) Vesicles

16. Glycosylation, which occurs in the Golgi apparatus, is best described by which of the following statements?
    (a) Addition of short chains of sugars to membranes in the Golgi
    (b) Addition of short chains of sugars to some proteins
    (c) Removal of sugar units from polysaccharides such as starch
    (d) None of the above

17. The vesicles that fuse with the *cis* face of the Golgi apparatus come from where?
    (a) Mitochondria
    (b) Other Golgi structures
    (c) Rough endoplasmic reticulum
    (d) Smooth endoplasmic reticulum

18. What is the protein that forms a coat around a virus called?
    (a) Bacteriophage
    (b) Capsid
    (c) Protomer
    (d) Virusoid

19. Which one of the following is not a feature of a prion?
    (a) A prion is made solely of protein
    (b) Infectious prions can convert the normal version into the infectious form
    (c) Normal versions of prions are found in the brains of mammals
    (d) The structures of the infectious and normal versions of a prion are indistinguishable

20. Tiny microfossils of structures resembling bacteria have been discovered in rocks of what age?
    (a) 3.4 million years
    (b) 34 million years
    (c) 340 million years
    (d) 3.4 billion years

21. The experiment carried out by Miller and Urey in 1952 resulted in synthesis of which type of biochemical compound from methane, ammonia, hydrogen and water vapor?
  (a) Amino acids
  (b) Nucleotides
  (c) RNA
  (d) Sugar

22. When did the first multicellular algae appear in the fossil record?
  (a) 3.4 billion years ago
  (b) 900 million years ago
  (c) 640 million years ago
  (d) 530 million years ago, during the Cambrian revolution

23. What is the mass extinction that resulted in the extinction of the dinosaurs called?
  (a) Cambrian revolution
  (b) Cretaceous–Tertiary event
  (c) Permian extinction
  (d) Triassic–Jurassic event

24. When did *Homo sapiens* first appear in Africa?
  (a) 4.4 million years ago
  (b) 2.5 million years ago
  (c) 195 000 years ago
  (d) At 18.00 hours on 22 October 4004 BC

Short answer questions

These do not require additional reading.

1. Describe the main differences in cell structure that are seen when prokaryotic and eukaryotic cells are examined with an electron microscope.

2. What are the similarities and differences between bacteria and archaea?

3. Explain what is meant by the term 'cell envelope' when referring to prokaryotic cell structure, and describe the components of the cell envelope.

4. Distinguish between the functional roles of bacterial flagella, pili and fimbriae.

5. Describe the structure of the nucleus of a eukaryotic cell. What similarities, if any, are there between the structures of a eukaryotic nucleus and prokaryotic nucleoid?

6. Compare the structures of mitochondria and chloroplasts, indicating components of the two organelle types that have similar functions. According to the endosymbiont theory, what is the origin of these organelles?

7. Outline the role of the Golgi apparatus.

8. Distinguish between the different types of capsid structure displayed by viruses.

9. What are the current theories regarding the processes by which the first polymeric biomolecules evolved on the early Earth?

10. How has evolution on the planet been affected by mass extinctions?

Self-study questions

These questions will require calculation, additional reading and/or internet research.

1. At one time the archaea were looked on as a primitive type of prokaryote that survived mainly in extreme environments such as hot springs and high-salt lakes. To what extent is this view supported by our current knowledge of bacteria and archaea?

2. It has been claimed that the rotary motor component of the bacterial flagellum is so complex, so specialized, and so unrelated to other structures in living cells, that it could not have arisen by evolution and provides evidence of intelligent design. Discuss this proposal.

3. The presence of DNA molecules in mitochondria and chloroplasts has led to the suggestion that these organelles are the relics of free-living bacteria that formed a symbiotic association with the precursor of the eukaryotic cell. But these DNA molecules contain, at most, a few hundred genes, compared to thousands of genes in a typical bacterium. Does this discrepancy in gene number suggest that the endosymbiont theory is incorrect?

4. Can viruses be considered a form of life?

5. In the early nineteenth century, biologists realized that structures as seemingly dissimilar as a human arm, a whale's flipper and the wing of a bird are all constructed from the same set of bones. Can this principle of homology be extended to biomolecules?

CHAPTER 3

Proteins

STUDY GOALS

After reading this chapter you will:

- understand that proteins are composed of amino acids, and know the general structure of an amino acid

- appreciate that the variability of the amino acid side-chains enables proteins with different biochemical properties to be constructed

- be able to discuss the key structural and chemical features of amino acids: enantiomeric pairs, ionization properties and polarity

- be aware of the range of modifications to amino acid structure that can occur after protein synthesis

- recognize the differences between the terms 'primary', 'secondary', 'tertiary' and 'quaternary' when referring to protein structure

- know the structure of the peptide group and the importance of the *psi* and *phi* bond angles in determining the conformation of a polypeptide

- be able to describe the α-helix and β-sheet secondary structures

- understand the structural features of fibrous and globular proteins, and be able to describe examples of both types

- recognize that quaternary structure involves the association of two or more polypeptides, and be able to describe the key features of examples of proteins with quaternary structure

- know how proteins fold and appreciate that the structures taken up are determined by the amino acid sequence

- be able to explain the link between the chemical variability of proteins and the range of different roles that proteins play in living organisms

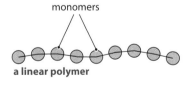
monomers

a linear polymer

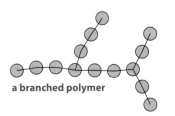

a branched polymer

Figure 3.1 **Linear and branched polymers.**

In the previous chapter we examined the diversity of living organisms and the important features of prokaryotic and eukaryotic cells. Now we must look more closely at the biomolecules within those cells.

There are four types of biomolecule: proteins, nucleic acids, lipids and polysaccharides. Each has a polymeric structure, built up by linking the monomeric units together in linear or branched chains (*Fig. 3.1*). The chemical features of the monomeric units are quite different, giving each type of biomolecule its own distinctive properties. As we will see, these properties underlie the specific roles that biomolecules play in living cells. We begin with proteins.

3.1　Proteins are made of amino acids

In proteins, the monomeric units are amino acids. These are linked together to form unbranched chains called **polypeptides**. Most polypeptides are a few hundred amino acids in length, although the shortest have less than 50 amino acids (these are more correctly called **peptides**) and the longest one known, a human muscle protein called titin, has 33 445 amino acids (*Table 3.1*).

Table 3.1. **Examples of human proteins**

Protein	Number of amino acids	Function
Sarcolipin	31	Calcium ion transport into muscle cells
Somatotropin	51	Growth hormone
Ribonuclease A	124	Breakdown of RNA molecules
Carbonic anhydrase	130	Removal of carbon dioxide from tissues
β-globin	146	Component of hemoglobin, which carries oxygen in the bloodstream
Myoglobin	154	Utilization of oxygen by muscle tissue
Tissue plasminogen activator	527	Part of the blood clotting system
Hsp70	641	Molecular chaperone, helps other proteins adopt their correct structures
Keratin type II	644	Component of hair and cytoskeleton
Type 1 collagen	1464	Component of tendons, ligaments and bones
Dystrophin	3685	Part of the internal skeleton of muscle cells
Titin	33 445	Structural component of muscle

3.1.1　Twenty different amino acids are used to make proteins

Table 3.2. **Amino acids**

Amino acid	Abbreviation Three-letter	One-letter
Alanine	Ala	A
Arginine	Arg	R
Asparagine	Asn	N
Aspartic acid	Asp	D
Cysteine	Cys	C
Glutamic acid	Glu	E
Glutamine	Gln	Q
Glycine	Gly	G
Histidine	His	H
Isoleucine	Ile	I
Leucine	Leu	L
Lysine	Lys	K
Methionine	Met	M
Phenylalanine	Phe	F
Proline	Pro	P
Serine	Ser	S
Threonine	Thr	T
Tryptophan	Trp	W
Tyrosine	Tyr	Y
Valine	Val	V

Polypeptides contain mixtures of 20 different amino acids (*Table 3.2*). Biochemists always use the common names for these amino acids, most of which were assigned when the individual amino acids were first discovered. Asparagine, for example, is named after asparagus, because it was first extracted from asparagus leaves, way back in 1806. Its full chemical name is 2-amino-3-carbamoylpropanoic acid. Each amino acid is also given a three-letter and one-letter abbreviation. The three-letter abbreviations are easy to remember as most are simply the first three letters of the full name, The exceptions are 'trp' for tryptophan (it is not 'try' because this might be confused with 'tyr' for tyrosine) and 'asn' and 'gln' for asparagine and glutamine ('asp' and 'glu' are used for aspartic acid and glutamic acid).

The one-letter abbreviations can be more difficult to learn. Eleven of these are the first letter of the full name (e.g. A for alanine), but nine are not, because there are not enough initial letters to go round. For two amino acids, the second letter of the name is used (R for arginine, and Y for tyrosine), but for the others the logic is more quirky. Phenylalanine is abbreviated to F because it sounds like it starts with that letter. Tryptophan is W, supposedly because a person with a lisp would call it twptophan. These abbreviations were assigned by Dr Margaret Oakley Dayhoff in the early 1960s, and have a very important purpose. A key feature of a protein is the order, or **sequence**, of amino acids in its polypeptide chain. A polypeptide 300 amino acids in length might therefore have the sequence methionine–glycine–alanine–leucine–glycine– followed by another 295 amino acids. If we wish to enter this sequence into a computer (for example, to compare it with the sequence of a related protein) then typing out the full names of the amino acids, or even the three-letter abbreviations (met–gly–ala–leu–gly–) would be very time-consuming. So we abbreviate the sequence to MGALG…, which can be typed in more quickly. This is exactly why Dayhoff devised the one-letter

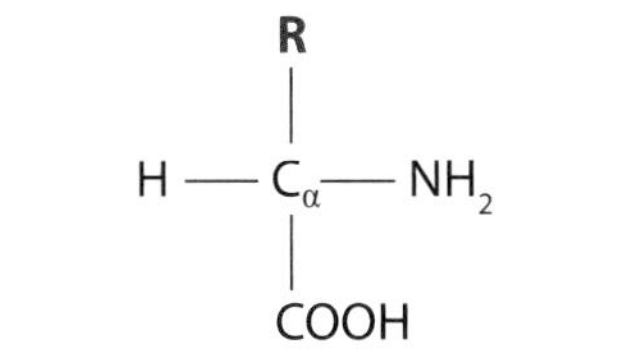

Figure 3.2 The general structure of an amino acid.
The central C is called the α-carbon.

abbreviations. She was one of the first **bioinformaticians** and the first person to use computers to study protein sequences.

Each amino acid has the same general structure (*Fig. 3.2*). This comprises a central carbon atom, called the α-carbon, to which four chemical groups are attached. These are a hydrogen atom (–H), a carboxyl group (–COOH), an amino group (–NH₂), and the **R group** or side-chain, which is different for each amino acid. The R groups vary greatly in complexity. For glycine, the R group is simply a hydrogen atom, whereas for phenylalanine, tryptophan and tyrosine they are large organic structures (*Fig. 3.3*). Note that proline has an unusual side-chain that includes the nitrogen of the amino group attached to the α-carbon. Because of this unusual structure, proline can introduce a kink into a polypeptide chain.

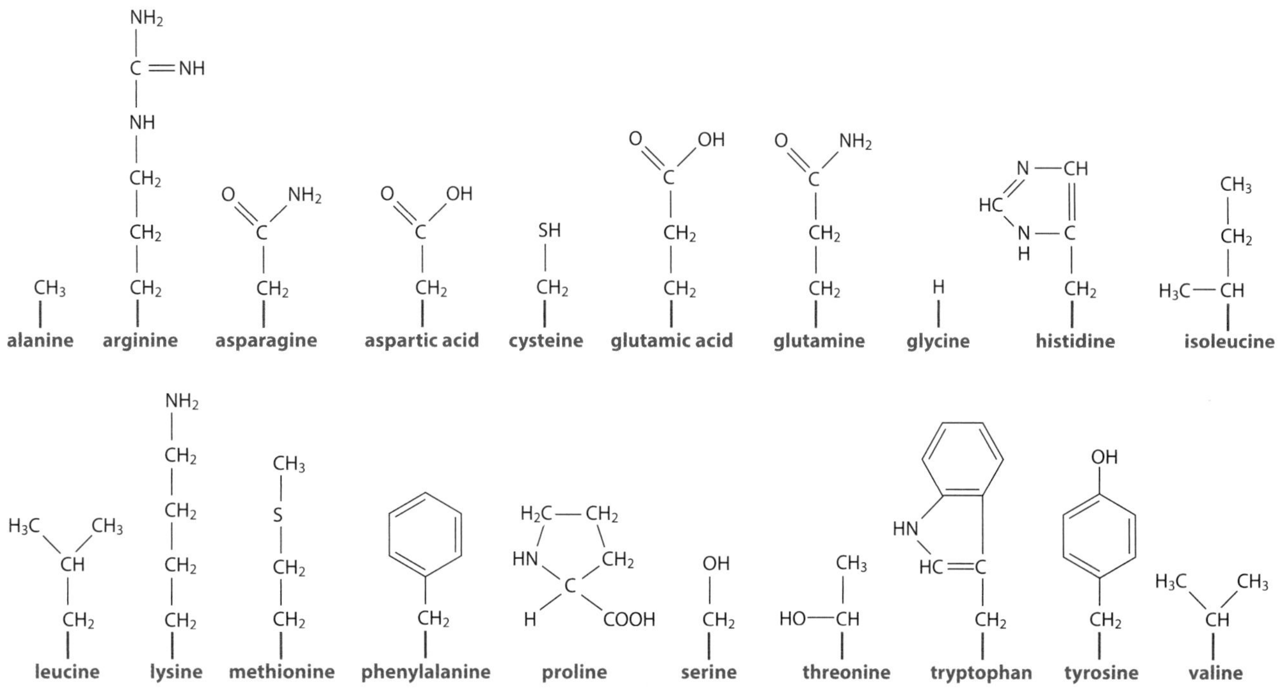

Figure 3.3 The structures of the amino acid R groups.
Note that the entire structure of proline is shown, not just its R group. This enables you to see the unusual structure of proline, in which the R group forms a bond not just to the α-carbon but also with the amino group attached to this carbon.

3.1.2 The biochemical features of amino acids

The differences between the side-chains mean that although all amino acids have the same general structure, each has its own specific chemical properties. This fact is of fundamental importance in biochemistry because it means that by combining amino acids together in different sequences, proteins with vastly different chemical features can be constructed. Later in this chapter we will examine how these different chemical features enable proteins to play a broad range of roles in living cells. First, we must understand the properties of amino acids in more detail.

There are L- and D-forms of each amino acid

The first feature of amino acids that we must consider is their precise structure. Although shown as a flat drawing in *Figure 3.2*, in reality each amino acid has a three-dimensional configuration. To understand this configuration we must consider the way in which chemical bonds are orientated around a carbon atom. Carbon has a **valency** of four, and so can form four single bonds. These bonds have a tetrahedral arrangement. This means that there are two versions of an amino acid, differing in the positioning of the four groups around the carbon (*Fig. 3.4A*). These two versions are mirror images and so are genuinely different, and it is not possible to go from one configuration to the other simply by rotating the molecule.

Figure 3.4 D- and L-isomers.
The D- and L-enantiomers of (A) an amino acid, and (B) glyceraldehyde, are shown. Note that each pair of enantiomers are mirror images and are not superimposable.

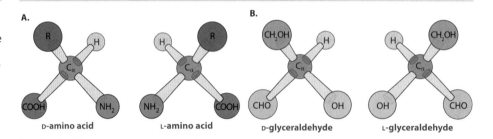

D-amino acid   L-amino acid   D-glyceraldehyde   L-glyceraldehyde

Two molecules that have identical chemical compositions but different structures are called **isomers**. If the isomers are mirror images, as with the alternative forms of an amino acid, then the molecules are said to be **optical isomers** or **enantiomers**.

The two enantiomers of an amino acid are called the L- and D-forms. The reason for using this designation is as follows. The first enantiomer to be studied was glyceraldehyde, which comprises a central carbon atom attached to a hydrogen, hydroxyl (–OH), formyl (–CHO) and hydroxymethyl (–CH₂OH) group (*Fig. 3.4B*). Like all pairs of enantiomers, the two versions of glutaraldehyde have identical chemical composition and special techniques have to be used to distinguish between them. One method is to shine plane-polarized light through a solution of the compound (*Fig. 3.5*). One enantiomer of glyceraldehyde rotates the plane of the light to the left, and the other rotates it to the right. The former was called laevo- or L-glyceraldehyde (*laevus* is Latin for 'left') and the latter dextro- or D-glyceraldehyde (*dexter* is Latin for 'right'). A carbon atom that is attached to four different groups, as in glyceraldehyde, is said to be **chiral**, from the Greek for 'hand'.

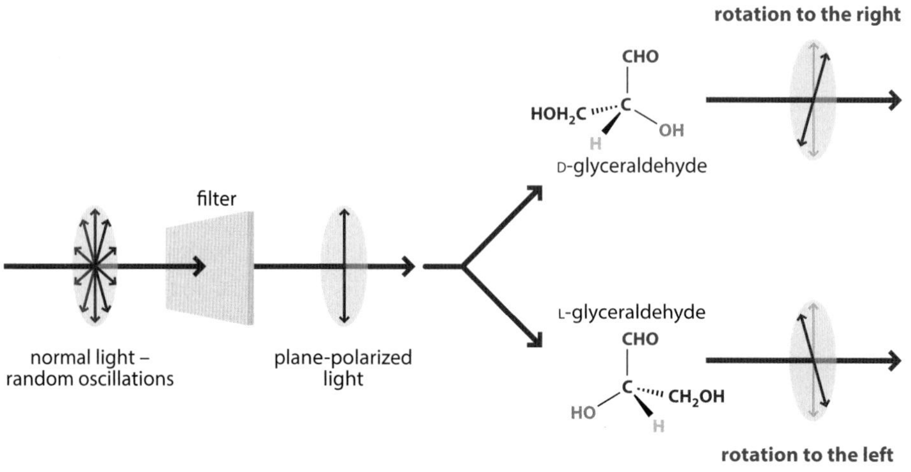

Figure 3.5 Distinguishing between the D- and L-forms of glyceraldehyde.
In normal light the waves oscillate randomly in all directions. Passing light through a special type of filter leaves only those oscillations in a single plane. When this plane-polarized light passes through a solution of an enantiomer the plane is rotated either to the right, as shown in the upper part of the diagram, or to the left, as shown below. D-glyceraldehyde causes rotation to the right, and L-glyceraldehyde causes rotation to the left.

rotation to the right

CHO
HOH₂C ''''C
            OH
       H
D-glyceraldehyde

filter

normal light –
random oscillations

plane-polarized
light

L-glyceraldehyde
CHO
    C '''''CH₂OH
HO
     H

rotation to the left

Box 3.1 Are there two versions of every amino acid?

The definition of a chiral molecule is one that is attached to four different groups and so has a non-superimposable mirror image. In this case the carbon is said to be a chiral carbon. Not all amino acids, however, have chiral carbon atoms because there is one in which the central α-carbon is not attached to four different groups. This is the simplest of the amino acids, glycine, whose R group is a hydrogen atom. In glycine the four groups attached to the α-carbon are a carboxyl group (–COOH), an amino group (–NH$_2$), a hydrogen atom (–H), and a second hydrogen atom (–H). Glycine is therefore not chiral and does not exist as a pair of optical isomers. Glycine is just glycine and there is no such thing as D- or L-glycine.

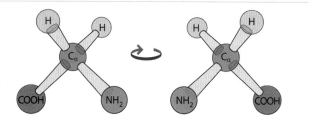

Glycine drawn in two orientations equivalent to the D- and L-amino acids shown in *Figure 3.4A*. These two glycine molecules are not optical isomers because one can be converted to the other by rotation about the vertical axis.

Because of the complicating effects of their side-chains, directly measuring the effect of an amino acid solution on the rotation of plane-polarized light is not a reliable way of distinguishing the D- and L-forms of an amino acid. Instead, the distinction is made by comparing the orientation of the groups around the α-carbon of the amino acid with their orientation around the chiral carbon of glyceraldehyde, as shown in *Figure 3.4*.

Although both L- and D-amino acids are found in living cells, only the L-forms are used to make proteins. There are only a few special exceptions to this rule, such as some small peptides found in bacterial cell walls, which contain one or more D-amino acids.

All amino acids have ionizable groups

The second feature of amino acids that we must consider is the presence, on all amino acids, of at least two ionizable groups. In chemistry, **ionization** is the conversion of an uncharged atom or molecule into a form with an electric charge, called an **ion**. Ionization can occur by adding or removing a proton or an electron. For example, a proton has a positive charge, so adding a proton to a molecule will result in an ionic version of that molecule with a positive charge of +1. Taking away a proton will leave an ion with a negative charge of –1.

The amino acid structure shown in *Figure 3.1* has two ionizable groups. The carboxyl group (–COOH) can lose a proton and so become a negative ion (–COO$^-$), and the amino group (–NH$_2$) can gain a proton and become a positive ion (–NH$_3^+$) (*Fig. 3.6*). A molecule that has two ionized groups but no net charge is called a **zwitterion**. In chemical terms, the presence of ionizable carboxyl and amino groups on an amino acid means that these compounds can act as both weak acids and weak bases. We refer to them as being **amphoteric**.

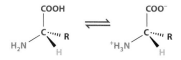

Figure 3.6 Ionization of an amino acid.

Whether or not the carboxyl and amino groups of an amino acid are ionized depends on the pH. *Figure 3.7* illustrates this point for glycine. We see that between about pH 4 and 8 all the glycine molecules are zwitterions, with both the carboxyl and amino groups ionized. The middle of this range is called the **isoelectric point** or **pI** and at this pH the molecules carry no electrical charge. Below pH 4, some of the molecules have regained a proton, converting their –COO$^-$ groups to –COOH. These molecules therefore have a positive charge. The pH at which there are an equal number of molecules with ionized and non-ionized carboxyl groups (at this point the carboxyl groups are said to be half-dissociated) is called the **pK_a** of the carboxyl group. At lower pH values molecules with the non-ionized version of the carboxyl group begin to predominate. Now we move to the other end of the pH scale. Above

---

Box 3.2 The ionization of water and the pH scale

Water is one of the molecules that can ionize. The chemical reaction can be described as:

$$H_2O \rightarrow H^+ + OH^-$$

In reality the H^+ ion, which is a proton, immediately combines with a second water molecule, to give a **hydronium ion** H_3O^+:

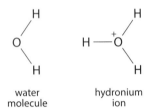

water   hydronium
molecule   ion

In pure water at 25°C about two in every 10^9 molecules are ionized. This corresponds to a hydronium ion concentration of 10^{-7} M.

Acids are compounds which release additional H^+ ions into a water solution. An example is hydrochloric acid, which ionizes to give a proton and a chloride ion:

$$HCl \rightarrow H^+ + Cl^-$$

Acids therefore increase the hydronium ion concentration of a solution.

Bases have the opposite effect, decreasing the hydronium ion concentration of a solution. Some bases do this directly by binding hydronium ions. Ammonia is an example, the combination between ammonia (NH_3) and a hydronium ion giving an ammonium ion (NH_4^+):

$$NH_3 + H_3O^+ \rightarrow NH_4^+ + H_2O$$

Others have an indirect effect on the hydronium ion concentration. For example, sodium hydroxide releases hydroxyl ions when it ionizes:

$$NaOH \rightarrow Na^+ + OH^-$$

These extra hydroxyl ions combine with hydronium ions to produce non-ionized water molecules:

$$H_3O^+ + OH^- \rightarrow 2H_2O$$

The **pH** of a solution is an inverse measure of its hydronium ion concentration:

$$pH = -\log_{10}[H_3O^+]$$

where $[H_3O^+]$ means 'concentration of hydronium ions'. Pure water, with its hydronium ion concentration of 10^{-7} M, therefore has a pH of 7. An acidic solution, with a higher hydronium ion concentration than pure water, has a pH less than 7. A basic solution, with a lower hydronium ion concentration, has a pH greater than 7.

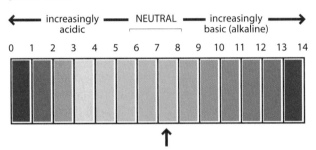

pH 7.4 – 'physiological pH'

The 'physiological pH', which is the pH of most tissues in the human body, is 7.4. A slight deviation from this physiological pH can have drastically harmful effects – if the pH of human blood moves outside of the range 6.9–7.9 then the result is either coma or death. There are many reasons why the pH of living tissues is so critical, as we will see as we progress further through this book. Perhaps most importantly, changes in the pH affect the stability of some types of chemical bond, including many of the bonds that are responsible for the three-dimensional structures of biomolecules, including proteins. Changing the pH can therefore result in protein structures becoming disrupted, preventing those proteins from performing their functions in the cell.

---

pH 8, some of the amino groups have lost their extra protons giving molecules with a negative charge. At the pK_a for the amino group, the number of molecules with and without ionized amino groups is the same, and at pH values above the pK_a the non-ionized version of this group predominates.

Most human and plant tissues have a pH of 7.4. Which ionized form of an amino acid is present at this 'physiological pH'? For the 'typical' amino acid glycine shown in *Figure 3.7*, pH 7.4 falls within the range at which the zwitterion predominates. This is true for all 20 of the amino acids, as indicated by the pK_a values for their carboxyl and amino groups (*Table 3.3*). These values are not all the same because they are affected by the structure of the side-chain, but they are all in the range 1.8–2.6 for the carboxyl group and 8.9–10.6 for the amino. This indicates that for all these amino acids the ionization patterns for the carboxyl and amino groups resemble those shown in *Figure 3.7*. But *Table 3.3* alerts us to a complication. Seven amino acids have side-chains that are also ionizable and which potentially could have a positive or negative charge at pH 7.4 (*Fig. 3.8*). Two of these amino acids, aspartic acid and glutamic acid,

Figure 3.7 Amino acid ionization at different pH values.
The graph shows the relative amounts of the three ionized versions of glycine at pH values from 0 to 14.

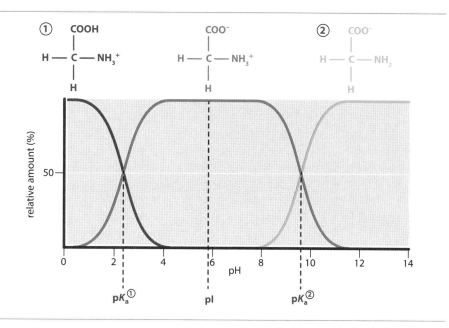

have carboxyl groups in their side-chains. These carboxyls have low pK_a values (3.86 and 4.07, respectively) and are fully ionized at pH 7.4. Aspartic acid and glutamic acid therefore have acidic properties in living tissues, as their names imply. This means that they can act as acceptors of protons in biochemical reactions.

Table 3.3. Amino acid pK_a values

Amino acid	pK_a		
	Carboxyl group	Amino group	Side-chain
Alanine	2.34	9.69	
Arginine	2.01	9.04	12.48
Asparagine	2.02	8.80	
Aspartic acid	2.10	9.82	3.86
Cysteine	2.05	10.25	8.00
Glutamic acid	2.10	9.47	4.07
Glutamine	2.17	9.13	
Glycine	2.35	9.78	
Histidine	1.82	9.17	6.00
Isoleucine	2.32	9.76	
Leucine	2.33	9.74	
Lysine	2.18	8.95	10.53
Methionine	2.28	9.21	
Phenylalanine	2.58	9.24	
Proline	2.00	10.60	
Serine	2.21	9.15	
Threonine	2.09	9.10	
Tryptophan	2.38	9.39	
Tyrosine	2.20	9.11	10.07
Valine	2.29	9.72	

basic at pH 7.4

acidic at pH 7.4

aspartic acid

glutamic acid

arginine

lysine

mix of ionized and non-ionized at pH 7.4

histidine

Figure 3.8 **Amino acids whose side-chains are ionized at pH 7.4.**

Arginine and lysine, on the other hand, both have positively charged side-chains at pH 7.4. They act as bases and can donate protons to other molecules during a biochemical reaction. The side-chains of cysteine and tyrosine also have ionizable groups, but these are largely non-ionized at pH 7.4. These two amino acids are therefore uncharged under physiological conditions. Histidine is the last in the list of amino acids with ionizable side-chains, and this one is interesting. At pH 7.4 there are significant amounts of both the ionized and non-ionized versions of the side-chain. A histidine within a protein molecule can therefore act as both a donor and acceptor of protons, a property that is exploited in a number of important biochemical reactions.

Some amino acids have polar side-chains

We have seen that some amino acids have distinctive chemical properties because their side-chains contain ionizable groups. A second distinctive feature of some R groups is their **polarity**.

Figure 3.9 **The polarity of (A) a hydroxyl group, and (B) a water molecule.**
The oxygen atom tends to attract electrons away from the hydrogen(s). The oxygen therefore becomes slightly electronegative (denoted by δ^-) and the hydrogen slightly electropositive (δ^+).

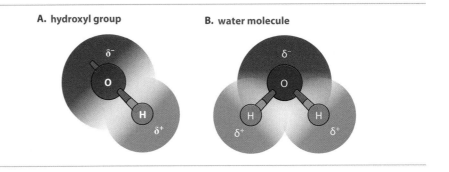

A. hydroxyl group

B. water molecule

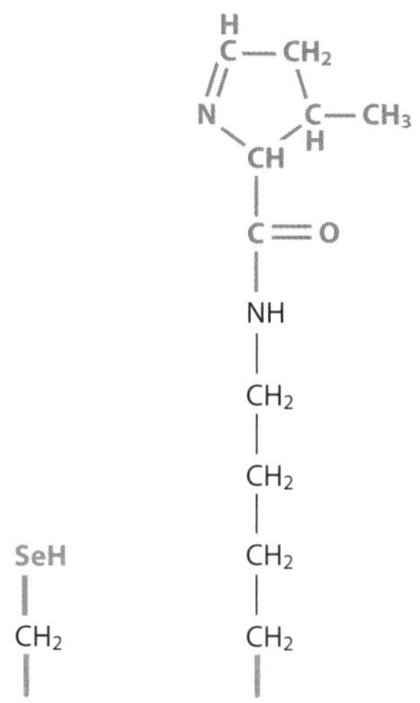

Figure 3.10 The structures of the R groups of selenocysteine and pyrrolysine.
The parts shown in brown indicate the differences between these amino acids and cysteine and lysine, respectively.

We will study the way in which the genetic code specifies the amino acid sequence of a protein in *Section 16.1*.

These modifications are described in *Section 16.3.2*.

Figure 3.11 Proline and 4-hydroxyproline.

Polarity arises if the electrons are not distributed evenly along an R group. Examples of polar amino acids are serine and threonine, both of which have side-chains containing a hydroxyl group (–OH). A hydroxyl group is polar because the oxygen atom tends to attract electrons away from the hydrogen. The oxygen therefore becomes slightly negative and the hydrogen slightly positive, setting up the polarity (*Fig. 3.9A*). It is important to appreciate that polarity is *not* the same as ionization. The hydroxyl group has not lost electrons, which it would need to do to become ionized. The number of electrons has remained the same, the difference lies in their distribution.

A water molecule is made up of two hydrogens attached to one oxygen. Again the oxygen attracts electrons from the hydrogens and becomes electronegative, so water is itself a polar molecule (*Fig. 3.9B*). Polar molecules like to associate together. This means that amino acids with polar side-chains are **hydrophilic**; they 'love water' and so are readily soluble. As well as serine and threonine, the polar amino acids include cysteine, which has a polar thiol group (–SH), and asparagine and glutamine, whose side-chains contain amides (–$CONH_2$).

The nonpolar amino acids are those whose side-chains have evenly distributed electrons. These are alanine, glycine, isoleucine, leucine, methionine, phenylalanine, proline, tryptophan, tyrosine and valine. Nonpolar compounds are **hydrophobic** or 'water fearing'. These amino acids tend not be on the surface of a protein, where they would be exposed to water. Instead they cluster within the protein away from water when the polypeptide folds into its three-dimensional structure.

3.1.3 Some amino acids are modified after protein synthesis

The 20 amino acids that we have studied so far are the ones that are specified by the **genetic code**. The genetic code is used by the cell to translate the information contained in its genes into the amino acid sequences of the proteins that it synthesizes. These 20 amino acids are therefore the ones that can be used during protein synthesis. In reality, the genetic code specifies 22 amino acids, but the other two are used very rarely and so are not usually included in the 'standard' set. These two uncommon amino acids are selenocysteine and pyrrolysine (*Fig. 3.10*). Proteins containing selenocysteine are found in most species, including humans, but pyrrolysine appears to be used only by the archaea.

After synthesis of a protein, some of its amino acids might become modified by the addition of new chemical groups. The simplest types of post-synthesis modification involve the addition of a small chemical group, such as a hydroxyl (–OH), methyl (–CH_3) or phosphate (–PO_4^{3-}), usually to the amino acid side-chain. These modifications increase the number of types of amino acids known in proteins to over 150. An amino acid that has been modified in this way will have slightly altered chemical properties, which might result in a subtle change in the function of the protein. Many modifications of this type are transient, the additional chemical group being removed equally easily, so the protein reverts to its original function. Amino acid modification is therefore a way of regulating the activity of a protein. In a few proteins, however, the modified amino acids are a permanent feature and are needed for the protein to adopt its correct three-dimensional structure. An example is the modified form of proline called 4-hydroxyproline (*Fig. 3.11*), which is present in collagen, a protein found in animal bones and tendons. We will examine the role of 4-hydroxyproline in collagen structure later in this chapter.

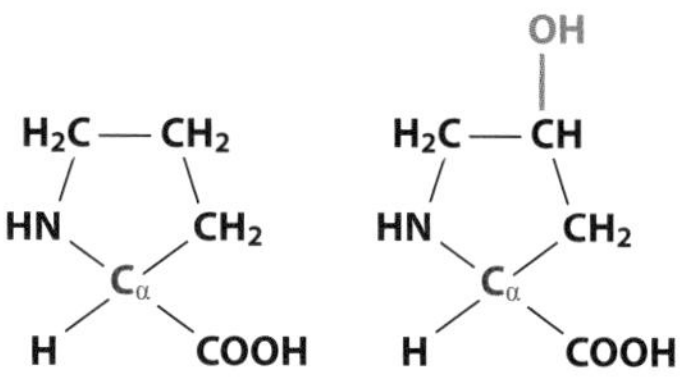

Box 3.3 Types of chemical bond

Chemical bonds are inherent and essential components of all of the molecular structures that are important in biochemistry.

- Chemical bonds hold together the atoms in a molecule such as an amino acid or a protein.
- Chemical bonds enable interactions to form between different parts of a polymeric molecule. This might result in the polymer adopting a helical or other conformation. Similar interactions might fold the polymer into a more complex, three-dimensional structure.
- Chemical bonds enable two or more molecules to bind to one another, resulting in, for example, a multisubunit protein.

We will encounter a variety of different types of chemical bond as we study the structures of proteins and other biomolecules. The most important of these bonds are described here.

Covalent bonds

All of the bonds contained within an amino acid are covalent ones, as are the bonds in the peptide linkage. Covalent bonds are also the predominant type of bonding in nucleic acids, lipids and polysaccharides. Covalent bonds are so common in biochemistry that if the word 'bond' is used without any adjective then you can assume that the bond is a covalent one.

Covalent bonds form when two atoms share electrons. If two atoms come close enough together then two or more pairs of electrons can become shared between the two atoms. In chemical terms, the shared electrons occupy **orbitals** of both atoms. The two atoms are held tightly together, forming the bond.

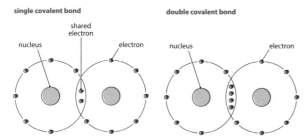

If one pair of electrons is shared then a **single bond** is formed. Both of the atoms can rotate about a single bond, changing the orientation of any other bonds that those atoms have formed. A **double bond**, on the other hand, involves two pairs of shared electrons, and does not allow any rotation.

The strength of a covalent bond depends on its **bond energy**, which is a measure of the amount of energy needed to break it. The strength depends on the identities of the atoms that are linked together, and double bonds are stronger than single ones. A single bond between two carbon atoms (C–C) has a bond energy of $348 \, kJ \, mol^{-1}$, whereas a carbon–carbon double bond (C=C) is 1.75× stronger, with an energy of $614 \, kJ \, mol^{-1}$. A C–H bond has an energy of $413 \, kJ \, mol^{-1}$, and the energy of a C–N bond is $308 \, kJ \, mol^{-1}$.

Electrostatic bonds

An electrostatic bond is an interaction between positively and negatively charged chemical groups. In proteins, they form between an amino acid with a positively charged side-chain (such as lysine or arginine) and one with a negatively charged side-chain (aspartic acid or glutamic acid). They have bond energies of $6–12 \, kJ \, mol^{-1}$, substantially less than covalent bonds. As well as stabilizing structures within proteins, electrostatic bonds are also important on the protein surface, where they hold together the various polypeptides of a multisubunit protein.

Hydrogen bonds

A hydrogen bond is an interaction that forms between the slightly electropositive hydrogen atom in a polar group and an electronegative atom, which might be in the same molecule or in a completely different one. The charge on the electropositive atom is designated δ^+ and on the electronegative one δ^-.

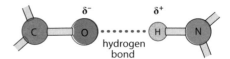

The hydrogen atom is shared between the two groups. The group to which it is more tightly linked (in this example, the -NH group) is called the 'donor' group, and the group to which it is less tightly linked (the -CO group in this example) is called the 'acceptor' group. Hydrogen bonds vary in strength depending on the atoms that are involved, but most are relatively weak. Those in biomolecules have energies between 8 and $29 \, kJ \, mol^{-1}$. Often a number of hydrogen bonds participate in the same interaction between two molecules or two parts of a molecule. The resulting structure can therefore be stable at physiological temperatures, even though the individual bonds are relatively weak. Examples of such structures are the α-helix and β-sheet of proteins, and the double helix of DNA.

van der Waals forces

These are weak attractions named after the Dutch physicist Johannes van der Waals (1837–1923), who first studied them in gases and liquids. They involve temporary electrical charges that occur because of random fluctuations in the distribution of electrons around an atom. Usually the electrons are distributed evenly, in which case the atom has no electrical charge. But by chance the electron cloud can become uneven, with more electrons on one side of the atom compared with the other. This results in a **dipole**, in which one side of the atom is slightly electropositive and the other side slightly electronegative.

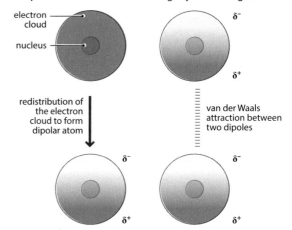

If two dipolar atoms are close enough together then they will attract one another, with a bond energy of about $2–4 \, kJ \, mol^{-1}$.

A van der Waals attraction will last only as long as the fluctuation in the electron clouds that give rise to the dipoles is maintained. However, within a biomolecule such as a protein, there will be so many dipolar chemical groups present at any one time that there are always pairs close enough together to stabilize the biomolecular structure. The identities of the dipole pairs will constantly change, but there will always be lots of them.

3.2   The primary and secondary levels of protein structure

Proteins are traditionally looked on as having four distinct levels of structure. These levels are hierarchical, the protein being built up stage by stage, with each level of structure depending on the one below it (*Fig. 3.12*).

- The **primary structure** is the sequence of amino acids in the polypeptide.

- The **secondary structure** refers to a series of conformations, including helices, sheets and turns, that can be adopted by different parts of the polypeptide.

- The **tertiary structure** is the overall three-dimensional configuration of the protein.

- A **quaternary structure** is the association between different polypeptides to form a multisubunit protein.

Figure 3.12 The four hierarchical levels of protein structure.

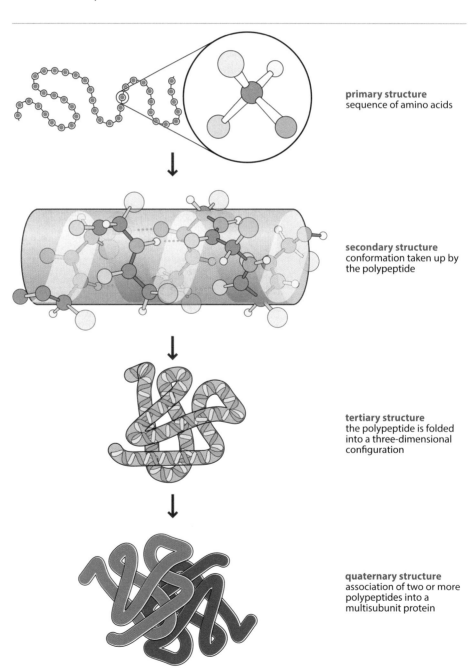

primary structure
sequence of amino acids

secondary structure
conformation taken up by the polypeptide

tertiary structure
the polypeptide is folded into a three-dimensional configuration

quaternary structure
association of two or more polypeptides into a multisubunit protein

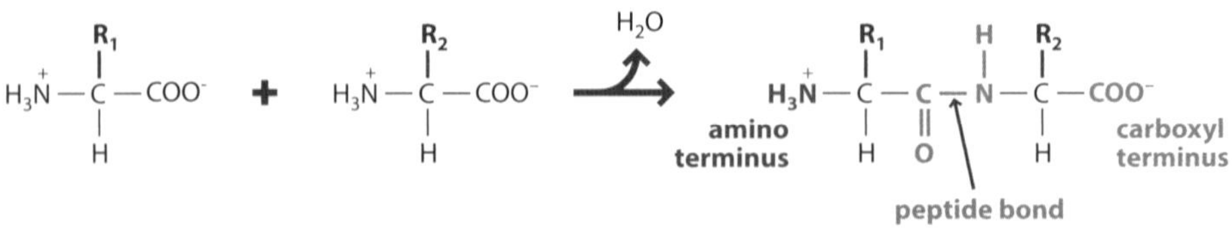

Figure 3.13 The chemical reaction that results in two amino acids becoming linked together by a peptide bond.

In this section we will study the first two of these levels of protein structure.

3.2.1 Polypeptides are polymers of amino acids

A polypeptide is built by linking amino acids together by **peptide bonds** (*Fig. 3.13*). Each peptide bond forms between the carboxyl and amino groups of adjacent amino acids, by a **condensation** reaction that expels a molecule of water. Note that this means that the two ends of the polypeptide are chemically distinct. One has a free amino group and is called the **amino, NH$_2$–,** or **N terminus.** The other has a free carboxyl group and is called the **carboxyl, COOH–,** or **C terminus**. A polypeptide therefore has a chemical direction that can be expressed as either N→C (left to right for the dipeptide shown in *Figure 3.13*) or C→N (right to left in *Figure 3.13*). Protein synthesis occurs in the N→C direction, which means that each new amino acid is added to the free carboxyl group of the growing polypeptide. We therefore use the N→C direction when we write out an amino acid sequence, or type it into a computer.

A **peptide group**, comprising the two α-carbons and the C, O, N and H atoms in between them, has a flat structure. In other words, all six atoms lie on the same plane (*Fig. 3.14A*). This flat structure is rigid because there is little opportunity for rotation around the peptide bond itself. Although drawn as a single bond, the peptide bond has some characteristics of a double bond, one of these being an inability to rotate.

Although the peptide bond cannot rotate, the bonds either side of it can. Rotation around these bonds does not alter the planar nature of the peptide group but does affect the polypeptide chain as a whole. Without any rotation, the polypeptide would be a rigid, linear chain. With these rotations, the polypeptide is able to bend and take up various secondary structural conformations. These rotations are therefore critical to the structure of the protein and we must understand them in detail before we look at the conformations that the polypeptide can adopt. The angle of rotation about the C$_\alpha$–C bond is called *psi* (ψ) and the angle about the N–C$_\alpha$ bond is *phi* (φ) (*Fig. 3.14B*).

A. the peptide group has a flat structure

B. the *psi* and *phi* angles

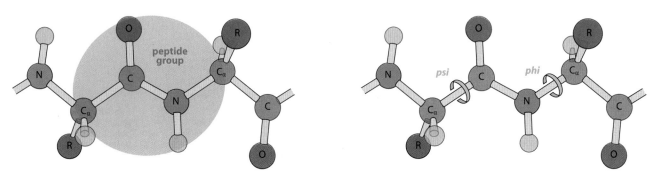

Figure 3.14 Important features of the peptide bond.

If two adjacent peptide groups are orientated in the same plane, then *psi* and *phi* are both 180°. If either bond is rotated clockwise (when looking towards the α-carbon from the other end of the bond) then the angle assigned to *psi* or *phi* increases. If the rotation is counterclockwise, then the angle decreases. The precise combination of angles for *psi* and *phi* either side of an α-carbon determines the conformation of the polypeptide at that point along its length.

It turns out that 77% of the possible combinations of *psi* and *phi* never occur, because of **steric effects**. These effects prevent two atoms from getting too close together and limit the possible conformations that any molecule can take up. The combinations of *psi* and *phi* that are allowed are shown by the **Ramachandran plot** (*Fig. 3.15*), named after G.N. Ramachandran, who led the team that first worked out the information summarized in this diagram, in 1963.

Figure 3.15 The Ramachandran plot.
The dark blue and red areas of the plot indicate the combinations of *psi* and *phi* that are possible without causing steric effects. The red areas are the most favorable regions, within which most combinations of bond angles found in real polypeptides are located. The types of secondary structure that result from the bond angles in the different regions of the plot are noted.

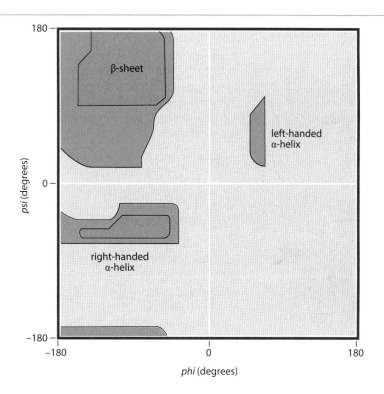

Box 3.4 The unusual characteristics of the peptide bond

PRINCIPLES OF CHEMISTRY

The peptide bond is usually drawn as a single bond, but it has some characteristics of a double bond. In particular, it is unable to rotate, contributing to the planarity of the peptide group. These properties arise from a process referred to as **resonance**, which involves the redistribution of electrons between adjacent atoms in a molecule. This can result in the replacement of a single bond with a double bond, and vice versa. The two resonance structures of the peptide bond are shown in the diagram on the right:

The continuous resonance between the two structures means that the peptide bond oscillates between single and double bond characteristics. The single bond predominates, but the double bond is sufficiently prevalent to prevent rotation.

3.2.2 Polypeptides can take up regular conformations

If one or both of the *psi* and *phi* angles either side of an α-carbon are different to 180°, then the polypeptide will change direction at that point. To illustrate this effect, we will study the two most common types of secondary structure found in proteins, called the α-helix and the β-sheet.

The α-helix is a common type of secondary structure

The α-helix was discovered by Linus Pauling, Robert Corey and Herman Branson in the late 1940s. At the time this was a rather odd type of discovery because it was not based entirely on experimental evidence. X-ray crystallography had suggested that a helical structure of some kind was a common feature of many proteins. Pauling decided to work out the structure of that helix by building models, initially just with a polypeptide chain drawn on a piece of paper. **Model-building** is still used to interpret data from X-ray crystallography, although today the models are built not with pieces of paper but with sophisticated computer programs.

The key feature that Pauling and his colleagues used to work out the details of the α-helix was the way in which the structure would be stabilized by hydrogen bonds forming between different parts of the polypeptide. In an α-helix, hydrogen bonds form between the CO of one peptide group and the NH of the peptide group four positions along the polypeptide (*Fig. 3.16*). The NH group is polar and so this hydrogen is electropositive, and the oxygen of the CO group is electronegative, providing the necessary conditions for hydrogen bond formation.

The α-helix has 3.6 amino acids per turn with the side-chains pointing outwards, an individual helix usually having 10–20 amino acids but sometimes as many as 40. It is possible to form both right- and left-handed α-helices, but almost all those in proteins are right-handed, which are slightly more stable. The *psi* and *phi* angles for a right-handed α-helix are –47° and –57°, respectively, within one of the favorable regions of the Ramachandran plot (see *Fig. 3.15*).

What determines whether or not an α-helix forms in a particular region of a polypeptide? The answer lies with the identities of the amino acids present in that stretch. The nature of the side-chain affects the ability of the bonds on either side of an α-carbon to take up the necessary *psi* and *phi* angles. Alanine is particularly suitable in this regard and acts as a 'helix former', promoting the folding of the polypeptide into an α-helix. Interactions between side-chains of different amino acids can also promote helix formation and stabilize a helix once it has formed. For such an interaction to occur, two side-chains have to be on the same face of the helix and therefore they must be 3–4 amino acids apart in the polypeptide. Electrostatic bonds between positively and negatively charged side-chains of amino acids spaced apart in this way often stabilize an α-helix.

Other amino acids are 'helix breakers' and either prevent a helix from forming or limit the length of one that does form. Proline is the main example of a helix breaker, the structure of its unusual side-chain (see *Fig. 3.3*) not allowing rotation about the N–C_α bond. The *phi* angle next to a proline is therefore invariant and cannot adopt the degree of rotation needed to form the α-helix. Often a proline is found at one or other end of an α-helix, marking the point where helix formation should terminate.

The β-sheet is another common secondary structure

The β-sheet was also predicted by Pauling and his colleagues following model-building experiments. As with an α-helix, a β-sheet is held together by hydrogen bonds between the CO and NH parts of different peptide groups. The distinction is that in a β-sheet the peptide groups that participate in hydrogen bonding are not close to one another in the polypeptide. In fact, their distance apart is immaterial. What is

We will examine how X-ray crystallography is used to study protein structure in *Section 18.1.3*.

○ hydrogen

● carbon

● nitrogen

● oxygen

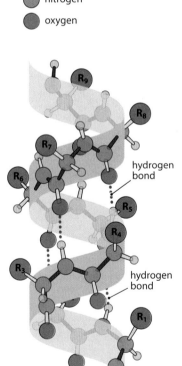

Figure 3.16 The α-helix.
The polypeptide chain is shown in outline. Hydrogen bonds occur between the CO of one peptide group and the NH of the peptide group four positions along the polypeptide.

Box 3.5 **What is the difference between a left-handed and a right-handed helix?**

The easiest way to answer this question is to imagine that the helix is a spiral staircase and you are climbing up that staircase. If the helix is right-handed then you will hold the outside rail with your right hand. If it is a left-handed helix then the outer rail will be adjacent to your left hand. The famous spiral staircase at the

Loretto Chapel in Santa Fe, New Mexico, which is claimed to have been constructed miraculously by St Joseph, is a left-handed helix. The equally famous spiral staircase at the Vatican Museums, designed by Giuseppe Momo in 1932, is right-handed.

---

important is that a series of hydrogen bonds forms between two parts of a polypeptide so that those segments are held together side by side (*Fig. 3.17*). Addition of more segments results in a sheet-like structure that can comprise 10 or more strands, each containing up to 15 amino acids. Within a **parallel β-sheet**, all the strands run in the same direction (N→C or C→N), whereas in the **antiparallel** version, adjacent strands run in opposite directions. A mixture of the two is also possible in a single sheet. The sheet itself might exhibit some degree of curvature in the form of a right-handed twist.

Stable hydrogen bonds will not form between adjacent strands if the polypeptides are fully extended, where the two rotation angles *psi* and *phi* are both 180°. Instead, there has to be some rotation of the bonds around the α-carbons so that *psi* is about 113° and *phi* about −119° in a parallel sheet, or 135° and −139° in an antiparallel one. These rotations give the polypeptide a zigzag shape and a sheet a pleated appearance (see *Fig. 3.17*). The side-chains point outwards at right angles to the plane of the sheet.

There is little interaction between the side-chains of different amino acids in individual or separate strands which means that, unlike with an α-helix, there are few rules regarding which amino acids can or cannot participate in a β-sheet. Proline is

---

Box 3.6 **Predicting the secondary structure of a polypeptide from its amino acid sequence**   **RESEARCH HIGHLIGHT**

It is easier to work out the amino acid sequence of a protein than its three-dimensional structure, especially as an amino acid sequence can be deduced from the DNA sequence of a gene, using the rules of the genetic code. DNA sequencing is relatively easy, as we will discover in *Section 19.2*, whereas the methods needed to determine the three-dimensional structure of a protein, such as X-ray crystallography and nuclear magnetic resonance (NMR) spectroscopy (*Section 18.1.3*), are more difficult and time-consuming. This means that there are a substantial number of proteins whose amino acid sequences are known but their three-dimensional structures are uncharacterized. In the cell the amino acid sequence specifies the three-dimensional structure of the protein. So is there any way in which we can predict that three-dimensional structure simply by examining the amino acid sequence?

Biochemists have attempted to devise rules for predicting protein structure since the 1960s. The early methods concentrated on trying to deduce the positions of α-helices in a polypeptide chain, making use of theoretical information about which amino acids should be helix formers and which helix breakers, along with actual knowledge of the frequencies at which different amino acids were present in helices in the small number of proteins whose structures were actually known at that time. In this way it proved possible to identify the positions of α-helices with 60–70% accuracy. A similar approach allowed β-sheets to be identified with slightly less confidence.

amino acid sequence

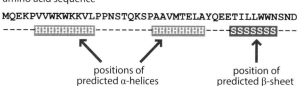

positions of predicted α-helices   position of predicted β-sheet

These methods only enable the secondary structure of a protein to be deduced with any degree of certainty. Predicting the way in which the polypeptide, containing its α-helices and β-sheets, folds up into its three-dimensional tertiary structure proved much more difficult. Gradually, the number of proteins whose structures were known increased to the stage where comparisons between the actual structures of related proteins with similar sequences could be made. Then it became possible to devise computer programs that would compare a new amino acid sequence with all the known protein structures, identify entire proteins or parts of proteins with similar sequences, and then use the structures of those proteins to predict the structure taken up by the new amino acid sequence. Even today this method is still not entirely accurate, but it provides a rapid way of identifying the important structural features of a protein before the results of a full X-ray crystallography or NMR analysis are obtained.

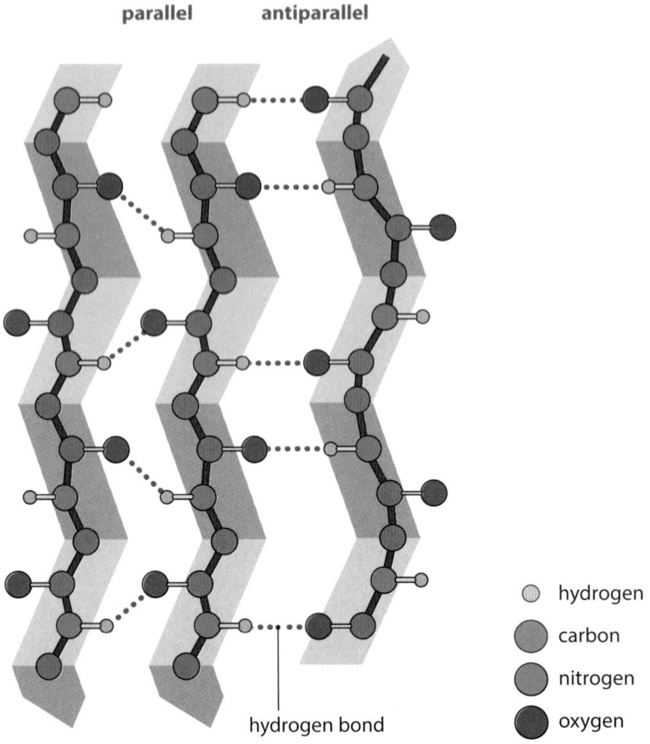

Figure 3.17 The β-sheet.
The polypeptide chains are shown in outline with the R groups omitted. The right-hand and middle strands form an antiparallel β-sheet, in which the polypeptides run in opposite directions. The middle and left-hand strands form a parallel sheet. Note the pleated appearance of the sheets.

parallel   antiparallel

hydrogen bond

○ hydrogen
● carbon
● nitrogen
● oxygen

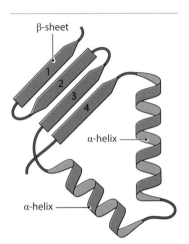

Figure 3.18 A typical combination of β-sheet and α-helices.
In this example there is a four-strand β-sheet. Strands 1–3 form an antiparallel sheet linked by two short hairpin turns. Strands 3 and 4 form a parallel sheet with a lengthy linking sequence containing two α-helices.

once again unfavored and if present is likely to be restricted to one of the edge strands, and amino acids with larger side-chains tend to be located towards the middle of a sheet. There are also very few rules regarding the number of amino acids between the end of one strand and the start of the next. The minimum is usually four because this number of amino acids is needed to execute a hairpin turn of the polypeptide. But the intervening segment can be much longer and can contain other structural motifs such as α-helices or even β-strands participating in a second, separate sheet (*Fig. 3.18*).

3.3   Fibrous and globular proteins

Proteins can broadly be divided into two types, **fibrous** and **globular**. Globular proteins are often soluble and perform a multitude of different functions in living cells. When we study their structure in the next section we will see that they comprise α-helices and β-sheets folded into complex three-dimensional **tertiary structures**. Fibrous proteins are insoluble and usually have more specialized structural roles. These proteins are not folded into tertiary structures. Instead the secondary structure is their highest level of organization.

3.3.1   Fibrous proteins: keratin, collagen and silk

Keratin, collagen and silk are three examples of fibrous proteins. Keratin is present in the hair, horns, nails and skin of animals. The protein is made up of two polypeptides, each of which is composed almost entirely of a slightly compacted version of the α-helix, with 3.5 rather than 3.6 amino acids per turn. This compaction gives the right-handed α-helix a left-handed **superhelix** conformation (*Fig. 3.19*). The configuration of

Figure 3.19 Keratin.
The individual polypeptides adopt compacted α-helix structures giving the strand a superhelix conformation. This diagram shows two polypeptides coiled around one another.
Image from *Essential Biochemistry* by Pratt *et al.* with permission from John Wiley and Sons, Inc.

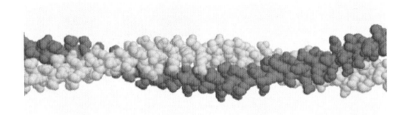

the superhelix is such that two keratin polypeptides can coil around one another, held together by weak bonds called **van der Waals forces**, and possibly by **disulfide bonds**, which are covalent bonds that form between cysteine residues occupying adjacent positions in the two polypeptides (*Fig. 3.20*). The resulting structure, technically called a **coiled coil**, can form fibrils with other coiled coils, which associate further to form microfilaments with high tensile strength, meaning that they are difficult to break by pulling at the ends. The helical conformation of the keratin polypeptide is therefore directly responsible for the physical properties of hair and other structures in which the protein is found.

Collagen also has a helical structure but one that presents new features that we have not encountered so far. A collagen polypeptide has a relatively simple primary structure made up of many repeats of the sequence glycine–X–Y, where X is frequently proline and Y the modified version of proline called 4-hydroxyproline (see *Fig. 3.11*). The repeat is therefore synthesized as glycine–proline–proline, with the second proline in the series converted to 4-hydroxyproline after the polypeptide has been made. The high proline content confers a left-handed helical structure on the polypeptide, with 3.3 amino acids per turn, partly because there cannot be any rotation about the $N–C_{\alpha}$ bond of a proline, and partly because the proline side-chains repel one another and try to be as far apart as possible. Three of these helical polypeptides then coil around

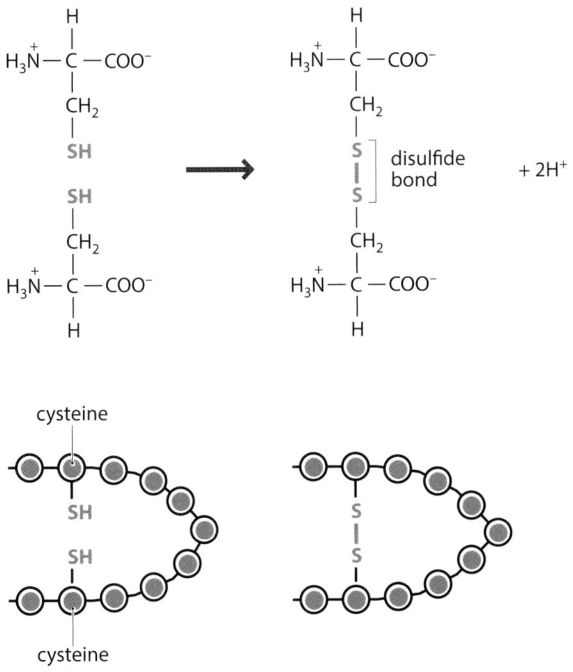

Figure 3.20 Disulfide bonds.
The upper drawing shows the chemical structure of a disulfide bond. Below is the effect that formation of a disulfide bond can have on the structure of a polypeptide.

Figure 3.21 Collagen.
Collagen is a triple helix of three polypeptides, each of which has a left-handed helical conformation. Image from *Essential Biochemistry* by Pratt *et al.* with permission from John Wiley and Sons, Inc.

one another to make a right-handed **triple helix** (*Fig. 3.21*). The structure requires every third amino acid in each of the polypeptides to be placed close to the central part of the triple helix. This is why every third amino acid is glycine. This amino acid, with its very small side-chain, is the only one that can fit in. The triple helix is held together by hydrogen bonding between the NH of a glycine in one polypeptide and the CO of a peptide group in one of the other two polypeptides. As in keratin, groups of collagen triple helices come together to form fibrils that provide collagen with the strength it needs to play its structural role in connective tissues including bones and tendons.

Silk is rather different. This fiber is produced by various insects and exploited by humans to make fine fabrics. The fibrous component of silk is the protein called fibroin, but this does not have a helical structure. Instead, each fibroin polypeptide has a high glycine and alanine content, which enables it to form extensive β-sheets, which layer on top of one another, with very close packing because of the small size of the glycine and alanine side-chains. The individual β-sheets provide tensile strength, but the layers of sheets are held together less tightly. This means that silk fibers are both strong and flexible.

Box 3.7 Using collagen structure to identify extinct animals

RESEARCH HIGHLIGHT

Collagen is one of the most important proteins in vertebrates, being present in bones, tendons and other structural tissues. In bones, collagen fibrils make up about 20% of the dry weight and are embedded in the mineral matrix called bioapatite. Collagen is a very stable protein that is not easily broken down and hence is often preserved in bones after the animal dies. There are even reports of small amounts of collagen being identified in the fossil leg bone of a dinosaur from 68 million years ago.

Although collagen polypeptides have a regular structure made up mainly of glycine, proline and 4-hydroxyproline, there are sufficient differences between the collagen molecules of different species for fossil bones to be identified. The method is called **collagen fingerprinting**. Recently it has been used to show that camels once lived in the Arctic. Small fragments of bone, dated to 3.5 million years ago, were discovered on Ellesmere Island in the Canadian High Arctic. Collagen fingerprinting showed that they came from an extinct species of giant camel that lived in what is now the Arctic during the mid-Pliocene, a warm period in the Earth's history. The Arctic was still very cold at that time, with winter blizzards and deep snow, but there were also forests that provided food and shelter for large animals. Nonetheless, the discovery that camels lived there has shaken up ideas about the types of species that were present in North America in the period immediately before the Ice Ages.

Image of camels on Ellesmere Island reproduced with permission from the artist Julius Csotonyi.

3.3.2 Globular proteins have tertiary and possibly quaternary structures

Globular proteins have spherical rather than elongated fibrous structures, and most are soluble in water. They have diverse biochemical roles, and equally diverse structures. In fact a major goal of biochemistry over the last 20 years has been to identify common structural features within different globular proteins, and to relate those features to the functions of the individual proteins.

The most important structural difference between a globular protein and a fibrous one is that the former displays at least one, and possibly two, higher levels of organization. These are called the **tertiary** and **quaternary structures**, and understanding them is the key to understanding the importance of globular proteins in biochemistry.

A. carbonic anhydrase   **B. myoglobin**   **C. concanavalin A**

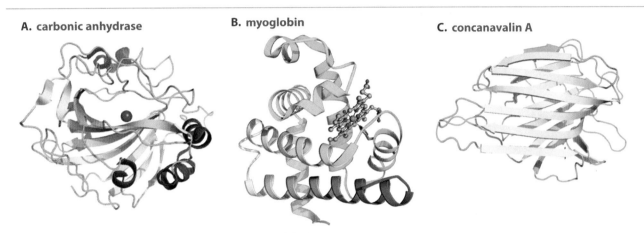

Figure 3.22 **Three globular proteins.**
(A) Carbonic anhydrase. This protein comprises a ten-stranded β-sheet (yellow) surrounded by five α-helices (pink and red). It also contains a zinc atom, shown in blue. (B) Myoglobin. The secondary structures are all α-helices. It also contains a heme molecule. (C) Concanavalin A. The secondary structure is entirely β-sheet.
(A) Reproduced with permission from University of Maine by Raymond Fort Jr (http://chemistry.umeche. maine.edu/CHY431.html). (B) Reproduced with permission from Science Photo Library. (C) Reproduced from Wikipedia under a Creative Commons license.

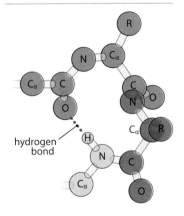

Figure 3.23 **A β-turn.**
The four amino acids involved in the turn are shown in different colors. For clarity, the hydrogen atoms are left out, except for the one that participates in the hydrogen bond.

The tertiary level of structure is the three-dimensional configuration of a protein

The tertiary structure of a globular protein results from folding the secondary structural components of the polypeptide into a three-dimensional configuration. For most proteins, the secondary structural components comprise a mixture of α-helices and β-sheets. An example is the enzyme carbonic anhydrase, which has a ten-stranded β-sheet surrounded by five α-helices (*Fig. 3.22A*). A few proteins have more uniform tertiary structures. Myoglobin, for example, is made up of eight α-helices and no β-sheet (*Fig. 3.22B*), and concanavalin A consists entirely of β-sheet (*Fig. 3.22C*). Whatever the combination, the secondary structural components are linked by less organized segments of polypeptide which might, nonetheless, include structures that cause the path of the polypeptide to change direction in a specific way. An example is the β-turn, which comprises four amino acids, often including glycine and proline. These four amino acids execute a 180° turn, held in place by hydrogen bonding between the CO of the first peptide group and the NH of the third (*Fig. 3.23*). This type of turn often connects pairs of strands in a β-sheet, and tends to be located at or near the surface of a globular protein.

Although the structures of globular proteins are very diverse, certain common features can be identified. The most consistent of these features is the distribution of the 'water-fearing' and 'water-loving' parts of the polypeptide chain. In most globular proteins, all of the nonpolar amino acid side-chains are located inside the structure. This is exactly what we expect because these side-chains are hydrophobic and so will tend to become buried within the protein when it folds into its tertiary structure. Similarly, the polar and charged parts of the polypeptide will usually be on the surface so that they can make contact with water molecules, presuming the protein is in an aqueous environment such as the inside of a cell. The polar parts of a polypeptide include not just amino acid side-chains, but also the CO and NH groups of the peptide linkages, unless these have formed hydrogen bonds, as in an α-helix or β-sheet. Peptide groups that are not participating in hydrogen bonding will therefore tend to be on the surface of the protein. These various forces dictate the way in which the polypeptide folds into its tertiary structure. Once folded, the structure will be stabilized by various interactions, such as van der Waals forces, and possibly by formation of disulfide bridges between cysteine amino acids.

Other common features of globular proteins are seen when combinations of secondary structural units are examined. Some combinations of units, folded in a particular way, are seen in many different proteins. The most frequent of these **motifs** is the **βαβ loop**, which is made up of two parallel strands of a β-sheet separated by an α-helix (*Fig. 3.24*). A second example is the **αα motif**, in which two α-helices lie side by side in antiparallel directions in such a way that their side-chains intermesh. Each type of motif is found in a range of different proteins, with diverse functions, suggesting that the motifs are structural units rather than functional ones.

In some larger globular proteins, the tertiary structure is split into separate segments called **domains**, usually linked by short segments of unstructured polypeptide. The domains might have identical or similar structures, as is the case for the four domains of the mammalian cell surface protein called CD4. In other proteins the domains are different in structure, each possibly contributing a different part of the overall function of the protein.

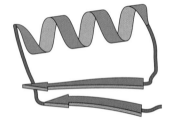

Figure 3.24 The βαβ loop.

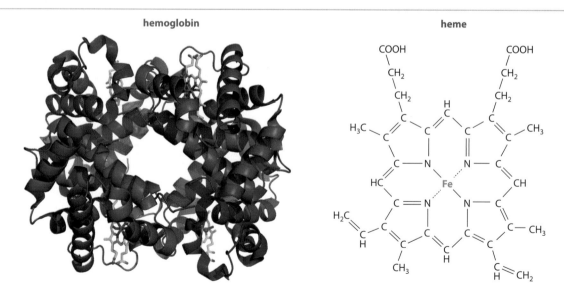

hemoglobin

heme

Figure 3.25 Hemoglobin.
This protein is a tetramer of two identical α subunits and two identical β subunits. The heme groups are shown in green in the protein structure. Heme is an organic compound containing an iron atom, which reversibly binds oxygen, enabling hemoglobin in red blood cells to carry oxygen from the lungs to other parts of the body.
(A) Haemoglobin image by Zephyris reproduced from Wikipedia under a CC BY-SA license.

Box 3.8 **An example of a protein with a mixture of domains**

Human tissue plasminogen activator (TPA), which is involved in blood clotting, is a good example of a multidomain protein. TPA has five domains:

- Two identical 'kringle' structures, which enable TPA to bind to other proteins, and also to lipids that act as mediators in the blood clotting process. Each kringle structure is a large loop stabilized by three disulfide bonds.
- A 'finger' module, which is a small β-sheet structure that binds fibrin, a fibrous protein found in blood clots.
- A growth factor module, made up of three loops held in place by two disulfide bonds. This module enables TPA to stimulate cell proliferation as part of the wound healing response.
- A large protease domain, comprising a β-sheet and α-helix.

The function of the protease domain is to convert an inactive protein called plasminogen into its active form, called plasmin. The protease does this by cutting a single peptide bond within the plasminogen polypeptide. Plasmin breaks down unused fibrin, ensuring that the clot does not spread into the bloodstream.

All of these domains are also found, with very similar structures, in other proteins. Kringle and finger domains are common in proteins involved in blood clotting, and growth factor domains are found in several proteins that stimulate cell growth. One of these, epidermal growth factor, is made up simply of a single growth factor domain.

Quaternary structure is the association of polypeptides into multisubunit proteins

The quaternary level of protein structure involves the association of two or more polypeptides, each folded into its tertiary structure, into a multisubunit protein. Not all proteins form quaternary structures, but it is a feature of many proteins with complex functions. Some quaternary structures are held together by disulfide bonds between the different polypeptides, resulting in a stable multisubunit protein that cannot easily be broken down to its component parts. Other quaternary structures comprise looser associations of subunits stabilized by relatively weak interactions such as hydrogen bonding. These proteins can revert to their component polypeptides, or change their subunit composition, according to the functional requirements of the cell.

Hemoglobin is an example of a protein with a quaternary structure. This is the protein in vertebrate red blood cells that carries oxygen from the lungs to other tissues in the body. It is a tetramer of four polypeptides, comprising two identical α subunits and two identical β subunits (*Fig. 3.25*). The polypeptides are called globins so the subunits are α-globins and β-globins. Each globin has an attached heme group, a non-protein compound that binds oxygen. The quaternary structure is stabilized by hydrogen and electrostatic bonds between the globin subunits.

Large quaternary structures are formed by the proteins that make up the coats, or capsids, of viruses. The capsid of tobacco mosaic virus (TMV), for example, is made up of 2130 identical subunits. Each subunit is a small globular protein, comprising 158 amino acids folded into a tertiary structure that includes four α-helices. The subunits are arranged into a tightly packed helical structure with 16.3 subunits per turn, which encloses the RNA genome of the virus. The TMV capsid is, in effect, a single multisubunit quaternary protein (*Fig. 3.26*). TMV is an example of a filamentous virus but the same principle applies to the capsids of icosahedral viruses. Human poliovirus has an icosahedral capsid with 20 faces. Each face is made up of 12 polypeptide subunits, three each of VP1, VP2, VP3 and VP4. The capsid as a whole therefore has 240 subunits, 60 of each of the four VP subunits.

individual subunit
(16.3 per turn)

total length
130 turns

Figure 3.26 The tobacco mosaic virus capsid.

3.4   Protein folding

A fundamental notion regarding globular proteins is that the secondary and tertiary structures are specified by the amino acid sequence of the polypeptide. In other words, a particular amino acid sequence will fold into just one, and no other, tertiary structure. This was first demonstrated by experiments carried out in the 1950s and has led to detailed models for the folding process and for the role in the cell of **molecular chaperones**, proteins that help other proteins to fold.

3.4.1 Small proteins fold spontaneously into their correct tertiary structures

The notion that the amino acid sequence contains all the information needed to fold the polypeptide into its correct tertiary structure derives from experiments carried out by Christian Anfinsen in the 1950s. He worked with ribonuclease, a small protein of 124 amino acids, whose tertiary structure is a mixture of α-helices and β-sheet and includes four disulfide bonds between cysteine amino acids in different parts of the polypeptide. Anfinsen used ribonuclease that had been purified from cow pancreas and resuspended in an aqueous buffer. The addition of urea, a compound that disrupts hydrogen bonding, resulted in a decrease in the activity of the enzyme, measured by testing its ability to cut up molecules of RNA into their monomeric subunits (*Fig. 3.27*). At the same time, the viscosity of the solution increased, indicating that the protein was being **denatured** by unfolding into an unstructured polypeptide chain.

The urea was then removed from the solution by **dialysis**. The viscosity decreased and the protein gradually regained its ability to cut RNA. The protein therefore refolds spontaneously when the denaturant is removed.

Urea does not disrupt disulfide bonds, so in the experiment just described these bonds remain intact. In a second experiment the urea was accompanied by a reducing agent, β-mercaptoethanol, which does break the disulfide bonds. The same result is obtained when the urea is removed from the solution: the protein activity still returns. This shows that the disulfide bonds are not critical to the protein's ability to refold, they merely stabilize the tertiary structure once it has been adopted.

Figure 3.27 Denaturation and spontaneous renaturation of ribonuclease.
The graph shows the changes in ribonuclease activity and solution viscosity that occur as the urea concentration is increased or decreased. As the urea concentration increases to 8 M, the protein becomes denatured by unfolding. Its activity decreases and the viscosity of the solution increases. When the urea is removed by dialysis, this small protein readopts its folded conformation. The activity of the protein increases back to the original level and the viscosity of the solution decreases.
Ribonuclease structure images reproduced from Wikipedia under a CC BY-SA 2.5 license.

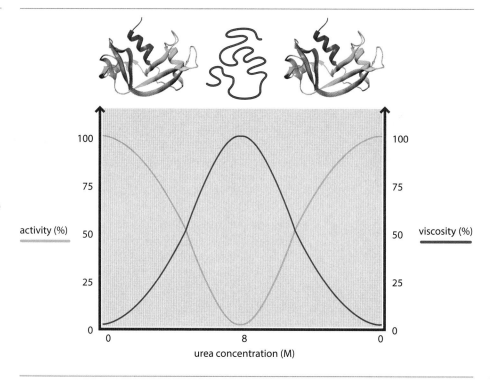

3.4.2 Protein folding pathways

Once it had become clear that proteins adopt their tertiary structures spontaneously, biochemists turned their attention to the folding process itself. It was quickly realized that the process cannot be random. It is not possible for a protein simply to explore all the possible conformations that it can take up until it finally hits on the correct one. This was made clear by Cyrus Levinthal in 1969, whose argument is as follows. The tertiary structure of a protein is set by the three-dimensional conformation of the polypeptide. This in turn is set by the *psi* and *phi* values for the bonds on either side of the α-carbons along its polypeptide chain. Remember that it is only by rotation around these bonds that the polypeptide can change direction. Levinthal argued that there would be at least three possible values for each *psi* and *phi* angle, which is almost certainly an under-estimate. This would mean that a polypeptide of 100 amino acids could adopt 3^{198} different conformations – this is about 10^{100}. Even if the protein could explore 10^{13} conformations per second (likely to be an *over*-estimate) this would mean that it would take about 10^{87} seconds for all conformations to be checked. This is a huge amount of time, longer even than the age of the universe (in fact, much longer). So proteins cannot find their correct tertiary structures just by a random search. This problem has been called **Levinthal's paradox**.

The folding process must be ordered in some way

Levinthal's paradox shows that the folding process must be ordered in some way. This had led biochemists to conclude that there is a **folding pathway** for each protein, each step involving just a small part of the polypeptide (*Fig. 3.28*). In this way the protein could find its way to its correct structure without having to test every possible conformation. These considerations, combined with experimental studies of protein folding, have led to the **molten globule** model. In this model, the initial step in folding is the rapid collapse of the polypeptide into a compact structure, with slightly larger dimensions than the final protein, driven by the desire of the hydrophobic amino acid side-chains to avoid water. Collapse into this molten globule might automatically fold some of the polypeptide into its α-helices and β-sheets. Because the globule is 'molten' it can change conformation rapidly, identifying additional folds so that the correct tertiary structure gradually emerges. For larger proteins, this step might involve construction of correctly folded subdomains which are then brought together to make the final tertiary structure. The whole process can take just a few seconds.

In thermodynamic terms, a decrease in randomness is accompanied by a reduction in **free energy** (see *Section 7.2.1*).

More sophisticated iterations of the molten globule and other models for protein folding imagine a **folding funnel** that the protein passes through, gradually taking up less random conformations until it reaches its final structure (*Fig. 3.29*). As the protein adopts an increasingly folded state, the funnel narrows because there are fewer options for the next steps towards the final structure. There are also side funnels into which the protein can be diverted, leading to an incorrect structure. If an incorrect structure is sufficiently unstable then partial or complete unfolding may occur, allowing the protein to return to the main funnel and pursue a productive route towards its correct conformation.

Figure 3.28 A protein folding pathway.

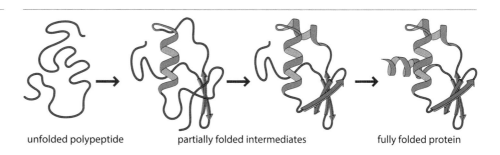

unfolded polypeptide   partially folded intermediates   fully folded protein

Box 3.9 Studying protein folding

How do biochemists study the way in which individual proteins fold? Anfinsen's experiments were revolutionary in their day, but that was 60 years ago and all he was able to do was measure the viscosity of his ribonuclease solutions, and the activity of the enzyme, to follow the unfolding and folding of the protein. He could not deduce any specific information about the folding pathway itself.

Biochemists today use three main approaches to study protein folding.

- For some proteins, it is possible to halt the folding pathway at certain points, and then use NMR to study the structure of the intermediate form directly. This provides very specific information about a folding pathway, but so far has only been used with a few proteins, and it may not be generally applicable because halting the pathway is not always possible.
- The degree of folding can be followed in real time by methods such as **circular dichroism**. In principle, this is the same as the approach used by Anfinsen, but with modern variations that make it much more informative. Circular dichroism measures the absorption of polarized light by a protein. Secondary structures such as α-helices and β-sheets absorb polarized light, so circular dichroism measures the rate at which these structures are formed. As with Anfinsen's experiments, this type of research is usually performed with proteins that have been denatured and are gradually reforming their folded structures. However, modern methods enable the renaturation process to be controlled much more carefully, so all the proteins in the solution begin to fold at exactly the same time. This means that the process is synchronized and hence much easier to study. It is even possible to study the folding of a single molecule that is initially held in a linear conformation with an **optical tweezer**, a laser device that can be used to manipulate individual molecules.
- The third approach is to change the amino acid sequence of the protein and see what effect this has on the folding pathway. The change in sequence is usually brought about by introducing a **mutation** into the gene coding for the protein (see *Section 19.1.2*). In this way the stage in the folding pathway at which a particular part of a protein adopts its structure can be identified. For example, imagine that we wish to test if a particular α-helix forms early in a folding pathway. To do this, we would change an amino acid predicted to be crucial for nucleating that helix into one that will prevent the helix from forming. If our hypothesis is correct, and the α-helix is indeed important in the early part of the folding pathway, then the altered protein should be unable to fold beyond that stage.

Figure 3.29 A folding funnel.
The top of the funnel is wide because the unfolded polypeptide can initially adopt any one of many initial intermediate structures. The funnel gradually narrows as the protein becomes more folded and its options for future folding are reduced. Gradually those options decrease and the protein becomes more completely folded and the fully folded protein emerges from the spout at the bottom. Side funnels lead to dead-ends. If the protein enters one of these side funnels then it has to partially unfold in order to return to the main funnel.

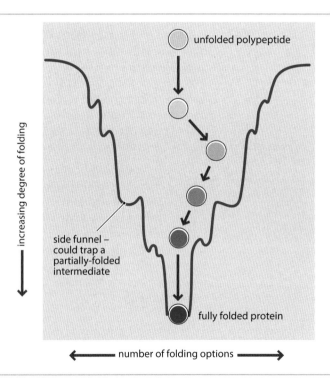

increasing degree of folding

unfolded polypeptide

side funnel – could trap a partially-folded intermediate

fully folded protein

number of folding options

In living cells, protein folding is aided by molecular chaperones

Experiments with purified proteins have been very useful in helping us to understand protein folding, but this type of *in vitro* work has two limitations. First, only smaller proteins with less complex structures fold spontaneously in the test tube. Larger proteins tend to get stuck as intermediate forms that are incorrectly folded but which are too stable to unfold to any significant extent. Secondly, the folding of a complete polypeptide might not be equivalent to folding *in vivo*, because a cellular protein might begin to fold before it has been fully synthesized.

Studies of protein folding in the cell have led to the discovery of proteins that help other proteins to fold. There are two types of these **molecular chaperones**. The first are the **Hsp70 proteins**. These bind to the hydrophobic regions of unfolded proteins, including proteins that are still being synthesized. They hold the protein in an open conformation and aid folding, presumably by modulating the association between those parts of the polypeptide which form interactions in the folded protein. Exactly how this is achieved is not understood, but it involves repeated binding and detachment of Hsp70 proteins.

The second type of molecular chaperone is the **chaperonins**, the main version of which is the **GroEL/GroES complex**. This is a multisubunit structure that looks like a hollowed-out bullet with a central cavity (*Fig. 3.30*). It is thought that an unfolded protein enters the cavity and emerges folded, possibly because the inside surface of the cavity changes from hydrophobic to hydrophilic in such a way as to promote the controlled burial of hydrophobic amino acids within the protein.

Figure 3.30 The GroEL/GroES chaperonin.
On the left is the view from the top, and on the right the view from the side. The GroES part of the structure is made up of seven identical protein subunits and is shown in gold. The GroEL components consist of 14 identical proteins arranged into two rings (shown in red and green), each containing seven subunits. The main entrance into the central cavity is through the bottom of the structure shown on the right.
Reproduced from Xu *et al.* (1997) *Nature*, 388:741 with permission from Macmillan Publishers Ltd.

3.4.3 Protein folding is one of the fundamental principles of biology

Although we do not fully understand the folding process, it is clear that the amino acid sequence of a polypeptide contains all the information needed for adoption of the higher levels of structure of a protein. The corollary is that by specifying a set of different amino acid sequences, an organism's genes can direct synthesis of proteins with different structures and hence different specific functions. These functions, when added together, constitute the phenomenon that we call 'life'.

The key to this interpretation of life is the functional diversity of proteins. Proteins with different amino acid sequences adopt quite different structures that have very different chemical properties and this enables different proteins to play a variety of roles in living systems. We have already seen how some amino acid sequences result in tough fibrous proteins, such as keratin and collagen, which give structure and rigidity to the framework of an organism. Other proteins have amino acid sequences that result in flexible structures. These **motor proteins** are able to change their shape, enabling organisms to move around. The muscle protein myosin is a motor protein, as is dynein in cilia and flagella.

Other types of proteins have quite different functions. Enzymes are proteins whose amino acid sequences enable them to catalyze biochemical reactions, such as those involved in metabolism. Other proteins have transport functions and carry compounds around the body. We have already studied hemoglobin, which carries oxygen from the lungs to other tissues. A second example is serum albumin, which transports fatty acids, which are the building blocks of lipids and are also used as energy sources.

Some proteins help to store molecules for future use by the organism. Examples include ovalbumin, which stores amino acids in egg white, and ferritin, which stores iron in the liver. A large group of proteins have protective functions, such as the immunoglobulins of mammals, which form complexes with foreign proteins and protect the body against infectious agents such as viruses and bacteria.

There are also **regulatory proteins** that control cellular and physiological activities. These include well-known hormones such as insulin, which regulates glucose metabolism in vertebrates, and the two growth hormones somatostatin and somatotropin. Although made inside a cell hormones are secreted so they can travel around the body and convey their regulatory messages to other cells. Other regulatory proteins work entirely within the cells in which they are synthesized, possibly responding to signals from extracellular hormones. Examples are components of the MAP kinase pathway, which regulate activities such as cell division in response to external signals.

All of these diverse functions are specified by the chemical properties of individual proteins, which in turn are specified by their three-dimensional structures and hence by their amino acid sequences. The adoption of those correct three-dimensional structures by protein folding is therefore one of the fundamental cornerstones of biology.

Further reading

Bragulla HH and Homberger DG (2009) Structure and functions of keratin proteins in simple, stratified, keratinized and cornified epithelia. *Journal of Anatomy* **214**, 516–59.

Covington AK, Bates RG and Durst RA (1985) Definition of pH scales, standard reference values, measurement of pH and related terminology. *Pure and Applied Chemistry* **57**, 531–42. *Everything that you will ever need to know about this subject.*

Eisenberg D (2003) The discovery of the α-helix and β-sheet, the principal structural features of proteins. *Proceedings of the National Academy of Sciences USA* **100**, 11207–10.

Jungck JR (1985) Margaret Oakley Dayhoff, "harnessing the computer revolution". *The American Biology Teacher* **47**, 9–10. *A review of the work of one of the first bioinformaticians.*

Klug A (1999) The tobacco mosaic virus particle: structure and assembly. *Philosophical Transactions of the Royal Society of London*, series B **354**, 531–5.

Mayer MP (2013) Hsp70 chaperone dynamics and molecular mechanism. *Trends in Biochemical Sciences* **38**, 507–14.

Pauling L and Corey RB (1951) The pleated sheet, a new layer configuration of polypeptide chains. *Proceedings of the National Academy of Sciences USA* **37**, 251. *The first description of the β-sheet.*

Pauling L, Corey RB and Branson HR (1951) The structure of proteins: two hydrogen-bonded helical configurations of the polypeptide chain. *Proceedings of the National Academy of Sciences USA* **37**, 205–11. *The first description of the α-helix.*

Ramachandran GN, Ramakrishnan C and Sasisekharan V (1963) Stereochemistry of polypeptide chain configurations. *Journal of Molecular Biology* **7**, 95–9. *Describes the psi and phi angles and the Ramachandran plot.*

Römer L and Scheibel T (2008) The elaborate structure of spider silk. *Prion* **2**, 154–61.

Rost B (2001) Protein secondary structure prediction continues to rise. *Journal of Structural Biology* **134**, 2014–18.

Rybczynski N, Gosse JC, Harington R, Wogelius RA, Hidy AJ and Buckley M (2013) Mid-Pliocene warm-period deposits in the High Arctic yield insight into camel evolution. *Nature Communications* **4**, 1550. *An Arctic camel identified by collagen fingerprinting.*

Shoulders MD and Raines RT (2009) Collagen structure and stability. *Annual Review of Biochemistry* **78**, 929–58.

Yébenes H, Mesa P, Muñoz IG, Montoya G and Valpoesta JM (2011) Chaperonins: two rings for folding. *Trends in Biochemical Sciences* **36**, 424–32.

Self-assessment questions

Multiple choice questions

Only one answer is correct for each question. Answers can be found on the website: www.scionpublishing.com/biochemistry

1. Titin, the longest known polypeptide, has how many amino acids?
   (a) 1464
   (b) 3685
   (c) 21 075
   (d) 33 445

2. Which amino acid is given the one-letter abbreviation 'A'?
   (a) Alanine
   (b) Arginine
   (c) Asparagine
   (d) Aspartic acid

3. Which amino acid has an unusual side-chain that includes the nitrogen of the amino group attached to the α-carbon?
   (a) Asparagine
   (b) Proline
   (c) Tryptophan
   (d) Tyrosine

4. The D- and L-forms of an amino acid are examples of what?
   (a) Enantiomers
   (b) Isomers
   (c) Optical isomers
   (d) All of the above

5. What is a molecule that has two ionized groups called?
   (a) Enantiomer
   (b) Hydrophile
   (c) Zwitterion
   (d) None of the above

6. Which one of the following statements regarding the isoelectric point of an amino acid is **incorrect**?
   (a) It is the pH at which an amino acid has no electrical charge
   (b) At the isoelectric point both the carboxyl and amino groups are ionized
   (c) It is a pH value greater than the pK_a of the amino group
   (d) For glycine, the isoelectric point is just below pH 6.0

7. Which two amino acids have positively charged side-chains at pH 7.4?
   (a) Arginine and lysine
   (b) Aspartic acid and glutamic acid
   (c) Cysteine and tyrosine
   (d) Histidine and proline

8. What is the name given to the type of chemical bond that forms between the slightly electropositive hydrogen atom in a polar group and an electronegative atom?
   (a) Covalent bond
   (b) Electrostatic bond
   (c) Hydrogen bond
   (d) van der Waals bond

9. Which one of the following statements is a feature of hydrophobic amino acids?
   (a) They are readily soluble
   (b) They are usually found on the surface of a protein
   (c) They have nonpolar side-chains
   (d) They often form hydrogen bonds with other hydrophobic amino acids

10. Which one of the following compounds is an example of a modified amino acid that is found in collagen?
   (a) 4-hydroxyproline
   (b) Pyrrolysine
   (c) Selenocysteine
   (d) Selenoproline

11. Which one of the following statements regarding a peptide bond is **incorrect**?
    (a) A peptide bond is able to rotate
    (b) A peptide bond is formed by a condensation reaction
    (c) A peptide bond forms between the carboxyl and amino groups of adjacent amino acids
    (d) A peptide bond is a single bond but due to resonance has some double bond characteristics

12. Because of steric effects, what proportion of the possible combinations of *psi* and *phi* bond angles never occur?
    (a) 7%
    (b) 57%
    (c) 77%
    (d) All combinations of *psi* and *phi* are possible

13. An α-helix is stabilized by what type of interactions?
    (a) Covalent bonds between cysteine amino acids
    (b) Hydrogen bonds between complementary amino acids
    (c) Hydrogen bonds between peptide groups four positions along the polypeptide
    (d) Hydrophobic interactions between peptide groups four positions along the polypeptide

14. A β-sheet is stabilized by what type of interactions?
    (a) Covalent bonds between proline amino acids which mark the start and end points of the β-sheet
    (b) Hydrogen bonds between complementary amino acids
    (c) Hydrogen bonds between two parts of a polypeptide so that those segments are held together side by side
    (d) Hydrophobic interactions between different parts of the β-sheet

15. A collagen polypeptide forms what type of secondary structure?
    (a) α-helix
    (b) β-sheet
    (c) Double helix
    (d) Left-handed helix

16. Which one of the following statements regarding silk fibroin is **incorrect**?
    (a) Fibroin forms extensive β-sheets
    (b) Fibroin has a high glycine and alanine content
    (c) It has a closely packed structure
    (d) The fibroin polypeptide forms a triple helix which gives it tensile strength

17. The structure in which two α-helices lie side by side in antiparallel directions in such a way that their side-chains intermesh is called what?
    (a) αα motif
    (b) βαβ loop
    (c) β turn
    (d) CD4 domain

18. Which of these proteins has a quaternary structure?
    (a) Carbonic anhydrase
    (b) Concanavalin A
    (c) Hemoglobin
    (d) Myoglobin

19. The tobacco mosaic virus capsid is made up of how many subunits?
    (a) 158
    (b) 240
    (c) 2130
    (d) 5200

20. The unfolding of a protein is called what?
    (a) Denaturation
    (b) Dialysis
    (c) Oxidation
    (d) Renaturation

21. What does the molten globule model for protein folding state?
    (a) Because the globule is molten it can change conformation rapidly
    (b) Collapse into a molten globule might automatically fold some of the polypeptide into its α-helices and β-sheets
    (c) The initial step in folding is the rapid collapse of the polypeptide into a compact structure
    (d) All of the above statements are part of the molten globule model

22. Hsp70 proteins are examples of what?
    (a) Chaperonins
    (b) Molecular chaperones
    (c) Molten globules
    (d) Motor proteins

23. The GroEL/GroES complex is a type of what?
    (a) Chaperonin
    (b) Hsp70 protein
    (c) Molten globule
    (d) Motor protein

24. Which one of the following is an example of a storage protein?
    (a) Dynein
    (b) Ferritin
    (c) Insulin
    (d) Keratin

Short answer questions

These do not require additional reading.

1. Draw the general structure of an amino acid and indicate the chemical groups that participate in formation of peptide bonds.

2. Distinguish between the L- and D-forms of an amino acid and explain how the two configurations are identified experimentally.

3. Define the term pK_a and explain why some amino acids have two pK_a values and others have three. How do these pK_a values affect the chemical properties of different amino acids?

4. Describe the differences between covalent, electrostatic and hydrogen bonds.

5. Explain how the Ramachandran plot enables combinations of the *psi* and *phi* bond angles that give rise to different polypeptide configurations to be identified.

6. Distinguish between the structures of the α-helix and β-sheet.

7. Describe the structure of the fibroin protein and explain how this structure enables silk to be both strong and flexible. To what extent is the structure of fibroin typical of that of other fibrous proteins?

8. Using examples, explain what is meant by the terms 'tertiary' and 'quaternary' with regard to the structure of a globular protein.

9. Summarize the molten globule model for protein folding.

10. Describe the roles of Hsp70 proteins and chaperonins in protein folding.

Self-study questions

These questions will require calculation, additional reading and/or internet research.

1. The Henderson–Hasselbalch equation defines the relationship between pH and pK_a as:

$$pH = pK_a + \log \frac{[A^-]}{[HA]}$$

where $[A^-]$ and $[HA]$ are the concentrations of the ionized and non-ionized forms of a chemical group, respectively. Explain how the Henderson–Hasselbalch equation relates to the ionization graph for glycine shown in *Fig. 3.7*.

2. Draw a graph showing the relative amounts of the different ionized versions of arginine at different pH values. The relevant pK_a values are 2.01 for the carboxyl group, 9.04 for the amino group, and 12.48 for the side-chain.

3. Most proteins denature at temperatures above approximately 50°C because of the disruptive effects that heat has on the chemical bonds that stabilize secondary and tertiary structures. However, some bacteria live at high temperatures, for example in hot springs, and their proteins retain their tertiary structures at temperatures up to 90°C. Speculate on the nature of the structural innovations that might enable a protein to withstand such high temperatures.

4. A protein with a molecular mass of 380 kDa is treated with β-mercaptoethanol. The molecular mass is measured again and now found to be 190 kDa. Provide an explanation for these results.

5. Does the existence of molecular chaperones contradict the statement that the amino acid sequence of a polypeptide contains the information needed to fold that polypeptide into its correct tertiary structure?

CHAPTER 4
Nucleic acids

STUDY GOALS

After reading this chapter you will:

- appreciate the importance of DNA as the store of biological information in living cells

- understand that nucleic acids are polynucleotides and be able to describe the basic structure of a nucleotide

- recognize that the variability between different nucleotides is conferred by the structure of the nitrogenous base

- be able to describe the structure of the phosphodiester bond and to explain how this bond results in the two ends of a polynucleotide being chemically distinct

- know the key features of the double helix structure for DNA, and understand what is meant by 'base pairing' and why base pairing is of fundamental importance in biology

- know that there are different versions of the double helix and be able to describe the structural variations between them

- be aware that RNA molecules often form intramolecular base pairs and be able to describe examples of base-paired RNA molecules

- understand how chemical modification can increase the variability of nucleotides in an RNA molecule

- recognize that DNA molecules are much longer than the chromosomes in which they are contained

- understand how this packaging problem is solved by association of DNA with histones and by higher levels of organization such as the 30 nm chromatin fiber

Nucleic acids are the second type of biomolecule that we will study. There are two types of nucleic acid in living cells, **deoxyribonucleic acid** or **DNA**, and **ribonucleic acid** or **RNA**. DNA is the store of genetic information in all cellular life forms and many viruses. RNA is the store of genetic information in some viruses but, more importantly, in all organisms it acts as the intermediate between DNA and the synthesis of proteins.

4.1   The structures of DNA and RNA

The structures of DNA and RNA are very similar, so we can deal with them together. We must ask two questions. First, what is the molecular structure of a DNA or RNA polymer? Secondly, what are the three-dimensional structures of these polymers in living cells? Answering the second of these questions will introduce us to the **double helix**, the famous structure of DNA, whose discovery by James Watson and Francis Crick in 1953 was the most important breakthrough in biology during the twentieth century.

4.1.1   Polynucleotide structure

Nucleic acids are polymeric molecules made up of monomeric units called **nucleotides**. The nucleotides are linked together to form **polynucleotide** chains that can be thousands of units in length for RNA, or millions for DNA.

Nucleotides are the monomeric units of a nucleic acid molecule

A nucleotide is made up of three distinct components: these are a sugar, a nitrogenous base and a phosphate group (*Fig. 4.1*).

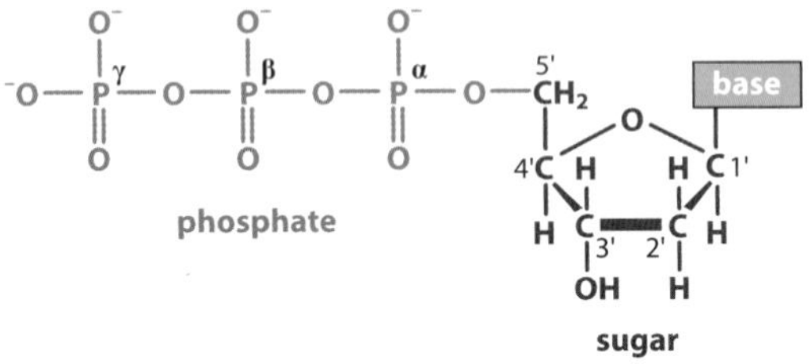

Figure 4.1 The components of a DNA nucleotide.

The sugar component of the nucleotide is a **pentose**. A pentose sugar has five carbon atoms which, in a nucleotide, are numbered from 1′ to 5′. The dash is called the 'prime' and the numbers are called 'one-prime', 'two-prime', etc. The prime is used to distinguish the carbon atoms in the sugar from the carbon and nitrogen atoms in the nitrogenous base, which are numbered 1, 2, 3, and so on. RNA nucleotides contain the pentose called ribose, and DNA contains 2′-deoxyribose. The name indicates that in 2′-deoxyribose the ribose structure has been altered by replacement of the hydroxyl group (–OH) attached to carbon atom number 2′ with a hydrogen group (–H) (*Fig. 4.2*).

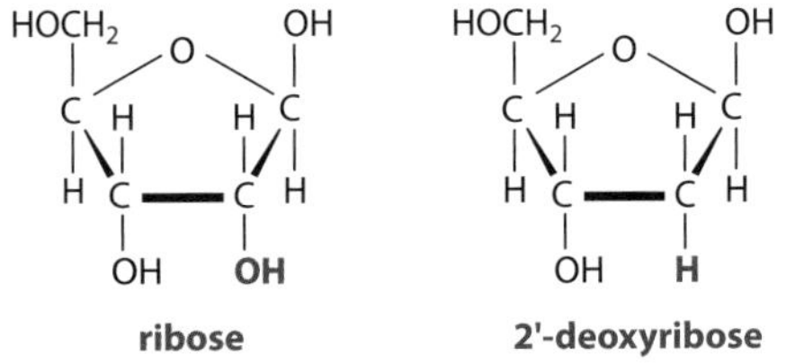

We will study the structure of pentose and related sugars in *Section 6.1.1.*

Figure 4.2   Ribose and 2′-deoxyribose.
The difference between these two sugars is the identity of the group attached to the 2′-carbon. This group is a hydroxyl group for ribose, and a hydrogen atom for 2′-deoxyribose.

The second part of a nucleotide is the nitrogenous base. These are single- or double-ring structures that are attached to the 1′-carbon of the sugar. In DNA any one of four different nitrogenous bases can be attached at this position. These are **adenine** and **guanine**, which are double-ring **purines**, and **cytosine** and **thymine**, which are single-ring **pyrimidines**. Three of these – adenine, guanine and cytosine – are also found in RNA, but the fourth, thymine, is replaced with a different pyrimidine called **uracil**. The structures of all five bases are shown in *Figure 4.3*. The base is joined to the sugar by a **β-N-glycosidic bond** attached to nitrogen number 1 of the pyrimidine or number 9 of the purine.

A molecule comprising the sugar joined to a base is called a **nucleoside**. This is converted into a nucleotide by attachment of a phosphate group to the 5′-carbon of the sugar. Up to three individual phosphates can be attached in series. These phosphate groups are designated α, β and γ, with the α-phosphate being the one attached directly to the sugar (see *Fig. 4.1*).

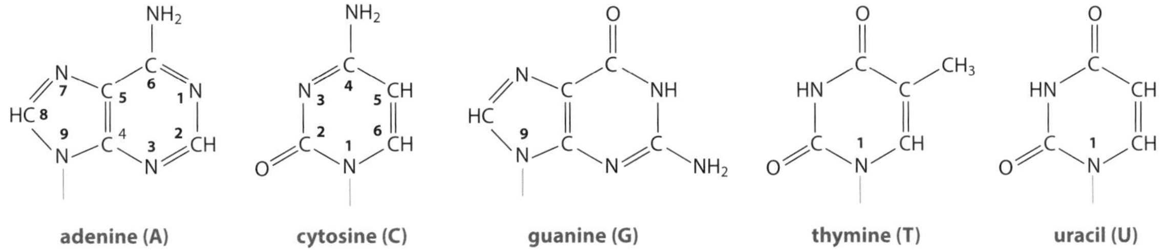

Figure 4.3 **The five nitrogenous bases found in DNA and RNA.**

Table 4.1. **The nucleotides present in nucleic acid molecules**

Nucleotide	Base component	Abbreviations		Found in
		3-letter	1-letter	
2'-deoxyadenosine 5'-triphosphate	Adenine	dATP	A	DNA
2'-deoxyguanosine 5'-triphosphate	Guanine	dGTP	G	DNA
2'-deoxycytidine 5'-triphosphate	Cytosine	dCTP	C	DNA
2'-deoxythymidine 5'-triphosphate	Thymine	dTTP	T	DNA
Adenosine 5'-triphosphate	Adenine	ATP	A	RNA
Guanosine 5'-triphosphate	Guanine	GTP	G	RNA
Cytidine 5'-triphosphate	Cytosine	CTP	C	RNA
Uridine 5'-triphosphate	Uracil	UTP	U	RNA

The full names of the nucleotides are listed in *Table 4.1*. Usually we refer to them by their abbreviations, dATP, dGTP, dCTP and dTTP for DNA, and ATP, CTP, GTP and UTP for RNA. If we are writing out the sequence of nucleotides in a DNA or RNA molecule then we use the one-letter abbreviations, which are A, C, G and T for DNA, and A, C, G and U for RNA. Using the same abbreviations for both sets of nucleotides rarely causes any confusion because the presence of Ts or Us in the sequence indicates whether it is DNA or RNA. For example, the sequence ATCGAGCGACGT is clearly DNA.

Nucleotides are joined by phosphodiester bonds

The next step in building up the structure of a nucleic acid molecule is to link the individual nucleotides together to form a polymer. This polymer is called a polynucleotide and is made by attaching one nucleotide to another through the phosphate groups.

The structure of a DNA trinucleotide, a short DNA molecule comprising three individual nucleotides, is shown in *Figure 4.4*. An RNA polynucleotide has the same structure except of course that RNA nucleotides are used. The nucleotide monomers are linked together by joining the α-phosphate group, attached to the 5'-carbon of one nucleotide, to the 3'-carbon of the next nucleotide in the chain. Normally a polynucleotide is built up from nucleoside triphosphate subunits, so during polymerization the β- and γ-phosphates are cleaved off. The hydroxyl group attached to the 3'-carbon of the second nucleotide is also lost. The resulting linkage is called a **phosphodiester bond**, 'phospho' indicating the presence of a phosphorus atom, and 'diester' referring to the two ester (C–O–P) bonds in each linkage. To be precise, we should call this a 3'-5' phosphodiester bond, so that there is no confusion about which carbon atoms in the sugar participate in the bond.

An important feature of a polynucleotide is that the two ends of the molecule are not the same. This is clear from an examination of *Figure 4.4*. The top of this polynucleotide ends with a nucleotide in which the triphosphate group attached to the

We will study the details of polynucleotide synthesis in *Section 14.1.2.*

Figure 4.4 The structure of a DNA trinucleotide.

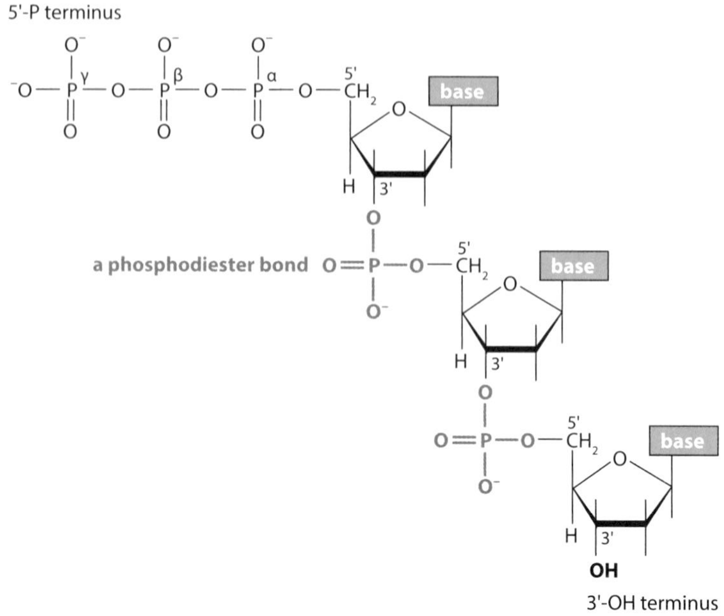

5'-P terminus

a phosphodiester bond

3'-OH terminus

5'-carbon has not participated in a phosphodiester bond and the β- and γ-phosphates are still in place. This end is called the **5' end** or **5'-P terminus**. At the other end of the molecule the unreacted group is the 3'-hydroxyl. This end is called the **3' end** or **3'-OH terminus**.

The chemical distinction between the two ends means that polynucleotides have a direction, which can be looked on as 5'→3' (down in *Fig. 4.4*) or 3'→5' (up in *Fig. 4.4*). The difference between the ends also means that the reaction needed to extend a DNA or RNA polymer in the 5'→3' direction is different from that needed to make a 3'→5' extension. In living cells, polynucleotides are always extended in the 5'→3' direction, by adding nucleotides to the free 3' end. No enzymes capable of catalyzing the chemical reaction needed to make DNA or RNA in the opposite direction, 3'→5', have ever been discovered.

There is apparently no limitation to the number of nucleotides that can be joined together to form a polynucleotide. RNA molecules containing several thousand nucleotides are known, and the DNA molecules in chromosomes are much longer, sometimes several million nucleotides in length. In addition, there are no chemical restrictions on the order of the different nucleotides in a DNA or RNA molecule.

4.1.2   DNA and RNA secondary structures

Both DNA and RNA molecules adopt secondary structures due to chemical interactions between different polynucleotides, or different parts of a single polynucleotide. In DNA, this secondary structure is the famous double helix discovered by James Watson and Francis Crick in 1953. The double helix is a complicated structure but the key facts about it are not too difficult to understand.

The features of the double helix

In the double helix the two polynucleotides are arranged in such a way that their sugar–phosphate 'backbones' are on the outside of the helix, and their bases are on the inside (*Fig. 4.5*). The bases are stacked on top of each other rather like a pile of plates or the steps in a spiral staircase. The two polynucleotides are **antiparallel**, meaning that they run in different directions, one being orientated in the 5'→3' direction, and

Figure 4.5 The double helix structure of DNA.
On the left the double helix is drawn with the sugar–phosphate backbone of each polynucleotide shown as a gray ribbon with the base pairs in green. On the right the chemical structure for three base pairs is shown.

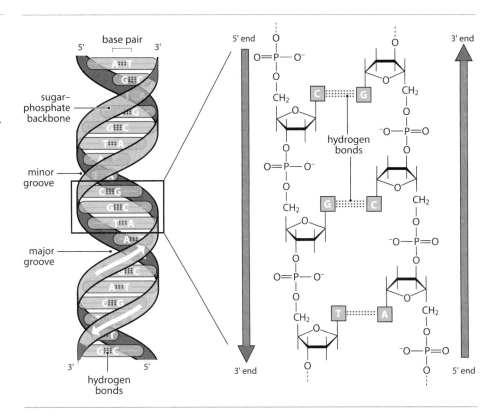

the other in the 3'→5' direction. The polynucleotides must be antiparallel in order to form a stable helix, and molecules in which the two polynucleotides run in the same direction are unknown in nature. The double helix is right-handed, but is not absolutely regular. Instead, two grooves spiral along the length of the helix. One of these grooves is relatively wide and deep and is called the **major groove**, and the other is narrow and less deep and is called the **minor groove**. These two grooves are clearly visible in *Figure 4.5*.

The helix is stabilized by two types of chemical interactions. The first of these are hydrogen bonds, which form between the bases that are adjacent to one another in the two strands of the helix. This **base pairing** can only occur between an adenine on one strand and a thymine on the other strand, or between a cytosine and a guanine (*Fig. 4.6*). These are the only pairs that are possible, partly because of the geometries of the nucleotide bases and the relative positions of the atoms that are able to participate in hydrogen bonds, and partly because the pairing must be between a purine and

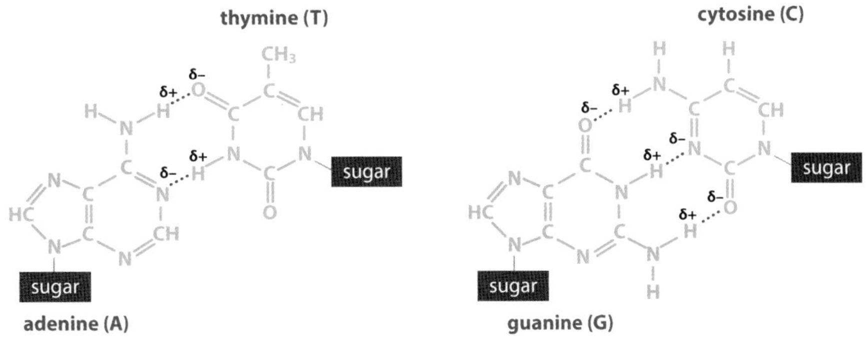

Figure 4.6 Base pairing.
The hydrogen bonds are indicated by dotted red lines. Note that a G–C base pair has three hydrogen bonds whereas an A–T base pair has just two.

PRINCIPLES OF CHEMISTRY

Box 4.1 Base stacking

The base stacking that occurs within the double helix is due to attraction between the aromatic rings of the nucleotide bases. These attractions are greatest when adjacent rings lie in the same plane, but are slightly displaced vertically, as occurs in a helical structure.

The underlying nature of the attraction involved in base stacking is unclear. The phenomenon is sometimes called pi stacking, based on the assumption that it involves p electrons, which are associated with double and triple bonds. Base stacking was originally thought to arise because of interactions between the p electrons in adjacent aromatic rings. This assumption is now being questioned and the possibility that base stacking involves a type of electrostatic interaction is being explored.

---

a pyrimidine; a purine–purine pair would be too big to fit within the helix, and a pyrimidine–pyrimidine pair would be too small. Because of the base pairing, the sequences of the two polynucleotides in the helix are **complementary**, the sequence of one polynucleotide matching the sequence of the other (see *Fig. 4.5*).

The second type of interaction holding the double helix together is called **base stacking**. This involves attractive forces between adjacent base pairs and adds stability to the double helix once the strands have been brought together by hydrogen bonding.

Both base pairing and base stacking are important in holding the two polynucleotides together, but base pairing has added significance because of its biological implications. The limitation that A can only base-pair with T, and G can only base-pair with C, means that DNA replication can result in two perfect copies of a parent molecule through the simple expedient of using the sequences of the pre-existing strands (the "templates") to dictate the sequences of the new strands (*Fig. 4.7*). This is **template-dependent DNA synthesis** and it is the system used by almost all of the enzymes that make new DNA molecules in the cell.

> The role of template-dependent DNA synthesis in replication of a double helix will be described in *Section 14.1.2*.

The double helix exists in several different forms

The double helix shown in *Figure 4.5* is referred to as the **B-form** of DNA. Its characteristic features are a helical diameter of 2.37 nm, a rise of 0.34 nm per base pair, and a pitch (i.e. distance taken up by a complete turn of the helix) of 3.4 nm, corresponding to ten base pairs per turn. The DNA in living cells is thought to be predominantly in this B-form, but we now know that DNA molecules are not entirely uniform in structure. This is mainly because each nucleotide in the helix has the flexibility to take up slightly different molecular shapes. To adopt these different conformations, the relative positions of the atoms in the nucleotide must change slightly. There are a number of possibilities, the most important being:

- Rotation around the β-*N*-glycosidic bond, which changes the orientation of the base relative to the sugar (*Fig. 4.8*), and influences the relative positioning of the two polynucleotides.

Figure 4.7 The role of complementary base pairing during DNA replication.
The limitation that A can only base-pair with T, and G can only base-pair with C, means that template-dependent DNA synthesis results in two perfect copies of a parent double helix. The polynucleotides of the parent double helix are shown in gray, and the newly synthesized strands are in pink.

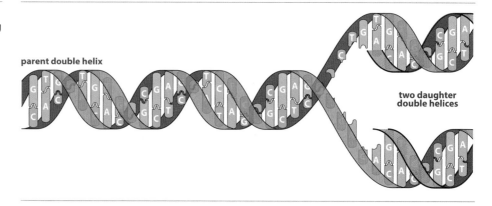

parent double helix

two daughter double helices

Box 4.2 The discovery of the double helix

The discovery of the double helix, by James Watson and Francis Crick of Cambridge University in the UK, in 1953, was the most important breakthrough in twentieth century biology. In the years leading up to 1953, it was shown that genes are made of DNA. One of the key properties of a gene is its ability to replicate, so copies can be passed to daughter cells during cell division and to the offspring during reproduction. So if genes are made of DNA then a DNA molecule must be able to replicate. Before the double helix structure was known, this replication process was a complete mystery, but once the double helix was revealed, with the two strands held together by complementary base pairing, the replication process became obvious.

When Watson and Crick began their work, the structures of the nucleotides, and the way these are linked together to form a polynucleotide, were known. What was not known was the structure of DNA in a living cell. Was it a single polynucleotide, or were there two or more polynucleotides in a DNA molecule? One way to address this question is to measure the density of DNA in the semi-crystalline fibers obtained when a DNA solution is mixed with salt. Several measurements of DNA fiber density had been reported, but they did not agree. Some measurements suggested that there were three polynucleotides in a single molecule, others suggested two. Linus Pauling, who had previously worked out the α-helix and β-sheet polypeptide conformations, devised an incorrect triple helix structure. Watson and Crick decided there were more likely to be two strands in a DNA molecule.

Studying DNA structure by X-ray diffraction analysis

In a DNA fiber, the individual molecules are oriented in a regular array. This means that their structure can be studied by **X-ray diffraction analysis**. In this technique, the fiber is bombarded with X-rays, some of which are deflected by the atoms in the DNA molecule. X-ray-sensitive photographic film placed across the beam reveals a series of spots, called the X-ray diffraction pattern. From the positions and intensities of features in the pattern, information on the structure of DNA can be deduced.

X-ray diffraction patterns were obtained from DNA fibers by Rosalind Franklin of King's College, London. The patterns showed that DNA is a helix and also revealed some of its dimensions. A periodicity of 0.34 nm indicated the spacing between individual

base pairs, and another periodicity of 3.4 nm gave the distance needed for a turn of the helix. By building models, Watson and Crick showed that if a helix with these dimensions contained just two polynucleotides, then the sugar–phosphate backbones had to be on the outside of the molecule, the strands had to be antiparallel, and the helix must be right-handed. This was the only way the various atoms could be spaced out appropriately.

The importance of Chargaff's base ratios

Franklin came very close to solving the double helix structure, but Watson and Crick were the ones who completed the puzzle, because they realized that the two strands must be held together by complementary base pairing. Erwin Chargaff at Columbia University in the USA had published data showing the amounts of each of the four nucleotides in DNA from different sources. To do this, he treated DNA extracts with weak acid to break the molecules into their component nucleotides. He then separated each nucleotide by **paper chromatography**. In this method, a mixture of nucleotides is placed at one end of a paper strip, and an organic solvent, such as *n*-butanol, allowed to soak along the strip. As the solvent moves it carries the nucleotides with it, but at different rates depending on how strongly each nucleotide is absorbed by the paper matrix. Each nucleotide therefore forms a different spot on the filter paper. After extracting the nucleotides from the spots, ultraviolet spectrophotometry was used to determine the relative amounts of each nucleotide in the sample.

paper chromatography

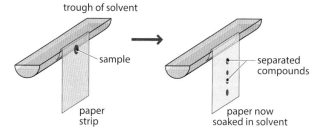

trough of solvent

sample

separated compounds

paper strip

paper now soaked in solvent

typical results obtained by Chargaff

Human cells		*Escherichia coli*	
base ratio		**base ratio**	
A : T	1.00	A : T	1.09
G : C	1.00	G : C	0.99

These experiments revealed a simple relationship between the proportions of the nucleotides in any one sample of DNA. The relationship is that the number of adenines equals the number of thymines (A = T), and the number of guanines equals the number of cytosines (G = C).

This relationship was the key to solving the double helix structure. Watson realized, on the morning of Saturday 7 March 1953, that the base pairs formed by adenine–thymine and guanine–cytosine have almost identical shapes. These pairs would fit neatly inside the double helix giving a regular spiral with no bulges. And if these were the only pairs that were allowed then the amount of A would equal the amount of T, and G would be the same as C. Everything fell into place and the greatest mystery of biology – how a gene can replicate – had been solved.

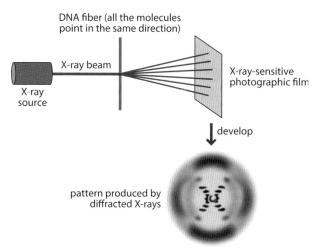

DNA fiber (all the molecules point in the same direction)

X-ray beam

X-ray source

X-ray-sensitive photographic film

develop

pattern produced by diffracted X-rays

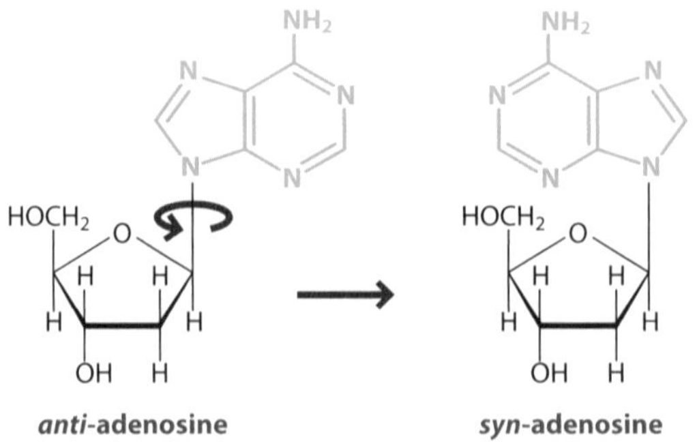

Figure 4.8 The structures of *anti*- and *syn*-adenosine.
The two structures differ in the orientation of the base relative to the sugar component of the nucleotide. Rotation around the β-*N*-glycosidic bond converts one form into the other. The three other nucleotides also have *anti*- and *syn*-conformations.

anti-adenosine   *syn*-adenosine

• **Sugar pucker**, which refers to the three-dimensional shape of the sugar, which can adopt either the C2′-endo or C3′-endo configuration. These configurations affect the conformation of the sugar–phosphate backbone.

Conformational changes within individual nucleotides can therefore lead to major changes in the overall structure of the helix. It has been recognized since the 1950s that changes in the dimensions of the double helix occur when fibers containing DNA molecules are exposed to different relative humidities. For example, the modified

Box 4.3 Sugar pucker

Sugar pucker occurs because the ribose sugar does not have a planar structure. When viewed from the side, one or two of the carbon atoms are either above or below the plane of the sugar. In the C2′-endo configuration the 2′-carbon is above the plane and the 3′-carbon slightly below, and in the C3′-endo configuration the 3′-carbon is above the plane and the 2′-carbon below. Because the 3′-carbon participates in the phosphodiester bond with the adjacent nucleotide, the two pucker configurations have different effects on the conformation of the sugar–phosphate backbone.

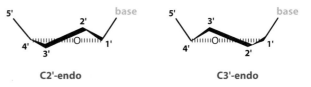

C2′-endo   C3′-endo

Figure 4.9 The A-, B- and Z-forms of the double helix.
The major and minor grooves on each molecule are indicated by 'M' and 'm', respectively.
Reproduced from http://en.wikipedia.org/wiki/Z-DNA under a Creative Commons license.

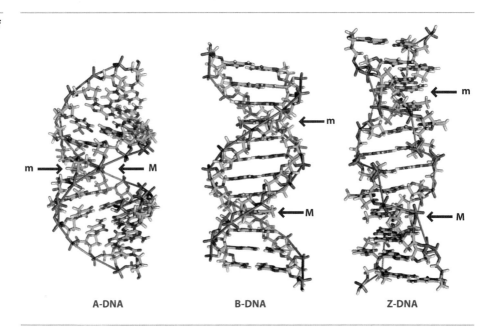

A-DNA   B-DNA   Z-DNA

Table 4.2. **Features of the different conformations of the DNA double helix**

Feature	A-form	B-form	Z-DNA
Type of helix	Right-handed	Right-handed	Left-handed
Base orientation	*anti*	*anti*	Mixture
Sugar pucker	C3'-endo	C2'-endo	Mixture
Number of base pairs per turn	11	10	12
Distance between base pairs (nm)	0.23	0.34	0.38
Distance per complete turn (nm)	2.5	3.4	4.6
Diameter (nm)	2.55	2.37	1.84
Major groove	Narrow, deep	Wide, deep	Flat
Minor groove	Shallow	Narrow, shallow	Narrow, deep

version of the double helix called the **A-form** (*Fig. 4.9*) has a diameter of 2.55 nm, a rise of 0.23 nm per base pair and a pitch of 2.5 nm, corresponding to 11 base pairs per turn (*Table 4.2*). Like the B-form, A-DNA is a right-handed helix and the bases are in the *anti*-conformation relative to the sugar. The main difference lies with the sugar pucker, the sugars in the B-form being in the C2'-endo configuration, and those in A-DNA in the C3'-endo configuration. This change in configuration alters the conformations of the sugar–phosphate backbones, with the result that in A-DNA the major groove is deeper than in the B-form, and the minor groove shallower and broader.

A third type, **Z-DNA**, is more strikingly different. In this structure the helix is left-handed, not right-handed as it is with A- and B-DNA. The sugar–phosphate backbone adopts an irregular zigzag conformation, with one of the two grooves virtually non-existent but the other very narrow and deep (*Fig. 4.9*). Z-DNA is more tightly wound with 12 bp per turn and a diameter of only 1.84 nm (*Table 4.2*). Z-DNA is known to occur in regions of a double helix that contain repeats of the motif GC (i.e. the sequence of each strand is ...GCGCGCGC...). In these regions each G nucleotide has the *syn* and C3'-endo conformations, and each C is *anti* and C2'-endo. Z-DNA is thought to form in cellular DNA adjacent to segments of B-DNA that have become slightly unwound, as occurs when a gene is being copied into RNA. Unwinding results in torsional stress and this might be relieved to some extent by forming the more compact Z version of the helix (*Fig. 4.10*).

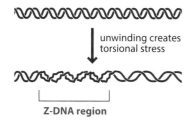

unwinding creates torsional stress

Z-DNA region

Figure 4.10 A possible function for Z-DNA in the cell.
Regions of Z-DNA might form adjacent to an under-wound segment of B-DNA, in order to relieve the torsional stress that is created.

RNA molecules often have intramolecular base pairs

The nucleotides in RNA molecules can also form base pairs, the rules being that A base pairs with U, and G with C. Some RNA double helices are known, but the pairing is not usually between different polynucleotides. Instead, a typical RNA molecule adopts a folded structure held together by intramolecular base pairs, ones that form between nucleotides in the same molecule. To illustrate this, we will look at the structure of **transfer RNA** or **tRNA**, a type of RNA present in all organisms and involved in the synthesis of proteins.

We will study the role of tRNA in protein synthesis in *Section 16.1.2*.

Box 4.4 **Units of length for nucleic acid molecules**

The **base pair (bp)** is the unit of length for a double-stranded DNA molecule.

- 1000 bp = 1 kilobase pair (1 kb)
- 1 000 000 bp = 1000 kb = 1 megabase pair (1 Mb)

Many natural DNA molecules are over 1 Mb in length. The single DNA molecule in human chromosome 1, for example, is 247 Mb long.

Most RNA molecules are single-stranded and so their lengths are described simply as 'nucleotides'. Very few RNA molecules are longer than a few thousand nucleotides, so terms such as 'kilonucleotides' are rarely used.

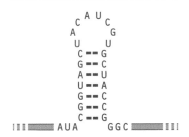

Figure 4.11 The base-paired cloverleaf structure of a tRNA molecule.
The four base-paired structures are shown in different colors. Small dots indicate regions where the number of nucleotides varies in different tRNAs. The olive green nucleotides form a fourth loop, but this loop has no base pairing.

The way in which miRNAs regulate gene expression is described in *Section 17.2.1*.

```
        A U
      C     C
    A         G
    U         U
      C == G
      G == C
      A == U
      U == A
      G == C
      G == C
      C == G
|||    AUA    GGC    |||
```

Figure 4.12 A typical RNA stem–loop structure.

Transfer RNA molecules are relatively small, mostly between 74 and 95 nucleotides in length. Each organism synthesizes a number of different tRNAs, each in multiple copies. However, virtually every tRNA molecule in every organism can be folded into a similar base-paired structure referred to as the **cloverleaf** (*Fig. 4.11*). The cloverleaf has four 'leaves' radiating from a central core. Three of these leaves are **stem–loop** structures formed by bending the polynucleotide chain back on itself, with a short stretch of base pairing making up the stem that holds the conformation together (*Fig. 4.12*). For this to happen, the nucleotide sequences in the two parts of the stem must be complementary. To produce a complex structure such as the cloverleaf, the components of these pairs of complementary sequences must be arranged in a characteristic order within the RNA sequence.

Stem–loops are found in many different types of RNA, with stems ranging from just three or four base pairs up to 100 or so. In longer stems, not all of the pairs of bases have to be complementary, because a few **mismatches** can be tolerated without destabilizing the structure. Hydrogen bonding can also occur between G and U nucleotides, resulting in nonstandard base pairs, and some structures are slightly irregular due to the pairing missing out one or more nucleotides on one side of the stem. All of these features are illustrated in *Figure 4.13*, which shows the structure of a microRNA (miRNA), a type of RNA that regulates gene expression.

The loop also contributes to the stability of the overall structure. The loop must contain at least three nucleotides, as this is the minimum needed in order for the polynucleotide to execute a 180° turn. The four nucleotide sequence 5'–UUCG–3' is particularly common because the resulting structure, called the **tetraloop**, is relatively stable due to strong base stacking that forms within this sequence. Larger loops tend to lack base stacking and so are less stable.

Although the cloverleaf is a convenient way of drawing the structure of a tRNA, it is only a representation and, in the cell, tRNAs have a different three-dimensional structure. This structure has been determined by X-ray crystallography and is shown in *Figure 4.14*. The base pairs in the stems of the cloverleaf are still present in the three-dimensional structure, but several additional base pairs form between nucleotides in different loops, which appear widely separated in the cloverleaf. This folds the molecule into a compact L-shaped conformation. The same is true of other RNA molecules – we can draw them in two dimensions as neat structures with stems and loops, but in reality their three-dimensional conformations are much more complex.

Figure 4.13 The structure of a human miRNA.
The miRNA adopts a stem–loop structure. Within the stem there are two mismatch positions (shown in green), and one example of irregular pairing (purple). There are also three nonstandard G–U base pairs.

4.1.3 RNAs display a diverse range of chemical modifications

Transfer RNAs display a second common feature of RNA molecules that we should consider at this point. Some of the nucleotides in a tRNA are altered after synthesis of the polynucleotide by various types of chemical modification. This means that these molecules, as well as many other RNAs, contain many more than the four standard nucleotides that we have encountered so far.

Figure 4.14 The three-dimensional structure of a tRNA.
The different parts of the tRNA are colored in the same way as in *Figure 4.11*.

We will study the chemical modification of RNA in more detail in *Section 15.2.3*

The following are the commonest types of modification (*Fig. 4.15*):

- **Methylation**, which involves the addition of one or more methyl groups (–CH₃) to the base or sugar component of the nucleotide. An example is the conversion of guanine to 7-methylguanine.

- **Deamination**, which is the removal of an amino group (–NH₂). Deamination converts adenine into hypoxanthine, and guanine to xanthine. It can also convert cytosine to uracil. Neither thymine nor uracil have an amino group and so cannot be deaminated.

- **Thio substitution**, in which an oxygen atom is replaced with a sulfur. An example of a thio-substituted base is 4-thiouracil, which results from sulfur substitution of uracil.

- **Base rearrangement**, in which the positions of atoms in the purine or pyrimidine ring becoming changed. The commonest example is the conversion of uracil to pseudouracil.

- **Double bond saturation**, which involves conversion of a double bond in the base to a single bond. This can also happen with uracil, to give dihydrouracil.

Over 50 types of chemical modification have been discovered so far in different types of RNA. The enzymes that carry out these modifications are thought to recognize particular nucleotide sequences or base-paired structures in an RNA molecule, or possibly a combination of both, and so modify only the appropriate nucleotides. The reasons for many of these modifications are unknown, although roles have been assigned to some specific cases. In tRNA, some of the modified nucleotides are recognized by the enzymes that attach an amino acid to the 3′ end of the molecule. This reaction is central to the intermediate role that tRNA plays during protein synthesis. The correct amino acid has to be attached to the correct tRNA, and the modifications within the tRNA are thought to provide some of the specificity that ensures that this happens.

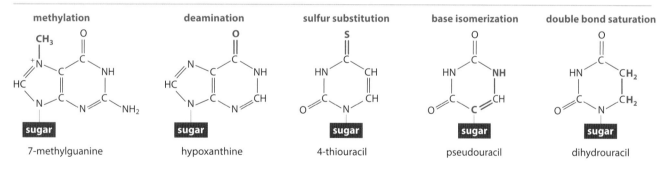

Figure 4.15 Examples of chemically modified bases occurring in RNA molecules.
The differences between these modified bases and the standard ones from which they are derived are shown in red.

4.2 Packaging of DNA

As we have seen, DNA molecules can be millions of nucleotides in length. The human genome, for example, comprises 24 double-stranded DNA molecules. The shortest of these is 47 Mb and the longest 247 Mb. If we bear in mind that in the B-form of DNA, the rise per base pair is 0.34 nm, then a quick calculation tells us that a 47 Mb DNA molecule has a length of 47 000 000 × 0.34 nm, which is equal to 1.6 cm. In fact, the average length of the 24 human DNA molecules is over 4 cm. Each one is contained in a chromosome which, during cell division, adopts a compact structure that is just a few micrometres (μm) in length. There must be a highly organized packaging system to fit such lengthy DNA molecules into such small structures.

4.2.1   Nucleosomes and chromatin fibers

The research that underlies our understanding of the way in which DNA is packaged into chromosomes began many years before the structure of DNA was known. Towards the end of the nineteenth century, cytologists discovered a component of the nucleus that stained deeply with certain types of dye. They called this material **chromatin**, 'chroma' being the Greek word for 'color'. The term 'chromosome' comes from the same root, literally meaning 'colored body'.

Chromatin was later shown to be a complex of DNA and protein. The proteins provide the packaging system that enables a lengthy DNA molecule to be squeezed into a tiny chromosome.

Histones are DNA-binding proteins

The protein component of chromatin mainly consists of **histones**. These are a family of quite short proteins, 100–220 amino acids in length, each with a relatively high content of basic amino acids (*Table 4.3*).

Table 4.3. Histones

Histone	Number of amino acids	Basic amino acid content
H1	194–346	30%
H2A	130	20%
H2B	126	22%
H3	136	23%
H4	103	25%
'Number of amino acids' refers to the human histones. H1 is a family of histones, including H1a–H1e, H1°, H1t and H5. 'Basic amino acid content' is the amount of lysine, histidine and arginine in each protein.		

Figure 4.16 The results of a nuclease protection experiment with purified chromatin.

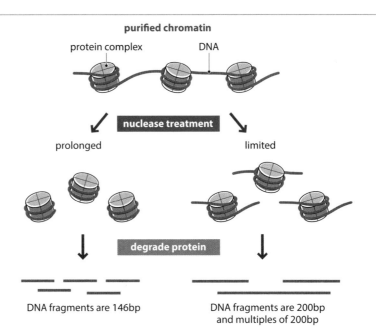

The way in which histone proteins are associated with the DNA in chromatin was first studied by **nuclease protection** experiments. In this procedure a DNA–protein complex is treated with an **endonuclease**, an enzyme that cuts phosphodiester bonds. An example is deoxyribonuclease I, or DNase I, which can be purified from cow pancreas. DNase I cuts DNA at any internal phosphodiester bond, so prolonged treatment breaks the DNA down into its constituent nucleotides. However, the endonuclease has to gain access to the DNA in order to cut it. If part of the DNA is masked ('protected') by attachment to a protein, then the enzyme will not be able to reach it. These protected regions are therefore unaffected by endonuclease treatment, and can be recovered intact after the enzyme has been inactivated and the attached proteins removed from the DNA.

Treatment of chromatin with a nuclease, under different conditions, enables two key features of the arrangement of histones to be deduced (*Fig. 4.16*):

- Prolonged nuclease treatment gives fragments of DNA 146 bp in length. This result suggests that each histone, or group of histones, is closely associated with a segment of DNA of this length.

- Limited nuclease treatment, intended to cut just a few of the phosphodiester bonds in the DNA, gives fragments of approximately 200 bp and multiples thereof. From this result we can conclude that the histones are associated with the DNA in a regular fashion, each histone or group of histones spread out at intervals roughly 200 bp apart.

Confirmation of the second of these two deductions was provided by electron microscopic studies of chromatin. The complex was revealed as a **beads-on-a-string** structure, with each bead of protein spaced out approximately 200 bp apart along the DNA molecule. The beads are called **nucleosomes**. Each nucleosome contains eight histone proteins, two each of histones H2A, H2B, H3 and H4, forming a barrel-shaped **core octamer**. The DNA is wound twice around the outside of the nucleosome, with 146 bp closely associated with the core octamer, and each nucleosome separated by 50–70 bp of unprotected **linker DNA** (*Fig. 4.17A*).

As well as the histones in the core octamer, there is an additional histone, H1, which is attached to the outside of the nucleosome. Structural studies suggest that this histone acts as a clamp, preventing the coiled DNA from detaching from the nucleosome (*Fig. 4.17B*). We now know that histone H1 is not a single protein but a

A. nucleosomes

linker DNA
(50–70 bp)

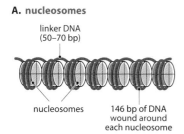

nucleosomes   146 bp of DNA
                  wound around
                  each nucleosome

B. the role of histone H1

octamer of
histones

histone H1

Figure 4.17 **Nucleosomes.**
(A) Nucleosomes form beads on a string of DNA. (B) The linker histone.

Box 4.5 **DNA packaging in bacteria**

Bacteria also have to package their DNA into a relatively small space. The *Escherichia coli* nucleoid contains a single, circular molecule of 4639 kb, corresponding to a contour length (i.e. circumference) of approximately 1.6 mm. In comparison, an *E. coli* cell is about 1 μm by 2 μm. The DNA molecule is folded up tightly by **supercoiling**, which occurs when additional turns are introduced into the DNA double helix (positive supercoiling) or if turns are removed (negative supercoiling). A circular DNA molecule responds to supercoiling by winding around itself to form a more compact structure. Supercoiling is therefore an ideal way of packaging a circular molecule into a small space.

The supercoiled bacterial DNA is attached to a protein core from which loops, each containing 10–100 kb of DNA, radiate out into the cell. Various packaging proteins have been identified. The most abundant of these is called HU, which forms a tetramer around which approximately 60 bp of DNA becomes wound. There are some 60 000 HU proteins per *E. coli* cell, enough to

cover about one-fifth of the DNA molecule, but it is not known if the tetramers are evenly spaced along the DNA or restricted to the core region of the nucleoid.

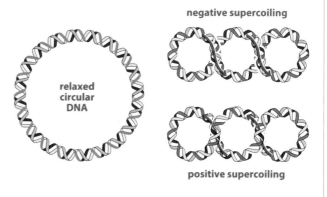

group of proteins, all closely related to one another and now collectively called the **linker histones**. In vertebrates these include the histones that are referred to as H1a–H1e, H1°, H1t and H5.

Higher levels of DNA packaging

The beads-on-a-string conformation reduces the length of a DNA molecule by about one-sixth, so a linear molecule of 4 cm would now be reduced to an effective length of 0.67 cm. This is still much longer than the length of a chromosome. Clearly there are higher levels of DNA packaging.

We now believe that the beads-on-a-string structure is an unpacked form of chromatin that occurs only infrequently in the nuclei of living cells. Electron microscopy of less compact chromatin preparations indicates that most of the nuclear DNA is in the form of the **30 nm fiber**, so named because it has a diameter of approximately 30 nm. The exact way in which the nucleosomes are arranged in the 30 nm fiber is not known, but the 'solenoid' model (*Fig. 4.18*) is the structure that most researchers favor. The individual nucleosomes within the 30 nm fiber could be held together by interactions between the linker histones, or the attachments might involve the core histones, whose N-terminal regions extend outside of the nucleosome (*Fig. 4.19*). The latter hypothesis is attractive because chemical modification of these N-terminal regions results in the 30 nm fiber opening up, enabling genes contained within it to be activated.

The 30 nm fiber reduces the length of the beads-on-a-string structure by about seven times, so a linear DNA molecule that began as 4 cm in length will now take up about 1 mm. The additional levels of packaging that reduce this length to the size of a chromosome are not particularly well understood. One possibility is that histones in different loops of the 30 nm fiber interact with one another to draw the structure into more compact conformations. The more condensed levels of packaging only occur

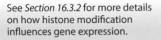

See *Section 16.3.2* for more details on how histone modification influences gene expression.

Figure 4.18 The solenoid model for the 30 nm chromatin fiber.
The nucleosomes have been left out of the drawing in order to show the way in which the DNA molecule is coiled within the fiber. The left-hand drawing shows the view from the side, and the right-hand drawing is the view along the axis of the fiber.
Reprinted with permission from Macmillan Publishers Ltd: *Nature Structural and Molecular Biology*, 12:6, © 2005.

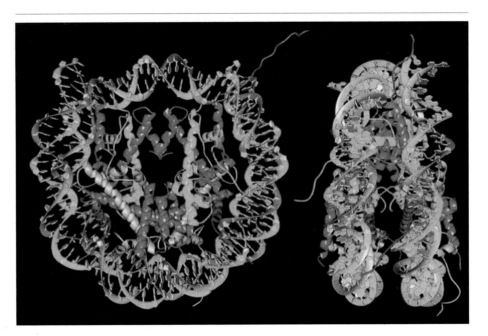

Figure 4.19 The structure of the nucleosome core octamer.
The view on the left is looking down from the top of the barrel-shaped octamer, and the view on the right is from the side. The two strands of the DNA double helix wrapped around the octamer are shown in brown and dark green. The histones are colored as follows: H2A, yellow; H2B, red; H3, blue; H4, bright green. The N-terminal tails of the histone proteins can be seen protruding from the core octamer.
Reprinted with permission from Macmillan Publishers Ltd: *Nature*, 389:251, © 1997.

when the cell is dividing, and the metaphase chromosomes become visible. These are the structures, visible with the light microscope, that have the appearance generally associated with the word 'chromosome' (*Fig. 4.20*). During **interphase**, the period between cell divisions, the DNA is less compact, much of it existing as the 30 nm fiber.

Figure 4.20 Human metaphase chromosomes.
The metaphase chromosomes appear only during cell division, when each chromosome takes up its most compact conformation. There are two copies of each of chromosomes 1–22, as well as an X and a Y.
Reproduced from www.contexo.info/DNA_Basics/chromosomes.htm.

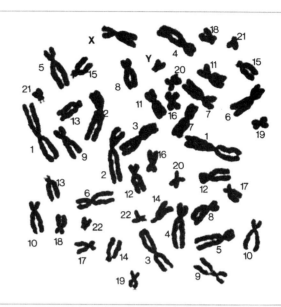

Further reading

Altona C and Sundaralingam M (1972) Conformational analysis of the sugar ring in nucleosides and nucleotides: a new description using the concept of pseudorotation. *Journal of the American Chemical Society* **94**, 8205–12. *Information on sugar pucker.*

Björk GR, Ericson JU, Gustafsson C, et al. (1987) Transfer RNA modification. *Annual Review of Biochemistry*, **56**, 263–87. *Information on modified nucleotides in tRNAs.*

Clark BFC (2001) The crystallization and structural determination of tRNA. *Trends in Biochemical Science* **26**, 511–14. *Determination of the three-dimensional structure of a tRNA.*

Cutter AR and Hayes JJ (2015) A brief review of nucleosome structure. *FEBS Letters* **589**, 2914–22.

Hagerman PJ (1991) RNA 'tetraloops': living in *syn*. *Current Biology* **1**, 50–2.

Harshman SW, Young NL, Parthun MR and Freitas MA (2013) HI histones: current perspectives and challenges. *Nucleic Acids Research* **41**, 9593–609.

Holley RW, Apgar J, Everett GA, et al. (1965) Structure of a ribonucleic acid. *Science*, **147**, 1462–5. *The discovery of the cloverleaf structure of tRNA.*

Rich A and Zhang S (2003) Z-DNA: the long road to biological function. *Nature Reviews Genetics* **4**, 566–72.

Robinson PJJ and Rhodes D (2006) Structure of the '30 nm' chromatin fibre: a key role for the linker histone. *Current Opinion in Structural Biology* **16**, 336–43. *Reviews models for the structure of the 30 nm fiber.*

Watson JD (1968) *The Double Helix*. Atheneum, London. *The most important discovery of 20th century biology, written as a soap opera.*

Watson JD and Crick FHC (1953) Molecular structure of nucleic acids: a structure for deoxyribose nucleic acid. *Nature*, **171**, 737–8. *The scientific report of the discovery of the double helix structure of DNA.*

Yakovchuk P, Protozanova E and Frank-Kamenetskii MD (2006) Base-stacking and base-pairing contributions into thermal stability of the DNA double helix. *Nucleic Acids Research* **34**, 564–74.

Self-assessment questions

Multiple choice questions

Only one answer is correct for each question. Answers can be found on the website: www.scionpublishing.com/biochemistry

1. A hydroxyl group is attached to which carbon of the deoxyribose sugar in DNA?
   (a) 1′
   (b) 2′
   (c) 3′
   (d) 4′

2. Which two purines are found in DNA molecules?
   (a) Adenine and cytosine
   (b) Adenine and guanine
   (c) Adenine and thymine
   (d) Cytosine and thymine

3. What is the link between the nitrogenous base and the sugar component of a nucleotide called?
   (a) Base pair
   (b) β-*N*-glycosidic bond
   (c) Hydrogen bond
   (d) Phosphodiester bond

4. What is the nitrogenous base that is found in DNA but not RNA called?
   (a) Adenine
   (b) Guanine
   (c) Thymine
   (d) Uracil

5. In a polynucleotide, the link between adjacent nucleotides forms between which pair of carbons?
   (a) 1′ and 2′
   (b) 1′ and 3′
   (c) 1′ and 5′
   (d) 3′ and 5′

6. Which one of the following techniques was not used in work that led to discovery of the double helix structure of DNA?
   (a) Model building
   (b) Nuclear magnetic resonance spectroscopy
   (c) Paper chromatography
   (d) X-ray diffraction analysis

7. Which one of these does the pairing between the two strands of the DNA helix involve?
   (a) Covalent bonds
   (b) Ionic interactions
   (c) Hydrogen bonds
   (d) Hydrophobic forces

8. Which one of these statements is **correct** with regard to base stacking?
   (a) It gives rise to the *anti-* and *syn-*conformations of adenosine
   (b) It occurs because the ribose sugar does not have a planar structure
   (c) It only occurs between A and T, and G and C bases
   (d) It results from attractions between the aromatic rings of the nucleotide bases

9. What causes the C2′-endo or C3′-endo conformations of a nucleotide to arise?
   (a) Base pairing
   (b) Base stacking
   (c) Complementarity between the polynucleotides
   (d) Sugar pucker

10. Which type of DNA forms a left-handed helix with 12 bp per turn and a diameter of only 1.84 nm?
    (a) A-form
    (b) B-form
    (c) Z-DNA
    (d) None of the above

11. When referring to 1 000 000 base pairs of DNA, which abbreviation is used?
    (a) kb
    (b) Mb
    (c) Gb
    (d) DNA molecules of 1 000 000 base pairs do not exist

12. Which one of the following statements is **incorrect** with regard to tRNA?
    (a) Each organism synthesizes a number of different tRNAs
    (b) G–T base pairs can occur in a tRNA
    (c) Most tRNAs are between 74 and 95 nucleotides in length
    (d) Most tRNA molecules can be folded into a cloverleaf structure

13. The four nucleotide sequence 5′–UUCG–3′ forms a relatively stable structure called what?
    (a) MicroRNA
    (b) Stem–loop
    (c) Tetraloop
    (d) tRNA

14. Which of the following is not a common type of chemical modification seen in tRNA molecules?
    (a) Base rearrangement
    (b) Deamination
    (c) Double bond saturation
    (d) Phosphorylation

15. How long is a 47 Mb DNA molecule?
    (a) 6 µm
    (b) 1.6 mm
    (c) 6 mm
    (d) 1.6 cm

16. What type of enzyme was used in the experiments which showed that proteins are associated with DNA in chromatin?
    (a) Endonuclease
    (b) Exonuclease
    (c) Pancrease
    (d) Protease

17. What is the name of the proteins that make up the 'beads' of the beads-on-a-string structure for chromatin?
    (a) Histones
    (b) Nucleases
    (c) Nucleosomes
    (d) Octamers

18. What are the vertebrate proteins called H1a–H1e, H1°, H1t and H5?
    (a) Core octamer proteins
    (b) Linker histones
    (c) Nucleosomes
    (d) Types of HU protein

19. Which one of the following statements is **correct**?
    (a) During interphase much of the DNA is in the form of the 30 nm chromatin fiber
    (b) The 30 nm chromatin fiber does not contain nucleosomes
    (c) The 30 nm chromatin fiber reduces the length of the beads-on-a-string structure by about six times
    (d) The cloverleaf is a popular model for the structure of the 30 nm chromatin fiber

20. In the bacterial nucleoid, the DNA is folded into a compact structure by what process?
    (a) Denaturation
    (b) Histone modification
    (c) Polymerization
    (d) Supercoiling

Short answer questions

These questions do not require additional reading.

1. Draw the structure of a nucleotide, including the numbering of the various carbon, nitrogen and phosphate atoms.

2. Explain why the structure of the phosphodiester bond means that the two ends of a polynucleotide are chemically distinct.

3. Describe the key features of the double helix structure for DNA, highlighting those features that are particularly important for the role of DNA as a store of biological information.

4. Describe the different types of experimental data that were used by Watson and Crick in their work that led to discovery of the double helix structure, and summarize the specific contribution that each of these sets of data made in understanding the details of the helix structure.

5. Distinguish between the terms 'base pairing' and 'base stacking' and describe the role of both types of interaction in the structure of the double helix.

6. List the key features of A-, B- and Z-forms of DNA.

7. Write a short essay on the role of intramolecular base pairing in RNA structure, with a focus on the structure of transfer RNA.

8. Summarize the different types of chemical modification that are displayed by nucleotides in RNA molecules.

9. Describe the structure of the nucleosome.

10. Outline our current knowledge on the structure of the 30 nm chromatin fiber and of higher levels of DNA packaging.

Self-study questions

These questions will require calculation, additional reading and/or internet research.

1. Discuss why the double helix gained immediate universal acceptance as the correct structure for DNA.

2. Discuss the reasons why polypeptides can take up a large variety of structures whereas polynucleotides cannot.

3. A tRNA has the nucleotide sequence 5'–GGGCGUGUGGCGUAGUCGGUAGCGCGCU CCCUUAGCAUGGGAGAGGUCUCCGG UUCGAUUCCGGACUCGUCCACCA–3'. Draw the cloverleaf structure that this tRNA could adopt.

4. An RNA molecule 75 nucleotides in length can form two stem–loops. One of these structures has a stem that is 15 bp in length and (including the loop) comprises nucleotides 15–51. The second structure forms a 9 bp stem–loop made up of nucleotides 40–64. The stems of these two structures have similar GC contents and contain no mismatches or G–U base pairs. As the two structures overlap it is not possible for both of them to form at the same time. Under normal circumstances, which of the two stem–loops would you expect to form? In a cell, what event(s) could result in the other stem–loop forming?

5. Discuss the impact that the presence of nucleosomes is likely to have on the expression of individual genes.

CHAPTER 5

Lipids and biological membranes

STUDY GOALS

After reading this chapter you will:

- appreciate that lipids comprise a diverse group of compounds with a variety of biochemical roles

- be able to describe the basic structure of a fatty acid

- understand the differences between saturated and unsaturated fatty acids, and appreciate why certain fatty acids are needed as part of the healthy human diet

- know how fatty acids are combined to produce triacylglycerols, and know how soaps and waxes are derived from triacylglycerols

- understand the key structural features of fatty acid derivatives such as glycerophospholipids and sphingolipids

- be able to describe the structures of terpenes, sterols and steroids, and know the importance of these compounds in biology

- be familiar with the structures and functions of the eicosanoids and lipid vitamins

- understand how the amphiphilic properties of certain lipids enable these lipids to form membrane bilayers

- be able to describe the fluid mosaic model of membrane structure and be aware of the importance of lipid rafts within that structure

- know the differences between integral and peripheral membrane proteins and be able to describe examples of both types

- be able to distinguish the different ways in which compounds can pass through membranes, with or without the aid of a transport protein

- know in outline how receptor proteins transmit extracellular signals across a cell membrane

Lipids are a broad group of compounds that include fats, oils, waxes, steroids and various resins. They have diverse functions in nature, two of which are particularly important in biochemistry. The first of these is energy storage, with catabolism of fats and oils providing the greater part of the energy requirements for many types of organism, not least animals such as humans. This is, of course, why we get 'fat' when our nutritional intake of certain lipids exceeds the energy requirement of our lifestyle. The second important function of lipids is as structural components of membranes. Again this involves all species because membranes are ubiquitous in nature, and although membrane composition varies in different types of organism, all membranes contain lipids of one type or another.

Energy storage and membrane structure are the two most important roles of lipids, but by no means the only ones. Waxes are secreted onto the surfaces of the leaves and fruits of plants where they protect against dehydration and attack by small predators such as insects, and some animals and birds secrete waxes and other lipids which have similar protective functions on fur and feathers. Many important hormones are lipids, as are the vitamins A, D, E and K. One single group of lipids, the **terpenes**, is the largest class of natural products and includes approximately 25 000 different compounds, synthesized mainly by plants. These compounds have a variety of functions including disease resistance, signaling and protection against predator attack, and are also involved in important physiological processes such as photosynthesis.

5.1   Lipid structures

Most lipids are hydrophobic and lipophilic. In other words, they are insoluble in water but soluble in organic solvents such as acetone and toluene. Beyond that, it is difficult to make general statements about their structures and chemical properties. These are as diverse as the functions of the different types of lipid. However, many of the most important lipids are **fatty acids** or derivatives of fatty acids. These derivatives include lipids that store energy as well as the lipids that are found in biological membranes. We therefore begin with this important family of compounds.

5.1.1   Fatty acids and their derivatives

Although much smaller than proteins and nucleic acids, fatty acids also have a polymeric structure. In the protein and nucleic acid polymers that we studied in the previous two chapters, the monomeric units were themselves complex molecules: amino acids for proteins and nucleotides for DNA and RNA. A fatty acid polymer is much less complicated, being a simple **hydrocarbon** chain of between four and 36 carbons with their attached hydrogen atoms (*Fig. 5.1*).

Fatty acids are hydrocarbon polymers

In chemical terms, fatty acids are a type of carboxylic acid. This is a compound made up of a central carbon atom attached by a double bond to an oxygen atom to form a carbonyl group (C=O), by a single bond to a hydroxyl group (–OH) and by a single bond to an R group that is different in each carboxylic acid (*Fig. 5.2A*). The general formula is therefore R–COOH, with the COOH being called a carboxyl group. The simplest carboxylic acids are familiar natural products, such as formic acid (where the R group is a hydrogen atom, giving H–COOH), which is present in ant bites and bee stings, and the acetic acid of vinegar (CH₃–COOH), but of course these are water

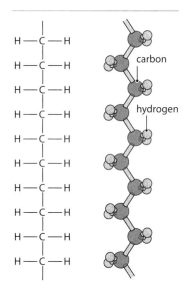

Figure 5.1 Part of a hydrocarbon chain.
The left drawing shows a simple representation of part of a hydrocarbon chain, and the drawing on the right shows the relative positions of the carbon and hydrogen atoms. Remember that the four bonds around a carbon atom have a tetrahedral arrangement.

Figure 5.2 Fatty acid structure.

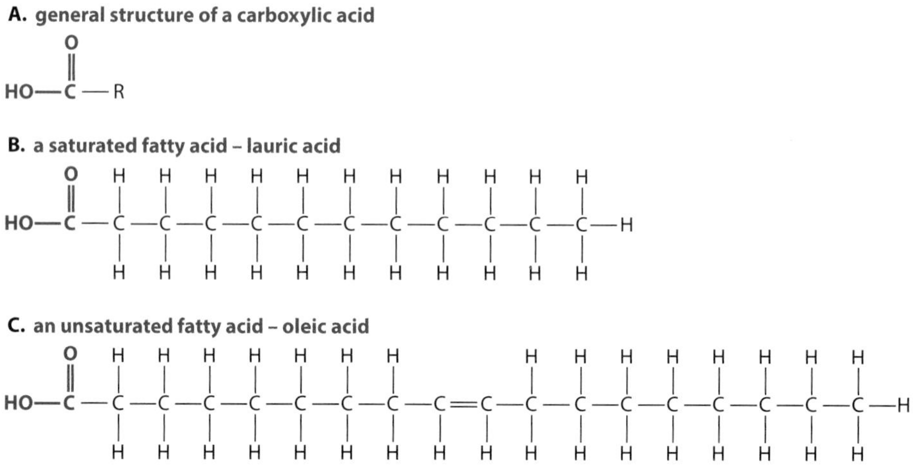

A. general structure of a carboxylic acid

B. a saturated fatty acid – lauric acid

C. an unsaturated fatty acid – oleic acid

Figure 5.3 The configurations of a saturated and unsaturated fatty acid.
The presence of a double bond introduces a kink into the hydrocarbon chain. Carbon atoms are shown in dark gray, hydrogens in light gray, and oxygens are red.

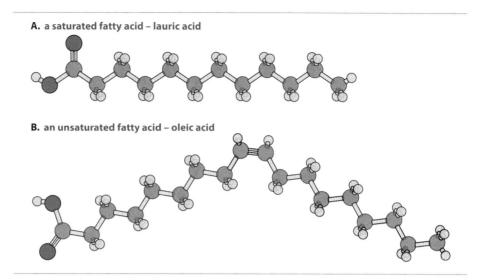

A. a saturated fatty acid – lauric acid

B. an unsaturated fatty acid – oleic acid

soluble compounds and not lipids. In a fatty acid the R group is the hydrocarbon chain. Because of this highly hydrophobic R group, fatty acids are mostly insoluble in water, but readily soluble in many organic solvents.

Fatty acids are divided into two classes depending on the structure of the hydrocarbon chain. If all the links between adjacent carbons are single bonds, which means that every carbon in the polymer chain carries two hydrogen atoms, then the fatty acid is said to be **saturated** (*Fig. 5.2B*). If, on the other hand, there are one or more pairs of carbons linked by double bonds, then the fatty acid is **unsaturated** (*Fig. 5.2C*).

The absence of double bonds means that the hydrocarbon chain of a saturated fatty acid has a linear structure (*Fig. 5.3A*). These linear molecules are able to pack together closely. One consequence of this tight packing is that most of the saturated fatty acids have melting points above 40°C and hence are fatty solids at room temperature. The presence of a double bond introduces a kink into the hydrocarbon chain (*Fig. 5.3B*), preventing the molecules of an unsaturated fatty acid from forming such closely associated arrays. Unsaturated fatty acids therefore have lower melting points and most are oily liquids at room temperature.

> The way in which fatty acids are synthesized will be described in *Section 12.1.1*.

Although fatty acids have the potential for vast diversity, not all possible structures are found in the natural world. Most fatty acids have an even number of carbons, reflecting their biochemical mode of synthesis, which involves linking together two-carbon units. If the fatty acid is unsaturated, then only rarely are there more than four double bonds in the hydrocarbon chain, and there are preferred positions for these bonds, immediately after the 9th, 12th or 15th carbon of the chain (*Fig. 5.4*).

Figure 5.4 The numbering system for the carbons in a fatty acid.
Double bonds often occur immediately after the 9th, 12th or 15th carbon of the chain.

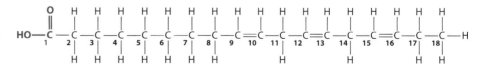

Some examples of fatty acids are given in *Table 5.1*. Each of these has a common name, often reflecting the main natural source of the compound. Lauric acid, for example, is extracted from the seed pods of the laurel tree. Different fatty acids are also distinguished by a useful nomenclature based on the formula $M:N(\Delta^{a,b,\cdots})$. In this formula, M is the number of carbons in the chain and N is the number of double

Table 5.1. Fatty acids

Structural formula	Name
Saturated	
12:0	Lauric acid (dodecanoic acid)
14:0	Myristic acid (tetradecanoic acid)
16:0	Palmitic acid (hexadecanoic acid)
18:0	Stearic acid (octadecanoic acid)
20:0	Arachidic acid (eicosanoic acid)
22:0	Behenic acid (docosanoic acid)
24:0	Lignoceric acid (tetracosanoic acid)
Monounsaturated	
16:1(Δ^9)	Palmitoleic acid
18:1(Δ^9)	Oleic acid
Polyunsaturated	
18:2($\Delta^{9,12}$)	Linoleic acid
18:3($\Delta^{9,12,15}$)	α-linolenic acid
18:3($\Delta^{6,9,12}$)	γ-linolenic acid
20:4($\Delta^{5,8,11,14}$)	Arachidonic acid

bonds. Lauric acid has 12 carbons and no double bonds and so is described as 12:0. If one or more double bonds are present then the ($\Delta^{a,b,\cdots}$) component is included, with a, b, …, indicating the number(s) of the carbons immediately preceding the double bond(s). Oleic acid, the major component of olive oil, is 18:1(Δ^9), because it has an 18-carbon chain and one double bond immediately after carbon number 9. Linoleic

---

Box 5.1 Structural notation for fatty acids

The M:N($\Delta^{a,b,\cdots}$) notation used in the text to describe fatty acid structure is based on the standard system in which the carbon of the carboxyl group is designated number 1, as shown in *Figure 5.4*. An alternative nomenclature, called the **omega (ω) system**, designates the carbon at the methyl end of the hydrocarbon chain as number 1.

the omega (ω) numbering system

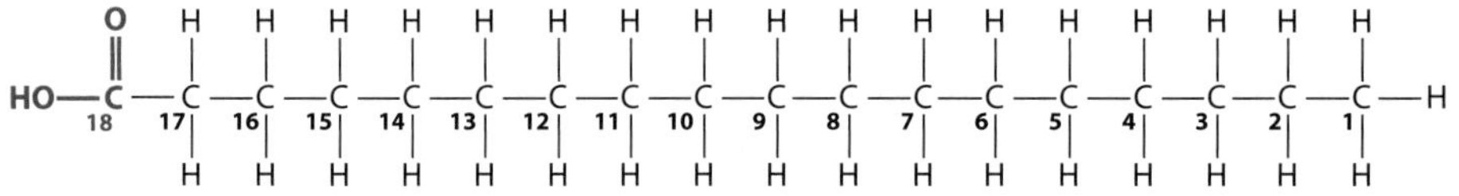

In this system, oleic acid is 18:1ω9, the double bond being immediately after the ninth carbon from the methyl end. Linoleic acid is 18:2ω6,ω9 and is part of the omega-6 family of fatty acids, those whose first double bond is located immediately after the sixth carbon from the methyl end.

It is also important to distinguish whether the carbons either side of a double bond are in the *cis* or *trans* configuration, because this affects the shape taken up by the hydrocarbon chain.

cis configuration

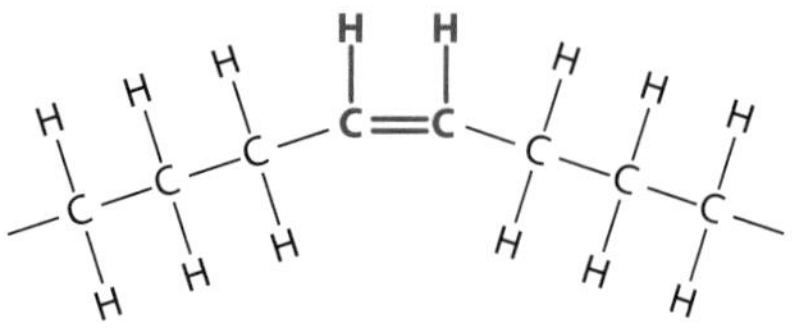

trans configuration

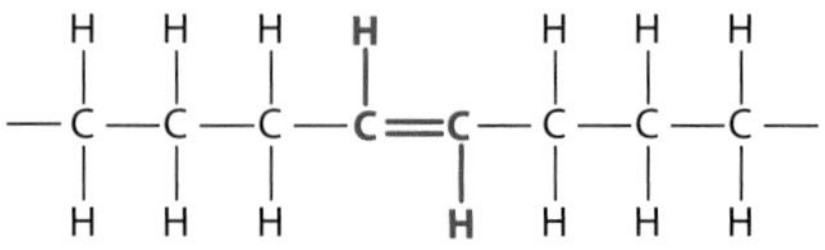

The *cis* configuration introduces a kink into the chain, whereas the *trans* configuration does not. Oleic acid is 18:1(*cis*-Δ^9), and γ-linolenic acid is 18:3(*cis,cis,cis*-$\Delta^{6,9,12}$). If all the bonds are in the *cis* configuration, the notation 'all-cis' is used: γ-linolenic acid would be called *all-cis*-9,6,12-octadecatrienoic acid.

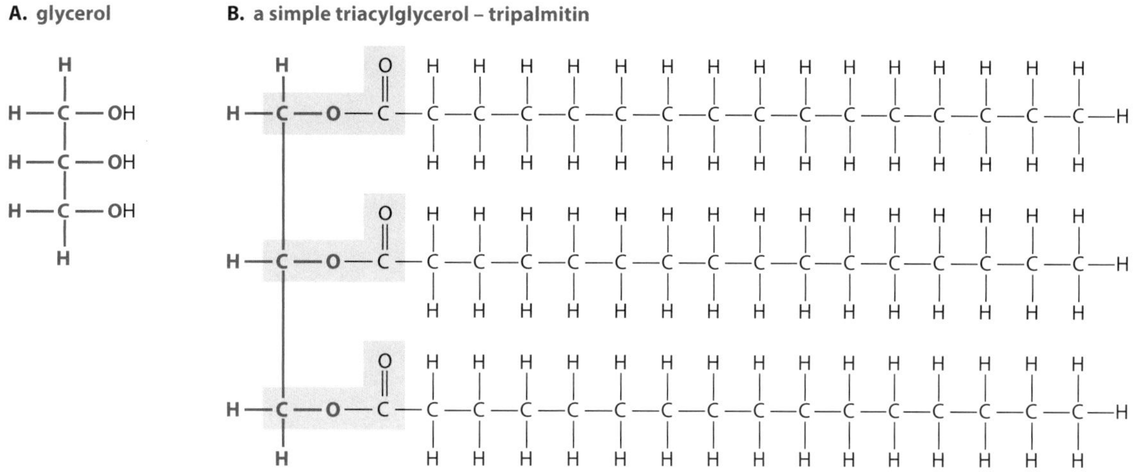

A. glycerol

B. a simple triacylglycerol – tripalmitin

Figure 5.5 **Triacylglycerol structure.**
The structure of glycerol is shown in part A, and that of a simple triacylglycerol in part B. The ester linkages between the glycerol unit and the three fatty acids are shaded.

acid, found in various plant oils, is 18:2($\Delta^{9,12}$), so has 18 carbons and two double bonds, these located after carbons 9 and 12.

Triacylglycerols are important energy storage compounds in eukaryotes

Most natural fats and oils are mixtures of fatty acids and derivatives of these compounds called **triacylglycerols** or **triglycerides**. The first of these names is more helpful because it tells us that these lipids are made up of three fatty acids attached to a glycerol molecule. Glycerol is a small organic compound with three hydroxyl groups (*Fig. 5.5A*). In a triacylglycerol, each of these hydroxyl groups acts as the attachment point for a fatty acid. The attachment is via the carboxyl group of the fatty acid, and results in an ester linkage (*Fig. 5.5B*).

In some triacylglycerols the three fatty acid chains are identical, examples being tripalmitin which has three 16:0 chains, and triolein which contains three 18:1(Δ^9) chains (*Table 5.2*). These are called **simple triacylglycerols**. Examples are known in nature but they are less common than the **complex triacylglycerols**, in which the chains are different fatty acids. As with the free fatty acids, fully saturated triacylglycerols have relatively high melting points and some are fats at room temperature. Those with one or more unsaturated chains are usually oils.

Triacylglycerols are important energy storage compounds for most animals and many plants. Animals have specialized fat storage cells called **adipocytes**, which are present in white and brown fat tissue. White fat cells contain a single droplet of fat and oil, whereas brown cells have multiple membrane-bound droplets (*Fig. 5.6*). White fat cells are the ones that increase in size and number if a person becomes obese. In

Table 5.2. **Triacylglycerols**

Fatty acid composition	Name
Simple triacylglycerols	
12:0, 12:0, 12:0	Trilaurin
16:0, 16:0, 16:0	Tripalmitin
18:0, 18:0, 18:0	Tristearin
18:1(Δ^9), 18:1(Δ^9), 18:1(Δ^9)	Triolein
Complex triacylglycerols	
18:1(Δ^9), 18:1(Δ^9), 16:0	Component of olive oil

Brown fat tissue   White fat tissue

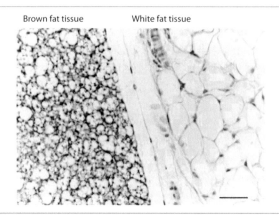

$$CH_3(CH_2)_{14}-\overset{\displaystyle O}{\overset{\|}{C}}-O-CH_2-(CH_2)_{28}-CH_3$$

palmitic acid   triacontanol

Figure 5.7 Beeswax.

triacylglycerol

glycerol   soap

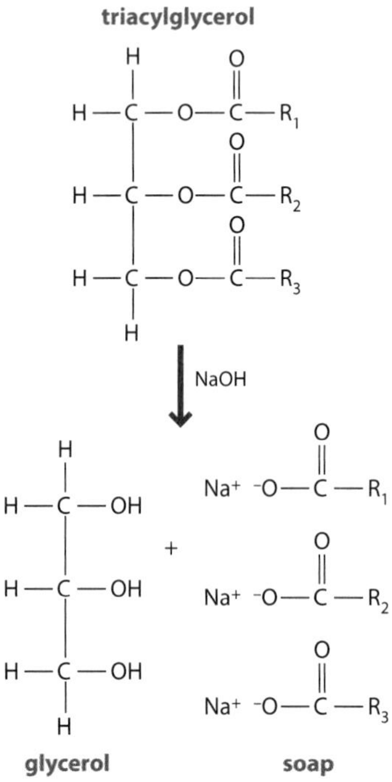

Figure 5.8 Formation of a soap.
The fatty acid hydrocarbon chains are denoted R_1, R_2 and R_3.

plants, triacylglycerols are stored in seeds and provide energy that is used by the new seedling after germination.

The triacylglycerols and fatty acids stored in plant seeds are also the components of the vegetable oils that we use in cooking and as nutrients. The beneficial and harmful effects of different types of plant lipid, as well as others that we obtain from the animal components of our diet, are widely debated. Perceived wisdom at the moment is that saturated fats, ones in which the fatty acid chains have no double bonds and which are major constituents of meat fats and milk, are bad for you, in particular by increasing the risk of cardiovascular disease. Polyunsaturated fats, which have multiple double bonds in their side-chains, are generally looked on as good, but there is debate about which ones are best. These polyunsaturated fats are more common in plants and fish oils, and are believed not only to reduce the risk of heart disease and stroke, but to protect against cancer, rheumatoid arthritis, autism and various other disorders, and even to increase brainpower in young children.

Waxes and soaps are derivatives of triacylglycerols

Triacylglycerols are not the only important derivatives of fatty acids. Fatty acids also form products when reacted with long chain alcohol compounds. An alcohol is any compound with the general structure R–CH$_2$–OH. The simplest alcohol is methanol (H–CH$_2$–OH) and the next most complex is ethanol from fermented and distilled products (CH$_3$–CH$_2$–OH). The alcohols that react with fatty acids have much longer R groups, such as triacontanol, whose formula is CH$_3$–(CH$_2$)$_{28}$–CH$_2$–OH. These alcohols form an ester linkage with the carboxyl group of a fatty acid. The product of esterification between triacontanol and palmitic acid (the 16:0 fatty acid) is beeswax (*Fig. 5.7*), which is made by worker bees and forms the honeycomb in which the new colony is raised. Waxes generally have higher melting points than fatty acids or triacylglycerols, mostly in the range 60–100°C.

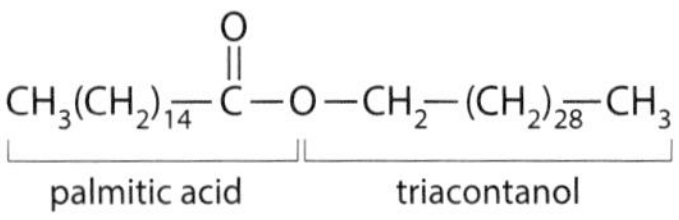

Soaps are also fatty acid derivatives, formed by heating a triacylglycerol with an alkali such as sodium hydroxide. This process is called **saponification**. The treatment breaks the ester linkages and converts the triacylglycerol back to its component fatty acids, which form salts with the cation of the alkali, sodium in the case of sodium hydroxide (*Fig. 5.8*). The first soaps were made with animal fats, over 4000 years ago. These have been supplemented in recent centuries by fine soaps made with vegetable oils, such as 'Castile soap' which comes from olive oil, and with liquid soaps based on olive, pine and palm oils.

In a soap, the presence of the cation increases the hydrophilic properties of the carboxyl end of the fatty acid, meaning that this end of the structure has an affinity for water. The hydrocarbon chain of the fatty acid remains lipophilic. A soap molecule is therefore an **amphiphile**, a type of compound with both water-liking and water-fearing

Box 5.2 **Essential fatty acids**

Humans and other mammals can synthesize a range of fatty acids (see *Section 12.1.1*) but are unable to create double bonds between the third and fourth, or sixth and seventh, carbons from the methyl end of the hydrocarbon chain.

This means that humans cannot make members of the omega-3 and omega-6 groups of polyunsaturated fatty acids. These fatty acids include α-linolenic acid (18,3ω3,ω6,ω9), a member of the omega-3 group, and linoleic acid (18:2ω6,ω9) and γ-linolenic acid (18:3ω6,ω9,ω12), which are omega-6 fatty acids. Omega-3 and omega-6 fatty acids are precursors for other important lipids, including arachidonic acid and the eicosanoid hormones. Linolenic and linoleic acid are therefore essential fatty acids that humans must obtain from their diet. They are obtained principally from green vegetables and various types of vegetable oil and, unless the diet is generally unhealthy, a deficiency in these fatty acids is unlikely to occur.

Humans can convert linolenic and linoleic acid into arachidonic acid and the eicosanoid precursors, but this conversion is not very efficient. Nutritionists therefore recommend that humans also acquire arachidonic acid and the eicosanoid precursors in their diet. Red meat, poultry and eggs provide arachidonic acid, but the eicosanoid precursors must be obtained from oily fish. As the latter does not include canned tuna (from which the oil is removed during the canning process), many people are at risk of eicosanoid deficiency. These factors underlie the huge popularity of omega-3 and omega-6 dietary supplements, though it should be noted that only those supplements that include a fish oil component will provide the full range of essential and semi-essential fatty acids.

properties (*Fig. 5.9*). Because of their amphiphilic nature, soaps can form aggregates called **micelles**. These are spheres with the carboxyl groups on the surface and the hydrocarbon chains embedded within the structure, away from the surrounding water (*Fig. 5.10*). The cleansing properties of soap are due to its ability to take the water-insoluble compounds that constitute 'dirt' out of solution by trapping them within the micelle.

soap molecule

dirt enclosed within micelle

micelle

Figure 5.10 **Soap molecules can form micelles.**

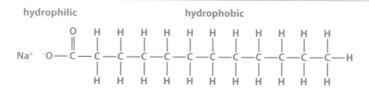

Figure 5.9 **A soap molecule is an amphiphile.**

Glycerophospholipids and sphingolipids are amphiphilic lipids

Soaps are not the only type of fatty acid derivative with amphiphilic properties. Two important classes of lipid, called **glycerophospholipids** and **sphingolipids**, are also amphiphiles. These are the lipids that are found in membranes.

A glycerophospholipid resembles a triacylglycerol, but one of the fatty acids is replaced by a hydrophilic group attached to the glycerol component by a phosphodiester bond (*Fig. 5.11*). This hydrophilic group is referred to as the 'head group' because it is located at the head of the molecule and it is the part to which the two fatty acid chains are attached. The simplest glycerophospholipid is **phosphatidic acid**, in which the head group is a hydrogen atom. Others are more complex, such as **phosphatidylserine**, in which the head group is a serine amino acid. **Phosphatidylglycerol** is particularly

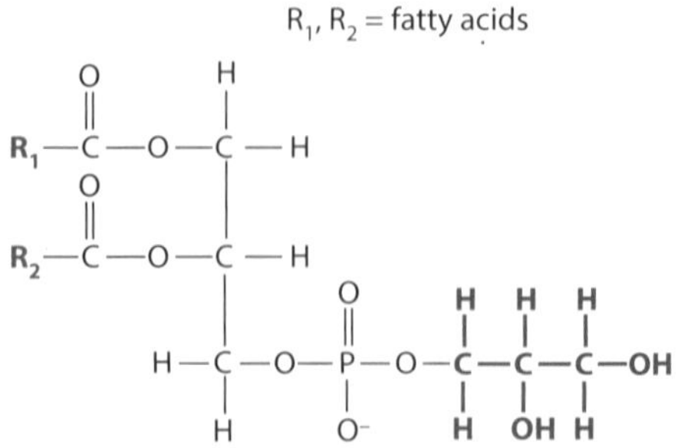

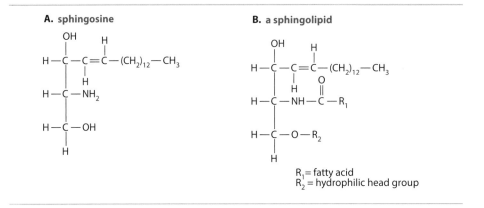

Figure 5.12 **Phosphatidylglycerol.**

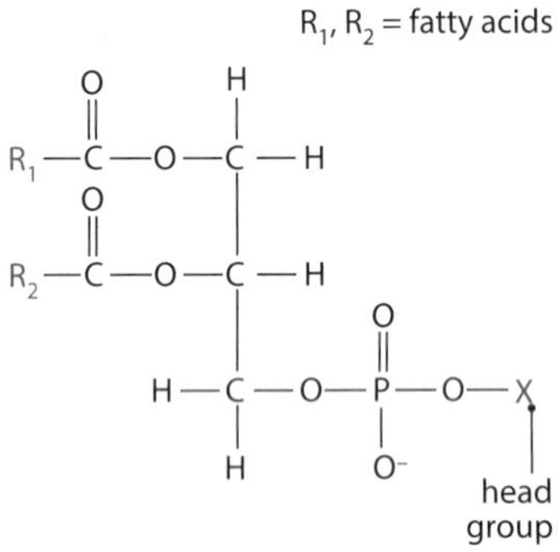

R₁, R₂ = fatty acids

Figure 5.11 **The general structure of a glycerophospholipid.**
R₁ and R₂ are the two fatty acid hydrocarbon chains and X is the hydrophilic head group.

important because it has a glycerol head group, which can be further modified to give additional structures (*Fig. 5.12*).

Sphingolipids have similar shapes to glycerophospholipids but a different chemical structure. The basic unit of a sphingolipid is **sphingosine**, a long chain hydrocarbon derivative with an internal hydroxyl group (*Fig. 5.13*). In a sphingolipid, a hydrophilic head group is attached to the last carbon of the chain, and a fatty acid to the second last. The molecule therefore has a hydrophilic head group and two hydrophobic tails, the tails being the fatty acid and the bulk of the sphingosine component. The head group is either a phosphate-containing compound such as phosphocholine, a simple sugar such as glucose, or a more complex sugar structure. A sphingolipid carrying a simple sugar head group is called a **cerebroside**, and those with complex sugars are called **gangliosides**.

5.1.2 Diverse lipids with diverse functions

Now we move away from the fatty acids and their derivatives to other types of lipid. We will encounter a diversity of structures and, importantly, a diversity of functions.

Terpenes are widespread in the natural world

First we will examine the terpenes. These are the most diverse of all types of natural product with over 25 000 different compounds known. Most terpenes are made by plants, and many are specific for a single species or a small group of species. The resins secreted by trees and other plants largely comprise terpenes, and these compounds are important components of products such as adhesives, varnishes, and some types of perfume.

Terpenes are hugely variable compounds, but they are all based on a small hydrocarbon called **isoprene** (*Fig. 5.14A*). Different terpenes are distinguished by the number of isoprene units that they contain, which can be anything from one in the

Figure 5.13 **Sphingosine and a sphingolipid.**

A. sphingosine

B. a sphingolipid

R₁ = fatty acid
R₂ = hydrophilic head group

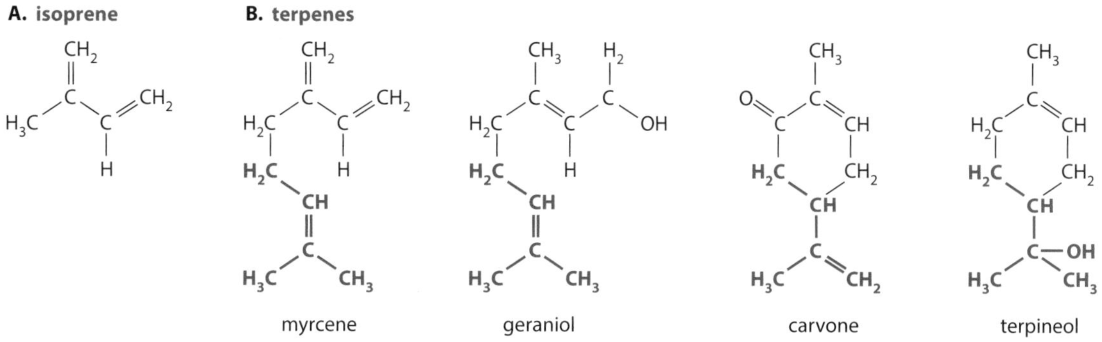

A. isoprene　　　**B. terpenes**

myrcene　　　geraniol　　　carvone　　　terpineol

Figure 5.14 Terpenes.
(A) Isoprene, which is the basic unit in terpene structure. (B) Four monoterpenes, each consisting of two modified isoprene units. In each molecule, the two units are shown in different colors.

hemiterpenes and two in the monoterpenes, up to hundreds in the polyterpenes. The latter include thick resinous substances such as rubber and gutta-percha. The different chain lengths account for one part of the great diversity of these compounds, but immense additional variability arises from the vast range of structural derivatives that exist for each class of terpene. Consider, for example, a few common monoterpenes, each consisting of two isoprene units (*Fig. 5.14B*). With myrcene and geraniol, fragrant chemicals obtained from the oils of bay and rose plants, respectively, the structures are relatively simple and the underlying isoprene units are easily identified. This becomes less easy when the derivatization has given rise to a terpene with a hydrocarbon ring component, as is the case with carvone from caraway and terpineol from pine oil.

Box 5.3 **Polyterpenes**

A polyterpene is a long-chain polymeric compound made up of many isoprene units. Examples are rubber and gutta-percha. Rubber, which is obtained from various trees native to South America and Africa, is *cis*-1,4-polyisoprene, with individual molecules containing 10000–200000 isoprene units. Gutta-percha comes from trees of the *Palaquium* genus, which are found in southeast Asia. Gutta-percha is *trans*-1,4-polyisoprene. Rubber and gutta-percha therefore differ only in the orientation

of the groups around the carbon–carbon double bonds present in the polymeric backbone.

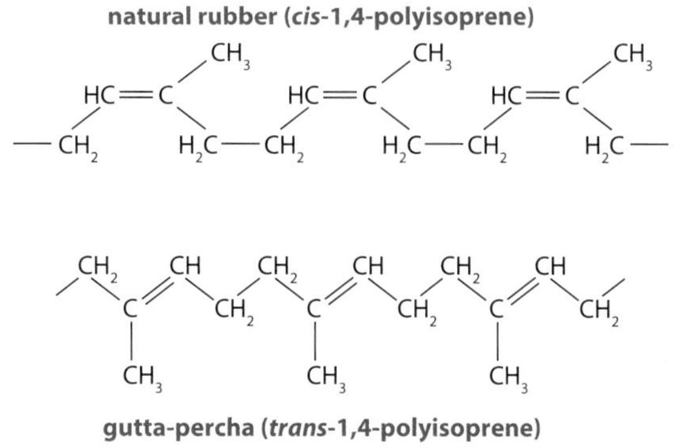

Rubber and gutta-percha are types of **latex**, tree exudates that are secreted primarily in response to wounding and thought to protect the tree from attack by herbivores. Some latexes contain toxic chemicals but the defense function is also provided in part by the sticky nature of the exudate, which prevents insects and other small herbivores from accessing the damaged part of the tree.

Latex can be collected, allowed to coagulate and then dried. Rubber produced in this way has many useful properties but remains sticky and is brittle at low temperatures. **Vulcanization** results in crosslinks being formed between individual chains, improving elasticity and giving greater mechanical stability. Most rubber products in everyday use, such as car tires, hoses and bowling balls, are made from vulcanized material. Gutta-percha is more elastic than non-vulcanized rubber, and is biologically inert. It has been used as an electrical insulator in extreme environments, including the early transatlantic telegraph cables. During the last century, these natural products have been replaced, in part, by synthetic alternatives, such as plastics and synthetic rubber made from petrochemicals.

natural rubber (*cis*-1,4-polyisoprene)

gutta-percha (*trans*-1,4-polyisoprene)

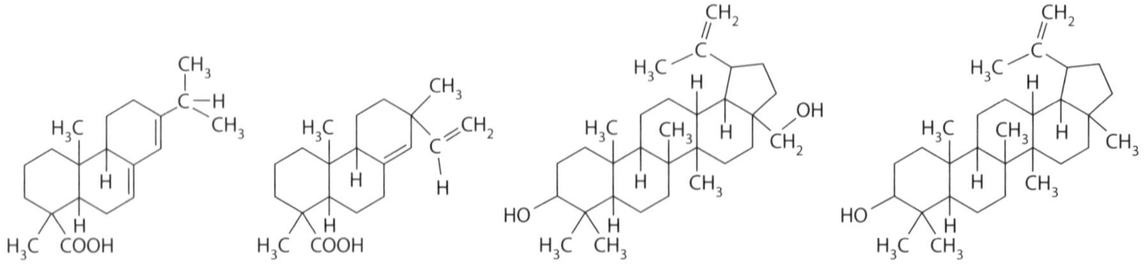

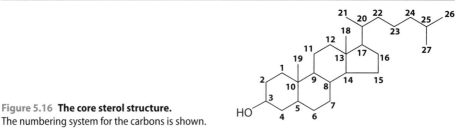

Figure 5.15 Important terpenes from tree resins.

A particular set of terpenes were among the first biological products used by prehistoric people. These are the resin terpenes of pine, spruce and birch trees. The resins of pine and spruce are largely composed of diterpenes, compounds containing four isoprene units. The two most important are abietic acid and pimaric acid (*Fig. 5.15*). These two compounds are closely related, both comprising three six-membered hydrocarbon rings derived from a four-unit isoprene backbone. The resin of birch bark contains betulin and lupeol, triterpenes containing five ring structures. The production of tars and pitches from tree resin by heating wood to high temperatures under anoxic conditions was being carried out 10 000 years ago, and possibly much earlier. This is well before metalworking and represents the beginning of the chemical industry. The tars were used for a range of purposes, notably as adhesives to attach stone arrowheads to wooden shafts. Today, betulin and related compounds have clinical applications as anti-inflammatory agents. They may also have been used for this purpose by some prehistoric groups.

Sterols and steroids are derivatives of terpenes

The complex ring derivatives of terpenes that we have just examined lead us towards the next type of lipid that we will consider. The **sterols** are formed by cyclization of squalene, which is a triterpene comprising six isoprene units. The core sterol structure produced by squalene cyclization has four hydrocarbon rings, three of which have six carbons each and one of which has five (*Fig. 5.16*).

We will examine the complex series of reactions that result in sterol synthesis in *Section 12.3.1*.

Figure 5.16 The core sterol structure.
The numbering system for the carbons is shown.

Sterols are another major lipid constituent of cell membranes. Like other membrane components, sterols are amphiphiles, having a hydrophilic head group provided by the hydroxyl group attached to carbon number 3 and, in most cases, a hydrophobic hydrocarbon chain, comprising some or all of carbons 20–27, as an R group at the other end of the molecule. **Cholesterol**, the best known animal sterol, is a typical example of this type of lipid, with an 8-member hydrocarbon R group comprising six carbons in a chain, with the two others forming short branches (*Fig. 5.17*). The equivalent compound in plants is **stigmasterol**, whose R group is similar in size to that of cholesterol but with a slightly different hydrocarbon configuration. As well as these membrane constituents, some sterols have hydrophilic R groups and are readily

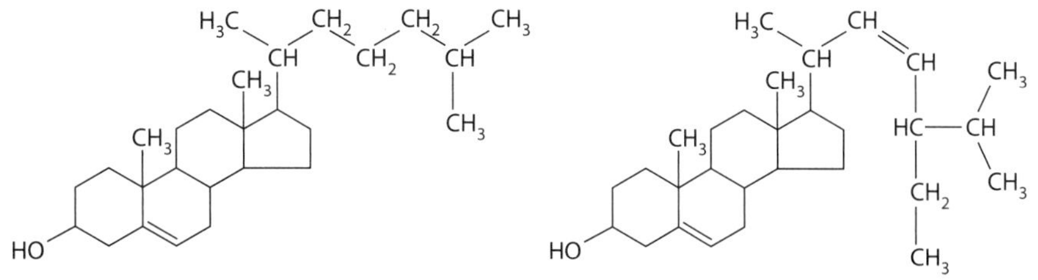

Figure 5.17 **Cholesterol and stigmasterol.**

cholesterol   stigmasterol

soluble in water. These include the **bile acids**, which have side-chains that terminate in a carboxyl group, the simplest example being **cholic acid**. Derivatives of cholic acid such as **glycocholate** and **taurocholate** are synthesized in the liver and secreted into the small intestine where they help to emulsify fats in the diet and hence aid their breakdown.

The **steroids**, which themselves are another large class of lipids, are sterol derivatives. The basic steroid unit is identical to that of sterols except that the hydroxyl attached to the C_3 carbon is replaced with a different chemical group. Because this group is variable in steroids, the sterols are, strictly speaking, a subclass of steroids, and the two names are occasionally used interchangeably. The R group possessed by a steroid is usually hydrophilic and these molecules are water soluble. They include a number of important hormones in humans and other mammals, including the male and female sex hormones (*Table 5.3*). Anabolic steroids, notorious in our modern world, include **testosterone** and other natural hormones which have roles in the regulation of bone and muscle synthesis.

Table 5.3. **Steroid hormones**

Type	Examples	Site of synthesis	Function
Glucocorticoids	Cortisol, cortisone	Adrenal cortex	Various effects on metabolism
Mineralocorticoids	Aldosterone	Adrenal cortex	Regulate the body's salt and water balance
Estrogens	Estrone, estradiol, estriol	Adrenal cortex, gonads	Female sex hormones
Androgens	Testosterone	Adrenal cortex, gonads	Male sex hormones
Progestins	Progesterone	Ovaries, placenta	Control menstrual cycle and pregnancy

Eicosanoids and lipid vitamins

Having covered most of the important classes of natural lipids, we will now briefly consider two final types, both of which have important biological functions.

First, the **eicosanoids** are compounds derived from the $20:4(\Delta^{5,8,11,14})$ fatty acid arachidonic acid. Eicosanoids are synthesized from arachidonic acid molecules released from membrane glycerophospholipids in response to hormone stimulation, and they themselves have hormone-like activity, controlling a number of biological processes including reproduction and the pain response. The common painkillers aspirin and ibuprofen act by preventing formation of certain types of eicosanoid. They are not true hormones because they stay within the tissues in which they are synthesized, rather than circulating to distant parts of the body in the bloodstream. Examples are **prostaglandin** and the **thromboxanes**.

Vitamins A, D, E and K are lipids (*Fig. 5.18*). Vitamins A, E and K are related to terpenoids and vitamin D has a steroid structure, but with one of the hydrocarbon rings broken open. Each vitamin is a group of related compounds:

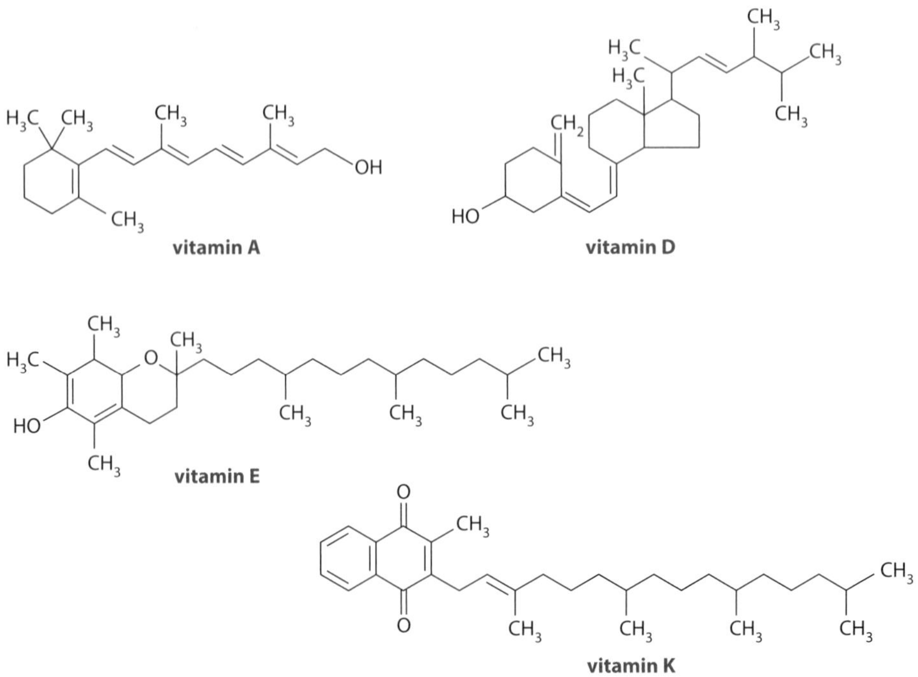

Figure 5.18 Structures of vitamins A, D, E and K.
Each vitamin is a family of related molecules. The versions shown here are retinol (the most common form of vitamin A in the diet), ergocalciferol (vitamin D_2), tocopherol (vitamin E), and phylloquinone (vitamin K_1).

vitamin A

vitamin D

vitamin E

vitamin K

- Vitamin A has a variety of functions but most notably includes retinol, which is needed for synthesis of the photoreceptor proteins, rhodopsin and iodopsin, in the retina of the eye. This role has led to the myth that eating lots of carrots (which contain vitamin A) will help you see in the dark.

Box 5.4 **Prostaglandins**

The prostaglandins are a family of eicosanoid compounds, each with a five-carbon aromatic ring and a pair of hydrocarbon tails.

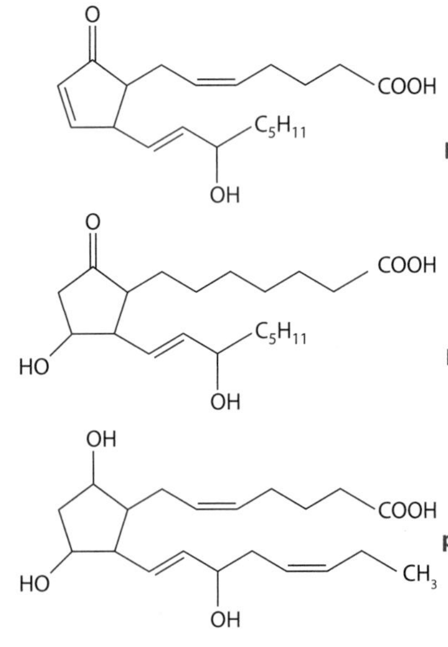

prostaglandin A_2

prostaglandin E_1

prostaglandin $F_{3\alpha}$

Prostaglandins are transported out of the cells in which they are synthesized, and then bind to receptor proteins on the surfaces of other cells in the same tissue. Binding to the receptor activates a series of biochemical events inside the target cells. We will study the general features of cell surface receptor proteins and how binding of a signaling molecule such as a prostaglandin influences events inside the cell later in this chapter. There are at least ten different types of prostaglandin receptor, which bind different groups of prostaglandin. Because different receptors stimulate different intracellular events, prostaglandins are able to control a diverse array of biochemical and physiological functions, including vasodilation, blood clotting, inflammation, ovulation and secretion of gastric acid.

Although prostaglandins are found in animals, a related compound, called jasmonic acid, is synthesized by plants.

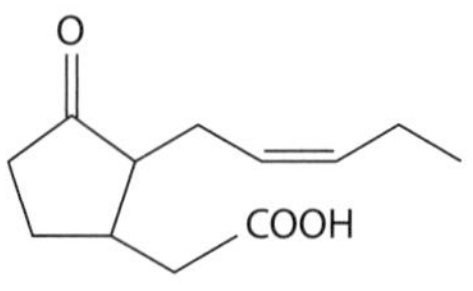

jasmonic acid

Jasmonic acid is also a type of signaling compound, and is involved in control of various plant processes, such as flowering, leaf abscission and the response to wounding.

- Vitamin D is obtained from the diet and is also synthesized in the skin in response to sunlight. Among its functions is the development of healthy bone, a deficiency in vitamin D being associated with the childhood bone disease called rickets. Recent cases of rickets have been ascribed to the overuse of sunblock.

- Vitamin E is involved in prevention of oxidative damage within cells. This group of vitamins is common in vegetables so dietary deficiencies rarely occur, but vitamin E uptake from the gastrointestinal tract can be affected by genetic disorders and if untreated leads to defects in the nervous system.

- Vitamin K is common in many leafy vegetables and is one reason why cabbage is good for you. Vitamin K is needed for correct functioning of the blood clotting response.

5.2 Biological membranes

Membranes are fundamental to all living systems, acting as selectively permeable barriers that control the movement of molecules into and out of cells, and also into and out of the organelles within cells. Membranes contain lipids, proteins and, in some cases, carbohydrates. First, we will look at membrane structure, and then we will explore how membranes act as selective barriers.

5.2.1 Membrane structure

The relative amounts of lipid, protein and carbohydrate in a membrane are variable, depending on the function of the membrane (*Table 5.4*). The inner mitochondrial membrane, for example, has a protein content approximately 50% greater than that of the outer mitochondrial membrane. This difference reflects the fact that the proteins of the energy-generating electron transport chain are located in the inner mitochondrial membrane. There is also variability in the types and relative proportions of the glycerophospholipids, sphingolipids, sterols and other lipids that are present. The first question we must ask is how these various types of lipids associate with one another to form a membrane.

A membrane is a lipid bilayer

The key feature shared by glycerophospholipids, sphingolipids and sterols is that each of these types of lipid is an amphiphile. Remember that an amphiphilic molecule has both hydrophobic and hydrophilic components. The hydrophobic tails of glycerophospholipids and sphingolipids, being water fearing, prefer being embedded in a lipid-rich environment away from water. Soap achieves this by forming a micelle (see *Fig. 5.10*), but soap molecules have just a single hydrophobic tail whereas each glycerophospholipid or sphingolipid molecule has two tails. With two tails, these molecules cannot form spherical micelles: there is simply not enough room inside

Table 5.4. **Compositions (by weight) of different membranes in human cells**

Membrane	Lipid					Protein	Carbohydrate
	Total	GPP	Sphingolipids	Sterols	Others		
Erythrocyte plasma membrane	43%	19%	8%	10%	6%	49%	8%
Liver cell plasma membrane	36%	23%	7%	6%	0%	54%	10%
Endoplasmic reticulum	28%	17%	1%	1%	9%	62%	10%
Outer mitochondrial membrane	45%	41%	0%	0%	3%	55%	0%
Inner mitochondrial membrane	22%	20%	0%	0%	2%	78%	0%

GPP, glycerophospholipids.

the micelle to fit all the tails. Instead, glycerophospholipids and sphingolipids protect their hydrophobic regions from water by aggregating into a **bilayer** (*Fig. 5.19*). Their hydrophobic tails are embedded inside the bilayer away from the surrounding water, and their hydrophilic head groups are positioned on the upper and lower surfaces. A biological membrane is therefore a lipid bilayer, made up of glycerophospholipids, sphingolipids and other amphiphilic lipids such as sterols.

Figure 5.19 A lipid bilayer.

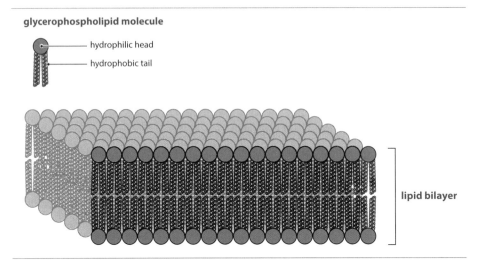

glycerophospholipid molecule

— hydrophilic head
— hydrophobic tail

lipid bilayer

The individual lipids in a membrane do not form strong attachments to one another. They are held in place predominantly by the desire of their hydrophobic tails to get away from the aqueous environment found within the cell and in the spaces between cells. These hydrophobic effects are strong enough to hold the two sheets of the bilayer close to one another, forming a stable structure that has both elasticity and flexibility. A membrane can be stretched by 2–4% without breaking and can be curved into the spherical vesicles and tubes that make up the internal membranous architecture of eukaryotic cells.

It is difficult for a lipid molecule to move from one side of the membrane to the other, because this would involve its hydrophilic head group passing through the bilayer, but there is little restriction on the movement of a lipid molecule within its own layer (*Fig. 5.20*). Each monolayer of a membrane can therefore be looked on as a two-dimensional fluid, within which the lipids are constantly moving. This is the basis of the **fluid mosaic model** for membrane structure, first proposed by Singer and Nicholson in 1972. Studies with artificial membranes suggest that an individual lipid molecule can move at rates approaching $2\,\mu m\,sec^{-1}$. This is fast enough to make an entire circuit of the outer membrane of a eukaryotic cell in less than a minute. The rate of diffusion depends on several factors, including the structure of the lipid. Lipids with longer hydrophobic tails form closer associations with one another and hence

Figure 5.20 Lateral and transverse movement of a lipid in a membrane. There are few restrictions on lateral movement, but transverse movement is less frequent because this involves passage of the hydrophilic part of the lipid through the membrane.

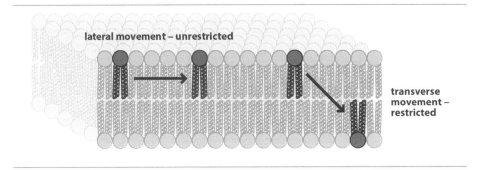

lateral movement – unrestricted

transverse movement – restricted

make less fluid membranes. Conversely, the presence of one or more double bonds in a tail structure introduces a kink that destabilizes the association, resulting in a more fluid membrane.

Membranes also contain proteins

The lipid bilayer is the key structural feature of a biological membrane, but membranes are not solely made of lipid. Most also contain proteins. The 'mosaic' part of the fluid mosaic model for membrane structure refers to these proteins. Proteins are present in much smaller numbers than the lipid molecules but, being larger in size, make a significant contribution to the overall mass of the membrane. For example, the plasma membranes of most human cells are made up of more protein than lipid by weight, but there are 50 times as many lipid molecules.

We distinguish between two types of membrane protein depending on the strength of their attachment to the lipid bilayer. First, there are **integral membrane proteins**, which form a tight attachment and can only be removed from the membrane by disrupting the structure of the bilayer. Experimentally this is achieved by treating a cell extract containing the membrane with a **detergent** such as sodium dodecyl sulfate (*Fig. 5.21*). Detergents are themselves fatty acid derivatives, similar to soaps, with a hydrophobic tail and strongly hydrophilic head group. Their hydrophobic tails penetrate the lipid bilayer and, if present in sufficient quantities, the detergent molecules dilute out the membrane lipids to the extent that the resulting mixture forms a micelle. The lipid bilayer therefore breaks down and the integral membrane proteins are released. In contrast, **peripheral membrane proteins** make looser attachments with the membrane and can be removed from the extract simply by gentle washing, without a detergent and hence without disrupting the bilayer.

Figure 5.21 Sodium dodecyl sulfate.

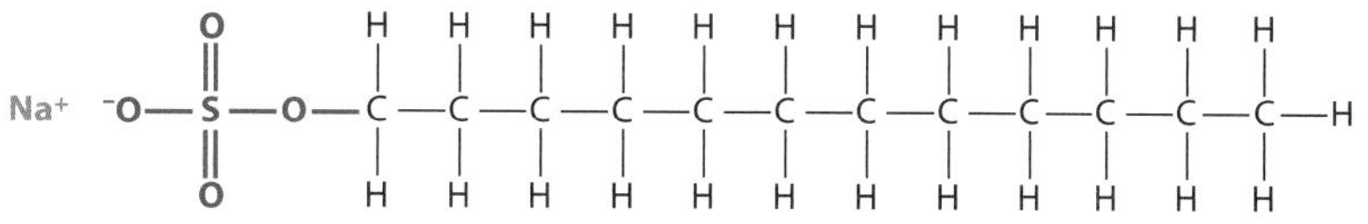

Many, but not all, integral membrane proteins span the entire lipid bilayer, with hydrophobic amino acid residues of the protein interacting with the fatty acyl groups of the lipids at the membrane core. Some of these **transmembrane proteins** have a barrel-like structure, the walls of the barrel made up of β-sheet (*Fig. 5.22*). With others, one or more α-helices span the membrane. The internal face of a transmembrane protein often makes attachments, possibly transient ones, with peripheral membrane proteins. Other peripheral proteins make direct attachments to one or other side of the lipid bilayer, either by an α-helix or other structure that penetrates part way into the

Figure 5.22 Three common types of integral membrane protein.

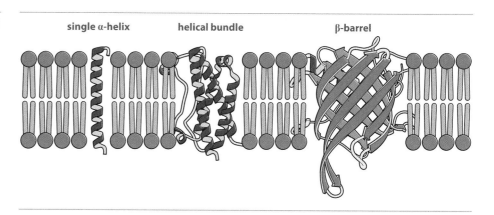

single α-helix   helical bundle   β-barrel

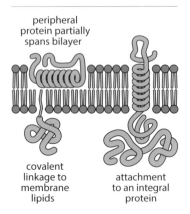

peripheral
protein partially
spans bilayer

covalent
linkage to
membrane
lipids

attachment
to an integral
protein

Figure 5.23 Different ways in which peripheral proteins make attachments with a membrane.

bilayer, or by a covalent attachment with the membrane lipids (*Fig. 5.23*). The latter are also called **lipid-linked proteins** and, because of their covalent attachment, are classified by some biochemists as integral membrane proteins, even though most are located on the membrane surface rather than being integrated into the lipid bilayer.

The fluid mosaic model envisages the proteins floating in a sea of lipids, but surely this would cause a problem for the functioning of some of these proteins? If two or more membrane proteins have to work together to carry out their biochemical role, as we know is often the case, then it would be highly inefficient if those proteins randomly moved around in their membrane. If that were allowed, then the biochemical function would only occur if and when those proteins happened to float into proximity with one another. When the fluid mosaic model was first proposed in the 1970s it was presumed that there must be relatively stable domains in a membrane where sets of proteins that work together could be co-located. These domains are now called **lipid rafts**, and are believed to be small areas of the bilayer, 10–100 nm in diameter, containing a high proportion of those types of lipid that form tight associations with one another (*Fig. 5.24*). Sterols are particularly common in lipid rafts, because a sterol molecule fits neatly into the space between two unsaturated glycerophospholipids, stiffening the membrane structure. A lipid raft therefore has more stability than the membrane as a whole, and so it floats within the rest of the less structured bilayer.

Figure 5.24 A lipid raft.

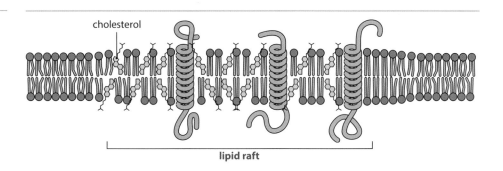

cholesterol

lipid raft

5.2.2 Membranes as selective barriers

Very few compounds can pass through a lipid bilayer. To pass directly through the internal hydrophobic region of a membrane a molecule must be small and nonpolar. Other compounds can only cross a membrane with the aid of one of the integral proteins. A biological membrane is therefore a **selective barrier**, allowing some compounds to pass, but preventing others. This means that the plasma membrane

---

Box 5.5 **The carbohydrate component of a membrane**

Most plasma membranes and some internal membranes have a carbohydrate component (see *Table 5.4*). The carbohydrates are located on the extracellular surface of a plasma membrane and are held in place by covalent attachments to membrane lipids and proteins, forming **glycolipids** and **glycoproteins**, respectively.

We are already familiar with the membrane glycolipids as these are sphingolipids with sugar head groups, also known as cerebrosides and gangliosides. The names of these glycolipids indicate that both are common in neuronal cell membranes

in the brain. In a glycoprotein, the carbohydrate component is a short chain of sugars attached to a serine, threonine or asparagine amino acid. We will examine glycoproteins in more detail in *Section 6.1.3*.

The carbohydrate coating on the extracellular surface of a cell has a protective role and also aids cell–cell recognition. The latter underlies interactions between groups of cells during tissue development, and also enables foreign cells to be recognized and destroyed as part of the body's defense against infection.

is able to regulate the internal chemical composition of the cell. Additionally, the presence of a membrane around an organelle enables the internal environment of that organelle to be different from that of the cytoplasm within which it resides.

Some compounds that are unable to pass across the membrane are still able to influence events within a cell by a process called **signal transduction**. These compounds are regulatory molecules that bind to transmembrane **receptor proteins**, which respond by initiating a series of intracellular biochemical reactions. Some growth factors, for example, bind to receptor proteins in order to induce cell division.

We will now examine the role of membranes in transport and signal transduction in more detail.

Transport processes that depend on diffusion do not require energy

Those molecules relevant to biochemistry that are able to pass through a lipid bilayer without the aid of a transport protein are water, some gases (such as oxygen, nitrogen and carbon dioxide) and a small number of organic molecules, including urea and ethanol. These molecules pass through a membrane simply by diffusion, at a rate that is proportional to the difference between the concentrations of the compound on the two sides of the membrane.

Those biochemicals that are unable to pass through the lipid bilayer include amino acids and sugars, as well as ions such as Na^+ and K^+. These molecules are transported across the membrane by integral membrane proteins. The simplest of these transport processes is **facilitated diffusion**, in which the protein, called a **uniporter**, moves its substrate from the side of the membrane at which the substrate concentration is higher to that at which the concentration is lower (*Fig. 5.25A*). No energy is required other than that inherent in the concentration gradient.

An example of facilitated diffusion is the transport of glucose into mammalian erythrocytes, by the **erythrocyte transporter protein**, also called GLUT1. This is a typical transmembrane protein, made up of twelve α-helices, each of which spans the erythrocyte plasma membrane. Binding of a molecule of glucose to the part of GLUT1 exposed on the outer surface of the erythrocyte results in a conformational change in the protein, which moves the glucose into a channel contained within the protein structure (*Fig. 5.25B*). This channel leads to the inside of the erythrocyte, so the glucose molecule can now pass across the membrane without encountering the impenetrable hydrophobic part of the lipid bilayer.

The glucose transport process is reversible, so if the glucose concentration within the erythrocyte exceeds that in the surrounding blood plasma, then glucose will be

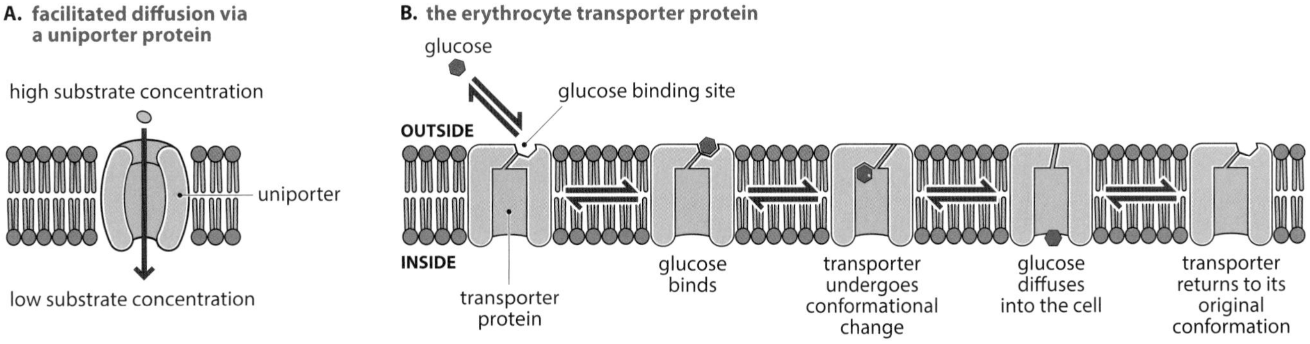

A. facilitated diffusion via a uniporter protein

high substrate concentration

uniporter

low substrate concentration

B. the erythrocyte transporter protein

glucose

glucose binding site

OUTSIDE

INSIDE

transporter protein

glucose binds

transporter undergoes conformational change

glucose diffuses into the cell

transporter returns to its original conformation

Figure 5.25 Facilitated diffusion across a membrane via a uniporter protein.
(A) The general mode of action of a uniporter. (B) Transport of glucose into an erythrocyte via the erythrocyte transporter protein.

Table 5.5. **Internal and external ionic concentrations for a typical mammalian cell**

Ion	Internal concentration (mM)	External concentration (mM)
K+	140	5
Na+	10	145
Ca+	4	110
Cl-	0.0001	5

Box 5.6 **Voltage-gated ion channels and nerve impulses**

The Na+/K+ ATPase helps to maintain the **membrane potential**, which is the electric charge across the membrane. Because the ATPase moves only two K+ ions into the cell for every three Na+ ions that it moves out, an electric potential is set up across the membrane, with the concentration of positively charged ions outside the cell being greater than that inside. The inside of the cell therefore has a negative voltage, usually between –40 and –80 mV compared with the cell exterior.

With most cells, the membrane potential does not vary over time. Nerve cells are an exception, because these have transmembrane proteins called **voltage-gated ion channels**, which can change their conformation in response to the electrical charge. In the activated conformation, the protein opens a channel that allows either Na+ or K+ ions to flow unhindered across the membrane by diffusion down their concentration gradients.

Opening of Na+ channels results in depolarization of the membrane because Na+ ions enter the cell balancing out the intracellular negative charge. In fact, the influx of Na+ ions is so rapid that the neutral point is overshot and the inside of the cell acquires a net positive charge, changing from approximately –60 mV to +40 mV in a millisecond. The positive internal charge now stimulates opening of K+ channel proteins, so K+ ions now leave the cell. This rapidly restores the negative intracellular charge.

For a short period after closing, the Na+ channels are insensitive to the membrane potential. This means that the same channels do not immediately reopen once the intracellular negative charge has been restored. This is the basis to the transmission of a nerve impulse along the axon of a neuron. The axon is a long, thin cylindrical structure and the nerve impulse travels from the main cell body to the far end of the axon. The directionality is set by the inability of this wave of depolarization or **action potential** to move back towards the cell body, because this would mean reopening Na+ channels that are temporarily quiescent.

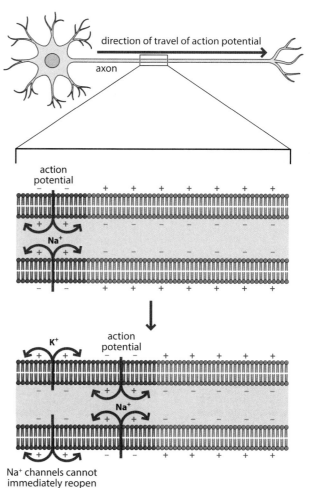

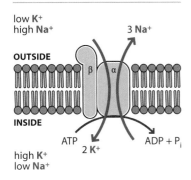

low K+
high Na+

OUTSIDE

INSIDE

3 Na+

ATP   ADP + P$_i$
    2 K+

high K+
low Na+

Figure 5.26 Active transport across the plasma membrane by the Na+/K+ ATPase.
The ATPase is a dimer of two proteins, called α and β.

We will study how hydrolysis of ATP releases energy for active transport and other biochemical processes in *Section 8.1.1.*

moved out of the cell. This situation rarely, if ever, arises, because the glucose released into the cell is immediately metabolized to produce energy. The internal glucose concentration therefore remains low, and glucose is continually transported into the erythrocyte, providing the cell with a continual energy supply. Uniporters similar to GLUT1 are present in the plasma membranes of most mammalian cells and enable various sugars and amino acids to be transported down their concentration gradients.

Active transport processes require energy

Processes based on diffusion are able to transport molecules across a membrane so long as the direction of transport is down a concentration gradient. However, there are also instances where a cell or organelle must transport molecules or ions against a concentration gradient. This is most important in maintenance of the correct ionic balance within a cell. Mammalian cells, for example, maintain a high internal concentration of K+, compared with the extracellular environment, and lower internal concentrations of Na+, Ca2+ and other ions (*Table 5.5*). To maintain these differentials, the cell must move ions across its plasma membrane against the concentration gradient, pumping K+ ions into the cell, and pumping Na+ and Ca2+ ions out. This is called **active transport**, and it requires energy.

The energy for active transport can be obtained in two different ways. The first is by hydrolysis of ATP, converting this nucleotide to ADP and inorganic phosphate, and the second is by coupling the transport of one ion against a concentration gradient with the movement of a second ion down a gradient.

RESEARCH HIGHLIGHT

Box 5.7 The biochemistry of cystic fibrosis

Cystic fibrosis (CF) affects approximately 8000 people in the UK and 30 000 in the USA. The main symptom of the disease is the build-up of mucus in the lungs, which must continually be cleared in order to avoid the respiratory tract becoming blocked. The disease also affects the pancreas, liver, kidneys and intestine. There is no cure and death usually results from lung failure or infection, but advances in care of CF patients mean that the life expectancy for a child born with the disease in the 2010s is over 50 years.

Cystic fibrosis is caused by a defect in a single protein. This is the **cystic fibrosis transmembrane regulator (CFTR)**, which is a member of the ABC group of transporters. CFTR is specifically responsible for transport of Cl– ions out of cells, but unlike most ABC transporters it does not do this by an active transport process. Instead, ATP binding induces a conformational change in the protein, opening a channel that allows Cl– ions to flow down the electrochemical gradient from inside to outside the cell. Disruption of this transport function in the epithelial cells of the lungs results in a change in the ionic balance of the fluid that covers the internal surfaces of the respiratory tract. This fluid becomes more viscous, resulting in accumulation of mucus, and less efficient in its ability to protect the lungs from bacterial infection.

What exactly are the defects in the CFTR protein that give rise to CF? In the majority of CF patients, the CFTR protein lacks a single phenylalanine amino acid at the 508th position of the polypeptide – which usually has 1480 amino acids in total. This alteration is called ΔF508, indicating a loss (Δ) of a phenylalanine

(F) from position 508. The loss of this amino acid prevents CFTR from folding correctly and the misfolded protein is degraded before insertion into the plasma membrane. This version of CF is therefore due to an absence of the protein and the concomitant loss of its Cl– transport function. A second type of CF, much less common in the population as a whole, is called G551D, meaning that a glycine (G) that is normally present at position 551 in the polypeptide is replaced by an aspartic acid (D). This change does not affect the way the protein folds and CFTR inserts into the plasma membrane in the correct way. However, the process by which the Cl– channel opens and closes is affected, with the channel no longer opening when ATP is bound by the protein. The channel is not permanently closed (the normal protein is able to transport a small amount of Cl– ions even when ATP is absent), but the substantially reduced activity of the defective CFTR protein means that the disease symptoms arise.

Understanding the biochemical basis to the different types of CF is important in attempting to design therapies for the disease. For example, knowing that patients with the G551D alteration have CFTR proteins in their plasma membranes, but that the Cl– channels in these proteins are closed most of the time, means that this type of CF might respond to treatment with a CFTR potentiator. This is a compound that binds directly to CFTR proteins and induces channel opening. Patients with the ΔF508 version of the disease would clearly not respond to this treatment, because their plasma membranes lack CFTR proteins. For these individuals, a type of **gene therapy** might be applicable, possibly by introducing the gene for the correct version of CFTR into the lung tissue by inhalation via an aspirator.

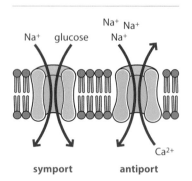

Figure 5.27 The roles of symporter and antiporter proteins.

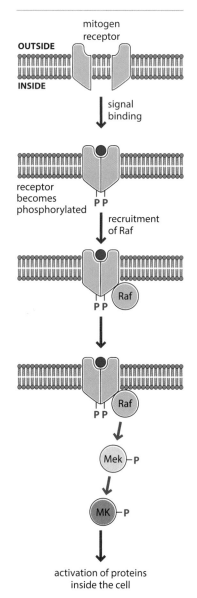

Figure 5.28 The MAP kinase signal transduction pathway.

Hydrolysis of ATP releases energy which some enzymes can utilize to drive biochemical reactions, such as the movement of an ion across a membrane against a concentration gradient. There are two important types of ATP-dependent transport protein:

- **P-type pumps**. With these, the phosphate released by ATP hydrolysis forms a transient attachment with the transport protein. An important example is the mammalian **Na⁺/K⁺ ATPase**, which maintains the high potassium and low sodium ion concentration within the cell. For every ATP that it uses, the ATPase moves two K⁺ ions into the cell, and three Na⁺ ions out (*Fig. 5.26*).

- **ATP-binding cassette (ABC) transporters**, which transport a variety of small molecules across membranes, mostly into cells, although a few examples of exporters are known. Humans have 48 different ABC transporters, each thought to be specific for a different compound or group of related compounds. Some other species have many more types – perhaps as many as 150 in plants.

The second way in which energy can be obtained for active transport is by coupling the transport of a molecule or ion against a concentration gradient with the movement of a second ion, usually H^+ or Na^+, down a gradient. The energy released by movement of the ion down its gradient is harnessed to drive the active component of the coupled transport system (*Fig. 5.27*). If the transport protein is a **symporter**, then both substrates move in the same direction across the membrane. An example is the mammalian **Na⁺/glucose transporter**. This protein links the diffusion of sodium from the gut into cells lining the intestine with the uptake by the same cells of dietary glucose. In contrast, an **antiporter** moves the two components of the coupled system in different directions. Many cells use an antiporter to maintain their low calcium ion content, for example, an **Na⁺/Ca²⁺ exchange protein** uses the energy from diffusion of three Na⁺ ions into the cell to drive the export of a single Ca²⁺ ion.

Receptor proteins transmit signals across cell membranes

Many extracellular compounds are unable to enter a cell because they are too hydrophilic to penetrate the lipid membrane and the cell lacks a specific transport mechanism for their uptake. In some cases, these compounds are still able to influence events inside the cell by the process called signal transduction. Extracellular compounds that fall into this category include hormones such as insulin and glucagon, which control the body's use of carbohydrate and fat, and compounds called **cytokines**, which regulate many cellular activities including cell division.

Attachment of one of these compounds to the external surface of a transmembrane receptor protein results in a conformational change, often dimerization of the receptor, with two subunits combining to form a single structure. This is possible because of the liquid nature of the cell membrane, which allows lateral movement of membrane proteins, even when these are contained in a lipid raft, enabling the two subunits of a receptor to associate and disassociate in response to the presence or absence of the extracellular compound. The change in the structure of the receptor induces a biochemical event within the cell, such as attachment of phosphate groups to a cytoplasmic protein. This phosphorylation initiates a series of reactions that bring about the changes in cellular activity stimulated by the hormone or cytokine.

The **MAP kinase system** is a typical example of a signal transduction pathway. 'MAP' stands for 'mitogen activated protein', indicating that this particular cell surface receptor responds to the binding of a mitogen, a type of regulatory molecule that specifically stimulates cell division. Binding of the mitogen results in dimerization of the receptor protein, which is accompanied by each subunit becoming phosphorylated (*Fig. 5.28*). Phosphorylation stimulates attachment of a protein called Raf to the internal side of the receptor. Once attached, Raf adds a phosphate to a third protein called Mek, which in turn adds a phosphate to the MAP kinase. Addition of the phosphate activates

An important example of the role of cAMP in metabolic control occurs during the 'fight or flight' response of mammals (see Section 11.1.2).

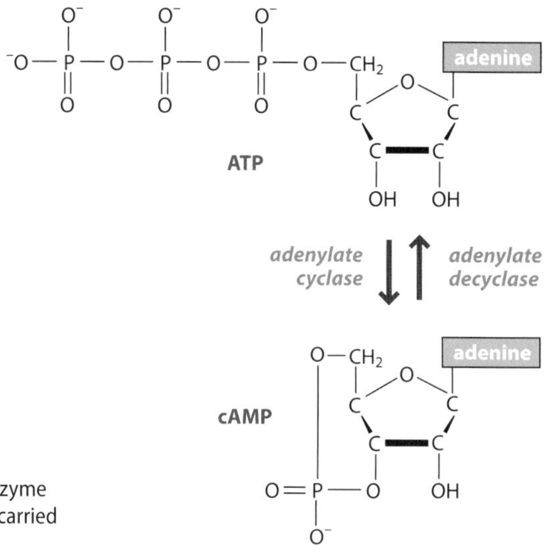

ATP

cAMP

adenylate cyclase | *adenylate decyclase*

Figure 5.29 Cyclic AMP.
Cyclic AMP is synthesized from ATP by the enzyme adenylate cyclase. Conversion back to ATP is carried out by adenylate decyclase.

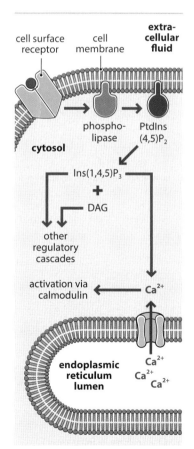

Figure 5.30 Induction of the calcium second messenger system.
Abbreviations: DAG, 1,2-diacylglycerol; Ins(1,4,5)P$_3$, inositol-1,4,5-trisphosphate; PtdIns(4,5)P$_2$, phosphatidylinositol-4,5-bisphosphate.

MAP kinase, which moves away from the membrane and activates other proteins elsewhere in the cell. Some of these proteins are enzymes responsible for catalyzing key biochemical pathways, others are regulatory proteins that switch particular sets of genes on and off. Binding of the mitogen to the receptor protein therefore initiates a cascade of events that result in diverse biochemical changes occurring within the cell. The MAP kinase pathway is used by vertebrate cells, but equivalent pathways, using intermediates similar to those identified in mammals, are known in other organisms.

Other signal transduction systems do not involve the direct transfer of the signal along a cascade of proteins, but instead utilize a less direct way of influencing biochemical activities within the cell. Binding of the extracellular compound – the 'first messenger' – to the membrane receptor induces a transient increase in the internal concentration of a **second messenger**. The spike in the second messenger concentration causes a rapid change in enzyme activities, leading to the desired change in cellular activity.

Important second messengers include the nucleotides **3′,5′-cyclic AMP** (**cAMP**) and **3′,5′-cyclic GMP** (**cGMP**). These are synthesized from ATP and GTP, respectively, by enzymes called **cyclases**, and converted back to ATP and GTP by **decyclases** (*Fig. 5.29*). Some cell surface receptors have guanidylate cyclase activity, and so convert GTP to cGMP, but most receptors in this family work indirectly by influencing the activity of cytoplasmic cyclases and decyclases. These cyclases and decyclases determine the cellular levels of cGMP and cAMP, which in turn control the activities of various target enzymes.

Other second messengers influence the cytoplasmic Ca^{2+} concentration by activating calcium transport proteins located in the endoplasmic reticulum. The concentration of Ca^{2+} within the lumen of the endoplasmic reticulum is higher than in the rest of the cell, so opening these channels allows Ca^{2+} to flow into the cytoplasm. In this system, the first messenger induces its cell surface receptor to activate a phospholipase enzyme, which cleaves **phosphatidylinositol-4,5-bisphosphate** (**PtdIns(4,5)P$_2$**), a lipid component of the cell membrane, into **inositol-1,4,5-trisphosphate** (**Ins(1,4,5)P$_3$**) and **1,2-diacylglycerol** (**DAG**). Ins(1,4,5)P$_3$ activates the calcium transport proteins (*Fig. 5.30*). The Ca^{2+} ions that are released into the cytoplasm bind to and activate a protein called **calmodulin**, which regulates a variety of additional enzymes, bringing about the change in biochemical activity. Additionally, Ins(1,4,5)P$_3$ and DAG can also initiate regulatory cascades.

Further reading

Atlas D (2014) Voltage-gated calcium channels function as Ca^{2+}-activated signaling receptors. *Trends in Biochemical Sciences* **39**, 45–52.

Bobadilla JL, Macek M, Fine JP and Farrell PM (2002) Cystic fibrosis: a worldwide analysis of *CFTR* mutations – correlation with incidence data application to screening. *Mutation Research* **19**, 575–606.

Claypool SM and Koehler CM (2012) The complexity of cardiolipin in health and disease. *Trends in Biochemical Sciences* **37**, 32–41. *A detailed review of the role of one particular type of glycerophospholipid.*

Dennis EA and Norris PC (2015) Eicosanoid storm in infection and inflammation. *Nature Reviews Immunology* **15**, 511–23. *The latest findings regarding the physiological role of eicosanoids.*

Kusumi A, Suzuki KGN, Kasai RS, Ritchie K and Fujiwara TK (2011) Hierarchical mesoscale domain organization of the plasma membrane. *Trends in Biochemical Sciences* **36**, 604–15. *Describes different levels of protein association in membranes.*

Lee AG (2011) Biological membranes: the importance of molecular detail. *Trends in Biochemical Sciences* **36**, 493–500. *Discusses the interactions between lipids and proteins in membranes.*

Nicholson GA (2014) The fluid-mosaic model of membrane structure: still relevant to understanding the structure, function and dynamics of biological membranes after more than 40 years. *Biochimica et Biophysica Acta* **1838**, 1451–66.

Schengrund C-L (2015) Gangliosides: glycerophospholipids essential for normal neural development and function. *Trends in Biochemical Sciences* **40**, 397–406.

Seifert R (2015) cCMP and cUMP: emerging second messengers. *Trends in Biochemical Sciences* **40**, 8–15.

Simopoulos AP (2008) The importance of the omega-6/omega-3 fatty acid ratio in cardiovascular disease and other chronic diseases. *Experimental Biology and Medicine* **233**, 674–88.

Singh B and Sharma RA (2015) Plant terpenes: defense responses, phylogenetic analysis, regulation and clinical applications. *3 Biotech* **5**, 129–51.

ter Beek J, Guskov A and Slotboom DJ (2014) Structural diversity of ABC transporters. *Journal of General Physiology* **143**, 419–35.

Self-assessment questions

Multiple choice questions

Only one answer is correct for each question. Answers can be found on the website: www.scionpublishing.com/biochemistry

1. Which one of these statements describes the typical lipid?
   (a) Hydrophilic and lipophilic
   (b) Hydrophilic and lipophobic
   (c) Hydrophobic and lipophilic
   (d) Hydrophobic and lipophobic

2. Which one of the following statements is **incorrect** with regard to fatty acids?
   (a) Fatty acids are a type of carboxylic acid
   (b) Fatty acids are mostly insoluble in water, but readily soluble in many organic solvents

   (c) The polymeric part of a fatty acid is a hydrocarbon chain
   (d) All of the above statements are correct

3. Which one of the following statements is **incorrect** with regard to unsaturated fatty acids?
   (a) Low melting points and hence are oily liquids at room temperature
   (b) The hydrocarbon chain contains at least one double bond
   (c) They form linear molecules that are able to pack together closely
   (d) All of the above statements are correct

4. Which one of these compounds is an example of an omega-6 fatty acid?
   (a) Lauric acid
   (b) Linoleic acid
   (c) α-Linolenic acid
   (d) Palmitic acid

5. What is the difference between a simple and complex triacylglycerol?
   (a) In a simple triacylglycerol the three fatty acids are identical, in a complex one they are different
   (b) In a simple triacylglycerol the three fatty acids are saturated, in a complex one they are unsaturated
   (c) In a simple triacylglycerol the three fatty acids have 16:0 chains, in a complex one they have $18:1(\Delta^9)$ chains
   (d) None of the above statements are correct

6. What is the compound formed by heating a triacylglycerol with an alkali such as sodium hydroxide called?
   (a) Oil
   (b) Soap
   (c) Sphingolipid
   (d) Wax

7. What is the difference between glycerophospholipids and other triacylglycerols?
   (a) A glycerophospholipid contains sphingosine
   (b) A glycerophospholipid is an alkali derivative of a triacylglycerol
   (c) A glycerophospholipid is the product of esterification between a triacylglycerol and triacontanol
   (d) In a glycerophospholipid, one of the fatty acids is replaced by a hydrophilic group attached to the glycerol component by a phosphodiester bond

8. Which of these compounds is **not** an example of a glycerophospholipid?
   (a) Phosphosphingosine
   (b) Phosphatidic acid
   (c) Phosphatidylglycerol
   (d) Phosphatidylserine

9. What is a ganglioside?
   (a) A sphingolipid carrying a complex sugar head group
   (b) A sphingolipid carrying a simple sugar head group
   (c) A sphingolipid that lacks sphingosine
   (d) A sphingolipid with an amino acid head group

10. Terpenes are based on a small hydrocarbon compound called what?
    (a) Abietic acid
    (b) Isoprene
    (c) Monoterpene
    (d) Squalene

11. What is the composition of the core sterol structure?
    (a) Four hydrocarbon rings, three of which have six carbons each and one of which has five carbons
    (b) Four hydrocarbon rings, three of which have five carbons each and one of which has six carbons

    (c) Four hydrocarbon rings, three of which have six carbons each and one of which has five carbons
    (d) Four hydrocarbon rings, two of which have six carbons each and two of which have five carbons

12. Which one of the following compounds is **not** a type of steroid hormone?
    (a) Androgens
    (b) Eicosanoids
    (c) Glucocorticoids
    (d) Progestins

13. What is the disease that results from vitamin D deficiency called?
    (a) Rackets
    (b) Rickets
    (c) Rockets
    (d) Pickets

14. Which of these types of eukaryotic membrane has the highest protein content?
    (a) Endoplasmic reticulum
    (b) Inner mitochondrial membrane
    (c) Outer mitochondrial membrane
    (d) Plasma membrane

15. The model for membrane structure proposed by Singer and Nicholson in 1972 is called what?
    (a) Fluid mosaic model
    (b) Lipid bilayer model
    (c) Lipid raft model
    (d) Tethered protein model

16. Which one of the following statements is **correct** with regard to an integral membrane protein?
    (a) Can only be removed from the membrane by disrupting the structure of the bilayer
    (b) Forms a tight attachment with the membrane
    (c) Most span the entire lipid bilayer
    (d) All of the above statements are correct

17. A barrel-like structure, with the walls of the barrel made up of β-sheet, is a typical feature of what?
    (a) Lipid-linked protein
    (b) Peripheral membrane protein
    (c) Transmembrane protein
    (d) None of the above

18. What is the erythrocyte transporter protein an example of?
    (a) Antiporter
    (b) P-type pump
    (c) Symporter
    (d) Uniporter

19. What is the mammalian Na^+/K^+ ATPase protein an example of?
    (a) Antiporter
    (b) P-type pump
    (c) Symporter
    (d) Uniporter

20. What is the mammalian Na⁺/glucose transporter protein an example of?
- **(a)** Antiporter
- **(b)** P-type pump
- **(c)** Symporter
- **(d)** Uniporter

21. What is the Na⁺/Ca²⁺ exchange protein an example of?
- **(a)** Antiporter
- **(b)** P-type pump
- **(c)** Symporter
- **(d)** Uniporter

22. What is the commonest cystic fibrosis mutation called?
- **(a)** ΔF508
- **(b)** G542X
- **(c)** G551D
- **(d)** N1303K

23. In the MAP kinase system, what are individual proteins in the cascade activated by?
- **(a)** Addition of Ca²⁺
- **(b)** Addition of a lipid group
- **(c)** Methylation
- **(d)** Phosphorylation

24. Which one of the following statements is **incorrect** with regard to calmodulin?
- **(a)** Activated by Ca²⁺ ions
- **(b)** Influenced by cleavage of phosphatidylinositol-4,5-bisphosphate
- **(c)** Is an integral membrane protein
- **(d)** Regulates a variety of enzymes

Short answer questions

These questions do not require additional reading.

1. Using examples, distinguish between the structures of saturated and unsaturated fatty acids.

2. Explain what is meant by the terms 'omega-3' and 'omega-6' when referring to fatty acid structure. Why do we require these types of fatty acid in our diet?

3. Describe the structures of (a) a simple triacylglycerol, (b) a complex triacylglycerol, (c) a soap, and (d) a wax.

4. Outline the key features of glycerophospholipids, sphingolipids and eicosanoids.

5. Explain why sterols and steroids are classified as terpene derivatives.

6. Describe the way in which certain lipids associate together to form a membrane bilayer.

7. Outline the key features of the fluid mosaic model for membrane structure. What is a 'lipid raft'?

8. Describe the distinguishing features of integral and peripheral membrane proteins.

9. Summarize, with examples, the different ways in which proteins aid the transport of small molecules across membrane bilayers.

10. Describe the various intracellular events that can be stimulated by attachment of an extracellular signaling compound to a transmembrane receptor protein.

Self-study questions

These questions will require calculation, additional reading and/or internet research.

1. A few years ago, coconut oil was looked on as unhealthy because of its high content of saturated fatty acids. Now coconut oil is a 'superfood' claimed to confer benefits ranging from glossier hair to resistance to disease. Evaluate the reasons for this change of opinion regarding the value of coconut oil in the diet.

2. Explain why many commercial products for unblocking sinks contain an alkali.

3. Identify the isoprene units in these terpenes.

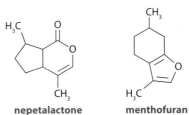

farnesol

ngaione

chrysanthemic acid

nepetalactone   **menthofuran**

4. What methods might be used to identify those parts of a transmembrane protein that are exposed on the outer surface of the plasma membrane of an animal cell?

5. The MAP kinase cascade is an example of a biochemical pathway in which phosphorylation is used to change the activity of a protein. Phosphorylation is also used to regulate protein activity in other contexts, especially in the regulation of enzymes in metabolic pathways such as glycolysis. Usually, the amino acid that is phosphorylated is one of serine, threonine, tyrosine or histidine. Explain why addition of a phosphate group to one of these amino acids can have a significant effect on the activity of a protein.

CHAPTER 6
Carbohydrates

STUDY GOALS

After reading this chapter you will:

- understand the differences between the terms 'monosaccharide', 'disaccharide', 'oligosaccharide' and 'polysaccharide'

- be able to distinguish between the structures of aldose and ketose monosaccharides of different chain lengths, including both the linear and ring forms of those compounds

- understand the various stereoisomeric terms relevant to carbohydrate structure: enantiomers, diastereomers, epimers and anomers

- know how monosaccharides are linked together to form disaccharides and longer chain structures

- be aware of the importance of oligosaccharides as side-chains attached to some types of protein

- be able to distinguish between homopolysaccharides and heteropolysaccharides, and to give examples of both types

- be aware of the important roles played by polysaccharides as energy stores and as structural components of plant and animal tissues

The carbohydrates are the fourth type of polymeric biomolecule found in living cells. The polymeric carbohydrates include starch and glycogen, which are important energy stores in plants and animals, respectively. Other carbohydrates have structural roles, the best known example being cellulose, which provides plants with part of their structural rigidity. Chitin, which is found in insect exoskeletons, is also a polymeric carbohydrate, as are several of the compounds that form the extracellular matrix between cells in animal tissues.

6.1   Monosaccharides, disaccharides and oligosaccharides

A carbohydrate is, strictly speaking, any compound made up of carbon, hydrogen and oxygen, with the hydrogen and oxygen in a 2:1 ratio as in water. The carbohydrates of most importance in biochemistry are called saccharides, a term derived from the Latin word for sugar (saccharum). Starch and the other polymeric carbohydrates are **polysaccharides**, and their monomeric units are **monosaccharides**. The monosaccharides, which include compounds such as glucose and galactose, are very important in their own right because these are the primary sources of the energy used to power cellular processes. **Disaccharides**, which contain two linked monosaccharide units, include important naturally occurring sugars such as sucrose and lactose. Short polymeric carbohydrates, called **oligosaccharides**, are important for a different reason, because these form side-chains on some proteins.

With carbohydrates we must therefore focus our attention not only on the large polymeric molecules but also on their constituent monosaccharides and the shorter chain di- and oligosaccharides. The easiest way to understand the relationships between all of these different compounds is to start small with monosaccharides and gradually work our way up to the largest molecules such as starch and cellulose.

6.1.1   Monosaccharides are the building blocks of carbohydrates

Monosaccharides include a number of familiar compounds, such as ribose and 2'-deoxyribose, found in RNA and DNA nucleotides, and glucose, the substrate for glycolysis, the central energy-generating pathway for most organisms.

The two simplest monosaccharides are glyceraldehyde and dihydroxyacetone

A monosaccharide is a carbohydrate with at least three carbon atoms, one of which is attached to an oxygen group (=O) and the others to hydroxyl groups. This definition allows for two quite distinct types of molecule, depending on which of the carbons carries the oxygen group. To understand this important point, we will look at the structures of the two simplest monosaccharides, glyceraldehyde and dihydroxyacetone, both of which have three carbon atoms (*Fig. 6.1*). In glyceraldehyde, the oxygen group is attached to one of the terminal carbons. This forms a formyl group (–CHO), which is the characteristic feature of compounds called **aldehydes** (indeed the –CHO group is commonly called an aldehyde group). Glyceraldehyde is therefore an aldehyde sugar or **aldose**, or more specifically an **aldotriose** because it has three carbon atoms.

Dihydroxyacetone, on the other hand, has the oxygen attached to the central carbon. The resulting C=O structure is characteristic of a **ketone**, so dihydroxyacetone is a ketone sugar or **ketose**. Dihydroxyacetone is therefore a **ketotriose** because it has three carbon atoms.

We encountered glyceraldehyde in *Section 3.1.2* when we studied the optical isomers or enantiomers of amino acids. Although shown as a flat structure in *Figure 6.1*, in reality glyceraldehyde has a tetrahedral configuration. This configuration comprises a central carbon atom attached to –H, –OH, –CHO and –CH₂OH groups (*Fig. 6.2*). As with an amino acid, there are two ways in which these four groups can be arranged around the central, chiral carbon. The two arrangements are mirror images of one another, one being the dextro-enantiomer and the other the laevo-enantiomer. The two forms – D-glyceraldehyde and L-glyceraldehyde – have identical chemical properties and differ only in their effect on plane-polarized light (see *Fig. 3.5*).

Dihydroxyacetone is different. It does not have a chiral carbon and so does not form enantiomers. Not only is it different from glyceraldehyde in this regard, it is different from all other monosaccharides. This is because all monosaccharides with four or more carbons, whether aldoses or ketoses, have at least one chiral carbon, and most have more than one. We must now examine the structural complications that arise from the presence of these multiple chiral centers.

Most monosaccharides exist as enantiomers and diastereomers

We will now move up a level and consider the **aldotetroses**, the aldehyde sugars that contain four carbon atoms. These compounds are **diastereomers**, meaning that they have more than one chiral carbon. The aldotetroses have two chiral carbons which means that there are four possible configurations, comprising two pairs of enantiomers. The compounds are called D- and L-erythrose and D- and L-threose (*Fig. 6.3*). Note that the structures of erythrose and threose are not themselves mirror images because the relative positioning of the hydroxyl groups is different. Erythrose and threose are therefore different compounds with distinct chemical properties.

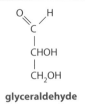

glyceraldehyde

dihydroxyacetone

Figure 6.1 Glyceraldehyde and dihydroxyacetone.

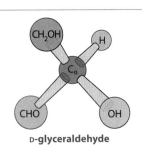

D-**glyceraldehyde**

L-**glyceraldehyde**

Figure 6.2 D-glyceraldehyde and L-glyceraldehyde.

Figure 6.3 Aldotetroses. The aldehyde group is shown in blue and the asymmetric center for each molecule is in red. The numbering of the carbon atoms is indicated on the left.

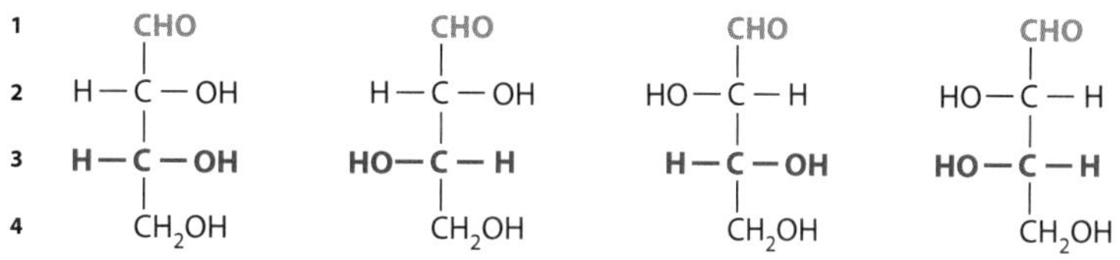
1
2
3
4
CHO
H—C—OH
H—C—OH
CH_2OH
CHO
H—C—OH
HO—C—H
CH_2OH
CHO
HO—C—H
H—C—OH
CH_2OH
CHO
HO—C—H
HO—C—H
CH_2OH

D-erythrose **L-erythrose** **D-threose** **L-threose**

At the next level are the five-carbon **aldopentoses** (*Fig. 6.4*). Each of these has three chiral carbons, so there are four pairs of enantiomeric aldopentoses in all. These include three sugars that are common in nature – ribose (present in nucleotides), arabinose and xylose. Arabinose is the only monosaccharide that occurs in nature predominantly as the L-enantiomer. For all of the others, the D-form is more common. Then there are eight **aldohexose** enantiomer pairs, which include glucose, mannose and galactose. We could continue up through the aldoheptoses, aldooctoses, aldononoses and aldodecoses, with 7, 8, 9 and 10 carbons, respectively, but these compounds are less common in nature and, as biochemists, we do not need to be so concerned about them.

Box 6.1 Representations of monosaccharide structures

Because of the tetrahedral arrangement of groups around a carbon atom, the structure of a monosaccharide cannot be depicted accurately when drawn on a flat piece of paper. The representation called the **Fischer projection** aims to solve this problem. When a compound is drawn as a Fischer projection, bonds drawn horizontally from the central carbon are ones that, in the tetrahedral arrangement, would project above the plane of the paper, and bonds drawn vertically are ones that would project below the plane of the paper.

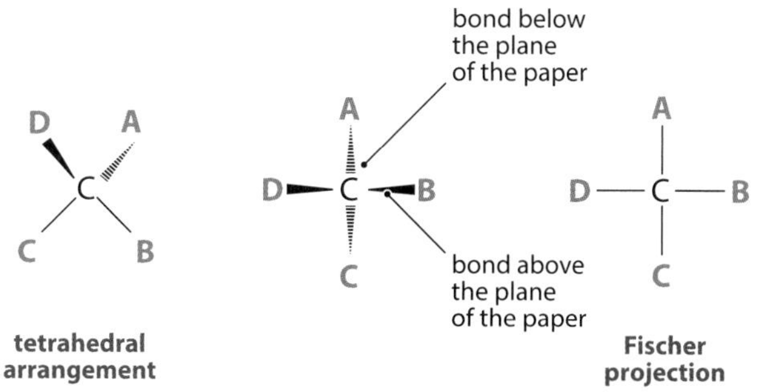
bond below the plane of the paper
bond above the plane of the paper
D
A
C
B
tetrahedral arrangement
Fischer projection

The two enantiomers of glyceraldehyde are therefore drawn like this:

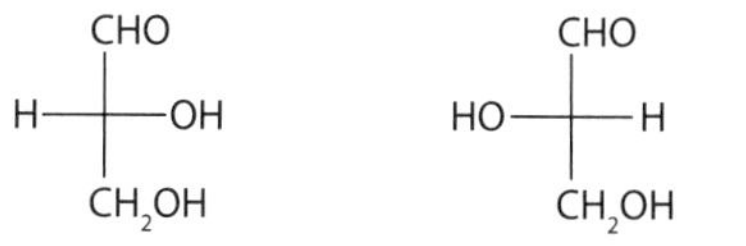
CHO
H
OH
CH_2OH
CHO
HO
H
CH_2OH

D-glyceraldehyde **L-glyceraldehyde**

When a monosaccharide has two chiral carbons, the D- and L-enantiomers are identified from the arrangement of groups around the carbon that is furthest from the aldehyde group if the sugar is an aldose, or the ketone group if it is a ketose. This carbon is referred to as the asymmetric center. The D-enantiomer is depicted with the hydroxyl on the right of this carbon, and the L-enantiomer is depicted with the hydroxyl on the left. The D- and L-versions of the aldotetrose erythrose are therefore drawn thus:

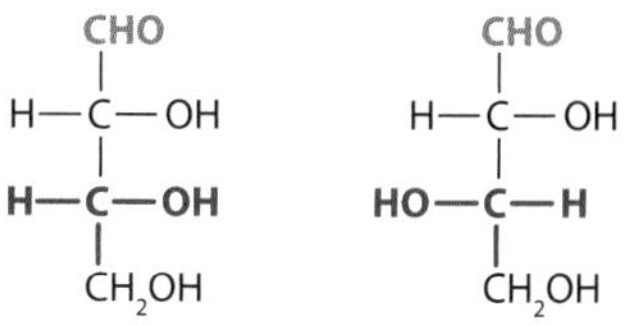
CHO
H—C—OH
H—C—OH
CH_2OH
CHO
H—C—OH
HO—C—H
CH_2OH

D-erythrose **L-erythrose**

In these drawings the aldehyde group is shown in blue and the asymmetric center in red.

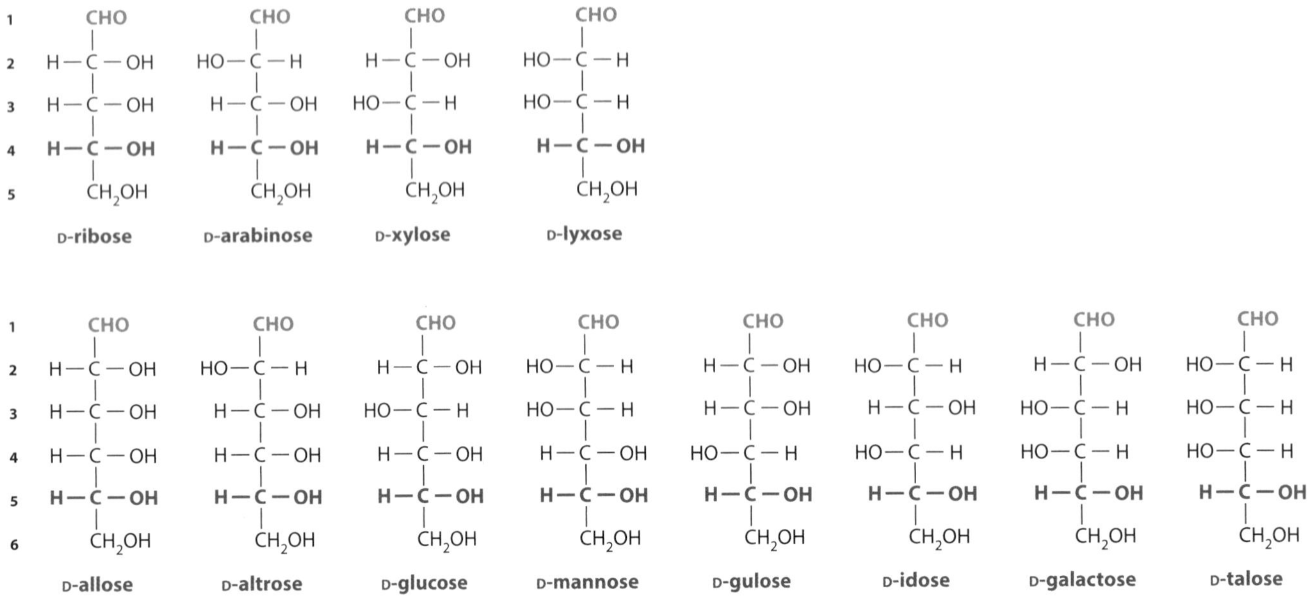

Figure 6.4 **Aldopentoses and aldohexoses.**
The aldehyde group is shown in blue and the asymmetric center for each molecule is in red. The numbering of the carbon atoms is indicated on the left. Each compound also has an L-enantiomer that is not shown here.

We must, however, consider the equivalent series of ketose sugars with 4–6 carbons. These are slightly less complicated because at each level we have one fewer chiral carbons. So there is just one pair of **ketotetroses**, called D- and L-erythrulose, two **ketopentoses**, ribulose and xylulose, and four **ketohexoses** (*Fig. 6.5*). This last set includes fructose which, like glucose, is an important dietary sugar obtained from many fruits and vegetables.

Some monosaccharides also exist as ring structures

As well as the linear structures that we have considered so far, monosaccharides with five or more carbons can also form ring or cyclic molecules. We are already aware of this because the form of ribose, a five-carbon sugar, present in nucleotides is a ring molecule (see *Fig. 4.1*), not the linear chain shown in *Figure 6.4*. The cyclic forms of five-carbon sugars like ribose are known as **furanoses** because of their structural similarity with the unrelated organic compound furan. The cyclic version of ribose

---

Box 6.2 **The different types of isomers relevant to carbohydrate structure**

It is easy to get confused by the different types of isomerism displayed by carbohydrates. Here is a summary of the important terms. Remember that, by definition, isomers are compounds that have the same chemical composition.

- **Stereomers** are isomers in which the atoms are joined together in the same sequence, but which differ in the arrangement of atoms around one or more asymmetric centers such as a chiral carbon. All of the types of isomer listed below are categories of stereoisomer.
- **Enantiomers** are isomers whose structures are mirror images of one another. D-glyceraldehyde and L-glyceraldehyde are enantiomers.

- **Diastereomers** are compounds that have two or more chiral carbons. Erythrose and threose are diastereomers.
- **Epimers** are diastereomers that differ in structure at just one of their chiral carbons. D-glucose and D-galactose are epimers.
- **Anomers** are cyclic monosaccharides that differ only in the arrangement of groups around the anomeric carbon; for an aldose this is carbon 1, and for a ketose it is carbon 2. Aldose examples are α-D-glucopyranose and β-D-glucopyranose.

Figure 6.5 The ketose sugars with 4–6 carbons.
The ketone group is shown in blue and the asymmetric center for each molecule is in red. The numbering of the carbon atoms is indicated on the left. Each compound also has an L-enantiomer that is not shown here.

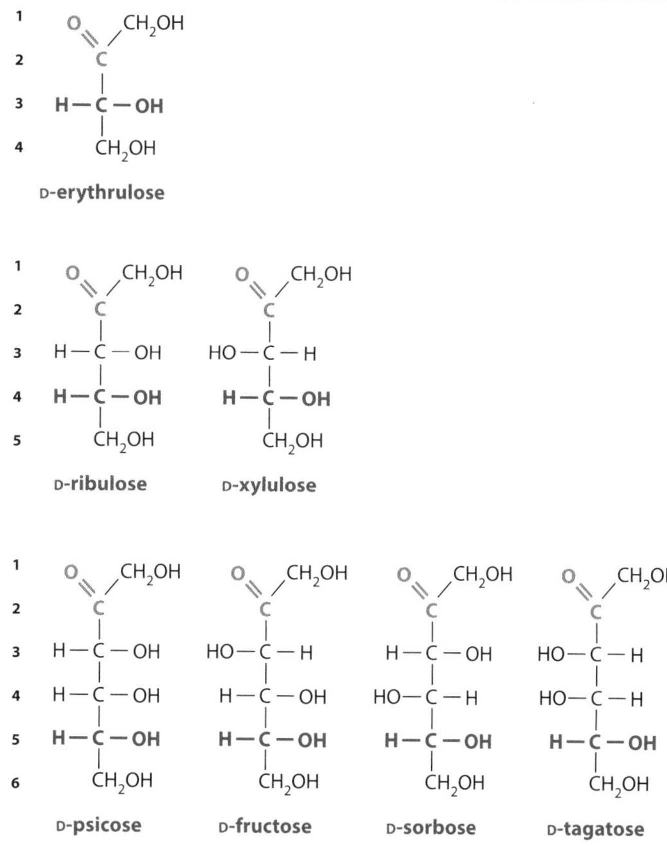

D-erythrulose

D-ribulose   D-xylulose

D-psicose   D-fructose   D-sorbose   D-tagatose

is formed by reaction between the aldehyde group at carbon number 1 with the hydroxyl group at carbon 4 (*Fig. 6.6*). A similar reaction can occur for the aldohexoses such as glucose. The cyclic form of an aldohexose is called a **pyranose** because it has a structural similarity with the organic compound called pyran. The specific chemical name for the ring form of glucose is therefore glucopyranose.

Formation of the ring provides further opportunities for creating variations in the structure of a monosaccharide. The cyclic form of D-glucose, for example, has α and β versions which differ in the positioning of the hydroxyl group attached to carbon 1. This is the carbon originally present in the aldehyde group of the linear form and which participated in the chemical reaction that led to cyclization (*Fig. 6.7*) The two structures, α-D-glucopyranose and β-D-glucopyranose, have slightly different optical properties but otherwise are chemically identical. They are called **anomers**. In solution, the α form can readily convert into β, and vice versa, so solutions of D-glucose contain a mixture of the two types, usually with a small amount of the linear form also present. This interconversion is called **mutarotation**.

Figure 6.6 Formation of the cyclic version of ribose.
The ring is formed by reaction between the aldehyde group at carbon number 1 and the hydroxyl group at carbon 4.

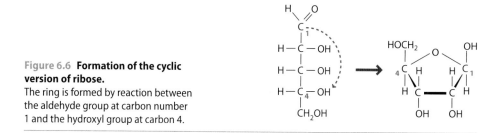

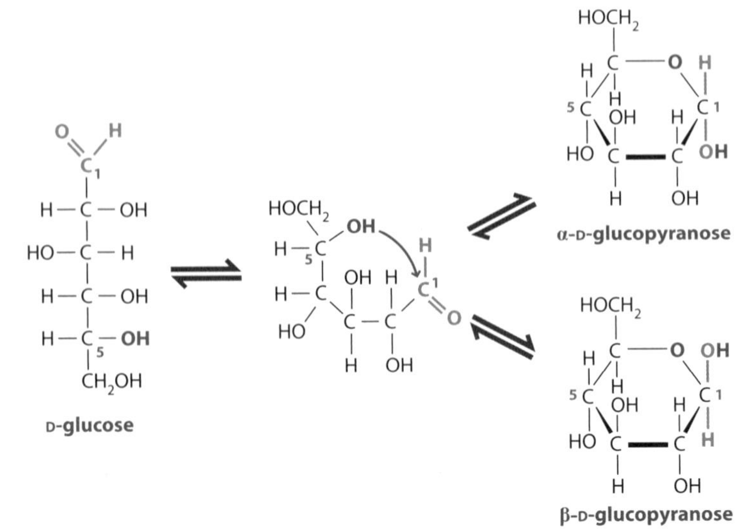

Figure 6.7 Formation of the two anomers of glucose.
The ring is formed by reaction between the aldehyde group at carbon number 1 (blue) and the hydroxyl group at carbon 5 (red). This can result in two different orientations of the groups around carbon 5, giving the α and β anomeric forms. The cyclization reaction is reversible and there is an equal chance of either anomer being formed, so solutions of D-glucose contain a mixture of the two anomers along with some linear molecules.

The 5- and 6-carbon ketose monosaccharides can also form ring structures. Reaction is between the ketone group at carbon 2 and the hydroxyl at carbon 5 to give a five-membered furanose ring (*Fig. 6.8*). The cyclic derivative of fructose is therefore called fructofuranose. As with glucopyranose, there are α and β anomeric forms.

6.1.2 Disaccharides are made by linking together pairs of monosaccharides

Now we move on to the way in which the cyclic monosaccharide units are linked together to form longer chain carbohydrates. The simplest of these are the **disaccharides**, which comprise just two monosaccharide units joined together. Some disaccharides are very common in nature (*Table 6.1*). These include sucrose, which is the type of sugar, obtained from sugar cane or beet, that some of us put in our coffee. Sucrose is made up of glucose and fructose units. Lactose, from milk, has glucose and galactose units, and maltose from malt barley has two glucose units.

The link between the two monosaccharide units in a disaccharide is called an **O-glycosidic bond**. This type of bond is formed between pairs of hydroxyl groups, one on each monosaccharide, so a great deal of variability is possible. In maltose, which

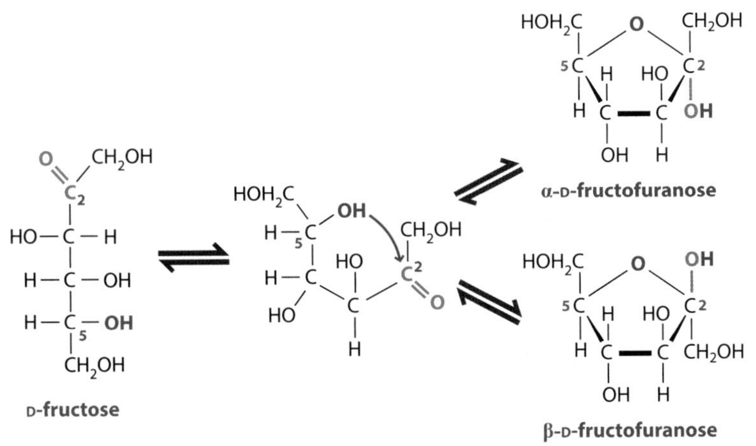

Figure 6.8 Formation of the cyclic form of fructose.
The ring is formed by reaction between the ketone group at carbon number 2 (blue) and the hydroxyl group at carbon 5 (red). As with glucose, two anomers can be formed.

Table 6.1. **Examples of disaccharides**

Name	Component sugars	Description
Sucrose	Glucose + fructose	From sugar cane and sugar beet
Lactose	Glucose + galactose	Milk sugar
Maltose	Glucose + glucose	Malt sugar, from germinating cereals
Trehalose	Glucose + glucose	Made by plants and fungi
Cellobiose	Glucose + glucose	Breakdown product of cellulose

contains two glucose units, the link is between the hydroxyls attached to carbon 1 of one glucopyranose and carbon 4 of the second unit (*Fig. 6.9*). The link is therefore denoted as '1→4'. Carbon 1 is the anomeric carbon so we must also distinguish if the bond involves the α or β version. In maltose, it is α, so the correct chemical name of this disaccharide is α-D-glucopyranosyl-(1→4)-D-glucopyranose.

In maltose, the second glucopyranose is free to interconvert between its α and β anomers because its carbon number 1 is not involved in the glycosidic bond. This is not always the case. Trehalose, for example, is also made up of two glucose units,

---

Box 6.3 Some humans have recently evolved the ability to digest milk

RESEARCH HIGHLIGHT

Milk and milk products form such an important part of the modern western diet that it is surprising to learn that our ancestors were unable to digest milk. Milk is rich in lactose, but lactose can only be used as a nutrient after it has been broken down into its constituent monosaccharides, glucose and galactose. This reaction is catalyzed by the enzyme lactase, which is present in the epithelial cells lining the small intestine. Most mammals only synthesize lactase during the early weeks of their lives, when they are dependent on the mother's milk for most of their nutrients. After weaning, lactase is no longer made and the ability to break down lactose, and hence to digest milk, gradually disappears. The majority of adult mammals are therefore lactose intolerant because they do not synthesize lactase.

Today, about 65% of the adult human population is lactose intolerant. If a lactose intolerant person drinks milk, or consumes dairy products from which the lactose has not been removed, then they experience stomach cramps and lower intestinal issues. However, as the map opposite shows, many people from northern Europe, part of Africa, the Middle East and southern Asia (and those who trace their ancestries to these populations) are able to digest milk. This is because they continue to produce lactase even after weaning. This is referred to as **lactose tolerance** or **lactase persistence**.

The lactase enzyme has an identical structure in both lactose tolerant and intolerant individuals, so we have to look beyond the protein to find the cause of lactase persistence. The difference lies with the DNA sequence in the region adjacent to the gene. Changes in this sequence alter the expression pattern of the gene for lactase synthesis, so that the normal regulatory process, which switches the gene off after weaning, does not operate. The gene remains active all the time, and lactase is synthesized into adulthood.

Comparisons between the DNA sequences of this regulatory region in different people enable some aspects of the evolution

of lactase persistence to be probed. These studies suggest that the crucial sequence changes that gave rise to lactase persistence in Europeans occurred between 7500 and 12 500 years ago. The timing coincides with the results of other investigations, which have detected traces of dairy lipids absorbed into fragments of cooking and storage pots from 9000 years ago. The origin of lactase persistence therefore appears to coincide with the first extensive use of milk, as we would expect. The DNA sequence changes are thought to have occurred in early farming populations in the Balkans or central Europe, and then to have spread gradually into northern Europe. This is consistent with our expectation that, once dairying became widespread, lactase persistence would have become highly advantageous.

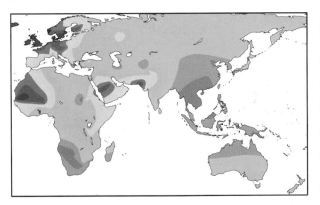

high ◼◼◼◼◼◼◼◼◼◼◼ low

frequency of lactase persistence

Image reproduced from *BMC Evolutionary Biology*, 2010; **10**:36.

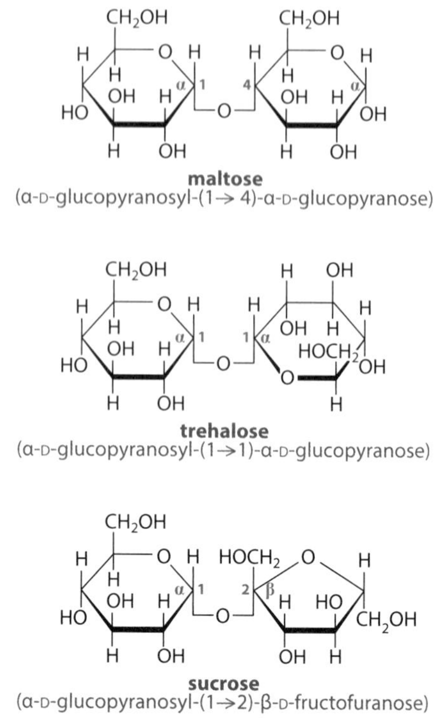

maltose
(α-D-glucopyranosyl-(1→ 4)-α-D-glucopyranose)

trehalose
(α-D-glucopyranosyl-(1→1)-α-D-glucopyranose)

sucrose
(α-D-glucopyranosyl-(1→2)-β-D-fructofuranose)

Figure 6.9 Three disaccharides.
The α anomer of maltose is shown.

but with this disaccharide the bond is between the pair of number 1 carbons (see *Fig. 6.9*). Trehalose is therefore α-D-glucopyranosyl-(1→1)-α-D-glucopyranose. Of course, the two monosaccharide units do not always have to be glucose. Sucrose is α-D-glucopyranosyl-(1→2)-β-D-fructofuranose, the link being between the α version of the anomeric carbon 1 of glucose and the β version of carbon 2 (the anomeric carbon) of fructose (see *Fig. 6.9*).

6.1.3 Oligosaccharides are short monosaccharide polymers

A disaccharide is the shortest type of **oligosaccharide**, a short polymeric carbohydrate made up of 2–20 monosaccharide units. Oligosaccharides made up entirely of fructose or xylose are synthesized by some plants, and a galactose oligosaccharide, with some glucose units, is present in human milk. These carbohydrates are attracting attention because of possible health benefits; oligogalactose, for example, is being linked with protection of infants against gastrointestinal infections.

Figure 6.10 Protein glycosylation.
The attachment of a glycan to a polypeptide chain by (A) *O*-linked glycosylation, and (B) *N*-linked glycosylation. The *O*-linked glycan can be attached to the –OH group of serine or threonine; the *N*-linked glycan attaches to the –NH₂ group of asparagine. The sugars in this diagram are *N*-acetylgalactosamine (GalNAc) and *N*-acetylglucosamine (GlcNAc).

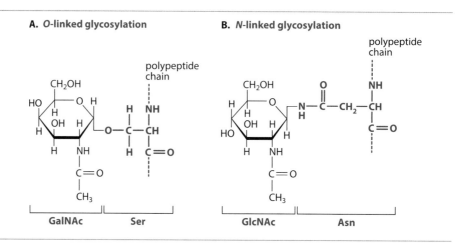

A. *O*-linked glycosylation

B. *N*-linked glycosylation

GalNAc Ser

GlcNAc Asn

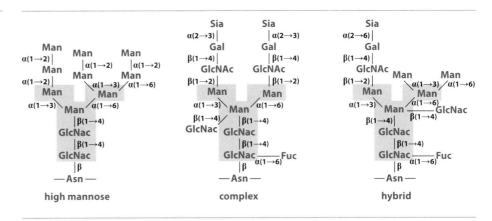

high mannose
complex
hybrid

Figure 6.11 **Typical *N*-linked glycans.**
All three *N*-linked glycan families have the same five-sugar core (shaded), whereas most *O*-linked glycans are smaller, with up to four sugars. Abbreviations: Fuc, fucose; Gal, galactose; GalNAc, *N*-acetylgalactosamine; GlcNAc, *N*-acetylglucosamine; Man, mannose; Sia, sialic acid.

In biochemistry, oligosaccharides are also important because they form side-chains that are attached to some types of protein. This is called **glycosylation** and it occurs within the Golgi apparatus of eukaryotic cells, after the polypeptide chains have been assembled. The oligosaccharide structures, or **glycans**, are attached either to the hydroxyl group of a serine or threonine amino acid, or to the amino group of asparagine (*Fig. 6.10*). These two versions are called ***O*-linked** and ***N*-linked glycosylation**, respectively.

Glycans are variable in structure, but typical examples are shown in *Figure 6.11*. An important feature is that some of the monosaccharide units are not the regular aldose and ketose structures that we have studied so far. They include modified monosaccharides in which one or more of the hydroxyls have been replaced by other chemical groups. Glucosamine, for example, is derived from glucose by replacement of the hydroxyl attached to carbon 2 with an amino group. Further addition of an acetyl group converts glucosamine to *N*-acetylglucosamine (see *Fig. 6.10*). The *N*-acetyl version of galactose is formed in a similar way, and *N*-acetylneuraminic acid, also called sialic acid, is derived from the nine-carbon ketose sugar neuraminic acid.

We do not fully understand the function of protein glycosylation. With some proteins, the attached glycans appear to act as zip codes, directing the protein to a particular compartment within the cell. Glycosylation also stabilizes proteins, slowing down the rate at which they are degraded by protease enzymes. The length of time that some protein hormones remain active in the bloodstream is significantly affected by glycosylation, and in some cases the glycan is essential for the activity of the hormone.

Another interesting function of glycosylation is protection against freezing. Many species of fish from the polar regions contain an antifreeze protein that is multiply glycosylated with *O*-linked β-galactosyl-(1→3)-α-*N*-acetylgalactosamine units. The resulting proteins circulate in the bloodstream and are thought to bind to small ice crystals, preventing them from growing and damaging the tissues inside the fish.

6.2 Polysaccharides

Now we move up the scale to the largest polymeric carbohydrates. Polysaccharides are made up of cyclic monosaccharide units linked by glycosidic bonds. The chains can be linear or branched, and the monosaccharide units can be identical or mixed. If all the units are the same, then the compound is a **homopolysaccharide**, and if mixed then it is a **heteropolysaccharide**.

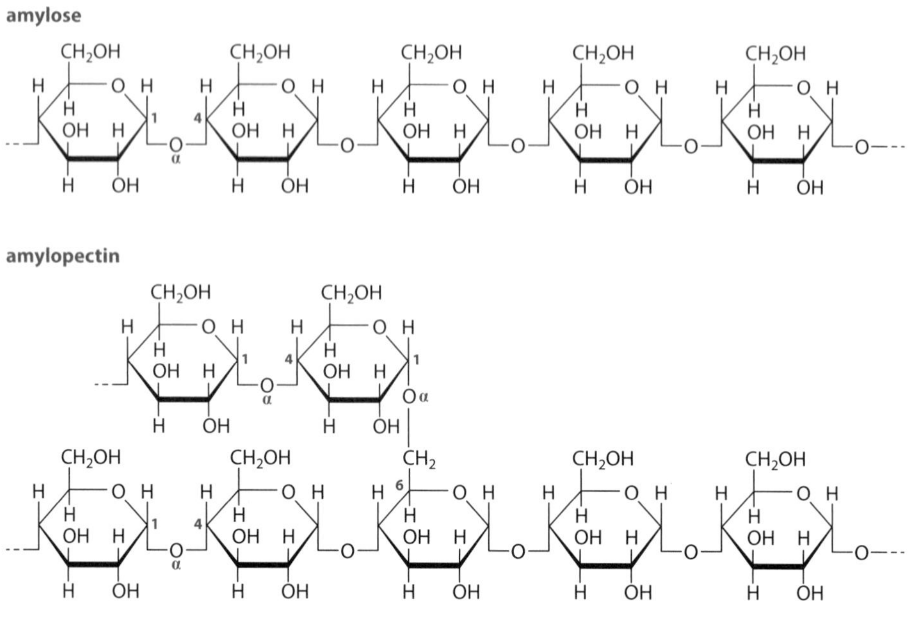

Figure 6.12 The polymeric structures of amylose and amylopectin.
Amylose is a linear polymer of D-glucose units linked by α(1→4) glycosidic bonds. Amylopectin has a branched structure made up of α(1→4) chains and α(1→6) branch points.

Figure 6.12 The polymeric structures of amylose and amylopectin. Amylose is a linear polymer of D-glucose units linked by α(1→4) glycosidic bonds. Amylopectin has a branched structure made up of α(1→4) chains and α(1→6) branch points.

6.2.1 Starch, glycogen, cellulose and chitin are homopolysaccarides

Starch is a homopolysaccharide made entirely of D-glucose units. There are two types of starch molecule, called **amylose** and **amylopectin**. The difference between the two is that amylose is a linear polymer of D-glucose units linked by α(1→4) glycosidic bonds, whereas amylopectin has a branched structure made up of α(1→4) chains and α(1→6) branch points, the branches occurring every 24–30 units along each linear chain (*Fig. 6.12*). All plants synthesize both amylose and amylopectin, with the latter usually being the predominant form.

Polysaccharides are variable in length. Amylose chains contain anything from a few hundred to over 2500 glucose units, and amylopectin spans a similar range but with a higher upper limit, possibly 6000 units in the largest molecules. The factors that determine the sizes of individual molecules are not well understood.

The starch polysaccharides can take up various conformations, the most stable of which gives the α(1→4) chain a fairly tight curvature. Amylose and amylopectin therefore have compact coiled structures that enable the molecules to be packed tightly into spherical granules, such as those found in the cells of the photosynthetic parts of a plant.

Starch is a storage polysaccharide, the monosaccharide units being utilized for energy generation by cleaving them from the ends of the amylose and amylopectin

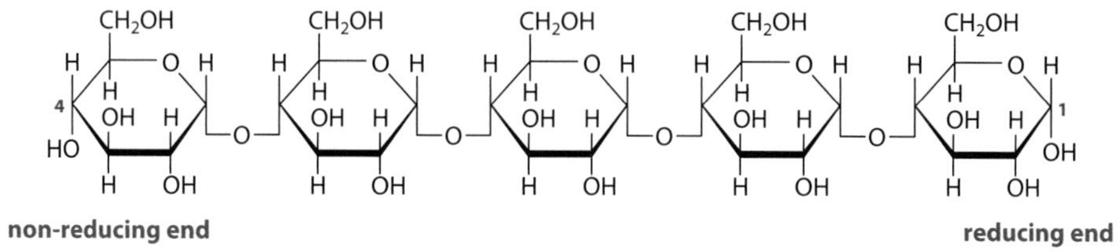

Figure 6.13 The non-reducing and reducing ends of a starch polysaccharide.

non-reducing end   reducing end

Box 6.4 **The reducing and non-reducing ends of a starch molecule**

The reducing end of a starch molecule is the end that terminates with the anomeric carbon. This terminal glucose unit can open up into its linear configuration, re-forming the aldehyde group.

reducing end of a starch polysaccharide

The aldehyde group acts as a reducing agent, which means it can donate electrons to other compounds. It can, for example, reduce cupric ions (Cu^{2+}) to cuprous ions (Cu^{+}), which is the basis of the Fehling's and Benedict's tests for **reducing sugars** – sugars which in their linear form have reducing activity.

The non-reducing end of a starch polymer, on the other hand, terminates with the non-anomeric carbon at position 4. This terminal glucose cannot open up into its linear configuration, and so cannot acquire reducing activity.

Starch is not the only sugar with reducing and non-reducing ends. The same will be true for any di-, oligo- or polysaccharide that terminates with a reducing sugar with a free anomeric carbon. Each of the simple aldose monosaccharides that we have studied (*Figs 6.3* and *6.4*) are reducing sugars because of the activity of the aldehyde group. The ketoses (*Fig. 6.5*) do not themselves have reducing activity. However, when in the linear configuration, a ketose exists in equilibrium with its equivalent aldose (e.g. fructose is in equilibrium with glucose). This equilibrium arises because of an isomerization that converts the ketose into the aldose, and vice versa.

D-**fructose**   **enediol intermediate**   D-**glucose**

The interconversion is favored by high pH and does not occur readily under physiological conditions, but a solution of a ketose sugar will always contain a small proportion of its aldose partner. A ketose is therefore not itself a reducing sugar, but a solution of a ketose sugar will display some reducing activity.

molecules. Only the 'non-reducing' ends are cut back, these being the ends that terminate with a free hydroxyl group at the non-anomeric carbon number 4 (*Fig. 6.13*). Amylose, being a linear chain, has one reducing and one non-reducing end, but in amylopectin the branches give rise to a molecule with many non-reducing ends that can be utilized concurrently. The non-reducing ends are attacked by β-amylase and other enzymes that release individual glucose molecules, possibly along with maltose disaccharides and maltotriose – the trisaccharide comprising three glucose units linked by α(1→4) bonds.

Glycogen, the main storage polysaccharide of animals, has a similar structure to amylopectin but with more frequent branch points. Cellulose is also a linear D-glucose homopolysaccharide, but with β(1→4) linkages (*Fig. 6.14A*). This subtle difference compared with amylose results in a molecule with completely different properties. The most stable conformation of cellulose is a straight chain, rather than the coil of the amylose molecule, with individual cellulose chains being able to line up side by side and attach to one another by hydrogen bonding. This gives rise to the rigid networks that play an important role in plant cell walls. Chitin has a very similar structure, being a β(1→4) linked homopolymer of *N*-acetylglucosamine, the acetyl-amino modification of glucose (*Fig. 6.14B*). The acetyl-amino groups participate in additional hydrogen bonding, so chitin molecules pack together even more tightly, giving a more rigid structure than cellulose. Chitin is present in the exoskeletons of insects and other arthropods such as crustaceans.

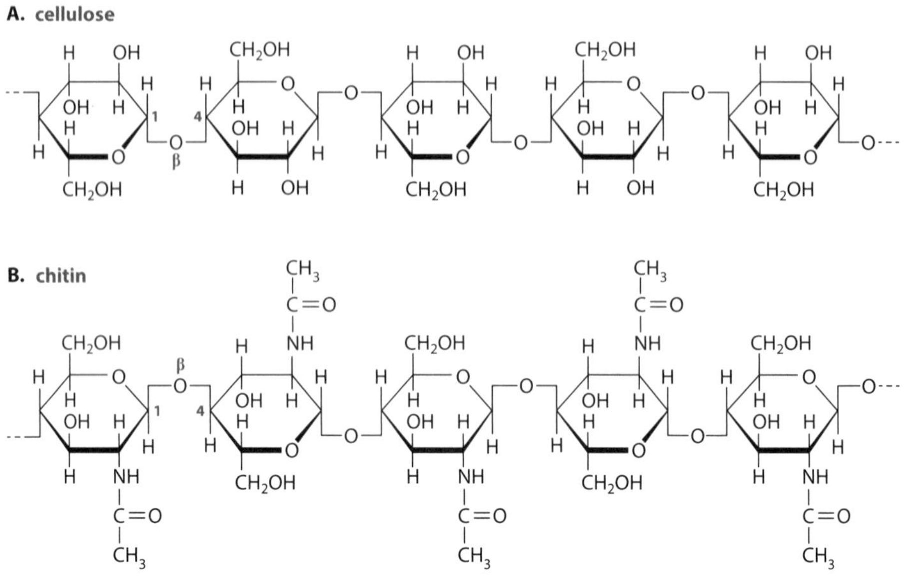

Figure 6.14 The polymeric structures of (A) cellulose and (B) chitin.
Cellulose is a linear polymer of D-glucose units linked by β(1→4) glycosidic bonds. Chitin is also a linear polymer but of N-acetylglucosamine units linked by β(1→4) glycosidic bonds.

A. cellulose

B. chitin

6.2.2 Heteropolysaccharides are found in the extracellular matrix and in bacterial cell walls

Heteropolysaccharides are also common in living organisms. In animals, they are important components of the extracellular matrix, which is the mixture of carbohydrates, proteins and other compounds that fills the spaces between cells and gives structure to tissues and organs. A good example of an extracellular heteropolysaccharide is **hyaluronic acid**, which is found in the synovial fluid of joints and is also the major component of the vitreous humor, the jelly-like substance within the eyeball. Hyaluronic acid is made up of alternating N-acetylglucosamine and D-glucuronic acid units, the latter being a glucose molecule whose carbon 6 has been converted into a carboxyl group (*Fig. 6.15*). The polymers, which are unbranched, contain up to 100 000 monomeric units and have lubricating properties that underlie their role in the fluid between joints.

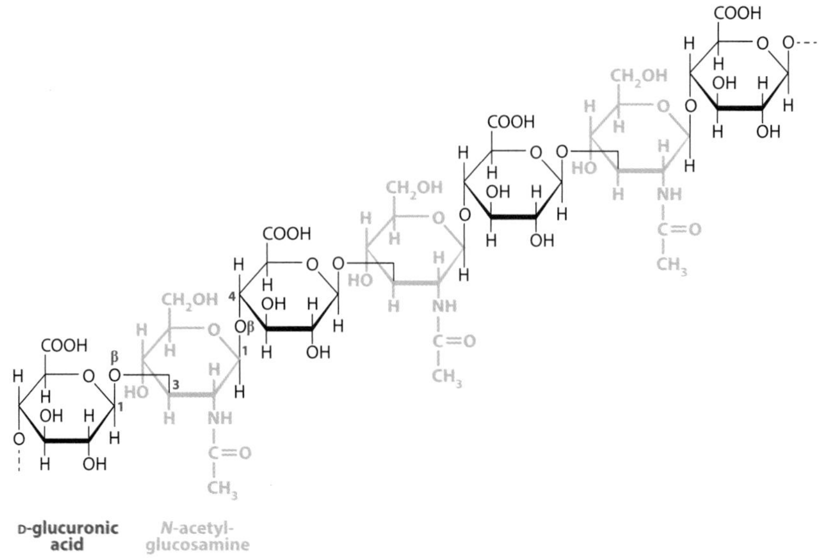

Figure 6.15 The polymeric structure of hyaluronic acid.
Hyaluronic acid has alternating D-glucuronic acid and N-acetylglucosamine units with β(1→3) and β(1→4) linkages.

D-glucuronic acid   N-acetyl-glucosamine

Hyaluronic acid is a member of a broader group of extracellular matrix polysaccharides called **glycosaminoglycans**. In most of these compounds, the repeating unit is a disaccharide made up of *N*-acetyl and carboxylated monosaccharide units. Some also have attached sulfate groups which, like the carboxylated units, have a negative charge. These charges, distributed along the length of the polysaccharide, naturally repel one another. The molecule is therefore forced into an extended rod-like conformation that can form cross-links with proteins to build up a matrix structure that gives rigidity to tissues and organs.

Finally, we move away from eukaryotes to an important example of a heteropolysaccharide that is found in bacteria. This is the polysaccharide component of **peptidoglycan**, which is the major constituent of the bacterial cell wall. The peptidoglycan polysaccharide has alternating *N*-acetylglucosamine and *N*-acetylmuramic acid units, the latter being obtained from *N*-acetylglucosamine by reaction of carbon 3 with lactic acid (*Fig. 6.16*). Individual polysaccharide molecules are attached to one another by short peptides, forming a huge matrix that encloses the entire bacterium. In effect the cell wall is a single giant molecule. Lysozyme, an enzyme present in tears, saliva and mucus, provides protection against bacterial infections by breaking the β(1→4) links between the peptidoglycan monosaccharide units in the bacterial cell wall. Penicillin has a similar protective effect by inhibiting synthesis of the peptide cross-links and so preventing bacterial growth. Archaea also have peptidoglycan cell walls, but in these prokaryotes the units are *N*-acetylglucosamine and an unusual highly modified monosaccharide called *N*-acetyltalosaminuronic acid that is found only in archaea.

A. *N*-acetylmuramic acid

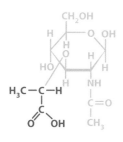

B. cross-linked structure of peptidoglycan

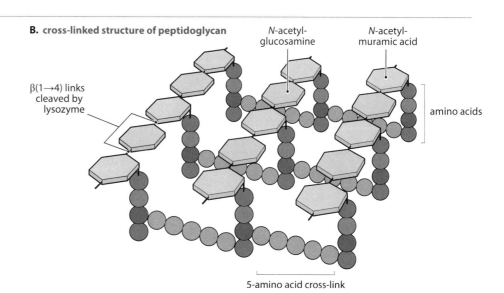

Figure 6.16 Peptidoglycan.
(A) The structure of *N*-acetylmuramic acid. (B) Peptidoglycan is made from polysaccharides cross-linked to one another by short peptide chains. The β(1→4) links in the polysaccharide are cleaved by lysozyme.

Further reading

Bang JK, Lee JH, Murugan RN, *et al.* (2013) Antifreeze peptides and glycopeptides, and their derivatives: potential uses in biotechnology. *Marine Drugs* **11**, 2013–41.

Itan Y, Powell A, Beaumont MA, Burger J and Thomas MG (2009) The origins of lactase persistence in Europe. *PLoS Computational Biology* **5**(8): e1000491.

Martínez JP, Falomir MP and Gozalbo D (2014) Chitin: a structural biopolysaccharide with multiple applications. *eLS* DOI: 10.1002/9780470015902.a0000694.pub3

Moremen KW, Tiemeyer M and Nairn AV (2012) Vertebrate protein glycosylation: diversity, synthesis and function. *Nature Reviews Molecular Cell Biology* **13**, 448–62.

Mouw JK, Ou G and Weaver VM (2014) Extracellular matrix assembly: a multiscale deconstruction. *Nature Reviews Molecular Cell Biology* **15**, 771–85. *Describes the components and organization of the extracellular matrix.*

Pérez S and Bertoft E (2010) The molecular structures of starch components and their contribution to the architecture of starch granules: a comprehensive review. *Starch* **62**, 389–420.

Vollmer W, Blanot D and de Pedro MA (2008) Peptidoglycan structure and function. *FEMS Microbiology Reviews* **32**, 149–67.

Self-assessment questions

Multiple choice questions

Only one answer is correct for each question. Answers can be found on the website: www.scionpublishing.com/biochemistry.

1. What is glyceraldehyde?
   (a) An aldehyde sugar
   (b) An aldotriose
   (c) A three-carbon carbohydrate
   (d) All of the above

2. Which one of these compounds does not have a chiral carbon?
   (a) Dihydroxyacetone
   (b) Erythrose
   (c) Glyceraldehyde
   (d) Threose

3. Which one of these sugars is an aldopentose?
   (a) Erythrose
   (b) Galactose
   (c) Glucose
   (d) Ribose

4. Which one of these sugars is not an aldohexose?
   (a) Galactose
   (b) Glucose
   (c) Mannose
   (d) Ribulose

5. Which of these statements best describes enantiomers?
   (a) Compounds that have two or more chiral carbons
   (b) Stereomers that differ in structure at just one of their chiral carbons
   (c) Cyclic monosaccharides that differ only in the arrangement of groups around the anomeric carbon
   (d) Isomers whose structures are mirror images of one another

6. Which of these statements best describes anomers?
   (a) Compounds that have two or more chiral carbons
   (b) Diastereomers that differ in structure at just one of their chiral carbons
   (c) Cyclic monosaccharides that differ only in the arrangement of groups around the anomeric carbon
   (d) Isomers whose structures are mirror images of one another

7. What are erythrose and threose examples of?
   (a) Anomers
   (b) Diastereomers
   (c) Enantiomers
   (d) Epimers

8. What are D-glucose and D-galactose examples of?
   (a) Anomers
   (b) Diastereomers
   (c) Enantiomers
   (d) Epimers

9. What is conversion between the α and β anomers of glucose called?
   (a) Cyclization
   (b) Epimerization
   (c) Glycosylation
   (d) Mutarotation

10. Which one of the following statements is **incorrect** with regard to lactase persistence?
    (a) It arose in Europeans between 7500 and 12 500 years ago
    (b) It enables an adult to digest lactose
    (c) It is also called lactose intolerance
    (d) It results from a change in the expression pattern of the lactase gene

11. What is the link between the two monosaccharide units in a disaccharide called?
    (a) Ester bond
    (b) *N*-glycosidic bond
    (c) *O*-glycosidic bond
    (d) Phosphodiester bond

12. What is the **correct** chemical name for maltose?
    (a) α-D-glucopyranosyl-(1→4)-D-glucopyranose
    (b) β-D-glucopyranosyl-(1→4)-D-glucopyranose
    (c) α-D-glucopyranosyl-(1→6)-D-glucopyranose
    (d) β-D-glucopyranosyl-(1→4)-D-glucopyranose

13. Sucrose is a disaccharide made up of which two compounds?
    (a) Glucose and galactose
    (b) Glucose and fructose
    (c) Glucose and ribose
    (d) Two glucose units

14. In proteins, *O*-linked glycans are attached to which two amino acids?
    (a) Asparagine and threonine
    (b) Glycine and leucine
    (c) Glycine and isoleucine
    (d) Serine and threonine

15. Which one of the following is **not** a modified monosaccharide found in glycans?
    (a) *N*-acetylglucosamine
    (b) Cellobiose
    (c) Glucosamine
    (d) Sialic acid

16. What is the branched form of starch called?
    (a) Amylopectin
    (b) Amylose
    (c) Hyaluronic acid
    (d) Maltotriose

17. The branch points in starch are made up of what type of linkage?
    (a) α(1→4)
    (b) α(1→6)
    (c) β(1→4)
    (d) β(1→6)

18. What does the reducing end of a starch molecule terminate with?
    (a) Carbon number 4
    (b) Carbon number 6
    (c) The anomeric carbon
    (d) The non-anomeric carbon

19. Chitin is a homopolysaccharide of which sugar?
    (a) *N*-acetylglucosamine
    (b) Galactose
    (c) Glucose
    (d) Sialic acid

20. Which of the following is an example of a heteropolysaccharide?
    (a) Cellulose
    (b) Glycogen
    (c) Hyaluronic acid
    (d) Starch

Short answer questions

These questions do not require additional reading.

1. What is the difference between an aldose and ketose sugar?

2. Draw the structures of the aldotetrose family of monosaccharides, indicating the positions of the chiral carbons and identifying which structures form enantiomer pairs.

3. Compare the structures of the linear and cyclic versions of ribose.

4. Giving examples, describe four types of stereoisomerism that are displayed by monosaccharides.

5. Draw the structures of maltose, trehalose and sucrose and explain how the full chemical names for these disaccharides (e.g. α-D-glucopyranosyl-(1→4)-D-glucopyranose) describe these structures.

6. By drawing structures, distinguish the key features of *O*- and *N*-linked glycans.

7. What are the differences between the structures of amylose and amylopectin?

8. Explain why the two ends of a starch molecule are called the reducing and non-reducing ends.

9. Describe the structures of glycogen, cellulose and chitin.

10. Write a short essay on the biological roles of heteropolysaccharides.

Self-study questions

These questions will require calculation, additional reading and/or internet research.

1. Identify the chiral carbons and the asymmetric center in each of the following monosaccharides:

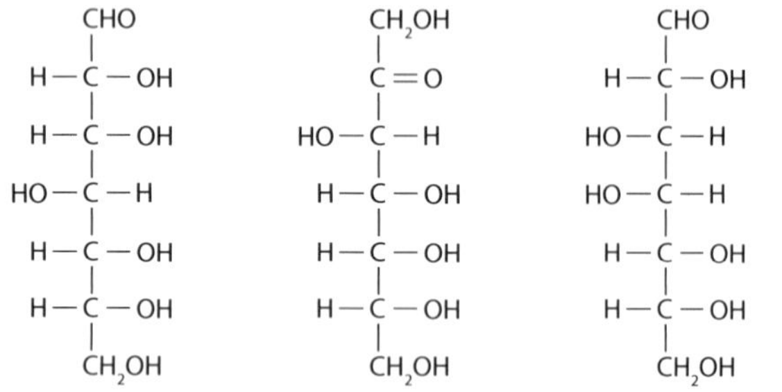

```
      CHO              CH₂OH             CHO
       |                 |                |
  H — C — OH            C = O        H — C — OH
       |                 |                |
  H — C — OH       HO — C — H       HO — C — H
       |                 |                |
 HO — C — H        H — C — OH       HO — C — H
       |                 |                |
  H — C — OH        H — C — OH       H — C — OH
       |                 |                |
  H — C — OH        H — C — OH       H — C — OH
       |                 |                |
     CH₂OH            CH₂OH            CH₂OH
```

  D-**glucoheptose**   D-**sedoheptulose**   D-**mannoheptose**

2. Draw diagrams illustrating the conversion of the linear forms of (a) galactose, and (b) tagatose into their α and β anomers.

3. Draw the structures of the following disaccharides:
   - Isomaltose: α-D-glucopyranosyl-(1→6)-α-D-glucopyranose
   - Sophorose: β-D-glucopyranosyl-(1→2)-α-D-glucopyranose
   - Turanose: α-D-glucopyranosyl-(1→3)-α-D-fructofuranose
   - Melibiose: α-D-galactopyranosyl-(1→6)-β-D-glucopyranose
   - Rutinose: α-L-rhamnopyranosyl-(1→6)-β-D-glucopyranose

4. Which of the following disaccharides are reducing sugars: maltose, trehalose, sucrose, lactose?

5. Amylase is a family of enzymes that break down starch molecules in different ways. The α-amylase can break α(1→4) bonds within a starch polymer, whereas β-amylase cuts the second α(1→4) bond from each non-reducing end, yielding maltose disaccharide molecules. What will be the end result of prolonged treatment of amylopectin with (a) α-amylase, (b) β-amylase, and (c) a combination of both enzymes?

CHAPTER 7

Enzymes

STUDY GOALS

After reading this chapter you will:

- appreciate that enzymes play key roles in biochemistry as catalysts of biochemical reactions

- be able to describe examples of both protein and RNA enzymes

- understand the roles of cofactors in enzyme-catalyzed reactions, and be able to give examples of different types of cofactor

- know how enzymes are classified according to the type of biochemical reaction that they catalyze

- be able to describe the free energy changes that occur during a biochemical reaction, distinguishing between exergonic and endergonic reactions

- understand how an enzyme affects the rate of a biochemical reaction by lowering the free energy of the transition state

- know the differences between the thermodynamic terms represented by ΔG and $\Delta G\ddagger$, and be able to explain the biochemical relevance of these terms

- be aware of how energy coupling enables the energy released by an exergonic reaction to drive an endergonic reaction

- understand the general ways in which temperature and pH affect the rate of an enzyme-catalyzed reaction

- be able to understand the effect of substrate concentration on reaction rate, and to explain the meaning of the terms V_{max} and K_m

- understand how the Lineweaver–Burk plot provides a graphical means of assigning V_{max} and K_m to an enzyme-catalyzed reaction

- be able to distinguish between an irreversible and reversible enzyme inhibitor, giving examples of both types

- know the different effects that competitive and non-competitive inhibition have on the kinetics of an enzyme-catalyzed reaction

We now move on to examine the major biochemical reactions in living organisms. These **metabolic** reactions are divided into two groups:

- **Catabolism**, which is the part of metabolism that is devoted to the breakdown of compounds in order to generate energy.

- **Anabolism**, which refers to those biochemical reactions that build up larger molecules from smaller ones.

The biochemical reactions that make up the metabolic activity of a cell are diverse and involve a large variety of compounds. These compounds include not just proteins, nucleic acids, lipids and carbohydrates, but also many smaller molecules that act as substrates, intermediates and products in the reactions that generate energy and participate in synthesis of the larger biomolecules. We will meet the most important of these compounds as we progress through the next seven chapters of this book.

Figure 7.1 **Enzymes catalyze the steps in a biochemical pathway.**
The first three steps in glycolysis are shown. This is the first stage of the energy-generating pathway of living cells. Each step is catalyzed by a different enzyme.

With such a vast range of molecules, and with so many different biochemical reactions serving so many diverse purposes, we might imagine that finding common themes would be difficult. In fact there is one common theme that unifies and underlies all of metabolism, and that is the role of **enzymes**. These are proteins, or very occasionally RNA molecules, that catalyze the individual steps in a biochemical pathway (*Fig. 7.1*). By responding to signals from inside and outside of the cell, enzymes also set the rates at which individual biochemical reactions occur. In this way they coordinate the overall metabolic activity of the cell, and ensure that this activity is appropriate for the cell's environment if it is a unicellular organism, or for its specialized function if the cell is part of a multicellular organism.

We therefore begin Part II of this book by studying enzymes and the way they work.

7.1   What is an enzyme?

The first scientific recognition of what we now refer to as an enzyme was made in 1833. In that year the French chemists Anselm Payen and Jean-François Persoz prepared an aqueous extract from malt, the germinated cereal grains used in brewing. They treated the extract with alcohol and obtained a milky precipitate. This precipitate had the ability to convert starch into sugar, an ability that was lost when the preparation was heated. They called the activity 'diastase', from the Greek for 'separation', because they looked on the activity that they observed as the 'separation' of the sugar from the starch.

Payen and Persoz's work was well ahead of its time, and they had little idea what diastase actually was. Looking back with today's knowledge, we realize that the milky precipitate that they obtained when they added alcohol to their malt extract was made up largely of proteins, which are insoluble in alcohol. We also know that the enzymatic activity of a protein is determined by its tertiary structure, and so is lost when the protein is heated and becomes denatured. Therefore, looking back, we can understand that diastase is a protein, which we now call **amylase**. But we also realize that Payen and Persoz's milky precipitate contained many different proteins, including various other enzymes, as well as a variety of other alcohol-insoluble compounds that are present in malt extract.

It took almost another century before biochemical techniques advanced to the stage where all the proteins and other molecules in an extract could be separated and individual enzymes obtained in pure form. The first person to do this was James Sumner of Cornell University, who in 1926 purified **urease** from jack beans and showed that this enzyme, which converts urea into carbon dioxide and ammonia, is a protein. During the next 10 years, Sumner and other biochemists purified several other enzymes and showed that each one was a protein. By 1946, when Sumner received the Nobel Prize, the fact that enzymes are proteins had become scientific dogma.

7.1.1   Most enzymes are proteins

Like many scientific dogmas, the one saying that enzymes are proteins turned out only to be partly correct. The vast majority of enzymes are indeed proteins, but a few are RNA molecules.

Examples of enzymes made of protein

First we will look at a few examples of the many enzymes that are indeed proteins. Protein enzymes come in all sizes, the smallest ones having fewer than 150 amino acids. **Ribonuclease A**, which we met in *Section 3.4.1* when we studied the denaturation and renaturation of proteins, is a typical small enzyme, with 124 amino acids. The biochemical reaction that it catalyzes is the conversion of a polymeric RNA molecule into two shorter molecules, by cutting one of the internal phosphodiester bonds (*Fig. 7.2A*). Repeated rounds of this reaction will eventually break the RNA

A. the reaction catalyzed by ribonuclease A

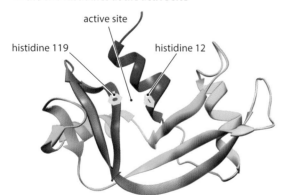

B. the two histidines at the active site

C. RNA bound to the enzyme

Figure 7.2 Ribonuclease A.
(A) The biochemical reaction catalyzed by ribonuclease A. The reaction requires a molecule of water, and results in a cut being made between a phosphodiester bond and the 5′-carbon of the adjacent nucleotide. (B) A representation of the enzyme structure showing the positions of the two histidine amino acids that flank the active site. (C) A model of the enzyme (blue) with an RNA (green) bound to the active site. This computer generated model shows the actual shape of the enzyme, based on the radii and relative positioning of all of the atoms in the tertiary structure.
Image (B) reproduced from Wikipedia under a CC BY-SA 2.5 license; (C) reprinted with permission from *Journal of Physical Chemistry*, 114:7371. © 2010 American Chemical Society.

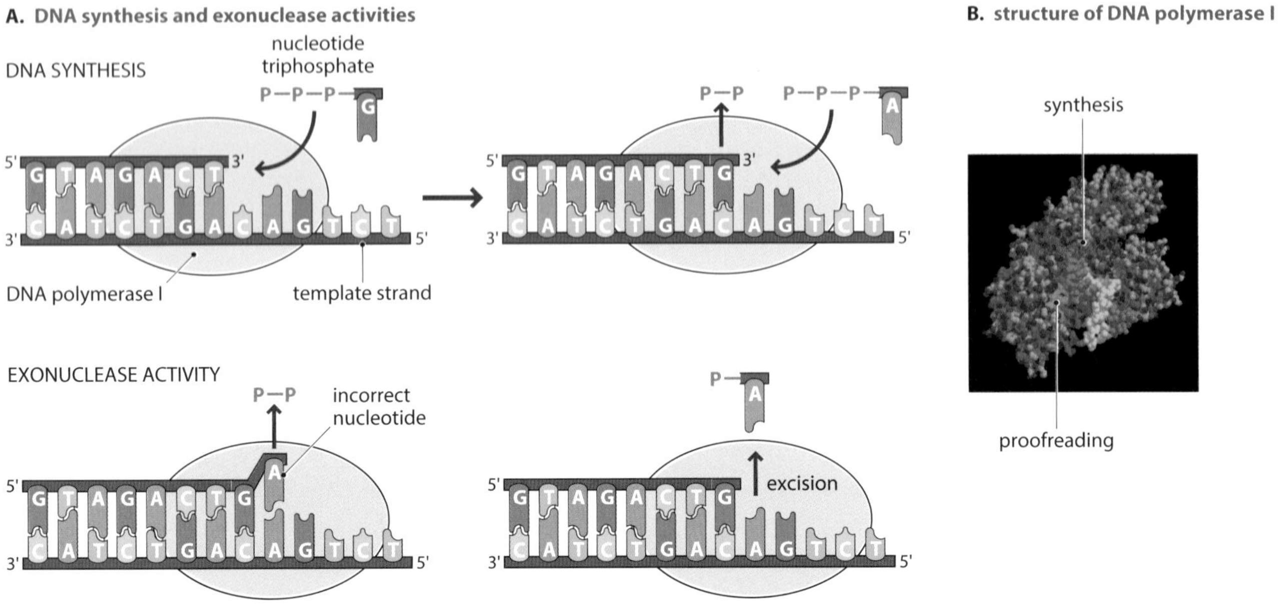

A. DNA synthesis and exonuclease activities

B. structure of DNA polymerase I

DNA SYNTHESIS

nucleotide
triphosphate

P—P—P—[G]

5' [G][T][A][G][A][C][T] 3'
3' [C][A][T][C][T][G][A][C][A][G][T][C][T] 5'

DNA polymerase I   template strand

P—P   P—P—P—[A]

5' [G][T][A][G][A][C][T][G] 3'
3' [C][A][T][C][T][G][A][C][A][G][T][C][T] 5'

EXONUCLEASE ACTIVITY

P—P   incorrect
        nucleotide
        [A]

5' [G][T][A][G][A][C][T][G]
3' [C][A][T][C][T][G][A][C][A][G][T][C][T] 5'

P—[A]

↑ excision

5' [G][T][A][G][A][C][T][G]
3' [C][A][T][C][T][G][A][C][A][G][T][C][T] 5'

synthesis

proofreading

Figure 7.3 The DNA synthesis and error correction activities of *E. coli* DNA polymerase I.
(A) DNA polymerase I synthesizes a new DNA strand by adding nucleotides to the 3'-end of the polynucleotide that is being made. The enzyme uses nucleoside triphosphates, with two of the phosphates being released when a nucleotide is added. An exonuclease activity enables the enzyme to remove a nucleotide that has been added in error. (B) The polymerase and exonuclease activities are specified by different parts of the DNA polymerase I protein. In this picture, a short piece of DNA is shown attached to the enzyme, with the template strand colored purple and the newly synthesized strand in green.
Image (B) produced by David S. Goodsell from The Scripps Research Institute shows the DNA polymerase I from *Escherichia coli*.

down into its constituent nucleotides. When we examine the tertiary structure of ribonuclease A we see a mixture of α-helices and β-sheets that form a U-shape. Two histidine amino acids, located at positions 12 and 119 in the polypeptide (we refer to these amino acids as 'his-12' and 'his-119'), flank the **active site**, the position where the biochemical reaction takes place (*Fig. 7.2B*). The RNA enters the active site and a phosphodiester bond is cut by a chemical reaction that involves the two histidines (*Fig. 7.2C*). Once cut, the two shorter RNA molecules are released from the enzyme.

Ribonuclease A, like most small enzymes, catalyzes a single, clearly defined biochemical reaction. Some larger enzymes have more complex activities, with different parts of the protein catalyzing different reactions. The enzyme called **DNA polymerase I**, from the bacterium *Escherichia coli*, is a good example. Like ribonuclease A, DNA polymerase I is a single polypeptide, but is much longer with a total of 928 amino acids. A DNA polymerase is an enzyme that makes a new DNA molecule by joining together individual nucleotides. The biochemical reaction is therefore synthesis of a phosphodiester bond. DNA polymerase I does this in a template-dependent manner, reading the sequence of nucleotides in an existing DNA strand (the template strand) to determine the sequence of the new polynucleotide, in accordance with the base-pairing rules. Nucleotides are added, one by one, to the 3'-end of the polynucleotide that is being made (*Fig. 7.3A*). DNA polymerase activity is specified by the amino acids between positions 521 and 928 of the DNA polymerase I polypeptide (*Fig. 7.3B*). Copying of the template is very accurate but every now and then, perhaps once for every 9000 nucleotides that are added, the polymerase makes a mistake and attaches an incorrect nucleotide. To correct the error,

See *Section 14.1.2* for further details regarding the actions of DNA polymerases.

the enzyme is able to break the phosphodiester bond that it has just made, releasing the incorrect nucleotide. This error correction is a completely different biochemical reaction, which we refer to as an **exonuclease** activity, removing a nucleotide from the end of a polynucleotide. This exonuclease is specified by amino acids 324–517. So DNA polymerase I is a multifunctional enzyme, with its different activities performed by different parts of its polypeptide.

Other multifunctional enzymes have multiple subunits, each one responsible for a different enzymatic reaction. An example, again in *E. coli*, is the enzyme **tryptophan synthase**. As its name implies, tryptophan synthase is involved in the anabolic pathway that results in synthesis of the amino acid tryptophan. This pathway begins with the aromatic compound called **chorismate**, and has branches leading to phenylalanine and tyrosine as well as the one leading to tryptophan. The last two steps in the tryptophan branch are catalyzed by tryptophan synthase. In the first of these reactions, indole-3-glycerol phosphate is converted to indole by removing the glycerol phosphate side-chain attached to carbon number 3, and in the second reaction a serine is attached at this position to form tryptophan (*Fig. 7.4A*). Tryptophan synthase is made up of four subunits, two called α and two β. An α-subunit has a barrel-like structure made up of an eight-stranded β-sheet surrounded by eight α-helices. A molecule of indole-3-glycerol phosphate enters this barrel and attaches to a glutamic acid and an aspartic acid at positions 49 and 60, respectively, in the α polypeptide. It is then cleaved to produce indole and 3-glyceraldehyde phosphate. The 3-glyceraldehyde phosphate molecule is ejected and the indole passed down a tunnel that leads to the active site of the β-subunit, some 2.5 nm away (*Fig. 7.4B*). A serine molecule is now attached to the indole, giving tryptophan. Channeling of the intermediate in a two-step biochemical reaction, from one subunit to another in a multifunctional enzyme, is a common way of ensuring that the intermediate does not diffuse away from the enzyme, and immediately enters the next step of the pathway.

Some enzymes are RNA molecules

Until the early 1980s it was believed that all enzymes were proteins. This rule was overturned when the first **ribozyme** was discovered, by Sidney Altman of Yale University and Thomas Cech of the University of Colorado. A ribozyme is an enzyme that is made of RNA.

> The synthesis pathways for phenylalanine, tyrosine and tryptophan are described in *Section 13.2.1*.

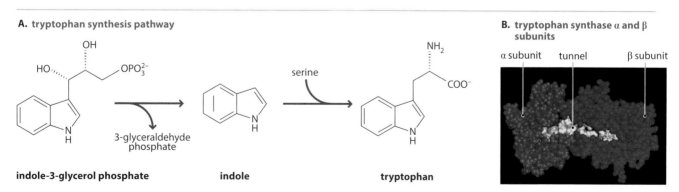

A. tryptophan synthesis pathway

indole-3-glycerol phosphate

3-glyceraldehyde phosphate

indole

serine

tryptophan

B. tryptophan synthase α and β subunits

α subunit   tunnel   β subunit

Figure 7.4 **Tryptophan synthase.**
(A) The last two steps of the biosynthetic pathway that results in synthesis of tryptophan. (B) The structure of the α- and β-subunits of *E. coli* tryptophan synthase, showing the tunnel between the two subunits along which indole travels. Note that the tryptophan synthase enzyme has four subunits, two α and two β. Two simultaneous reactions can therefore occur, one in each of the αβ dimers.
Image (B) reproduced from *Essential Biochemistry* by Pratt *et al.* with permission from John Wiley and Sons, Inc.

Many ribozymes work in conjunction with proteins to carry out their enzymatic function but, in all the examples known at present, the catalytic activity itself is specified by the RNA component. A good example is the bacterial enzyme **ribonuclease P**. This is a different enzyme to the protein ribonuclease A that we studied above. Rather than making random cuts in an RNA molecule, ribonuclease P has a more specialized function, making single cuts at specific positions in a small number of cellular RNA molecules. These molecules include tRNAs, which are initially made as precursor RNAs that are longer than the mature tRNAs that assist in protein synthesis. The pre-tRNA molecules are processed by ribonuclease processing enzymes that remove the additional stretches of polynucleotide to either side of the cloverleaf structure. One of these processing enzymes is ribonuclease P, which makes a single cut at the 5'-end of the mature tRNA (*Fig. 7.5A*).

We will look at the role of ribonuclease P in tRNA processing in *Section 15.2.1*.

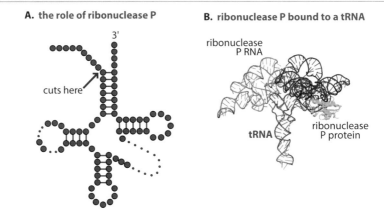

A. the role of ribonuclease P

cuts here

3'

B. ribonuclease P bound to a tRNA

ribonuclease P RNA

tRNA

ribonuclease P protein

Figure 7.5 Ribonuclease P.
(A) The role of ribonuclease P. (B) Ribonuclease P bound to a tRNA. The RNA component of ribonuclease P is shown in blue/gray, the protein component in green, and the tRNA is in red/gray.
Image (B) is reproduced by permission from Macmillan Publishers Ltd: *Nature*, 2010; 468:784, © 2010.

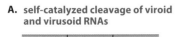

A. self-catalyzed cleavage of viroid and virusoid RNAs

genomes linked head to tail

self-catalyzed cleavage

individual linear genomes

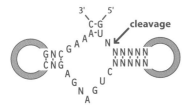

B. the hammerhead structure

3'   5'
C-G
A-U
G N C A A N cleavage
C N G N N N N N
A N N N N N
G U
N G C
A

Figure 7.6 Self-catalyzed cleavage of linked genomes during replication of viroids and virusoids.
(A) The replication pathway. (B) The hammerhead structure, which forms at each cleavage site and which has enzymatic activity. N, any nucleotide.

Theories about the origins of the first cells, which are thought to have contained self-replicating RNA molecules, were described in *Section 2.2.1*.

Ribonuclease P has two subunits. One of these is a protein of about 120 amino acids and the other is an RNA molecule of about 400 nucleotides. The RNA forms intramolecular base pairs which folds the molecule into a three-dimensional structure that binds and cuts the pre-tRNA (*Fig. 7.5B*). In the bacterium, both the RNA and protein subunits are required in order for the enzyme to cut the precursor tRNA, but experiments *in vitro* have shown that the catalytic activity resides entirely in the RNA subunit. The protein subunit is thought to stabilize the interaction between the enzyme and the pre-tRNA, but cannot itself carry out the ribonuclease activity.

Many of the ribozymes that have been discovered so far are ribonucleases. They are thought to have originated in the **RNA world**, the very early stage in evolution before DNA and proteins existed and when all biochemical systems were thought to have been centered on RNA. According to the theories, these early RNA molecules included some that carried genes. In today's world, the genes of all cellular organisms are contained in DNA molecules, but a few viruses still have genomes made of RNA. These include the **virusoids** and **viroids**, whose genomes are RNA molecules some 200–400 nucleotides in length. The replication process for some virusoids and viroids results in a series of genome copies joined end-to-end in a single long RNA molecule. This RNA molecule is a ribozyme, and is able to cut itself up, releasing the individual genome copies, by a self-catalyzed reaction (*Fig. 7.6A*). As with ribonuclease P, the enzymatic activity is contained within a three-dimensional structure that is formed by intramolecular base pairs. This structure is not the same in all virusoids and viroids, but a common type is the **hammerhead** (*Fig. 7.6B*).

7.1.2   Some enzymes require cofactors

We have seen how certain amino acids in an enzyme play a central role in the biochemical reaction catalyzed by that enzyme. For example, in ribonuclease A the two histidines at positions 12 and 119 in the polypeptide participate in the biochemical reaction that results in breakage of phosphodiester bonds. Similarly, glu-49 and asp-60 in the α-subunit of tryptophan synthase are involved in the cleavage of indole-3-glycerol phosphate to indole and 3-glyceraldehyde phosphate. With these and many other enzymes, the catalytic activity is provided solely by chemical groups present on particular amino acids within the protein. This is not always the case though, and many enzymes require additional ions or molecules, called **cofactors**, in order to perform their catalytic function.

The commonest cofactors are metal ions, which are required by about one-third of all known enzymes (*Table 7.1*). These ions include Cu^{2+}, Fe^{2+}, Fe^{3+}, Mg^{2+}, Mn^{2+}, Ni^+ and Zn^{2+}. They form attachments at specific sites within the enzyme, usually at or close to the active site so that they can participate directly in the catalytic process. An enzyme that contains a metal ion is called a **metalloenzyme**, and their ubiquity in human cells is the reason why we require trace elements in our diet.

Table 7.1. **Examples of cofactors**

Cofactor	Enzymes requiring the cofactor
Metal ion cofactors	
Cu^{2+}	Cytochrome oxidase
Fe^{2+}, Fe^{3+}	Catalase, nitrogenase
Mg^{2+}	Hexokinase
Mn^{2+}	Arginase
Ni^+	Urease
Zn^{2+}	Carboxypeptidase, carbonic anhydrase
Organic cofactors (coenzymes)	
NAD^+	Oxidoreductases
$NADP^+$	Fatty acid synthesis enzymes
FAD	Succinate dehydrogenase
FMN	NADH dehydrogenase
Coenzyme A	Fatty acid synthesis enzymes
Ascorbic acid	Antioxidative defense enzymes

Other cofactors are organic compounds. Many of these are derived from dietary vitamins. These vitamins include:

- **Niacin** (vitamin B_3), which is the precursor of the cofactors **nicotinamide adenine dinucleotide** (**NAD⁺**) and **nicotinamide adenine dinucleotide phosphate** (**NADP⁺**) (*Fig. 7.7A*). NAD^+ and $NADP^+$ are required by several enzymes involved in energy generation and anabolism, respectively.

- **Riboflavin** (vitamin B_2) is the precursor of **flavin adenine dinucleotide** (**FAD**) and **flavin mononucleotide** (**FMN**) (*Fig. 7.7B*). These have a similar function to NAD^+ and $NADP^+$.

- **Pantothenic acid** (vitamin B_5) is converted into **coenzyme A** (*Fig. 7.7C*), which is involved in energy generation as well as lipid metabolism.

- **Ascorbic acid** (vitamin C; *Fig. 7.7D*) is a cofactor in its own right, in particular in enzymatic reactions that modify amino acids on collagen polypeptides, enabling the polypeptides to form the triple helix structure that collagen adopts in

A. NAD⁺ and NADP⁺

B. FAD and FMN

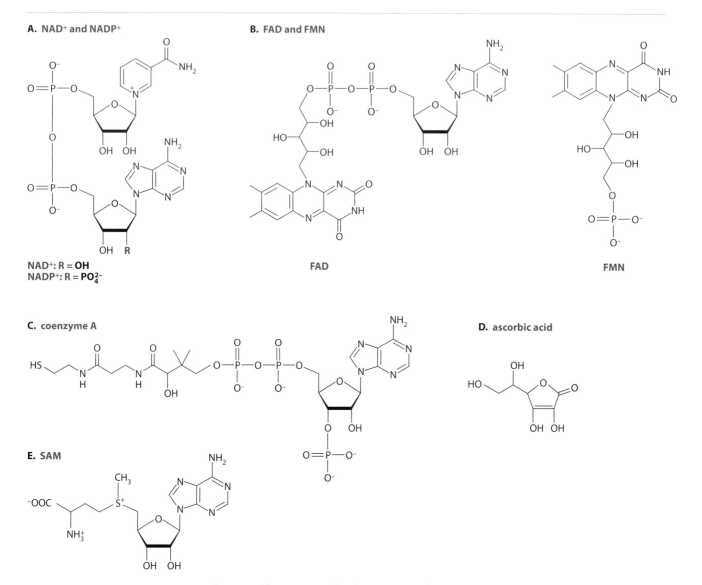

NAD⁺: R = **OH**
NADP⁺: R = **PO₄²⁻**

FAD

FMN

C. coenzyme A

D. ascorbic acid

E. SAM

Figure 7.7 **The structures of various organic cofactors.**

connective tissues such as tendons, ligaments, cartilage and bones. When there is a deficiency of vitamin C in the diet, the action of these enzymes is limited and collagen triple helices cannot be formed correctly. The defective collagen fibrils have a severe effect on the connective tissues, leading to the disease called scurvy.

Other cofactors for human proteins can be made in our cells. These cofactors include the modified amino acid **S-adenosyl methionine** or **SAM** (*Fig. 7.7E*).

Organic compounds that are cofactors for enzymatic reactions are sometimes called **coenzymes**. Some coenzymes participate in catalysis by forming a transient attachment to the enzyme they are assisting. Others form a permanent or semi-permanent linkage and so become an inherent part of the enzyme's structure. The latter are called **prosthetic groups**. Most metal ion cofactors are an inherent part of the enzyme structure and these are also classed as prosthetic groups.

Box 7.1 **Metalloproteins and metalloenzymes**

Metal ions are present in many different types of **metalloprotein**, including ones that have no catalytic function and hence are not enzymes. In most metalloproteins, the metal ion is held in place by one or more **coordinate bonds**. A coordinate bond is a special type of covalent bond that forms between the metal ion and the nitrogen, oxygen or sulfur-containing group present in the side-chains of amino acids such as histidine, cysteine and glutamic acid. For example, the Zn^{2+} ion in carboxypeptidase (a protein-degrading enzyme secreted by the pancreas) is held in place by coordinate bonds to two histidines and one glutamic acid, and also makes an attachment to a water molecule, as shown in the diagram to the right.

The resulting structure is called a **coordination sphere**, with the metal ion forming the **coordination center**.

In some metalloproteins, the metal ion forms part of a non-protein organic compound, the commonest example being the iron-containing porphyrin called heme (see *Fig. 3.25*). A heme molecule is made up of four pyrrole subunits linked together in a circle, with an Fe^{2+} ion coordinated in the center. Heme is found in hemoglobin, the oxygen-carrying protein of red blood cells (see *Section 3.3.2*), and in the cytochrome proteins, which are components of the electron transport chain (see *Section 9.2.2*).

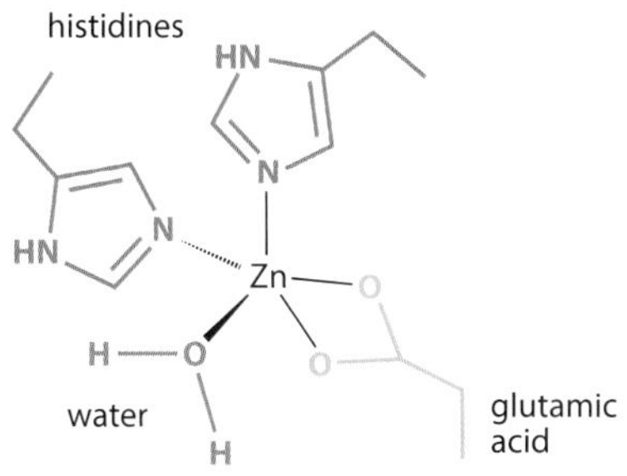

An enzyme plus its cofactor is called the **holoenzyme**. When the cofactor is absent what is left is known as the **apoenzyme**. Finally, we should note that it is not just protein enzymes that require cofactors. Some ribozymes, including ribonuclease P, need Mg^{2+} ions in order to carry out their enzymatic reactions.

7.1.3 Enzymes are classified according to their function

How many different enzymes are there? There are billions if we consider each species separately, and so, for example, count the amylase enzyme of barley malt, the one discovered by Payen and Persoz in 1833, as different from the amylase enzymes of other cereals such as wheat or rye. But how many are there if we only consider the different enzyme *activities* known in nature, and so count all these amylases, whether from barley, wheat or rye, as a single enzyme? Using this criterion, there are about 3200 different enzymes.

Over the years, various ways of classifying all these different enzymes have been proposed. Today, everybody uses a single standard scheme which was first agreed by the International Union of Biochemistry and Molecular Biology in 1961. In this classification, all 3200 types of enzyme are initially divided into six broad groups:

- Enzyme Commission group 1 (EC 1) comprises the **oxidoreductases**. These are enzymes that catalyze oxidation or reduction reactions.

- EC 2 are the **transferases**, enzymes that transfer a chemical group from one compound to another.

- EC 3 are **hydrolases**, which carry out hydrolysis reactions in which a chemical bond is cleaved by the action of water.

- EC 4 are **lyases**, enzymes that break chemical bonds by processes other than oxidation and hydrolysis.

- EC 5 comprises enzymes that rearrange the atoms within a molecule. This rearrangement is called **isomerization** and the enzymes are called **isomerases**.

- EC 6 are the **ligases**, which join molecules together.

Box 7.2 Oxidation and reduction reactions

The oxidation and reduction reactions catalyzed by oxidoreductase enzymes are important in many areas of biochemistry. In particular, oxidation and reduction underlie the biochemical processes that release energy from carbohydrates and lipids, as we will see in *Chapters 8* and *9*.

In chemistry, oxidation is often defined as the gain of oxygen by a substance, and reduction is the loss of oxygen. For example, when copper oxide is heated with magnesium metal, the following reaction occurs:

$$CuO + Mg \rightarrow Cu + MgO$$

The magnesium metal gains an oxygen and so is oxidized, whereas the copper oxide loses oxygen and is reduced. Most oxidation and reduction reactions are linked in this way, with the oxygen being lost by one compound and gained by another. We therefore refer to them as **redox reactions**.

In biochemistry, we usually look on redox reactions in a slightly different way. Rather than focusing on the oxygen transfer, we consider the gain and loss of electrons that occurs during the reaction. For example, leaving out the oxide part, we can rewrite the reaction between copper oxide and magnesium as:

$$Cu^{2+} + Mg \rightarrow Cu + Mg^{2+}$$

This representation of the reaction shows us that:
- The oxidation part of the reaction involves loss of two electrons by magnesium, converting the Mg atom to an Mg^{2+} ion.
- The reduction involves gain of two electrons by the Cu^{2+} ion, converting it to a Cu atom.

There is a useful mnemonic:

OIL RIG

Oxidation Is Loss, Reduction Is Gain.

Each of these groups is further subdivided in such a way that every individual enzyme has its own four-part **EC number**. For example, the amylase discovered by Payen and Persoz is referred to as EC 3.2.1.2. This number indicates the precise nature of the enzyme's activity (*Fig. 7.8*):

- EC <u>3</u> tells us that amylase is a member of EC group 3, being a hydrolase that uses water to break a chemical bond. This is the underlying nature of the biochemical reaction that results in the conversion of starch to sugar, as observed by Payen and Persoz with their malt extract.

- EC 3.<u>2</u> indicates that amylase is a **glycosidase** enzyme, one that breaks glycosidic bonds. A glycosidic bond is any bond that links two sugar units or a sugar and another molecule. This part of the EC number distinguishes amylase from other types of hydrolase that work on, for example, ester bonds (which form EC Group 3.1 and include the ribonucleases which break the phosphodiester bonds in RNA) or peptide bonds (EC Group 3.4, this group containing the proteases which break peptide bonds in proteins).

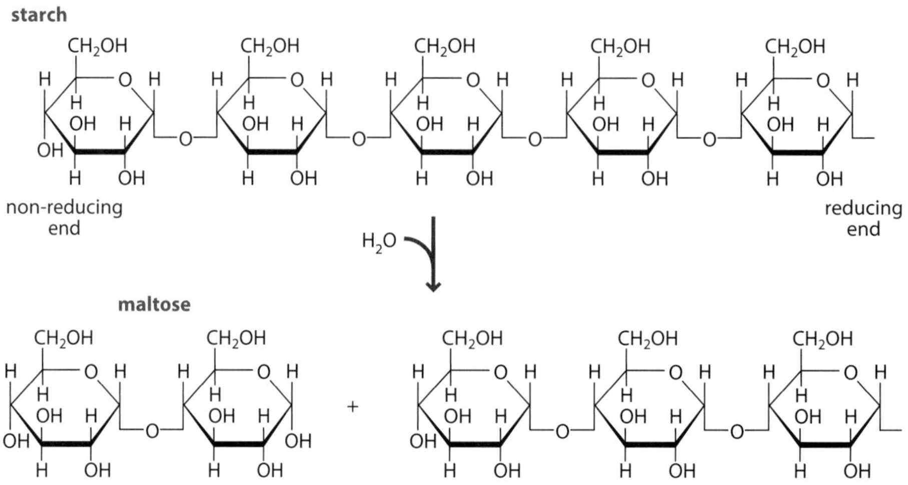

Figure 7.8 The β-amylase reaction (EC 3.2.1.2).

- EC 3.2.<u>1</u> indicates that amylase is a member of a subgroup of glycosidases that hydrolyze *O*- or *S*-glycosidic bonds. These are glycosidic bonds in which the linkage between the sugar and the second molecule includes an oxygen or sulfur atom. The second subgroup at this level, 3.2.<u>2</u>, comprises enzymes that work on *N*-glycosidic bonds, in which the linkage is via a nitrogen atom. Enzymes in subgroup 3.2.2 therefore include ones that break the bond between the sugar and base in a nucleotide.

- EC 3.2.1.<u>2</u> is the specific number for β-amylase, the enzyme which hydrolyzes the α(1→4) *O*-glycosidic bonds between glucose units in starch, glycogen and related polysaccharides, in such a way as to release maltose disaccharide units from the non-reducing ends of the polymers.

There are 135 enzymes altogether in the EC 3.2.1 subgroup, all of them hydrolases that break *O*- or *S*-glycosidic bonds between pairs of sugar units or between a sugar and another molecule. The first in the list, EC 3.2.1.1, is α-amylase, which also hydrolyzes α(1→4) *O*-glycosidic bonds in starch and other α(1→4) glucose polysaccharides, but cleaves these bonds randomly within the polymeric chains, rather than just at the non-reducing ends. This type of amylase is found in the salivary and pancreatic secretions of humans and other mammals, and enables us to digest oligo- and polysaccharides in our diet.

The distinction between α- and β-amylase illustrates the importance of the EC classification scheme. It enables us not only to distinguish between different enzymes in a single organism, but also to recognize **homologous** enzymes – ones with identical functions – from different organisms. The β-amylases are considered to be homologous regardless of which type of cereal they are obtained from. There are also enzymes with β-amylase activity in bacteria, and these are also assigned to EC 3.2.1.2. The α-amylases are looked on as a separate type of enzyme because the biochemical reaction that they catalyze is different. They therefore have a different EC number, and again that number is used for α-amylases from any species.

7.2   How enzymes work

Now that we know what enzymes are and how they are classified according to their biochemical activity, we can move on to study the way in which enzymes work in living cells. First, we will examine the fundamental property of an enzyme, which is its ability to act as a biological catalyst.

7.2.1   Enzymes are biological catalysts

We have already established that an enzyme is a biological catalyst. A catalyst is a substance that increases the rate of a chemical reaction but is not itself used up as a result of that reaction. An enzyme is no different from any other type of catalyst except that the reaction that it catalyzes is a biochemical one.

To understand how enzymes work we must therefore study some of the principles of catalysis. In order to do this, we will examine the events that occur during a biochemical reaction, in particular those events relating to the **energetics** of the reaction.

Most biochemical reactions result in a change in free energy

We will look at a typical biochemical reaction, catalyzed by a transferase enzyme (a member of EC group 2), and resulting in transfer of a chemical group from one compound to another (*Fig. 7.9*). The chemical equation for this reaction could be written as:

$$A-R + B \rightarrow A + B-R$$

Figure 7.9 Transferase reactions.
(A) The general formula for a transferase reaction. (B) An example of a transferase reaction, occurring in the pentose phosphate pathway (see *Section 11.3*). Transaldolase catalyzes the transfer of dihydroxyacetone from sedoheptulose 7-phosphate to glyceraldehyde 3-phosphate.

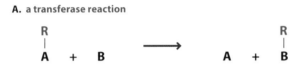

A. a transferase reaction

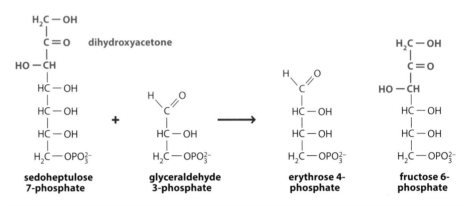

B. the reaction catalyzed by transaldolase

sedoheptulose 7-phosphate + glyceraldehyde 3-phosphate → erythrose 4-phosphate + fructose 6-phosphate

In this equation, the chemical group 'R' is transferred from compound 'A' to compound 'B'. The starting compounds, A–R and B, are the **substrates** for the reaction, and A and B–R are the **products**.

The chemical equation that we have just written summarizes the reaction that takes place, but it does not tell us a great deal about what actually happens. To begin to delve more deeply into the nature of the reaction we need to consider whether it results in a change in **free energy**. The Gibbs free energy, referred to as G, is a very useful thermodynamic function invented by the American scientist Josiah Willard Gibbs in 1873. In simple terms, it is a measure of the energy content of a 'system'. A system with a low energy content and therefore a low value for G is more stable than one with a higher energy content and higher G value. In our typical biochemical reaction, the two 'systems' are the substrates A–R + B and the products A + B–R.

When considering a biochemical reaction, what interests us is not so much the actual values for G for the substrates and products but the difference between these values. We use the Greek letter Δ (an upper case 'delta') to denote the difference between two values for G. The change in free energy that occurs during a biochemical reaction is therefore expressed as ΔG.

If the ΔG for a reaction is negative, then the free energy of the products is less than that of the substrates (*Fig. 7.10A*). This reaction can occur spontaneously, releasing an amount of energy equivalent to the ΔG value. This is called an **exergonic** reaction. Some chemical reactions have high negative ΔG values and are highly exergonic. The reaction of sodium with water, a favorite of high school chemistry classes, is an example. The products are sodium hydroxide, hydrogen gas and a great deal of energy released as heat, so much that the hydrogen ignites and the sodium metal speeds around the surface of a beaker of water emitting flames. Many biochemical reactions have negative ΔG values and hence release energy when they occur. This is the basis of catabolism, the part of metabolism that results in the generation of energy.

What if the free energy of the products is greater than that of the substrates (*Fig. 7.10B*)? In this case, ΔG is positive and the reaction is energy-requiring or **endergonic**. In biochemistry, many anabolic reactions are endergonic. These are the reactions that result in synthesis of larger products from smaller molecules. Endergonic reactions cannot occur spontaneously because they always need an input of energy.

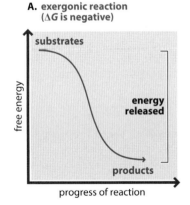

A. exergonic reaction (ΔG is negative)

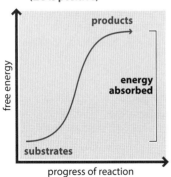

B. endergonic reaction (ΔG is positive)

Figure 7.10 The difference between an exergonic and endergonic reaction.

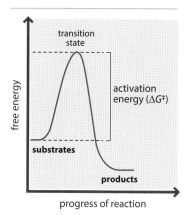

Figure 7.11 The transition state and its energetic implications.
The activation energy is needed to push the reaction over the barrier represented by the transition state.

There is an energy barrier between substrates and products

We have seen how the difference in the free energy values for the substrates and products of a chemical reaction are expressed as a ΔG value. The next issue that we must consider is the free energy of any intermediate structures that are formed during the course of a reaction. By 'intermediate structure' we do not mean actual compounds such as those that are formed during a multistep process, and which might be individually purified. Instead we mean a structure formed part way during a single reaction.

To clarify this important point, we will look more closely at the typical transferase reaction from the previous section. We drew the equation for this reaction as:

$$A–R + B \rightarrow A + B–R$$

If we think more carefully about this reaction then we will realize that it probably involves an intermediate structure that is formed at the very instant that group R is being passed from a molecule of A to a molecule of B. To indicate the existence of this structure we should therefore redraw the equation as:

$$A–R + B \rightarrow A \cdots R \cdots B \rightarrow A + B–R$$

Here, we use dotted lines '$\cdots$' to indicate that the bonds connecting group R to molecules A and B are in the process of being broken (in the case of $A \cdots R$) or formed ($R \cdots B$). This intermediate structure will be very unstable and hence will have a high free energy content. We call the point in the reaction pathway at which this structure is formed the **transition state**.

The existence of a transition state means that when considering the free energy changes that occur during a reaction, we must look beyond the ΔG value obtained simply by comparing the substrates and products. We must also consider a second ΔG, between the substrates and the transition state (*Fig. 7.11*). We call this free energy difference the **activation energy** or $\Delta G\ddagger$. Often $\Delta G\ddagger$ is much greater than ΔG and forms a significant barrier that must be overcome in order for the reaction to proceed. It is the existence of this energy barrier that limits the rate of most biochemical reactions.

Enzymes lower the free energy of the transition state

Now we begin to understand how an enzyme, or any other type of catalyst, can speed up the rate of a chemical reaction. A catalyst has no influence on the free energy values of the substrates and products and hence does not change ΔG. Instead, a catalyst reduces $\Delta G\ddagger$, usually by stabilizing the intermediate structure formed at the transition state (*Fig. 7.12*). A catalyst therefore decreases the size of the energy barrier that has to be crossed in order to convert the substrates into the products. Because the energy barrier is easier to cross, the rate of the reaction increases.

How does an enzyme stabilize the transition state of the reaction that it catalyzes? The answer is by reducing **entropy**. In thermodynamics, entropy is a measure of the degree of disorder of a system. Entropy contributes to the free energy value, so by reducing the entropy of the transition state an enzyme reduces the value of $\Delta G\ddagger$. This might sound complicated but in fact the underlying principle is quite straightforward. Entropy is reduced by bringing order to a system. In our favorite biochemical reaction

$$A–R + B \rightarrow A \cdots R \cdots B \rightarrow A + B–R$$

the 'system' at the start of the reaction is $A–R + B$. There might be 100 molecules of A–R in the cell, and 100 molecules of B, all floating quite close together in the aqueous cytoplasm. But the A–R and B molecules have no natural attraction for one another and, in the absence of the enzyme, diffuse in random directions according to chance collisions with other molecules (*Fig. 7.13A*). Occasionally a molecule of A–R might encounter a molecule of B in just the correct manner needed to form the

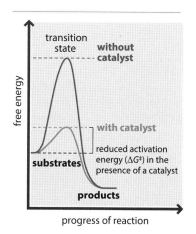

Figure 7.12 A catalyst reduces the free energy of the transition state.

Box 7.3 **Reversible reactions**

Many biochemical reactions are reversible. This means that the products of the reaction can react with one another to re-form the substrates. A reversible transferase reaction would therefore be made up of two parts, which we call the **forward** and **reverse reactions**:

forward reaction $X\text{–}R + Y \rightarrow X + Y\text{–}R$
reverse reaction $X + Y\text{–}R \rightarrow X\text{–}R + Y$

Usually, we combine these two part-reactions into a single equation, using a bidirectional arrow to indicate that the overall reaction is reversible:

$$X\text{–}R + Y \rightleftharpoons X + Y\text{–}R$$

For a reversible reaction, the rate at which the products are formed is counterbalanced by the rate at which the products are converted back to the substrates. As the reaction proceeds, an equilibrium point is reached. After this point, the number of substrate molecules that are converted into products during a particular time period equals the number of substrate molecules that are re-formed by the reverse reaction. The forward and reverse reactions continue to take place, but the relative concentrations of substrate and product no longer change.

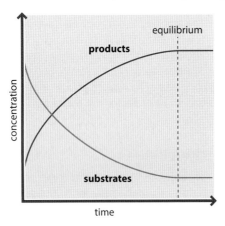

The ratio of products to substrates at the equilibrium point depends on the relative rates of the forward and reverse reactions. If the forward reaction is much more rapid than the reverse, then the substrate concentration at the equilibrium point will be small. On the other hand, if there is little difference between the forward and reverse reaction rates, then the substrates and products will have similar concentrations at the equilibrium point. Most reversible reactions are ones where the difference between the free energy values for the substrates and

products is relatively small, so in energetic terms the forward reaction is only slightly favored over the reverse one.

What effect will addition of an enzyme have on a reversible reaction? It is important to recognize that the enzyme does *not* affect the equilibrium point. This is because the forward and reverse reactions have the same transition state, even though the $\Delta G^{\ddagger}$ values for the two reactions are different.

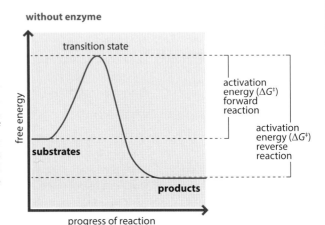

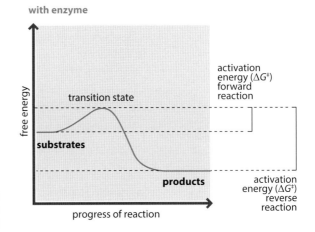

Addition of an enzyme lowers the free energy of the transition state, and hence has an equivalent effect on the $\Delta G^{\ddagger}$ values for both the forward and reverse reactions. The enzyme therefore increases the rates of the forward and reverse reactions, but does not affect the equilibrium between the substrates and products.

transition state structure and enable the conversion to $A + B\text{–}R$ to occur. But because of the disorder of the system (its high entropy) these chance encounters are few and far between. The rate of the reaction – the conversion of $A\text{–}R + B$ to $A + B\text{–}R$ – is very slow.

Now we will introduce into the system the transferase enzyme that catalyzes this reaction. The enzyme reduces entropy by binding one molecule of $A\text{–}R$ and one

Figure 7.13 An enzyme stabilizes the transition state of a reaction. (A) In a random mixture of reactants, chance collisions are needed in order for the transition state to form. (B) By binding the reactants, the enzyme lowers the entropy of the system, increasing the rate of conversion of reactants to products.

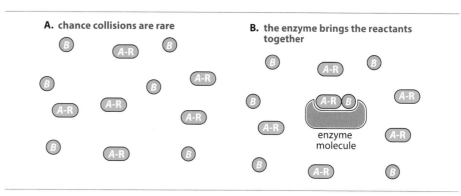

A. chance collisions are rare

B. the enzyme brings the reactants together

enzyme molecule

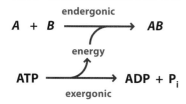

Figure 7.14 Energy coupling. In this example, the free energy required by the endergonic reaction, in which reactants A and B are combined to form product AB, is provided by hydrolysis of ATP to ADP and inorganic phosphate (P_i).

We will learn more about how ATP stores energy in *Section 8.1.1*.

molecule of B in precisely the correct relative positioning and orientation needed for these two substrates to form the transition state structure (*Fig. 7.13B*). By binding the substrates in this way the enzyme introduces order into the system, reducing the entropy and hence also reducing the value of $\Delta G\ddagger$. The rate of conversion of A–R + B to A + B–R is therefore increased.

An enzyme therefore lowers the energy barrier between the substrates and the transition state, but does not alter the free energy values for the substrates and products. In thermodynamic terms, the enzyme reduces $\Delta G\ddagger$ but has no effect on ΔG. If the reaction is exergonic, with a negative ΔG, then it will proceed with the release of energy. But what if the reaction is endergonic, with the products having a higher free energy content than the substrates? Endergonic reactions require an input of energy in order to reach completion. Many enzymes obtain this energy by coupling the endergonic reaction with a second reaction that generates energy. This is called **energy coupling**. Often the second reaction is hydrolysis of the nucleotide ATP, which gives ADP and inorganic phosphate (*Fig. 7.14*). The free energy released by this exergonic reaction is used by the enzyme to drive the coupled endergonic reaction.

Box 7.4 **The specificity of substrate binding**

Most enzymes have high specificity for their substrates and are able to distinguish between molecules with very similar structures, binding only those that are the correct substrates for the reaction that the enzyme catalyzes. What is the basis to this specificity?

During the early years of biochemistry, Emil Fischer suggested that the interaction between a substrate and an enzyme is similar to the way in which a key fits into a lock. According to this **lock and key model**, the binding pocket on the surface of an enzyme has a shape that precisely matches that of its substrate.

Specificity is achieved because only the substrate, and no other compound, has the shape needed to fit into the binding pocket.

A more recent suggestion is that the enzyme binding site is not a rigid structure, but instead has some flexibility. Attachment of the substrate induces a change in the binding pocket, so that the substrate becomes more precisely enclosed by the enzyme. Only the correct substrate can induce the necessary change to the structure of the binding pocket, increasing the specificity of the enzyme for the substrate. This is called the **induced fit model**.

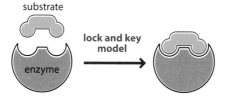

substrate

lock and key model

enzyme

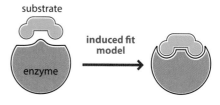

substrate

induced fit model

enzyme

7.2.2 Factors influencing the rate of an enzyme-catalyzed reaction

Our study of the principles of catalysis has shown us that an enzyme increases the rate of a reaction by reducing the energy barrier between the substrates and the products. Without catalysis, most biochemical reactions occur at a negligible rate and product synthesis is extremely slow. When catalyzed by their enzymes, the same reactions proceed much more rapidly and the products are produced at rates sufficient to satisfy the requirements of the cell. This does not mean, however, that a biochemical reaction has only two possible rates: either switched off because no enzyme is present, or switched on because it is being catalyzed by its enzyme. The rate of the catalyzed reaction is affected by various factors which together determine precisely how rapidly the products are made at any given time. We must now examine these factors and look at the effects that each one has on the rate of an enzyme-catalyzed biochemical reaction.

Temperature and pH affect the rates of enzyme-catalyzed reactions

All chemical reactions are affected by heat, occurring more rapidly at higher temperatures. This is because heating results in an increase in thermal energy, which causes the substrates to move about more rapidly, increasing the frequency with which they come into contact with one another. In thermodynamic terms, the addition of thermal energy helps to push the substrates over the energy barrier of the transition state. In this regard, enzyme-catalyzed reactions are no different to any other chemical reaction, and occur more rapidly when the temperature is increased. However, this is only true at relatively low temperatures, because at higher temperatures a second factor comes into play. This is the effect that temperature has on the stability of chemical bonds, in particular the relatively weak hydrogen bonds that hold together the secondary structures within a protein molecule. As the temperature rises, hydrogen bonds break and the secondary structure of the protein unfolds (denatures) causing a loss of enzyme activity. High temperature therefore denatures proteins in the same way as a chemical denaturant such as urea. The rate of a typical enzyme-catalyzed reaction therefore gradually increases as the temperature is raised, reaching an optimum beyond which the activity declines, possibly quite rapidly because small additional temperature increments cause a relatively large disruption to the enzyme's structure (*Fig. 7.15A*).

In vertebrates, most enzymes have a **temperature optimum** of around 37°C, this being the temperature within the tissues of these warm-blooded animals. The enzymes

> We examined the effects of urea on protein activity in *Section 3.4.1.*

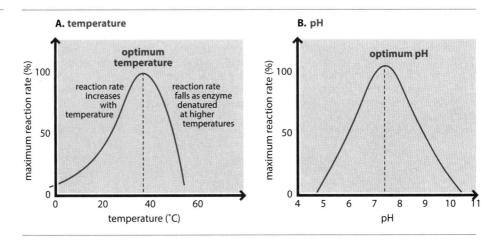

Figure 7.15 The effect of (A) temperature and (B) pH on the rate of an enzyme-catalyzed reaction.

of many of the bacteria that live in or on the bodies of vertebrates have a similar optimum temperature, but bacteria that live in other environments might be quite different in this regard. Thermophilic bacteria, which live naturally in hot springs, are a good example. The temperatures within hot springs can approach boiling point, so the proteins in these bacteria must be able to withstand high temperatures. The **thermostable** enzymes present in these bacteria typically have temperature optima of 75–80°C.

Proteins will also denature at extreme pH values, but the effect of pH on a reaction rate is rather more subtle than a simple disruption of the enzyme's structure. Often the amino acids at the active site of an enzyme are ones that have ionizable side-chains, these side-chains participating in the enzymatic reaction in some way. If we recall how the ionizable groups of an amino acid are affected by the pH then we will immediately appreciate how critical a pH change can be to the rate of an enzymatic reaction. Most enzymes have **pH optima** of 6.8–7.4, matching the physiological pH found in living cells and tissues (*Fig. 7.15B*). As with the effects of temperature, there are exceptions. Pepsin, one of the enzymes that breaks down proteins in the stomach, has a pH optimum of about 2.0, reflecting the highly acid environment in which this enzyme has to function.

Substrate concentration has an important effect on reaction rate

Although pH and temperature have important effects on the rates of enzyme-catalyzed reactions, most organisms have mechanisms for ensuring that the pH within their cells stays constant, and vertebrates also control their internal temperatures so that these rarely vary far from 37°C. So, temperature and pH are not themselves important determinants of the actual rate at which individual biochemical reactions take place. Of much greater importance is the availability of the substrates for the reaction.

To illustrate the important effects of substrate concentration we will consider the simplest type of biochemical reaction, in which there is a single substrate and a single product. This is the type of reaction catalyzed by the isomerase enzymes of EC group 5. We could write this reaction as:

$$S \rightarrow P$$

where S is the substrate, and P is the product. Imagine that we have purified this isomerase enzyme and mixed it with its substrate. The enzyme begins to catalyze the conversion of the substrate into the product, and we are following the reaction by measuring the amount of product that is present at successive time intervals. We plot the results as a graph and see the pattern shown in *Figure 7.16*. The shape of this curve tells us that initially the reaction proceeds at a linear rate, which we refer to as the **initial velocity** or V_0. Gradually, however, the rate of the reaction decreases until at some point the graph forms a plateau because no additional product is being produced, indicating that the reaction rate is now zero.

The simplest explanation for the leveling out of the curve shown in *Figure 7.16* is that product synthesis stops when all the substrate has been used up. This is not a complete explanation, but for the time being it is enough of an answer to suit our purpose. The important point is that the graph shows that the reaction rate gradually slows down as the amount of substrate decreases. In other words, the reaction rate is dependent on the substrate concentration.

The effect of substrate concentration reveals features of an enzyme's mode of action

The relationship between substrate concentration and reaction rate forms the basis of **enzyme kinetics**. This is an important subject because we can use the kinetics of an enzyme-catalyzed reaction to make deductions about the way that the enzyme works.

The effect of pH on ionization of amino acids was described in *Section 3.1.2.*

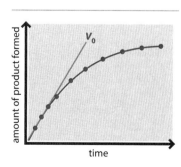

Figure 7.16 The time course for a typical enzyme-catalyzed reaction. The initial velocity (V_0) is shown as an extrapolation of the linear part of the reaction.

Box 7.5 Exploiting thermostable enzymes in biofuel production

Purified enzymes have been used in industrial processes for decades. Examples include chymosin (also called rennin), a protease obtained from the stomach lining of calves, which is used in cheese-making, and invertase from yeast, which breaks sucrose into glucose and fructose and is used to make syrup for the production of toffee and other candy. In recent years, biotechnologists have started to explore the possible applications of thermostable enzymes in industrial processes that involve high temperatures, ones that would denature most proteins. An example is in the production of **biofuel** from plant material.

Biofuels are becoming increasing attractive in the search for greener types of energy, ones that are not derived from fossil fuels and which generate fewer pollutants. Various biofuels are being produced in different parts of the world, but the most widely used type is based on ethanol obtained by breakdown of carbohydrates from plant material. Production of this biofuel involves the initial conversion of the plant's cellulose into

glucose, followed by the breakdown of the glucose into ethanol and carbon dioxide. We will study the latter pathway in more detail in *Section 8.2.2*.

Conversion of cellulose into glucose is achieved by adding an enzyme preparation called cellulase to the plant material. Cellulase is a mix of different enzymes, the most important being:
- An endoglucanase, which breaks internal β-glycosidic bonds in cellulose, breaking the polymer into smaller fragments.
- A cellobiohydrolase, which removes cellobiose units sequentially from the ends of the fragments created by endoglucanase treatment. Cellobiose is a disaccharide comprising two glucoses linked by a β(1→4) bond.
- β-glucosidase, which cleaves the β(1→4) bond, converting cellobiose into glucose.

These three enzymes therefore work together to release glucose from cellulose.

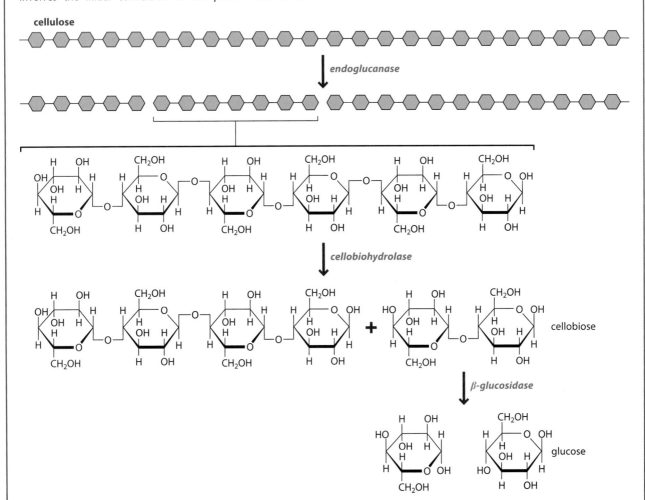

The cellulases currently used in biofuel production are obtained from fungi, and are not heat-resistant. They are therefore denatured at temperatures above 60°C. This complicates the industrial process because the plant material has to be heated

to 75°C in order to release the cellulose content from other plant biopolymers such as lignin, which cannot be broken down into glucose. The conventional technology therefore requires two stages, each carried out in a different bioreactor. In the first stage the plant material is heated to liberate its cellulose and, in the second, cellulase is added to the cooled extract to convert the cellulose to glucose. Having two stages lengthens the time needed to complete the process and, importantly, increases the overall cost.

A thermostable cellulase would therefore reduce the costs of this type of biofuel production by enabling both the release and breakdown of cellulose to be carried out as a single-stage process. Suitable thermostable enzymes do not appear to be common among thermophilic bacteria, but a few examples are known and these are being investigated as alternatives to the fungal enzymes. The main question is whether the costs of growing the thermophilic bacteria and extracting their enzymes will make their use uneconomic.

Another possibility is to use **protein engineering** to increase the heat stability of a fungal cellulase. Protein engineering involves making changes to the amino acid sequence of a protein, using techniques that we will study in *Box 19.2*. The intention would be to change the amino acid sequences of a set of fungal cellulose-degrading enzymes in order to make these enzymes more heat resistant. The problem with this approach is that we do not yet understand exactly why a thermostable enzyme is able to withstand high temperatures without being denatured. A variety of structural innovations have been identified in different thermostable enzymes that might explain their heat tolerance, but none of these features are present in all enzymes of this type. The innovations include a more compact conformation, with a relatively high percentage of the polypeptide folded into α-helices and β-sheets, rather than unfolded in loops and turns. Often the different secondary structural units are held together, and attached to one another, by a higher number of hydrogen bonds and van der Waals interactions than would be present in a non-thermostable protein. The surface features of a thermostable protein are also likely to be important, as these will dictate how the protein interacts with the surrounding water molecules, which in turn will influence the ease with which the protein is unfolded at higher temperatures. Even if the key features of a thermostable enzyme can be identified, it will be difficult to work out what changes should be made to the amino acid sequence of a non-thermostable enzyme in order to bring about these types of structural change.

Because it is difficult to predict what amino acid changes should be made to fungal cellulase enzymes, biotechnologists are exploring a different type of protein engineering called **directed evolution**. In this approach, random changes are made to the amino acid sequence of a protein, and the resulting variants then tested to see which ones have improved properties. For biofuel production, we would make random changes to one of the fungal cellulase enzymes, and then test each new variant to identify any that, by chance, displayed increased heat resistance. The increase might be small, but if a variant was then subjected to further rounds of random alteration, with the most heat-resistant versions carried forward at each stage, then eventually we might obtain a cellulase with sufficient thermostability to be used in a single-stage biofuel production process.

First, we need a consistent way of comparing the rate of a reaction at different substrate concentrations. The experiment we depicted in *Figure 7.16* indicates how we can do this. To compare reaction rates at different substrate concentrations we simply set up a series of experiments with identical amounts of enzyme but different amounts of substrate, and measure the V_0 for each of these substrate concentrations (*Fig. 7.17A*). Plotting these results on a graph will give a hyperbolic curve (*Fig. 7.17B*). This curve reveals two key parameters relating to the activity of the enzyme:

- The maximum velocity or V_{max} is achieved when the curve eventually reaches a plateau. This parameter indicates the maximum rate at which the enzyme can carry out the reaction.

- The K_m or **Michaelis constant** is the substrate concentration at which the reaction rate is half of the maximum value (i.e. $0.5 \times V_{max}$). This gives us a numerical value for K_m for any enzyme. But what does K_m tell us about an enzyme? The K_m is a measure of the stability of the enzyme–substrate complex, or to be more precise the 'affinity' of the enzyme for its substrate. This is a reciprocal relationship; a low K_m indicates high affinity and a high K_m indicates low affinity.

The precise relationship between the substrate concentration, V_{max} and K_m, was first worked out by Leonor Michaelis and Maud Menten in 1913. The **Michaelis–Menten equation** states that:

$$V_0 = \frac{V_{max} \times [S]}{K_m + [S]}$$

In this equation, the square brackets indicate 'concentration of', so [S] refers to the substrate concentration.

Figure 7.17 Comparing the rate of an enzyme-catalyzed reaction at different substrate concentrations. (A) Measuring the V_0 at different substrate concentrations (2.5 mM, 5 mM and 10 mM substrate). (B) Using the V_0 value to calculate the V_{max} and K_m for the enzyme.

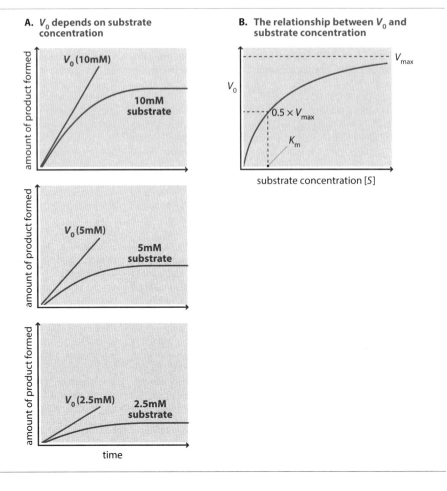

A. V_0 depends on substrate concentration

B. The relationship between V_0 and substrate concentration

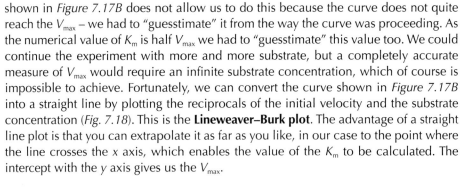

How do we measure the V_{max} and K_m of an enzyme experimentally? The graph shown in *Figure 7.17B* does not allow us to do this because the curve does not quite reach the V_{max} – we had to "guesstimate" it from the way the curve was proceeding. As the numerical value of K_m is half V_{max} we had to "guesstimate" this value too. We could continue the experiment with more and more substrate, but a completely accurate measure of V_{max} would require an infinite substrate concentration, which of course is impossible to achieve. Fortunately, we can convert the curve shown in *Figure 7.17B* into a straight line by plotting the reciprocals of the initial velocity and the substrate concentration (*Fig. 7.18*). This is the **Lineweaver–Burk plot**. The advantage of a straight line plot is that you can extrapolate it as far as you like, in our case to the point where the line crosses the x axis, which enables the value of the K_m to be calculated. The intercept with the y axis gives us the V_{max}.

7.2.3   Inhibitors and their effects on enzymes

To complete our study of the way that enzymes work, we must end this chapter by examining how enzymes are affected by **inhibitors**. An inhibitor is a compound that interferes with the activity of an enzyme, reducing its catalytic rate. There is a vast range of compounds that affect the activities of different enzymes in this way, but we can place all of these compounds into two broad groups, depending on whether or not their inhibitory action can be reversed. We will look first at **irreversible inhibitors**, whose effects are permanent.

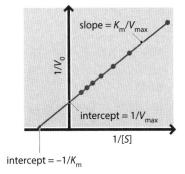

Figure 7.18 The Lineweaver–Burk plot.

Box 7.6 **The Michaelis–Menten equation**

The Michaelis–Menten equation is central to the study of enzymes and it is important that we understand how it is derived. The Michaelis–Menten equation is based on the following concept of enzyme catalysis.

$$E + S \underset{k_2}{\overset{k_1}{\rightleftharpoons}} ES \overset{k_3}{\rightarrow} E + P$$

In this scheme, the enzyme E combines with its substrate S to form an enzyme–substrate complex ES. The ES complex can dissociate again to form E + S, or can proceed to form E and the product P. The symbols k_1, k_2 and k_3 are **rate constants**, which describe the rates associated with each step of the process. We assume that there is no significant rate for the backward reaction of $E + P \rightarrow ES$.

According to this model, the concentration of the enzyme–substrate complex, which we represent by [ES], remains approximately constant until nearly all the substrate is used. This means that the rate of synthesis of ES equals the rate of its consumption over most of the course of the reaction. In other words, [ES] maintains a **steady state**.

We know that the initial velocity (V_0) at low substrate concentrations is directly proportional to the concentration of substrate, [S], while at high substrate concentrations the velocity becomes independent of [S], eventually reaching its maximum value, V_{max}. The Michaelis–Menten equation describes the hyperbolic curve obtained when (V_0) is plotted against [S] (see *Fig. 7.17B*). The equation is:

$$V_0 = \frac{V_{max} \times [S]}{K_m + [S]}$$

In deriving the equation, Michaelis and Menten defined a new constant, K_m, the Michaelis constant:

$$K_m = \frac{k_2 + k_3}{k_1}$$

K_m is therefore equal to the rate of breakdown of ES ($k_2 + k_3$) divided by its rate of formation (k_1). This means that the K_m of an enzyme indicates the stability of the ES complex. However, for many enzymes k_2 is much greater than k_3. If this is the case then K_m becomes dependent on the relative values of k_1 and k_2, which are the rate constants for ES formation and dissociation, respectively. Under these circumstances, the K_m becomes a measure of the degree of affinity of an enzyme for its substrate:

• If an enzyme has a weak affinity for its substrate then k_2 (dissociation of ES into E and S) will be predominant over k_1 (association of E and S to form ES). The K_m value will therefore be high.
• Conversely, an enzyme with a strong affinity for its substrate will have a low K_m, because for this enzyme k_1 will be predominant over k_2.

Finally, we will examine what happens if we take the reciprocal of the Michaelis–Menten equation. This would give us:

$$\frac{1}{V_0} = \frac{K_m + [S]}{V_{max} [S]} = \frac{K_m}{V_{max}} \frac{1}{[S]} + \frac{1}{V_{max}}$$

This is the equation presented by Hans Lineweaver and Dean Burk in 1934. It tell us that a plot of $1/V_0$ against $1/[S]$ will give us a straight line. The slope of this line will equal K_m/V_{max}, the intercept with the y axis ($1/[S] = 0$) will indicate the value of $1/V_{max}$, and the intercept with the x-axis ($1/V_0 = 0$) will give $-1/K_m$. This graph is called the Lineweaver–Burk plot (see *Fig. 7.18*).

An irreversible inhibitor causes a permanent reduction in an enzyme's activity

Most irreversible inhibitors are compounds that alter the active site of an enzyme in such a way that the enzyme is no longer able to bind to its substrate. Often the inhibitor compound simply forms a covalent bond with one of the amino acids at the active site, blocking the active site so that the substrate cannot enter. This is usually a permanent, irreversible change, because the inhibitor can only be removed from the active site by cleaving the covalent bond that now attaches it to the enzyme's polypeptide chain. Those amino acids with hydroxyl (–OH) or sulfydryl (–SH) groups in their side-chains are often the targets of irreversible inhibitors, so enzymes with serine, threonine, tyrosine or cysteine at their active sites are particularly susceptible to this type of inhibition.

Diisopropyl fluorophosphate (DIFP) is an example of an irreversible inhibitor. DIFP reacts with many compounds that contain a hydroxyl group, including serine (*Fig. 7.19*). Attachment of DIFP to a serine side-chain at an active site is likely to block entry of the substrate, and will also prevent the serine side-chain from playing its role in the biochemical reaction catalyzed by the enzyme. The activity of the enzyme molecule to which DIFP is attached is therefore inhibited, totally and irreversibly. DIFP inhibits many proteases (enzymes that cleave peptide bonds and so break polypeptides into

Figure 7.19 The reaction between diisopropyl fluorophosphate (DIFP) and serine.

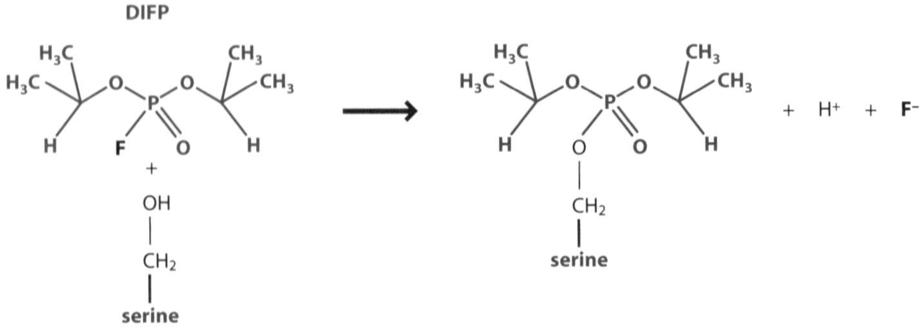

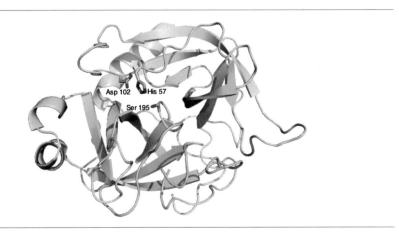

Figure 7.20 Chymotrypsin.
The chymotrypsin active site comprises a 'catalytic triad' of three amino acids: a serine, a histidine and an aspartic acid. Reaction between the serine and DIFP results in irreversible inhibition of chymotrypsin. Image reproduced from Wikimedia under a CC BY-SA 3.0 license.

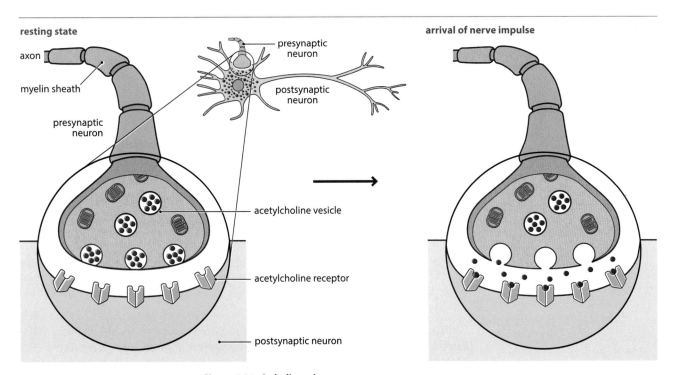

Figure 7.21 A cholinergic synapse.
Arrival of a nerve impulse stimulates release of the neurotransmitter acetylcholine from the presynaptic neuron. Binding of the acetylcholine molecules to receptor proteins on the surface of the postsynaptic neuron results in transmission of the impulse across the synapse. Immediately after transmission, the acetylcholine molecules are broken down by acetylcholinesterase, so the synapse returns to its resting state.

amino acids) because many of these enzymes have a serine at their active site. An example is **chymotrypsin** (*Fig. 7.20*), which is secreted by the pancreas and is involved in the digestive breakdown of proteins in the duodenum.

Diisopropyl fluorophosphate also inhibits the enzyme called **acetylcholinesterase**, which is present in nerve cells and degrades acetylcholine. Acetylcholine is a neurotransmitter, a compound that passes nerve impulses across the **synapses** between adjacent nerve cells or **neurons** (*Fig. 7.21*). Once a nerve impulse has passed, the neurotransmitter must be broken down, otherwise the nerve cells continue to signal to one another. Inhibition of acetylcholinesterase by DIFP therefore disrupts the nervous system by preventing acetylcholine being broken down in those synapses in which it acts as the neurotransmitter. DIFP is notorious as a component of some types of nerve gas.

Before we move on we must make a careful distinction between irreversible inhibition and the more general inactivation of enzyme activity caused by heat, pH and those chemicals that act as denaturants. Both types of event have the same result, this being a substantial or complete loss of enzyme activity. The difference is that heat, pH and chemical denaturants are non-specific in their action. They affect all enzymes because their mode of action is to disrupt the non-covalent chemical bonds that stabilize a protein's three-dimensional structure. An inhibitor, on the other hand, has a specific effect on a single enzyme, or group of similar enzymes, with which it is able to react because of the structure of the active site. The same compound will be unable to react with other enzymes, whose active sites have different structures, and hence will display no inhibitory effect with those other enzymes.

Reversible inhibition can be competitive or non-competitive

A **reversible inhibitor** is a compound whose inhibitory effects can be reversed, at least to some extent, by the presence of the substrate. Different types of reversible inhibition can be distinguished, the most common being described as **competitive reversible inhibition** and **non-competitive reversible inhibition**.

In competitive reversible inhibition, the inhibitor binds to the active site, but not in a permanent fashion as is the case with an irreversible inhibitor. Instead, the reversible inhibitor forms only relatively weak non-covalent attachments with the amino acids in the active site. Because the attachment is not via covalent bonds, it is possible for the enzyme substrate to displace the inhibitor. The substrate and inhibitor therefore *compete* for access to the active site. This means that the rate at which the enzymatic reaction proceeds depends on the relative amounts of substrate and inhibitor that are present. With a relatively large amount of inhibitor, the reaction rate will be slow, but this inhibition can be overcome by increasing the substrate concentration (*Fig. 7.22A*). This relationship has a specific effect on the kinetics of the reaction. The V_{max} of the reaction is unchanged, because the enzyme is still capable of achieving its maximum catalytic activity, if enough substrate is added to completely displace the inhibitor. However, the K_m is increased because the presence of the inhibitor decreases the affinity of the enzyme for its substrate. Whether or not a reversible inhibitor is acting in this competitive manner can therefore be determined by examining its effect on the Lineweaver–Burk plot for the enzyme-catalyzed reaction (*Fig. 7.22B*). Presence of the inhibitor will not change the intercept of the plot with the y axis, which corresponds to the V_{max}, but does move the position of the intercept with the x axis, which gives the K_m.

A **non-competitive reversible inhibitor** does not compete directly with the substrate, usually because it binds to some other part of the enzyme, away from the active site. This is called **allosteric inhibition** and the binding position for the inhibitor is called the **allosteric site**. Binding of the inhibitor to the allosteric site will still cause an alteration in the active site, and hence affect substrate binding, but this will be

A. the effect of substrate concentration

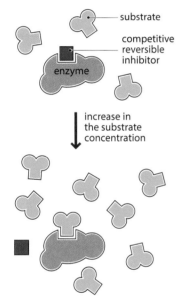

B. the effect on V_{max} and K_m

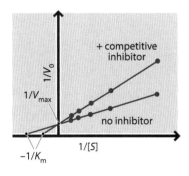

Figure 7.22 Competitive reversible inhibition.
(A) The substrate and inhibitor compete for access to the active site, so the inhibition can be overcome by increasing the substrate concentration. (B) The effect on the V_{max} and K_m of the reaction, as revealed by the Lineweaver–Burk plot.

A. the effect of inhibitor binding

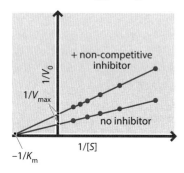

B. the effect on V_{max} and K_m

Figure 7.23 Non-competitive reversible inhibition.
(A) Binding of the inhibitor to some other part of the enzyme results in an alteration in the active site, and hence affects substrate binding. (B) The effect on the V_{max} and K_m of the reaction, as revealed by the Lineweaver–Burk plot.

by changing the enzyme's structure in some way rather than by entering the active site (*Fig. 7.23A*). Increasing the amount of substrate will increase the reaction rate, but there is no direct competition between substrate and inhibitor, because only the former is able to enter the active site. The kinetics of a non-competitively inhibited reaction are therefore different to the kinetics resulting from competitive inhibition. The V_{max} of the reaction is reduced, because addition of substrate does not displace the inhibitor, which means that there will always be some inhibitory effect however much substrate is added. Because substrate binding is not affected by the inhibitor, the K_m, which indicates the affinity of the enzyme for its substrate, is unaltered. Once again, the Lineweaver–Burk plot gives a diagnostic result from which this type of inhibition can be identified (*Fig. 7.23B*).

Allosteric inhibition is important in the regulation of metabolic pathways

Reversible inhibition is an important part of the natural control processes by which the metabolic pathways within a cell are regulated so that the correct amounts of the end products are synthesized. Many pathways are controlled by a type of **feedback regulation**, in which the end product controls the rate of its own synthesis by acting as a reversible inhibitor of one of the enzymes that catalyzes an early step in the pathway (*Fig. 7.24A*). Usually the structure of the product of a pathway is quite different from that of the substrate for the first step, so the product is unable to enter the active site of the first enzyme and cannot exert competitive reversible inhibition. This type of control is therefore almost always exerted by an allosteric effect.

Feedback regulation usually operates at the **commitment step** of a pathway. This is the first step in the pathway that produces an intermediate that is unique to that pathway, so synthesis of this intermediate only affects the pathway in question and has no effect on any other part of the metabolic network of the cell. In energy terms, this is the most economic strategy, because it means that no energy is wasted in synthesizing intermediates that are not needed.

Figure 7.24 Feedback inhibition of a biochemical pathway.
(A) Regulation of a linear pathway. The end product Z controls the rate of its own synthesis by acting as a reversible inhibitor of enzyme E_1, which catalyzes an early step in the pathway. (B) Regulation of a branched pathway. End product Y controls the rate of its own synthesis by acting as a reversible inhibitor of enzyme E_3, and end-product Z regulates its synthesis by acting on enzyme E_5. If there are sufficient amounts of both Y and Z then intermediate C accumulates, which inhibits enzyme E_1, at the commitment step for the entire branched pathway.

A. feedback regulation of a linear pathway

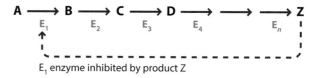

B. feedback regulation of a branched pathway

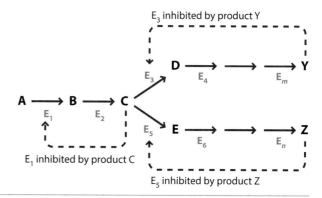

Box 7.7 **Allosteric enzymes**

Allosteric inhibition is one facet of a broader set of regulatory processes that are mediated by binding of an **effector** molecule to an enzyme. Many effectors have a negative impact on the activity of the target enzyme, as we have seen with non-competitive reversible inhibition, but other effectors have positive effects, stimulating enzyme activity when attached to an allosteric site. An **allosteric enzyme** is any enzyme whose activity is influenced by allosteric effectors, irrespective of whether the effect is to stimulate (activate) the enzyme (**positive allosteric control**) or inhibit it (**negative allosteric control**).

Positive allosteric control is used by some enzymes to enhance their sensitivity to small changes in the availability of a substrate. With these enzymes, the binding of a substrate molecule to one active site induces a conformational change that facilitates substrate binding at other active sites in the enzyme. Substrate binding is therefore **cooperative**. Two models have been proposed to explain this effect. Both assume that the allosteric enzyme is a multisubunit protein:

- The **concerted model** was first proposed by Jacques Monod, Jeffries Wyman and Jean-Pierre Changeux. In this model, each subunit of the enzyme can take up either of two conformations. One of these is a 'tensile' conformation which has low affinity for the substrate, whereas the other is a 'relaxed' conformation, which has higher substrate affinity. When there is no substrate, each subunit is in the tensile conformation. Binding of a substrate molecule to one of the subunits induces the immediate conversion of all the subunits to their relaxed conformations. Binding of the first substrate molecule therefore increases the affinity of the enzyme for other substrate molecules.

- The **sequential model**, first proposed by Daniel Koshland, also assumes that there are tensile and relaxed conformations. The difference is that in this model the binding of the first substrate molecule only influences the substrate affinity of neighboring subunits, rather than all the subunits of the enzyme.

concerted model

sequential model

■ tensile conformation, no substrate bound
□ tensile conformation, substrate bound
● relaxed conformation, no substrate bound
◎ relaxed conformation, substrate bound

Studies of a number of allosteric enzymes suggest that neither model is precisely correct, with most enzymes responding to binding of the initial substrate molecule in a manner that is intermediate between the predictions of the concerted and sequential processes.

Feedback regulation is particularly useful if a metabolic pathway contains branches, the initial substrate being converted into more than one end product. Allosteric inhibition can then switch off one branch of the pathway when that particular end product is present in adequate amounts, so that the substrate is directed entirely towards synthesis of the product of the second branch (*Fig. 7.24B*). If the end products of both branches of a pathway are present in sufficient amounts, then the intermediate immediately before the branchpoint accumulates. This intermediate might be able to inhibit an earlier commitment step so now the entire pathway is shut down. We will encounter several examples of feedback regulation when we examine individual metabolic pathways in the next few chapters.

Further reading

Atkins P (2010) *The Laws of Thermodynamics: a very short introduction.* Oxford University Press, Oxford.

Cleland WW (1963) The kinetics of enzyme-catalyzed reactions with two or more substrates or products. II. Inhibition: nomenclature and theory. *Bichimica et Biophysica Acta* **67**, 173–87.

Čolović MB, Krstić DZ, Lazarević-Pašt TD, Bondžić AM and Vasić VM (2013) Acetylcholinesterase inhibitors: pharmacology and toxicology. *Current Neuropharmacology* **11**, 315–35.

Cornish–Bowden A (2014) Current IUBMB recommendations on enzyme nomenclature and kinetics. *Perspectives in Science* **1**, 74–87. *Describes the EC classification for enzyme nomenclature, and also gives extensive details of enzyme kinetics.*

Cornish–Bowden A (2014) Understanding allosteric and cooperative interactions in enzymes. *FEBS Journal* **281**, 621–52.

Hashim OH and Adnan NA (1994) Coenzyme, cofactor and prosthetic group – ambiguous biochemical jargon. *Biochemical Education* **22**, 93–4. *Discusses the confusion that has arisen regarding the precise meaning of these terms.*

Jimenez RM, Polanco JA and Luptak A (2015) Chemistry and biology of self-cleaving ribozymes. *Trends in Biochemical Sciences* **40**, 648–61.

Johnson KA and Goody RS (2011) The original Michaelis constant: translation of the 1913 Michaelis–Menten paper. *Biochemistry* **50**, 8264–9.

Koshland DE (1995) The key–lock theory and the induced fit theory. *Angewandte Chemie* **33**, 23–4. *Models for enzyme–substrate binding.*

Kumar S and Nussinov R (2001) How do thermophilic proteins deal with heat? *Cellular and Molecular Life Sciences* **58**, 1216–33.

Lineweaver H and Burk D (1934) The determination of enzyme dissociation constants. *Journal of the American Chemical Society* **56**, 658–66.

Yennamalli RM, Rader AJ, Kenny AJ, Wolt JD and Sen TZ (2013) Endoglucanases: insights into thermostability for biofuel applications. *Biotechnology for Biofuels* **6**: 136.

Self-assessment questions

Multiple choice questions

Only one answer is correct for each question. Answers can be found on the website: www.scionpublishing.com/biochemistry.

1. What was the first enzyme to be shown to be a protein?
   - **(a)** Amylase
   - **(b)** Diastase
   - **(c)** Ribonuclease A
   - **(d)** Urease

2. The active site of ribonuclease A contains two copies of which amino acid?
   - **(a)** Glycine
   - **(b)** Histidine
   - **(c)** Isoleucine
   - **(d)** Leucine

3. Which one of the following statements is **correct** with regard to tryptophan synthase?
   - **(a)** An intermediate in the biochemical reaction is channeled between two subunits of the enzyme
   - **(b)** It is a dimer of two identical subunits
   - **(c)** It possesses an exonuclease activity
   - **(d)** It uses chorismate as its substrate

4. What is an RNA enzyme called?
   - **(a)** Ribosome
   - **(b)** Ribozyme
   - **(c)** Transfer RNA
   - **(d)** This is a trick question, all enzymes are made of protein

5. Which is the metal ion cofactor in cytochrome oxidase?
   (a) Cu^{2+}
   (b) Fe^{2+}
   (c) Mg^{2+}
   (d) Zn^{2+}

6. Riboflavin (vitamin B_2) is the precursor of which organic cofactors?
   (a) Coenzyme A
   (b) FAD and FMN
   (c) NAD^+ and $NADP^+$
   (d) S-adenosyl methionine

7. What is the term used to describe the combination of an enzyme with its cofactor?
   (a) Apoenzyme
   (b) Holoenzyme
   (c) Multisubunit enzyme
   (d) Ribozyme

8. Which one of the following statements is **correct** with regard to redox reactions?
   (a) Both oxidation and reduction result in gain of electrons
   (b) Both oxidation and reduction result in loss of electrons
   (c) Oxidation is loss of electrons, reduction is gain
   (d) Reduction is loss of electrons, oxidation is gain

9. What are enzymes with identical functions from different organisms called?
   (a) Allosteric enzymes
   (b) Homologous enzymes
   (c) Isozymes
   (d) Paralogous enzymes

10. What is the term used to describe an enzymatic reaction that releases energy?
    (a) Endergonic
    (b) Energy coupled
    (c) Exergonic
    (d) Reversible

11. What term is used to denote the energy difference between the substrates of an enzymatic reaction and the transition state?
    (a) ΔG
    (b) $\Delta G\ddagger$
    (c) $\Delta G'$
    (d) $\Delta G^{0'}$

12. Which one of the following statements is **incorrect**?
    (a) An enzyme changes the ΔG values for substrates and products
    (b) An enzyme increases the reaction rate
    (c) An enzyme reduces the free energy of the transition state
    (d) None of the above statements is incorrect

13. Which thermodynamic term is a measure of the degree of disorder of a system?
    (a) Chaos
    (b) Enthalpy
    (c) Entropy
    (d) Free energy

14. The lock and key and induced fit models refer to which aspect of enzyme behavior?
    (a) Cooperative substrate binding
    (b) Irreversible inhibition
    (c) Reduction of the free energy of the transition state
    (d) Specificity of substrate binding

15. Which one of the following statements is **incorrect** regarding thermostable enzymes?
    (a) They are able to withstand high temperatures without denaturing
    (b) They are obtained from thermophilic bacteria
    (c) They have a temperature optimum of 75–80°C
    (d) All of the above statements are incorrect

16. What is the term used to denote the substrate concentration at which the rate of an enzymatic reaction is half of the maximum value?
    (a) k_1
    (b) K_m
    (c) $[S]$
    (d) V_0

17. In the Lineweaver–Burk plot, what does the intercept with the x axis give?
    (a) K_m
    (b) $\dfrac{1}{V_{max}}$
    (c) $-\dfrac{1}{K_m}$
    (d) $\dfrac{1}{K_m}$

18. Diisopropyl fluorophosphate (DIFP) is an example of what type of enzyme inhibitor?
    (a) Allosteric
    (b) Competitive
    (c) Irreversible
    (d) Non-competitive

19. What is the name of the enzyme, inhibited by DIFP, that is involved in transmission of nerve impulses?
    (a) Acetylcholinesterase
    (b) Chymotrypsin
    (c) Neuraminidase
    (d) Synapsase

20. In which type of inhibition does V_{max} stay the same, but K_m is increased?
    (a) Competitive reversible
    (b) Irreversible
    (c) Non-competitive reversible
    (d) The scenario described never occurs

21. In which type of inhibition is V_{max} reduced, but K_m stays the same?
  (a) Competitive reversible
  (b) Irreversible
  (c) Non-competitive reversible
  (d) The scenario described never occurs

22. An allosteric site is the part of an enzyme that does what?
  (a) Binds an inhibitor or other effector molecule
  (b) Binds the substrate
  (c) Binds the product prior to its release by the enzyme
  (d) Channels an intermediate between two subunits of the enzyme

23. What is the name given to the first step in a metabolic pathway that produces an intermediate that is unique to that pathway?
  (a) Allosteric step
  (b) Commitment step
  (c) Concerted step
  (d) Cooperative step

24. The concerted and sequential models refer to which aspect of enzyme behavior?
  (a) Cooperative substrate binding
  (b) Irreversible inhibition
  (c) Reduction of the free energy of the transition state
  (d) Specificity of substrate binding

Short answer questions

These questions do not require additional reading.

1. Compare the structures of ribonuclease A, DNA polymerase I and tryptophan synthase, in each case explaining how the structure relates to the enzymatic activity.

2. What is unusual about the structure of ribonuclease P?

3. Giving as many examples as possible, describe the main categories of enzyme cofactor.

4. Outline the EC enzyme classification system.

5. Explain what is meant by the term 'free energy' and describe the free energy differences between the substrates and products of exergonic and endergonic biochemical reactions.

6. Why is the free energy of the transition state central to any discussion of enzyme-catalyzed reactions?

7. How does substrate concentration affect the rate of an enzyme-catalyzed reaction?

8. Describe how the Lineweaver–Burk plot is derived from the Michaelis–Menten equation, and draw examples of the Lineweaver–Burk plots expected in the presence or absence of (a) a competitive reversible inhibitor, and (b) a non-competitive reversible inhibitor.

9. Explain why diisopropyl fluorophosphate disrupts the transmission of nerve impulses.

10. Define the term 'allosteric inhibition' and describe why allosteric inhibition is important in the control of metabolic pathways.

Self-study questions

These questions will require calculation, additional reading and/or internet research.

1. The existence of ribozymes is looked upon as evidence that RNA evolved before proteins and therefore at one time, during the earliest stages of evolution, all enzymes were made of RNA. Assuming that this hypothesis is correct, explain why some ribozymes persist to the present day.

2. Identify the EC numbers for (a) ribonuclease A, (b) DNA polymerase I, and (c) tryptophan synthase.

3. The rate constants for the reactions catalyzed by two different enzymes are given below. Calculate the K_m for each enzyme and identify which one has the strongest affinity for its substrate.

	k_1	k_2	k_3
Enzyme A	$5 \times 10^6\,M^{-1}\,sec^{-1}$	$2 \times 10^3\,sec^{-1}$	$5 \times 10^2\,sec^{-1}$
Enzyme B	$2 \times 10^7\,M^{-1}\,sec^{-1}$	$5 \times 10^3\,sec^{-1}$	$2 \times 10^2\,sec^{-1}$

4. Explain why k_1, the rate constant for formation of the enzyme–substrate complex, is expressed as $M^{-1} sec^{-1}$, whereas k_2 and k_3, which are the rate constants for breakdown of the enzyme–substrate complex, are expressed as sec^{-1}.

5. The initial velocity was measured for an enzyme-catalyzed reaction at different substrate concentrations, with and without the presence of two different inhibitors. Using the data in the table opposite, determine V_{max} and K_m values for the enzyme with and without the inhibitors, and identify the type of inhibition that is occurring in each case.

Substrate concentration	Initial velocity ($\mu M\ sec^{-1}$)		
(mM)	no inhibitor	inhibitor 1	inhibitor 2
1.0	2.0	1.1	1.0
2.0	3.3	2.0	1.7
5.0	5.9	4.0	3.0
10.0	7.7	5.9	4.0
20.0	10.0	8.3	5.0

CHAPTER 8

Energy generation: glycolysis

STUDY GOALS

After reading this chapter you will:

- understand the role of activated carrier molecules in energy storage and be able to describe the most important of these carriers

- know that the pathway for biochemical energy generation involves two stages, and be able to list the amounts of ATP, NADH and $FADH_2$ that are produced during each stage

- be able to describe the steps of the glycolysis pathway, knowing the substrates, products and enzymes involved at each step

- recognize that ATP is used up at an early stage in glycolysis, but that subsequent steps lead to a net gain in ATP

- know how some organisms carry out glycolysis in the absence of oxygen, and be able to describe the ways by which these organisms re-oxidize the NADH molecules resulting from glycolysis

- be able to describe how sugars other than glucose enter the glycolytic pathway

- understand the importance of the step catalyzed by phosphofructokinase as a key control point in glycolysis, and be able to explain how the activity of this enzyme is regulated by ATP, AMP, citrate and hydrogen ions

- be able to describe how fructose 6-phosphate also regulates phosphofructokinase and how this regulatory effect is responsive to the amount of glucose in the blood

- appreciate the importance of hexokinase and pyruvate kinase as additional control points in glycolysis

The metabolic reactions which provide the energy that a cell needs in order to carry out its physiological activities and to grow and divide are vitally important. Humans and other animals obtain their energy by breakdown of organic molecules which they ingest as food. Carbohydrates (in particular glucose), lipids and amino acids can all be utilized as energy sources.

In this chapter and the next we will examine how the free energy contained in the chemical bonds of a glucose molecule is released and utilized by the cell. We will discover that the process is a multistep metabolic pathway involving a variety of enzymes and a series of intermediate compounds that are formed during the gradual breakdown of glucose. It is important that we study the individual steps in the pathway, but equally important that we do not lose sight of the overall purpose of the pathway as a whole. We will therefore begin with an overview of the process, so that throughout the next two chapters we will be aware of the broader context for the individual reactions being studied.

8.1   An overview of energy generation

The complete breakdown of a molecule of glucose yields six molecules of carbon dioxide and six of water:

$$C_6H_{12}O_6 + 6O_2 \rightarrow 6CO_2 + 6H_2O$$

Oxygen is used up during the reaction, so in chemical terms the process is an oxidation.

Glucose oxidation is a highly exergonic reaction, yielding 2870 kJ of energy for every mole of glucose that is broken down. In biochemical terms, this is a substantial amount of energy; a typical endergonic enzyme-catalyzed reaction requires only about 10 kJ of energy to convert a mole of substrate into a mole of product. The cell therefore breaks glucose down gradually, releasing smaller units of energy at different stages of the process. These packets of energy are stored in **activated carrier molecules**.

Box 8.1 **Units of energy**   PRINCIPLES OF CHEMISTRY

In biochemistry, quantities of energy are expressed as **kilojoules per mole**, written as **kJ mol^{-1}**. Kilojoules and moles are standard SI units that are defined as follows:
• A kilojoule is 1000 **joules**, a joule being the work done by a force of one newton when its point of application moves through a distance of one meter in the direction of the force.

• A mole is the amount of a substance that contains as many atoms, molecules, ions, or other elementary units as the number of atoms in 0.012 kg of carbon-12.

The complete oxidation of glucose yields 2870 kJ mol^{-1} of energy. In other words, the ΔG for this reaction is -2870 kJ mol^{-1}, the negative value indicating that this is an exergonic reaction (see *Section 7.2.1*).

8.1.1   Activated carrier molecules store energy for use in biochemical reactions

In *Section 7.2.1* we learnt that the energy needed to drive an endergonic biochemical reaction is often obtained by hydrolysis of a molecule of ATP. ATP is an example of an activated carrier molecule, which is a molecule that acts as a temporary store of the free energy released by breakdown of glucose and other organic compounds.

ATP is the most important biological energy carrier, with a typical human cell containing approximately 10^9 molecules of ATP, which in some cells is completely used up (and replaced by new ATP molecules) once every few minutes. Hydrolysis of ATP releases 30.84 kJ mol^{-1} of energy and results in ADP and inorganic phosphate (*Fig. 8.1*).

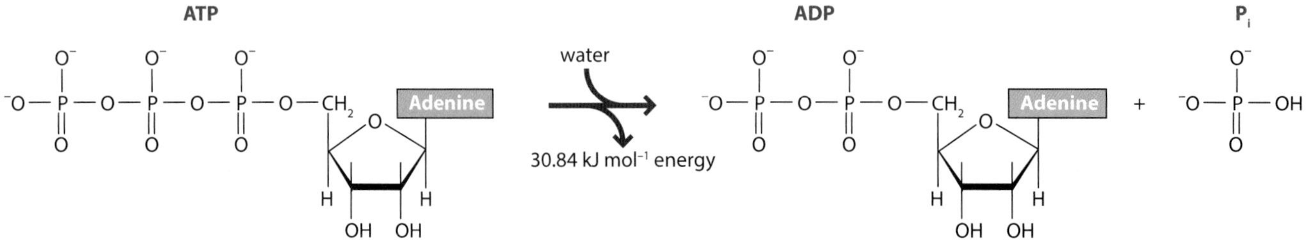

Figure 8.1 Hydrolysis of ATP.
Abbreviations: ATP, adenosine 5′-triphosphate; ADP, adenosine 5′-diphosphate; P$_i$, inorganic phosphate.

The phosphate–phosphate linkage that is broken during ATP hydrolysis is sometimes called a 'high-energy' bond, but this is a confusing description and not a correct interpretation of the source of the energy released when ATP is hydrolyzed. The energy that is released does not come directly from the splitting of a phosphate–phosphate bond, and is certainly not the bond energy for that linkage. Instead, the free energy arises, as in all chemical reactions, because of the ΔG between the reactants and products. In this case, the ΔG between the reactants (ATP and water) and the products (ADP and inorganic phosphate) is relatively large because of differences between the resonance (distribution of electrons) and solvation (interaction with water) properties of ATP and ADP. Because of these resonance and solvation differences, ADP is, in thermodynamics terms, a more ordered system than ATP, and so has a lower free energy content. The conversion of ATP to ADP therefore releases energy.

ATP may be the most important activated carrier molecule in living cells, but it is by no means the only compound of this type. A second type of nucleotide, GTP, also acts as an energy carrier, particularly during the reactions that result in synthesis of proteins.

We will look at the role of GTP in protein synthesis in *Section 16.2.2*.

Some enzyme cofactors are also activated carrier molecules. These include NAD^+ and $NADP^+$, each of which can carry energy in the form of a pair of electrons and a proton (H^+ ion), converting the molecules into their reduced forms referred to as NADH and NADPH. The chemical equations for reduction of NAD^+ and $NADP^+$ are therefore:

$$NAD^+ + H^+ + 2e^- \rightarrow NADH$$

$$NADP^+ + H^+ + 2e^- \rightarrow NADPH$$

See *Figure 7.7A* for the structures of NADH and NADPH, and *Figure 7.7B* for FAD and FMN.

Reversal of these reactions releases the stored energy. NADH acts as an energy carrier between different components of the energy-generating pathway, as we will see below, whereas NADPH is mainly used in anabolic reactions leading to synthesis of large organic molecules from smaller ones.

FAD and FMN act in a similar way, but reacting with two protons rather than one:

$$FAD + 2H^+ + 2e^- \rightarrow FADH_2$$

$$FMN + 2H^+ + 2e^- \rightarrow FMNH_2$$

Both FAD and FMN, like NAD^+, are involved in the energy-generating pathway.

8.1.2 Biochemical energy generation is a two-stage process

The series of reactions that release the energy contained in a glucose molecule in incremental steps, transferring it to ATP molecules, can be described as a two-stage process (*Fig. 8.2*). The first stage is called **glycolysis**. Each six-carbon glucose molecule is broken down to two molecules of the three-carbon sugar called **pyruvate**. Glycolysis does not require oxygen and so can occur in all cells of all organisms. However, it releases less than 7% of the total free energy content of glucose. This released energy is used to synthesize two molecules of ATP. In addition, two molecules of NADH are made for every molecule of glucose that is metabolized.

The TCA cycle is also called the citrate cycle or the Krebs cycle, after Hans Krebs who described it in 1937. We will study the TCA cycle in *Section 9.1* and the electron transport chain in *Section 9.2*.

The second stage of the process requires oxygen, and therefore only occurs under aerobic conditions in cells capable of carrying out **respiration**. This stage comprises two linked pathways. First, the **tricarboxylic acid** (**TCA**) **cycle** completes the breakdown of the pyruvate molecules resulting from glycolysis. Before entering the TCA cycle, pyruvate is converted into acetyl CoA, generating another molecule of NADH. The TCA cycle then breaks down the **acetyl CoA**, with each molecule of acetyl CoA yielding one molecule of ATP, in addition to three of NADH and one of $FADH_2$. Acetyl CoA is also obtained from the breakdown of storage fats, which means that the TCA cycle can utilize energy from this other energy store. The second pathway

Figure 8.2 The two stages of the biochemical energy-generating process.

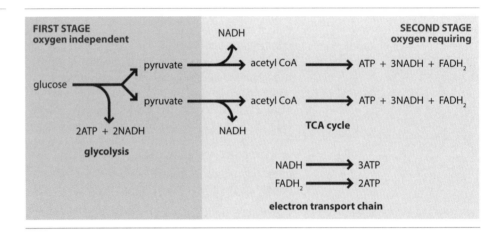

is the **electron transport chain**, which uses the energy contained in the NADH and FADH$_2$ molecules to synthesize another three molecules of ATP per NADH and two per FADH$_2$.

In summary therefore, each molecule of glucose yields 38 molecules of ATP:

• Glycolysis results in eight of these – two molecules of ATP made directly during the glycolysis pathway and six more from the NADH molecules that glycolysis generates.

• Another six ATP molecules are obtained from the two NADH molecules produced when the two pyruvates resulting from glycolysis are converted into two acetyl CoA molecules.

• Two further ATP molecules are made during the TCA cycle, one from each of the two acetyl CoA molecules.

• The final 22 ATPs are generated from the NADH and FADH$_2$ molecules that also result from the TCA cycle.

The 38 ATP molecules corresponds to $38 \times 30.84 = 1173\,kJ\,mol^{-1}$ of energy. This is only 41% of the total energy contained in glucose. What happens to the remainder? It is lost as heat, which in warm-blooded creatures such as humans helps to maintain our body temperature.

8.2 Glycolysis

As we have seen, the first stage of the process that releases energy from glucose is called glycolysis. In the remainder of this chapter we will consider this process from four angles. First, we will look in detail at the steps in the glycolytic pathway, in particular highlighting those that result in transfer of energy to an ATP molecule. Secondly, we will study the role of glycolysis in anaerobic organisms – those that cannot respire and therefore depend on glycolysis as their principal energy source. Thirdly, we will ask how sugars other than glucose enter the pathway, and finally we will examine how glycolysis is regulated so that the amount of glucose that is consumed is appropriate for the energy needs of the cell.

8.2.1 The glycolytic pathway

The glycolysis pathway is shown in outline in *Figure 8.3*. We will run through the individual steps in the pathway and then look in more detail at the key features.

Figure 8.3 Glycolysis in outline.
The names of the enzymes that catalyze the steps in the pathway are given in italics.

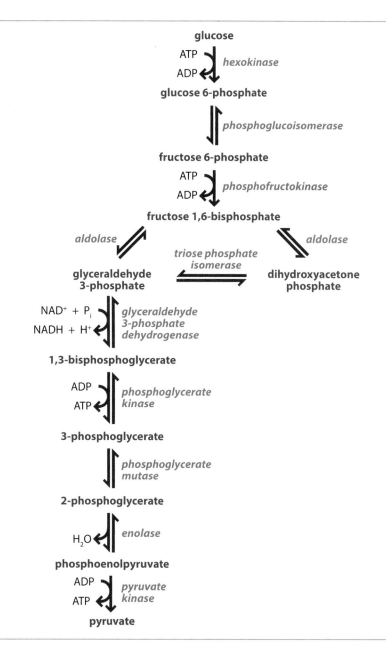

Glycolysis converts one molecule of glucose into two of pyruvate

The individual steps in glycolysis are as follows.

Step 1. To begin the pathway, glucose is phosphorylated by ATP to form glucose 6-phosphate and ADP. The reaction is catalyzed by the enzyme **hexokinase**.

Step 2. Glucose 6-phosphate is converted to fructose 6-phosphate by **phosphoglucoisomerase**.

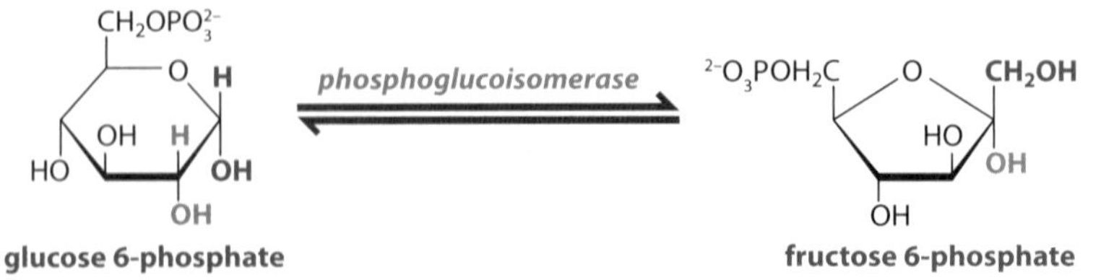

glucose 6-phosphate   **fructose 6-phosphate**

Glucose 6-phosphate is an aldose sugar whereas fructose 6-phosphate is a ketose. The conversion of one to the other is therefore an isomerization reaction, which is more easily visualized by looking at its effect on the linear versions of the two compounds.

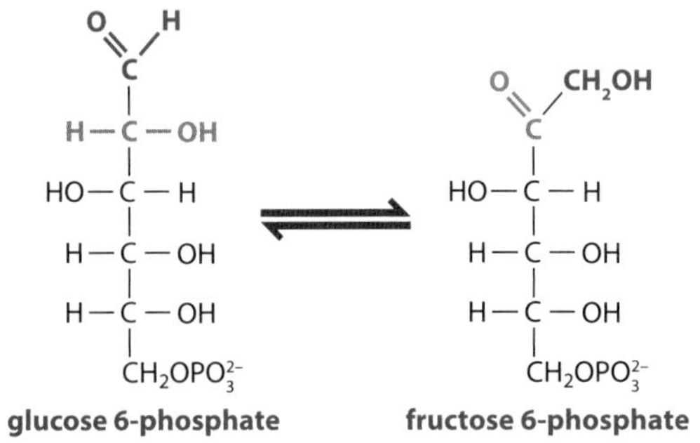

glucose 6-phosphate   **fructose 6-phosphate**

Step 3. Fructose 6-phosphate is phosphorylated by ATP to form fructose 1,6-bisphosphate and ADP. The enzyme catalyzing this step is **phosphofructokinase**.

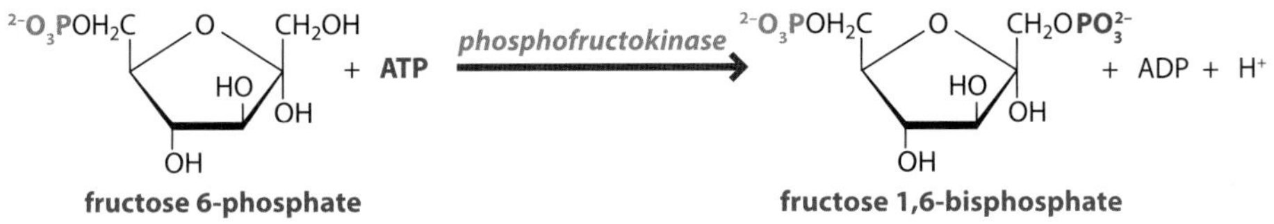

fructose 6-phosphate   **fructose 1,6-bisphosphate**

Step 4. **Aldolase** splits fructose 1,6-bisphosphate, which is a six-carbon sugar, into two three-carbon compounds. These compounds are glyceraldehyde 3-phosphate and dihydroxyacetone phosphate.

dihydroxyacetone phosphate

fructose 1,6-bisphosphate

glyceraldehyde 3-phosphate

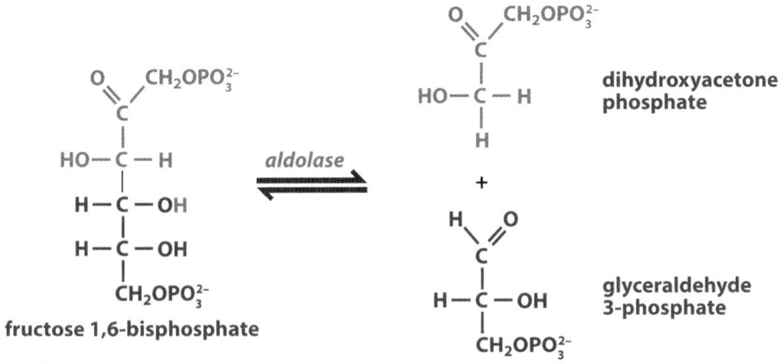

Step 5. Dihydroxyacetone phosphate cannot itself be used in the remainder of the glycolytic pathway. It is therefore converted into glyceraldehyde 3-phosphate by an isomerization reaction catalyzed by **triose phosphate isomerase**.

dihydroxyacetone phosphate

triose phosphate isomerase

glyceraldehyde 3-phosphate

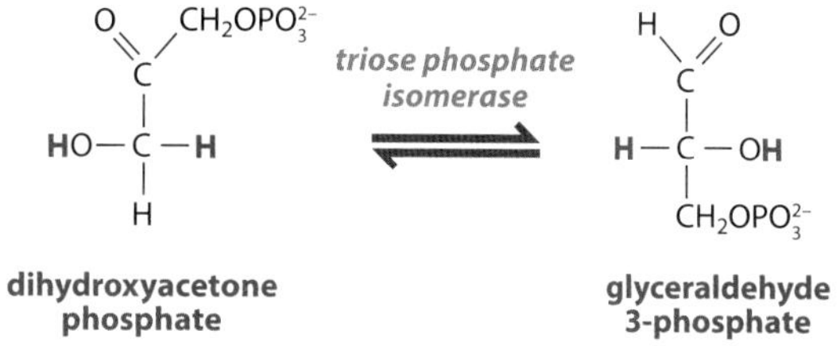

Step 6. Glyceraldehyde 3-phosphate is converted to 1,3-bisphosphoglycerate. The reaction is catalyzed by **glyceraldehyde 3-phosphate dehydrogenase** and uses inorganic phosphate (P_i) and NAD^+. It generates one molecule of NADH and hence is the first step in the pathway at which some of the energy content of the original glucose molecule becomes stored in an activated carrier.

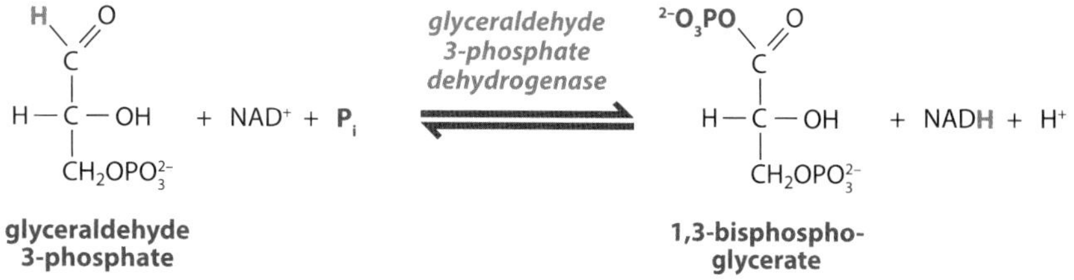

glyceraldehyde 3-phosphate $+ NAD^+ + P_i$

glyceraldehyde 3-phosphate dehydrogenase

1,3-bisphospho-glycerate $+ NADH + H^+$

Step 7. **Phosphoglycerate kinase** catalyzes the transfer of a phosphate group from 1,3-bisphosphoglycerate to ADP, generating ATP and 3-phosphoglycerate.

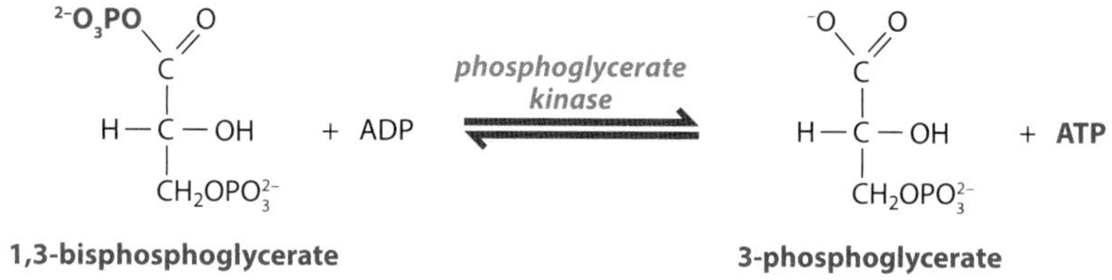

1,3-bisphosphoglycerate $+$ ADP

phosphoglycerate kinase

3-phosphoglycerate $+$ **ATP**

Step 8. 3-Phosphoglycerate is converted to 2-phosphoglycerate by **phosphoglycerate mutase**. This reaction moves the phosphate group present in 3-phosphoglycerate to a different carbon atom within the same molecule.

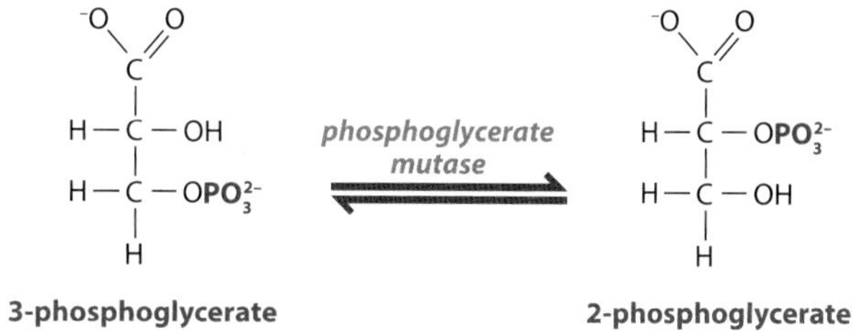

3-phosphoglycerate

phosphoglycerate mutase

2-phosphoglycerate

Step 9. **Enolase** catalyzes the removal of water from 2-phosphoglycerate, giving phosphoenolpyruvate.

2-phosphoglycerate   *enolase*   phosphoenolpyruvate

Step 10. In the final reaction of the pathway, **pyruvate kinase** catalyzes the transfer of the phosphate group from phosphoenolpyruvate to ADP to form ATP and pyruvate.

phosphoenolpyruvate   ADP + H⁺   **ATP**   *pyruvate kinase*   pyruvate

Glycolysis uses up ATP in order to make more ATP

The glycolysis pathway can be divided into two phases, the first phase comprising steps 1–5 and culminating in synthesis of glyceraldehyde 3-phosphate, and the second phase made up of steps 6–10, when glyceraldehyde 3-phosphate is metabolized into pyruvate. The first phase does not generate ATP. In fact, the reverse is true, because two molecules of ATP are needed to convert one molecule of glucose (which has no phosphate groups) into two molecules of glyceraldehyde 3-phosphate (both of which have a single phosphate). It is only in phase 2 of glycolysis that ATP molecules are produced, two for every molecule of glyceraldehyde 3-phosphate, and hence four for each starting molecule of glucose (*Fig. 8.4*). The pathway as a whole therefore achieves a net gain of two ATPs per glucose molecule. This net gain can be increased to eight in respiring organisms where glycolysis is linked to the electron transport chain, because now the energy contained in the NADH molecule generated in step 6 can be used to synthesize three further ATPs. Again, we double this number to get the number of ATPs obtained from a single starting glucose molecule.

The two ATPs that are used in the first phase of the pathway are recovered in the final step, when phosphoenolpyruvate is converted to pyruvate. Why use up two ATPs during the first phase of glycolysis when they are simply recovered again at the end of the pathway? There are two reasons. First, the initial phosphorylation ensures that glucose continues to flow into the cell. Remember that in *Section 5.2.2* we learnt how the GLUT1 uniporter transports glucose across the plasma membrane. Glucose transport is an example of facilitated diffusion, so to be transported into the cell the internal glucose concentration must be lower than that outside of the cell. Conversion of glucose to glucose 6-phosphate, which is not a substrate for the GLUT1 uniporter, immediately or soon after transport ensures that the internal glucose concentration remains low (*Fig. 8.5*). In effect, the phosphorylation traps the energy source within the cell so it is not lost if the external glucose concentration drops.

The second reason for the initial phosphorylations is that these favor the reactions occurring during steps 6 and 7. These two steps result in conversion of glyceraldehyde 3-phosphate to 3-phosphoglycerate, generating one molecule of ATP and one of NADH. These steps therefore make up the critical part of the pathway because

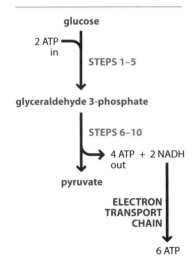

Figure 8.4 The energy balance of the glycolytic pathway.

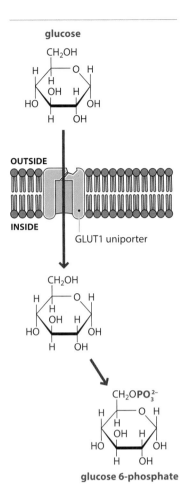

glucose

OUTSIDE

INSIDE

GLUT1 uniporter

glucose 6-phosphate

Figure 8.5 Phosphorylation traps glucose within the cell.
Immediately after transport into the cell, glucose is converted into glucose 6-phosphate. The latter cannot pass back through the GLUT1 uniporter and so remains inside the cell even if the external glucose concentration drops.

this is when the net energy gain is achieved. Glyceraldehyde 3-phosphate, as its names implies, is an aldehyde, and 3-phosphoglycerate is a type of carboxylic acid. The conversion from one to the other is an oxidation reaction. The first of the two enzymes involved in this conversion, glyceraldehyde 3-phosphate dehydrogenase, uses inorganic phosphate as the source of the oxygen to carry out the oxidation, generating 1,3-bisphosphoglycerate (see step 6, above). The displaced hydrogen is used to reduce NAD^+ to NADH, capturing a portion of the energy released by the oxidation reaction. The 1,3-bisphosphoglycerate molecule is immediately passed to the second enzyme, phosphoglycerate kinase, which transfers the phosphate group to ADP, converting the latter to ATP (step 7, above). These reactions would be possible with glyceraldehyde as the substrate instead of glyceraldehyde 3-phosphate, but with glyceraldehyde the energy barrier that would have to be surmounted to bring the oxidation to completion is greater. The increased energy input needed to oxidize glyceraldehyde would reduce the overall energy balance of these two steps to the extent that insufficient energy would be left over to generate either the ATP or NADH molecules. The two phosphorylations in the first phase of glycolysis, by reducing the energy barrier, therefore make it possible for the energy released during the oxidation to drive the production of ATP and NADH.

8.2.2 Glycolysis in the absence of oxygen

Glycolysis does not require the presence of molecular oxygen and so can take place under anaerobic conditions. As glycolysis results in a net yield of ATP molecules, a cell operating under anaerobic conditions is able to generate energy, even though it is unable to utilize the additional energy contained in the NADH molecules generated in step 6. This is a disadvantage, but the main problem that an anaerobic cell faces is that if these NADH molecules are not re-oxidized then its supply of NAD^+ may become low. As NAD^+ is a substrate for step 6 of glycolysis, a shortage of this compound would, amongst other things, cause glycolysis to stall at this point, before the pathway has reached the steps where the net gain in ATP is achieved. As we will see, different species have evolved different strategies for converting the NADH back to NAD^+.

In exercising muscles, pyruvate is converted to lactate

In animals, oxygen can become limiting in muscles after a period of prolonged exercise. The TCA cycle and electron transport chain are then unable to work rapidly enough to regenerate all the NAD^+ needed to maintain glycolysis at its maximum rate. To alleviate this problem, some of the pyruvate that now accumulates in the muscle cells is converted to lactate by the enzyme **lactate dehydrogenase**.

pyruvate   NADH + H⁺   NAD⁺   **lactate**

lactate dehydrogenase

This conversion of pyruvate to lactate is a reduction and hence can be coupled to the oxidation of NADH to NAD^+, ensuring that the cell's supply of the latter remains sufficient for glycolysis to continue.

What happens to the lactate that is now being produced? Lactate cannot be metabolized into any other useful compounds and so the only way to get rid of it is to convert it back to pyruvate. This reverse reaction can also be catalyzed by lactate dehydrogenase, but of course would consume NAD^+ molecules in the muscle. Instead, the lactate is transported from the anaerobic muscle environment, via the bloodstream, to the liver, whose cells are unaffected by exercise and will still be

---

Box 8.2 **Biochemical synthesis of ATP**

ATP is the most important activated carrier molecule and reactions that result in its synthesis are critical for maintaining the energy supply available to the cell. There are two ways in which ATP can be generated, by **substrate-level phosphorylation** and by **oxidative phosphorylation**.

In substrate-level phosphorylation, the phosphate used to generate ATP from ADP comes from a phosphorylated intermediate, which is one of the *substrates* of the reaction. We can denote this intermediate as $R-OPO_3^{2-}$, where 'R' is the sugar component of the compound:

$$R-OPO_3^{2-} \xrightarrow{\text{ADP \quad ATP}} R-OH$$

The energy released when the phosphate group is detached from the intermediate is conserved and used to drive the transfer

of this group to ADP. Steps 7 and 10 of glycolysis are both substrate-level phosphorylations.

In oxidative phosphorylation, an ATP synthase enzyme synthesizes ATP from ADP and inorganic phosphate:

$$ADP + P_i \longrightarrow ATP$$

The energy needed to drive this reaction is obtained by oxidation of NADH or $FADH_2$. We will study the process in detail in *Section 9.2.3*.

Respiring cells obtain most of their ATP by oxidative phosphorylation. Substrate-level phosphorylation is more important in tissues that are suffering from oxygen starvation and in organisms that live in natural environments that lack oxygen.

---

operating in an aerobic environment. The lactate is then oxidized back to pyruvate by lactate dehydrogenase.

The pyruvate in the liver could now enter the TCA cycle, but usually this does not occur. This is because the liver is able to generate sufficient energy for its own needs without this pyruvate boost. Instead, a process called **gluconeogenesis** converts the pyruvate back to glucose, which is then passed into the bloodstream for use by other tissues. If the period of exercise that initiated this process is prolonged and severe, then its maintenance may be dependent on the muscle cells accessing this new supply of glucose. The combination of glycolysis and lactate production in muscle cells linked to regeneration of pyruvate and glucose in the liver is called the **Cori cycle** (*Fig. 8.6*). The cycle, and the exercise it is supporting, cannot continue indefinitely because there is a net loss of ATP. This is because gluconeogenesis consumes six ATP molecules for every pyruvate that is converted back to glucose, and only two of these ATPs are recovered when the glucose is converted back to pyruvate via glycolysis.

> Gluconeogenesis will be described in *Section 11.2*.

Figure 8.6 The Cori cycle.
Lactate synthesized in exercising muscle is transported to the liver, where it is converted to pyruvate by lactate dehydrogenase, and then to glucose by the gluconeogenesis pathway. During periods of extreme exercise, the glucose can be sent back to the muscle in order to maintain glycolysis in the muscle cells.

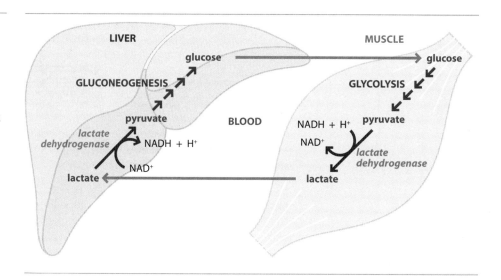

Box 8.3 **Aerobes and anaerobes**

Microorganisms can be classified according to their requirement for oxygen:

- An **obligate aerobe** must have oxygen in order to grow. These organisms obtain the bulk of their ATP from oxidative phosphorylation, hence their requirement for oxygen. Most fungi and algae are obligate aerobes, as are many bacteria.
- A **facultative anaerobe** is able to use oxygen to make ATP, but can also grow in the absence of oxygen. The yeast *Saccharomyces cerevisiae* is a typical facultative anaerobe. If no oxygen is available then yeasts can use fermentation to regenerate NAD⁺, enabling them to continue to obtain ATP by substrate-level phosphorylation during glycolysis.
- An **obligate anaerobe** never uses oxygen. Indeed oxygen is lethal to many organisms of this type, because they are

unable to detoxify compounds such as superoxide (O_2^-) and hydrogen peroxide (H_2O_2), which then accumulate in their cells causing oxidative damage to enzymes and membranes. Some obligate anaerobes convert pyruvate from glycolysis into lactate or some other compound whose synthesis enables the NAD⁺ used in glycolysis to be regenerated. Others regenerate NAD⁺ and possibly synthesize ATP via modified versions of the electron transport chain, in which a compound other than oxygen is used as the final electron acceptor. For example, *Desulfobacter* uses sulfate as its final electron acceptor, and is therefore a type of sulfate-reducing bacterium.

Yeast converts pyruvate to alcohol and carbon dioxide

Various microorganisms, including the yeast *Saccharomyces cerevisiae*, can live in natural environments that lack oxygen. These species are therefore called **facultative anaerobes**, to distinguish them from **obligate aerobes**, which are organisms that must have oxygen in order to grow. When oxygen is available, yeast carries out the full energy-generation pathway including the TCA cycle and electron transport chain. But when the oxygen supply falls below a certain level, the TCA cycle and electron transport chain are temporarily switched off, and the yeast cell relies only on glycolysis for provision of ATP.

Under anaerobic conditions, yeast regenerates the NADH resulting from glycolysis by a two-step process called **alcoholic fermentation** (*Fig. 8.7*).

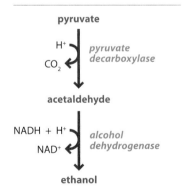

Figure 8.7 **Alcoholic fermentation.**

Step 1. Pyruvate is converted to acetaldehyde by **pyruvate decarboxylase**.

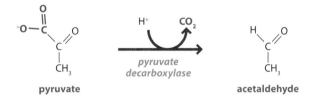

Step 2. Acetaldehyde is converted to ethanol by **alcohol dehydrogenase**.

The products of alcoholic fermentation are therefore NAD⁺, ethanol and carbon dioxide.

For yeast cells, the purpose of alcoholic fermentation is to regenerate NAD⁺ for use in glycolysis. For humans, the commercial importance of the pathway is synthesis of the ethanol byproduct, the use of yeast to produce this compound representing the earliest example of prehistoric biotechnology. Alcoholic beverages are made by allowing yeast to carry out alcoholic fermentation of sugar contained in natural products such as grapes, to produce wine, and barley, to produce beer. The archaeological record

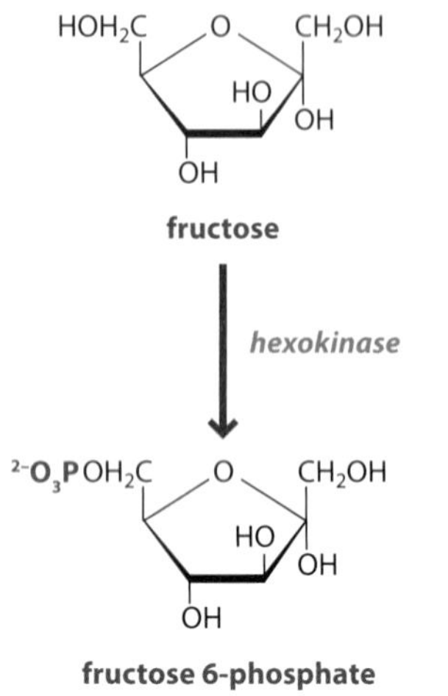

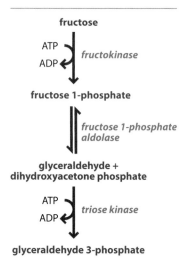

Figure 8.8 The conversion of fructose to fructose 6-phosphate catalyzed by hexokinase.

suggests that a type of rice wine was being made in China about 9000 years ago, and at about the same time beer was being brewed in Mesopotamia. The carbon dioxide produced during alcoholic fermentation is exploited during bread-making; addition of yeast to the flour generates carbon dioxide that causes the resulting dough to rise and produces nice bouncy bread. Certain chemicals such as baking soda (sodium bicarbonate) that release carbon dioxide during baking can be used instead of yeast. Agents such as yeast and baking soda which cause bread dough to rise are called leavening agents and the resulting bread is called leavened bread. The ancient Egyptians were making bread leavened with yeast 2500 years ago, and there is evidence that this type of bread-making was being practiced in Greece some 1000 years earlier than that. Today, the production of alcoholic beverages and bread are important industries worldwide.

8.2.3 Glycolysis starting with sugars other than glucose

Glucose is one of three sugars, fructose and galactose being the others, that can be absorbed into the bloodstream during digestion. Having considered glucose, we must now look at how these other sugars enter the glycolytic pathway.

Fructose has two routes into glycolysis, which are used in different tissues

Fructose is common in the human diet, being present in many fruits and most root vegetables, and is also the major sugar in honey. Sucrose is a disaccharide of fructose and glucose and so, after digestion, is another major dietary source of fructose.

In most tissues, the presence of fructose rather than glucose does not cause any difficulty because hexokinase, which catalyzes the conversion of glucose to glucose 6-phosphate, can also use fructose as a substrate (*Fig. 8.8*). The resulting fructose 6-phosphate then enters step 3 of glycolysis.

A difficulty arises in liver cells, because these make use of an alternative means of phosphorylating glucose, using the enzyme **glucokinase** instead of hexokinase. Glucokinase does not recognize fructose as a substrate, so in these cells fructose must enter glycolysis via a different route. This is achieved by the **fructose 1-phosphate pathway** (*Fig. 8.9*). This pathway has three steps:

Step 1. **Fructokinase** phosphorylates fructose, converting it into fructose 1-phosphate.

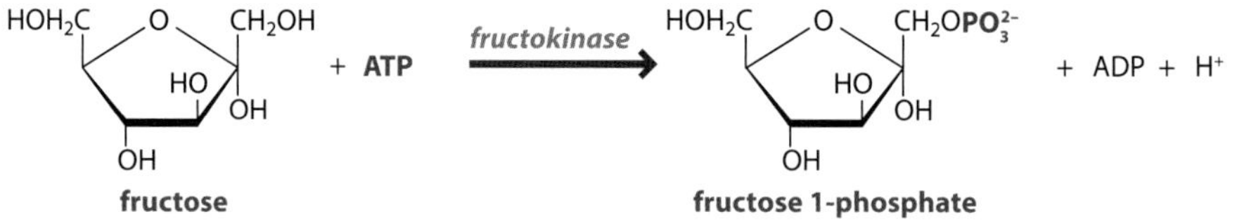

fructose + **ATP** →(fructokinase) **fructose 1-phosphate** + ADP + H$^+$

Step 2. **Fructose 1-phosphate aldolase** splits fructose 1-phosphate into glyceraldehyde and dihydroxyacetone phosphate.

fructose 1-phosphate →(fructose 1-phosphate aldolase) glyceraldehyde + dihydroxyacetone phosphate

Figure 8.9 The fructose 1-phosphate pathway.

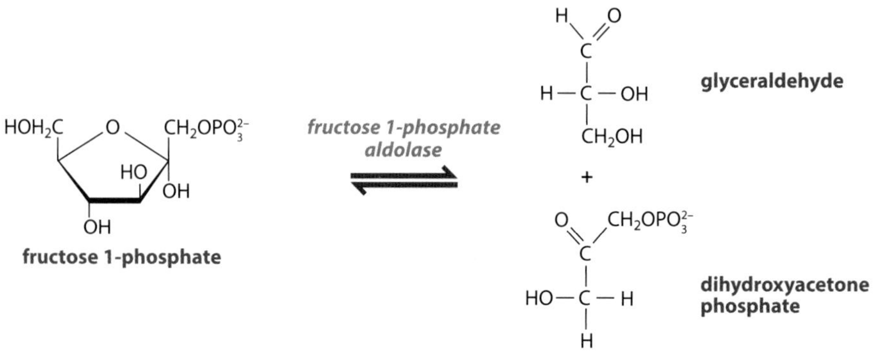

The dihydroxyacetone enters the glycolytic pathway at step 5 and is converted into glyceraldehyde 6-phosphate by triose phosphate isomerase.

Step 3. **Triose kinase** phosphorylates glyceraldehyde to give glyceraldehyde 3-phosphate.

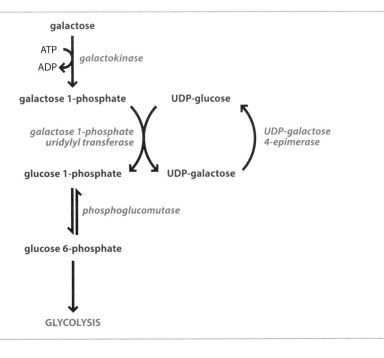

glyceraldehyde   **glyceraldehyde 3-phosphate**

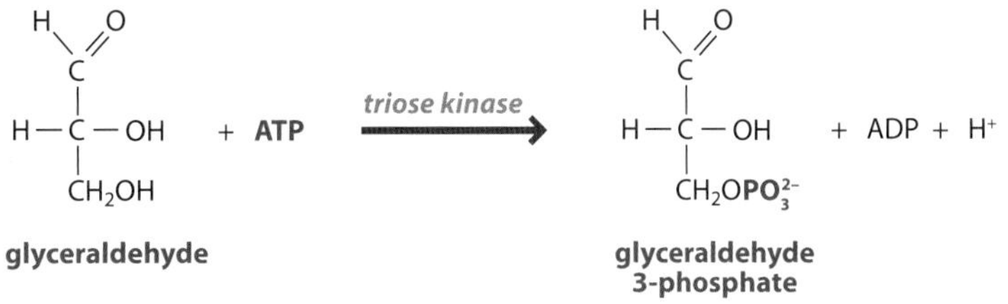

The fructose 1-phosphate pathway therefore yields two molecules of glyceraldehyde 3-phosphate, just like the first phase of the standard glycolysis pathway.

Galactose is converted to glucose before use in glycolysis

Galactose is less common than glucose and fructose in fruits and vegetables, though it is present in sugar beet. In the human diet, its main source is milk and dairy products, which contain lactose, a disaccharide of galactose and glucose. Babies are able to split lactose into its component sugars using the enzyme lactase and the resulting glucose and galactose are then absorbed into the bloodstream. Humans who display lactase persistence, in which lactase is still active in adulthood, are also able to metabolize lactose in this way.

The molecular structures of galactose and glucose differ only in the arrangement of the –H and –OH groups around carbon number 4 (*Fig. 8.10*). Conversion of one to the other therefore requires an isomerization reaction, specifically the type of isomerization called epimerization, in which chemical groups are rearranged around a chiral carbon. The **galactose–glucose interconversion pathway** (*Fig. 8.11*) carries out this epimerization in four steps.

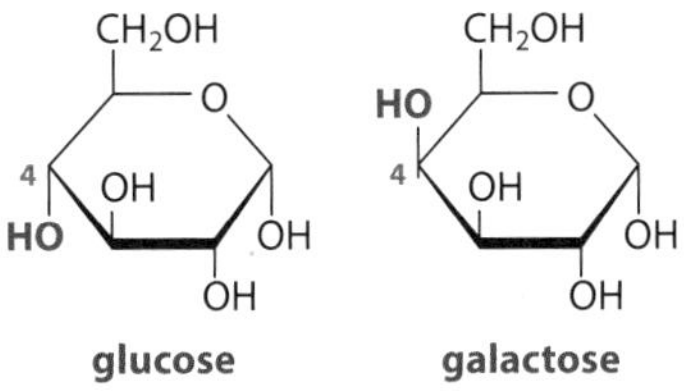

See *Box 6.3* for details regarding the evolution of lactase persistence in the human population.

Figure 8.10 Glucose and galactose are epimers.
The two sugars differ only in the arrangement of the –H and –OH groups around carbon number 4.

Figure 8.11 The galactose–glucose interconversion pathway.

Step 1. **Galactokinase** phosphorylates galactose to galactose 1-phosphate.

galactose   galactose 1-phosphate

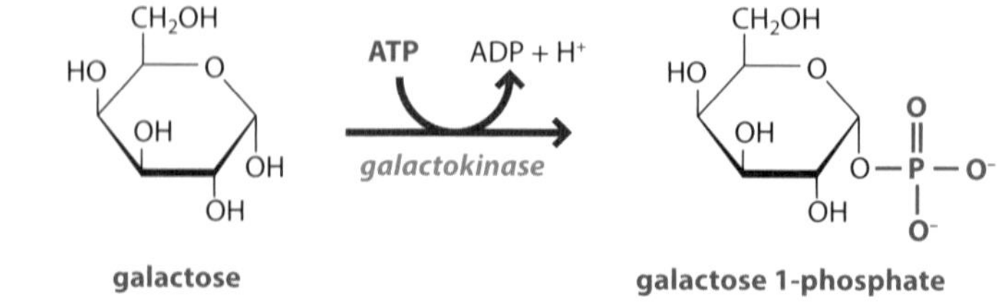

Step 2. **Galactose 1-phosphate uridylyl transferase** transfers a uridine group from UDP–glucose to galactose 1-phosphate. This gives one molecule of UDP–galactose and one of glucose 1-phosphate.

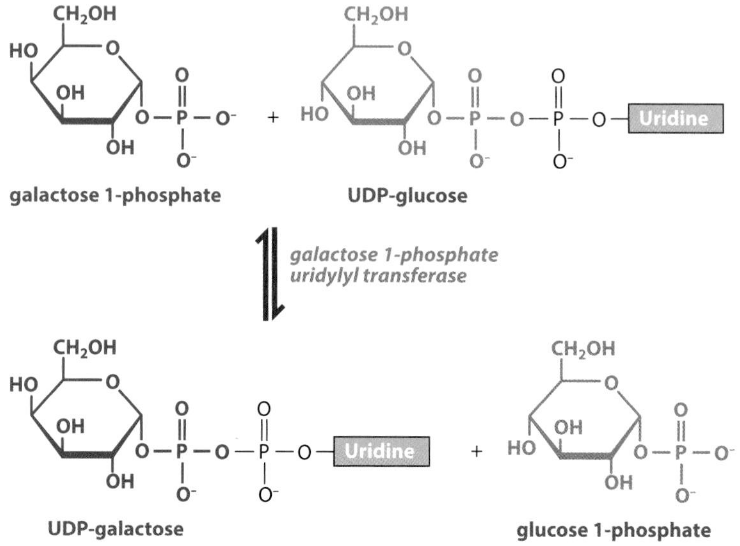

galactose 1-phosphate   UDP-glucose

galactose 1-phosphate uridylyl transferase

UDP-galactose   **glucose 1-phosphate**

Step 3. **UDP–galactose 4-epimerase** converts UDP–galactose to UDP–glucose. This step therefore regenerates the UDP–glucose molecule used in step 2.

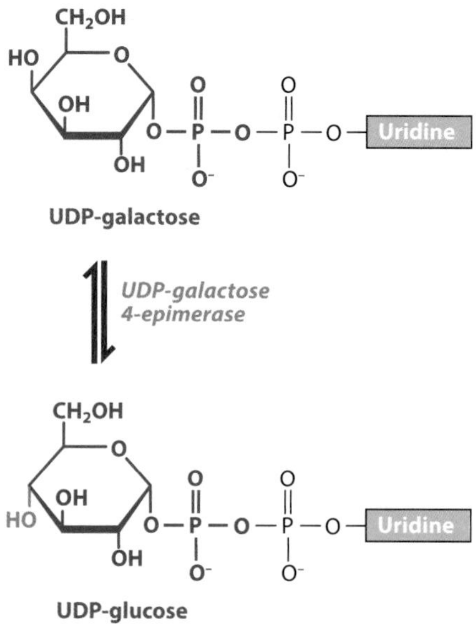

UDP-galactose

UDP-galactose 4-epimerase

UDP-glucose

Step 4. **Phosphoglucomutase** repositions the phosphate group on the glucose 1-phosphate molecule formed in step 2.

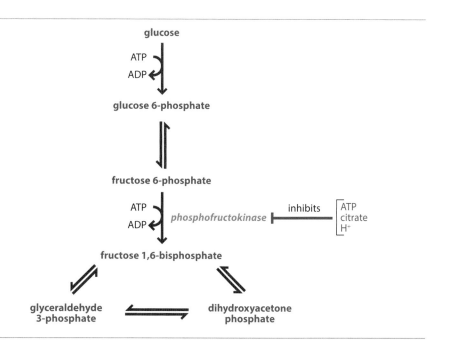

glucose 1-phosphate   glucose 6-phosphate

This reaction produces glucose 6-phosphate, which enters at step 2 of the standard glycolytic pathway.

8.2.4 Regulation of glycolysis

The final aspect of glycolysis that we must consider is how the pathway is regulated. Glycolysis has two main roles: it degrades glucose to generate ATP and it produces intermediates that act as precursors for biosynthetic pathways, such as those involved in the synthesis of fatty acids. Glycolysis must therefore be regulated to ensure that these two roles are fulfilled.

The conversion of fructose 6-phosphate to fructose 1,6-bisphosphate is the main control point in glycolysis

The main control point in the glycolytic pathway is step 3, when fructose 6-phosphate is phosphorylated by ATP to form fructose 1,6-bisphosphate and ADP. In eukaryotes, the enzyme that catalyzes this step, phosphofructokinase, is inhibited by three of the later products of glycolysis (ATP, citrate, and hydrogen ions) enabling the pathway as a whole to be regulated in response to different physiological conditions (*Fig. 8.12*).

The most straightforward of the inhibitory effects on phosphofructokinase is that brought about by ATP. Clearly, if the cell is lacking ATP then it needs to increase the rate at which glycolysis is taking place and, conversely, when ATP is abundant the pathway should be slowed down. This is achieved by ATP attaching to the surface

Figure 8.12 Phosphofructokinase is the main control point in glycolysis. The first five steps of glycolysis are shown, with the step catalyzed by phosphofructokinase highlighted.

We studied the way in which an allosteric inhibitor regulates enzyme activity in *Section 7.2.3*.

of the phosphofructokinase enzyme. This attachment is away from the active site of the enzyme, where ATP also binds in order to participate in the phosphorylation catalyzed by the enzyme (*Fig. 8.13*). ATP therefore acts as an allosteric regulator of phosphofructokinase. AMP competes with ATP for attachment to the allosteric site, and reverses the inhibitory effect of ATP. The rate of the phosphofructokinase reaction, and hence the flow of metabolites through the subsequent steps of the glycolytic pathway, is therefore regulated in response to the relative amounts of AMP and ATP in the cell. When ATP is plentiful the rate goes down so that the ATP pool does not become over-abundant, and when ATP is scarce the rate increases so that the cell's ATP supplies are replenished.

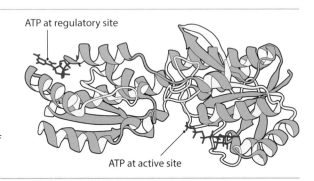

ATP at regulatory site

ATP at active site

Figure 8.13 ATP binding at the active and regulatory sites of phosphofructokinase.
Phosphofructokinase is a tetramer of four identical subunits, one of which is shown in this drawing.

Citrate affects phosphofructokinase activity by promoting the binding of ATP to the allosteric site on phosphofructokinase. Increased levels of citrate therefore result in decreased phosphofructokinase activity, so glycolysis slows down. This is logical because citrate is one of the intermediates of the TCA cycle, and its accumulation in the cell would indicate that the energy-generating pathway as a whole is over-active. There is, however, some doubt about the role of citrate inhibition in living cells. Its effect on phosphofructokinase has been studied in the test tube, but in the cell it is possible that any excess of citrate resulting from the TCA cycle is immediately used as a source of acetyl CoA for fatty acid synthesis. If this is the case then citrate is unlikely to accumulate sufficiently to have a significant effect on phosphofructokinase activity.

We will study the link between the TCA cycle and fatty acid synthesis in *Section 12.1.1*.

Hydrogen ions also inhibit phosphofructokinase, again by increasing the allosteric effect of ATP. This means that at low pH values, phosphofructokinase activity is reduced and glycolysis slows down. Why should pH be an important influence on the rate of glycolysis? The answer lies with the accumulation of lactate that occurs in active muscle tissue. Excess amounts of lactate can damage muscle tissue and also

---

Box 8.4 **Why is phosphofructokinase regulated by AMP and not ADP?**

The product of ATP hydrolysis is ADP, not AMP, and so it might appear logical that ADP would be the positive regulator of phosphofructokinase. AMP plays this role because the level of ADP is not always an accurate indication of the energy requirements of the cell. This is because ADP can be directly converted to ATP by **adenylate kinase**.

$$\text{ADP} + \text{ADP} \underset{}{\overset{\textit{adenylate kinase}}{\rightleftharpoons}} \text{ATP} + \text{AMP}$$

This conversion, which occurs when ATP is used up rapidly, decreases the amount of ADP in the cell under conditions when ATP is required. ADP could not therefore act as the positive regulator of phosphofructokinase, because its concentration drops when ATP is needed. On the other hand, the AMP produced by adenylate kinase supplements the very small amount of this compound that is usually present in the cell. The resulting large increase in AMP concentration enables AMP to act as the positive regulator of phosphofructokinase. It out-competes ATP for occupancy of the allosteric site, so the enzyme is stimulated to increase the flow of metabolites through the glycolytic pathway and hence increase the synthesis of ATP.

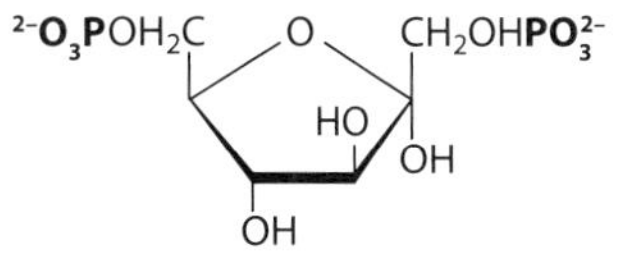

fructose 1,6-bisphosphate

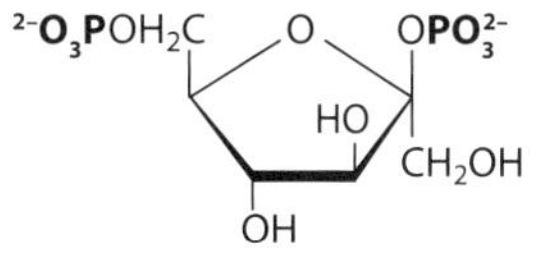

fructose 2,6-bisphosphate

Figure 8.14 The structures of fructose 1,6-bisphosphate and fructose 2,6-bisphosphate.

bring about acidosis, when the blood pH falls to dangerously low levels. Inhibition of phosphofructokinase by hydrogen ions means that, at low pH values, glycolysis is slowed down so less lactate is produced and its dangerous effects are ameliorated. This is one of the reasons why excessive exercise cannot be continued indefinitely. Although muscle cells can switch to anaerobic respiration when their oxygen supply becomes limiting, and some of the lactate that is then produced can be transported to the liver and converted back to pyruvate and glucose, at some point the rate of accumulation of lactate will defeat the body's attempts to adapt, and energy production will begin to decline because of the hydrogen ion effect.

Substrate availability also regulates phosphofructokinase activity

So far we have examined how the activity of phosphofructokinase can be inhibited by the products of glycolysis, so that the flow of metabolites through the pathway is increased or decreased depending on how much ATP, citrate or lactate is being produced. Phosphofructokinase is also regulated by the amount of substrate that is present. This stimulatory effect is mediated by fructose 2,6-bisphosphate, a phosphorylated sugar with a structure slightly different to the fructose 1,6-bisphosphate that is produced by phosphofructokinase activity (*Figure 8.14*).

Fructose 2,6-bisphosphate is synthesized from fructose 6-phosphate by an enzyme called **phosphofructokinase 2**. This is a different enzyme to the phosphofructokinase involved in glycolysis, but it catalyzes a similar reaction.

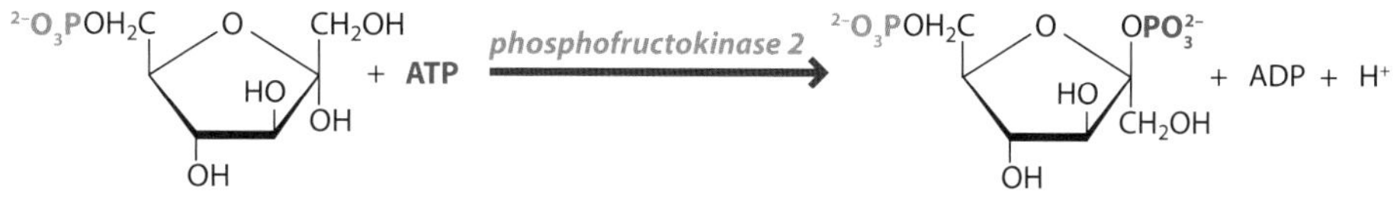

fructose 6-phosphate   fructose 2,6-bisphosphate

The only difference between the activities of the two types of phosphofructokinase is the number of the carbon to which the phosphate group is attached. Phosphofructokinase 2 attaches this phosphate to carbon number 2, whereas phosphofructokinase uses carbon 1.

The reverse reaction, converting fructose 2,6-bisphosphate back to fructose 6-phosphate is catalyzed by **fructose bisphosphatase 2**.

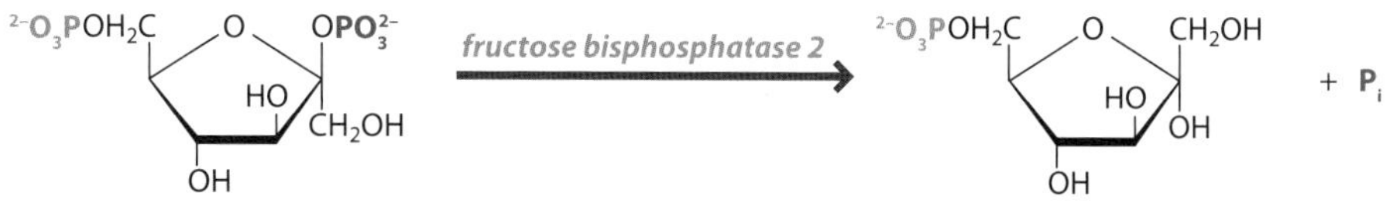

fructose 2,6-bisphosphate   fructose 6-phosphate

Although phosphofructokinase 2 and fructose bisphosphatase 2 are different enzyme activities, both are catalyzed by the same protein. The activity of this protein is regulated by fructose 6-phosphate in two separate ways (*Figure 8.15*):

• Fructose 6-phosphate *stimulates* the phosphofructokinase 2 activity, and hence promotes its own conversion into fructose 2,6-bisphosphate.

• Fructose 6-phosphate *inhibits* the fructose bisphosphatase 2 activity, and hence reduces its synthesis from fructose 2,6-bisphosphate.

The net result of these two complementary regulatory activities is that fructose 6-phosphate exerts self-control over its concentration in the cell. When the amount of fructose 6-phosphate increases, the excess is converted into fructose 2,6-bisphosphate rather than proceeding down the glycolytic pathway with the possible over-production of ATP. If the level of fructose 6-phosphate falls, then more can be obtained from the fructose 2,6-bisphosphate pool, so the rate of glycolysis is maintained.

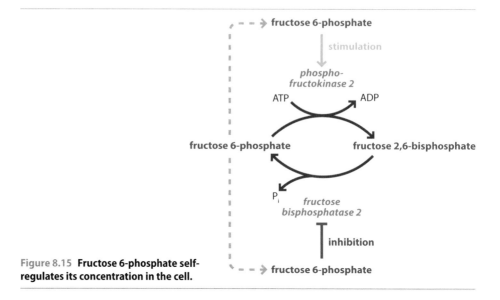

Figure 8.15 Fructose 6-phosphate self-regulates its concentration in the cell.

Fructose 6-phosphate is the product of glucose, in steps 1 and 2 of glycolysis, so the regulation exerted by fructose 6-phosphate and fructose 2,6-bisphosphate is responsive to the amount of glucose available to the cell. Glucose also has its own more direct effect on this regulatory network. When the amount of glucose in the bloodstream falls, the hormone called **glucagon** is released by the pancreas. Glucagon triggers a series of reactions that result in modification of the phosphofructokinase 2/fructose bisphosphatase 2 protein. The modified protein has an increased fructose bisphosphatase 2 activity and reduced phosphofructokinase 2 activity (*Figure 8.16*). This means that more fructose 2,6-bisphosphate is converted to fructose 6-phosphate, maintaining the rate of glycolysis even though the availability of glucose has become low.

We have examined the phosphofructokinase 2/fructose bisphosphatase 2 control system in some detail, not just because it is a key aspect of glycolysis regulation, but also because this system illustrates the exquisite complexity and fitness-for-purpose that regulatory networks can display in living organisms. The direct and indirect influences of glucose and fructose 6-phosphate on this bifunctional enzyme, which is not itself an integral part of the glycolysis pathway, enables the rate of glycolysis to be precisely set so that the most efficient use is made of the amount of substrates that are available to the cell.

Figure 8.16 Glucagon increases fructose bisphosphatase 2 activity and reduces phosphofructokinase 2 activity.

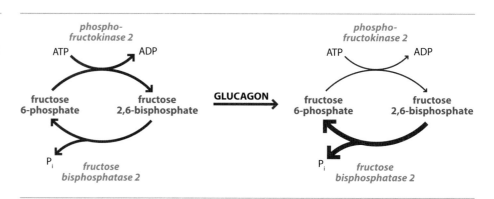

Box 8.5 Control of fructose 6-phosphate levels by glucagon

The process by which glucagon controls the fructose 6-phosphate content of the cell provides a typical example of a signal transduction pathway (*Section 5.2.2*). Glucagon is the extracellular signaling compound, and the signal from cell membrane to target enzyme is transduced via a second messenger system involving cAMP.

The first stage of any signal transduction pathway is the attachment of the extracellular signaling compound to the outside of the cell. Glucagon binds to the glucagon receptor, which is a transmembrane protein with seven α-helices forming a barrel-shaped structure that spans the cell membrane.

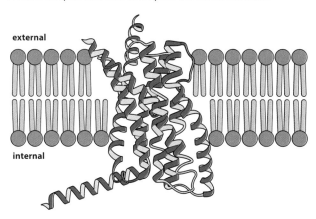

Because of their structure, this class of receptor is called **seven-transmembrane-helix** or **7TM proteins**.

Binding of glucagon to the outer surface of the receptor induces a conformational change in the positioning of the loops on the internal side of the protein. This in turn activates a **G protein** that is associated with the receptor. A G protein is a small protein that

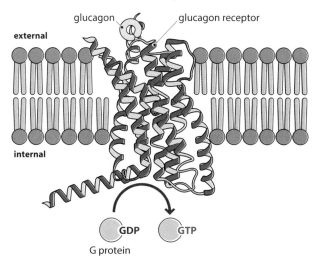

binds either a molecule of GDP or GTP. When GDP is bound the G protein is inactive. The change in conformation of the glucagon

receptor causes the GDP to be replaced by GTP, converting the G protein to its active form. Because it works via a G protein the glucagon receptor is called a **G-protein coupled receptor**.

Once activated, the G protein interacts with adenylate cyclase which, like the receptor protein, is attached to the cell membrane with its active site on the internal side. The interaction with the G protein changes the conformation of adenylate cyclase, changing it to its active form, which now converts ATP to cAMP (see *Figure 5.29*). The increased cellular level of cAMP in turn activates **protein kinase A**, which adds a phosphate group to one of the serines in the phosphofructokinase 2/fructose bisphosphatase 2 enzyme. This phosphorylation is the modification that increases the fructose bisphosphatase 2 activity and reduces the phosphofructokinase 2 activity.

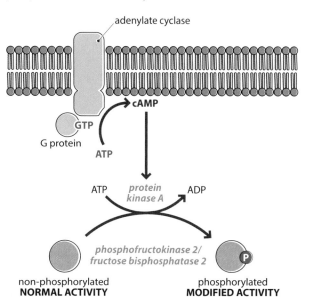

When the blood glucose level increases, glucagon is no longer released by the pancreas and the signal transduction pathway is switched off. A **phosphatase** enzyme now removes the phosphate from phosphofructokinase 2/fructose bisphosphatase 2, so this enzyme reverts back to its alternative state, with reduced fructose bisphosphatase 2 activity and increased phosphofructokinase 2 activity.

As well as glycolysis, glucagon also controls other metabolic pathways that result in the increase or depletion of cellular glucose levels. An example that we will study later is the synthesis and degradation of glycogen, which is the polymeric store of glucose reserves in animals. In response to glucagon, protein kinase A phosphorylates glycogen synthase which, as its name implies, is one of the key enzymes involved in glycogen synthesis. Phosphorylation inactivates glycogen synthase. Protein kinase A has the converse effect on glycogen phosphorylase, activating this enzyme, which breaks down glycogen (*Section 11.1.2*). Glycogen reserves are therefore utilized in order to replenish the blood sugar level.

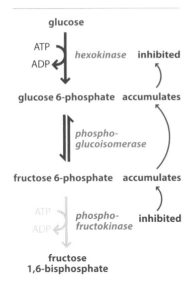

Figure 8.17 Control over the entry of glucose into glycolysis.
When phosphofructokinase is inhibited, fructose 6-phosphate and glucose 6-phosphate accumulate. The latter inhibits hexokinase so that additional glucose does not enter glycolysis.

These alternative uses of glucose will be described in *Section 11.1.1* (glycogen synthesis) and in *Section 11.3* (the pentose phosphate pathway).

See *Section 13.2.1* for details of amino acid synthesis.

Hexokinase and pyruvate kinase are also control points in glycolysis

Although phosphofructokinase is the main control point in glycolysis, two other enzymes have important roles in regulating the pathway. These are hexokinase and pyruvate kinase, which catalyze the first and last steps in the pathway, respectively.

Hexokinase is inhibited by its product, glucose 6-phosphate. Step 2 of glycolysis, in which glucose 6-phosphate is converted to fructose 6-phosphate by phosphoglucoisomerase is a reversible reaction. This means that there is a balance between the amounts of glucose 6-phosphate and fructose 6-phosphate in the cell, the relative amounts kept in equilibrium by the reversible nature of their interconversion. So when phosphofructokinase, the enzyme for step 3 of the pathway, is inhibited, and fructose 6-phosphate accumulates, glucose 6-phosphate also accumulates. The latter then inhibits hexokinase, the enzyme responsible for its synthesis, so that additional glucose does not enter the pathway until it is needed (*Figure 8.17*).

Why is phosphofructokinase the main control point for glycolysis and not hexokinase, the first enzyme in the pathway? The answer is because glucose 6-phosphate is not only used as a substrate for glycolysis. Some glucose 6-phosphate is used to synthesize glycogen, and some is also used by the pentose phosphate pathway, which generates the NADPH needed for synthesis of fatty acids and other biomolecules. If hexokinase was the main control point for glycolysis, then the availability of glucose 6-phosphate would be subject to a regulatory process that did not take account of the requirements of these other metabolic pathways (*Figure 8.18*). Phosphofructokinase is therefore the first commitment step in glycolysis, and hence is the main control step.

Pyruvate kinase, catalyzing the last step in glycolysis, can be looked on as regulating the junction between this pathway and the TCA cycle, into which pyruvate enters prior to its complete breakdown into carbon dioxide and water. Pyruvate kinase is activated by fructose 1,6-bisphosphate and inhibited by ATP, exactly as we might anticipate from what we have learnt so far about the respective effects of substrates and products on the control of glycolysis. Pyruvate kinase is also inhibited by glucagon, and so is a second site at which this hormone achieves a slow-down of glycolysis when blood glucose levels are low. Finally, pyruvate kinase is inhibited by the amino acid alanine. Amino acids are one of several types of biomolecule that are made from intermediates synthesized during the TCA cycle. A relatively high amount of alanine indicates that the cell has a plentiful supply of these biomolecules, reducing the need for pyruvate to be fed into the TCA cycle.

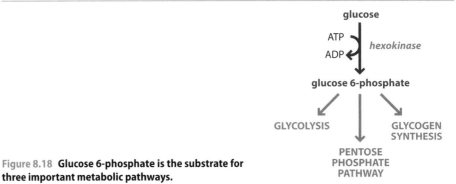

Figure 8.18 Glucose 6-phosphate is the substrate for three important metabolic pathways.

Further reading

Authier F and Desbuquois B (2008) Glucagon receptors. *Cellular and Molecular Life Sciences* **65**, 1880–99.

Guo X, Li H, Ku H, et al. (2012) Glycolysis in the control of blood glucose homeostasis. *Acta Pharmaceutica Sinica* **2**, 358–67.

Lenzen S (2014) A fresh view of glycolysis and glucokinase regulation: history and current status. *Journal of Biological Chemistry* **289**, 12189–94.

Li X-B, Gu J-D and Zhou G-H (2015) Review of aerobic glycolysis and its key enzymes – new targets for lung cancer therapy. *Thoracic Cancer* **6**, 17–24. *Discusses the Warburg effect (see Box 9.1) and how this might be exploited in the design of cancer drugs.*

Müller M, Mentel M, van Hellemond JJ, et al. (2012) Biochemistry and evolution of anaerobic energy metabolism in eukaryotes. *Microbiology and Molecular Biology Reviews* **76**, 444–95.

Scrutton MC and Utter MF (1968) The regulation of glycolysis and gluconeogenesis in animal tissues. *Annual Review of Biochemistry* **37**, 249–302.

Sola-Penna M, Da Silva D, Coelho WS, Marinho-Carvalho MM and Zancan P (2010) Regulation of mammalian muscle type 6-phosphofructo-1-kinase and its implication for the control of metabolism. *IUBMB Life* **62**, 791–806.

Self-assessment questions

Multiple choice questions

Only one answer is correct for each question. Answers can be found on the website: www.scionpublishing.com/biochemistry.

1. How much energy is produced by the complete oxidation of 1 mole of glucose?
   (a) 287 cal
   (b) 287 kJ
   (c) 2870 cal
   (d) 2870 kJ

2. What are the activated carrier molecules synthesized during glycolysis?
   (a) ATP and NADH
   (b) ATP and $FADH_2$
   (c) ATP, NADH and $FADH_2$
   (d) ATP and NADPH

3. One NADH molecule can generate how many ATPs when entered into the electron transport chain?
   (a) 1
   (b) 2
   (c) 3
   (d) 4

4. Glycolysis results in a net gain of how many ATP molecules?
   (a) 2
   (b) 4
   (c) 6
   (d) 8

5. Which enzyme catalyzes the first step in the glycolysis pathway?
   (a) Aldolase
   (b) Enolase
   (c) Hexokinase
   (d) Phosphoglucoisomerase

6. Which compound is split to give one molecule of glyceraldehyde 3-phosphate and one of dihydroxyacetone phosphate?
   (a) Fructose 1,6-bisphosphate
   (b) Fructose 2,6-bisphosphate
   (c) Fructose 6-phosphate
   (d) Glucose 6-phosphate

7. Which compound is converted into pyruvate by the enzyme pyruvate kinase?
   (a) Acetyl CoA
   (b) 2-Phosphoglycerate
   (c) 3-Phosphoglycerate
   (d) Phosphoenolpyruvate

8. What is the production of ATP by phosphoglycerate kinase called?
   (a) Activation
   (b) Kinasing
   (c) Oxidative phosphorylation
   (d) Substrate-level phosphorylation

9. In exercising muscle cells, what is excess pyruvate converted into?
   (a) Acetyl CoA
   (b) Alcohol and carbon dioxide
   (c) Lactate
   (d) Phosphoenolpyruvate

10. Which one of the following statements is **correct** with regard to the Cori cycle?
    (a) Acetyl CoA is used as a substrate
    (b) Lactate from muscles is transported to the liver where it is converted to glucose
    (c) It is responsible for alcohol production by yeast
    (d) It results in a net gain of ATP molecules

11. What is *Saccharomyces cerevisiae* an example of?
    (a) Facultative anaerobe
    (b) Obligate aerobe
    (c) Obligate anaerobe
    (d) None of the above

12. What are the two enzymes involved in alcoholic fermentation?
    (a) Lactate dehydrogenase and alcohol dehydrogenase
    (b) Lactate dehydrogenase and pyruvate decarboxylase
    (c) Pyruvate decarboxylase and alcohol dehydrogenase
    (d) Pyruvate decarboxylase and triose kinase

13. To be used in glycolysis, fructose is first converted to which of the following?
    (a) Fructose 1,6-bisphosphate
    (b) Fructose 1-phosphate
    (c) Fructose 6-phosphate
    (d) Glucose

14. Which of the following is UDP–glucose involved in?
    (a) Entry of fructose into glycolysis
    (b) Galactose–glucose interconversion pathway
    (c) Regulation of glycolysis
    (d) The Cori cycle

15. The main control point in glycolysis is the step that results in synthesis of what?
    (a) Fructose 1,6-bisphosphate
    (b) Fructose 6-phosphate
    (c) Glucose 6-phosphate
    (d) Pyruvate

16. Which one of the following is **not** an inhibitor of phosphofructokinase?
    (a) ADP
    (b) ATP
    (c) Citrate
    (d) Hydrogen ions

17. Which compound regulates phosphofructokinase activity in response to substrate availability?
    (a) Fructose 1,6-bisphosphate
    (b) Fructose 2,6-bisphosphate
    (c) Glucose 6-phosphate
    (d) Glucose 1,6-bisphosphate

18. The glucagon receptor protein is an example of what?
    (a) G-protein coupled receptor
    (b) Integral membrane protein
    (c) Seven-transmembrane-helix protein
    (d) All of the above

19. Hexokinase is inhibited by which one of these compounds?
    (a) ADP
    (b) Glucose
    (c) Glucose 1-phosphate
    (d) Glucose 6-phosphate

20. Regulation of pyruvate kinase involves which one of the following?
    (a) Activation by ATP and inhibition by fructose 1,6-bisphosphate
    (b) Activation by fructose 1,6-bisphosphate and inhibition by ATP
    (c) Activation by both ATP and fructose 1,6-bisphosphate
    (d) Inhibition by both ATP and fructose 1,6-bisphosphate

Short answer questions

These questions do not require additional reading.

1. Giving examples, describe the biochemical role of activated carrier molecules.

2. Explain in detail how a single molecule of glucose can yield 38 molecules of ATP.

3. Draw an outline of the glycolytic pathway, showing the substrates, products and enzymes for each step.

4. Give a detailed description of those steps of glycolysis that either consume or synthesize ATP. Based on your description, explain why glycolysis results in a net gain of two ATPs per glucose molecule.

5. What is the role of the GLUT1 uniporter in glycolysis?

6. Describe the special features of glycolysis in (a) exercising muscle, and (b) yeast cells growing under anaerobic conditions.

7. Outline how fructose and galactose are entered into the glycolytic pathway.

8. Describe why conversion of fructose 6-phosphate to fructose 1,6-bisphosphate is the main control point in glycolysis.

9. Describe how substrate availability regulates phosphofructokinase activity.

10. Outline the signal transduction pathway that enables glucagon to influence the intracellular level of fructose 6-phosphate.

Self-study questions

These questions will require calculation, additional reading and/or internet research.

1. Although glycolysis, the TCA cycle and electron transport chain can yield 38 molecules of ATP per molecule of glucose, it has been estimated that most cells can only generate 30–32 ATPs per glucose. What might be the reason(s) for this discrepancy?

2. Identify which carbon atom(s) in pyruvate correspond to carbons 1 and 4 of the glucose molecule that entered the glycolysis pathway. What assumption must be made in order to answer this question?

3. Arsenate ions (AsO_4^{3-}) are able to replace phosphate in many biochemical reactions, including the one catalyzed by glyceraldehyde 3-phosphate dehydrogenase. The resulting compound is unstable and immediately breaks down to give 3-phosphoglycerate. Describe the impact that arsenate will have on energy generation during glycolysis.

4. Pyruvate kinase deficiency (PKD) is estimated to affect 51 per million Caucasians. Individuals with this disorder present a range of symptoms, with anemia usually the most prevalent. Without treatment, patients can experience severe and possibly lethal complications, but if the anemia is managed, most individuals enjoy relatively good health. Those patients with a more mild form of PKD may not have any symptoms at all. Bearing in mind the essential role that pyruvate kinase plays in glycolysis, explain why PKD is not lethal in all patients, and why there are severe and mild forms of the disorder.

5. From the information provided about the Cori cycle (*Section 8.2.2* and *Fig. 8.6*), predict what would happen if glycolysis and gluconeogenesis operated simultaneously in the same cell.

CHAPTER 9

Energy generation: the TCA cycle and electron transport chain

STUDY GOALS

After reading this chapter you will:

- understand how ATP is generated by the combined action of the TCA cycle and electron transport chain

- be able to describe how pyruvate is converted to acetyl CoA, and recognize that pyruvate must be transported into the mitochondrion before this conversion can take place

- know the steps of the TCA cycle including the substrates, products and enzymes for each reaction, and in particular be able to indicate the steps at which ATP, NADH and $FADH_2$ are generated

- appreciate that the pyruvate dehydrogenase complex is the main target for control of the TCA cycle, and be able to explain how acetyl CoA, NADH, ATP and pyruvate regulate the activity of this enzyme complex

- understand the importance of differences in redox potential in the context of electron transfer and be able to explain why oxidation of a single NADH and $FADH_2$ molecule can yield multiple ATP molecules

- know the components of the electron transport chain and be able to indicate the entry points for NADH and $FADH_2$

- appreciate the way in which proton pumping establishes an electrochemical gradient

- be able to describe how the electrochemical gradient is used to synthesize ATP, and in particular know the structure of the F_0F_1 ATPase and how the components of this structure work together in ATP synthesis

- understand the role that ADP availability plays in regulation of electron flow along the electron transport chain

- be able to give examples of inhibitors and uncouplers of electron transport

- know how mitochondrial shuttles enable the energy contained in the NADH molecules synthesized during glycolysis to be used in ATP synthesis

In the second stage of the energy-generation pathway, the pyruvate produced by glycolysis is broken down to carbon dioxide and water, and ATP is produced. This second stage can be divided into two parts. The first part comprises the TCA cycle, which completes the breakdown of pyruvate, releasing energy in the form of one molecule of ATP, three of NADH and one of $FADH_2$ for every starting molecule of pyruvate. The second part is the electron transport chain, which oxidizes the NADH and $FADH_2$ molecules, yielding additional ATPs.

9.1   The TCA cycle

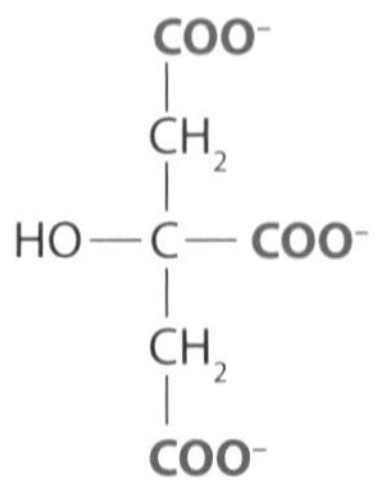

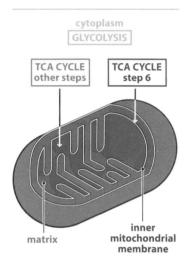

citrate

Figure 9.1   Citrate is a tricarboxylic acid.

The **tricarboxylic acid cycle** is so named because the first step in the pathway generates citric acid, which has three carboxyl groups and is therefore a tricarboxylic acid (*Fig. 9.1*). The process is also called the **citric acid cycle**, or sometimes the **Krebs cycle**, after Sir Hans Krebs, who was the first person to describe the cycle, in 1937.

To understand the TCA cycle we must ask three questions. First, how do the pyruvate molecules generated by glycolysis enter the cycle? Secondly, what reactions occur during the TCA cycle and what do they achieve, individually and collectively? Thirdly, how is the cycle regulated?

9.1.1   The entry of pyruvate into the TCA cycle

The end product of glycolysis is pyruvate, two molecules of this three-carbon sugar being generated for each molecule of six-carbon glucose that enters the pathway (see *Fig. 8.3*). Before entering the TCA cycle, these pyruvate molecules must be transported from one part of the cell to another.

The TCA cycle occurs inside mitochondria

The enzymes involved in the TCA cycle are located within the mitochondria of eukaryotic cells, most of them within the mitochondrial matrix, but with one enzyme, **succinate dehydrogenase**, attached to the inner mitochondrial membrane (*Fig. 9.2*). Glycolysis, on the other hand, takes place in the cytoplasm. This means that, before a pyruvate molecule can enter the TCA cycle, it must be transported from the cytoplasm to the inside of a mitochondrion.

Pyruvate carries a negative charge, which makes it impossible for the molecule to pass directly through the hydrophobic interior region of a membrane bilayer. For the outer mitochondrial membrane, the problem is solved by the presence of transmembrane proteins called **porins**, which have a barrel-like structure and form a channel across the membrane, through which charged molecules such as pyruvate can pass (*Fig. 9.3*). This occurs simply by diffusion.

Transport across the inner mitochondrial membrane is more difficult. The mitochondrial matrix is a specialized part of the cell carrying out important biochemical reactions that must be carefully regulated. The inner mitochondrial membrane is therefore a highly selective barrier that restricts access to the matrix to those specific compounds that are needed inside the mitochondrion. For this reason,

Figure 9.2   The cellular locations of glycolysis and the TCA cycle. Glycolysis takes place in the cytoplasm and the TCA cycle in the mitochondrion. Most of the steps of the TCA cycle occur in the mitochondrial matrix, but succinate dehydrogenase, which catalyzes step 6, is attached to the inner mitochondrial membrane.

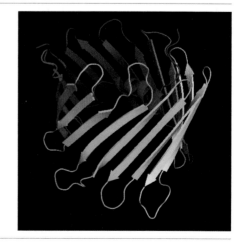

Figure 9.3   The structure of a porin. This is a porin from the outer mitochondrial membrane of human cells. It is made up of a barrel-like β-sheet which forms a pore through which metabolites such as pyruvate can pass. Reproduced from Wikipedia under a CC BY-SA 3.0 license; image by Plee579.

the inner mitochondrial membrane lacks porins, through which many different types of compound can freely pass, and instead contains more specialized transport proteins that import just one or a few individual compounds. The **mitochondrial pyruvate carrier** is one such protein. Only recently discovered, it appears to consist of multiple copies of two small protein subunits, each subunit being about 15 kDa in size. It is assumed that the assembled subunits form an integral membrane protein spanning the membrane, binding pyruvate on the outer surface and transporting it to the inner surface, where it is released into the mitochondrial matrix.

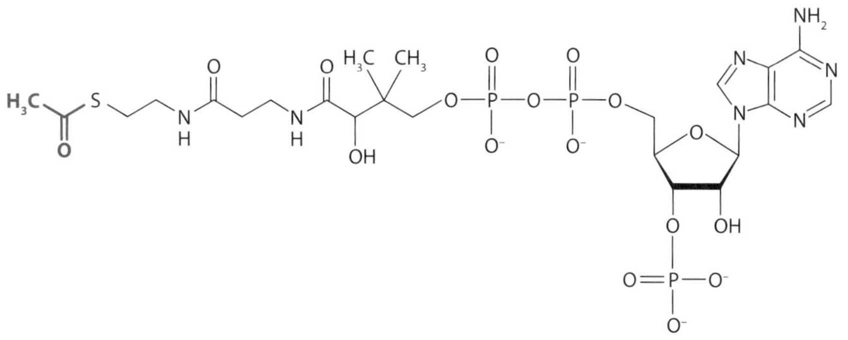

Figure 9.4 Acetyl CoA.
The acetyl group is highlighted in red.

Pyruvate is converted to acetyl CoA prior to entry into the TCA cycle

Once inside the mitochondrial matrix, pyruvate is converted into **acetyl CoA** (*Fig. 9.4*). Acetyl CoA comprises an acetyl group (CH_3CO-) attached to the enzyme cofactor called coenzyme A, which is derived from pantothenic acid (vitamin B_5). The reaction yields one molecule of carbon dioxide and converts one molecule of NAD^+ into NADH.

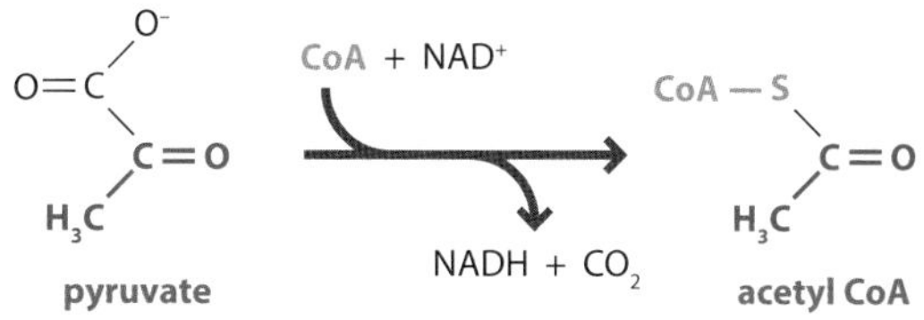

The production of NADH means that this is one of the important steps of the energy-generating process, during which energy is captured by a carrier molecule. When transferred to the electron transport chain, the molecule of NADH yields three ATPs, which equates to six ATPs for every starting molecule of glucose. This compares to the net eight ATPs generated per glucose for the whole of glycolysis.

The conversion of pyruvate to acetyl CoA is catalyzed by the **pyruvate dehydrogenase complex**, which is made up of three different enzymes that work together to bring about this complicated biochemical reaction. The roles of the three component enzymes are as follows:

- **Pyruvate dehydrogenase** binds pyruvate and converts it into acetate, with the release of carbon dioxide.

- **Dihydrolipoyl transacetylase** collects the acetyl from pyruvate dehydrogenase and attaches it to coenzyme A, forming acetyl CoA.

- **Dihydrolipoyl dehydrogenase** uses the pair of electrons, released during conversion of pyruvate to acetate, to generate the molecule of NADH.

The reaction is therefore a type of **oxidative decarboxylation**, the pyruvate substrate being oxidized (it loses a pair of electrons) and decarboxylated (loss of CO_2).

Box 9.1 Identification of the mitochondrial pyruvate carrier protein

The transport of pyruvate into the mitochondrion is a key stage in the energy-generation pathway for eukaryotic cells, being the link between the glycolysis reactions, which take place in the cytoplasm, and the TCA cycle, which occurs inside the mitochondria. With pyruvate transport being so important, it is perhaps surprising that the carrier protein was not identified until 2012, some 40 years after the search for it began.

The first question that was tackled, back in the 1970s, is whether pyruvate entry into the mitochondrion does actually require a specific transport protein. The undissociated form of pyruvic acid is able to diffuse across the membrane unaided, and at first it was thought that a carrier protein would be unnecessary. It was then shown that most of the pyruvate in the cell is in the ionic form, and hence has a net positive charge that prevents simple diffusion through a membrane. Studies of the kinetics of pyruvate uptake also suggested that transport is mediated by a carrier protein. The point was then proven by the discovery that a compound called α-cyano-4-hydroxycinnamate, which is structurally similar to pyruvate, blocks pyruvate transport into mitochondria. The effect of this inhibitor is a strong indication that pyruvate is not simply diffusing across the membrane and must be transported by a carrier protein of some kind.

pyruvic acid (undissociated)
can pass through inner
mitochondrial membrane

pyruvate (ionic)
cannot pass through inner
mitochondrial membrane

α-cyano-4-hydroxycinnamate
specific inhibitor of the
pyruvate carrier protein

During the 1980s, progress towards identifying the pyruvate carrier was achieved using liposome reconstitution assays. In this technique, partially purified proteins are mixed with lipids to form small vesicles called **liposomes**, which are made up of a lipid bilayer enclosing a small internal aqueous compartment. The biochemical activities of the liposomes are then studied in order to deduce the functions of the proteins. Liposomes able to take up pyruvate were reconstituted from bovine heart tissue and castor beans, but in both cases the initial protein mixture was complex, with at least six major components in the case of the castor bean preparation. With the methods available at the time, these initial results could not be followed up.

Subsequent attempts to identify the mitochondrial pyruvate carrier were hindered by what turned out to be erroneous assumptions about its likely structure. The plasma membrane also has a pyruvate carrier, for import of pyruvate into the cell, and this carrier protein is also sensitive to α-cyano-4-hydroxycinnamate inhibition. This might indicate that the plasma membrane and mitochondrial carriers are similar

proteins. Additional research on the plasma version then showed that the plasma membrane carrier is inhibited by other compounds that have no effect on the mitochondrial protein. The two must therefore be distinct and any knowledge obtained about the plasma protein would be of little use in identifying the mitochondrial carrier.

Eventually, in 2012, two research groups independently identified the mitochondrial pyruvate carrier. In both cases, the discovery was made by serendipity, neither group having set out specifically to find the elusive carrier protein. In one project, mitochondrial proteins of unknown function but with similar structures in humans, yeast and fruit flies were identified. The rationale was that any protein that has a similar structure in such different species must have an important biochemical role. Two of these proteins, subsequently called Mpc1 and Mpc2, were shown to be involved in the conversion of cytoplasmic pyruvate into acetyl CoA, which requires transport of the pyruvate into the mitochondrion. A key result was the demonstration that a mutation in Mpc1, which changes an aspartic acid into a glycine, produces a yeast strain that is resistant to inhibitors related to α-cyano-4-hydroxycinnamate.

Identification of the mitochondrial pyruvate carrier protein is important not just for our academic understanding of the link between glycolysis and the TCA cycle. The discovery also has profound implications for research into cancer. Back in 1927, Otto Warburg observed that most types of cancerous cell obtain their energy predominantly by glycolysis, the resulting pyruvate being converted to lactate in the cytoplasm, rather than being transported into the mitochondrion and being metabolized by the TCA cycle and electron transport chain.

The role of the pyruvate carrier in this 'Warburg effect' can now be studied much more directly than was possible before the structure of the protein was known. Understanding the role of the carrier in this key metabolic change might shed new light on the biochemical basis to cancer formation, and aid the search for therapies that prevent cancers developing.

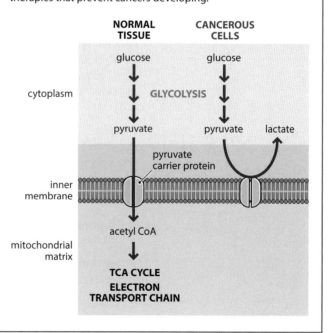

9.1.2   The steps of the TCA cycle

The TCA cycle, as its name implies, is a circular pathway, with one of the substrates, **oxaloacetate**, regenerated at the end of each round of the cycle (*Fig. 9.5*). The cycle comprises eight steps.

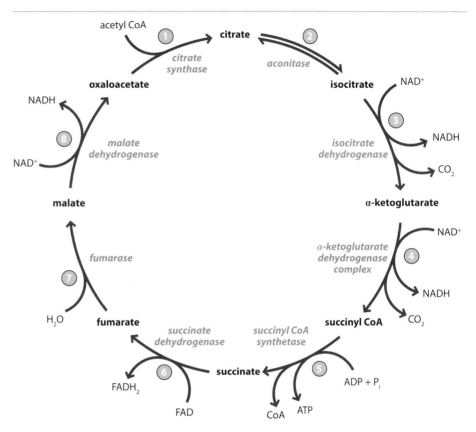

Figure 9.5 The TCA cycle.
The steps in the process, as described in the text, are indicated by the green circles.

Step 1. The acetate group carried by acetyl CoA, derived from pyruvate, is transferred to the four-carbon dicarboxylic acid called oxaloacetate. This produces a molecule of citrate, which is a six-carbon tricarboxylic acid. The reaction is catalyzed by **citrate synthase.**

Step 2. An isomerization reaction, catalyzed by **aconitase**, converts citrate to isocitrate.

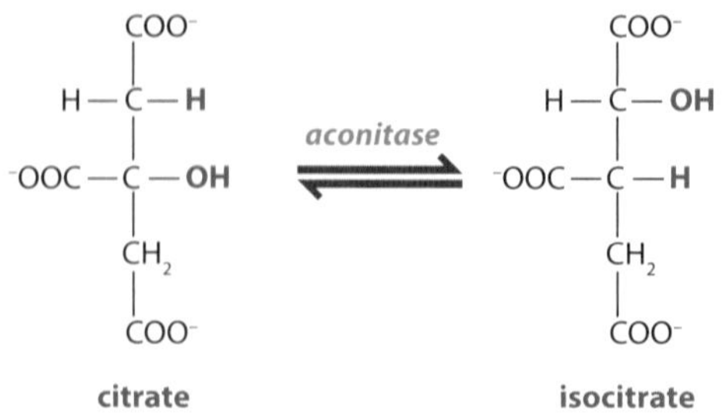

Step 3. **Isocitrate dehydrogenase** oxidizes isocitrate to α-ketoglutarate. This reaction releases carbon dioxide and enables a molecule of NAD^+ to be converted to NADH. It is therefore another example of an oxidative decarboxylation.

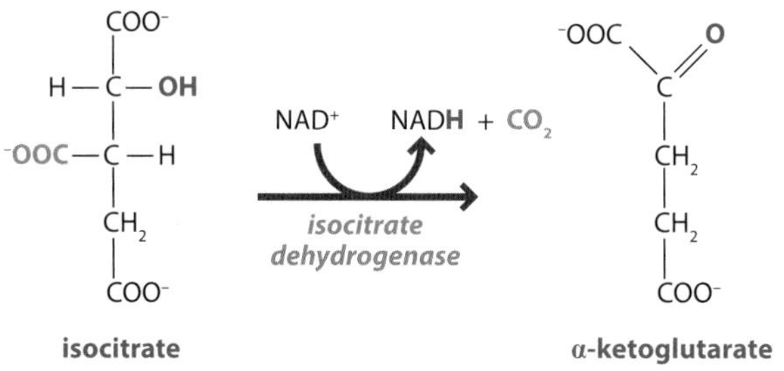

Step 4. Another NADH is generated by the oxidative decarboxylation of α-ketoglutarate to succinyl CoA, catalyzed by the **α-ketoglutarate dehydrogenase complex**, another combination of three different enzymes that work together to bring about a single biochemical reaction.

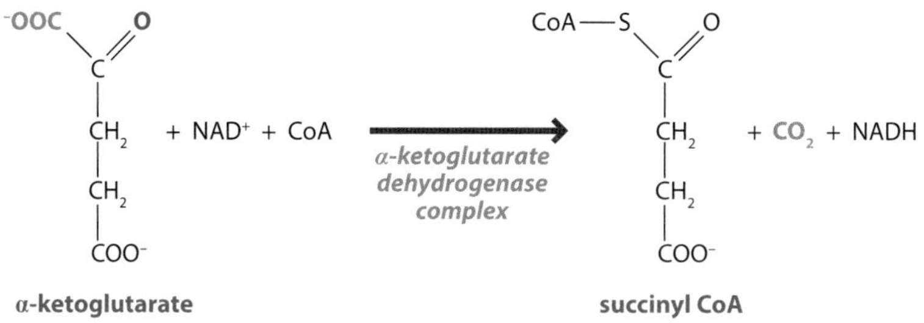

Step 5. **Succinyl CoA synthetase** converts succinyl CoA to succinate. The name of this enzyme indicates that it can also carry out the reverse reaction, in which succinyl CoA is synthesized from succinate. In the TCA cycle, the enzyme breaks the succinate–CoA linkage, releasing sufficient energy to phosphorylate a molecule of ADP, giving ATP.

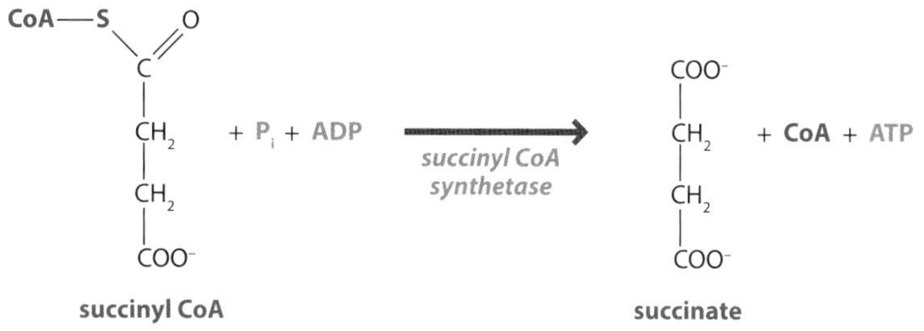

succinyl CoA + P$_i$ + ADP $\xrightarrow{\text{succinyl CoA synthetase}}$ **succinate** + **CoA** + **ATP**

The reaction can also generate GTP from GDP and hence is one way of producing this second type of nucleotide energy carrier. Because this is the second decarboxylation reaction of the TCA cycle, we now have a four-carbon compound. This means that the original pyruvate has been completely broken down, with all three carbons released as CO$_2$, one at the pyruvate dehydrogenase step, one by the isocitrate dehydrogenase reaction and the final one by the α-ketoglutarate dehydrogenase reaction. However, some of the energy released by pyruvate breakdown is still stored within the succinate molecule. This energy is utilized in the remaining steps of the cycle.

Step 6. **Succinate dehydrogenase** oxidizes succinate to fumarate, with the conversion of FAD to FADH$_2$.

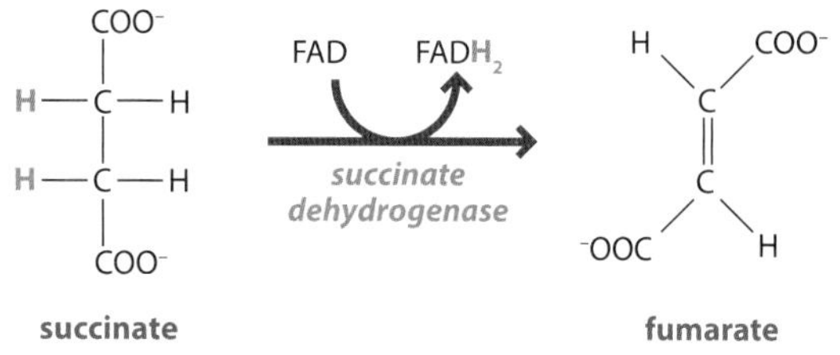

succinate $\xrightarrow[\text{dehydrogenase}]{\text{succinate}}$ fumarate (FAD → FADH$_2$)

Step 7. **Fumarase** converts fumarate to malate by a hydration reaction that requires the addition of a water molecule.

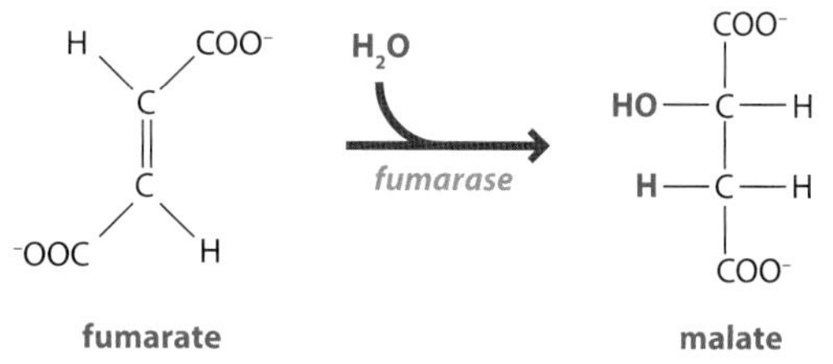

fumarate $\xrightarrow[\text{fumarase}]{\text{H}_2\text{O}}$ malate

Step 8. **Malate dehydrogenase** oxidizes malate to produce oxaloacetate, yielding another molecule of NADH.

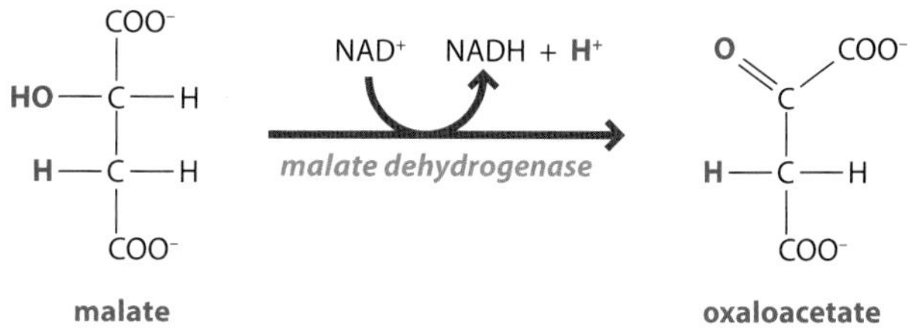

malate $\xrightarrow[\text{malate dehydrogenase}]{\text{NAD}^+ \rightarrow \text{NADH} + \text{H}^+}$ oxaloacetate

The last step regenerates oxaloacetate which was used at the start of the cycle. The cycle can therefore proceed into another round with a second molecule of acetyl CoA. The cycle has produced one molecule of ATP or GTP, as well as three NADH and one $FADH_2$. The electron transport chain can generate three additional ATPs from each NADH, and two ATPs from the $FADH_2$.

As well as its role in energy generation, the TCA cycle is also an important starting point for many biosynthetic pathways:

- Oxaloacetate is a starting point for the production of aspartate, other amino acids, purines and pyrimidines.

- Citrate is used as a source of acetyl CoA for fatty acid synthesis.

- α-Ketoglutarate is a substrate for glutamate, other amino acids and purines.

- Succinyl CoA is a starting point for the production of porphyrins such as heme and chlorophyll.

We will study these biosynthetic pathways later in book. See *Section 12.1.1* for fatty acid synthesis, *13.2.1* for amino acid synthesis and *13.2.3* for tetrapyrrole synthesis.

Box 9.2 Succinyl CoA synthetases

The substrate-level phosphorylation catalyzed by succinyl CoA synthetase is able to use both ADP and GDP as substrates. In most cells, both ATP and GTP are synthesized, but the relative amounts differ. In muscles, ATP is produced in large excess over GTP, but in liver cells GTP is synthesized in significantly larger amounts than ATP. It has been suggested that the balance of ATP to GTP synthesis depends on the overall metabolic activity of the cell. In muscles, especially during exercise, there is a high energy requirement, and in these tissues succinyl CoA synthetase helps to meet this need by synthesizing ATP. In liver, on the other hand, there is a smaller energy requirement which can be met without the additional two ATPs per glucose that can be produced by succinyl CoA synthetase. Instead, the enzyme uses GDP as the substrate, generating GTP, which is required as an energy supply for protein synthesis (*Section 16.2.2*).

The two activities of succinyl CoA synthetase are specified by two closely related but distinct enzymes. Both are called succinyl CoA synthetase because both catalyze the same biochemical reaction – the conversion of succinyl CoA to succinate accompanied by substrate-level phosphorylation of a nucleotide diphosphate. They have slightly different amino acid sequences which account for their different substrate specificities. The two versions of succinyl CoA synthetase are called **isozymes**. Many enzymes exist as isozymes, with different members of an isozyme family active in different tissues (as is the case with succinyl CoA synthetase) or at different developmental stages.

9.1.3   Regulation of the TCA cycle

Not surprisingly, in view of its pivotal role at the entry point, the pyruvate dehydrogenase complex is the main target for regulation of the TCA cycle. This enzyme complex is inhibited by its immediate products – acetyl CoA and NADH – and also by ATP. These compounds do not, however, exert their influence by allosteric inhibition. Instead, their effect is mediated by another pair of enzymes, **pyruvate dehydrogenase kinase** and **pyruvate dehydrogenase phosphatase**. These enzymes, respectively, add or remove a phosphate group to or from each of three serine amino acids in the pyruvate dehydrogenase enzyme. The phosphorylated version of the enzyme is inactive. Acetyl CoA, NADH and ATP stimulate the kinase, increasing the rate at which it phosphorylates and hence inactivates pyruvate dehydrogenase (*Fig. 9.6*).

The presence of pyruvate, on the other hand, increases the activity of pyruvate dehydrogenase. This is achieved by pyruvate inhibiting the kinase, allowing the phosphatase to dephosphorylate and hence activate pyruvate dehydrogenase. Phosphoenolpyruvate, one of the intermediates in glycolysis, also has a stimulatory effect on pyruvate dehydrogenase, but in this case by activating the phosphatase.

Figure 9.6 Control of pyruvate dehydrogenase by acetyl CoA, NADH, ATP and pyruvate.
Acetyl CoA, NADH and ATP stimulate the activity of pyruvate dehydrogenase kinase, which phosphorylates and thereby inactivates pyruvate dehydrogenase. Pyruvate inhibits the kinase.

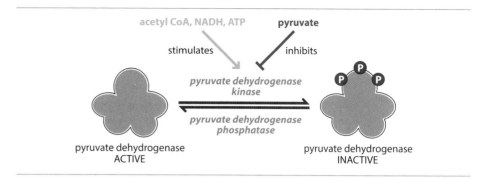

Within the TCA cycle itself there are three additional points at which products exert a feedback inhibition on the enzymes responsible for their synthesis (*Fig. 9.7*):

- Citrate synthase is inhibited by citrate and ATP.
- Isocitrate dehydrogenase is inhibited by NADH and ATP.
- α-Ketoglutarate dehydrogenase is inhibited by succinyl CoA and NADH.

The overall effect of the various regulatory processes is that the TCA cycle slows down when the cell has an adequate supply of stored energy, signaled by the accumulation of ATP and NADH, which inhibit the entry of acetyl CoA into the cycle and the progression of the cycle past the other three control points.

Figure 9.7 Control points within the TCA cycle.

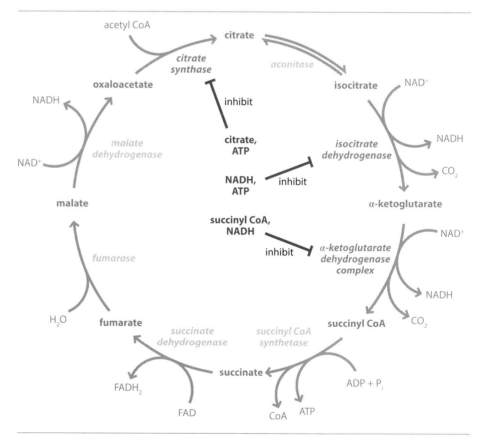

9.2   The electron transport chain and the synthesis of ATP

The complete breakdown of a single molecule of glucose via glycolysis and the TCA cycle yields sufficient energy to synthesize four ATP molecules, ten NADH and two FADH$_2$. The electron transport chain enables a further 34 ATPs to be generated from the energy that has been temporarily transferred to the NADH and FADH$_2$ molecules. This is achieved as follows:

- Eight of the NADH molecules are already located in the mitochondrion and these are used to produce 24 ATPs – three for each NADH.

- The two FADH$_2$ molecules, which are also present in the mitochondrion, generate an additional four ATPs – two for each FADH$_2$.

- Finally, there are two NADH molecules, originally produced by glycolysis, which are located in the cytoplasm. Because of the impermeability of the inner mitochondrial membrane, these two NADH molecules have to remain in the cytoplasm, but they are still able to generate six ATPs (three per NADH) by a process that makes indirect use of the electron transport chain.

We will now examine these three different ways in which ATP is synthesized by the electron transport chain – from NADH in the mitochondrion, from FADH$_2$, and from NADH in the cytoplasm. First, however, we will begin with our usual overview of the process.

9.2.1   Energy is released as electrons are passed along the electron transport chain

To understand how the electron transport chain generates ATP, we must first consider the energetics of the underlying biochemical reactions.

Oxidation of NADH and FADH$_2$ yields sufficient energy to make several ATP molecules

The chemical formula for the conversion of NADH to NAD$^+$ is:

$$NADH + H^+ + \tfrac{1}{2}O_2 \rightarrow NAD^+ + H_2O$$

This is an oxidation–reduction reaction: NADH is oxidized to NAD$^+$ and oxygen is reduced to water. The reaction therefore involves the transfer of two electrons from NADH to oxygen. This transfer can occur because NADH has a lower affinity for electrons than does oxygen, so the electrons move from NADH, the electron donor, to oxygen, the electron acceptor. During electron transfer from a donor to an acceptor, energy is released, the amount depending on the difference between the electron affinities of the donor and acceptor molecules.

We are already familiar with the use of the term ΔG to indicate the change in free energy that occurs during an enzyme-catalyzed reaction. When comparing different reactions, we measure ΔG under standard conditions, with each reactant present at equimolar amounts and the pH set at 7.0. This version of ΔG is called the **standard free energy change** or $\Delta G^{0\prime}$. For the oxidation of NADH, the value of $\Delta G^{0\prime}$ is $-220.2\,\mathrm{kJ\,mol^{-1}}$ (*Fig. 9.8*). Remember that a negative ΔG indicates an energy yield. The equivalent reaction for FADH$_2$, giving rise to FAD, has a $\Delta G^{0\prime}$ of $-181.7\,\mathrm{kJ\,mol^{-1}}$.

The $\Delta G^{0\prime}$ values for NADH and FADH$_2$ indicate the amounts of energy that we have available to make ATP molecules. How much do we actually need? The formula for the synthesis of ATP from ADP and inorganic phosphate is:

$$ADP^{3-} + HPO_4^{2-} + H^+ \rightarrow ATP^{4-} + H_2O$$

PRINCIPLES OF CHEMISTRY

Box 9.3 Redox potential

The affinity of a compound such as NADH for electrons is described by its **redox potential**, which is measured relative to the redox potential of hydrogen:
- A positive redox potential means that a compound has a higher affinity for electrons than hydrogen and so accepts electrons from hydrogen.
- A negative redox potential means the compound has a lower affinity for electrons than hydrogen and so donates electrons to H^+ ions, forming hydrogen.

In the reaction:

$$NADH + H^+ + \tfrac{1}{2}O_2 \rightarrow NAD^+ + H_2O$$

NADH is a strong reducing agent with a negative redox potential and has a tendency to donate electrons. Oxygen is a strong oxidizing agent with a positive redox potential and has a tendency to accept electrons. The electron transfer is therefore from NADH to oxygen.

The **standard redox potential (E_0')** of a compound is measured under standard conditions at pH 7 and expressed in volts. The standard free energy change of a reaction at pH 7 ($\Delta G^{0'}$) can be calculated from the change in redox potential ($\Delta E_0'$) of the substrates and products:

$$\Delta G^{0'} = -nF \, \Delta E_0'$$

where n is the number of electrons transferred and F is the Faraday constant ($96.485 \text{ kJ V}^{-1} \text{mol}^{-1}$). Because of the minus sign on the right hand side of this equation, a reaction with a positive $\Delta E_0'$ will have a negative $\Delta G^{0'}$, and so will be exergonic. For the transfer of electrons from NADH to oxygen, the value for $\Delta E_0'$ is 1.14 V, and $\Delta G^{0'}$ is $-220.2 \text{ kJ mol}^{-1}$.

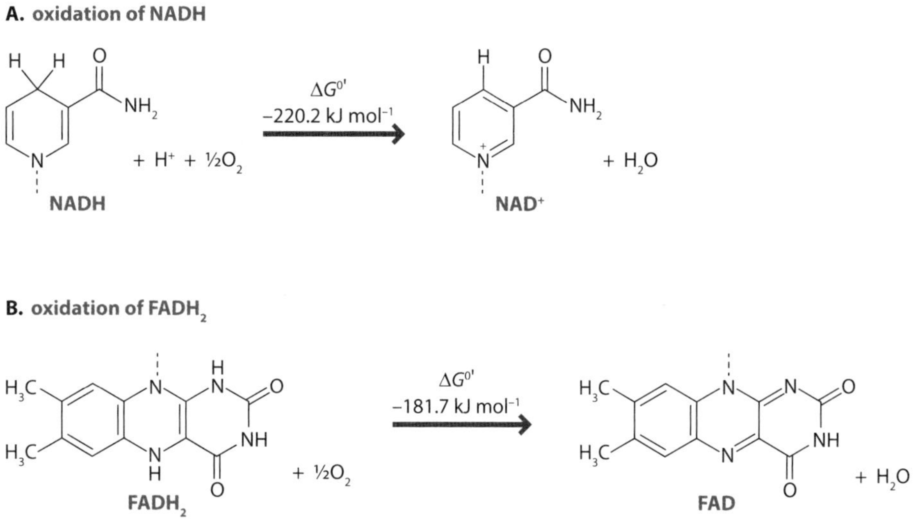

Figure 9.8 Oxidation of (A) NADH and (B) FADH$_2$.
See *Figure 7.7* for the complete structures of these compounds.

A. oxidation of NADH

$$\Delta G^{0'}$$
$$-220.2 \text{ kJ mol}^{-1}$$

NADH $+ H^+ + \tfrac{1}{2}O_2$ → NAD$^+$ $+ H_2O$

B. oxidation of FADH$_2$

$$\Delta G^{0'}$$
$$-181.7 \text{ kJ mol}^{-1}$$

FADH$_2$ $+ \tfrac{1}{2}O_2$ → FAD $+ H_2O$

This reaction is energy-requiring and therefore has a positive $\Delta G^{0'}$ of 30.6 kJ mol^{-1}.

The energy released by oxidation of NADH or FADH$_2$ is significantly larger than 30.6 kJ mol^{-1} and is therefore sufficient to synthesize multiple ATP molecules. In practice, not all of this energy can be utilized, some being lost as heat simply because of the inefficiency of the transfer. On average, oxidation of one NADH molecule can generate up to three ATPs and one FADH$_2$ can yield two ATPs. The linked reactions, in which oxidation of NADH or FADH$_2$ drives synthesis of ATP, are called **oxidative phosphorylations**.

The electron transfer from NADH or FADH$_2$ to oxygen occurs via intermediates

If, during the oxidation of NADH or FADH$_2$, the electrons were transferred directly to oxygen, then all of the available energy would be released in one burst. Some of this could be captured to drive synthesis of ATP, but it is likely that a large proportion would be lost. The transfer could become so inefficient that it might not be possible to make even a single ATP from the oxidation of one molecule of NADH or FADH$_2$.

To improve the efficiency, the transfer of electrons does not occur directly to oxygen, and this is where the electron transport chain comes in. The chain is made up of a series of compounds, arranged in order of increasing electron affinities as we move along the chain (*Fig. 9.9*). Each step in the transfer of electrons along the chain therefore releases a small quantum of energy, until at the end of the chain the electrons are donated to oxygen and the oxidation process is complete. It is the controlled, incremental release of energy along the electron transport chain that enables the oxidation of NADH and $FADH_2$ to drive synthesis of ATP.

Figure 9.9 The principle of the electron transport chain.
The chain is made up of a series of compounds arranged in order of increasing electron affinities as we move along the chain. Each step in the transfer of electrons along the chain releases a small quantum of energy.

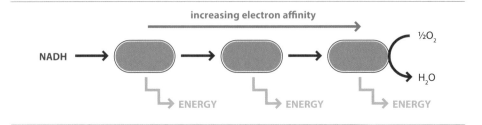

9.2.2   The structure and function of the electron transport chain

Now we must examine the components of the electron transport chain and understand exactly how electrons pass along the chain from an NADH or $FADH_2$ donor.

The components of the electron transport chain

The electron transport chain comprises four large structures, called Complexes I–IV, located in the inner mitochondrial membrane. Electrons from NADH enter the chain via Complex I, and those from $FADH_2$ enter Complex II (*Fig. 9.10*). Electron transfer from Complex I or II to Complex III requires an intermediate carrier molecule called **ubiquinone** or **coenzyme Q (CoQ)**, and transfer from Complex III to Complex IV is via **cytochrome *c***. From Complex IV, electrons are transferred to oxygen.

Figure 9.10 The structure of the electron transport chain.
The two intermediate carriers are coenzyme Q (CoQ) and cytochrome *c* (cytc).

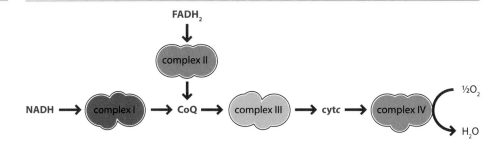

Each of the four complexes is a multisubunit protein, ranging from four subunits for Complex II to 44 for Complex I. But the proteins themselves are not the important parts of the complexes as far as the electron transfer is concerned. Some amino acids can act as electron acceptors and donors, but within polypeptide chains these amino acids lack the range of electron affinities needed to achieve the incremental energy release that underlies the electron transport chain. Instead, the electron binding components of the complexes are non-protein prosthetic groups. The identities of these are as follows:

- Complex I contains flavin mononucleotide (FMN) and eight **iron–sulfur** or **FeS clusters**. The latter consist of iron atoms coordinated with inorganic sulfur atoms and with the sulfur of a cysteine side-chain within one of the polypeptide subunits of the complex (*Fig. 9.11*). A polypeptide that contains an iron–sulfur cluster is called an **iron–sulfur** or **FeS protein**.

Box 9.4 **The location of the electron transport chain**

In eukaryotes the four complexes of the electron transport chain are located in the inner mitochondrial membrane. Remember that a membrane is a fluid mosaic of lipids and proteins (see *Section 5.2.1*). This means that if the four complexes and the intermediate carriers were independent structures then they would float around in the membrane, rather than being held next to one another in any particular order. This would mean that electron transfer could only occur when the members of a donor–acceptor pair (e.g. Complex I and CoQ) came into chance contact. This might place limitations on the efficiency of electron transfer. Even if the membrane has a high concentration of the components of the electron transport chain, chance collisions between the necessary pairs might be relatively uncommon.

The gradual realization that groups of membrane proteins that work together are often co-located, for example on lipid rafts, has led to a new model for the electron transport chain. Now we believe that Complexes I, III and IV (the ones required from NADH oxidation) and the intermediate carriers CoQ and cytochrome *c* are assembled together into structures called supercomplexes or **respirasomes**. Within a respirasome, the complexes are placed in a specific orientation, with electron transport between pairs of complexes mediated by the movement of CoQ and cytochrome *c* within the respirasome, as shown in the diagram on the right.

It is not yet clear if Complex II (the entry point for FADH$_2$) is also located with Complexes III and IV in a respirasome. There is evidence for a stable association between Complex II and III in yeast, but the evidence is not conclusive and has not yet been reproduced with mammalian cells.

Many prokaryotes also generate energy via the TCA cycle and an electron transport chain. The components of the complexes of the electron transport chain are slightly different to those in eukaryotes, and species that are obligate anaerobes use a compound other than oxygen as the terminal electron acceptor (see *Box 8.3*). Prokaryotes do not have mitochondria so where are these pathways located? The enzymes of the TCA cycle are present in the cytoplasm of a prokaryotic cell, and the components of the electron transport chain in the plasma membrane. The F$_0$F$_1$ ATPase, which synthesizes ATP, is also located in the plasma membrane, with protons flowing through it from the external side of the membrane back into the cell. As in eukaryotes, the individual components of a prokaryotic electron transport chain are thought to assemble into respirasomes.

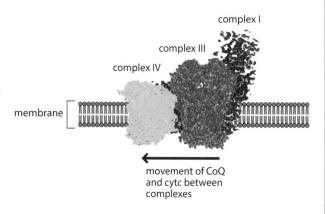

Image of complexes reproduced from *PNAS* 2011; 108(37): 15196 with permission.

- Complex II includes one FeS protein with three FeS clusters.

- Complex III has a single FeS protein, as well as three **cytochromes**. A cytochrome is a protein that contains one or more heme prosthetic groups. There are various types of cytochrome depending on the amino acid sequence of the polypeptide and the precise structure of the heme group. Those present in Complex III are called cytochrome b_{562} and cytochrome b_{566}, both of which have two heme groups, and cytochrome c_1 which has one heme.

Figure 9.11 Iron–sulfur clusters. The complexes of the electron transport chain contain both (A) 2Fe–2S and (B) 4Fe–4S clusters.

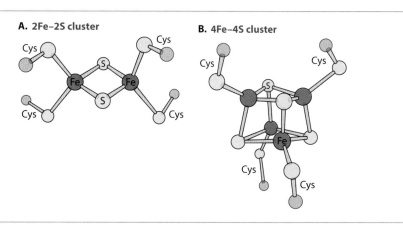

- Complex IV contains two other cytochromes, cytochrome a and cytochrome a_3, the former associated with two copper ions, which are denoted Cu_A, and the latter with a third copper ion, Cu_B.

The transfer of electrons along the electron transport chain

We have seen that the electron transport chain consists of four complexes with CoQ mediating electron transfer between Complexes I–III and II–III, and cytochrome c mediating transfer between III and IV (see *Fig. 9.10*). Now we will look at the detail within this pathway, first for NADH. We can divide the process into three steps, each corresponding to a different complex (*Fig. 9.12*).

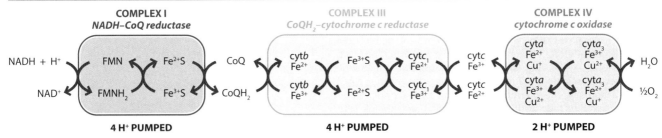

Figure 9.12 The electron transfers occurring within the electron transport chain.
Some steps are combined in order to reduce the complexity of the diagram. The cytochrome b cycle in Complex III in fact involves two cytochromes, cytochrome b_{562} and cytochrome b_{566}, and in Complex IV the electrons transfer from the Cu ion of cytochrome a to the Fe ion of this cytochrome, and then to the Cu and then Fe ions of cytochrome a_3.

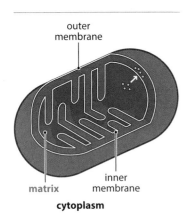

outer membrane

inner membrane

matrix

cytoplasm

Figure 9.13 Pumping of protons across the inner mitochondrial membrane.
Four H^+ ions, depicted as yellow dots, are transferred from the mitochondrial matrix to the space between the inner and outer membranes.

Step 1. The **NADH–CoQ reductase complex** (Complex I). NADH binds to Complex I, releasing two electrons which are passed to FMN, reducing FMN to $FMNH_2$. Each electron transfer is also accompanied by uptake of an H^+ ion from the aqueous mitochondrial matrix. The electrons are then passed to the FeS clusters in Complex I, reducing the Fe^{3+} (ferric) ions in these clusters to Fe^{2+} (ferrous) ions. At the same time, FMN is regenerated from $FMNH_2$. Finally, the electrons are passed from the FeS clusters to CoQ, yielding the reduced form of this compound, called **ubiquinol** (**$CoQH_2$**). The overall reaction for step 1 is therefore the transfer of one H^+ ion from NADH and one from the matrix to CoQ:

$$NADH + CoQ + H^+ \rightarrow NAD^+ + CoQH_2$$

This reaction has a $\Delta G^{0\prime}$ of $-69.5\,kJ\,mol^{-1}$, which is used to transfer four H^+ ions across the inner mitochondrial membrane, from the mitochondrial matrix to the space between the two mitochondrial membranes (*Fig. 9.13*). Complex I therefore acts as a **proton pump**.

Step 2. The **$CoQH_2$–cytochrome c reductase complex** (Complex III). $CoQH_2$ is lipid soluble and so can move from one site to another in the inner mitochondrial membrane. It can therefore transfer the electrons it is carrying to Complex III, the next component in the chain. Within Complex III the electrons are first transferred to the heme groups of cytochrome b_{562}, reducing the Fe^{3+} ion contained in the heme to Fe^{2+} (see *Fig. 9.12*). The electrons then flow to cytochrome b_{566}, then to the FeS cluster of Complex III, on to the heme group of cytochrome c_1, and finally to cytochrome c. We can therefore describe the overall reaction as:

$$CoQH_2 + 2cytc^{3+} \rightarrow CoQ + 2H^+ + 2cytc^{2+}$$

where $cytc^{3+}$ and $cytc^{2+}$ are the oxidized and reduced forms of cytochrome c, respectively. The energy released in step 2 is sufficient to pump four more H^+ ions across the inner mitochondrial membrane.

Step 3. The **cytochrome c oxidase complex** (Complex IV). Cytochrome c forms the link between Complexes III and IV. It transfers electrons initially to the Cu_A ions, reducing these from Cu^{2+} (cupric) to Cu^+ (cuprous) (see *Fig. 9.12*). The electrons then move via the heme of cytochrome a to the Cu_B ion, then to the heme of cytochrome a_3, and finally to oxygen, giving H_2O. The overall reaction is:

$$2cytc^{2+} + 2H^+ + \tfrac{1}{2}O_2 \rightarrow 2cytc^{3+} + H_2O$$

This reaction, which completes the oxidation of the original NADH molecule, releases more energy, sufficient to move two H^+ ions across the inner mitochondrial membrane.

All that remains is to return to the start of the electron transport chain and see how $FADH_2$ enters the pathway. We will call this step 1*.

Step 1.* The **succinate–CoQ reductase complex** (Complex II). At step 6 of the TCA cycle (see *Fig. 9.5*), succinate is oxidized to fumarate by succinate dehydrogenase. This enzyme is located in the inner mitochondrial membrane, closely linked to Complex II. The two electrons released by oxidation of succinate generate the $FADH_2$ molecule which immediately transfers its electrons to the FeS clusters of Complex II (*Fig. 9.14*). From the clusters the electrons pass to CoQ and then on to Complexes III and IV, exactly as during the pathway for NADH. Because of the close linkage between succinate dehydrogenase and Complex II, we usually denote the overall reaction for entry of $FADH_2$ into the electron transport chain as:

$$succinate + CoQ \rightarrow fumarate + CoQH_2$$

The reaction releases energy but not enough to transfer any H^+ ions across the inner mitochondrial membrane. This inability of Complex II to pump electrons across the membrane accounts for the difference in the number of ATPs that can be generated per molecule of $FADH_2$ compared to NADH.

Figure 9.14 The transfer of electrons from $FADH_2$ to Complex II of the electron transport chain.

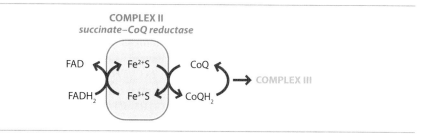

9.2.3   Synthesis of ATP

We have seen how electrons pass along the electron transport chain, with an incremental release of energy at each of the four complexes, and how the energy that is released is used to pump protons across the inner mitochondrial membrane. Now we must examine how this process gives rise to the synthesis of ATP molecules.

Proton pumping generates an electrochemical gradient

For many years, the way in which the transfer of electrons along the electron transport chain results in the synthesis of ATP was a mystery. Biochemists thought that the energy released by electron transfer must be stored in a high energy compound of some description, before being used to make ATP molecules. The problem was that efforts to detect this high energy compound were unsuccessful, even though it should be present in mitochondria in reasonably abundant amounts. Then, in 1961, Peter Mitchell proposed a radical new idea called the **chemiosmotic theory**. Mitchell proposed that the pumping of protons across the inner mitochondrial membrane creates an **electrochemical gradient** that drives synthesis of ATP.

When it was first proposed the chemiosmotic theory was highly controversial, not least because at that time it was not even known that H⁺ ions are indeed transferred across the inner mitochondrial membrane during the passage of electrons through Complexes I, III and IV of the electron transport chain. Now we recognize that the theory is indeed the correct answer to the puzzle of ATP synthesis, and Mitchell is looked on as one of the giants of twentieth century biochemistry.

An electrochemical gradient is formed when charged chemical ions are distributed unevenly, with a greater concentration at one location compared to a second location. The uneven distribution means that there is a charge differential between the two locations. As a consequence of the proton pumping that occurs along the electron transport chain, an electrochemical gradient is set up across the inner mitochondrial membrane (*Fig. 9.15A*). Proton pumping results in a greater concentration of H⁺ ions in the space between the inner and outer membranes compared to within the mitochondrial matrix.

There is a natural tendency for electrochemical gradients to revert to an equilibrium state, in our case this would mean the surplus protons in the space between the membranes moving back into the mitochondrial matrix, so that the charge becomes the same on either side of the inner mitochondrial membrane. To re-enter the mitochondrial matrix, protons pass through a multisubunit protein called the **F₀F₁ ATPase** (*Fig. 9.15B*). The passage of protons through the ATPase results in the synthesis of ATP. As electrons continue to pass down the electron transport chain from NADH and FADH₂, more and more protons are pumped out of the mitochondrial matrix, maintaining the gradient and hence maintaining the flow of protons back into the matrix via the ATPase. The ATPase therefore makes ATP in direct response to the oxidation of NADH and FADH₂.

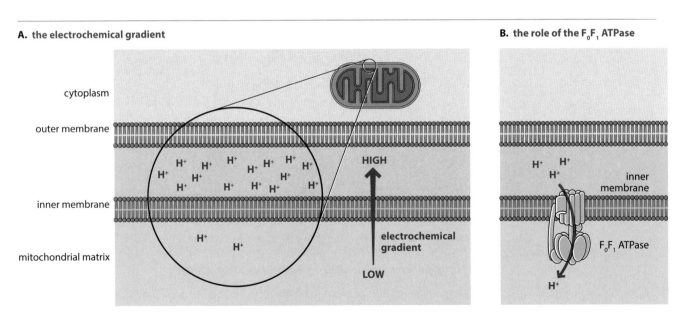

A. the electrochemical gradient

cytoplasm

outer membrane

HIGH

electrochemical gradient

LOW

inner membrane

mitochondrial matrix

B. the role of the F₀F₁ ATPase

inner membrane

F₀F₁ ATPase

Figure 9.15 Proton pumping sets up an electrochemical gradient across the inner mitochondrial membrane.
(A) The electrochemical gradient. (B) Protons re-enter the mitochondrial matrix via the F₀F₁ ATPase.

Box 9.5 **Why is a protein that makes ATP called an ATPase?**

The abbreviation ATPase denotes an ATP hydrolase, in other words an enzyme that breaks ATP down into ADP and inorganic phosphate. This is the reverse of the reaction catalyzed by the F_0F_1 ATPase, which should really be called an ATP synthase. Why therefore do we refer to it as an ATPase?

The answer lies in the series of experiments that led to discovery of the F_0F_1 ATPase. One of the lines of investigation involved isolation of mitochondria from living cells, followed by disruption of these mitochondria by sonication. This treatment resulted in formation of submitochondrial vesicles in which the spheres of the F_0F_1 structure point outward.

In 1960, the Austrian biochemist Efraim Racker showed that spheres which had been detached from these vesicles have ATPase activity, converting ATP to ADP. The spheres are the F_1 component of the complete multiprotein structure, so this activity was called the F_1 ATPase. It is only when attached to the F_0 part of the structure that the catalytic activity of the F_1 component acts in reverse, so the ATPase becomes an ATP synthase.

Because of its physiological role in making ATP, and to avoid confusion, the F_0F_1 ATPase is sometimes referred to as the F_0F_1 ATP synthase, but this is not its biochemically correct name. The designation F_0F_1 ATPase is more correct because it indicates that it is composed of the F_1 ATPase attached to the membrane-spanning F_0 unit.

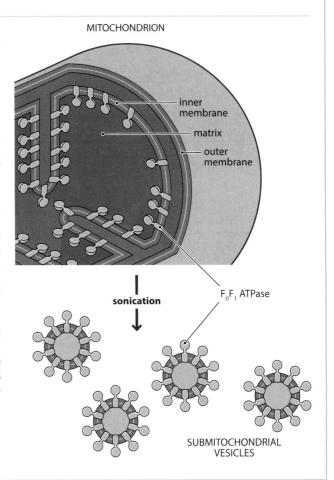

The F_0F_1 ATPase is a molecular motor driven by protons

We still need energy to generate ATP molecules, so the next question we have to address is how the ATPase obtains this energy. The answer lies in the structure of the remarkable F_0F_1 ATPase protein.

The F_0F_1 ATPase, as its name indicates, is made up of two components, referred to as F_0 and F_1, each of which is constructed from multiple protein subunits. The F_0 component comprises between 10 and 14 copies of subunit c, forming a barrel that spans the inner mitochondrial membrane (*Fig. 9.16*). This barrel is able to rotate within the membrane. Alongside the barrel, and also spanning the membrane, is a single copy of subunit a. As we will see in a moment, subunit a forms the channel through which protons pass on their way back to the mitochondrial matrix.

The F_1 component of the ATPase forms a ball-and-stick structure, the stick of which fits into the F_0 barrel, with the ball on the inner side of the membrane, within the mitochondrial matrix. The most important parts of this component are subunits α and β, six of which (three of each) make up the ball structure. The ball is attached to the γ subunit, which forms the stick that holds the F_1 component into the F_0 barrel. The γ subunit can rotate as the F_0 barrel rotates, but the $\alpha\beta$ ball cannot rotate, because it is held in place by a structure made up of two copies of subunit b and one of subunit δ.

The ability of the stick to rotate, but not the ball, is the key feature of the ATPase structure. As mentioned above, protons pass through the membrane via subunit a. This

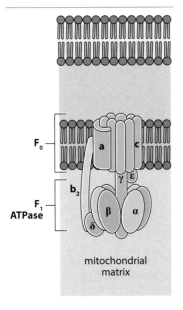

Figure 9.16 The F_0F_1 ATPase.
The different subunits are labeled. There are two subunits b adjacent to one another, only one of which is shown in this drawing.

causes the barrel of c subunits, and the γ stalk that is attached to the barrel, to rotate. Rotation of the γ subunit within the stationary αβ ball generates mechanical energy that the ATPase uses to make ATP from ADP and inorganic phosphate. The precise mechanism is not known but the current model is the **binding-charge mechanism**, which suggests that rotation of the γ subunit causes a cycle of conformational changes in the β subunits, resulting in a repeated series of ADP binding, phosphorylation, and release as ATP. Three ATP molecules can be generated per turn of the γ subunit, one for each of the β subunits in the stationary ball structure. The maximum speed of rotation has been estimated at over 300 revolutions per second.

The F_0F_1 ATPase is therefore a molecular motor, driven by protons and generating ATP molecules. It forms the end point of the energy-generation pathway, completing

---

Box 9.6 The rotation of the F_0F_1 ATPase

RESEARCH HIGHLIGHT

The central feature of the F_0F_1 ATPase is the ability of the F_0 barrel and the attached γ subunit of the F_1 ATPase to rotate, generating the mechanical energy that is used to convert ADP to ATP. Rotation of the F_0 barrel is driven by the passage of protons through subunit a. Exactly how the flow of protons is converted into the rotatory motion is unknown. Current models are based on the identification of those amino acids in subunits a and c that play the most critical roles in the link between proton flow and rotation. Mutation studies have shown that the most important of these is an aspartic acid at position 61 of subunit c. Conversion of this aspartic acid to asparagine completely abolishes rotation of the ATPase. As aspartic acid has a negatively charged side-chain, it can act as a proton acceptor. The implication is that the transient conversion of asp-61 to its protonated form, as protons pass through the ATPase, is the key to conversion of this proton flow to the rotatory motion of the F_0 barrel.

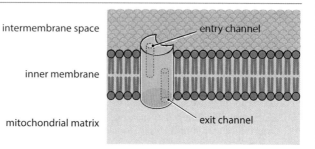

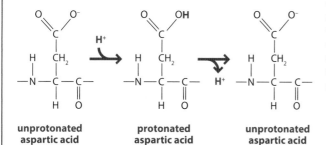

unprotonated aspartic acid   **protonated aspartic acid**   **unprotonated aspartic acid**

The c subunit is made up of a pair of α-helices running approximately parallel to one another, with asp-61 located in the middle of one of these helices. In the F_0F_1 ATPase, this helix lies close to a third helix, this one located within subunit a. According to one model, protonation of asp-61 causes a change in conformation of the c subunit helix. This conformational change causes the c subunit helix to push against the nearby helix of subunit a, forcing the barrel of c subunits to rotate.

A second model proposes a different mechanism for rotation of the F_0 barrel. The channel formed by subunit a, through which protons pass, is not continuous. Instead the channel is in two parts, with the entry channel from the intermembrane space slightly offset compared to the exit channel leading to the mitochondrial matrix.

The consequence of this offset is that a proton cannot pass through subunit a unless there is some means by which it jumps from the end of the entry channel to the start of the exit channel. From what is known about the structures of subunits a and c, it seems likely that as the F_0 barrel rotates, the asp-61 of each c subunit moves past the start of the exit channel of subunit a, and then past the end of the entry channel. The proposal is therefore that the asp-61 of a c subunit becomes protonated when it is adjacent to the entry channel. Complete rotation of the F_0 barrel results in that asp-61 becoming positioned next to the exit channel, at which point the proton detaches and moves on to the mitochondrial matrix.

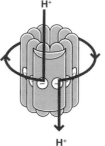

So protonation of asp-61 might form the bridge between the two channels, but how can protonation induce rotation? To understand this point we need to consider the effect of protonation on the properties of an aspartic acid molecule. Protonation neutralizes the negative charge on the aspartic acid side-chain. This converts the charged, hydrophilic amino acid into an uncharged and relatively hydrophobic one. Once protonated, the aspartic acid is therefore repulsed by the hydrophilic environment of the subunit a entry channel, and attracted by the hydrophobic environment of the membrane into which subunit a and the F_0 barrel are embedded. The combination of repulsion and attraction moves the protonated asp-61 away from the entry channel and towards the membrane environment. The F_0 barrel therefore rotates.

the events that release and capture in a readily useable form the energy contained in glucose and other dietary nutrients.

Control of ATP synthesis

When we studied glycolysis and the TCA cycle we were able to identify specific control points at which the flow of metabolites through the energy-generating pathway could be regulated. The rates at which the electron transport chain and the F_0F_1 ATPase operate are also affected by the availability of substrates, in particular ADP, but the control process is more subtle than the simple allosteric inhibition of a key enzyme.

The basis to this control process is a tight coupling that exists between the flow of electrons along the electron transport chain and ATP synthesis. This can be demonstrated with isolated mitochondria. When ADP is added the rate of oxygen utilization by the mitochondria increases, and stays at a high level until all the ADP has been converted into ATP. Oxygen utilization then declines (*Fig. 9.17A*). This type of experiment shows that the availability of ADP controls the flow of electrons along the electron transport chain. When ADP is available, and can be phosphorylated to make ATP, electrons flow along the chain. Conversely, when ADP is unavailable, there is no flow of electrons and oxygen utilization decreases (*Fig. 9.17B*). The chemiosmotic theory explains these observations in the following way:

- When ADP is available, the electron transport chain is active, resulting in the pumping of protons across the inner mitochondrial membrane and into the intermembrane space. The resulting electrochemical gradient drives the movement of protons through the F_0F_1 ATPase and back into the mitochondrial matrix, generating ATP and using up the available ADP.

- When the ADP is depleted, and it not possible to synthesize ATP, there is no movement of protons through the F_0F_1 ATPase. Without this movement,

Figure 9.17 The effect of ADP on oxygen consumption by isolated mitochondria.
(A) Addition of ADP to a preparation of isolated mitochondria increases the rate of oxygen utilization. (B) When ADP is available, electrons flow along the electron transport chain and oxygen is used up. When ADP is unavailable, there is no flow of electrons.

A. effect of ADP on oxygen utilization by isolated mitochondria

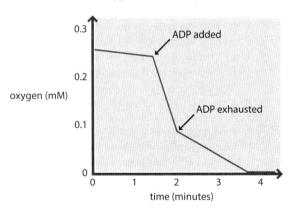

B. the explanation of the ADP effect

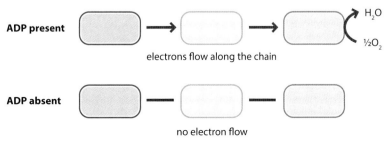

protons accumulate in the intermembrane space, increasing the electrochemical gradient. Very soon, this gradient becomes so steep that further proton pumping is impossible, because the energy available to pump an individual proton is insufficient to move that proton up the increased transmembrane concentration gradient. Because proton pumping stops, the flow of electrons through the electron transport chain also stops.

The regulation of the electron transport chain by ADP availability is called **respiratory** or **acceptor control**, the latter name reflecting the role of ADP, which 'accepts' phosphates to generate ATP. Inhibition of electron flow by respiratory control leads to an increase in the levels of NADH and $FADH_2$ and, further back in the energy-generating pathway, citrate accumulates (*Fig. 9.18*). As a consequence, both glycolysis and the TCA cycle are also inhibited.

Figure 9.18 Respiratory control.
The availability of ADP controls the flow of electrons through the chain. When there is no ADP, electron flow stops, leading to the accumulation of NADH, $FADH_2$ and citrate, and inhibition of glycolysis and the TCA cycle.

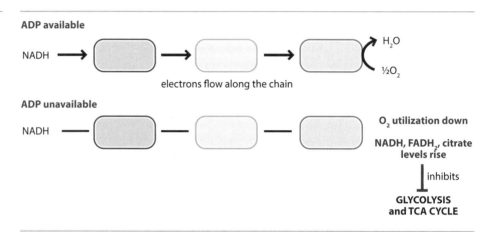

9.2.4 Inhibitors and uncouplers of the electron transport chain

Being such a vital part of the cell's biochemical activity, it is not surprising that most of the compounds that inhibit one or other step in the electron transport chain are potent toxins. Most notorious of these is cyanide, the favorite of murder mystery writers, which inhibits cytochrome c oxidase by binding to the iron within the heme group of cytochrome a_3, thereby terminating electron transport at Complex IV. Rotenone, used in pesticides, prevents electron transfer from Complex I to CoQ, and so inhibits NADH oxidation, but not that of $FADH_2$.

Other compounds interfere with the functioning of the electron transport chain by uncoupling the oxidation of NADH and $FADH_2$ from the production of ATP. **Uncouplers** are typically small lipid-soluble compounds that are able to bind H^+ ions. As they are lipid soluble, they can pass through a membrane, taking their bound protons with them. Compounds of this type are **ionophores**. In a mitochondrion, they uncouple electron transport from ATP synthesis because they transport pumped protons directly back into the mitochondrial matrix, so the ATPase channel is bypassed (*Fig. 9.19*). Energy is still released by the electron transport chain, but it dissipates as heat without generating any ATP. An example of an uncoupler is **2,4-dinitrophenol**, which was used as an aid to weight loss in the days when the toxic side-effects of drugs were less well recognized than they are today.

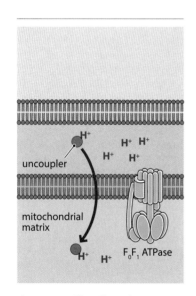

Figure 9.19 The effect of an uncoupler on ATP synthesis.
The uncoupler transports protons directly across the inner mitochondrial membrane, so the F_0F_1 ATPase is bypassed and ATP is not synthesized.

Uncoupling also occurs naturally under some physiological conditions. In brown adipose tissue, the inner mitochondrial membrane contains **thermogenin**, a proton transport protein that reverses the effects of proton pumping and increases the proportion of the energy generated by the electron transport chain that is released as heat. The heat thus produced enables brown adipose tissue to play a protective role in cold-sensitive organs in newborn animals and during hibernation.

9.2.5 Cytoplasmic NADH cannot gain access to the electron transport chain

One final detail remains before we complete our study of the energy-generation pathway. We have still not accounted for the NADH molecules produced during glycolysis. There are two of these for every starting molecule of glucose. Because glycolysis takes place in the cytoplasm, these two NADH molecules are outside of the mitochondrion. The inner mitochondrial membrane prevents the entry of NADH into the mitochondrial matrix, so the NADH produced by glycolysis cannot directly enter the electron transport chain. How can the energy carried by these NADH molecules be used to synthesize ATP?

The answer is to transfer the electrons from these NADH molecules to a molecule that can pass through the inner mitochondrial membrane. The subsequent oxidation of this **mitochondrial shuttle** molecule generates NADH or $FADH_2$ inside the mitochondrial matrix, which can then enter into the electron transport chain (*Fig. 9.20*).

The malate–aspartate shuttle operates in most mammalian tissues

In humans, the most important mitochondrial shuttle is the **malate–aspartate shuttle**, which involves malate, oxaloacetate and aspartate. In the cytoplasm, NADH is used by the enzyme **malate dehydrogenase** to reduce oxaloacetate to malate (*Fig. 9.21*). Malate is then transported across the inner mitochondrial membrane by a carrier protein specific for malate and α-ketoglutarate. Once inside the mitochondrial matrix, the malate molecules are oxidized back to oxaloacetate by the mitochondrial version of malate dehydrogenase, catalyzing the reverse reaction to its cytoplasmic counterpart. This oxidation is coupled to the conversion of mitochondrial NAD^+ to NADH, which then enters the electron transport chain.

To complete the cycle, the oxaloacetate produced in the mitochondrion must be transported back to the cytoplasm. The inner mitochondrial membrane does not, however, contain a transporter protein capable of working with oxaloacetate. Instead, the oxaloacetate molecules are converted to aspartate, which can cross the inner mitochondrial membrane, thanks to the presence of a glutamate–aspartate carrier protein. The oxaloacetate to aspartate conversion is a **transamination** reaction, in which a carbonyl group on oxaloacetate is replaced by an amino group (*Fig. 9.22*). The

Figure 9.20 The mode of action of a mitochondrial shuttle.

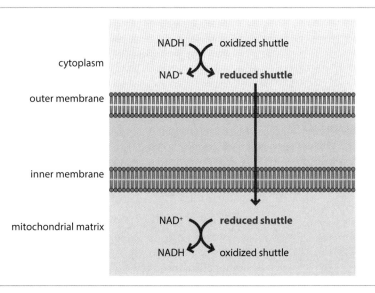

Figure 9.21 The malate–aspartate shuttle.
Only the inner mitochondrial membrane is shown because malate and aspartate pass freely through porins in the outer membrane.

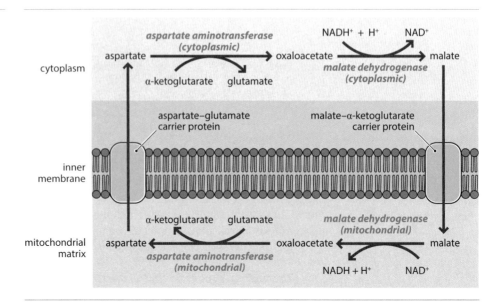

enzyme responsible for this transamination, **aspartate aminotransferase**, synthesizes aspartate in the mitochondrion, and then regenerates oxaloacetate by the reverse reaction in the cytoplasm.

The transamination in the mitochondrion uses an amino group from glutamate, which converts glutamate to α-ketoglutarate. The α-ketoglutarate also has to be transported to the cytoplasm so that it can act as an amino acceptor when oxaloacetate is regenerated, with the glutamate that results being shuttled back into the mitochondrion so the cycle can continue.

The glycerol 3-phosphate shuttle is restricted to brown adipose tissue

The second mitochondrial shuttle system in humans is the **glycerol 3-phosphate shuttle**, which occurs only in brown adipose tissue. In the cytoplasm, **glycerol 3-phosphate dehydrogenase** converts dihydroxyacetone phosphate to glycerol 3-phosphate, with transfer of electrons from NADH (*Fig. 9.23*). Glycerol 3-phosphate does not require a transport protein because the mitochondrial version of glycerol 3-phosphate dehydrogenase is located on the outer surface of the inner mitochondrial membrane. The mitochondrial enzyme regenerates dihydroxyacetone phosphate, the electrons being transferred to an FAD group that is linked to the enzyme. The FADH$_2$ that results enters the electron transport chain at Complex II, and the dihydroxyacetone phosphate diffuses back to the cytoplasm.

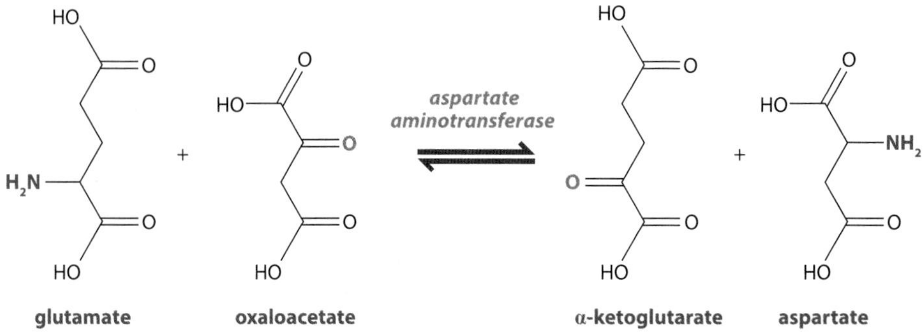

Figure 9.22 The transamination catalyzed by aspartate aminotransferase.

glutamate   oxaloacetate   α-ketoglutarate   aspartate

Figure 9.23 **The glycerol 3-phosphate shuttle.**

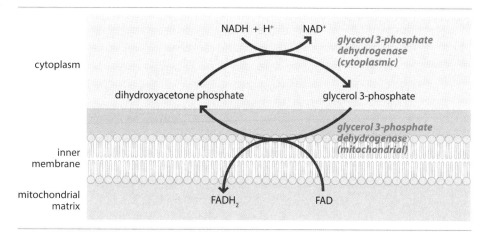

Because FADH$_2$ is the direct donor into the electron transport chain, only two ATP molecules are generated rather than the three that result from mitochondrial NADH. But energy cannot simply disappear, which means that the glycerol 3-phosphate shuttle produces a greater amount of heat than is the case with the malate–aspartate shuttle. As with the uncoupling of the electron transport chain, the heat generated by the glycerol 3-phosphate shuttle contributes to the protective role of brown adipose tissue in cold-sensitive parts of the body and during hibernation.

Box 9.7 **The smell of skunk cabbage**

Skunk cabbage, *Symplocarpus foetidus*, grows in wet areas of North America. As its name implies, it produces a perfume that is pungent rather than fragrant (it is rich in dimethyl disulfide), but which is highly attractive to the plant's pollinators, which include flies and bees.

The smell is not the only remarkable feature of skunk cabbage. It is one of only a very few plants that generate heat. It is thought that thermogenesis in skunk cabbage helps the seedling to heat the surrounding soil to 10–30°C above the ambient temperature, enabling it to germinate and flower very early in the season, even when the ground is still frozen. By flowering early the plant is able to exploit resources, including pollinators, that are not accessible to species that do not germinate until later in the season. The heat generated by the plant also diffuses its aromatic chemicals, helping to attract its pollinators.

Thermogenesis in skunk cabbage and other plants is due to a modification of the electron transport chain. Electrons from NADH and FADH$_2$ are transferred in the usual way to CoQ, the intermediate carrier between Complex I or II and Complex III. However, in thermogenic plants, CoQ does not donate its electrons to Complex III, but instead passes them directly to oxygen, mediated by a protein called the 'alternative oxidase'. This version of the electron transport chain is referred to as **cyanide-resistant respiration**, because it bypasses Complex IV, which contains cytochrome a_3. Cyanide inactivates this cytochrome by binding to its heme group, and hence has a severely toxic effect on organisms, such as humans, that are wholly dependent on the standard electron transport chain for their survival. Bypassing Complexes III and IV also means that ATP production is reduced in plants using the cyanide-resistant pathway. The energy that is not used to make ATP is instead lost as heat, resulting in the plant's thermogenic properties.

Image reproduced from *The Quantum Biologist*.

Further reading

Akram M (2014) Citric acid cycle and role of its intermediates in metabolism. *Cell Biochemistry and Biophysics* **68**, 475–8.

Anraku Y (1988) Bacterial electron transport chains. *Annual Review of Biochemistry* **57**, 101–32.

Bendell DS and Bonner WD (1971) Cyanide-insensitive respiration in plant mitochondria. *Plant Physiology* **47**, 236–45. *Skunk cabbage.*

Brand MD and Murphy MP (2008) Control of electron flux through the respiratory chain in mitochondria and cells. *Biological Reviews* **62**, 141–93.

Halestrap AP (2012) The mitochondrial pyruvate carrier: has it been unearthed at last? *Cell Metabolism* **16**, 141–3.

Kornberg H (2000) Krebs and his trinity of cycles. *Nature Reviews Molecular Cell Biology* **1**, 225–8.

Kühlbrandt W and Davies KM (2016) Rotary ATPases: a new twist to an ancient machine. *Trends in Biochemical Sciences* **41**, 106–16.

Mitchell P (1961) Coupling of phosphorylation to electron and hydrogen transfer by a chemi-osmotic type of mechanism. Nature **191**, 144–8. *The chemiosmotic theory.*

Palou A, Picó C, Bonet ML and Oliver P (1998) The uncoupling protein, thermogenin. *International Journal of Biochemistry and Cell Biology* **30**, 7–11.

Patel MS, Nemeria NS, Furey W and Jordan F (2014) The pyruvate dehydrogenase complexes: structure-based function and regulation. *Journal of Biological Chemistry* **289**, 16615–23.

Saier MH (1997) Peter Mitchell and his chemiosmotic theories. *ASM News* **63(1),** 13–21.

Sazanov LA (2015) A giant molecular proton pump: structure and mechanism of respiratory complex I. *Nature Reviews Molecular Cell Biology* **16**, 375–88.

Schell JC and Rutter J (2013) The long and winding road to the mitochondrial pyruvate carrier. *Cancer and Metabolism* **1,** 6.

Winge DR (2012) Sealing the mitochondrial respirasome. *Molecular and Cellular Biology* **32**, 2647–52.

Self-assessment questions

Multiple choice questions

Only one answer is correct for each question.
Answers can be found on the website:
www.scionpublishing.com/biochemistry.

1. What does each molecule of pyruvate yield during the TCA cycle?
   (a) One molecule of ATP, one of NADH and one of FADH$_2$
   (b) One molecule of ATP, three of NADH and one of FADH$_2$
   (c) Three molecules of ATP, three of NADH and one of FADH$_2$
   (d) Three molecules of ATP, one of NADH and one of FADH$_2$

2. The enzymes of the TCA cycle are located in which part(s) of the mitochondrion?
   (a) Inner mitochondrial membrane
   (b) Inner mitochondrial membrane and intermembrane space

   (c) Inner mitochondrial membrane and mitochondrial matrix
   (d) Mitochondrial matrix

3. How does pyruvate pass through the outer mitochondrial membrane?
   (a) Directly through the lipid bilayer
   (b) It is transported by the mitochondrial pyruvate carrier protein
   (c) Through a porin
   (d) None of the above processes is correct

4. How does pyruvate pass through the inner mitochondrial membrane?
   (a) Directly through the lipid bilayer
   (b) It is transported by the mitochondrial pyruvate carrier protein
   (c) Through a porin
   (d) None of the above processes is correct

5. The pyruvate dehydrogenase complex converts pyruvate into which compound?
- **(a)** Acetyl CoA
- **(b)** Citrate
- **(c)** Isocitrate
- **(d)** Oxaloacetate

6. ATP is generated by which enzyme(s) during the TCA cycle?
- **(a)** Isocitrate dehydrogenase
- **(b)** Succinate dehydrogenase
- **(c)** Succinyl CoA synthetase
- **(d)** All of the above

7. Which enzyme regenerates the oxaloacetate used up in the first step of the TCA cycle?
- **(a)** Aconitase
- **(b)** Fumarase
- **(c)** Malate dehydrogenase
- **(d)** Succinate dehydrogenase

8. Which of the intermediates in the TCA cycle can be used as a substrate for production of glutamate, other amino acids and purines?
- **(a)** Citrate
- **(b)** α-Ketoglutarate
- **(c)** Oxaloacetate
- **(d)** Succinyl CoA

9. Which group of compounds are stimulators of pyruvate dehydrogenase kinase?
- **(a)** Acetyl CoA, ATP and NADH
- **(b)** Acetyl CoA, ATP and pyruvate
- **(c)** Acetyl CoA, ATP and oxaloacetate
- **(d)** Acetyl CoA, NADH and oxaloacetate

10. Which compound is an inhibitor of pyruvate dehydrogenase kinase?
- **(a)** Acetyl CoA
- **(b)** ATP
- **(c)** Oxaloacetate
- **(d)** Pyruvate

11. What is the symbol used to denote the standard free energy charge?
- **(a)** G
- **(b)** $\Delta G^{\ddagger}$
- **(c)** $\Delta G'$
- **(d)** $\Delta G^{0\prime}$

12. What is the standard free energy charge for oxidation of NADH?
- **(a)** $-220.2\,cal\,mol^{-1}$
- **(b)** $220.2\,cal\,mol^{-1}$
- **(c)** $-220.2\,kJ\,mol^{-1}$
- **(d)** $220.2\,kJ\,mol^{-1}$

13. At which complex does NADH enter the electron transport chain?
- **(a)** Complex I
- **(b)** Complex II
- **(c)** Complex III
- **(d)** Complex IV

14. At which complex does $FADH_2$ enter the electron transport chain?
- **(a)** Complex I
- **(b)** Complex II
- **(c)** Complex III
- **(d)** Complex IV

15. Which of these complexes contains one or more FeS proteins?
- **(a)** Complex I
- **(b)** Complex II
- **(c)** Complex III
- **(d)** All of the above

16. How many protons are pumped for every NADH molecule that is oxidized by the electron transport chain?
- **(a)** 4
- **(b)** 8
- **(c)** 10
- **(d)** 12

17. Proton pumping involves movement of protons from where to where?
- **(a)** From the intermembrane space to the cytoplasm
- **(b)** From the mitochondrial matrix to the cytoplasm
- **(c)** From the intermembrane space to the mitochondrial matrix
- **(d)** From the mitochondrial matrix to the intermembrane space

18. Between 10 and 14 copies of which subunit of the F_0F_1 ATPase form a barrel that spans the inner mitochondrial membrane?
- **(a)** Subunit a
- **(b)** Subunit b
- **(c)** Subunit c
- **(d)** Subunit d

19. The rate of ATP synthesis is controlled by availability of which compound?
- **(a)** ADP
- **(b)** Citrate
- **(c)** NADH
- **(d)** Pyruvate

20. Which one of the following statements is **incorrect** with regard to a typical uncoupler of the electron transport chain?
- **(a)** Able to bind H^+ ions
- **(b)** Able to pass through the inner mitochondrial membrane
- **(c)** A small lipid-soluble compound
- **(d)** Stimulates over-production of ATP

21. Which one of the following compounds is an uncoupler of the electron transport chain?
- **(a)** Cyanide
- **(b)** Diisopropyl fluorophosphate
- **(c)** 2,4-Dinitrophenol
- **(d)** Glycerol 3-phosphate

22. Which one of the following statements is **correct** with regard to thermogenin?
  (a) Important for stimulating ATP synthesis during exercise
  (b) It is a proton transport protein that reverses the effects of proton pumping
  (c) Present in muscle
  (d) Synthesis is controlled by glucagon

23. During the malate–aspartate shuttle, which enzyme converts oxaloacetate to malate in the cytoplasm?

  (a) Aspartate aminotransferase
  (b) Malate dehydrogenase
  (c) Malate synthase
  (d) Oxaloacetate dehydrogenase

24. The glycerol 3-phosphate shuttle takes place in which type of tissue?
  (a) Brown adipose tissue
  (b) Liver
  (c) Muscle
  (d) White adipose tissue

Short answer questions

These questions do not require additional reading.

1. Describe how pyruvate is transported into mitochondria.

2. Outline the structure of the pyruvate dehydrogenase complex and explain why this enzyme plays a central role in energy generation. Your answer should include a description of the regulation of pyruvate dehydrogenase activity.

3. Draw an outline of the TCA cycle, showing the substrates, products and enzymes for each step.

4. Why is succinyl CoA synthetase able to synthesize both ATP and GTP?

5. Describe the importance of redox potential in operation of the electron transport chain.

6. Draw a diagram showing, in outline, the structure of the electron transport chain, marking the entry points for NADH and $FADH_2$, and indicating the number of protons that are pumped at each stage.

7. Give a detailed explanation of how proton pumping can give rise to ATP synthesis.

8. Describe the structure of the F_0F_1 ATPase and outline current theories regarding the operation of this molecular motor.

9. Explain how the flow of electrons along the electron transport chain is regulated by ADP availability. How does an uncoupler such as 2,4-dinitrophenol affect the operation of the electron transport chain?

10. How is the energy contained in the two NADH molecules synthesized during glycolysis used to synthesize ATP?

Self-study questions

These questions will require calculation, additional reading and/or internet research.

1. Why did it take so long to identify the mitochondrial pyruvate carrier protein?

2. Patients suffering from beriberi (vitamin B_1 or thiamine deficiency) often have high levels of pyruvate in their blood and urine. Explain this observation.

3. Albert Szent-Györgyi, who won the Nobel Prize in 1937, carried out a number of the early experiments that led to our understanding of the TCA cycle. In one experiment, Szent-Györgyi showed that addition of succinate to a pigeon breast muscle extract stimulated the production of CO_2. He observed that for every mole of succinate that was added, many additional moles of CO_2 were produced. In other words, the amount of CO_2 that was generated was greater than could be obtained by breakdown of the succinate that was added. What is the explanation of this observation?

4. Succinate dehydrogenase is inhibited by a variety of compounds, including the three-carbon compound malonate and various compounds that resemble quinones. Explain why such different types of compound can inhibit succinate dehydrogenase.

5. You have purified the different components of the electron transport chain and reconstituted these, in various combinations, in artificial membranes. For each of the following combinations, what would you predict to be the final electron acceptor if (a) NADH, or (b) succinate were added?
  - complex 1, complex 2, complex 3, complex 4, cytochrome *c*, ubiquinone
  - complex 1, complex 3, complex 4, cytochrome *c*, ubiquinone
  - complex 2, complex 3, complex 4, cytochrome *c*, ubiquinone
  - complex 1, complex 2, complex 3, complex 4, cytochrome *c*
  - complex 1, complex 2, complex 3, complex 4, ubiquinone
  - complex 1, complex 2, complex 3, cytochrome *c*, ubiquinone

CHAPTER 10
Photosynthesis

STUDY GOALS

After reading this chapter you will:

- understand the importance of photosynthesis as the primary source of the energy used by all living organisms

- be able to distinguish between the events occurring during the light and dark stages of photosynthesis

- appreciate that plants, algae and some types of bacteria are able to photosynthesize, and know the key differences between the processes in these types of organism

- understand the role of chlorophyll and other light-harvesting pigments in photosynthesis

- be able to describe the structures of the two photosystems, and be able to explain how light energy is captured by these photosystems

- know the structure of the photosynthetic electron transport chain and be able to explain how electron transport results in synthesis of NADPH and ATP

- be able to describe the special features of the cyclic photophosphorylation pathway, and explain what this pathway achieves

- understand the central role played by ribulose bisphosphate carboxylase in carbon fixation, and know how the activity of this enzyme is regulated

- know the key reactions of the Calvin cycle, including the substrates, products and enzymes for each step, and in particular be able to explain how the Calvin cycle results in conversion of carbon dioxide into glyceraldehyde 3-phosphate

- be able to describe the role of ferrodoxin in regulation of the Calvin cycle

- be able to describe how sucrose and starch are synthesized from glyceraldehyde 3-phosphate

- understand the difficulties caused by the oxygenation reaction catalyzed by ribulose bisphosphate carboxylase, and be able to describe how C4 and CAM plants circumvent these difficulties

The ultimate source of most of the energy used by living organisms is sunlight. The energy in sunlight is used directly by those organisms that are able to carry out **photosynthesis**. These organisms, which comprise plants, algae and some types of bacteria, are called **primary producers** or **autotrophs**, and make up less than 5% of all of the species on the planet. Other organisms utilize the energy from sunlight indirectly, either by consuming the primary producers, or even more indirectly via a **food chain** (see *Fig. 1.2*).

Photosynthesis is therefore a preliminary stage to the energy-generating pathway that we studied in the previous two chapters. During photosynthesis, the energy from sunlight is used to drive a series of endergonic reactions that result in the synthesis of carbohydrates from carbon dioxide and water. The resulting carbohydrates act as stores of energy that can subsequently be utilized, via glycolysis, the TCA cycle and the electron transport chain, either by the photosynthetic organism or by another organism that eats it.

10.1 An overview of photosynthesis

Before looking in detail at the biochemical basis to photosynthesis, we will examine what it achieves, and where in the cell it occurs.

10.1.1 Photosynthesis is the light-driven production of carbohydrates

Photosynthetic organisms use sunlight to synthesize carbohydrates. The overall reaction can be described as:

$$CO_2 + H_2O \xrightarrow{\text{sunlight}} (CH_2O)_n + O_2$$

In this equation, $(CH_2O)_n$ is the general formula for carbohydrate, and 'sunlight' above the arrow indicates that the reaction requires the input of energy from sunlight.

Photosynthesis is divided into two stages which are called the **light** and **dark reactions**. These are the traditional names and the ones we will use, though in reality we should call them the light-dependent and light-independent reactions. This is because the dark reactions (light-independent) do not actually have to occur in the dark, they simply do not require the input of sunlight. These two stages of photosynthesis have quite distinct outcomes (*Fig. 10.1*):

- During the **light reactions**, the energy from sunlight is harnessed to make ATP and NADPH.

- During the **dark reactions**, the energy contained in these ATP and NADPH molecules is used to drive the synthesis of carbohydrates from carbon dioxide and water.

The products of the dark reactions include glucose 1-phosphate and fructose 6-phosphate, which combine to make sucrose. Sucrose is the main energy supply for plants, and is transported from the photosynthetic tissues to other parts of the plant, where it is broken down to glucose and fructose, which enter into the glycolytic pathway. Excess glucose is polymerized into starch and stored by the plant for future use. Herbivores grazing on plants use the sucrose and stored starch for their own energy needs, beginning the food chain that culminates with the top carnivores in the ecosystem.

10.1.2 Photosynthesis occurs in specialized organelles

In plants and algae, photosynthesis takes place in special organelles called chloroplasts. Like mitochondria, chloroplasts have a permeable outer membrane and largely impermeable inner membrane, surrounding an internal matrix which in chloroplasts is called the stroma (*Fig. 10.2*). Unlike mitochondria, within the stroma there is a third membrane system, which forms interconnected structures called **thylakoids**. Each thylakoid is shaped like a disc, containing an internal space (the **thylakoid space**) completely surrounded by the thylakoid membrane. Individual thylakoids are stacked on top of one another like piles of plates, forming structures called **grana**. The light reactions of photosynthesis take place in the thylakoids and the dark reactions in the stroma.

There are other similarities between chloroplasts and mitochondria that we will encounter as we study the details of the light and dark reactions. One of these is that

LIGHT REACTIONS	DARK REACTIONS
energy in sunlight used to make ATP and NADPH	ATP and NADPH used to synthesize carbohydrate from carbon dioxide and water

Figure 10.1 The light and dark reactions of photosynthesis.

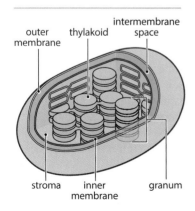

Figure 10.2 A chloroplast.
The structural components important in photosynthesis are labeled.

photosynthesis involves an electron transport chain, which is located in the thylakoid membrane. Movement of electrons along this chain results in proton pumping, building up an excess of H^+ ions in the space within the thylakoids. Protons flow from the thylakoid space back to the stroma through an ATPase that generates ATP molecules. So we can approach photosynthesis confident in the knowledge that we are already familiar with many of the underlying processes, albeit that these occur within a different architecture.

Many bacteria are also able to photosynthesize. These include the **cyanobacteria**, which used to be called the blue–green algae (the name was changed because algae are now looked on as exclusively eukaryotic organisms), and the **purple bacteria** and **green bacteria**. The names of these bacteria indicate that they are colored, a feature of photosynthetic organisms that we will deal with in a moment. Cyanobacteria have internal membrane structures equivalent to thylakoids, but in purple and green bacteria the light reactions take place in the plasma membrane, which in the purple bacteria is enfolded to form invaginations called **chromatophores** (*Fig. 10.3*).

Figure 10.3 Chromatophores.
Cells of a purple bacterium showing the membranous invaginations that are the sites of the photosynthetic light reactions.
Image reproduced from HS Pankratz and RL Uffen, MSU/Biological Photo Service with permission.

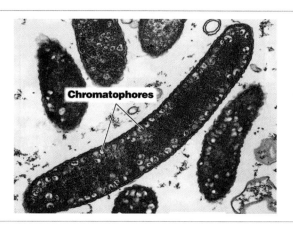

10.2 The light reactions

During the light reactions, sunlight is used to generate high energy electrons which release their energy incrementally by passing down the electron transport chain located in the thylakoid membrane. The energy that is released is used partly to establish the electrochemical gradient that drives ATP synthesis, and partly to convert $NADP^+$ to NADPH. We must therefore examine two aspects of the light reactions: the way in which sunlight is captured and used to generate high energy electrons, and the components and functioning of the electron transport chain and the ATPase.

10.2.1 Sunlight is harvested by photosynthetic pigments

In order to photosynthesize, plants and other photosynthetic organisms must absorb energy from sunlight. This is called **light harvesting** and depends on the presence of light-absorbing pigments in the photosynthetic tissues.

Chlorophyll is the main light harvesting pigment in plants

We all know that plants are green and many people are aware that because plants are green they are able to photosynthesize. The link between the color of a plant and its ability to photosynthesize is the presence in plant tissue of **chlorophyll**, a compound that absorbs light. Specifically, chlorophyll absorbs light at the two ends of the visible light spectrum, in the red–orange–yellow and blue–indigo–violet regions (*Fig. 10.4*).

Figure 10.4 The absorbance spectra of chlorophyll *a* and chlorophyll *b*.

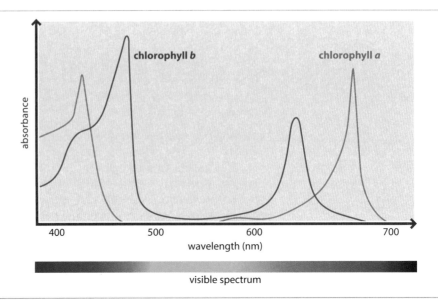

Figure 10.4 The absorbance spectra of chlorophyll *a* and chlorophyll *b*.

This means that green light (in the middle of the visible spectrum) is not absorbed and is reflected to our eyes making plants appear green. We can be thankful for this limitation in the light-harvesting capability of chlorophyll. If all parts of the light spectrum were absorbed then plants would be black and the world a sadder place.

Chlorophyll is a type of **porphyrin**, which is the same class of compound as the heme found in cytochrome proteins and in hemoglobin. In heme, the porphyrin contains an iron atom at its center, but in chlorophyll this atom is magnesium. There are several types of chlorophyll, with slightly different structures and different light absorption spectra. In plants, the two main types are chlorophyll *a* and *b* (*Fig. 10.5*). Some algae also contain chlorophyll *b*, along with types *c*1 and *c*2. In photosynthetic bacteria the equivalent compound is **bacteriochlorophyll** (*Table 10.1*).

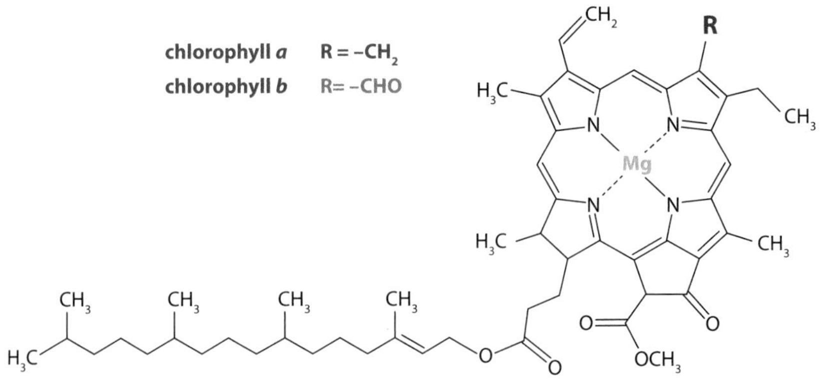

Figure 10.5 The structures of chlorophyll *a* and chlorophyll *b*.

chlorophyll *a*   R = –CH$_2$
chlorophyll *b*   R= –CHO

Table 10.1 Types of chlorophyll present in different organisms

Organism	Light-harvesting pigments
Plants, green algae	chlorophyll *a*, chlorophyll *b*
Red algae	chlorophyll *a*
Brown algae, diatoms	chlorophyll *a*, chlorophyll *c*1, chlorophyll *c*2
Some photosynthetic bacteria	bacteriochlorophyll

Box 10.1 Fall colors

The changes in leaf color from green to various shades of orange, red and yellow that occur during the fall are due to the presence in these leaves of accessory pigments. In most plant species, chlorophyll is the most abundant pigment and the leaves appear green. During the later weeks of summer, the chlorophyll content of the leaves begins to decline, whereas many of the accessory pigments remain stable until the leaves wither and drop from the trees. The colors of the accessory pigments are therefore revealed as the chlorophyll levels decrease, giving rise to the magnificent fall displays associated with the deciduous woodlands of temperate regions, such as the northern parts of North America and Europe.

Image shows an aerial view of a section of woodland in Alaska; image by Frans Lanting and reproduced with permission from Science Photo Library.

Plants also contain other light-absorbing compounds, called **accessory pigments**, including some that trap light in the green region of the spectrum and hence compensate in part for the limitation of chlorophyll in this respect. These accessory pigments include **carotenoids** such as **β-carotene** and **xanthophyll**, which are red and yellow, respectively (*Fig. 10.6*). Carotenoids are built up from isoprene units and hence are a type of lipid. Another type of carotenoid, **fucoxanthin**, is a brown-colored light-harvesting pigment found in brown algae, which include some seaweeds. Many photosynthetic bacteria also contain **phycobilins**, which are related to chlorophyll but lack a metal ion.

Figure 10.6 The structure of β-carotene and xanthophyll.

β-carotene   R = –H
xanthophyll   R= –OH

Solar energy is captured by chlorophyll located in two photosystems

Chlorophyll and the accessory carotenoid pigments are located in large, multisubunit protein complexes called **photosystems**, which are embedded in the thylakoid membrane. There are two different photosystems in plants, **photosystem I** and **photosystem II**. Although photosystem I is larger, the overall structures of the two protein complexes are very similar. Both types of photosystem are made of two components, called the **reaction center** and the **antenna complex**. The latter is rich in chlorophyll and accessory pigments and is laid out on the surface of the thylakoid membrane, exactly like an antenna, in order to capture energy from sunlight (*Fig. 10.7A*). The reaction center, as the name indicates, is located at the center of the antenna.

Energy from sunlight is initially captured by chlorophyll and carotenoid molecules in the antenna complex. When a photon of sunlight strikes the antenna complex, the energy contained in that photon can be transferred to a chlorophyll molecule. If the quantum of energy is of a particular size it can excite an electron in the chlorophyll molecule, so this electron gains energy and moves to a higher orbital in the electron

A. structure of a photosystem

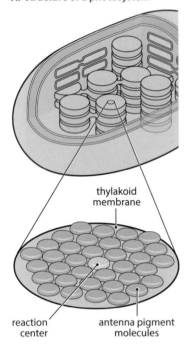

thylakoid
membrane

reaction
center

antenna pigment
molecules

B. channeling of energy to the reaction center

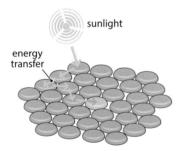

sunlight

energy
transfer

Figure 10.7 The structure and mode of action of a photosystem.
(A) A photosystem is a protein complex containing light-harvesting pigments. The antenna complex, shown in green, surrounds the reaction center. (B) Energy captured from sunlight is channeled through the antenna complex to the reaction center.

shell. The energy is then channeled to an adjacent chlorophyll molecule and then step by step to the reaction center (*Fig. 10.7B*). This energy transfer can occur in either of two ways:

- By **resonance energy transfer**, also called **exciton transfer**, in which the energy quantum is simply passed from one chlorophyll molecule to another, the recipient chlorophyll becoming excited and the donor chlorophyll returning to its ground state.

- By **direct electron transfer**, in which the high energy electron is itself transferred to the neighboring chlorophyll, in return for a low energy electron.

Accessory pigment molecules can also become excited by absorbance of light, at different wavelengths compared to the chlorophyll molecules, extending the energy-capturing capability of the antenna complex. Once excited, an accessory pigment passes its energy quantum to the nearest chlorophyll molecule, to begin the shuttling process along to the reaction center.

The reaction center contains a pair of closely adjacent chlorophyll molecules, usually slightly modified versions of chlorophyll *a*. The molecules in the reaction center of photosystem I absorb light most efficiently at a wavelength of 700 nm. Because of this property, the photosystem I reaction center is called **P700**. The photosystem II reaction center, for a similar reason, is called **P680**. Under normal conditions, the pair of chlorophylls in each reaction center are the ultimate acceptors of the energy quanta harvested from sunlight by the antenna pigments. However, at high light intensities, excess energy is transferred to carotenoids to protect the photosystems from damage.

10.2.2 Electron transport and photophosphorylation

The excited chlorophyll molecules in the P680 and P700 reaction centers participate in an electron transport chain that results in synthesis of NADPH and ATP. Because the process is light driven, the synthesis of ATP by this route is called **photophosphorylation**.

Electrons flow from photosystem II to NADPH

The P680 reaction center forms the start of the photosynthetic electron transport chain. This is the reaction center in photosystem II, which is so-named not because it works subsequent to photosystem I in electron transport, but because it was the second of the two photosystems to be discovered.

The excited version of P680, generated by energy passed along the antenna complex from a light-harvesting pigment, is designated P680*. The series of events in the electron transport chain, beginning with P680*, are as follows (*Fig. 10.8*).

Step 1A. A high energy electron is passed from P680* to **plastoquinone** (**PQ**). This is a lipid-soluble compound comprising a modified benzene ring (*Fig. 10.9*) that can move within the thylakoid membrane. Two electrons, plus a pair of H$^+$ ions, convert plastoquinone to its reduced form PQH$_2$.

$$P680^* + PQ + 2H^+ \rightarrow P680^+ + PQH_2$$

P680*, having donated an electron to PQ, now has one electron fewer than it should have. This form is denoted P680$^+$.

Step 1B. P680$^+$ replaces its missing electron with an electron from water to generate P680. Removal of four electrons from water, by four P680$^+$ molecules, results in the production of oxygen, one of the characteristic features of photosynthesis.

$$2H_2O \rightarrow 4e^- + 4H^+ + O_2$$

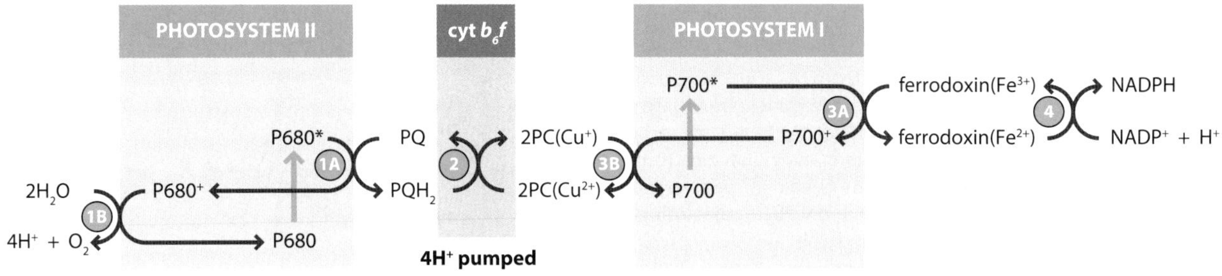

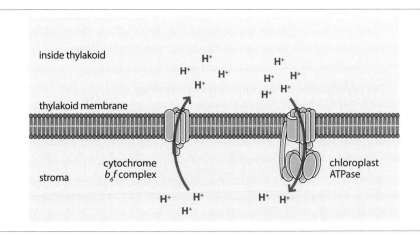

Figure 10.8 **Electron transfer from photosystem II to NADPH.**
The steps in the process, as described in the text, are indicated by the dark green circles. The vertical green arrows indicate the excitation of the P680 and P700 reaction centers. Abbreviations: PC, plastocyanin; PQ, plastoquinone.

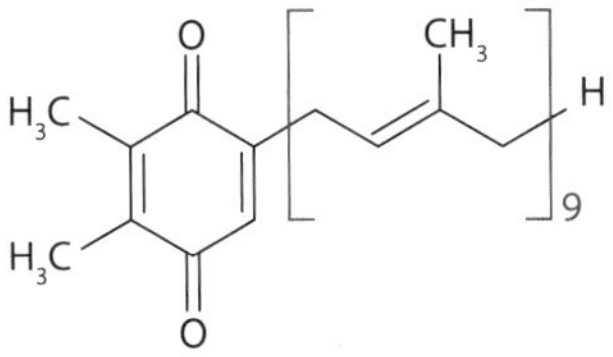

Figure 10.9 **Plastoquinone.**
The part of the structure enclosed in brackets is repeated nine times.

Step 2. Back on the electron transport chain, PQH_2 donates its electrons to the **cytochrome b_6f complex**. This complex comprises the two iron-containing cytochromes, cytochrome b_6 and cytochrome f, and also an iron–sulfur (FeS) protein. From this complex the electrons transfer to **plastocyanin (PC)**, a copper-containing protein.

$$PQH_2 + 2PC(Cu^{2+}) \rightarrow PQ + 2PC(Cu^+) + 2H^+$$

Step 3A. P700 is converted to its excited form P700* by acceptance of a high energy electron from the antenna complex. It donates an electron to **ferrodoxin**, a type of FeS protein, becoming the P700+ cation.

$$P700* + ferrodoxin(Fe^{3+}) \rightarrow P700^+ + ferrodoxin(Fe^{2+})$$

Step 3B. P700 is then regenerated by transfer of an electron to P700+ from the PC(Cu+) formed in *Step 2*.

Step 4. Now we make NADPH. Two electrons from the reduced form of ferrodoxin are used by **NADP reductase** to convert NADP+ to NADPH.

$$2ferrodoxin(Fe^{2+}) + NADP^+ + H^+ \rightarrow 2ferrodoxin(Fe^{3+}) + NADPH$$

Figure 10.10 **Proton pumping and ATP synthesis in the chloroplast.**
The cytochrome b_6f complex pumps protons across the thylakoid membrane and into the thylakoid. Exit of the protons through the ATPase results in synthesis of ATP.

Box 10.2 Atomic orbitals

The nucleus of an atom is surrounded by a cloud of negatively charged electrons. Although it is tempting to equate the movement of electrons around the nucleus with the orbit of planets around a star, the trajectories of individual electrons are much more complex and cannot be predicted with certainty. Rather than attempting to define individual trajectories, we identify the region of space around the nucleus in which a particular electron is likely to be found; this region is called an **orbital**.

The simplest atom is that of hydrogen, which has just a single electron. This electron occupies the 1s orbital. The nomenclature indicates that the orbital represents energy level 1 (the lowest energy level) and is spherical, with the nucleus at its center. The 2s orbital is also spherical, but is at a higher energy level, and so extends a greater distance from the nucleus. To move from the 1s to the 2s orbital, an electron must be **excited** – it must gain energy. If the electron subsequently moves back to the 1s orbital then it loses energy. There are also p orbitals, which are shaped like a pair of balloons, one on either side of the nucleus. The lowest energy level for a p orbital is 2. At higher energy levels the balloons become larger.

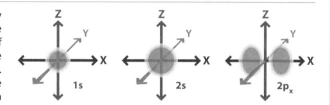

If two atoms come close enough together for their orbitals to overlap, then one or more electrons can be shared between the atoms. These atoms form a bond (see *Box 3.3*) and their shared electrons now occupy molecular orbitals which, like atomic orbitals, have characteristic shapes and can exist as different energy levels.

During the light-harvesting phase of photosynthesis, an electron within a chlorophyll molecule gains energy from a photon of sunlight, enabling it to jump from its molecular orbital to one with a higher energy level. Resonance energy transfer to an adjacent chlorophyll molecule results in the electron returning to its original orbital, with an electron in the second molecule now becoming excited. Alternatively, in direct electron transfer, the electron jumps directly to a higher level orbital in the adjacent chlorophyll molecule, with an electron from a lower orbital in this chlorophyll moving in the opposite direction.

Photosynthetic electron transport generates an electrochemical gradient

When we studied the mitochondrial electron transport chain, we discovered that movement of H^+ ions across the inner mitochondrial membrane creates an electrochemical gradient which is subsequently used to power ATP synthesis. Photophosphorylation (the synthesis of ATP during photosynthesis) is also driven by an electrochemical gradient set up during electron transport. The only significant difference is in the way that this gradient is generated.

One component of the photosynthetic electron transport chain, the cytochrome b_6f complex, acts as a proton pump in a similar manner to Complexes I, III and IV of the mitochondrial respiratory chain. The cytochrome b_6f complex moves H^+ ions from the stroma to the space within the thylakoids (*Fig. 10.10*). This is the only component of the photosynthetic electron transport chain that can actively transfer H^+ ions across the thylakoid membrane, but the gradient is supplemented by two other reactions occurring during electron transport:

Box 10.3 The role of carotenoid pigments in photoprotection

As well as contributing to light harvesting, the carotenoid pigments play an important role in **photoprotection**. This is a process that prevents the photosystems from becoming damaged at high light intensities. Under these conditions, excitation of electrons can occur at such high rates that not all of the absorbed energy can be channeled to the reaction centers. This situation is potentially harmful because the energy could be passed to oxygen molecules in the chloroplast, converting these into **singlet oxygen** (O^*), which in turn can give rise to **reactive oxygen species** such as hydrogen peroxide. Reactive oxygen species are potent oxidizing agents that can impair cellular function by damaging membranes and inactivating enzymes.

To avoid the production of singlet oxygen, the excess energy absorbed by chlorophyll at high light intensity is transferred to carotenoids present in or adjacent to the antenna complex. High light intensity induces the **xanthophyll cycle**, which results in chemical modification of certain carotenoids to give derivatives with energy quenching properties. One example is the synthesis of zeaxanthin from violaxanthin. Transfer of energy from an excited chlorophyll molecule to zeaxanthin enables the energy to be dissipated safely as heat. The technical name for this process is **non-photochemical quenching**.

Box 10.4 **The Z scheme**

The photosynthetic electron transport system is sometimes called the **Z scheme**. The name refers to the shape of a graph depicting the changes in redox potential that occur along the pathway that begins with the initial excitation of the P680 reaction center and ends with the reduction of NADP+ to NADPH by NADP reductase. Usually, redox potential is plotted on the y axis, as shown here, which means that the resulting graph is more N-shaped than Z-shaped. The graph shows an initial increase in redox potential caused by excitation of P680, followed by a gradual decline as electrons are transferred to plastoquinone, the cytochrome b_6f complex and then to plastocyanin. At this point the graph shows a second increase in redox potential with excitation of P700, followed by another decline as electrons flow to ferrodoxin and NADP reductase.

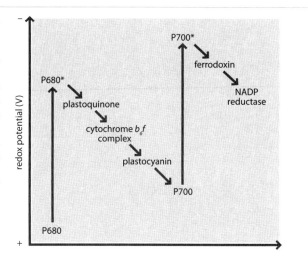

- The H+ ions released when water is oxidized by P680+ (*Step 1B* above) accumulate within the thylakoid, adding to the net positive charge on the internal side of the thylakoid membrane.

- The H+ ions that are consumed in order for NADP reductase to convert NADP+ to NADPH (*Step 4*) are taken from the stroma, contributing to the net negative charge on the stromal side of the thylakoid membrane.

Combined with the proton pumping of the cytochrome b_6f complex, the oxidation of water and reduction of NADP+ sets up an electrochemical gradient steep enough to generate ATP.

The generation of ATP from ADP and inorganic phosphate occurs exactly as in the mitochondrion. The chloroplast ATP synthase is structurally very similar to the mitochondrial F_0F_1 ATPase, acting as a molecular motor driven by the movement of protons from the thylakoid space to the stroma.

Photophosphorylation can be uncoupled from NADPH synthesis

The system we have looked at so far links the synthesis of ATP with that of NADPH. It is impossible in the scheme shown in *Figure 10.8* to generate the electrochemical gradient for photophosphorylation without reducing NADP+ to NADPH. This presents a problem in that there might be occasions when the chloroplast's supply of NADP+ is low, and hence it is unable to make NADPH at any significant rate. Yet there might still be a need for ATP synthesis. How can this problem be circumvented?

The answer is by switching to an alternative electron transport pathway, one that enables photophosphorylation to be uncoupled from NADPH synthesis. This is called **cyclic photophosphorylation** and is shown in *Figure 10.11*. In this system, the excited state of the P700 reaction center (P700*) passes high energy electrons to ferrodoxin, as in the non-cyclic pathway, but from ferrodoxin the electrons flow to the cytochrome b_6f complex, and from there to plastocyanin. The electrons are then passed back to P700+, to return the reaction center to its ground state. This means that the cytochrome b_6f complex proton pump is still operating and an electrochemical gradient can be set up, so ATP can still be made. By sending the electrons from plastocyanin back to P700+, rather than on to NADP reductase, no NADP+ is used up. Additionally, because the P680 photosystem is not involved, cyclic photophosphorylation does not generate oxygen.

Figure 10.11 Cyclic photophosphorylation. In this pathway, electrons flow from ferrodoxin to plastocyanin via the cytochrome b_6f complex, so proton pumping occurs and ATP can be synthesized. Plastocyanin regenerates P700 from P700+, which means that no NADPH synthesis occurs.

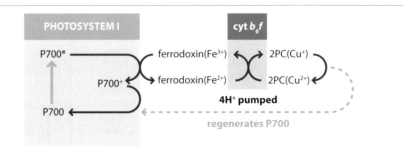

10.3 The dark reactions

In the second stage of photosynthesis the NADPH and ATP made by the light reactions are used to synthesize carbohydrate from carbon dioxide and water. This converts an inorganic form of carbon (CO_2) into an organic form (carbohydrate) and is therefore called **carbon fixation**. It occurs in the stroma and the end products are sucrose and starch.

10.3.1 The Calvin cycle

Central to the dark reactions is a cyclic pathway called the **Calvin cycle**, named after Melvin Calvin of Berkeley University who, with his colleagues Andrew Benson and James Bassham, first described the pathway in a series of landmark papers published between 1949 and 1953. The Calvin cycle generates one molecule of the three-carbon sugar glyceraldehyde 3-phosphate from three molecules of carbon dioxide. The energy required to achieve this carbon fixation is provided by six molecules of NADPH and nine of ATP. The overall reaction is therefore:

$$3CO_2 + 6NADPH + 9ATP \rightarrow \text{glyceraldehyde 3-phosphate} + 6NADP^+ + 9ADP + 8P_i$$

where P_i is, as usual, inorganic phosphate. The glyceraldehyde 3-phosphate that is produced is then used to make sucrose and starch.

Box 10.5 **Photosynthesis in bacteria**

Photosynthesis is not restricted to the chloroplasts of plants and algae. Several types of bacteria are also able to convert light energy into chemical energy.

In the cyanobacteria, the process is very similar to that occurring in chloroplasts. Cyanobacteria have internal membranous structures that resemble thylakoids, and capture light with a blue pigment called phycocyanin. They have photosystem I and II reaction centers and an electron transport chain almost identical to that of plants and algae. Because of these similarities, it has been suggested that chloroplasts are descended from cyanobacteria that formed a symbiotic relationship with the precursor of the eukaryotic cell, at an early stage in evolution (see *Section 2.1.3*).

Other photosynthetic prokaryotes include the purple and green bacteria, and members of the heliobacteria family.

These species contain one or more versions of the porphyrin called bacteriochlorophyll, which has a similar structure to the chlorophyll of chloroplasts, but absorbs light across a broader range of wavelengths. Energy is channeled to a single photosystem, linked to an electron transport chain comprising quinones and cytochromes, with protons pumped across the plasma membrane to generate the electrochemical gradient that enables ATP to be synthesized. One important difference is that these bacteria do not use water as the electron donor for regeneration of their photosystems. This means that they do not make oxygen and so carry out **anoxygenic photosynthesis**. The purple and green sulfur bacteria, as their names imply, both use hydrogen sulfide as their main electron donor, and so are characterized by the synthesis of elemental sulfur. With other species, hydrogen gas or ferrous ions act as the electron donor.

The starting point for the Calvin cycle is the reaction catalyzed by the enzyme called **ribulose bisphosphate carboxylase,** often abbreviated to **Rubisco**. This is the step that is directly responsible for the carbon fixation. We will begin by examining this reaction, before surveying the Calvin cycle as a whole and the way in which sucrose and starch are synthesized from glyceraldehyde 6-phosphate.

Carbon is fixed by ribulose bisphosphate carboxylase

The key step in the dark reactions is the one that is directly responsible for the fixation of carbon. In this step, one molecule of CO_2 combines with the five-carbon sugar ribulose 1,5-bisphosphate (*Fig. 10.12*). This gives an unstable six-carbon sugar, called 3-keto-2-carboxyarabinitol-1,5-bisphosphate, that immediately breaks down into two molecules of the triose 3-phosphoglycerate.

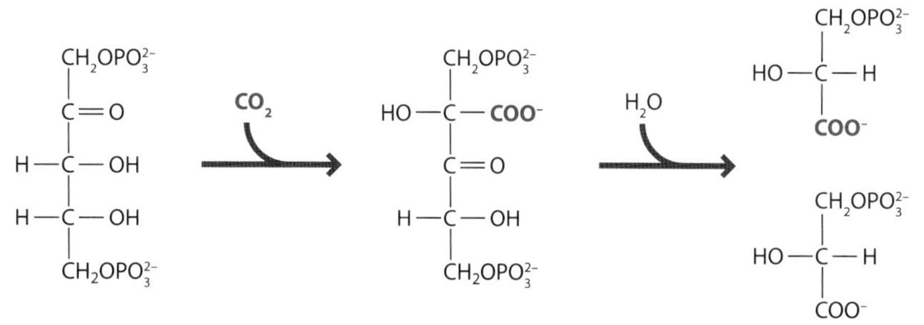

Figure 10.12 The reaction catalyzed by Rubisco.

ribulose 1,5-bisphosphate   3-keto-2-carboxyarabinitol-1,5-bisphosphate   2 × 3-phosphoglycerate

Ribulose bisphosphate carboxylase, the enzyme that catalyzes this reaction, has 16 subunits, eight of the large or L subunits and eight small or S subunits. Each of the large subunits contains an active site at which carbon dioxide can be combined with ribulose 1,5-bisphosphate. Part of the active site is contributed by a lysine that has been modified by addition of a carboxyl group to give the **carbamoyl** derivative of this amino acid (*Fig. 10.13*). This lysine binds a magnesium ion, Mg^{2+}, which plays the central role in bringing the reactants into the active site and placing these in the correct relative positions for the carbon fixation reaction to occur.

Rubisco works very slowly, carrying out only three reactions per second at 25°C. For this reason, a great deal of the enzyme must be present in each chloroplast in order to maintain the overall rate of carbon fixation at an acceptable level. This feature of the enzyme leads to some of the more amazing facts of biology. Rubisco makes up between 15 and 50% of the total protein in chloroplasts and is almost certainly the most abundant protein on the planet. One estimate is that 1000 kg of Rubisco is synthesized every second in the biosphere, and that there is the equivalent of 44 kg of Rubisco for every person alive.

The accuracy of these amazing facts about Rubisco is open to debate. What is more certain is that Rubisco is the main control point for the dark reactions, and in particular is responsive to the amount of sunlight, being less active when sunlight levels are low. This response ensures that synthesis of carbohydrate by the dark reactions is coordinated with the production of NADPH and ATP by the light ones.

In low light conditions, the active site of Rubisco is blocked. This can happen in two different ways. Newly synthesized Rubisco, in which the lysine at the active site has not yet been carbamoylated by carbon dioxide, binds ribulose 1,5-bisphosphate, but non-productively, meaning that no conversion to 5-phosphoglycerate occurs (*Fig. 10.14A*). When light levels rise, a second enzyme called **Rubisco activase** removes the ribulose 1,5-bisphosphate so lysine modification can occur and the productive binding of

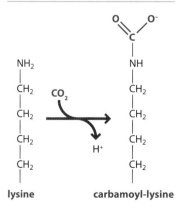

lysine   carbamoyl-lysine

Figure 10.13 Formation of the carbamoyl derivative of lysine.

Figure 10.14 Inhibition of Rubisco under low light conditions.
(A) Inhibition by non-productive binding of ribulose 1,5-bisphosphate to the unmodified lysine at the active site of newly-synthesized Rubisco, and (B) non-productive binding of 2-carboxy-ᴅ-arabitinol 1-phosphate to a lysine that has already been carbamoylated. Both types of inhibition can be reversed by Rubisco activase in response to increasing light intensity.

A. binding of ribulose 1,5-bisphosphate to unmodified lysine

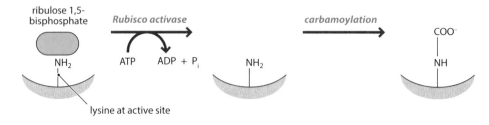

B. binding of 2-carboxy-ᴅ-arabitinol 1-phosphate to carbamoylated lysine

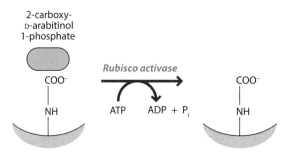

ribulose 1,5-bisphosphate can take place. Rubisco that already has modified lysines can be inactivated in a similar way, but not by ribulose 1,5-bisphosphate. Instead, the blocking compound is 2-carboxy-ᴅ-arabitinol 1-phosphate, a stable analog of the unstable six-carbon sugar produced as an intermediate of Rubisco activity (*Fig. 10.14B*). Again, Rubisco activase removes the bound 2-carboxy-ᴅ-arabitinol 1-phosphate at the appropriate time. Removal of either blocking compound by Rubisco activase requires conversion of ATP to ADP. The activity of the enzyme is therefore dependent on the amount of ATP that is present in the chloroplast stroma. If ATP is low, because the light intensity is low and little photophosphorylation is occurring, then Rubisco activase is inhibited and the blocking compounds are not removed from the Rubisco active sites. When light intensity increases and ATP levels rise, Rubisco activase is switched on and the blocking compounds are removed.

A second regulatory regime for Rubisco centers on the availability of magnesium ions. These are an essential cofactor for Rubisco activity, Mg^{2+} being required for substrate binding at the active site. The Mg^{2+} content of the stroma, in which Rubisco is located, is influenced by the proton gradient set up by the light reactions. When

Figure 10.15 Magnesium ions enter the stroma in response to the electrochemical gradient set up by proton pumping.
It is assumed that the thylakoid membrane contains a transport protein for movement of Mg^{2+} across the membrane, but this protein has not yet been isolated.

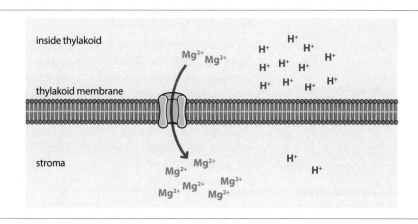

there is a net negative charge in the stroma compared to the thylakoid space, Mg^{2+} ions move from the thylakoid space to the stroma, ensuring a continual supply for Rubisco activity (*Fig. 10.15*). Conversely, when the light reactions are less active there will be a smaller net negative charge in the stroma, resulting in reduced import of Mg^{2+} and hence a limitation on Rubisco. Magnesium ion availability therefore provides a second means of linking Rubisco activity with that of the light reactions.

The steps of the Calvin cycle

As described above, the Calvin cycle results in the synthesis of one molecule of glyceraldehyde 3-phosphate from three of carbon dioxide. The cycle as a whole is shown in *Figure 10.16*. Note that in the Calvin cycle it is important to consider the **stoichiometry** at each step. This refers to the number of molecules of each reactant that are used and the number of molecules of each product that are made. The stoichiometry is important because we need to fix three molecules of carbon

Figure 10.16 The Calvin cycle.
The steps in the process, as described in the text, are indicated by the green circles. The abbreviations C3, C5 and C6 indicate the number of carbons in the different sugar compounds.

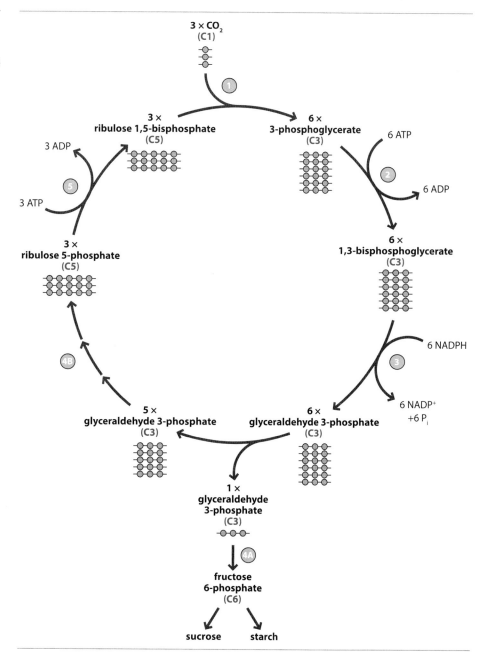

dioxide, requiring three parallel Rubisco reactions, in order eventually to produce one molecule of glyceraldehyde 3-phosphate.

The individual steps of the Calvin cycle are as follows:

Step 1. Rubisco carries out the carbon fixation step. This step was illustrated in *Figure 10.12.*

Step 2. The 3-phosphoglycerate molecules are phosphorylated to 1,3-bisphosphoglycerate by **phosphoglycerate kinase**, using up one molecule of ATP for each molecule of 3-phosphoglycerate.

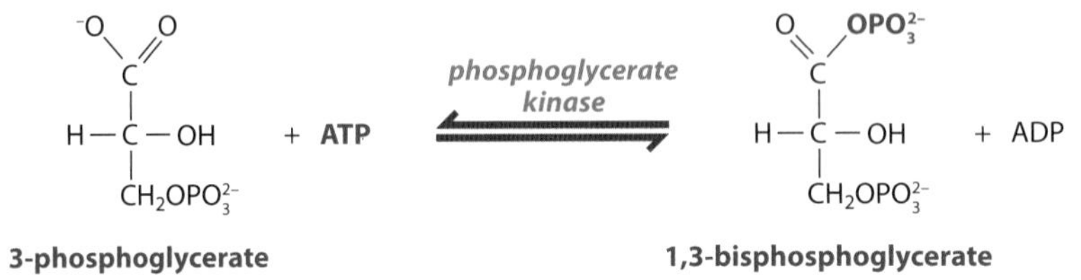

3-phosphoglycerate   **1,3-bisphosphoglycerate**

Step 3. **Glyceraldehyde 3-phosphate dehydrogenase** converts 1,3-bisphosphoglycerate to glyceraldehyde 3-phosphate, with loss of inorganic phosphate. Energy is provided by NADPH.

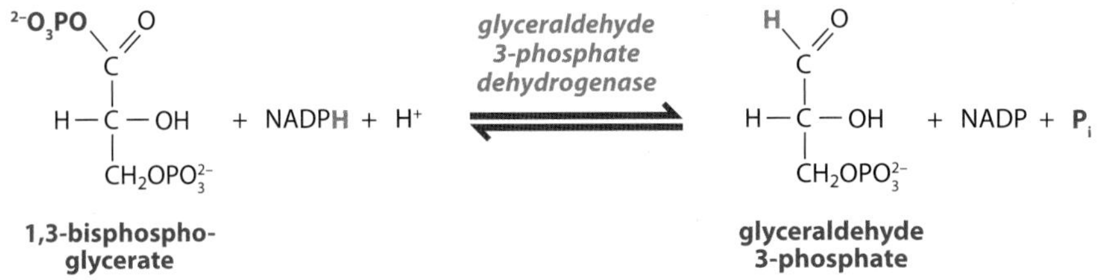

1,3-bisphospho-glycerate   **glyceraldehyde 3-phosphate**

Phosphoglycerate kinase and glyceraldehyde 3-phosphate dehydrogenase are also involved in glycolysis, catalyzing steps 7 and 6 of that pathway, respectively. The reactions in the Calvin cycle are therefore the reverse of that part of glycolysis, the only difference being that in glycolysis the energy released at step 6 is stored as NADH, and in the Calvin cycle the energy required for step 3 is provided by NADPH.

Step 4A. One molecule of glyceraldehyde 3-phosphate is diverted out of the Calvin cycle as the product that is subsequently used in production of sucrose and starch.

Step 4B. The other five molecules of glyceraldehyde 3-phosphate are converted to three molecules of the five-carbon sugar ribulose 5-phosphate. This is, in reality, a complex series of individual steps involving eight intermediate sugars and catalyzed by an aldolase, a transketolase and five other enzymes. Ignoring the intermediates, the reaction can be summarized as:

glyceraldehyde 3-phosphate   **ribulose 5-phosphate**

Step 5. Ribulose 5-phosphate is phosphorylated to ribulose 1,5-bisphosphate by **ribulose 5-phosphate kinase**. ATP is consumed.

$$
\begin{array}{ccc}
\text{O} \diagdown \text{CH}_2\text{OH} & & \text{O} \diagdown \text{CH}_2\text{OPO}_3^{2-} \\
\text{C} & & \text{C} \\
\text{H} - \text{C} - \text{OH} & \xrightarrow{\text{ATP} \quad \text{ADP}} & \text{H} - \text{C} - \text{OH} \\
\text{H} - \text{C} - \text{OH} & \textit{ribulose} & \text{H} - \text{C} - \text{OH} \\
\text{CH}_2\text{OPO}_3^{2-} & \textit{5-phosphate kinase} & \text{CH}_2\text{OPO}_3^{2-} \\
\textbf{ribulose} & & \textbf{ribulose} \\
\textbf{5-phosphate} & & \textbf{1,5-bisphosphate}
\end{array}
$$

We have now regenerated the ribulose 1,5-bisphosphate needed for another round of the cycle.

Ferrodoxin from the light reaction pathway regulates the Calvin cycle

We have already seen how the carbon fixation step catalyzed by Rubisco is regulated indirectly by sunlight, via Rubisco activase and the magnesium ion content of the stroma. The subsequent flow of metabolites around the Calvin cycle is regulated by controlling the activities of some of the other enzymes involved in the pathway. These enzymes include:

- Phosphoglycerate kinase and glyceraldehyde 3-phosphate dehydrogenase, catalyzing *Steps 2* and *3*, respectively.

- Ribulose 5-phosphate kinase, catalyzing *Step 5*.

- Two enzymes, fructose 1,6-bisphosphatase and sedoheptulose-1,7-bisphosphatase, that are involved in the complex series of reactions that make up *Step 4B*.

The catalytic activity of each of these enzymes can be inhibited by formation of an additional disulfide bond. These bonds are not themselves at the active sites of their enzymes, but their presence disrupts the active site such that the catalytic reaction can no longer take place (*Fig. 10.17A*). Control over the presence or absence of these disulfide bonds is therefore a means of regulating the activities of the enzymes.

Ferrodoxin, the electron carrier immediately downstream of P700* in the light reactions, indirectly influences the presence of these inhibitory disulfide bonds in the Calvin cycle enzymes. The reduced version, ferrodoxin(Fe^{2+}), which is produced when

Figure 10.17 Regulation of Calvin cycle enzymes by disulfide bond formation.
(A) With some enzymes, formation of an additional disulfide bond results in inactivation. (B) Disulfide bond formation is mediated by ferrodoxin(Fe^{2+}), which reduces thioredoxin by breaking a disulfide bond. The reduced form of thioredoxin can in turn break the inhibitory disulfide bond on the target enzyme.

A. inactivation of an enzyme by disulfide bond formation

B. the role of thioredoxin in enzyme inactivation

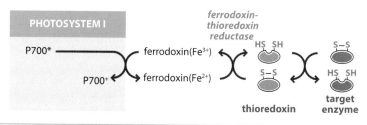

the light reactions are active, in turn converts the protein **thioredoxin** to its reduced form, by a reaction catalyzed by **ferrodoxin–thioredoxin reductase** (*Fig. 10.17B*). Reduced thioredoxin cleaves the inhibitory disulfide bonds, ensuring that the activity of the Calvin cycle enzymes is coordinated with that of the light reactions. The action of thioredoxin has to be continuous, because other enzymes are constantly attempting to re-form the disulfide bonds. This means that as soon as the light reactions slow down, bond formation occurs and the Calvin cycle also slows down.

10.3.2 Synthesis of sucrose and starch

In the final set of dark reactions, the glyceraldehyde 3-phosphate molecules produced by the Calvin cycle are used to synthesize fructose 6-phosphate and glucose 6-phosphate, which can then be further metabolized to generate sucrose and starch. Sucrose is transported around the plant and used as the immediate energy supply, and the formation of starch provides a means of storing the energy surplus.

Sucrose is synthesized in the cytoplasm

The majority of the glyceraldehyde 3-phosphate molecules produced by the Calvin cycle are used to make sucrose. The first step in this process occurs in the chloroplast, where some of the glyceraldehyde 3-phosphate is converted to dihydroxyacetone phosphate by the enzyme triose phosphate isomerase (*Fig. 10.18*). This reaction is the reverse of step 5 of glycolysis. The glyceraldehyde 3-phosphate and dihydroxyacetone phosphate molecules then move out of the chloroplast and into the cytoplasm. Transfer across the inner chloroplast membrane requires a special membrane-spanning transport protein, but both compounds can pass directly though the more permeable outer membrane.

In the cytoplasm, aldolase joins together glyceraldehyde 3-phosphate and dihydroxyacetone phosphate to form fructose 1,6-bisphosphate, which **fructose 1,6-bisphosphatase** converts to fructose 6-phosphate, and phosphoglucoisomerase rearranges to give glucose 6-phosphate. Again these reactions reproduce, in reverse direction, a part of the glycolysis pathway, in this case steps 4–2.

There are now two steps that we do not see in glycolysis. The phosphate group in glucose 6-phosphate is moved from carbon number 6 to carbon number 1, to give

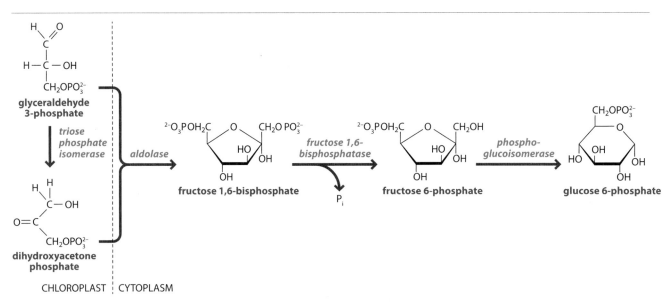

Figure 10.18 The first steps in the synthesis of sucrose.

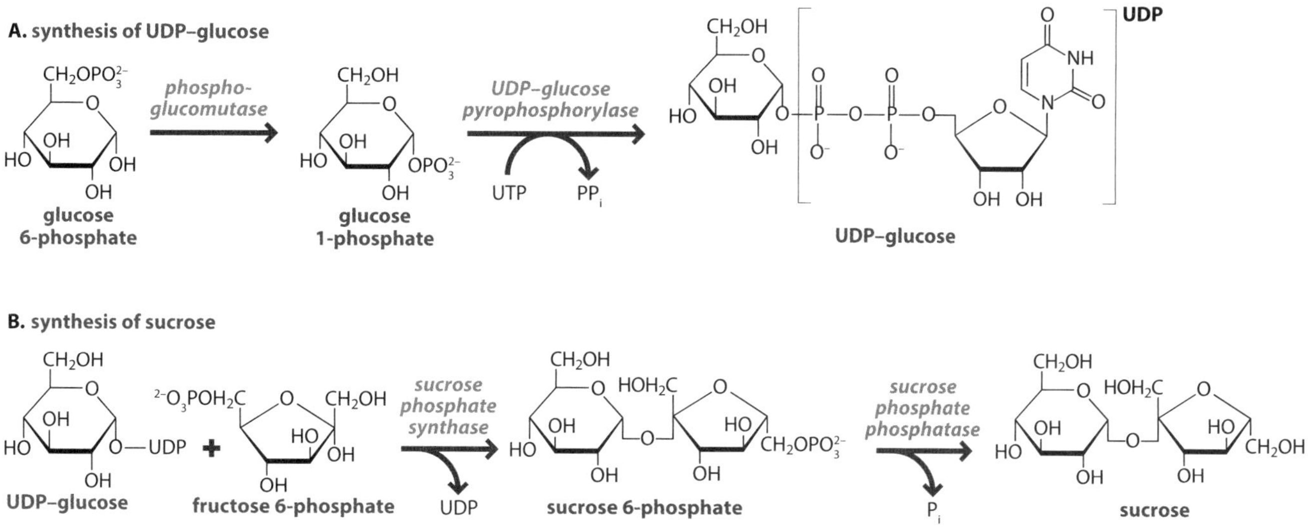

A. synthesis of UDP–glucose

B. synthesis of sucrose

Figure 10.19 **The final two stages in the synthesis of sucrose.**

glucose 1-phosphate (*Fig. 10.19A*). This reaction is catalyzed by **phosphoglucomutase**. Then the phosphate is replaced with uridine diphosphate (UDP) by **UDP–glucose pyrophosphorylase**. This results in an activated form of glucose that can now be linked to the fructose 6-phosphate formed earlier in the pathway to produce the disaccharide sucrose 6-phosphate, which is then dephosphorylated to give the end product, sucrose (*Fig. 10.19B*). The enzymes for these two steps are **sucrose phosphate synthase** and **sucrose phosphate phosphatase**, respectively.

Starch is synthesized in chloroplasts and amyloplasts

Starch is also synthesized from an activated glucose intermediate, though this is usually **ADP–glucose**, occasionally CDP– or GDP–glucose, but never UDP–glucose. The ADP–glucose is made in the stroma of the chloroplast, from glucose 1-phosphate that is produced by a parallel reaction pathway to that occurring in the cytoplasm. ADP–glucose molecules are then added to the ends of a growing starch molecule by **starch synthase** (*Fig. 10.20*). This facet of biochemistry has proved controversial in recent years due to the presence of two distinct ends on a starch molecule, described as reducing and non-reducing, depending on whether the chain terminates with the anomeric carbon 1 (reducing end) or non-anomeric carbon 4 (non-reducing end). For many years it was thought that starch synthase only adds glucose units to the reducing end, but now it has been shown that the enzyme can also catalyze transfer to the non-reducing end. There is also a **starch branching enzyme** that can synthesize the $\alpha(1\rightarrow6)$ links that result in the branched structure of the amylopectin version of starch (see *Fig. 6.12*).

These events, occurring in the chloroplast, are termed **transient** or **transitory starch synthesis**. The starch reserves that are formed are short-lived, most being used up to supply energy for the plant during the following period of darkness. **Stored starch synthesis**, in contrast, takes place not in the chloroplasts but in related structures called **amyloplasts** which are present in roots, developing seeds and storage organs such as fruits and tubers. The glucose used to produce stored starch is transported to these tissues, from the sites of photosynthesis within the leaves and other aerial parts of the plant.

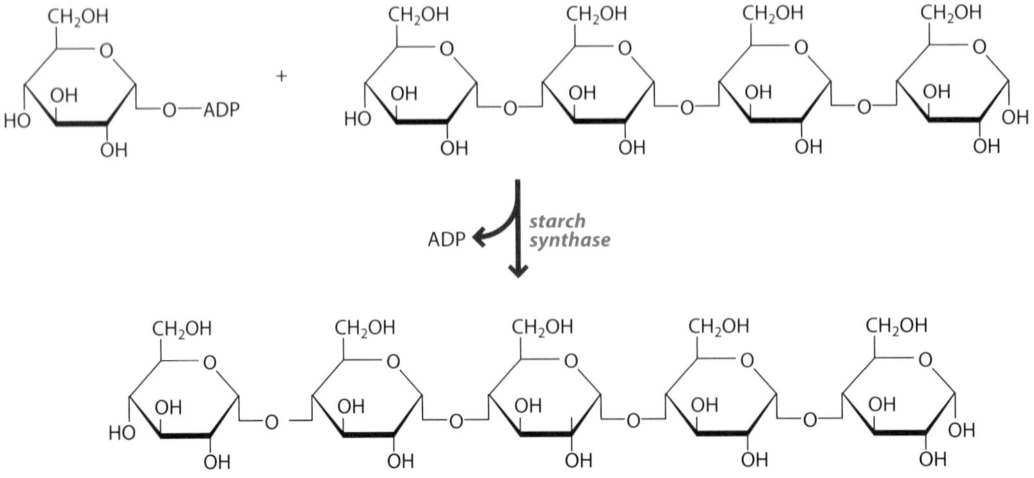

Figure 10.20 Synthesis of starch.
The addition of a glucose unit to the non-reducing end of a starch polymer is shown.

10.3.3 Carbon fixation by C4 and CAM plants

The main function of Rubisco is to add carbon dioxide to ribulose 1,5-bisphosphate producing two molecules of 3-phosphoglycerate. But the enzyme can also carry out an alternative reaction in which it adds oxygen instead of carbon dioxide, resulting in synthesis of just one 3-phosphoglycerate molecule, along with a molecule of 2-phosphoglycolate (*Fig. 10.21*). Formation of 2-phosphoglycolate is a problem for the plant because this compound inhibits some of the enzymes of the Calvin cycle. A lengthy series of reactions, referred to as **photorespiration** because they result in the generation of carbon dioxide, converts 2-phosphoglycolate into 3-phosphoglycerate, but the pathway uses up ATP and NADH and so is energetically wasteful.

Oxygen and carbon dioxide compete for the Rubisco active site, with approximately 75% of the enzyme reactions using carbon dioxide and the remainder going along the photorespiration route. This ratio can be tolerated by most plants, but some species live in environments which enhance the problem caused by the ability of Rubisco to use oxygen as well as carbon dioxide. These species, which make up the **C4** and **CAM** groups of plants, have evolved alternative ways of delivering carbon dioxide to Rubisco.

Figure 10.21 The first stage of photorespiration, resulting in synthesis of 2-phosphoglycolate.

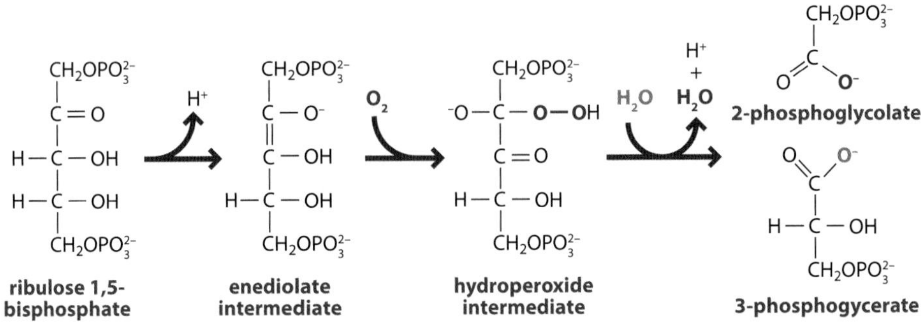

Box 10.6 Increasing the photosynthetic capability of crop plants

Plant biologists are not certain why Rubisco carries out the alternative oxygen reaction. Possibly the enzyme evolved at a stage in the Earth's geological prehistory when the oxygen content of the atmosphere was very low, so that primordial Rubisco did not need to discriminate between oxygen and carbon dioxide. Whatever the explanation, the outcome is that on today's Earth photosynthesis is an inherently inefficient process. In the natural environment, most plants are able to tolerate the loss of approximately 25% of Rubisco function to the oxygenation reaction. The situation is different in agricultural settings, where the inefficiency of photosynthesis is looked on as a significant factor limiting the productivity of crop species. Plant biotechnologists are therefore searching for ways of improving Rubisco activity in the hope that this will help to increase the amount of plant protein that can be obtained per acre of cultivated land.

One strategy for improving Rubisco activity in crops is based on the discovery that the Rubisco enzymes of cyanobacteria operate at a higher catalytic rate than those in plants. The cyanobacterial enzymes are therefore able to synthesize 3-phosphoglycerate more rapidly than their eukaryotic counterparts. If genetic engineering (see *Section 19.3.1*) could be used to transfer the genes for the large and small subunits of Rubisco from a cyanobacterium into a plant, then the engineered plant would possibly photosynthesize at a higher rate and be more productive. To test this hypothesis, the genes for the *Synechococcus elongates*

Rubisco enzyme have been transferred into tobacco. The photosynthetic activity of the engineered plants was tested by homogenizing leaf segments in an extraction buffer containing different amounts of sodium bicarbonate.

The results of this assay (graph at bottom left) reveal that the rate of carbon dioxide fixation in the engineered plants is up to three times that of normal tobacco, showing that the cyanobacterial enzyme retains its higher catalytic rate even when expressed in the new host. However, this is far from the end of the project. When grown into mature plants, the engineered tobacco varieties develop more slowly than normal plants and do not display increased productivity. This outcome was not unexpected. It is known that in cyanobacteria there is a trade-off between Rubisco activity and specificity. In other words, the high catalytic activity comes with a lesser degree of discrimination between oxygen and carbon dioxide. This means that although cyanobacterial Rubisco works more rapidly than the plant enzyme, a greater proportion of the reactions that it catalyzes are the less productive oxygenations. Cyanobacteria mitigate this problem by locating their Rubisco enzymes in structures called **carboxysomes**. These are protein spheres that also contain carbonic anhydrase, an enzyme that converts bicarbonate ions (resulting from the dissolution of carbon dioxide in water) into carbon dioxide. Within the cyanobacterium, bicarbonate diffuses into a carboxysome, where it is converted into carbon dioxide. The local CO_2 concentration within the carboxysome is therefore relatively high, so the majority of the reactions catalyzed by the Rubisco enzymes are carboxylations, simply because there is much more carbon dioxide than oxygen available.

To utilize fully the efficiency of cyanobacterial photosynthesis in engineered crop plants, it will therefore be necessary not only to transfer the genes for Rubisco into the crop, but also genes for the carboxysome proteins and for the cyanobacterial carbonic anhydrase. That is a much more challenging proposition, but it has already been shown that carboxysome proteins can be synthesized in engineered plants, and that these proteins assemble into structures resembling the carboxysomes present in cyanobacteria. Use of the cyanobacterial Rubisco system to boost the productivity of crops might therefore be a future possibility.

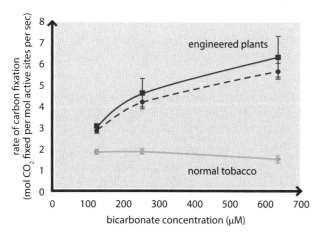

C4 plants shuttle fixed carbon between different types of cell

The 75:25 ratio for carbon dioxide versus oxygen utilization by Rubisco depends on the relative amounts of carbon dioxide and oxygen in the atmosphere, which of course is the same in all parts of the world. In hot climates, however, plants have to conserve water, and one way they do this is temporarily to close the stomata on their leaves (*Fig. 10.22*). As well as reducing water loss, closure of the stomata also prevents gas exchange between the inner parts of the plant and the outside environment. As a consequence the trapped air supply becomes depleted in carbon dioxide as photosynthesis continues, favoring photorespiration.

To avoid this increase in photorespiration, some plants restrict the Calvin cycle reactions to a specialized tissue made up of **bundle sheath cells**, which is surrounded

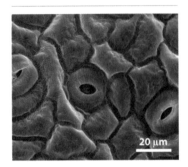

Figure 10.22 Open and closed stomata on a cherry leaf.
Image reproduced from www.deviantart.com.

by **mesophyll cells** (*Fig. 10.23*). In the mesophyll cells, carbon dioxide is fixed by the enzyme **phosphoenolpyruvate carboxylase**, the carbon dioxide being added to phosphoenolpyruvate to give oxaloacetate. Oxygen has no effect on the fixation of carbon dioxide by phosphoenolpyruvate carboxylase, so this reaction occurs at high efficiency even when the carbon dioxide levels are low. The oxaloacetate that is produced is converted to malate by **NADP-linked malate dehydrogenase**, and the malate transported from the mesophyll cells to the bundle sheath cells. There, the **NADP-linked malate enzyme** releases the carbon dioxide, which is collected by Rubisco and re-fixed to give 3-phosphoglycerate, initiating a round of the Calvin cycle. This direct delivery of carbon dioxide to Rubisco means that in the bundle sheath cells the Calvin cycle can proceed at maximal rate despite the low ambient level of carbon dioxide within the rest of the plant tissue.

To complete this carbon dioxide shuttle system, pyruvate, which is the second product of the NADP-linked malate enzyme reaction, is moved back to the mesophyll cells, where it is phosphorylated by **pyruvate-P$_i$ dikinase**, regenerating the phosphoenolpyruvate used in the initial carbon fixation. Species which use this shuttle are called C4 plants, because the carbon dioxide is initially fixed as the four-carbon oxaloacetate. Those species in which Rubisco carries out the initial carbon fixation, giving rise to three-carbon 3-phosphoglycerate molecules, are called **C3 plants**. There are about 7500 C4 plant species, representing less than 5% of the total number of terrestrial plant species, common examples being maize, sorghum and various grasses.

CAM plants use temporal control over carbon fixation

Some tropical plant species, including cacti and orchids, use the same biochemical pathway as C4 plants to deliver pre-fixed carbon dioxide to Rubisco, but do so without the need to shuttle malate between cells. Instead the two parts of the cycle occur at different times of day. During the night, when temperatures are low, the stomata of these plants are open, enabling carbon dioxide to enter. This carbon dioxide is

Figure 10.23 The C4 pathway.
A section through a leaf of finger millet (a typical C4 plant) shows bundle sheath cells surrounded by mesophyll cells. In the C4 pathway the enzymes catalyzing each step are as follows: 1, phosphoenolpyruvate carboxylase; 2, NADP-linked malate dehydrogenase; 3, NADP-linked malate enzyme; 4, pyruvate-P$_i$ dikinase. Note that the pathway has an energetic cost because step 4 uses a molecule of ATP. The scale bar is 50 μm.
Photograph reproduced from *Plant and Cell Physiology* 2009; **50**: 1736, with permission from Oxford University Press.

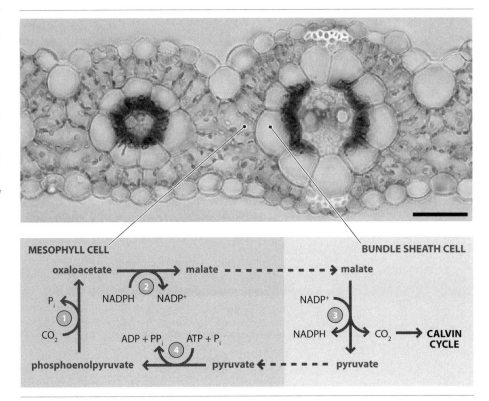

immediately fixed to give malate, which is stored in vacuoles inside the plant cells. During the day, when the stomata are closed, the malate is shuttled to the chloroplasts where the carbon dioxide is released, to be re-fixed by Rubisco.

Species using this system are called CAM plants. 'CAM' stands for 'crassulacean acid metabolism', but you will search biochemistry textbooks in vain for a description of the structure and properties of crassulacean acid. Instead the name refers to acid metabolism in the Crassulaceae family of plants, in which this process, with malic acid at its center, was first discovered.

Further reading

Allen JF (2003) Cyclic, pseudocyclic and noncyclic photophosphorylation: new links in the chain. *Trends in Plant Science*, **8**, 15–19.

Andersson I and Backlund A (2008) Structure and function of Rubisco. *Plant Physiology and Biochemistry* **46**, 275–91.

Cogdell RJ, Isaacs NW, Howard TD, McLuskey K, Fraser NJ and Prince SM (1999) How photosynthetic bacteria harvest solar energy. *Journal of Bacteriology* **181**, 3869–79.

Denning-Adams B and Adams WW (1992) Photoprotection and other responses of plants to high light stress. *Annual Review of Plant Physiology and Plant Molecular Biology* **43**, 599–626.

Law CJ, Roszak AW, Southall J, Gardiner AT, Isaacs NW and Cogdell RJ (2004) The structure and function of bacterial light-harvesting complexes. *Molecular Membrane Biology* **21**, 183–91.

Lin MT, Occhialini A, Andralojc PJ, Parry MAJ and Hanson MR (2014) A faster Rubisco with potential to increase photosynthesis in crops. *Nature* **513**, 547–50.

Nelson N and Ben-Shem A (2004) The complex architecture of oxygenic photosynthesis. *Nature Reviews Molecular and Cell Biology* **5**, 971–82. *The photosynthetic electron transport chain.*

Portis AR, Li C, Wang D and Salvucci ME (2008) Regulation of Rubisco activase and its interaction with Rubisco. *Journal of Experimental Botany* **59**, 1597–604.

Tetlow IJ and Emes MJ (2014) A review of starch-branching enzymes and their role in amylopectin biosynthesis. *IUBMS Life* **66**, 546–58.

Vinyard DJ, Ananyev GM and Dismukes GC (2013) Photosystem II: the reaction center of oxygenic photosynthesis. *Annual Review of Biochemistry* **82**, 577–606.

Yamori W, Hikosaki K and Way DA (2014) Temperature response of photosynthesis in C3, C4 and CAM plants: temperature acclimation and temperature adaptation. *Photosynthesis Research* **119**, 101–17.

Self-assessment questions

Multiple choice questions

Only one answer is correct for each question. Answers can be found on the website: www.scionpublishing.com/biochemistry.

1. Which of the following terms is a synonym for 'primary producer'?
   (a) Amphitroph
   (b) Antitroph
   (c) Autotroph
   (d) Auxotroph

2. Where is the chloroplast electron transport chain located?
   (a) Inner chloroplast membrane
   (b) Stroma
   (c) Thylakoid membrane
   (d) Thylakoid space

3. Which of the following types of bacteria are able to photosynthesize?
   (a) Cyanobacteria

(b) Green bacteria

(c) Purple bacteria

(d) All of the above

4. Which are the main light-harvesting pigments of plants and green algae?

(a) Chlorophyll *a* and chlorophyll *b*

(b) Chlorophyll *a* and chlorophyll *c*

(c) Chlorophyll *b* and chlorophyll *c*

(d) Chlorophyll *d* and chlorophyll *f*

5. Which one of the following compounds is not an accessory pigment?

(a) β-Carotene

(b) Ferrodoxin

(c) Fucoxanthin

(d) Xanthophyll

6. What is the process by which an energy quantum is passed from one chlorophyll molecule to another called?

(a) Direct electron transfer

(b) Orbital transfer

(c) Photonics

(d) Resonance energy transfer

7. What is the photosystem I reaction center called?

(a) P600

(b) P680

(c) P690

(d) P700

8. What is the photosystem II reaction center called?

(a) P600

(b) P680

(c) P690

(d) P700

9. The xanthophyll cycle is involved in which process?

(a) Light harvesting

(b) Photophosphorylation

(c) Photoprotection

(d) Xanthophyll synthesis

10. What is the intermediate compound between photosystems II and I in the photosynthetic electron transport chain called?

(a) Ferrodoxin

(b) Plastocyanin

(c) Plastoquinone

(d) Xanthophyll

11. Which one of the following statements regarding cyclic photophosphorylation is **incorrect**?

(a) From ferrodoxin the electrons flow to the cytochrome b_6f complex

(b) Oxygen is generated

(c) The cytochrome b_6f complex proton pump still operates

(d) The excited state of the P700 reaction center passes high energy electrons to ferrodoxin

12. Rubisco combines one molecule of CO_2 with which five-carbon sugar?

(a) Ribulose 1,2-bisphosphate

(b) Ribulose 1,5-bisphosphate

(c) Ribulose 1,6-bisphosphate

(d) Ribulose 2,5-bisphosphate

13. The active site of Rubisco contains a lysine that has been modified by what process?

(a) Carbamoylation

(b) Methylation

(c) Oxidation

(d) Phosphorylation

14. What is the enzyme that controls Rubisco activity called?

(a) Phosphoribulokinase

(b) Rubisco activase

(c) Rubisco regulase

(d) This is a trick question because Rubisco activity is controlled solely by substrates and products

15. What is the enzyme that uses ATP to make ribulose 1,5-bisphosphate called?

(a) Glyceraldehyde 3-phosphate dehydrogenase

(b) Phosphoglycerate kinase

(c) Ribulose 5-phosphate kinase

(d) Phosphoglucomutase

16. What is the compound that activates Calvin cycle enzymes by cleaving the inhibitory disulfide bonds in those enzymes' molecules?

(a) Oxidized ferrodoxin

(b) Reduced ferrodoxin

(c) Oxidized thioredoxin

(d) Reduced thioredoxin

17. Sucrose is synthesized in what part of a plant cell?

(a) Amyloplasts

(b) Chloroplasts

(c) Cytoplasm

(d) Mitochondria

18. What is the activated intermediate used in synthesis of starch?

(a) ADP–glucose

(b) ADP–fructose

(c) UDP–glucose

(d) UDP–fructose

19. Stored starch synthesis takes place in what part of a plant cell?

(a) Amyloplasts

(b) Chloroplasts

(c) Cytoplasm

(d) Mitochondria

20. Which one of the following statements regarding C4 plants is **incorrect**?

(a) Calvin cycle reactions take place in bundle sheath cells

(b) Carbon dioxide is fixed in mesophyll cells

(c) Carbon dioxide is fixed by phosphoenolpyruvate carboxylase

(d) Carbon fixation and the Calvin cycle occur at different times of the day

Short answer questions

These questions do not require additional reading.

1. Draw a diagram showing the internal structure of a chloroplast and indicate where in the chloroplast the different stages of photosynthesis take place.

2. Distinguish between the roles of chlorophyll and accessory pigments in photosynthesis.

3. Describe how the antenna complex of a photosystem captures light energy and channels this energy to the reaction center.

4. Distinguish between the role of photosystems I and II in photophosphorylation.

5. What is cyclic photophosphorylation and why is it important?

6. Describe the reaction catalyzed by Rubisco and outline how the activity of this important enzyme is regulated.

7. Draw an outline of the Calvin cycle, showing the substrates, products and enzymes for each of the important steps.

8. Describe how ferrodoxin regulates the activity of TCA cycle enzymes.

9. Distinguish between the biochemical pathways that result in synthesis of (a) sucrose, and (b) starch in plant cells.

10. What is the influence on photosynthesis of the ability of Rubisco to use oxygen as a substrate, and how is this problem solved by C4 and CAM plants?

Self-study questions

These questions will require calculation, additional reading and/or internet research.

1. 'Red tides' are caused by proliferations of dinoflagellates and other algae in seawater. Sketch the expected absorbance spectrum for the major light-harvesting pigments in these algae.

2. It has been claimed, though not substantiated, that a diet with a high content of plants that produce zeaxanthin can confer protection against some types of age-related macular degeneration, a disease of the eye. Provide a possible explanation for this claim.

3. The thylakoid membrane is more permeable to Mg^{2+} and Cl^- than is the inner mitochondrial membrane. As a consequence, the movement of protons across the thylakoid membrane during the photosynthetic electron transport process is accompanied by a certain amount of movement of Mg^{2+} and Cl^- across the membrane. Discuss the impact that the movement of these ions will have on the way in which the electrochemical gradient that drives ATP synthesis is set up in chloroplasts.

4. In *Section 9.2.4* the effects of various inhibitors of the mitochondrial electron transport chain were described. What would you predict to be the effect of (a) cyanide and (b) 2,4-dinitrophenol on photophosphorylation?

5. If a C3 and C4 plant are grown side by side in a sealed container, in pots containing an adequate supply of nutrients and moisture, and under appropriate lighting conditions, the C3 plant will gradually die whereas the C4 plant will continue to grow. Explain this observation.

CHAPTER 11
Carbohydrate metabolism

STUDY GOALS

After reading this chapter you will:

- be able to describe the pathway for synthesis and degradation of glycogen

- understand how glycogen metabolism is regulated by hormonal and allosteric control

- know what a 'futile cycle' is and why cells must avoid futile cycles

- understand the role of gluconeogenesis as a source of glucose during starvation and excessive exercise

- be able to describe the steps in the gluconeogenesis pathway

- know the positions at which various substrates enter gluconeogenesis

- understand how gluconeogenesis and glycolysis are coordinated

- know the various roles of the pentose phosphate pathway

- be able to distinguish between the oxidative and non-oxidative phases of the pentose phosphate pathway

- be able to describe the steps in the pentose phosphate pathway

In the next three chapters we will look at the metabolic pathways that are relevant to the major types of biomolecule in living cells: carbohydrates, lipids, and the nitrogen-containing compounds which include proteins and nucleic acids. We begin in this chapter with carbohydrates.

We have, in fact, already covered a large part of the information relevant to carbohydrate metabolism. This is because glycolysis and the TCA cycle are metabolic pathways involved in the breakdown of carbohydrates, and the dark reactions of photosynthesis make up a metabolic pathway that synthesizes carbohydrate. There are, however, three other important aspects of carbohydrate metabolism that we have not studied so far. These are:

- **Glycogen metabolism**, comprising the pathways that synthesize and break down glycogen in animal cells.

- **Gluconeogenesis**, which synthesizes glucose from precursor compounds that are not carbohydrates.

- The **pentose phosphate pathway**, which has various functions, but importantly is the cell's major source of NADPH.

11.1 Glycogen metabolism

Animals store glucose as polymeric glycogen, in a manner similar to the storage of glucose as starch in plants. The main glycogen stores are in the muscle and liver, the glycogen in muscles being used as a source of energy for muscular activity, and the stores in the liver providing energy for most other parts of the body. So we must examine the pathways by which glycogen is synthesized and broken down in liver and muscle cells and, importantly, also look closely at the regulatory processes that control the cycle of glycogen synthesis and degradation. These control processes are vital because they are responsible for ensuring that muscle tissue and organs such as the brain receive an unbroken and adequate energy supply.

11.1.1 Synthesis and degradation of glycogen

Glycogen is a branched polysaccharide made up of $\alpha(1{\rightarrow}4)$ chains and $\alpha(1{\rightarrow}6)$ branch points, the branches occurring approximately every ten glucose units along each linear chain (*Fig. 11.1*). Its structure is therefore similar to that of the amylopectin version of starch, and the processes for glycogen synthesis and degradation are very similar to those for amylopectin.

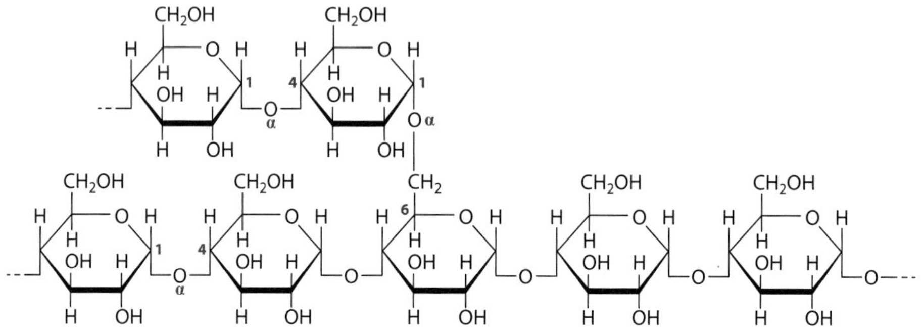

Figure 11.1 The polymeric structure of glycogen.
Glycogen has the same structure as amylopectin, but with more frequent branch points.

Glycogen is synthesized from activated glucose

When we studied starch biosynthesis in chloroplasts, we discovered that the polymer is built up from activated glucose intermediates, predominantly ADP–glucose. The same is true for glycogen synthesis, although in this case the activated molecules are UDP–glucose, which are synthesized from glucose 1-phosphate and UTP by UDP–glucose pyrophosphorylase. This is the same as the reaction that makes UDP–glucose for sucrose synthesis (see *Fig. 10.19A*). Activated glucose units are then added to the non-reducing ends of the growing glycogen molecule by **glycogen synthase**, in a manner analogous to starch synthesis (see *Fig. 10.20*).

A second difference between starch and glycogen synthesis is that whereas starch synthase can initiate a starch polymer by joining together the first few glucose units, glycogen synthase can only extend an existing molecule. If an existing molecule is required, then how does glycogen synthesis ever get started? The answer is provided by a separate enzyme, called **glycogenin**, which is a dimer of identical subunits, each of which make a **primer** consisting of at least eight glucose units in a linear chain. The non-reducing ends of these two primers provide the starting points for subsequent glucose polymerization by a pair of glycogen synthase enzymes. The primers remain attached to the glycogenin protein, and it appears that the glycogen synthases that are extending the glycogen molecule are only fully active when they are able to make physical contact with the glycogenin at the core of the molecule. This means that when the glycogen molecule reaches a certain size, and contact between the

Figure 11.2 Cross-section of a glycogen molecule.
The dimer of glycogenin proteins that prime glycogen synthesis are seen at the center of the molecule. The size of the molecule is limited by the need for glycogen synthase to retain contact with glycogenin.

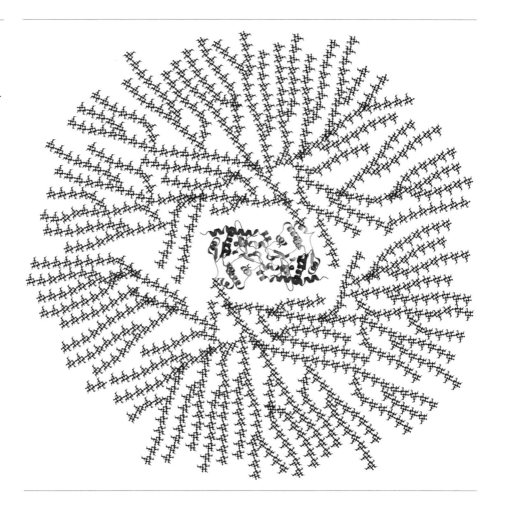

glycogen synthases and glycogenin is lost, further polymerization stops. The size of the glycogen molecule is therefore limited by the need for contact between glycogenin and glycogen synthase (*Fig. 11.2*).

Glycogen is a branched polymer, with the **glycogen branching enzyme** forming the α(1→6) linkages at the branch points. UDP–glucose is not a substrate for the branching enzyme, which instead breaks off a chain of seven glucose units from one of the growing ends of the glycogen molecule, reattaching these at an internal position, with an α(1→6) linkage, to create the branch (*Fig. 11.3*). During the transfer, the detached chain forms a transient covalent bond with an aspartic acid within the glycogen branching enzyme, so there is no chance of the chain becoming lost.

Glycogen degradation provides glucose for glycolysis

The pathway for degradation of glycogen is, in essence, the reverse of its synthesis. Glucose units are removed one by one from the non-reducing ends of the glycogen molecule by the enzyme **glycogen phosphorylase**. As the enzyme name indicates, phosphate is involved, the enzyme adding inorganic phosphate to each glucose unit that it removes, creating glucose 1-phosphate.

Glycogen phosphorylase can break only the α(1→4) links and stops when it approaches an α(1→6) linkage. This leaves a short chain of 4–6 glucose units still attached to the α(1→6) linkage (*Fig. 11.4*). These are dealt with by the **glycogen debranching enzyme**, in a process similar, in reverse, to the synthetic activity of the branching enzyme. First, the glycogen debranching enzyme cleaves the α(1→4) link

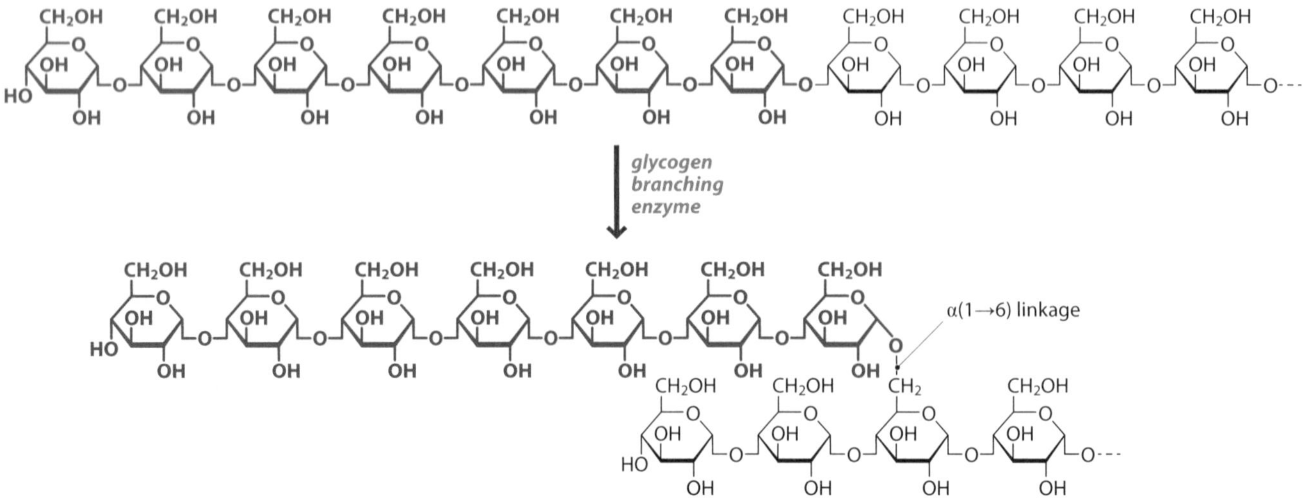

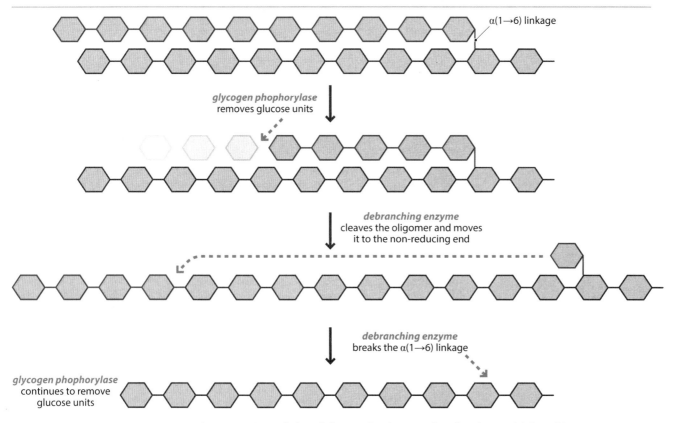

Figure 11.3 **The role of the glycogen branching enzyme.**

between the first and second glucose units away from the branch point. This leaves one glucose attached to the glycogen molecule via a α(1→6) linkage, and a short oligomer of 3–5 units still connected by their α(1→4) links. The debranching enzyme then attaches this oligomer to one of the non-reducing ends of the glycogen molecule, where they are depolymerized by glycogen phosphorylase. Finally, the debranching enzyme cleaves the α(1→6) linkage holding the last glucose to the main chain. The branch is now completely degraded.

Figure 11.4 **Degradation of glycogen by glycogen phosphorylase and debranching enzyme.**

The glycogen debranching enzyme therefore has two distinct catalytic activities:

- A **transferase**, which transfers the oligomer of glucose units from the branch to a second non-reducing end.

- An **α(1→6) glucosidase**, which removes the glucose at the actual branch point.

Because the second of these activities is a glucosidase and not a phosphorylase, the glucose unit released by cleavage of the α(1→6) link by the debranching enzyme is released as glucose itself and not glucose 1-phosphate.

The purpose of glycogen breakdown is to provide glucose that can enter the glycolysis pathway and hence be used to generate energy. The glucose 1-phosphate resulting from glycogen breakdown is therefore converted to glucose 6-phosphate by **phosphoglucomutase** (*Fig. 11.5*). In muscle cells, which use their own glycogen stores as energy supplies, glucose 6-phosphate then enters step 2 of glycolysis (see *Fig. 8.3*). In liver cells, the glycogen stores are for more general use around the body rather than as a local supply of energy. In liver, the glucose 6-phosphate molecules are therefore converted to glucose by **glucose 6-phosphatase**. The glucose is then passed into the bloodstream which transports it to those tissues that are dependent on the liver as their source of energy.

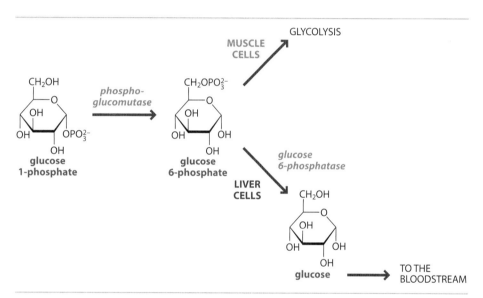

Figure 11.5 Fates of glucose 1-phosphate in different tissues.

11.1.2 Control of glycogen metabolism

The balance between the synthesis and degradation of glycogen is the primary determinant of the amount of glucose that animals have available for energy generation. Glycogen metabolism must therefore be tightly controlled. This is achieved by a combination of hormonal and allosteric control. The hormonal control principally involves:

- **Insulin** and **glucagon**, which together ensure that the blood glucose level is maintained within its normal range of 75–110 mg dl^{-1}, only rising temporarily immediately after a meal.

- **Epinephrine**, also called **adrenaline**, which increases the general availability of glucose as part of the 'fight or flight' response that this hormone coordinates.

Hormonal control acts broadly throughout the body. Allosteric control, on the other hand, acts more locally within muscle and liver cells in response to the levels of metabolites such as glucose 6-phosphate, AMP and ATP.

Box 11.1 **Blood sugar**

The blood sugar level is a measure of the concentration of glucose in the blood, usually expressed as $mg\,dl^{-1}$. In a healthy adult, the reciprocal effects of hormones such as glucagon and insulin maintain the blood sugar level within a range of $75-110\,mg\,dl^{-1}$, with the higher levels occurring immediately after a meal.

Measurements of blood sugar are informative in clinical settings because levels above or below the norm (**hyperglycemia** and **hypoglycemia**, respectively) are indicative of various disease states. The commonest cause of hyperglycemia is **diabetes mellitus**, which includes Type I diabetes, where the pancreas fails to produce sufficient insulin to control blood sugar levels, and Type II diabetes, where insulin is produced but the target cells become less sensitive to its presence. Both can be treated with insulin injections, but the dosage and times of application must be carefully controlled. Indeed, the commonest cause of hypoglycemia is over-administration of insulin in the treatment of diabetes, although hypoglycemia can also be a symptom of liver or kidney disease.

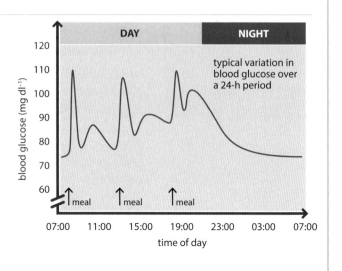

typical variation in blood glucose over a 24-h period

Hormonal control involves regulation of glycogen phosphorylase and glycogen synthase

The two key targets for regulation of glycogen metabolism are glycogen phosphorylase and glycogen synthase, these being the two enzymes that play the central roles in glycogen degradation and synthesis, respectively. Both of these enzymes are dimers consisting of two identical subunits, and both exist in two forms called *a* and *b*. In simple terms, the *a* form of the enzyme is always active, but the *b* form is usually inactive. Conversion between the *a* and *b* forms is brought about by the addition of a single phosphate group to one amino acid in each subunit. However, the effect of this addition is very different in the two enzymes (*Fig. 11.6*):

- Adding the phosphate group *activates* glycogen phosphorylase, in other words converts glycogen phosphorylase *b* into glycogen phosphorylase *a*. This is carried out by **phosphorylase kinase**.

- Adding the phosphate group *inactivates* glycogen synthase, and so converts glycogen synthase *a* into glycogen synthase *b*. This is done by **protein kinase A**.

The same enzyme, **protein phosphatase**, is responsible for the reverse reactions, removing the phosphate from both enzymes, inactivating glycogen phosphorylase and activating glycogen synthase.

Figure 11.6 Regulation of glycogen phosphorylase and glycogen synthase by phosphorylation – the addition of a phosphate group.

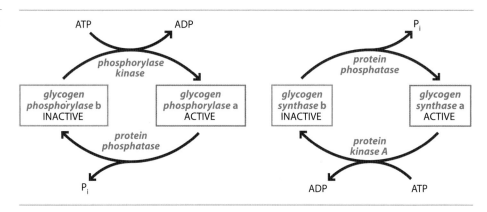

As part of the 'fight or flight' response, epinephrine increases the glucose available to the body by reducing glycogen synthesis and stimulating glycogen degradation. It achieves these effects via a signal transduction pathway that begins with attachment of the hormone to the **β-adrenergic receptor** protein on the surface of the cell. The receptor protein undergoes a conformational change that activates adenylate cyclase, increasing the intracellular level of cAMP, which, in turn, activates protein kinase A (*Fig. 11.7*). Once activated, protein kinase A does two things:

- It acts directly on glycogen synthase, inactivating it and switching off glycogen synthesis.

- It acts indirectly on glycogen phosphorylase, by adding a phosphate group to phosphorylase kinase, activating this enzyme, which then converts inactive glycogen phosphorylase *b* into active glycogen phosphorylase *a*.

The net result is an increase in glycogen degradation, resulting in higher levels of glucose in the liver cells and more glucose 6-phosphate in muscle cells. The additional glucose 6-phosphate in muscle is metabolized by glycolysis, releasing extra energy, so the animal can either fight or run away quickly. In the liver, the glucose is passed via the bloodstream to other organs such as the brain. Once the crisis has passed, the epinephrine levels fall and the receptor reverts to its original conformation. Adenylate cyclase is no longer stimulated and the cAMP levels fall as this compound is converted back to AMP. Protein kinase A is, in turn, deactivated and protein phosphatase removes the phosphates from glycogen synthase and phosphorylase kinase, restoring the original balance between glycogen synthesis and degradation.

Figure 11.7 The signal transduction pathway by which epinephrine reduces glycogen synthesis and stimulates glycogen degradation. Note that the scheme shown here involves a cascade of effects. Binding of one molecule of epinephrine results in synthesis of many cAMP molecules, each of which activate many protein kinase A enzymes, etc. This means that a small amount of hormone can have a major effect on cellular glycogen metabolism.

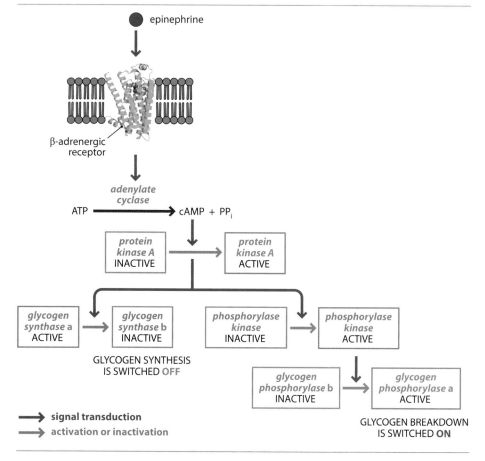

Box 11.2 **Avoiding a futile cycle**

As well as controlling the amount of glucose that is available for energy generation, the processes that regulate glycogen metabolism also prevent the occurrence of a **futile cycle**. A futile cycle is possible when there are two metabolic pathways that run

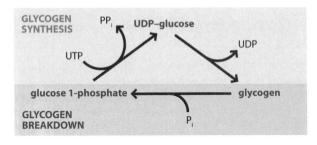

in reverse opposite directions. With glycogen metabolism, a futile cycle would operate if a cell was synthesizing glycogen from glucose at the same time that it was breaking other glycogen molecules back down into glucose. The futile glycogen cycle would waste energy, because UTP would be consumed during glycogen synthesis, but not recovered during glycogen breakdown.

There are several examples of possible futile cycles. Later in this chapter we will see how gluconeogenesis converts pyruvate to glucose, which is the reverse of the glycolytic pathway. As we have seen with glycogen metabolism, the enzymes that are unique to either glycolysis or gluconeogenesis are carefully regulated to ensure that just one pathway operates at any particular time, rather than both at once (see *Fig. 11.16*).

Glucagon, which is part of the physiological process that prevents blood sugar levels from falling below an acceptable level, acts in the same way as epinephrine, but on a continual basis rather than in response to a sudden period of stress. Insulin has the opposite and complementary effect, preventing blood sugar levels from getting too high by ensuring that excess glucose is converted into glycogen. It does this by binding to liver cells and activating an **insulin-responsive protein kinase**, whose target is protein phosphatase (*Fig. 11.8*). So by activating the kinase, which activates the phosphatase, insulin activates glycogen synthase and inactivates glycogen phosphorylase. This combination of events results in increased glycogen synthesis and reduced glycogen degradation, lowering the amount of glucose that enters the blood. The balance between the effects of glucagon and insulin therefore maintains the blood glucose level within its normal range.

Allosteric control also acts on glycogen phosphorylase and glycogen synthase

As well as the broad controls over glycogen metabolism exerted by epinephrine, glucagon and insulin, the relative levels of AMP and ATP in muscle cells can also influence the activity of glycogen phosphorylase and glycogen synthase. This means that the energy supply in muscle cells can quickly be turned up during periods of exercise, and turned down again when the exercise has ended and there is less need for energy consumption.

This allosteric control process acts on the *b* versions of glycogen phosphorylase and glycogen synthase, which are usually inactive. In muscle (but not in liver), the *b* version of glycogen phosphorylase is activated by high levels of AMP, but this is

Box 11.3 **Control of glycogen metabolism by calcium**

The activity of glycogen phosphorylase is also influenced by the concentration of calcium ions in a muscle cell. Calcium is another activator of glycogen phosphorylase and so increases glycogen breakdown.

Why do Ca^{2+} ions have this effect? The answer lies with the role of Ca^{2+} in muscle activity. When a muscle cell is stimulated by a nerve impulse, Ca^{2+} ions are released from a specialized type of smooth endoplasmic reticulum called the **sarcoplasmic reticulum**,

resulting in a tenfold increase in the Ca^{2+} concentration in the cytoplasm. Calcium binds to and causes a conformational change in one of the three polypeptides that make up the **troponin** protein. This conformational change initiates the series of events that results in muscle contraction. Muscle contraction requires energy and by having a direct stimulatory effect on glycogen phosphorylase, the elevated Ca^{2+} ion concentration increases the cell's supply of glucose at precisely the time when the energy requirement increases.

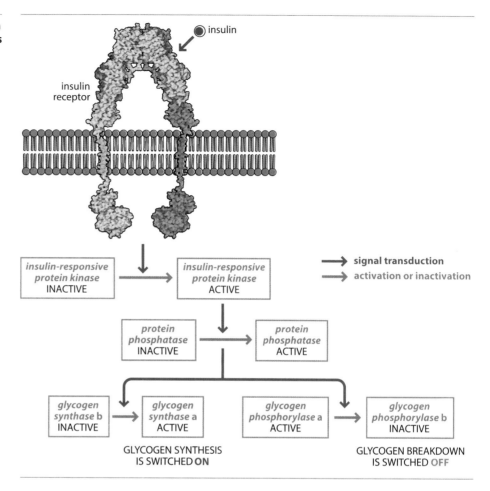

Figure 11.8 The signal transduction pathway by which insulin stimulates glycogen synthesis and reduces glycogen degradation.
Insulin receptor image produced by David Goodsell and reproduced from the Protein Data Bank (doi: 10.2210/rcsb_pdb/mom_2015_2).

opposed by the high levels of ATP and glucose 6-phosphate found in resting muscle. This means that during rest, glycogen phosphorylase *b* is inactive, but during periods of exercise, when ATP and glucose 6-phosphate are depleted and AMP levels rise, phosphorylase *b* is stimulated to release more glucose 6-phosphate from glycogen, increasing the available energy supply (*Fig. 11.9*). As you would anticipate, the control over glycogen synthase acts in the opposite manner. During periods of rest, a high concentration of glucose 6-phosphate can stimulate glycogen synthase *b* to increase synthesis of glycogen. When exercise begins, the depletion of glucose 6-phosphate causes glycogen synthase *b* to become inactive, so glucose remains available for use in glycolysis and is not diverted away into glycogen synthesis. Glycogen degradation to generate more energy is also stimulated by calcium during muscle contraction.

Allosteric control of glycogen metabolism has a different basis to hormonal control and does not involve addition and removal of phosphate groups to the *b* forms of

---

Box 11.4 **Allosteric control of glycogen metabolism in liver cells**

In liver, glycogen phosphorylase *b* is not activated by AMP. This means that, contrary to the situation in muscle, glycogen degradation in the liver is not responsive to the energy status of the cell. Instead, glycogen phosphorylase *a* is inhibited by glucose. This fits with the role of the glycogen stores in liver cells, these stores being used to maintain the level of glucose in the blood. As glucose levels rise, glycogen phosphorylase *a* is inhibited so glycogen breakdown in the liver is switched off. When the glucose level falls the process is reversed and glycogen breakdown is switched on again.

Figure 11.9 Allosteric control of glycogen synthesis and breakdown in exercising and resting muscle cells.
Abbreviation: G6P, glucose 6-phosphate.

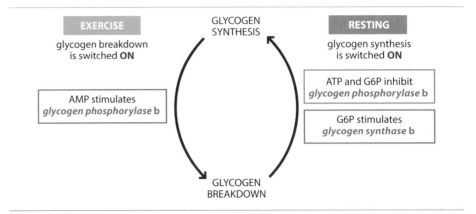

glycogen phosphorylase and glycogen synthase. Instead, binding of the allosteric compound to glycogen phosphorylase causes a slight repositioning of the two enzyme subunits relative to one another, resulting in a transition between a 'relaxed' active state and a 'tense' inactive conformation. In the tense conformation, substrates cannot enter the active site of the enzyme. The basis to allosteric control of glycogen synthase has not been worked out in as much detail, but the two enzymes have very similar structures and it is likely that glycogen synthase has similar relaxed and tense conformations.

11.2 Gluconeogenesis

Gluconeogenesis is a pathway for the synthesis of glucose from various non-carbohydrate precursors. It enables the body to maintain a supply of glucose for energy generation during extreme conditions, such as starvation or excessive exercise, when sources of carbohydrate can run low. The liver can only store enough glycogen to supply the brain with energy for up to 12 hours of starvation. After 12 hours, the liver uses gluconeogenesis to maintain the brain in operational mode for as long as possible. Gluconeogenesis is clearly a very important biochemical process and we must therefore examine the steps in the pathway and its regulation.

11.2.1 The gluconeogenesis pathway

Gluconeogenesis is usually looked on as a multistep pathway that begins with pyruvate and ends with glucose. In reality, several of the non-carbohydrate compounds that can be used as substrates for gluconeogenesis are not metabolized via pyruvate, but instead enter the pathway at a later step. We will therefore begin by examining the pyruvate to glucose component of gluconeogenesis, and then look at the entry points for the most important of the various substrates for the pathway.

The conversion of pyruvate to glucose is partly glycolysis in reverse

The reason why we tend to look on gluconeogenesis as a pyruvate to glucose pathway is because this enables us to make a direct comparison with glycolysis, which has the opposite outcome of converting glucose to pyruvate. This often leads to the suggestion that gluconeogenesis is glycolysis in reverse, but this is only partly true. To understand why, we need to make a distinction between those steps in glycolysis which are readily reversible, and those which are not (*Fig. 11.10*). We see that in the middle of the glycolysis pathway there are a series of reactions, beginning with fructose 1,6-bisphosphate and ending with phosphoenolpyruvate, each of which

Figure 11.10 The reversible reactions in glycolysis.
Those reactions that are highlighted with colored arrows have ΔG values close to zero, which means that the equilibrium between the reactants and products is close to 1:1, and the same enzyme can catalyze both the forward and reverse reactions.

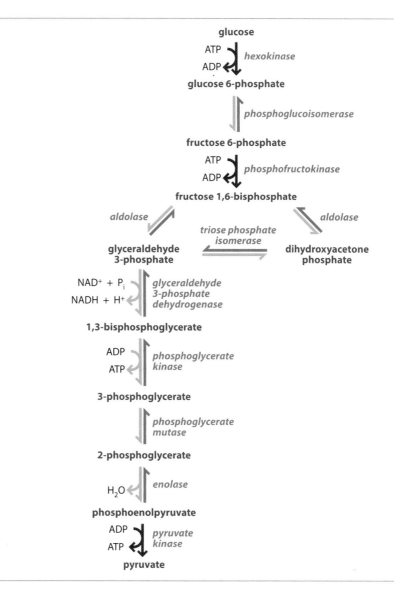

has a ΔG value close to zero. At these steps the equilibrium between the reactants and products is close to 1:1, and the same enzyme can catalyze both the forward and reverse reactions. These reactions are also utilized, in the reverse direction, in gluconeogenesis.

'Glycolysis in reverse' therefore enables us to achieve that part of the gluconeogenesis pathway that converts phosphoenolpyruvate to fructose 1,6-bisphosphate. However, there are three steps in glycolysis which have highly negative ΔG values when proceeding forwards (i.e. in the direction followed during glycolysis). Under cellular conditions these reactions are therefore irreversible. Two of these reactions come at the start of glycolysis, when glucose is converted to glucose 6-phosphate by hexokinase, and fructose 6-phosphate is converted to fructose 1,6-bisphosphate by phosphofructokinase. The third irreversible reaction is the last step in glycolysis, from phosphoenolpyruvate to pyruvate, catalyzed by pyruvate kinase. If these three reactions can only proceed in the glycolysis direction, then how can we achieve their reverse outcomes in gluconeogenesis?

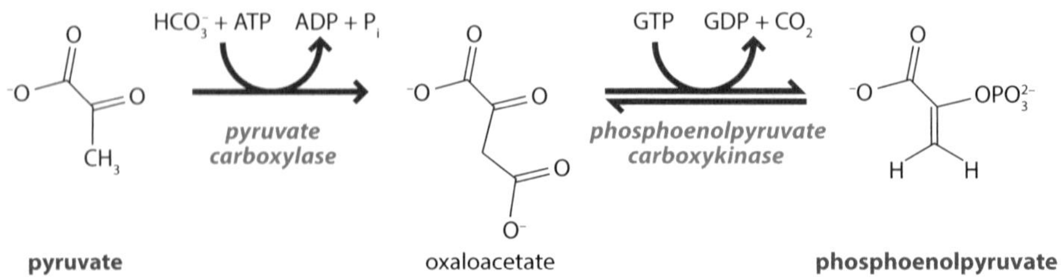

Figure 11.11 Synthesis of phosphoenolpyruvate from pyruvate during gluconeogenesis.

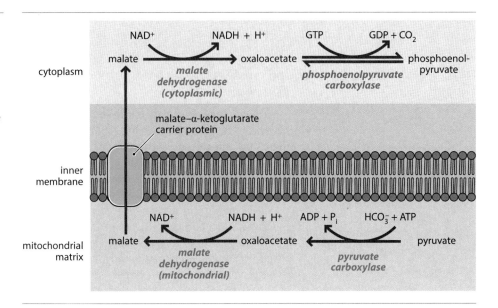

pyruvate   oxaloacetate   phosphoenolpyruvate

The answer is to bypass these steps by using alternative reactions with different chemistries. This means that the process that achieves the outcome $B \rightarrow A$ uses different chemical steps to those involved when $A \rightarrow B$. The two equations, $B \rightarrow A$ and $A \rightarrow B$, are now different chemical reactions, not the forward and reverse versions of the same reaction. Because they are different reactions they have different ΔG values. So now there is the possibility that although $A \rightarrow B$ has a highly negative ΔG and is irreversible, $B \rightarrow A$ has a manageable ΔG and can proceed when catalyzed by the appropriate enzyme. This is what happens in gluconeogenesis.

The irreversible reactions in glycolysis are bypassed in gluconeogenesis

First we will consider step 1 in the pyruvate to glucose part of gluconeogenesis, when pyruvate is converted to phosphoenolpyruvate. In glycolysis, pyruvate kinase transfers the phosphate group from phosphoenolpyruvate to ADP to form pyruvate and ATP. In gluconeogenesis, phosphoenolpyruvate is generated from pyruvate and a phosphate group by a two-part reaction, catalyzed by **pyruvate carboxylase** and **phosphoenolpyruvate carboxykinase**. The phosphate group is donated by GTP and energy is obtained by hydrolysis of ATP. The overall reaction is:

pyruvate + GTP + ATP + H$_2$O → phosphoenolpyruvate + GDP + ADP + P$_i$ + 2H$^+$

This equation hides a variety of important events that occur during the course of this reaction. Oxaloacetate is formed as an intermediate, which is why the two enzymes are called carboxylases – they add and remove carboxyl (–COOH) groups (*Fig. 11.11*). The carboxyl group is donated from carbon dioxide which exists in solution as bicarbonate (HCO$_3^-$) ions, and is initially bound to a **biotin** (vitamin B$_7$) molecule that is attached to pyruvate carboxylase as a prosthetic group.

Figure 11.12 Malate transport during gluconeogenesis.
The oxaloacetate formed by pyruvate carboxylase is converted to malate by the mitochondrial malate dehydrogenase enzyme and then transported out of the mitochondrial matrix via the malate–α-ketoglutarate carrier protein (see *Fig. 9.21*). The malate is then converted back to oxaloacetate by the cytoplasmic version of malate dehydrogenase.

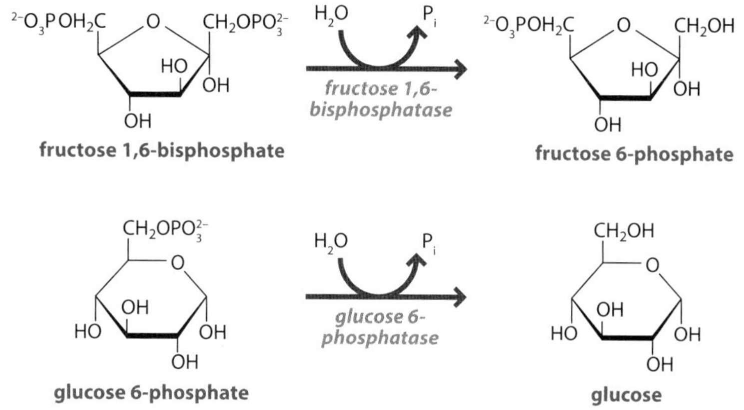

Figure 11.13 **The reactions catalyzed by fructose 1,6-bisphosphatase and glucose 6-phosphatase during gluconeogenesis.**

fructose 1,6-bisphosphate — *fructose 1,6-bisphosphatase* → fructose 6-phosphate

glucose 6-phosphate — *glucose 6-phosphatase* → glucose

Pyruvate carboxylase is a mitochondrial enzyme, so this part of the reaction occurs in the mitochondrial matrix. The oxaloacetate that is formed then has to be transported out of the mitochondrion into the cytoplasm, where the remainder of gluconeogenesis takes place, including the decarboxylation of oxaloacetate to give phosphoenolpyruvate. This raises a difficulty because oxaloacetate cannot cross the inner mitochondrial membrane. Instead, it has to be converted to malate, which can be exported via the malate transport protein (*Fig. 11.12*). Once in the cytoplasm, the malate is reconverted to oxaloacetate, which is then decarboxylated by phosphoenolpyruvate carboxykinase to form phosphoenolpyruvate.

Now we move to the end of the gluconeogenesis pathway and to the two reactions catalyzed during glycolysis by hexokinase and phosphofructokinase. These two enzymes transfer phosphate groups from ATP to their substrates. During gluconeogenesis, the energetically unfavorable transfer of the phosphates back to ADP is not attempted. Instead the phosphate groups are removed by hydrolysis, yielding inorganic phosphate as one of the products (*Fig. 11.13*). The enzymes are **fructose 1,6-bisphosphatase** and glucose 6-phosphatase. The second of these steps takes place in the lumen of the endoplasmic reticulum. The glucose that is produced is either enclosed in vesicles that fuse with the plasma membrane, or transported back into the cytoplasm and then across the plasma membrane by glucose transporter proteins. The exported glucose is then taken to the brain and other tissues where it is needed. When the energy crisis passes, gluconeogenesis terminates in the synthesis of glucose 6-phosphate, which is used to rebuild the glycogen stores in the cytoplasm of the liver cell.

Box 11.5 **The energy budget for gluconeogenesis**

The following steps in the gluconeogenesis pathway require energy:
- Pyruvate to oxaloacetate (catalyzed by pyruvate carboxylase) uses 1 ATP
- Oxaloacetate to phosphoenolpyruvate (catalyzed by phosphoenolpyruvate carboxykinase) uses 1 GTP
- 3-Phosphoglycerate to 1,3-bisphosphoglycerate (catalyzed by phosphoglycerate kinase) uses 1 ATP

In addition, glyceraldehyde 3-phosphate dehydrogenase, when operating in the gluconeogenesis direction, uses up one molecule of cytoplasmic NADH, which could otherwise be used to generate three ATPs via the electron transport chain (see *Section 9.2.5*).

Two pyruvate molecules are required to make one of glucose, so we have to double the above numbers to calculate the energy needed to make a single glucose molecule by gluconeogenesis. The total is therefore 10 ATP + 2 GTP. In contrast, glycolysis has a net yield of just two ATPs. The cell must therefore avoid the glycolysis–gluconeogenesis futile cycle (see *Box 11.2*) because this would waste a substantial amount of energy.

The main substrates for gluconeogenesis enter the pathway at different places

Although pyruvate is looked on as the starting point for gluconeogenesis, the primary substrates for the pathway are other non-carbohydrate compounds, which are present in cells and are sacrificed as starvation or intense exercise make the body more and more desperate for sources of energy.

The main substrates for gluconeogenesis are lactate, amino acids and triacylglycerols. Lactate is produced in muscles by anaerobic respiration when oxygen runs short, and so is available during heavy exercise, one of the stresses that might make gluconeogenesis necessary. Lactate is therefore transported from the muscles to the liver, where it is directly converted to pyruvate by lactate dehydrogenase (see *Fig. 8.6*). The pyruvate can then enter gluconeogenesis as described above (*Fig. 11.14*).

Figure 11.14 The entry points for the main substrates for gluconeogenesis.
The gluconeogenesis pathway is shown as 'glycolysis in reverse' with the enzymes that are unique to gluconeogenesis in orange. The entry points for lactate, amino acids and triacylglycerols are indicated.

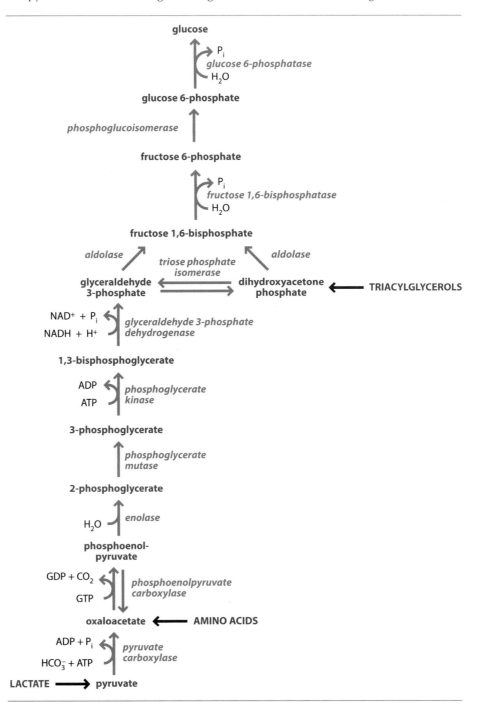

Amino acids for gluconeogenesis are obtained from the diet or, in more extreme cases, from breakdown of proteins, largely ones in muscle. Most of the 20 standard amino acids can be converted to oxaloacetate and so enter gluconeogenesis at the step catalyzed by phosphoenolpyruvate carboxykinase. For some amino acids, the pathway to oxaloacetate is direct, but for others one of the intermediates in the TCA cycle is initially formed. The enzymes of the TCA cycle then convert the intermediate into oxaloacetate.

Finally, triacylglycerols are broken into their component fatty acids and glycerol. The fatty acids cannot be utilized in gluconeogenesis, but glycerol can be metabolized to glycerol 3-phosphate and then to dihydroxyacetone phosphate by the action of glycerol kinase and glycerol 3-phosphate dehydrogenase, respectively (*Fig. 11.15*). Dihydroxyacetone phosphate then enters gluconeogenesis at the relatively late stage immediately before fructose 1,6-bisphosphate synthesis.

Figure 11.15 Synthesis of dihydroxyacetone phosphate from glycerol.
This conversion enables triacylglycerol breakdown products to enter gluconeogenesis.

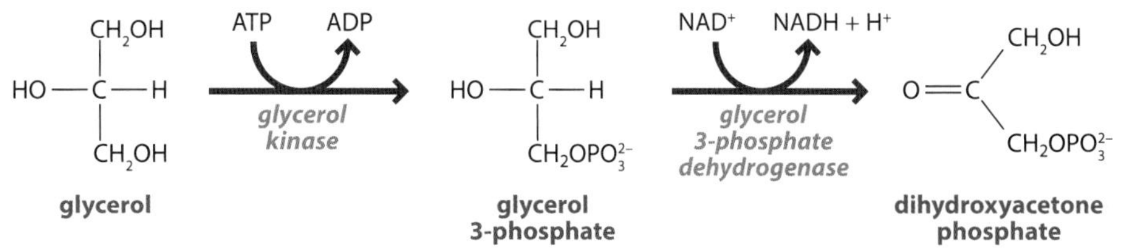

11.2.2 Regulation of gluconeogenesis

Gluconeogenesis must be regulated in such a way that the pathway is switched on when there is an energy crisis, to which the liver must respond by synthesizing glucose to send to the brain. Regulation of gluconeogenesis therefore requires coordination with glycolysis, to ensure that the latter is inhibited when gluconeogenesis is in operation.

The coordination between gluconeogenesis and glycolysis is achieved by a reciprocal regulation process acting on phosphofructokinase and fructose 1,6-bisphosphatase. These are the enzymes responsible for the interconversion of fructose 6-phosphate and fructose 1,6-bisphosphate in glycolysis and gluconeogenesis, respectively. When we studied the regulation of glycolysis we learned that phosphofructokinase is stimulated by AMP and inhibited by ATP and citrate. The rate of glycolysis therefore increases when energy supplies are low, signaled by high AMP, and decreases when there are adequate energy reserves, indicated by high ATP and citrate. AMP and citrate have the complementary effects on gluconeogenesis. AMP inhibits fructose 1,6-bisphosphatase so gluconeogenesis is shut down when glycolysis must be operational, and citrate stimulates fructose 1,6-bisphosphate so gluconeogenesis can occur when glycolysis is inhibited (*Fig. 11.16*).

Figure 11.16 Reciprocal control of glycolysis and gluconeogenesis.

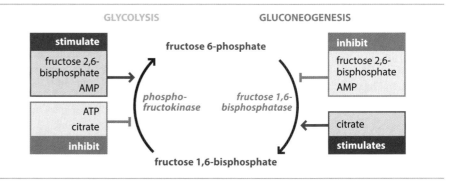

The regulatory molecule fructose 2,6-bisphosphate (see *Section 8.2.4*) coordinates glycolysis and gluconeogenesis in a similar manner. This molecule stimulates phosphofructokinase and inhibits fructose 1,6-bisphosphatase. During starvation, glucagon is released into the bloodstream. One of the effects of this increase in glucagon is the breakdown of fructose 2,6-bisphosphate, enabling gluconeogenesis to predominate over glycolysis.

11.3   The pentose phosphate pathway

The pentose phosphate pathway, which is also called the **hexose monophosphate shunt** or the **phosphogluconate pathway**, has three main roles:

- It is a major source of NADPH, which is used as an energy carrier during important biosynthetic reactions such as those that result in the synthesis of fatty acids and sterols.

- One of the intermediates in the pathway is ribose 5-phosphate, which is a precursor for nucleotide synthesis and for synthesis of the amino acids histidine and tryptophan.

- Another intermediate is erythrose 5-phosphate, which is a precursor for the synthesis of phenylalanine, tryptophan and tyrosine.

We will study these biosynthetic pathways in the next two chapters: fatty acid synthesis in *Section 12.1.1*, sterol synthesis in *Section 12.3*, amino acid synthesis in *Section 13.2.1*, and nucleotide synthesis in *Section 13.2.2*.

The pentose phosphate pathway takes place in the cytoplasm, in particular in those tissues which synthesize fatty acids or steroid hormones, such as the mammary glands, the adrenal cortex and adipose tissue.

11.3.1   The oxidative and non-oxidative phases of the pentose phosphate pathway

The pentose phosphate pathway can be divided into two stages. The first of these is an oxidative phase that begins with glucose 6-phosphate and produces ribose 5-phosphate plus two molecules of NADPH. This is followed by a non-oxidative or synthetic stage that yields a variety of sugars with 3–7 carbons.

The oxidative phase results in NADPH synthesis

The overall reaction for the oxidative phase of the pentose phosphate pathway is:

glucose 6-phosphate $+ 2NADP^+ + H_2O \rightarrow$ ribose 5-phosphate $+ 2NADPH + 2H^+ + CO_2$

The pathway is summarized in *Figure 11.17*. In detail, the steps are as follows:

Step 1. Glucose 6-phosphate is oxidized to 6-phosphoglucono-δ-lactone by **glucose 6-phosphate dehydrogenase**. The reaction produces one molecule of NADPH.

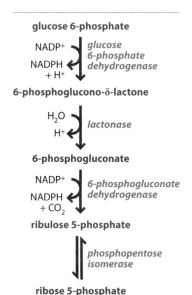

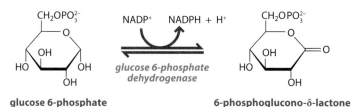

Figure 11.17 The oxidative phase of the pentose phosphate pathway.

Step 2. **Lactonase** hydrolyzes 6-phosphoglucono-δ-lactone to 6-phosphogluconate.

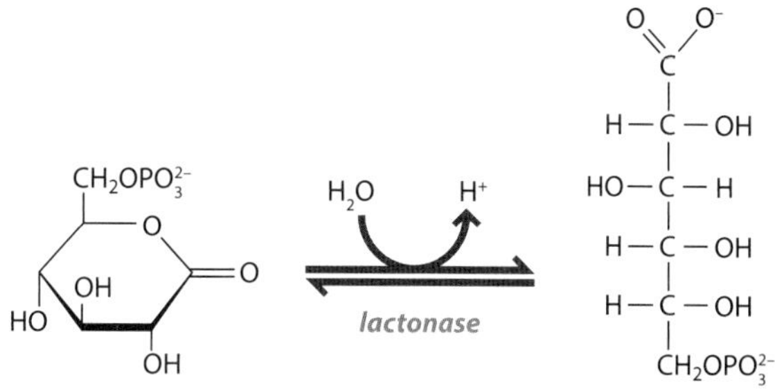

6-phosphoglucono-δ-lactone   **6-phosphogluconate**

Step 3. An oxidative decarboxylation catalyzed by **6-phosphogluconate dehydrogenase** converts 6-phosphogluconate (a 6-carbon sugar) to ribulose 5-phosphate (a 5-carbon sugar), yielding another molecule of NADPH.

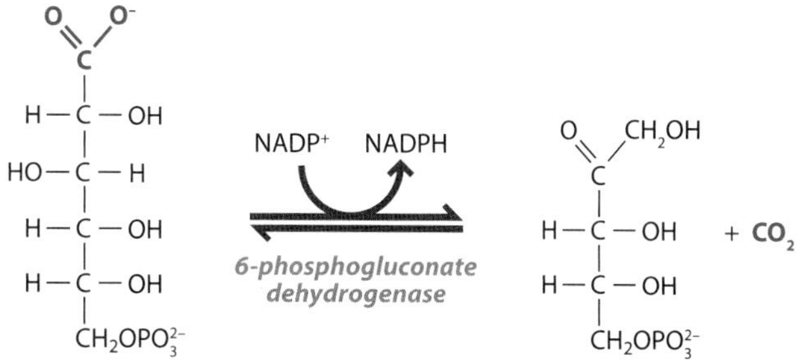

6-phosphogluconate   **ribulose 5-phosphate**

Step 4. Ribulose 5-phosphate undergoes isomerization to ribose 5-phosphate, catalyzed by **phosphopentose isomerase**.

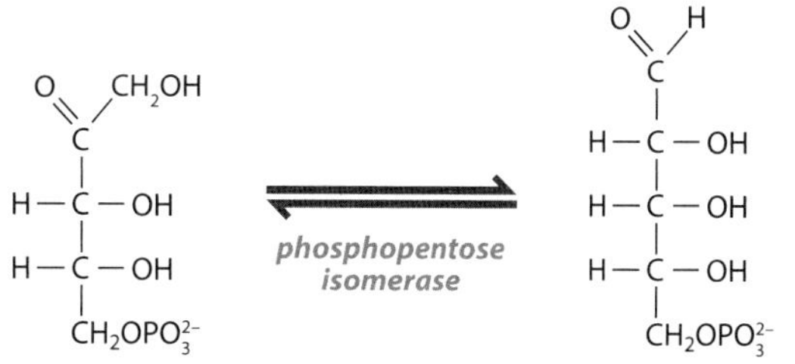

ribulose 5-phosphate   **ribose 5-phosphate**

At the end of this stage of the pentose phosphate pathway we have achieved two of its main objectives: the production of NADPH for fatty acid and sterol biosynthesis, and the synthesis of ribose 5-phosphate for use in making nucleotides and amino acids.

The non-oxidative phase yields a variety of products

If the ribose 5-phosphate generated by the oxidative phase of the pentose phosphate pathway is not all used up in synthesis of nucleotides and amino acids, then the non-oxidative phase of the pathway can operate. This second phase gives rise to a variety of carbohydrate intermediates and end products (*Fig. 11.18*).

Figure 11.18 The non-oxidative phase of the pentose phosphate pathway.
The abbreviations C3, C4, C5, C6 and C7 indicate the number of carbons in the different sugar compounds.

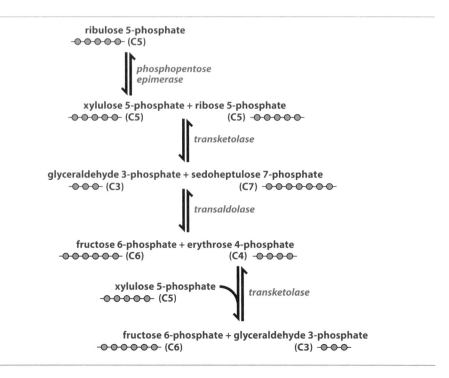

Step 5. Ribulose 5-phosphate (from *Step 3*) is converted into its isomer xylulose 5-phosphate by the enzyme **phosphopentose epimerase**.

Step 6. Now **transketolase** transfers two carbons and attached side groups from xylulose 5-phosphate to ribose 5-phosphate. This gives the triose glyceraldehyde 3-phosphate and the 7-carbon sugar sedoheptulose 7-phosphate.

Step 7. **Transaldolase** catalyzes another intermolecular transfer to convert glyceraldehyde 3-phosphate and the 7-carbon sugar sedoheptulose 7-phosphate into fructose 6-phosphate (a 6-carbon sugar) and erythrose 4-phosphate (4 carbons).

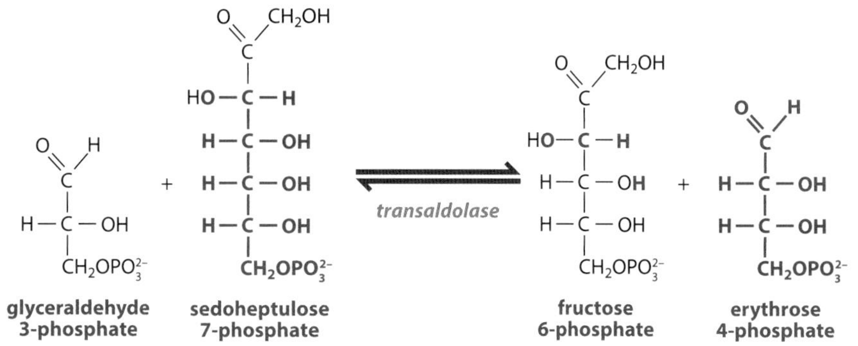

glyceraldehyde 3-phosphate   sedoheptulose 7-phosphate   *transaldolase*   fructose 6-phosphate   erythrose 4-phosphate

Step 8. Finally, transketolase catalyzes the reaction between the erythrose 4-phosphate with another molecule of xylulose 5-phosphate (from *Step 5*) which produces another molecule of fructose 6-phosphate and regenerates glyceraldehyde 3-phosphate.

erythrose 4-phosphate   xylulose 5-phosphate   *transketolase*   fructose 6-phosphate   glyceraldehyde 3-phosphate

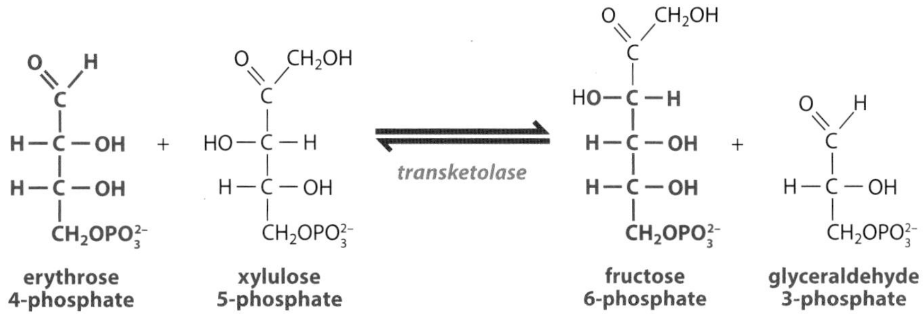

The non-oxidative phase of the pathway has therefore achieved the following:

2 xylulose 5-phosphate + ribose 5-phosphate →
  2 fructose 6-phosphate + glyceraldehyde 3-phosphate

The intermediates include erythrose 4-phosphate, which can be used to synthesize aromatic amino acids. The products are fructose 6-phosphate and glyceraldehyde 3-phosphate, which are intermediates in glycolysis and so can enter that pathway. The transketolase and transaldolase reactions are reversible, which means that the non-oxidative part of the pentose phosphate pathway can operate backwards, using the glycolysis intermediates as substrates for synthesis of ribose 5-phosphate. This provides a way of making ribose 5-phosphate without generating NADPH, which is useful if the cell has an ample supply of NADPH but a shortage of ribose 5-phosphate for nucleotide and amino acid biosynthesis.

Box 11.6 Did Pythagoras ban the broad bean because of the pentose phosphate pathway?

Although best known for his thoughts on the square of the hypotenuse, the Greek philosopher Pythagoras made many contributions to mathematics, astronomy, medicine and other areas of early scientific endeavor. Of the various statements attributed to him, one of the more enigmatic is "Avoid broad beans". Scholars down through the ages have argued about the reasons for this edict. Possibly broad beans were looked on as harboring the souls of the dead, or they may have been considered unclean due to the resemblance between their shape and that of a certain part of the male anatomy, or possibly the resulting flatulence was considered inimical to philosophical thought. Whatever the reason, Pythagoras' aversion to broad beans was so great that he met his death when he refused to enter a field of beans when being pursued by assassins.

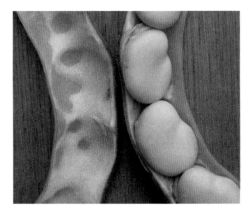

Image reproduced from Wikimedia under a CC BY-SA 3.0 license.

A recent theory is that Pythagoras' views on broad beans had a solid scientific basis, one involving the pentose phosphate pathway. About 400 million people worldwide suffer from **favism**, an ailment named after the broad bean (which is also called fava bean). Favism is characterized by breakdown of red blood cells after consumption of broad beans. The hemolytic response is due to the presence in broad beans of sugar derivatives – glycosides such as vicine – that act as reducing agents, converting oxygen inside the erythrocytes into reactive species such as hydrogen peroxide. These reactive species cause hemolysis by damaging the erythrocyte plasma membrane.

Individuals suffering from favism are unable to withstand the toxic effects of hydrogen peroxide because they do not synthesize enough of the reduced form of **glutathione**. This is a tripeptide, made up of glutamic acid, cysteine and glycine, with an unusual linkage between the first two amino acids. In the oxidized version, two tripeptides are attached to one another by a disulfide bond.

Hydrogen peroxide is detoxified by the enzyme **glutathione peroxidase**, which catalyzes the following reaction:

$$2GSH + H_2O_2 \rightarrow GSSG + 2H_2O$$

Regeneration of reduced glutathione is the role of **glutathione reductase**:

The continued functioning of reduced glutathione as an antioxidant therefore depends on the supply of NADPH in the erythrocyte being maintained. Only a few biochemical pathways generate NADPH, and all of these, with the exception of the pentose phosphate pathway, operate inside mitochondria. Erythrocytes lack mitochondria and so are entirely dependent on the pentose phosphate pathway for their supply of NADPH.

The biochemical basis to favism is depletion of the erythrocyte NADPH supply due to a defect in the pentose phosphate pathway. Favism is specifically **glucose 6-phosphate dehydrogenase (G6PD) deficiency**, and results from a mutation in the gene for this enzyme. The gene is on the X chromosome, which means that the disease is prevalent in males, who possess just a single X chromosome. In females, the presence of the correct version of the gene on the second X chromosome compensates to a certain extent for the defective activity of the mutant gene, so women only display a mild version of the disease.

The disease is more prevalent in the Mediterranean region, and would probably have been familiar to Pythagoras and his adherents, a possible explanation for the philosopher's ban on broad beans.

Further reading

Beutler E (2008) Glucose-6-phosphate dehydrogenase deficiency: a historical perspective. *Blood* **111**, 16–24.

Gerich JE, Meyer C, Woerle HJ and Stumvoll M (2001) Renal gluconeogenesis: its importance in human glucose homeostasis. *Diabetes Care* **24**, 382–91.

Jensen TE and Richter EA (2012) Regulation of glucose and glycogen metabolism during and after exercise. *Journal of Physiology* **590**,1069–76.

Jitrapakdee S, St Maurice M, Rayment I, Cleland WW, Wallace JC and Attwood PV (2008) Structure, mechanism and regulation of pyruvate carboxylase. *Biochemical Journal* **413**, 369–87.

Nuttal FQ, Ngo A and Gannon MC (2008) Regulation of hepatic glucose production and the role of gluconeogenesis in humans: is the rate of gluconeogenesis constant? *Diabetes/Metabolism Research and Reviews* **24**, 438–58.

Patra KC and Hay N (2014) The pentose phosphate pathway and cancer. *Trends in Biochemical Sciences* **39**, 347–54.

Roach PJ, Depaoli-Roach AA, Hurley TD and Tagliabracci VS (2012) Glycogen and its metabolism: some new developments and old themes. *Biochemical Journal* **441**, 763–87.

Smythe C and Cohen P (1991) The discovery of glycogenin and the priming mechanism for glycogen biogenesis. *European Journal of Biochemistry* **200**, 625–31.

Stein RB and Blum JJ (1978) On the analysis of futile cycles in metabolism. *Journal of Theoretical Biology* **72**, 487–522.

Van Schaftingen E and Gerin I (2002) The glucose-6-phosphatase system. *Biochemical Journal* **362**, 513–52.

Self-assessment questions

Multiple choice questions

Only one answer is correct for each question. Answers can be found on the website: www.scionpublishing.com/biochemistry

1. What is the structure of glycogen?
 (a) $\alpha(1\rightarrow6)$ chains and $\alpha(1\rightarrow4)$ branch points
 (b) $\alpha(1\rightarrow4)$ chains and $\beta(1\rightarrow6)$ branch points
 (c) $\alpha(1\rightarrow4)$ chains and $\alpha(1\rightarrow6)$ branch points
 (d) $\beta(1\rightarrow6)$ chains and $\alpha(1\rightarrow6)$ branch points

2. The predominant substrate for glycogen synthesis is what?
 (a) ADP–glucose
 (b) UDP–glucose
 (c) ADP–galactose
 (d) Glucose 1-phosphate

3. What is the role of glycogenin?
 (a) It is the enzyme that synthesizes glycogen
 (b) It makes the primer for glycogen synthesis
 (c) It makes the branch points in a glycogen molecule
 (d) It limits the size of a glycogen molecule by degrading excessively long branches

4. What is the enzyme that removes glucose units from the non-reducing ends of the glycogen molecule called?
 (a) Glycogen phosphorylase
 (b) Glycogeninase
 (c) Phosphoglucomutase
 (d) Glycogen debranching enzyme

5. Which of the following are the two hormones that maintain blood glucose levels within the normal range of $75–110\,mg\,dl^{-1}$?
 (a) Insulin and epinephrine
 (b) Insulin and adrenaline
 (c) Insulin and glucagon
 (d) Glucagon and epinephrine

6. Which of the following statements is **incorrect** regarding glycogen phosphorylase?
 (a) Adding a phosphate group activates glycogen phosphorylase
 (b) Glycogen phosphorylase is activated by phosphorylase kinase
 (c) Glycogen phosphorylase is activated by protein kinase A
 (d) The inactive form is glycogen phosphorylase *b*

7. Which of the following statements is **incorrect** regarding the β-adrenergic receptor protein?
   (a) It is the receptor protein for insulin
   (b) It undergoes a conformational change after binding the effector molecule
   (c) It activates adenylate cyclase
   (d) It is present in the plasma membrane

8. What is the role of AMP in allosteric control of glycogen metabolism?
   (a) AMP stimulates glycogen phosphorylase *a*
   (b) AMP stimulates glycogen phosphorylase *b*
   (c) AMP inhibits glycogen phosphorylase *b*
   (d) AMP stimulates glycogen synthase *b*

9. What is the role of glucose 6-phosphate in allosteric control of glycogen metabolism?
   (a) Glucose 6-phosphate stimulates glycogen phosphorylase *a*
   (b) Glucose 6-phosphate stimulates glycogen phosphorylase *b*
   (c) Glucose 6-phosphate inhibits glycogen phosphorylase *b*
   (d) Glucose 6-phosphate stimulates glycogen synthase *b*

10. Which enzymes catalyze the synthesis of phosphoenolpyruvate from pyruvate and a phosphate group during gluconeogenesis?
    (a) Pyruvate kinase and pyruvate carboxylase
    (b) Pyruvate kinase and phosphoenolpyruvate carboxykinase
    (c) Pyruvate carboxylase and phosphoenolpyruvate carboxykinase
    (d) Pyruvate kinase and enolase

11. Which mitochondrial transport protein is utilized during the synthesis of phosphoenolpyruvate from pyruvate and a phosphate group during gluconeogenesis?
    (a) The malate–α-ketoglutarate carrier protein
    (b) The aspartate–glutamate carrier protein
    (c) The oxaloacetate carrier protein
    (d) None of the above

12. Which of the following compounds is converted to pyruvate prior to entry into gluconeogenesis?
    (a) Lactate
    (b) Amino acids
    (c) Triacylglycerols
    (d) Glucose

13. Which of the following compounds is converted to oxaloacetate prior to entry into gluconeogenesis?
    (a) Lactate
    (b) Amino acids
    (c) Triacylglycerols
    (d) Glucose

14. Which of the following statements is **correct** regarding the regulation of gluconeogenesis?
    (a) Citrate stimulates phosphofructokinase
    (b) AMP inhibits phosphofructokinase
    (c) Fructose 2,6-bisphosphate inhibits phosphofructokinase
    (d) ATP inhibits phosphofructokinase

15. Which of the following statements is **incorrect** regarding the regulation of gluconeogenesis?
    (a) Citrate stimulates fructose 1,6-bisphosphatase
    (b) AMP inhibits fructose 1,6-bisphosphatase
    (c) Fructose 2,6-bisphosphate inhibits fructose 1,6-bisphosphatase
    (d) ATP inhibits fructose 1,6-bisphosphatase

16. Which of the following is an alternative name for the pentose phosphate pathway?
    (a) Gluconeogenesis
    (b) Phosphogluconate pathway
    (c) Hexose diphosphate shunt
    (d) All of the above

17. Which of the following is not a role of the pentose phosphate pathway?
    (a) A major source of NADPH
    (b) A source of glucose for glycolysis
    (c) A source of ribose 5-phosphate
    (d) A source of erythrose 5-phosphate

18. Which of these enzymes is not involved in the oxidative phase of the pentose phosphate pathway?
    (a) Phosphopentose epimerase
    (b) Glucose 6-phosphate dehydrogenase
    (c) Phosphopentose isomerase
    (d) Lactonase

19. Which compound is the starting point for the non-oxidative phase of the pentose phosphate pathway?
    (a) Xylulose 5-phosphate
    (b) Sedoheptulose 7-phosphate
    (c) Fructose 6-phosphate
    (d) Ribulose 5-phosphate

20. Which philosopher banned broad beans, possibly because of the pentose phosphate pathway?
    (a) Socrates
    (b) Pythagoras
    (c) Aristotle
    (d) Captain Jack Sparrow

Short answer questions

These questions do not require additional reading.

1. Describe the pathway for glycogen synthesis. What limits the size of the glycogen molecule that is made?

2. Explain how a glycogen molecule is broken down and the monomeric glucose units entered into glycolysis.

3. Explain how insulin, glucagon and epinephrine regulate glycogen metabolism.

4. Describe how glycogen metabolism is regulated by allosteric control.

5. What is a 'futile cycle' and why must cells avoid futile cycles?

6. Draw an outline of the gluconeogenesis pathway and explain why gluconeogenesis is sometimes called 'glycolysis in reverse'.

7. On your outline from question 6, indicate the positions at which various substrates enter the gluconeogenesis pathway.

8. Describe how the gluconeogenesis and glycolysis pathways are coordinated.

9. List the various roles of the pentose phosphate pathway.

10. Describe the steps in the pentose phosphate pathway, making a clear distinction between the oxidative and non-oxidative phases.

Self-study questions

These questions will require calculation, additional reading and/or internet research.

1. The genetic disorder called von Gierke's disease is characterized by a deficiency in glucose 6-phosphatase. What will be the effect of this deficiency on the biochemical capability of the patient, and what symptoms would you predict?

2. Gluconeogenesis results in synthesis of the D-enantiomer of glucose. Identify the most likely step in the gluconeogenesis pathway at which chirality would be introduced into the product, and speculate about how synthesis of the D-enantiomer of this product might be achieved.

3. Identify the commitment step in the gluconeogenesis pathway. Explain your reasoning.

4. Individuals with glucose 6-phosphate dehydrogenase deficiency often display resistance to malaria. Provide an explanation for this observation.

CHAPTER 12
Lipid metabolism

STUDY GOALS

After reading this chapter you will:

- understand how acetyl CoA is moved out of the mitochondrion prior to its use in fatty acid synthesis

- be able to describe the sequential construction of a 16-carbon fatty acid

- know how longer fatty acids are made and how ones with double bonds are synthesized

- understand the differences between the processes for fatty acid synthesis in animal and plant cells

- appreciate the role of malonyl CoA synthesis as the main control point in fatty acid synthesis

- know how triacylglycerols are synthesized and broken down

- understand that triacylglycerol cleavage is the major control point in lipolysis

- be able to describe the steps in the pathway of fatty acid degradation

- know the energy yield for breakdown of a fatty acid

- appreciate the distinctive features of fatty acid breakdown in peroxisomes

- know how unsaturated fatty acids and those with odd numbers of carbons are broken down

- be able to describe in outline how cholesterol is synthesized

- know how the energy status of the organism and sterol content of the cell influence the rate of cholesterol synthesis

- be able to describe in outline how the important derivatives of cholesterol are synthesized

Lipids have a number of important functions in living cells. Fatty acids and triacylglycerols are the main energy stores in animals, and glycerophospholipids, cholesterol and various other lipid types are the main structural components of membranes. Cholesterol is also the precursor for **steroid hormones**, which include the **glucocorticoids**, **estrogens** and **progestogens**. In this chapter we will examine how the various types of lipids are synthesized, and how those that act as energy stores are broken down so that the energy they contain can be utilized by the cell.

12.1 Synthesis of fatty acids and triacylglycerols

In *Chapter 5* we learned that fatty acids are carboxylic acids with the general formula R–COOH, where R is a hydrocarbon chain of 5–36 carbon units and their attached hydrogens. Most of the carbons in the chain are linked by single bonds, but one or more double bonds are also present in certain types of fatty acid (see *Fig. 5.2*). Triacylglycerols are lipids that consist of three fatty acids attached to a glycerol

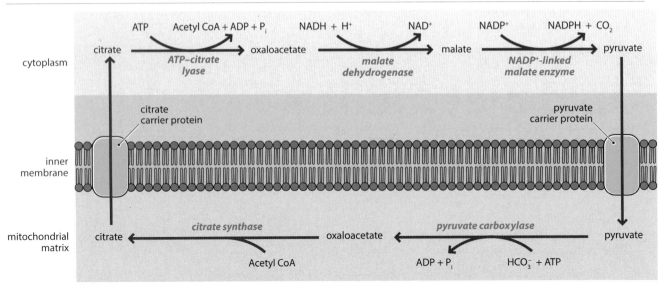

Figure 12.1 Shuttling of acetyl CoA from mitochondrion to cytoplasm.
Shuttling occurs by transferring the acetyl group of acetyl CoA to oxaloacetate, forming citrate, which is moved to the cytoplasm via the citrate carrier protein. Acetyl CoA is then recovered by converting citrate back to oxaloacetate. The remainder of the shuttle is required to maintain the oxaloacetate content of the mitochondrion.

molecule. First, we must examine how the fatty acid chains are built up, and then how these chains are joined to glycerol to make a triacylglycerol molecule.

12.1.1 Fatty acid synthesis

The hydrocarbon chain in a fatty acid is built up from two-carbon acetyl units donated by acetyl CoA. Clearly we must examine this polymerization process in some detail, but first we must deal with a problem that we have met previously, in different guises: acetyl CoA is synthesized in the mitochondrion and fatty acids are made in the cytoplasm. We therefore need to shuttle acetyl CoA from mitochondrion to cytoplasm, but we cannot do this directly because this molecule cannot cross the inner mitochondrial membrane.

Citrate shuttles acetyl units across the inner mitochondrial membrane

Acetyl CoA is synthesized from pyruvate by the pyruvate dehydrogenase complex, at the link between glycolysis and the TCA cycle (see *Section 9.1.1*). The pyruvate dehydrogenase complex operates inside the mitochondrion, whereas fatty acid synthesis occurs in the cytoplasm. Fatty acids themselves can be transported across the inner mitochondrial membrane, but acetyl CoA cannot cross this barrier.

This particular shuttling problem is solved by transferring the acetyl group of acetyl CoA to oxaloacetate, forming citrate (*Fig. 12.1*). This is the same reaction that

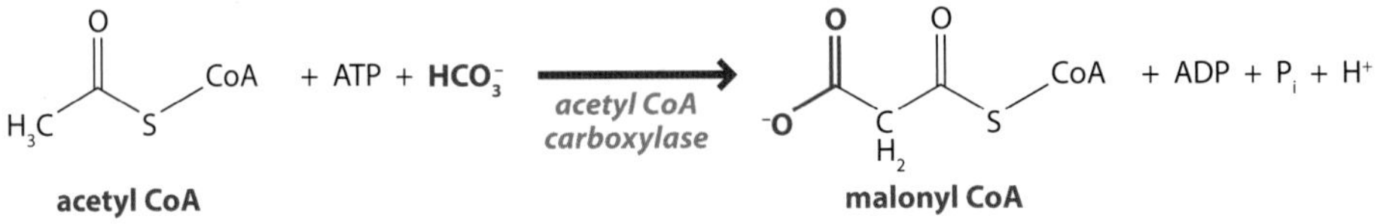

Figure 12.2 Synthesis of malonyl CoA.

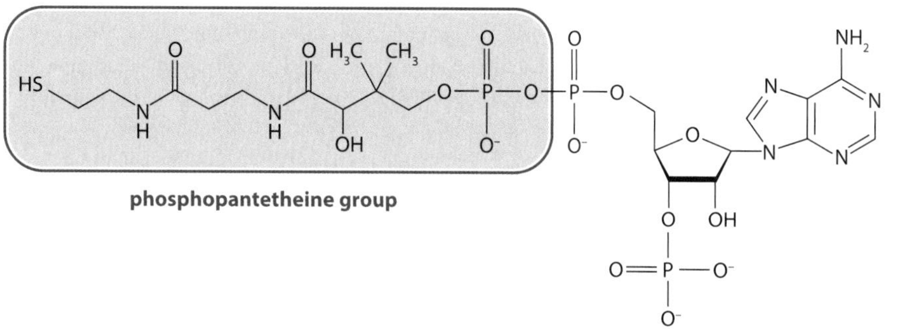

Figure 12.3 Structure of coenzyme A with the phosphopantetheine group highlighted.
A phosphopantetheine prosthetic group is also attached to a serine in the acyl carrier protein.

phosphopantetheine group

occurs during the first step of the TCA cycle, but rather than being utilized for energy generation, the citrate that is formed is transported out of the mitochondrion via the **citrate carrier** protein, which is located in the inner mitochondrial membrane. Once in the cytoplasm, the acetyl group is transferred back to coenzyme A, regenerating acetyl CoA or, more precisely, creating a cytoplasmic supply of this compound.

The oxaloacetate that is now located in the cytoplasm must be moved back into the mitochondrion. Once again there is a transport problem, because the inner mitochondrial membrane does not have a carrier protein for oxaloacetate. The oxaloacetate is therefore converted first to malate, by malate dehydrogenase, and then to pyruvate by an oxidative decarboxylation catalyzed by the $NADP^+$-linked malate enzyme. There are shuttle proteins for transporting both malate and pyruvate back into the mitochondrion, so why do we have this second step in which the malate is converted to pyruvate? The answer is that this second reaction also yields a molecule of NADPH. This is important because NADPH is needed to join together the acetyl units to form fatty acid chains. Oxidative decarboxylation of malate provides some of this NADPH, the remainder being supplied by the pentose phosphate pathway.

It is still necessary to replenish the oxaloacetate used up in the mitochondrion, so after re-entry into the mitochondrion, the pyruvate is converted back to oxaloacetate, by **pyruvate carboxylase**, which adds a carboxyl (–COOH) group obtained from dissolved bicarbonate ions.

Fatty acids are built up by sequential addition of two-carbon units

Fatty acid synthesis is an iterative process in which two-carbon units are added to the end of the growing hydrocarbon chain. The process is begun by the enzyme **acetyl CoA carboxylase**, which carboxylates acetyl CoA, giving malonyl CoA (*Fig. 12.2*). The acetyl and malonyl groups from these two compounds are then transferred to separate molecules of the **acyl carrier protein** (**ACP**). This is a small, non-enzymatic protein that contains a **phosphopantetheine** prosthetic group attached to a serine amino acid within its polypeptide chain. Phosphopantetheine is derived from vitamin B_5 and is also a component of coenzyme A (*Fig. 12.3*). The transfer simply involves the acetyl and malonyl units being detached from the phosphopantetheine structure in coenzyme A and reattached to the same structure in an ACP. The transfers are catalyzed by the **acetyl transacylase** and **malonyl transacylase** enzymes, respectively.

The synthesis phase of the process then begins, with four steps needed to add each two-carbon unit (*Fig. 12.4*):

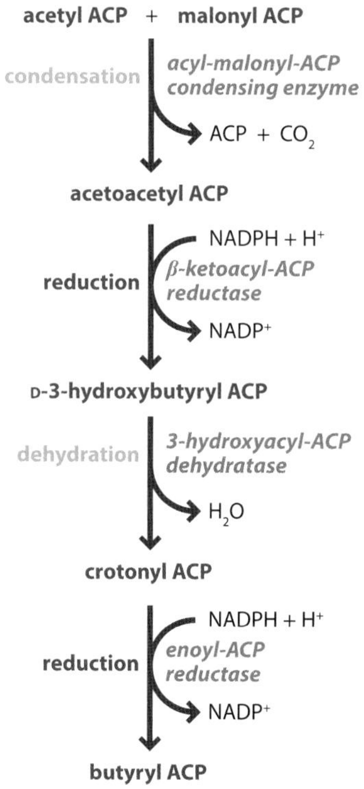

Figure 12.4 Outline of fatty acid synthesis.
Each two-carbon unit is added by a cycle of four reactions, involving a condensation, reduction, dehydration and second reduction step.

Step 1. Acetyl ACP and malonyl ACP are linked together to form acetoacetyl ACP. The ACP that was attached to the acetyl group is released. This is a condensation reaction – the type of reaction in which two molecules combine to form a larger one. It is catalyzed by the **acyl-malonyl-ACP condensing enzyme.**

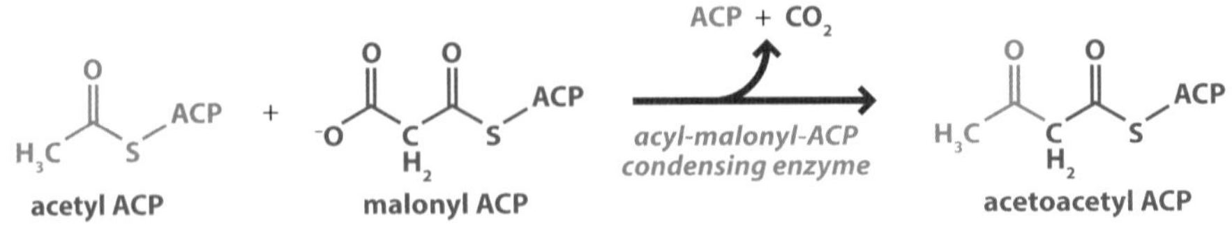

Step 2. Acetoacetyl ACP is reduced, giving D-3-hydroxybutyryl ACP. This reaction requires NADPH and is catalyzed by **β-ketoacyl-ACP reductase.**

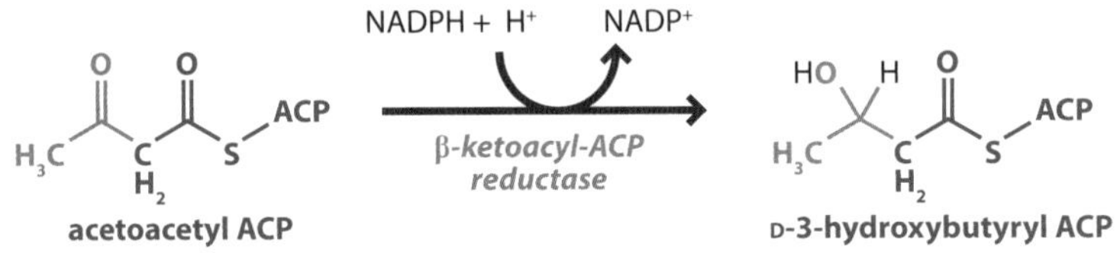

Step 3. A dehydration reaction (removal of water) converts D-3-hydroxybutyryl ACP to crotonyl ACP. The enzyme is **3-hydroxyacyl-ACP dehydratase.**

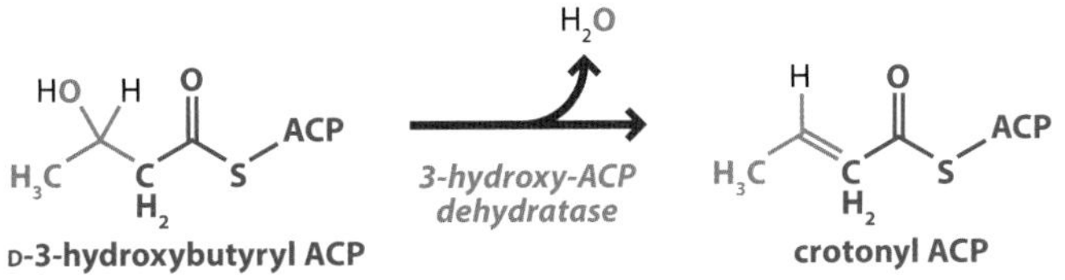

Step 4. A second reduction, using another molecule of NADPH, converts crotonyl ACP to butyryl ACP. This reaction is catalyzed by **enoyl-ACP reductase.**

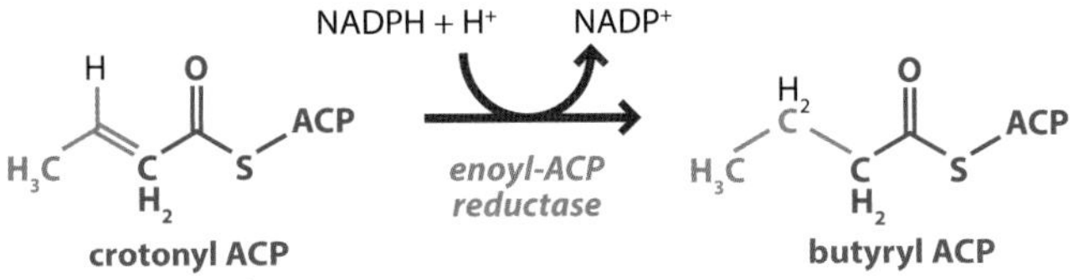

Butyryl ACP is a four-carbon fatty acid linked to ACP. We have therefore completed the first round of synthesis. The second round begins with *step 1*, but with butyryl ACP replacing acetyl ACP. The second round therefore yields a six-carbon fatty acid, which is used as the substrate for the third cycle, and so on, with each cycle adding on a two-carbon unit.

In animals, the synthesis cycle continues until the 16-carbon fatty acid palmitic acid is formed; it is still linked to ACP and so is more accurately described as palmitoyl ACP. The acyl-malonyl-ACP condensing enzyme cannot use palmitoyl ACP as a substrate, so at this stage chain elongation stops. To complete the synthesis, the fatty acid is cleaved from the carrier protein by a **thioesterase** enzyme (*Fig. 12.5*).

Making fatty acids other than palmitic acid

Palmitic acid is not the only fatty acid found in living cells. There are many other types with different chain lengths and some with one or more double bonds linking pairs of carbons. How are these other types made?

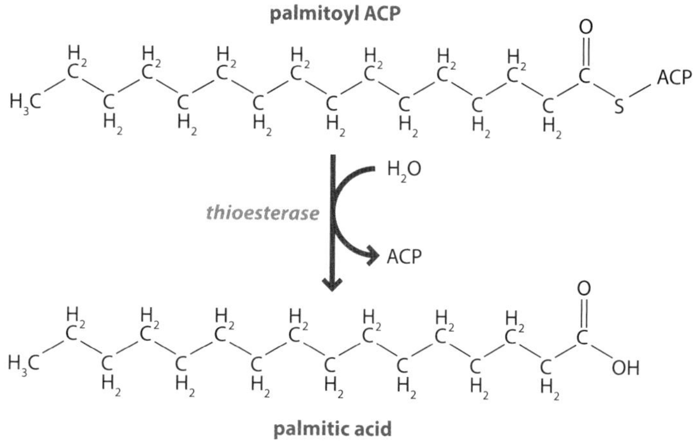

Figure 12.5 **Conversion of palmitoyl ACP to palmitic acid by thioesterase.**

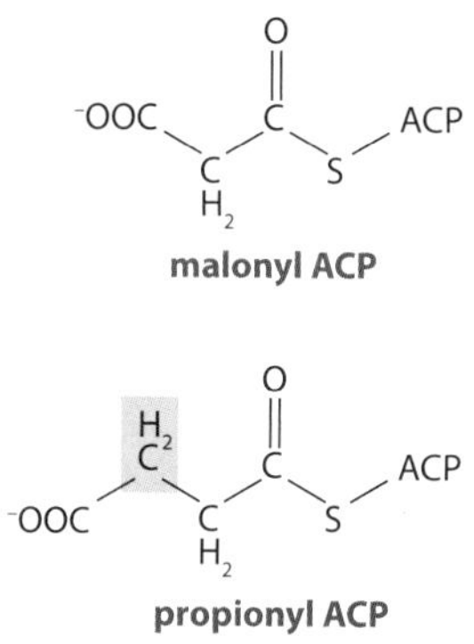

Figure 12.6 **The difference between malonyl and propionyl ACP.**

Most of the important fatty acids have an even number of carbons, but a few are odd numbered. These are still built up by sequentially adding two-carbon units, but the initial malonyl ACP, used at the very start of the process, is replaced by propionyl ACP (*Fig. 12.6*). Propionyl ACP is one carbon longer than malonyl ACP, so the result of the first round of synthesis is a five-carbon fatty acid. The synthesis then continues exactly as for even-numbered fatty acids, until the 15-carbon molecule, pentadecylic acid, is produced, which is cleaved from its carrier protein in the same way as palmitic acid.

Fatty acids with chain lengths greater than the 15 carbons of pentadecylic acid or 16 carbons of palmitic acid are synthesized by enzymes that are located on the outer surface of the smooth endoplasmic reticulum. To initiate this type of synthesis, pentadecylic acid or palmitic acid is attached to coenzyme A, rather than ACP, and elongated with acetyl units from malonyl CoA.

Each double bond is introduced into the hydrocarbon chain by an oxidation reaction, catalyzed by an enzyme complex comprising **NADH–cytochrome b_5 reductase**, **cytochrome b_5** and **desaturase**. This complex is also located on the outer surface of the smooth endoplasmic reticulum and uses the CoA derivative of the fatty acid as the substrate. Electrons are transferred through the enzyme complex in order to drive the oxidation (*Fig. 12.7*). Mammals have four different saturases, called Δ^4, Δ^5, Δ^6 and Δ^9, indicating that they are able to add double bonds immediately after the 4th, 5th, 6th and 9th carbons in a fatty acid. This means that mammals are not able to synthesize linoleic and the α- and γ-forms of linolenic acid, which have the structures denoted $18:2(\Delta^{9,12})$, $18:3(^{9,12,15})$ and $18:3(^{6,9,12})$, respectively. Mammals must therefore obtain linoleic and linolenic acid from their diet because these fatty acids are precursors for

---

Box 12.1 **The energy requirement for fatty acid synthesis**

The synthesis of a fatty acid molecule requires both ATP and NADPH. To calculate the extent of this energy requirement we must look closely at the synthesis pathway. Seven rounds of the synthesis cycle are needed to make one molecule of palmitate – the anion of palmitic acid. The basic equation for these seven synthesis cycles is:

acetyl CoA + 7malonyl CoA + 14NADPH + 20H+ →
palmitate + 7CO$_2$ + 14NADP+ + 8CoA + 6H$_2$O

To make this equation more comprehensive we must also consider how the seven malonyl CoA molecules are synthesized:

7acetyl CoA + 7CO$_2$ + 7ATP → 7malonyl CoA + 7ADP + 7P$_i$ + 14H+

Adding these two reactions together we see that that the overall equation for synthesis of one molecule of palmitate is:

8acetyl CoA + 7ATP + 14NADPH + 6H+ →
palmitate + 14NADP+ + 8CoA + 6H$_2$O + 7ADP + 7P$_i$

Figure 12.7 Introduction of a double bond into a fatty acid.
Electrons are transferred from NADH to the FAD component of NADH–cytochrome b_5 reductase, and then to the heme of cytochrome b_5, reducing this from Fe^{3+} to Fe^{2+}. The same reduction now occurs in the non-heme iron present in the desaturase, which oxidizes the bond targeted by the desaturase.

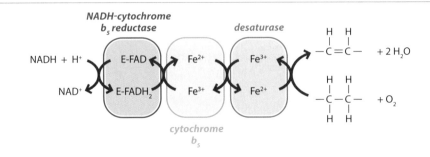

other important lipids, including arachidonic acid, from which eicosanoid hormones are made (see *Box 5.2*). Arachidonic acid has the formula $20:4(\Delta^{5,8,11,14})$, and mammals can make it from linoleic acid by adding further double bonds with the Δ^6 and Δ^5 desaturases, the two oxidations occurring, respectively, before and after addition of another two-carbon unit.

Animals are unable to make fatty acids with chain lengths shorter than the 15 carbons of pentadecylic acid. Plants on the other hand make a greater variety of fatty acids including some with fewer than 15 carbons, such as lauric acid (12 carbons), found in the seed pods of the laurel tree and also in coconut and palm oils, and myristic acid (14 carbons), from nutmeg. Plants are able to make these shorter structures because of a difference in the arrangement of the enzymes involved in fatty acid synthesis. In mammals, the six key enzyme activities (the transacylases, acyl-malonyl-ACP condensing enzyme, β-ketoacyl-ACP reductase, 3-hydroxyacyl-ACP dehydratase, enoyl-ACP reductase and thioesterase) are all contained within a single multifunctional protein. Within this protein, called **fatty acid synthase**, the six enzyme activities are specified by different parts of the polypeptide chain (*Fig. 12.8*).

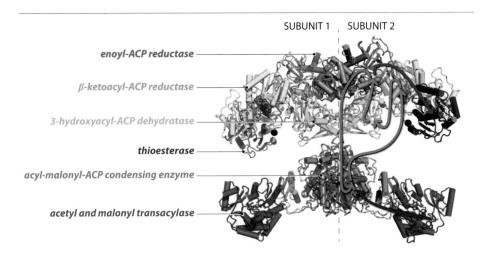

Figure 12.8 Mammalian fatty acid synthase.
The protein is a dimer of two identical subunits. In the left subunit, the approximate locations of the six enzyme activities involved in fatty acid synthesis are shown. On the right, the arrows indicate the shuttling of the growing fatty acid within this subunit. The initial step, involving acylation of acetyl and malonyl units, occurs at the bottom right, with the products then shuttled to the condensing enzyme (gray arrow). The cyclical condensation–reduction–dehydration–reduction reaction, which elongates the fatty acid chain, is depicted by the blue arrows. Once completed, palmitoyl ACP is shuttled to the thioesterase activity (red arrow) and the palmitic acid molecule is cleaved from its ACP carrier. In reality, the pathway might not be exactly as shown, because synthesis of an individual fatty acid may involve shuttling between enzyme activities on different subunits.
Main image reproduced from *Science* 2008;**321**:1315 with permission from the AAAS.

During synthesis of a fatty acid, the growing end of the hydrocarbon chain is moved from one active site to the next, the translocations aided in part by the flexibility of the arm formed by the phosphopantetheine group of the carrier protein, to which the fatty acid is attached. The growing fatty acid does not emerge from the fatty acid synthase complex until the 15- or 16-carbon stage has been reached, so shorter fatty acids are not made. In plants, there is no fatty acid synthase, the enzyme activities being provided by different proteins that form only a loose association. The growing fatty acid therefore has to shuttle between the individual enzymes. As in animals, these enzymes only make fatty acids up to 15 or 16 carbons in length, but because the growing structure is not synthesized within a single protein, shorter chain molecules can be released from the elongation cycle as and when they are needed.

Synthesis of malonyl CoA is the main regulatory step in fatty acid synthesis

Malonyl CoA has very few roles in the cell other than as a precursor for fatty acid synthesis, so it is not surprising that regulation over the synthesis pathway for fatty acids is exerted primarily by controlling the conversion of acetyl CoA into malonyl CoA. This is a good example of the regulation of a pathway being controlled at the first commitment step, which in this case is the formation of malonyl CoA. This regulatory regime ensures that fatty acid synthesis only occurs when the cell has a plentiful supply of energy, some of which it can afford to divert into storage in the form of fatty acids and triacylglycerols.

A plentiful supply of energy is signaled by a surplus of ATP and a deficiency of AMP. Under these conditions, the acetyl CoA carboxylase enzyme, which converts acetyl CoA to malonyl CoA, is fully active. If, on the other hand, the cellular ATP concentration goes down and AMP increases, then the carboxylase becomes inhibited by the AMP level. The regulation is mediated by an **AMP-activated protein kinase**, which phosphorylates a serine amino acid within the carboxylase polypeptide. This modification inactivates the carboxylase. When energy supplies are again high, the level of AMP declines and the kinase is inhibited by the high concentration of ATP (*Fig. 12.9*). No further phosphorylation of the carboxylase occurs and this enzyme is reactivated by **protein phosphatase 2A**, which removes the phosphate group from the serine.

It is also possible for the phosphorylated version of acetyl CoA carboxylase to be partially reactivated without removing the phosphate group. This effect is brought about by citrate, which is made directly from acetyl CoA at the start of the TCA cycle. A high level of citrate therefore indicates that there is a surplus of acetyl CoA, which

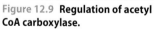

Figure 12.9 Regulation of acetyl CoA carboxylase.
It is not known how citrate activates phosphorylated acetyl CoA carboxylase, but in experimental systems activation is associated with a change in conformation of the enzyme from octameric spheres to polymeric filaments. Whether this conformation change is the cause of the enzyme activation, a result of the activation, or possibly an experimental artefact, has not yet been established.

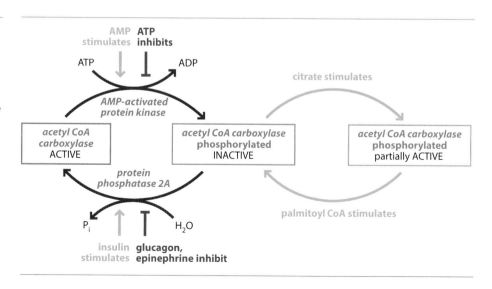

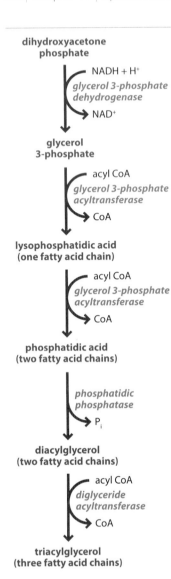

Figure 12.10 Outline of triacylglycerol synthesis.

can be used for fatty acid synthesis. Conversely, an abundance of palmitoyl CoA indicates that fatty acid levels are high and further synthesis is unnecessary. Palmitoyl CoA therefore switches off fatty acid synthesis by reversing the effect of citrate.

The phosphorylation state of acetyl CoA carboxylase is more generally affected by glucagon, epinephrine and insulin, hormones that regulate the level of glucose in the blood and hence monitor the energy status of the body as a whole. Glucagon and epinephrine respond to low energy levels by inhibiting protein phosphatase 2A, keeping the carboxylase in its inactive state so that fatty acid synthesis does not take place. Insulin has the opposite effect, increasing fatty acid synthesis when blood glucose levels are high, probably by activating the phosphatase.

12.1.2 Triacylglycerol synthesis

Triacylglycerol synthesis involves a short pathway that first converts dihydroxyacetone phosphate, one of the intermediates produced during glycolysis, into glycerol 3-phosphate, and then attaches the fatty acid side-chains one by one from acyl CoA substrates (*Fig. 12.10*).

Step 1. **Glycerol 3-phosphate dehydrogenase** reduces dihydroxyacetone phosphate, to give glycerol 3-phosphate.

Step 2. The first fatty acid is transferred from its coenzyme A carrier to carbon number 1 of glycerol 3-phosphate (the carbon carrying the phosphate being number 3). The enzyme is **glycerol 3-phosphate acyltransferase** and the resulting structure is called lysophosphatidic acid.

Step 3. The second fatty acid is attached to carbon 2, creating phosphatidic acid. Again this is catalyzed by glycerol 3-phosphate acyltransferase.

Phosphatidic acid is a simple type of glycerophospholipid and can also be used to synthesize the more complex glycerophospholipids found in membranes.

Step 4. **Phosphatidate phosphatase** removes the phosphate group on carbon 3 to give diacylglycerol.

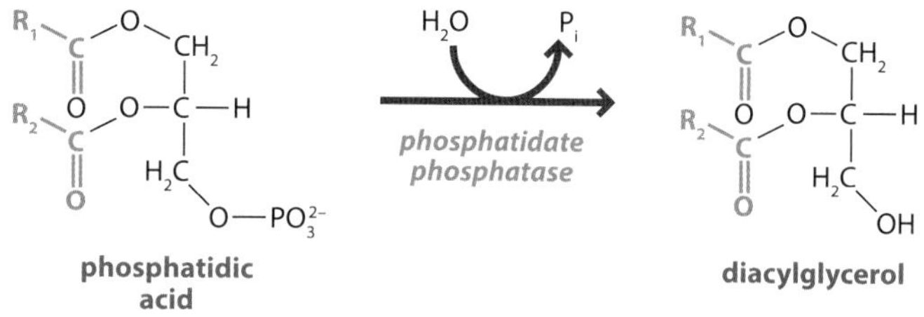

phosphatidic
acid

phosphatidate
phosphatase

diacylglycerol

Step 5. The third fatty acid is now added to complete the process.

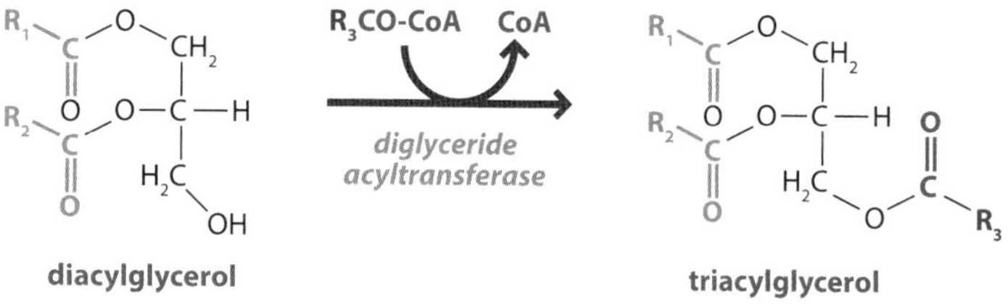

diacylglycerol

diglyceride
acyltransferase

triacylglycerol

The enzyme for this step, **diglyceride acyltransferase**, forms a complex with phosphatidate phosphatase, the pair of activities sometimes being referred to as **triacylglycerol synthetase**.

Triacylglycerol synthesis occurs mainly in liver cells, within the endoplasmic reticulum. The triacylglycerols are then combined with proteins to form **lipoprotein** complexes and transported to storage cells called **adipocytes** (found in adipose tissue and commonly known as fat cells), or sent to the muscles where they are broken down again to release energy.

12.2 Breakdown of triacylglycerols and fatty acids

We have seen how fatty acids are built up from acetyl CoA molecules, and how fatty acids are combined with glycerol 3-phosphate to make triacylglycerols. Now we must examine **lipolysis**, the process by which triacylglycerols and fatty acids are broken down again in order to release the energy that they contain.

12.2.1 Breakdown of triacylglycerols into fatty acids and glycerol

The first step in lipolysis is the conversion of triacylglycerols into their fatty acid and glycerol components. Although a fairly straightforward process, this step is important because it is the major control point for the entire lipolysis pathway.

Triacylglycerols are cleaved by lipases

Triacylglycerols are broken down by **lipase** enzymes which cleave the three fatty acid chains from the glycerol component of the molecule (*Fig. 12.11*). Individual lipases are specific for the different carbons within the glycerol, so a combination of three enzymes, with different specificities, is needed to remove all three of the fatty acids from a single triacylglycerol.

In mammals, lipases produced in the pancreas are secreted into the intestine, where they remove the fatty acids from triacylglycerols consumed in the diet. The fatty acids are then taken up by the cells lining the inside of the intestine. However, the fatty acids

Box 12.2 **Lipoproteins**

Triacylglycerols, phospholipids and cholesterol are relatively insoluble in aqueous solutions such as blood and lymph. They are therefore transported around the body as components of multimolecular structures called **lipoproteins**. A lipoprotein is a micelle-like particle that consists of a spherical lipid monolayer, made of amphiphilic lipids with various embedded proteins, surrounding a hydrophobic core containing triacylglycerol and cholesterol molecules.

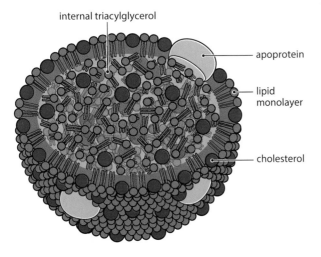

internal triacylglycerol

apoprotein

lipid monolayer

cholesterol

The protein components, which are called **apolipoproteins** or **apoproteins**, are recognized by cell surface receptors and therefore ensure that the lipoproteins are taken up by the correct tissue.

There are three groups of lipoproteins:
- **Chylomicrons** are the largest lipoproteins, with a molecular mass of >400 kDa. They transport dietary triacylglycerols and cholesterol from the intestines to other tissues. After leaving the intestine, they first move to muscle and adipose tissue, where the triacylglycerols are broken down to fatty acids and monoacylglycerols by the action of **lipoprotein lipase**. This enzyme is located on the outside of muscle and adipose cells

and is activated by **apoC-II**, one of the apoproteins on the chylomicron surface. The fatty acids and monoacylglycerols are taken up by the tissues, and either used for energy production or converted back to triacylglycerol for storage. As their triacylglycerol content is depleted, the chylomicrons shrink and form cholesterol-rich **chylomicron remnants**. These are transported to the liver where they bind to a cell surface receptor and are taken up by endocytosis.
- **Very low density lipoproteins (VLDLs), intermediate density lipoproteins (IDLs)** and **low density lipoproteins (LDLs)** are all related to one another. VLDLs are synthesized in the liver and transport a variety of lipids to muscle and adipose tissue. As with chylomicrons, the triacylglycerols in VLDLs are broken down by lipoprotein lipase and the released fatty acids are taken up by the muscle or adipose cells. The VLDL remnants remain in the blood as IDLs, and then lose most of their apoprotein content to become LDLs. The LDLs are then taken up by various cell types and their contents metabolized by enzymes located in the cellular organelles called **lysosomes**.
- **High density lipoproteins (HDLs)** have the opposite function to that of LDLs in that they transport cholesterol back to the liver. HDLs are synthesized in the blood mainly from lipids and apoproteins resulting from the degradation of other lipoproteins. HDLs then extract cholesterol from cell membranes and move it to the liver. The liver then disposes of the excess cholesterol as **bile acids** (see *Sections 5.1.2* and *12.3.2*).

Blood cholesterol level is associated with the risk of cardiovascular disease because the deposition of cholesterol and other lipids from LDLs onto the inner surfaces of blood vessels is thought to promote **atherosclerosis** ('hardening of arteries'). White blood cells accumulate in the cholesterol deposits, which can lead to inflammation and an eventual blockage. A blockage in a coronary artery can lead to **myocardial infarction** or **heart attack**, which is the most common cause of death in Western industrialized countries. Blood clots in cerebral arteries cause stroke, while those in peripheral blood vessels in the limbs can lead to gangrene. LDLs are therefore looked on as 'bad' cholesterol, in contrast to HDLs which scavenge lipids from the blood and are therefore 'good' cholesterol.

are not utilized as an energy source in this tissue, but instead are converted back into triacylglycerols which are released, as large lipoproteins called **chylomicrons**, into the lymphatic system and the bloodstream. These newly formed triacylglycerols are now transported to muscle cells for use as an energy supply, or to adipose tissue for storage (*Fig. 12.12*).

triacylglycerol   diacylglycerol   monoacylglycerol   glycerol

lipase   *lipase*   *lipase*

Figure 12.11 Breakdown of a triacylglycerol by a series of lipase enzymes.
R1, R2 and R3 are the fatty acid hydrocarbon chains.

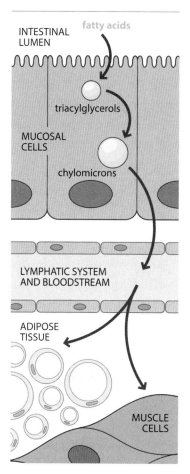

Figure 12.12 Transport of fatty acids from the intestine to muscle and adipose tissue.
Fatty acids are absorbed by the intestinal mucosal cells, converted to triacylglycerols and packaged into chylomicrons, which are then transported via the lymphatic system and bloodstream to muscle cells and adipose tissue.

Different lipases are responsible for breakdown of triacylglycerols in adipocytes – the fat cells in adipose tissue. The fatty acids and glycerol molecules produced by these lipases are passed to the bloodstream and transported to the muscles and other active tissues. Within these tissues the fatty acids are further degraded by the energy generating process called **β-oxidation**, which we will study in the next section. The glycerol is taken to the liver, where it is converted into glycerol 3-phosphate and then dihydroxyacetone phosphate (*Fig. 12.13*). The two steps of this short pathway are catalyzed by **glycerol kinase** and glycerol 3-phosphate dehydrogenase, respectively, the second step being a reversal of step 1 of the triacylglycerol synthesis pathway. Dihydroxyacetone phosphate is an intermediate in glycolysis, so both the fatty acid and glycerol components of the triacylglycerol are utilized for energy generation.

Triacylglycerol cleavage is the major control point in lipolysis

The β-oxidation pathway by which the individual fatty acids are broken down does not contain any major control points, the rate of fatty acid oxidation depending only on the availability of fatty acids. This means that the amount of energy that is being generated from stored lipids at any particular time is determined almost entirely by the rate at which triacylglycerols are being converted into fatty acids. Lipase activity is therefore the key event that regulates the entire lipolysis pathway.

The activity of the lipases in the adipocytes is responsive to the levels of various hormones in the bloodstream. Glucagon, epinephrine, **norepinephrine** and **adrenocorticotropic hormone** each stimulate lipase activity, increasing the rate at which fatty acids are released from triacylglycerols. Insulin has the opposite effect, inhibiting the lipases and reducing triacylglycerol breakdown. All five hormones act in the manner with which we have now become familiar: they indirectly regulate the activity of a protein kinase which, in turn, regulates the activity of the lipases (*Fig. 12.14*). The phosphorylated version of a lipase is the active enzyme, and the dephosphorylated version is inactive. Glucagon, epinephrine, norepinephrine and adrenocorticotropic hormone stimulate the protein kinase by activating adenylate cyclase and increasing the cellular cAMP concentration. In response to increased cAMP, the protein kinase phosphorylates the lipases, increasing triacylglycerol breakdown. This occurs when blood glucose levels are low, and energy is needed. Insulin, on the other hand, responds to high blood glucose levels by reducing the rate of lipolysis. Insulin therefore inactivates adenylate cyclase, so the cAMP level drops and the kinase is switched off. Insulin also stimulates the phosphatase that removes the phosphate groups added by the kinase and hence inactivates the lipases.

12.2.2 Breakdown of fatty acids

Traditionally, the mitochondrial matrix is looked on as the site of fatty acid breakdown. This was the conclusion of important experiments carried out by Eugene Kennedy and Albert Lehninger in 1949. Today we recognize that their work only gave us half the picture, and that some fatty acid degradation also occurs in the **peroxisomes**.

Figure 12.13 Utilization of glycerol resulting from triacylglycerol breakdown.

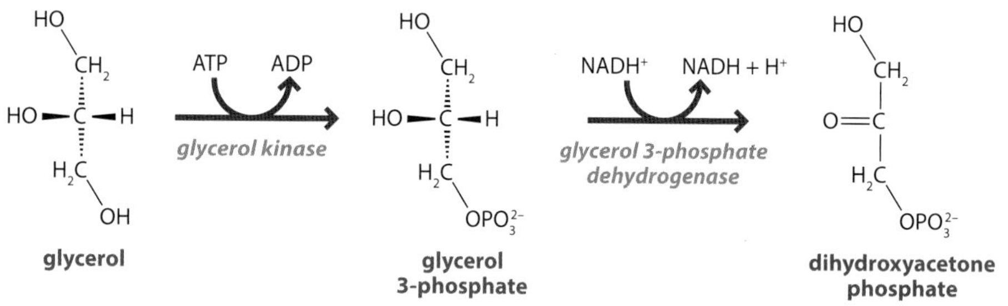

Figure 12.14 **Control of triacylglycerol breakdown.**
Abbreviation: ACTH, adrenocorticotropic hormone.

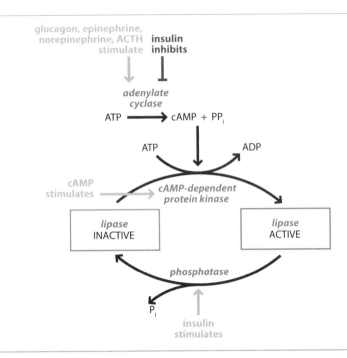

These are small organelles, surrounded by a single membrane, that are found in the cytoplasm of all eukaryotes. Peroxisomes were not discovered until 1954, and there was uncertainty about their biochemical roles for many years after this. Today we recognize that these roles include some aspects of fatty acid breakdown. Indeed, in plants and some other eukaryotes, fatty acid breakdown occurs exclusively in the peroxisomes – the mitochondria are not involved at all. In other eukaryotes, including animals such as humans, most fatty acids are degraded in the mitochondria, but longer chain molecules are initially broken down into shorter units in the peroxisomes, the partly degraded fatty acids then being moved to the mitochondria where the process is completed.

We will look first at the role of the mitochondria in fatty acid degradation, by considering how a molecule of palmitic acid is broken down in an animal cell. We will then examine the slightly different degradation process that takes place in peroxisomes. As with synthesis, we must also consider the special arrangements that are necessary to deal with double bonds and with fatty acids with odd numbers of carbons.

Fatty acids are converted to acyl forms before transport into the mitochondria

As a prelude to their breakdown, the fatty acids taken up from the bloodstream are attached to coenzyme A to form acyl CoA derivatives (*Fig. 12.15*). The enzyme that is responsible for this reaction is called **fatty acid thiokinase**, and the reaction requires energy that is provided by ATP hydrolysis. Note that the ATP is converted to AMP, which means that two of its 'high energy' bonds are cleaved, compared to just one

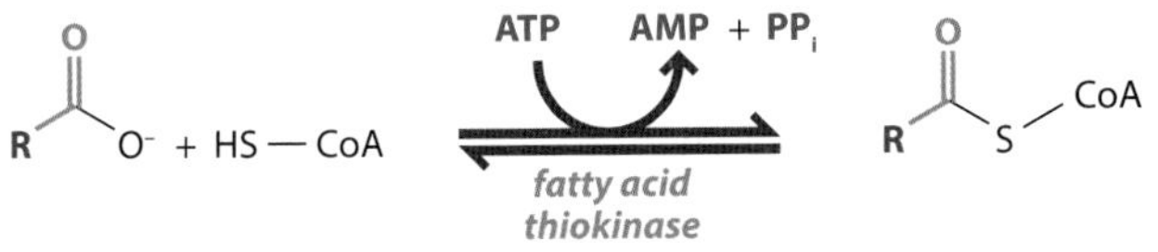

Figure 12.15 **Attachment of a fatty acid to CoA prior to breakdown.**
'R' indicates the fatty acid hydrocarbon chain.

Figure 12.16 **Shuttling of fatty acids across the inner mitochondrial membrane as carnitine derivatives.** This process operates only for unsaturated fatty acids and saturated ones longer than ten carbons. Short, saturated fatty acids can pass directly through the inner mitochondrial membrane.

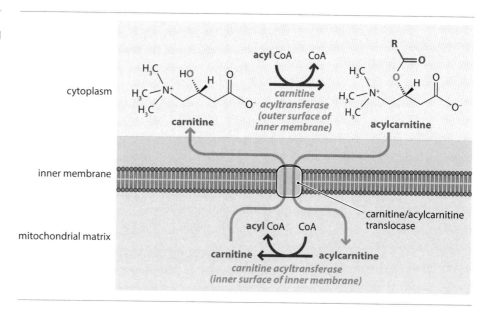

Figure 12.16 **Shuttling of fatty acids across the inner mitochondrial membrane as carnitine derivatives.** This process operates only for unsaturated fatty acids and saturated ones longer than ten carbons. Short, saturated fatty acids can pass directly through the inner mitochondrial membrane.

cleavage in most of the ATP-requiring reactions that have studied so far, in which the product is ADP.

Fatty acid thiokinase is attached to the outer mitochondrial membrane, so the 'activated' fatty acids are produced on the surface of the organelle into which they must be transported in order to be degraded. Shorter types of saturated fatty acid, up to 10 carbons in length and without any double bonds, are able to pass directly through the mitochondrial membranes and into the matrix as their CoA derivatives. In contrast, longer fatty acid CoAs and unsaturated ones of any length, cannot get across the inner mitochondrial membrane unaided. For these molecules, the CoA group is removed and replaced with **carnitine**. This is a small polar molecule, derived from lysine and methionine, that has achieved some notoriety as a dietary supplement especially for bodybuilding, due to its supposed ability to reduce body fat. In fact, carnitine is synthesized in adequate amounts for most purposes in the liver and kidneys, and is also a natural dietary component of red meat and milk, and there is no convincing evidence that dietary supplements can improve athletic performance.

The replacement of CoA by carnitine is catalyzed by **carnitine acyltransferase**, which is located on the outer surface of the inner mitochondrial membrane (*Fig. 12.16*). The acylcarnitine derivatives are then transported across the inner mitochondrial membrane by the **carnitine/acylcarnitine translocase**, an integral membrane protein that spans the membrane and hence can carry the acylcarnitine molecules into the mitochondrial matrix. Once inside, a second type of carnitine acyltransferase, attached to the inner side of the inner mitochondrial membrane, removes the carnitine group, replacing it with CoA and hence re-creating the original acyl CoA molecule. The carnitine released during this step is moved back across the inner mitochondrial membrane by the translocase, ready to participate in another cycle of fatty acid transport.

Fatty acids are broken down by sequential removal of acetyl CoA units

Inside the mitochondrial matrix, fatty acids are broken down by a series of reactions that result in the sequential removal of acetyl CoA units from the ends of the hydrocarbon chains. Two of these reactions are oxidations, and the process as a whole is sometimes referred to as the 'β-oxidation' pathway. The steps required for removal of an acetyl CoA unit from the end of an acyl CoA molecule are as follows (*Fig. 12.17*):

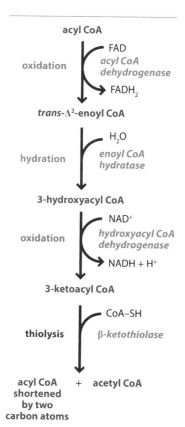

Figure 12.17 **Outline of fatty acid breakdown.**

Step 1. The single bond linking carbons 2 and 3 (the α and β carbons, respectively) is converted to a double bond. This is an oxidation reaction that generates a molecule of FADH₂ and is catalyzed by **acyl CoA dehydrogenase**.

The reaction results in the Δ² version of the fatty acid. In chemical terms, the fatty acid is now a type of enoyl compound (*Fig. 12.18*) and so is described as *trans-Δ²-enoyl* CoA. There are three types of acyl CoA dehydrogenase that are active with fatty acyl CoA chains of different lengths.

Step 2. A hydroxyl group is added to the β carbon, resulting in conversion of the double bond to a single bond. Water is used, and so this is a hydration reaction, catalyzed by **enoyl CoA hydratase**. The resulting compound is 3-hydroxyacyl CoA.

Step 3. A second oxidation, generating one molecule of NADH⁺, converts the hydroxyl to a carbonyl group. The enzyme is **hydroxyacyl CoA dehydrogenase** and the resulting version of the fatty acid is a 3-ketoacyl CoA.

Step 4. Finally, cleavage of the bond linking the α and β carbons removes an acetyl CoA unit, giving a new acyl CoA that is two carbons shorter than the one that entered this series of reactions. A second CoA molecule is needed to create the new acyl CoA. The enzyme that catalyzes this reaction is called **β-ketothiolase**.

This is a **thiolysis** reaction, the cleavage being driven by the thiol (–SH) group present at the end of the CoA molecule.

Subsequent rounds of the reaction cycle continue to remove additional acetyl CoA units. The final cycle, which begins with an acyl CoA comprising four carbons, yields two acetyl CoA molecules, completing the degradation of the original fatty acid.

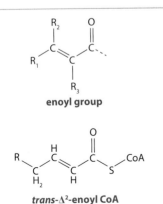

Figure 12.18 An oxidized fatty acid is an enoyl compound.

Box 12.3 **The Greek notation for fatty acid structure.**

The term 'β-oxidation' indicates that the bond which is cleaved is the one immediately before the β carbon of the hydrocarbon chain. As with the M:N($\Delta^{a,b,...}$) notation, we are counting along the fatty acid from the end that terminates with the carboxyl group (see *Fig. 5.4*). In the M:N($\Delta^{a,b,...}$) notation, the numbering (carbon 1, carbon 2, etc.) begins with the carbon of the carboxyl group, but the Greek notation (α carbon, β carbon, etc.) refers only to the hydrocarbon chain. The α carbon is therefore number 2 in the chain and the β carbon is number 3 as shown below.

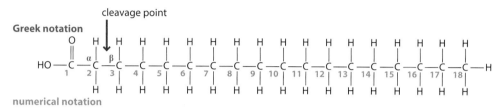

Breakdown of palmitic acid gives a net yield of 129 ATPs

The purpose of fatty acid breakdown is to release the energy contained in these molecules, making that energy available, in the form of ATP, to the tissues that require it. It is therefore important that we consider the energy budget for lipid breakdown, this budget representing the yield of ATP molecules minus any that are used up in the breakdown process. The energy budget will depend on the identity of the fatty acid that is being degraded, because longer chain molecules yield more acetyl CoA units and hence more energy than shorter ones. We will therefore consider the energy balance for palmitic acid, the common 16-carbon, unsaturated fatty acid.

The overall reaction for β-oxidation of palmitic acid is:

$$\text{Palmitoyl CoA} + 7\text{FAD} + 7\text{NAD}^+ + 7\text{CoA} + 7\text{H}_2\text{O} \rightarrow 8\text{acetyl CoA} + 7\text{FADH}_2 + 7\text{NADH} + 7\text{H}^+$$

From this equation we can see that there are three separate ways of generating ATP from the breakdown of palmitic acid:

- The NADH molecules can enter the electron transport chain, each one yielding three ATPs. Because we have seven NADH molecules we will obtain 21 ATPs by this route.

- The seven FADH$_2$ molecules can also enter the electron transport chain, each giving two ATPs, a total of 14 ATPs per molecule of palmitic acid.

- Each acetyl CoA molecule can enter the TCA cycle, yielding another 12 ATPs. With eight of these molecules, the acetyl CoAs can generate a total of 96 ATPs.

The energy yield per molecule of palmitic acid is therefore 21 + 14 + 96 = 131 ATPs. But we must take two ATPs away from this to get the final budget. This is because we used a molecule of ATP at the very start of the process, when the palmitic acid molecule was 'activated' by attachment to coenzyme A (see *Fig. 12.15*). If we used one molecule of ATP to activate one molecule of palmitic acid, then why do we have to take two ATPs away from the final yield? This is because, as we noted when we looked at the activation reaction, the ATP used up when palmitoyl CoA is synthesized is converted not to ADP but to AMP. Thus two of the 'high energy' bonds in the ATP were broken, not just one.

Breakdown of one molecule of palmitic acid therefore gives a net gain of 129 ATPs. Longer fatty acids will give more ATPs, and shorter ones will give fewer. In comparison, a single molecule of glucose will generate just 38 ATPs when taken through the entire

Box 12.4 **The glyoxylate cycle**

In animals most of the acetyl CoA produced from fatty acid degradation enters the TCA cycle, where it is utilized for energy generation, or is used in the synthesis of cholesterol and its derivatives. Plants and many microorganisms have greater flexibility and can also use acetyl CoA as the substrate for synthesis of four- and six-carbon sugars including glucose. This is achieved by the **glyoxylate cycle**, which in plants takes place in special organelles called **glyoxysomes**.

The glyoxylate cycle is similar to the TCA cycle, but bypasses the series of reactions from isocitrate to malate, which include the two decarboxylations that convert acetyl CoA to CO_2. These steps are replaced by two new ones, catalyzed by **isocitrate lyase**, which converts isocitrate into succinate and glyoxylate, and **malate synthase**, which combines the glyoxylate with a second molecule of acetyl CoA to form malate.

The product of the pathway is therefore succinate, the overall equation for a single cycle being:

$$2\text{acetyl CoA} + NAD^+ + 2H_2O \rightarrow \text{succinate} + 2\text{CoA} + NADH + 2H^+$$

Succinate can be used as the substrate for synthesis of a variety of carbohydrates. One possible route is to convert it to oxaloacetate via the TCA cycle and then to glucose by gluconeogenesis. This means that organisms which possess the glyoxylate cycle can convert fatty acids into glucose, which animals are unable to do.

In plants, the main role of the glyoxylate cycle is to convert part of the lipid stored in the germinating seed into carbohydrate, which the seedling uses for energy generation and for synthesis of structural polysaccharides such as cellulose. Storing carbon in seeds in the form of oil is more efficient than storing carbohydrate, because oils have a higher carbon content. Many bacteria also have the ability to convert acetate to acetyl CoA. When combined with the glyoxylate cycle, this function enables a bacterium to use acetate as its sole source of carbon.

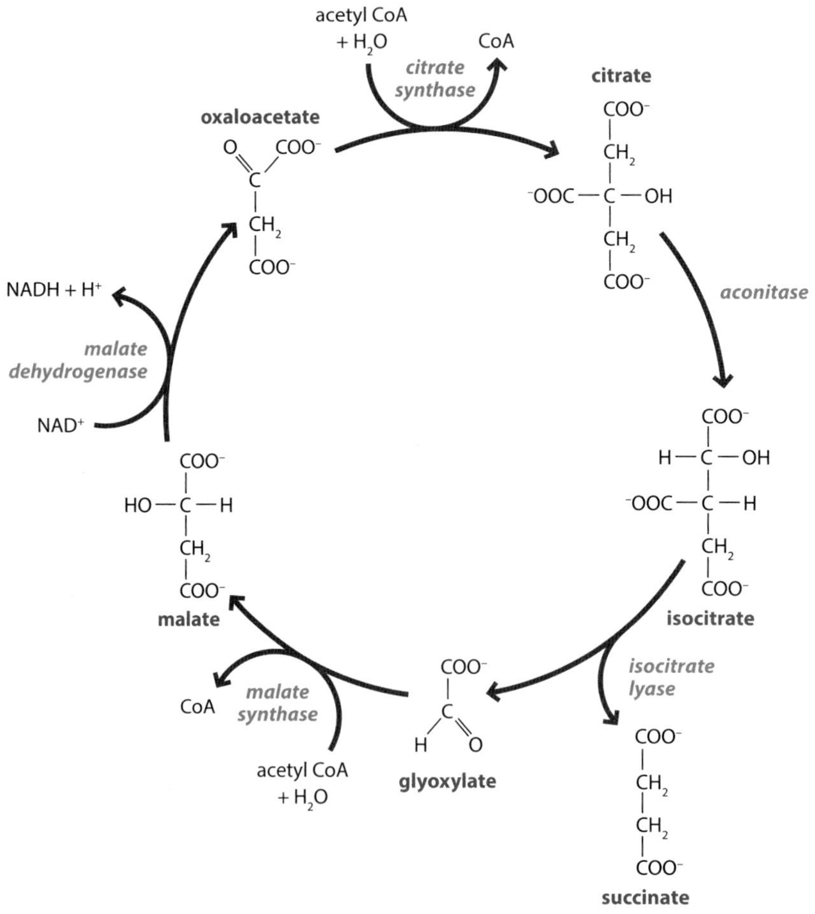

glycolysis and TCA pathways. This might explain why fats are such a good energy store in animals.

Fatty acid breakdown in peroxisomes

Now we return to the second site of fatty acid breakdown, the peroxisomes. The process for fatty acid breakdown in peroxisomes is very similar to that operating in the mitochondria, with one important difference. This concerns step 1 of the β-oxidation pathway, during which the single bond linking the α and β carbons is converted to a double bond. In mitochondria, this oxidation reaction is catalyzed by acyl CoA dehydrogenase and yields one molecule of $FADH_2$, which can subsequently generate two ATPs via the electron transport chain. In peroxisomes, this oxidation step is carried out in a different way. An acyl CoA dehydrogenase is still involved, but in peroxisomes this enzyme transfers electrons not to FAD, but to water, converting it into hydrogen peroxide (*Fig. 12.19*). The enzyme called **catalase** then converts the hydrogen peroxide back to water and oxygen, a reaction that can be linked to the oxidation of a variety of toxic compounds such as phenols and alcohol. Especially in liver, these detoxification reactions appear to be an important role of peroxisomes.

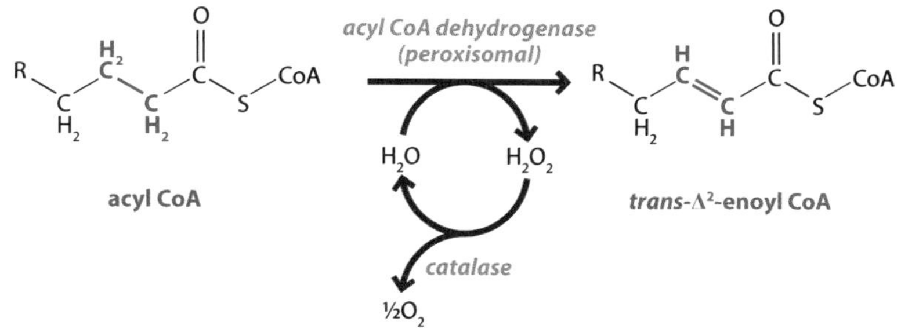

Figure 12.19 The first step of β-oxidation in peroxisomes. The peroxisomal acyl CoA dehydrogenase transfers electrons to water, generating hydrogen peroxide, which is converted back to water by catalase. The notation '½O_2' indicates that two molecules of hydrogen peroxide are needed to make a single molecule of oxygen. The reaction that catalase catalyzes is therefore $2H_2O_2 \rightarrow 2H_2O + O_2$.
'R' indicates the remainder of the fatty acid hydrocarbon chain.

The subsequent steps in the β-oxidation pathway are identical in both mitochondria and peroxisomes, with the resulting NADH and acetyl CoA exported out of the peroxisomes. However, a single round of β-oxidation in the peroxisomes generates fewer ATP molecules than the equivalent events in mitochondria. This is for two reasons. First, no $FADH_2$ is produced because of the difference in the initial oxidation step. Secondly, the NADH molecules cannot be entered directly into the electron transport chain because they cannot cross the inner mitochondrial membrane and hence cannot enter the mitochondrial matrix where the electron transport chain is located. To get across the inner mitochondrial membrane they must use either the malate–aspartate or glycerol 3-phosphate shuttles. Those that use the latter are regenerated within the mitochondrion as $FADH_2$, and hence yield fewer ATPs than would be the case if they remained as NADH.

These shuttles were described in *Section 9.2.5.*

In plants and some other organisms, such as yeast, all fatty acid breakdown takes place in the peroxisomes. In animals, only long chain fatty acids are dealt with in peroxisomes, being broken down into eight-carbon octanyl CoA molecules which are then transferred to the mitochondria, where their breakdown is completed. Some long chain fatty acids pass directly through the peroxisome membrane, and are attached to CoA inside the organelle. Others are activated on the outer surface of the peroxisome and transported inside via a peroxisomal membrane transporter protein.

Dealing with unsaturated fatty acids and those with odd numbers of carbons

As is the case with fatty acid synthesis, special mechanisms are needed to deal with the degradation of fatty acids that contain double bonds or have odd numbers of carbon atoms.

Box 12.5 The biochemistry of *Lorenzo's Oil*

Lorenzo's Oil is a 1992 film that depicts a dramatized version of the true story of Augusto and Michaela Odone's attempts to find a cure for the disease from which their son, Lorenzo, was suffering. The disease is **adrenoleukodystrophy (ALD)**, a genetic disorder resulting from a defect in the gene for the peroxisomal membrane transporter protein called **ALDP**, which transports long chain fatty acids into peroxisomes prior to their breakdown. As with favism (see *Box 11.6*), the gene for the membrane transporter is on the X chromosome, and ALD therefore affects boys more severely than girls. Accumulation of long chain fatty acids can be detected in most tissues, but the harmful effects are most manifest in the brain and adrenal cortex.

The name 'leukodystrophy' indicates that the disease results in breakdown of the myelin sheath that surrounds the axon of a neuron.

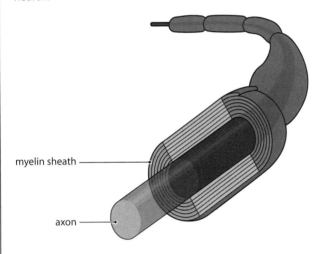

myelin sheath

axon

Demyelination results in a disruption of signal transmission along the affected axons, which initially leads to symptoms such as blurred vision and muscular weakness, but after progression results in a loss of brain function and eventual death.

Lorenzo's Oil is a 4:1 mixture of oleic and erucic acids which Augusto and Michaela Odone formulated after extensive discussions with scientists working on ALD and demyelination.

The dramatized account presents the Odones struggling against conventional and hidebound medical wisdom, but in reality they worked productively with the research community, and in particular funded a conference to bring together experts from all over the world, with the discussions at this conference stimulating new avenues for research. Among these was the possibility that consumption of Lorenzo's Oil might provide a therapy for ALD. Initial results were promising. The long chain fatty acid content of Lorenzo's blood serum returned to a level close to normal after treatment with the oil mixture. Lorenzo's neurological degeneration was slowed down, but not halted altogether, and he passed away aged 30 in 2008. The normal life expectancy for ALD diagnosed in childhood, as was the case with Lorenzo, is 3–10 years.

The mode of action of Lorenzo's Oil is not certain, but it may interfere with the synthesis pathway for long chain fatty acids. After treatment, the fatty acids do not accumulate because they are not synthesized in the first place. This may explain why Lorenzo's Oil appears to have its greatest positive effect with male children who possess the genetic defect but have not developed symptoms of ALD. With these asymptomatic children, consumption of the oil reduces their chance of becoming symptomatic by about one-half. Sadly, for children already expressing the disease, Lorenzo's Oil seems to provide no long term benefit. That is the conclusion of detailed studies that have been carried out since the oil was first introduced.

Other types of treatment to alleviate ALD are now being pursued. For those patients whose symptoms are limited to disruption of the adrenal cortex, the disease can be successfully managed by hormone replacement. The cerebral version, if diagnosed early enough, can be treated by **hematopoietic stem cell transplant**. In this method, the patient's own hematopoietic stem cells, from which all the blood cells develop, are replaced with cells from a non-ALD donor. After transplantation, the stem cells divide to produce blood cells that have the non-defective version of the ALD gene and which are therefore able to break down long chain fatty acids. Follow-up tests of patients who have undergone this treatment have shown that, providing the demyelination process has not progressed too far, transplantation can halt neurodegeneration.

Unsaturated fatty acids are degraded in the normal way until the double bond position is reached. What happens next depends on whether the double bond is at an odd-numbered position (for example, Δ^5 or Δ^9) or an even position (such as Δ^4 or Δ^6): if the double bond is at an odd-numbered position then successive rounds of β-oxidation will eventually result in this double bond being located between carbons 3 and 4 (*Fig. 12.20*). This molecule is not acted on by acyl CoA dehydrogenase at step 1 of the β-oxidation cycle. Instead, an isomerase enzyme moves the double bond one position closer to the CoA terminus, so it is now between the α and β carbons. This creates a *trans*-Δ^2-enoyl CoA which is processed in the normal way.

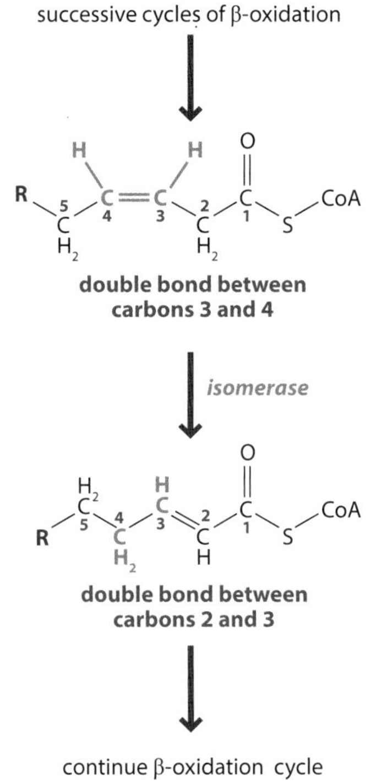

double bond between carbons 3 and 4

isomerase

double bond between carbons 2 and 3

continue β-oxidation cycle

successive cycles of β-oxidation

Figure 12.20 Breakdown of an unsaturated fatty acid with a double bond at an odd-numbered position.
'R' indicates the remainder of the fatty acid hydrocarbon chain.

If the double bond is at an even-numbered position then a slightly more complicated innovation is needed. After successive rounds of β-oxidation, this double bond will be located between carbons 4 and 5 (*Fig. 12.21*). Acyl CoA dehydrogenase oxidizes this molecule, introducing a second double bond between carbons 2 and 3, producing a Δ²,⁴-dienoyl CoA. The two double bonds are then reduced into one double bond, located between carbons 3 and 4, by **Δ²,⁴-dienoyl CoA reductase**. The isomerase can now convert this Δ³-enoyl CoA into the Δ² molecule, and again β-oxidation can continue as normal.

Those fatty acids with odd numbers of carbons are degraded by β-oxidation until the very final cycle, during which one molecule of acetyl CoA and one of propionyl CoA are produced. The latter can then be metabolized into succinyl CoA (*Fig. 12.22*), which is an intermediate in the TCA cycle. This pathway involves two enzymes, **propionyl CoA carboxylase**, which converts propionyl CoA into an intermediate compound called methylmalonyl CoA, and **methylmalonyl CoA mutase**, whose product is succinyl CoA.

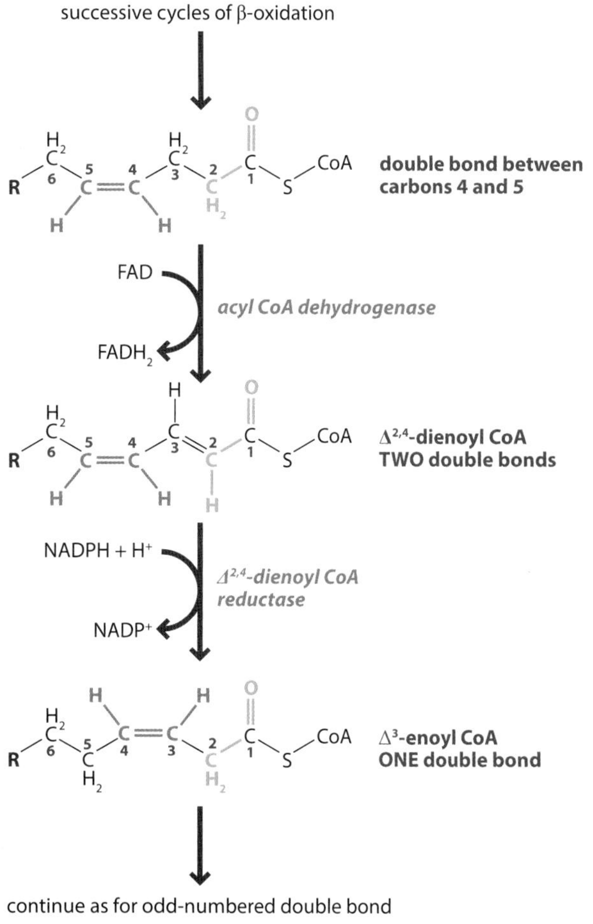

successive cycles of β-oxidation

double bond between carbons 4 and 5

FAD → FADH₂
acyl CoA dehydrogenase

Δ²,⁴-dienoyl CoA TWO double bonds

NADPH + H⁺ → NADP⁺
Δ²,⁴-dienoyl CoA reductase

Δ³-enoyl CoA ONE double bond

continue as for odd-numbered double bond

Figure 12.21 Breakdown of an unsaturated fatty acid with a double bond at an even-numbered position.
After the series of reactions shown here, breakdown of the fatty acid continues as shown in *Figure 12.20*. 'R' indicates the remainder of the fatty acid hydrocarbon chain.

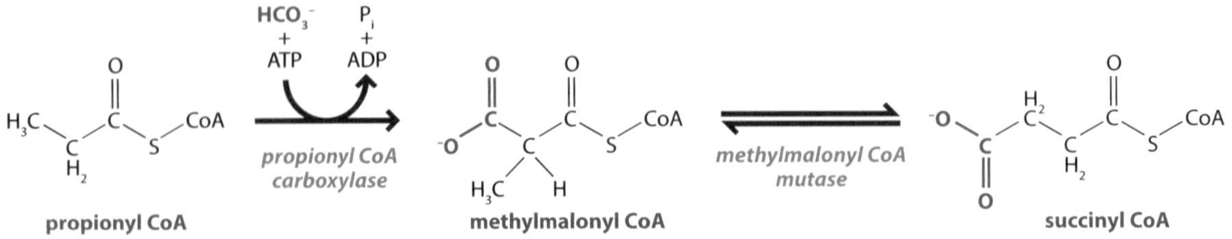

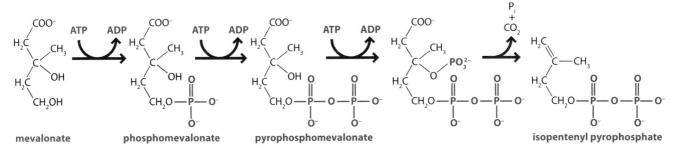

Figure 12.22 **Conversion of propionyl CoA to succinyl CoA.**

12.3   Synthesis of cholesterol and its derivatives

The third and final aspect of lipid metabolism that we must study is the synthesis of cholesterol and those compounds that are derivatives of cholesterol, including vitamin D and the steroid hormones.

12.3.1   Synthesis of cholesterol

Cholesterol is an important component of cell membranes. Animals obtain some of their cholesterol requirement from their diet, but are also able to make it, mainly in liver cells.

Figure 12.23 **The first stage of the cholesterol synthesis pathway.**
In this part of the pathway, isopentenyl pyrophosphate is synthesized from acetyl CoA.

A cholesterol molecule is made up of four hydrocarbon rings, three comprising six carbons each and one with five carbons, along with an 8-member hydrocarbon head group attached to the five-carbon ring (see *Fig. 5.17*). Amazingly, this complex structure is synthesized simply by linking together acetyl CoA molecules.

We will divide the overall synthesis pathway for cholesterol into three stages. The first stage will result in synthesis of **isopentenyl pyrophosphate**, which acts as the basic building block for the sterol structure. Stage 2 will end with squalene, a linear molecule that cyclizes to form the sterol family, and stage 3 will involve the cyclization of squalene and modification of the cyclic product to give cholesterol.

The first stage begins with the enzyme **thiolase**, which links together pairs of acetyl CoA molecules to give **acetoacetyl CoA** (*Fig. 12.23A*). One molecule of acetyl CoA then joins with one of acetoacetyl CoA to give the compound **3-hydroxy-3-methylglutaryl CoA** or **HMG CoA**. Reduction by **HMG CoA reductase** gives mevalonate, this reaction requiring two molecules of NADPH and releasing coenzyme A (*Fig. 12.23B*). A series of three ATP-requiring reactions then yields isopentenyl pyrophosphate (*Fig. 12.23C*).

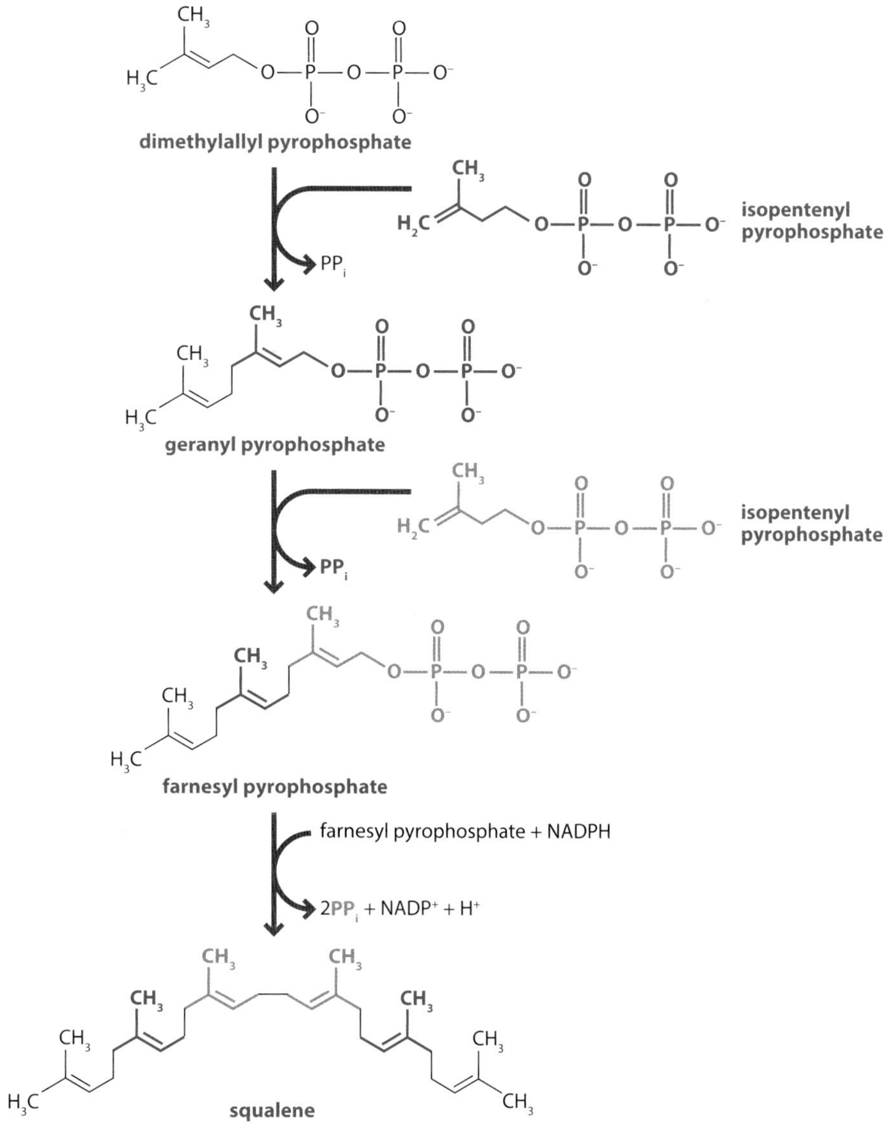

Figure 12.24 The second stage of the cholesterol synthesis pathway. Isopentenyl pyrophosphate is the substrate for a series of reactions that result in synthesis of squalene.

Now to stage 2. Some of the isopentenyl pyrophosphate molecules are isomerized into dimethylallyl pyrophosphate. The two compounds (isopentenyl pyrophosphate and dimethylallyl pyrophosphate) join together to give geranyl pyrophosphate, which then links to another molecule of isopentenyl pyrophosphate, resulting in farnesyl pyrophosphate (*Fig. 12.24*). Finally, a pair of farnesyl pyrophosphates are joined together, and we have squalene.

The final stage starts with oxidation of squalene, resulting in **squalene epoxide** (*Fig. 12.25*). This molecule cyclizes to give **lanosterol,** which has the four-ring structure characteristic of all sterols. Lanosterol is then converted to cholesterol by a short series of reactions which replaces three methyl groups with hydrogens, reduces one double bond to a single bond, and moves the position of a second double bond.

The commitment step in this lengthy pathway is the conversion of HMG CoA to mevalonate by HMG CoA reductase. The activity of this enzyme is regulated in various ways, both in response to the general energy status of the organism and the sterol content of individual cells. Energy depletion, indicated by high levels of AMP, results in inhibition of HMG CoA reductase, by phosphorylation carried out by AMP-activated protein kinase (*Fig. 12.26*). This is the same enzyme that phosphorylates acetyl CoA carboxylase in order to regulate fatty acid synthesis. The synthesis of both cholesterol and fatty acids is therefore co-regulated, with both pathways being switched off when the AMP levels are high and there is a general need for energy generation. When the energy needs have been satisfied, the increased amount of ATP inhibits the kinase and HMG CoA reductase is reactivated by a phosphatase. Cholesterol synthesis is therefore switched back on.

Cholesterol and other sterols also inhibit HMG CoA reductase activity. This regulatory process involves degradation of the protein, a mechanism for controlling enzyme activity that we have not previously encountered. The HMG CoA reductase protein has two domains, a membrane-binding domain, which attaches the enzyme to the endoplasmic reticulum, and the catalytic domain, which extends out into the cytoplasm. The membrane-binding domain changes conformation when the sterol content of the membrane in which it is embedded reaches a certain level. The conformational change exposes a lysine amino acid, which becomes modified by attachment of a small protein called **ubiquitin**. The attachment of ubiquitin to a protein acts as a signal for degradation of that protein, so the HMG CoA reductase enzyme is removed from the membrane and broken down into short peptides and eventually into individual amino acids. The enzyme activity declines for the simple reason that the protein is no longer present in the cell.

> We will examine the role of ubiquitin in protein degradation in *Section 17.2.2*.

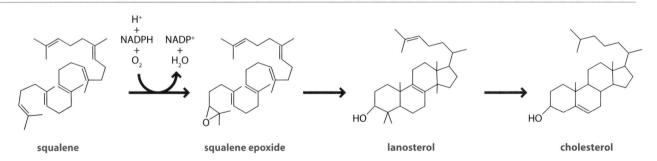

Figure 12.25 The third stage of the cholesterol synthesis pathway.
Squalene is circularized and additional modifications produce cholesterol.

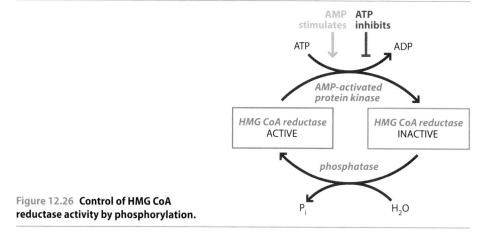

Figure 12.26 Control of HMG CoA reductase activity by phosphorylation.

12.3.2 Synthesis of cholesterol derivatives

A number of important sterols and steroids are cholesterol derivatives that are synthesized in the cell by modification of cholesterol molecules. We will consider three of the most important of these derivatives: the bile acids, **vitamin D** and the steroid hormones.

Figure 12.27 Synthesis of the bile acids glycocholate and taurocholate.

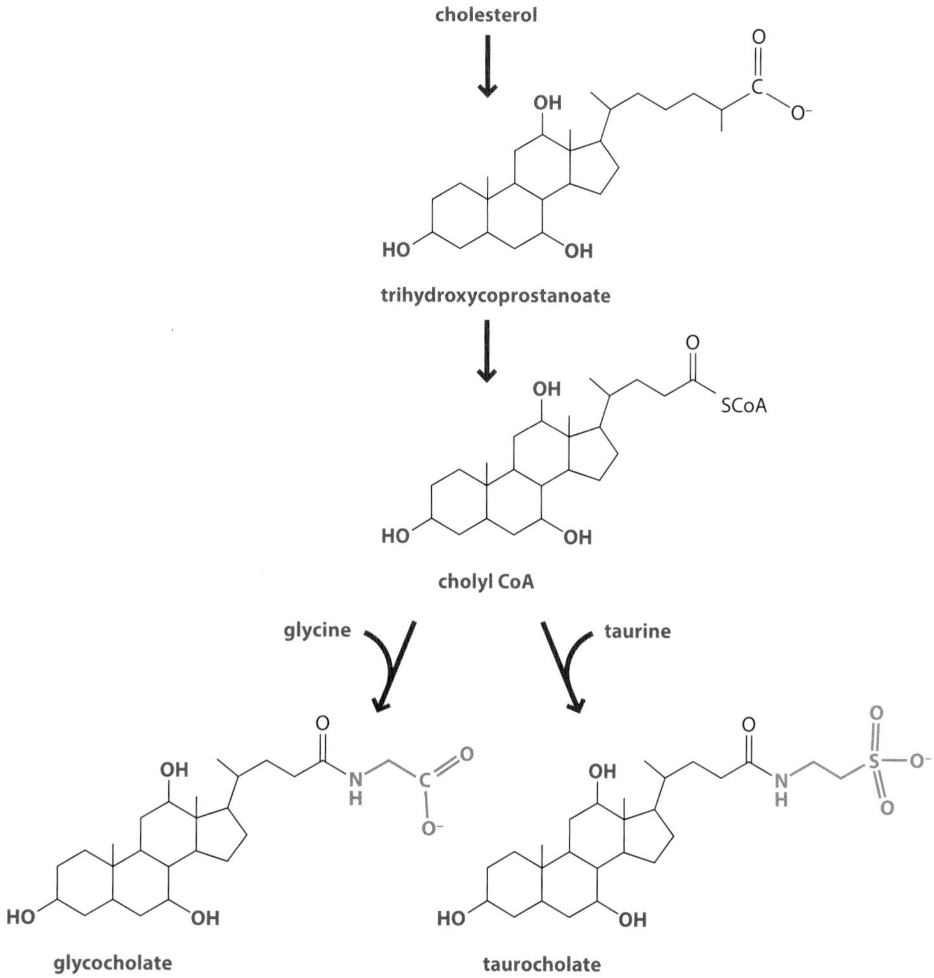

Bile acids are amphipathic molecules and hence effective detergents which are used to solubilize dietary lipids and the lipid-soluble vitamins A, D, E and K. They are also the main form in which excess cholesterol is excreted from the body. Their synthesis involves the initial activation of cholesterol by linkage with CoA, giving **cholyl CoA**. Reaction of cholyl CoA with the amino group of glycine gives **glycocholate**, and reaction with taurine, derived from cysteine, gives **taurocholate** (*Fig. 12.27*). These are the two main bile acids. After synthesis in the liver, glycocholate and taurocholate are stored in the gall bladder and then released into the small intestine.

Vitamin D is synthesized in the skin in response to ultraviolet radiation in sunlight. The modified form of cholesterol called 7-dehydrocholesterol undergoes photolysis, which opens up one of the six-carbon rings (*Fig. 12.28*). The initial product is **previtamin D$_3$**, which isomerizes to give **vitamin D$_3$** or **cholecalciferol**. Subsequent hydroxylation reactions take place in the liver and kidneys to produce **calcitriol** (1,25-dihydroxycholecalciferol), which is the active hormone. Vitamin D deficiency, which arises if exposure to sunlight is low and there is no compensatory increase in the vitamin D content of the diet, gives rise to **rickets** in children and **osteomalacia** in adults. Both disorders are characterized by a softening or weakening of the bones.

Finally, the steroid hormones are synthesized from **pregnenolone**, which is obtained from cholesterol. Modifications of pregnenolone to give the various members of the steroid hormone family are catalyzed by the **cytochrome P450** group of heme-containing enzymes. Examples of steroid hormones and their functions were given in *Table 5.3* and details of their synthesis are as follows (*Fig. 12.29*):

- **Progesterone** is synthesized from pregnenolone by oxidation of the keto group at carbon 3 and transfer of the double bond from between carbons 5 and 6 to between carbons 4 and 5.

- **Cortisol** is derived from progesterone by hydroxylations at carbons 11, 17 and 21.

- **Aldosterone** is also derived from progesterone, by hydroxylations at carbons 11 and 21, and oxidation of the methyl group at carbon 18.

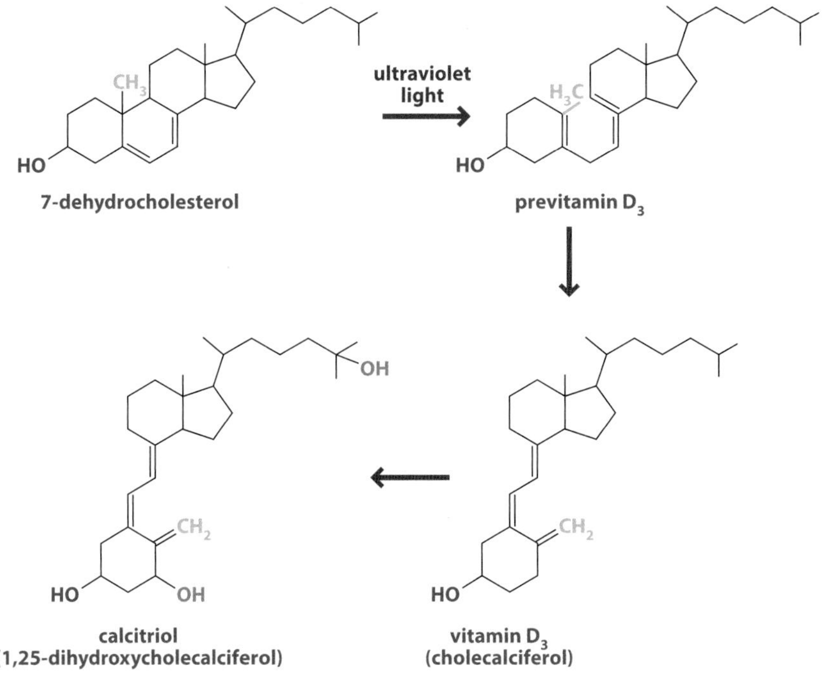

Figure 12.28 Synthesis of vitamin D.

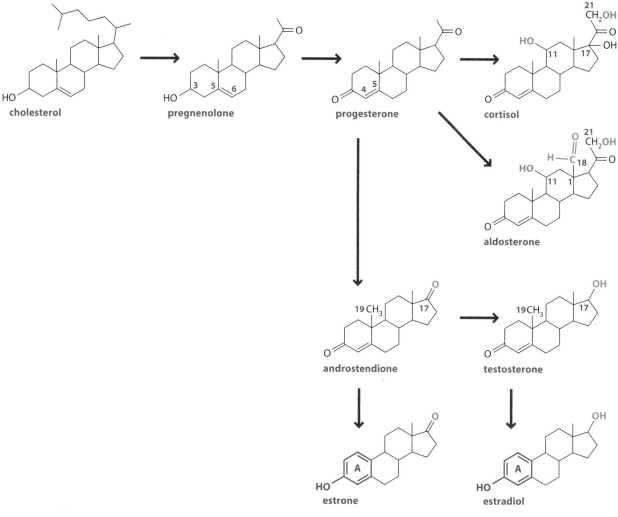

Figure 12.29 Synthesis of the steroid hormones.
For the numbering of the carbons, see *Figure 5.16*.

- **Testosterone** is synthesized from progesterone by hydroxylation of carbon 17, followed by removal of the side-chain containing carbons 20 and 21 (which is attached to carbon 17). This gives androstendione, which is converted to testosterone by reduction of the 17-keto group resulting from side-chain removal.

- The **estrogens** are derived from androstendione and testosterone, by removal of the methyl group at carbon 19, followed by rearrangement of the bonds in the A ring. Modification of androstendione in this way gives estrone and modification of testosterone gives **estradiol**.

Further reading

Beld J, Lee DJ and Burkart MD (2015) Fatty acid biosynthesis revisited: structure elucidation and metabolic engineering. *Molecular BioSystematics* **11**, 38–59.
Discusses approaches for engineering microorganisms for the commercial production of fatty acids for use as biofuel.

Byers DM and Gong H (2007) Acyl carrier protein: structure–function relationships in a conserved multifunctional protein family. *Biochemistry and Cell Biology* **85**, 649–62.

Chiang JYL (2009) Bile acids: regulation of synthesis. *Journal of Lipid Research* **50**, 1955–66.

Coleman RA and Lee DP (2004) Enzymes of triacylglycerol synthesis and their regulation. *Progress in Lipid Research* **43**, 134–76.

Ghayee HK and Auchus RJ (2007) Basic concepts and recent developments in human steroid hormone biosynthesis. *Reviews in Endocrine and Metabolic Disorders* **8**, 289–300.

Houten SM, Violante S, Ventura FV and Wanders RJA (2016) The biochemistry and physiology of mitochondrial fatty acid β-oxidation and its genetic disorders. *Annual Review of Physiology* **78**, 23–44.

Ikonen E (2008) Cellular cholesterol trafficking and compartmentalization. *Nature Reviews Molecular Cell Biology* **9**, 125–38.

Johnson MNR, Londergan CH and Charkoudian LK (2014) Probing the phosphopantetheine arm conformations of acyl carrier proteins using vibrational spectroscopy. *Journal of the American Chemical Society* **136**, 11240–3.

Kornberg H (2000) Krebs and his trinity of cycles. *Nature Reviews Molecular Cell Biology* **1**, 225–8. *The discovery of the glyoxylate cycle.*

Mansbach CM and Siddiqi SA (2010) The biogenesis of chylomicrons. *Annual Review of Physiology* **72**, 315–33.

Moser HW, Raymond GV, Lu S-E, *et al*. (2005) Follow-up of 89 asymptomatic patients with adrenoleukodystrophy treated with Lorenzo's Oil. *Archives of Neurology* **62**,1073–80. *Showing the beneficial effects of Lorenzo's Oil for asymptomatic ALD patients.*

Poirier Y, Antonenkov VD, Glumoff T and Hiltunen JK (2006) Peroxisomal β-oxidation – a metabolic pathway with multiple functions. *Biochimica et Biophysica Acta* **1763**, 1413–26.

Wakil SJ, Stoops JK and Joshi VC (1983) Fatty acid synthesis and its regulation. *Annual Review of Biochemistry* **52**, 537–79.

Wanders RJ, van Grunsven EG and Jansen GA (2000) Lipid metabolism in peroxisomes: enzymology, functions and dysfunctions of the fatty acid alpha- and beta-oxidation systems in humans. *Biochemical Society Transactions* **28**, 141–9.

Ye J and DeBose-Boyd RA (2011) Regulation of cholesterol and fatty acid synthesis. *Cold Spring Harbor Perspectives in Biology* **3**, a004754.

Self-assessment questions

Multiple choice questions

Only one answer is correct for each question. Answers can be found on the website: www.scionpublishing.com/biochemistry

1. In what form are acetyl units shuttled across the inner mitochondrial membrane prior to fatty acid synthesis?
   (a) Citrate
   (b) Oxaloacetate
   (c) Malate
   (d) Pyruvate

2. What is the name of the prosthetic group contained in the acyl carrier protein?
   (a) Vitamin B_5
   (b) Coenzyme Q
   (c) Phosphopantetheine
   (d) Riboflavin

3. Which of these reactions is **not** involved in the cycle of events that build up a saturated fatty acid?
   (a) Phosphorylation
   (b) Reduction
   (c) Condensation
   (d) Dehydration

4. What is the name of the enzyme that cleaves the completed fatty acid from its carrier protein?
   (a) Enoyl-ACP reductase
   (b) Thioesterase
   (c) ACP delipase
   (d) Acyl-malonyl-ACP condensing enzyme

5. Which compound replaces malonyl ACP during synthesis of a fatty acid with an odd number of carbons?
   (a) Propionyl ACP
   (b) Acetyl ACP
   (c) Acetoacetyl ACP
   (d) Palmitoyl ACP

6. Which complex of enzymes introduces double bonds into fatty acids?
   (a) NADH–cytochrome b_5 oxidase, cytochrome b_5 and desaturase
   (b) NADH–cytochrome b_5 reductase, cytochrome c and desaturase
   (c) NADH–cytochrome b_5 oxidase, cytochrome c and desaturase
   (d) NADH–cytochrome b_5 reductase, cytochrome b_5 and desaturase

7. Which of the following statements concerning regulation of fatty acid synthesis is **incorrect**?
   (a) Synthesis of malonyl CoA is the main regulatory step
   (b) Regulation is mediated by an AMP-activated protein kinase
   (c) AMP stimulates fatty acid synthesis
   (d) Fatty acid synthesis only occurs when the cell has a plentiful supply of energy

8. Which hormones inhibit fatty acid synthesis?
   (a) Glucagon and insulin
   (b) Insulin and epinephrine
   (c) Glucagon and epinephrine
   (d) None of the above

9. The cells that store fat in adipose tissue are called what?
   (a) Adipocytes
   (b) Lipoproteins
   (c) Sarcomeres
   (d) Apidocytes

10. Which of the following statements is **incorrect** with regard to triacylglycerol breakdown?
    (a) It is the major control point for the entire lipolysis pathway
    (b) The process is called β-oxidation
    (c) Triacylglycerols are broken down by lipase enzymes
    (d) A combination of three enzymes, with different specificities, is needed to remove all three of the fatty acids from a single triacylglycerol

11. Glycerol released by triacylglycerol degradation is converted into which intermediate in glycolysis?
    (a) Pyruvate
    (b) Dihydroxyacetone phosphate
    (c) Phosphoenolpyruvate
    (d) Glyceraldehyde 3-phosphate

12. Which group of hormones increase the rate of triacylglycerol breakdown?
    (a) Glucagon, insulin, norepinephrine and adrenocorticotropic hormone
    (b) Glucagon, insulin and adrenocorticotropic hormone
    (c) Insulin, norepinephrine and adrenocorticotropic hormone
    (d) Glucagon, epinephrine, norepinephrine and adrenocorticotropic hormone

13. How is energy provided for attachment of a fatty acid to coenzyme A, prior to fatty acid breakdown?
    (a) Hydrolysis of ATP to ADP
    (b) Hydrolysis of ATP to AMP
    (c) Hydrolysis of GTP to GDP
    (d) The process does not require energy

14. What is the name of the compound to which fatty acids are attached in order to pass through the inner mitochondrial membrane?
    (a) Carnine
    (b) Caprine
    (c) Carnithreonine
    (d) Carnitine

15. Which of these reactions is **not** involved in the cycle of events that break down a saturated fatty acid?
    (a) Thiolysis
    (b) Oxidation
    (c) Hydration
    (d) Reduction

16. How many ATPs are obtained by complete breakdown of palmitic acid?
    (a) 38
    (b) 92
    (c) 120
    (d) 129

17. Which of the following statements is **incorrect** concerning fatty acid breakdown in peroxisomes?
    (a) The enzyme called catalase is involved
    (b) The process does not occur in plants
    (c) In animals, only long chain fatty acids are broken down in peroxisomes
    (d) The resulting NADH and acetyl CoA are exported out of the peroxisomes

18. Which of the following statements is **incorrect** with regard to breakdown of unsaturated fatty acids?
    (a) Double bonds at odd- and even-numbered positions are dealt with in different ways
    (b) Acyl CoA dehydrogenase might introduce a second double bond
    (c) An isomerase enzyme might be needed to move the position of the double bond
    (d) Degradation of an unsaturated fatty acid always yields propionyl CoA

19. Which enzyme links together acetyl CoA units at the start of the cholesterol synthesis pathway?
    (a) Thiolase
    (b) HMG CoA reductase
    (c) Enolase
    (d) Acetoacetyl CoA synthase

20. What is the name of the linear molecule that cyclizes to form the sterol family?
    (a) Lanosterol
    (b) Squalene
    (c) Farnesyl pyrophosphate
    (d) HMG CoA

21. Which is the commitment step in cholesterol synthesis?
    (a) Formation of the sterol ring
    (b) Formation of acetoacetyl CoA
    (c) Conversion of squalene to lanosterol
    (d) Conversion of HMG CoA to mevalonate

22. How does cholesterol regulate the activity of HMG CoA reductase?
    (a) By stimulating protein kinase A, which phosphorylates HMG CoA reductase
    (b) By stimulating its breakdown
    (c) By inducing a conformation change which activates the enzyme
    (d) By influencing the amount of cAMP in the cell

23. Which of the following is **not** a derivative of cholesterol?
    (a) Bile acids
    (b) Vitamin D
    (c) Steroid hormones
    (d) Heme

24. Which of these compounds is **not** a steroid hormone?
    (a) Progesterone
    (b) Cortisol
    (c) Testosterone
    (d) Insulin

Short answer questions

These questions do not require additional reading.

1. Describe how acetyl CoA is moved out of the mitochondrion at the start of the fatty acid synthesis pathway.

2. Give a detailed description of the cycle of reactions that build up an unsaturated fatty acid. How does this process differ in animals and plants?

3. Explain how long chain fatty acids and those with double bonds are made in animals.

4. Outline the way in which fatty acid synthesis is regulated.

5. Describe how lipid degradation is regulated.

6. Explain how fatty acids are transported into the mitochondrion prior to their breakdown.

7. Describe the β-oxidation pathway for fatty acid breakdown.

8. Outline the process by which cholesterol is synthesized.

9. Describe how cholesterol synthesis responds to the overall energy status of the organism and to the sterol content of the cell.

10. Outline how the main derivatives of cholesterol are synthesized.

Self-study questions

These questions will require calculation, additional reading and/or internet research.

1. From the information in this and previous chapters, draw a diagram illustrating the hormonal control of energy storage and utilization in humans.

2. Why is it inadvisable to consume a high-sugar drink immediately prior to running a marathon?

3. What would be the net yield of ATP for β-oxidation of (A) 12:0 lauric acid; (B) 18:0 stearic acid; (C) 24:0 lignoceric acid?

4. The glyoxylate cycle is present in plants and many bacteria but is generally assumed to be absent in animals. Over the years, however, there have been occasional reports in the scientific literature suggesting that the cycle may exist in some animals, such as in the livers of newborn rats. Outline a research project designed to test the hypothesis that animals possess the glyoxylate cycle.

5. Statins are inhibitors of HMG CoA reductase. Explain why statins are used (A) to reduce the risk of cardiovascular disease, and (B) as a treatment for familial hypercholesterolemia.

CHAPTER 13

Nitrogen metabolism

STUDY GOALS

After reading this chapter you will:

- understand the importance of nitrogen fixation and nitrate reduction as pathways for incorporation of inorganic nitrogen into ammonia

- be able to describe the main types of nitrogen-fixing organisms and the key features of the symbiosis between leguminous plants and rhizobia

- understand how nitrogen is fixed by the nitrogenase complex

- know in outline the pathway for nitrate reduction

- understand the difference between a non-essential and essential amino acid, and be able to list the amino acids in each group

- be able to describe how ammonia is converted into glutamate and glutamine

- know in detail the pathways for synthesis of the other nine non-essential amino acids in humans

- be able to describe in outline the pathways for the synthesis of essential amino acids by bacteria

- understand how the branch points in some amino acid synthesis pathways are regulated by the end products

- be able to describe the salvage and *de novo* pathways for nucleotide synthesis

- understand how tetrapyrroles such as heme are synthesized

- know in outline how nucleotides and tetrapyrroles are broken down

- be able to describe how the nitrogen components of amino acids are converted to ammonia during breakdown of these compounds

- know how the carbon skeletons of amino acids are degraded and understand the difference between glucogenic and ketogenic amino acids

- be able to distinguish between the ways in which different organisms dispose of excess ammonia

- know the steps of the urea cycle and understand how the cycle is regulated

- know how some of the energy used by the urea cycle can be recovered by linkage with the TCA cycle

The nitrogen-containing compounds are the final group whose biosynthesis and degradation we must study. Principal among these biomolecules are the amino acids, the purine and pyrimidine bases found in nucleotides, and the **tetrapyrrole** compounds, which include the heme family of cofactors in addition to chlorophyll. The major part of this chapter will be devoted to the pathways by which these compounds are made and subsequently broken down again when they are no longer needed. These are not, however, the only aspects of nitrogen metabolism that we must consider. We must also examine the processes by which some plants and microorganisms convert inorganic nitrogen from the environment into ammonia. These processes are extremely important because the ammonia that they provide is the substrate for synthesis of all of the organic nitrogen-containing compounds present in living organisms.

13.1 Synthesis of ammonia from inorganic nitrogen

At the very beginning of this book we looked at the elemental composition of an adult human, and asked what would be needed to turn this mixture of elements into our favorite cult movie star. For the three most abundant elements (carbon, oxygen and hydrogen) part of the answer is photosynthesis, which converts inorganic sources of these elements into carbohydrates. These carbohydrates, as well as acting as sources of energy, provide the photosynthetic organisms, and those other organisms that consume them, with the organic carbon/oxygen/hydrogen groups needed to construct biomolecules such as proteins, lipids and nucleic acids. But the amino acid and nucleotide subunits of proteins and nucleic acids also contain nitrogen, the fourth most abundant element in human biochemistry. How is this nitrogen acquired?

The answer is that most plants and many microorganisms are able to synthesize ammonia from inorganic sources of nitrogen. The nitrogen contained in ammonia can then be used to construct all of the organic nitrogen-containing compounds that are needed by living organisms. As with carbon, oxygen and hydrogen, animals acquire nitrogen by consuming the plants responsible for its initial assimilation.

There are two pathways for the incorporation of inorganic nitrogen into ammonia (*Fig. 13.1*):

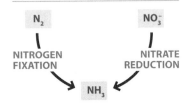

- **Nitrogen fixation**, in which the inorganic source is nitrogen gas from the atmosphere.

- **Nitrate reduction**, in which inorganic nitrate ions present in soil are used.

Figure 13.1 Two pathways for the incorporation of inorganic nitrogen into ammonia.

Nitrogen fixation is the more complex of these two pathways and the one that we will study first.

13.1.1 Nitrogen fixation

Nitrogen fixation is restricted to a small number of bacterial and archaeal species, called **diazotrophs**. Many of these species are free-living, which in a microbial context means that they live independently of other organisms rather than forming symbiotic relationships of any kind. In contrast, a small number of nitrogen-fixing bacteria form symbioses with the roots of certain types of plant. As part of this symbiosis, the plant acquires organic nitrogen compounds that result from the nitrogen-fixing activities of the bacteria. Before studying the biochemistry of nitrogen fixation we will look briefly at the details of the symbiosis.

Symbiotic bacteria fix nitrogen within root nodules

The symbiotic nitrogen-fixing bacteria fall into two main groups (*Fig. 13.2*):

- The **rhizobia** group, which includes members of several genera including *Rhizobium*, *Bradyrhizobium* and *Burkholderia*.

- Members of the genus *Frankia*, which is a type of **actinomycete** or filamentous bacterium.

Although the *Frankia* diazotrophs are less well studied, it is clear that the physiological and biochemical features of the symbiosis are very similar in both groups of bacteria. The main distinction is the identities of the plants that participate in the symbiosis. For the rhizobia, these plants are almost exclusively members of the Fabaceae family.

Figure 13.2 Two species of nitrogen-fixing bacteria.
(A) *Rhizobium trifolii* – image provided by Prof. Frank Dazzo from Michigan State University. (B) *Frankia* sp. – image provided by Prof. David Benson of the University of Connecticut.

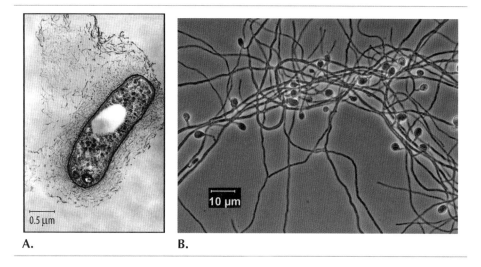

A.   B.

These are the **legumes**, the word 'legume' sometimes thought to mean 'nitrogen-fixing' but in fact referring to the structure of the fruit. The Fabaceae include important crops such as beans, peas, alfalfa, peanuts and clover. The *Frankia*, on the other hand, form symbioses with members of eight taxonomic families of plants, collectively called the **actinorhizal** plants, and including trees and shrubs such as alder and bayberry.

The nitrogen-fixing bacteria are initially free-living in soil, but they detect the presence of organic compounds called **flavonoids** that are secreted by the roots of a suitable host plant. The bacteria migrate towards the roots and in turn secrete **nod factors** (short oligosaccharides with a fatty acid side-chain) that alert the plant to

---

Box 13.1 Nitrogen fixation by symbiotic cyanobacteria

Various cyanobacteria also participate in symbiotic nitrogen fixation, with a diverse range of host species and involving partnerships different from the well-studied relationships between rhizobia and leguminous plants. Here are three particularly interesting examples:

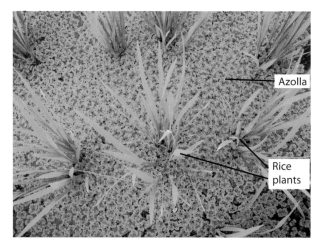

Image of Azolla in paddy fields reproduced from the Integrated Rural Development Organization website.

• **Azolla** is a small aquatic fern that lives on the surface of ponds and small lakes in many parts of the world. It forms a symbiosis with nitrogen-fixing cyanobacteria called *Anabaena*, the latter visible as filaments living inside cavities within the leaves of Azolla. In many parts of southwest Asia, Azolla is added to rice paddies as a 'green manure'. Some of the fixed nitrogen products become available to the rice plants, boosting crop productivity.

• Several **lichens** include a nitrogen-fixing component. All lichens are symbiotic organisms, one partner being a fungus and the other a photosynthetic bacterium or alga. A few species, such as *Lobaria* and *Peltigeria*, are tripartite, and have a cyanobacterium such as *Nostoc* as a third, nitrogen-fixing partner. Lichens of this type are important sources of fixed nitrogen in ecosystems such as redwood forests.

• The diatom *Rhopalodia gibba* has taken the symbiosis one step further. This unicellular organism contains intracellular structures, derived from cyanobacteria, which carry out nitrogen fixation. The relationship is a type of endosymbiosis, similar to that displayed by mitochondria and chloroplasts.

their presence. This two-way signaling prepares the two partners for initiation of the symbiosis. The bacteria enter the plant roots either via a modified root hair or simply by squeezing into the gaps between cells on the root surface. Infection induces plant cell division, forming a specialized structure call a **root nodule**, within which nitrogen fixation takes place (*Fig. 13.3*).

Figure 13.3 Nitrogen-fixing nodules on the roots of *Medicago italica*.
The root nodules appear pink because they contain leghemoglobin. Reproduced from Wikimedia Commons under a CC BY-SA 3.0 license.

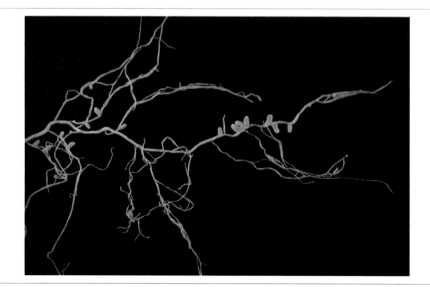

Inside the nodule, the bacteria enter the plant cells and differentiate into **bacteroids**, in effect becoming intracellular organelles similar in size to mitochondria. The plant provides the bacteroids with carbohydrates such as succinate and malate, which the bacteroids use as sources of energy, and the bacteroids provide the plant with ammonia. This is an example of the type of symbiosis called **mutualism**, a cooperative relationship of mutual benefit to both participating species.

Nitrogen fixation results in the reduction of nitrogen to ammonia

Nitrogen fixation is the reduction of atmospheric nitrogen to cellular ammonia. The reaction requires six electrons as N≡N is sequentially reduced to HN=NH, then $_2$HN–NH$_2$, and finally to two NH$_3$ molecules. The reduction is achieved by the bacterial **nitrogenase complex**, which comprises two enzymes:

- A **reductase**, which is a dimer of two identical subunits, each containing an FeS cluster. This is the same type of electron-binding prosthetic group that we first met when we studied the components of the electron transport chain.

- A **nitrogenase**, which is made up of four subunits, two of type α and two β, with a single **molybdenum–iron** or **MoFe center**. This is a type of electron-binding cofactor that we have not encountered before.

The electrons required for conversion of nitrogen to ammonia are donated by the reduced form of ferrodoxin, which is generated in the host plant's chloroplasts during photosynthesis. The electrons are passed, one by one, from ferrodoxin to the FeS clusters of the reductase, and then to the MoFe center of the nitrogenase, and finally to nitrogen (*Fig. 13.4*). The transfer of an electron between the reductase and nitrogenase requires energy, which is obtained by hydrolysis of ATP, with two ATPs needed to transfer one electron. Although only six electrons are needed for the reduction of one molecule of nitrogen to two molecules of ammonia, the nitrogenase complex is slightly inefficient, and on average wastes two electrons during each transformation.

Figure 13.4 Electron transfer during nitrogen fixation.

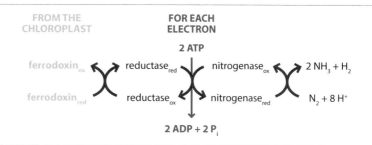

This means that eight electrons are needed and hence 16 ATPs. The overall reaction is very endergonic because of the stability of the triple bond in atmospheric nitrogen. The equivalent chemical reaction for reduction of nitrogen to ammonia is called the **Haber process**, and involves heating a mixture of nitrogen and hydrogen at 500°C under a pressure of 300 atmospheres (3.04×10^7 Pa) with an iron catalyst. With such extreme measures needed to achieve the reduction in the laboratory, it is remarkable that the reaction is possible at all under biological conditions.

The nitrogenase complex is irreversibly inactivated by oxygen and so must be protected from it. The necessary protection is provided by **leghemoglobin**, a protein with a structure very similar to the hemoglobin of blood, but with a higher affinity for oxygen binding. Both the protein component and the heme prosthetic group are made by the plant cell. Like hemoglobin, leghemoglobin is red, its presence revealed by the pinkish color that is seen when a root nodule is sliced open with a razor blade, and which is sometimes visible even in uncut nodules (see *Fig. 13.3*).

13.1.2 Nitrate reduction

Nitrate reduction is the second way in which inorganic nitrogen is converted into ammonia. Most plants and many bacteria are able to carry out this transformation.

In plants, nitrate utilization begins with the transport of nitrates from the soil into the roots. The absorbed nitrates are then moved to the shoots, which are the main sites of nitrate reduction, the roots being relatively inactive in this regard. The reduction pathway has two steps, the first occurring in the cytoplasm and resulting in conversion of nitrate (NO_3^-) to nitrite (NO_2^-), and the second in chloroplasts, reducing the nitrite to ammonia (*Fig. 13.5*).

The cytoplasmic conversion of nitrate to nitrite is catalyzed by **nitrate reductase**, an enzyme made up of two identical protein subunits in addition to FAD, heme and a molybdenum-containing cofactor. A pair of methionine amino acids at the active site of the enzyme stabilize an interaction between the molybdenum of the cofactor and a nitrate ion, which is reduced to nitrite, using electrons donated by either NADH or NADPH. The nitrite is then moved into the chloroplasts via a proton-dependent transporter, where it is converted to ammonia by **nitrite reductase**. This is a single-subunit protein with an FeS cluster and a **siroheme** prosthetic group, the latter being a modified version of heme found in several enzymes involved in the reduction of nitrogen and sulfur-containing compounds. As with nitrogen fixation, six electrons are needed to fully reduce one nitrite ion into ammonia. These are provided by ferrodoxin but, unlike nitrogen reduction, there is no energy requirement so no ATPs are consumed.

Figure 13.5 Reduction of nitrate to ammonia in plants.
The nitrate reduction step can use either NADH (as shown here) or NADPH as the electron donor.

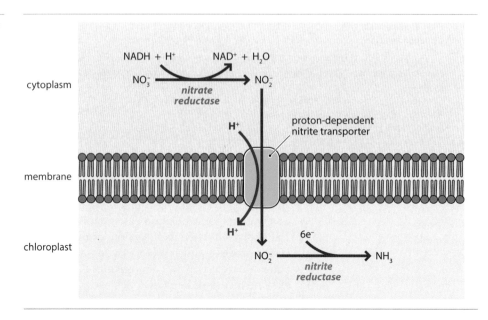

13.2 Synthesis of nitrogen-containing biochemicals

Nitrogen fixation and nitrate reduction provide living organisms with a source of ammonia which acts as the initial substrate for the metabolic pathways that result in synthesis of all the other nitrogen-containing biochemicals in living cells. Our next objective is to explore the pathways leading to the most important of these compounds: the amino acids, nucleotide bases, and tetrapyrroles.

13.2.1 Synthesis of amino acids

Table 13.1. **Essential and non-essential amino acids for humans**

Essential	Non-essential
histidine	alanine
isoleucine	arginine
leucine	asparagine
lysine	aspartate
methionine	cysteine
phenylalanine	glutamate
threonine	glutamine
tryptophan	glycine
valine	proline
	serine
	tyrosine

Human cells only have the necessary enzymes to make 11 of the 20 amino acids that are used in protein synthesis. The remaining nine are called the **essential amino acids** because it is essential that they are obtained from the diet (*Table 13.1*). Of the 11 (non-essential) amino acids that humans can synthesize, 10 are obtained via relatively short pathways that begin with one of the intermediates from the TCA cycle, glycolysis or the pentose phosphate pathway. The eleventh, tyrosine, is synthesized from phenylalanine.

Glutamate and glutamine are synthesized from α-ketoglutarate and ammonia

We begin with glutamate and glutamine, because these two amino acids are made directly from ammonia. The second substrate is α-ketoglutarate, one of the intermediates in the TCA cycle. The pathway has two steps:

Step 1. **Glutamate dehydrogenase** uses ammonia as the source of an amino group (–NH$_2$) that it adds to α-ketoglutarate. The resulting product is glutamate. The reaction is a reduction and uses a molecule of NADPH.

Box 13.2 **Synthesis of the correct enantiomer of glutamate**

Amino acids exist in two different forms, the L- and D-enantiomers, distinguished from one another by the relative positioning of the atoms around the chiral α-carbon (see *Fig. 3.4*). As with all amino acids, the L-form of glutamate is the predominant version in the cell and the enantiomer that is used in protein synthesis. The substrate for glutamate synthesis, α-ketoglutarate, is achiral and does not exist in L- and D-forms. The chirality of the amino acid is therefore set by glutamate dehydrogenase during the conversion of α-ketoglutarate into glutamate. How does the enzyme ensure that it exclusively makes L-glutamate?

The reaction catalyzed by glutamate dehydrogenase occurs in two steps. First, the amino group from ammonia is transferred to the carbon that will become the α-carbon in the amino acid.

At this stage of the reaction, the carbon we are interested in is still achiral. The key event is therefore the second stage, when a hydrogen is transferred from NADPH to the α-carbon. This transfer must be done in such a way as to establish the L-configuration around the α-carbon.

The critical factor that ensures that the L-configuration is generated is the positioning of NADPH relative to the aminated intermediate within the glutamate dehydrogenase active site. The orientation is such that the hydrogen transfer from NADPH to the α-carbon results in the L-configuration.

The architecture of the active site therefore sets the orientation of groups around the α-carbon. Because of the structure of its active site, glutamate dehydrogenase can only synthesize L-glutamate.

The orientation of groups around the α-carbon is maintained when glutamate is converted to glutamine by glutamine synthetase, and when glutamate is metabolized into proline and arginine (see *Fig. 13.6*). However, the α-carbons of other amino acids are not derived from glutamate and their configurations must be set up independently at the appropriate step in their biosynthetic pathways. This step is always the transamination, when an amino group is donated (from glutamate) to the α-carbon (e.g. step 2 of the pathway leading to serine, see *Fig. 13.7A*). Each of these transaminations is catalyzed by a similar enzyme, with the hydrogen donated not by NADPH but by a lysine present at the active site. The orientation of this lysine relative to the substrate is such that the L-configuration is always established around the α-carbon.

Step 2. **Glutamine synthetase** uses a second ammonia molecule to add a second amino group to glutamate, giving glutamine. Energy is provided by ATP.

In plants, the synthesis of glutamate and glutamine is linked with nitrogen fixation and nitrate reduction, occurring in the bacteroids of the root nodule and in the chloroplasts of nodular and nitrate-reducing cells. Humans and other organisms synthesize glutamate and glutamine from ammonium ions acquired in the diet or by breakdown of other nitrogen compounds.

Pathways for the other nine amino acids that are made by humans

Now we will look at the pathways that humans use to make the other nine non-essential amino acids.

Glutamate is the starting point for synthesis of two other amino acids, proline and arginine. First, glutamate is reduced to glutamate γ-semialdehyde via a phosphorylated intermediate. Glutamate γ-semialdehyde can then follow either of two pathways:

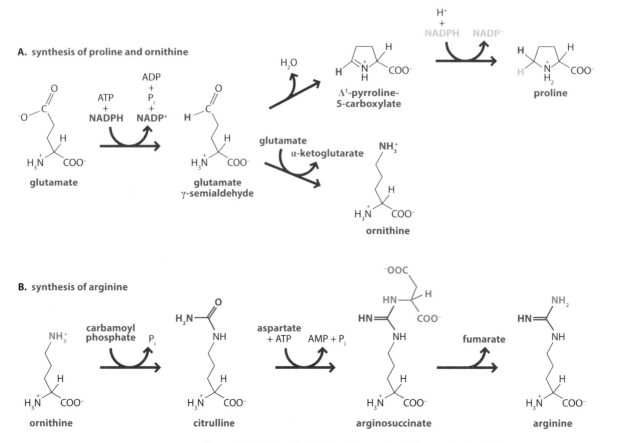

Figure 13.6 **Synthesis of (A) proline and ornithine, and (B) arginine.**

- To produce proline, glutamate γ-semialdehyde spontaneously cyclizes and the resulting compound (Δ¹-pyrroline 5-carboxylate) undergoes a second reduction (*Fig. 13.6A*).

- To produce arginine, glutamate γ-semialdehyde is converted to **ornithine** by a **transamination** reaction, in which a carbonyl (–C=O) group is replaced with an amino (–NH₂) group, the latter donated by a second molecule of glutamate. Ornithine can then be metabolized to arginine via a short pathway (*Fig. 13.6B*).

There is some debate as to whether the transamination of glutamate γ-semialdehyde is a major source of arginine in humans. The cyclization reaction that gives proline is very rapid, which means that there is unlikely to be much glutamate γ-semialdehyde available for transamination. In addition, humans can obtain arginine from the **urea cycle**, which we will study later in this chapter.

Serine, glycine and cysteine are each derived from 3-phosphoglycerate, an intermediate in glycolysis (*Fig. 13.7*). Again, this involves a transamination from glutamate, with serine initially being produced. Replacement of the –CH₂OH side-chain of serine with a hydrogen then gives glycine, and replacement of the hydroxyl component of the side-chain with a sulfhydryl (–SH) gives cysteine.

Transaminations also yield aspartate and alanine, by simple one-step reactions from oxaloacetate and pyruvate, respectively (*Fig. 13.8*). Asparagine is then obtained from aspartate by amidation, the –NH₂ group coming from glutamine.

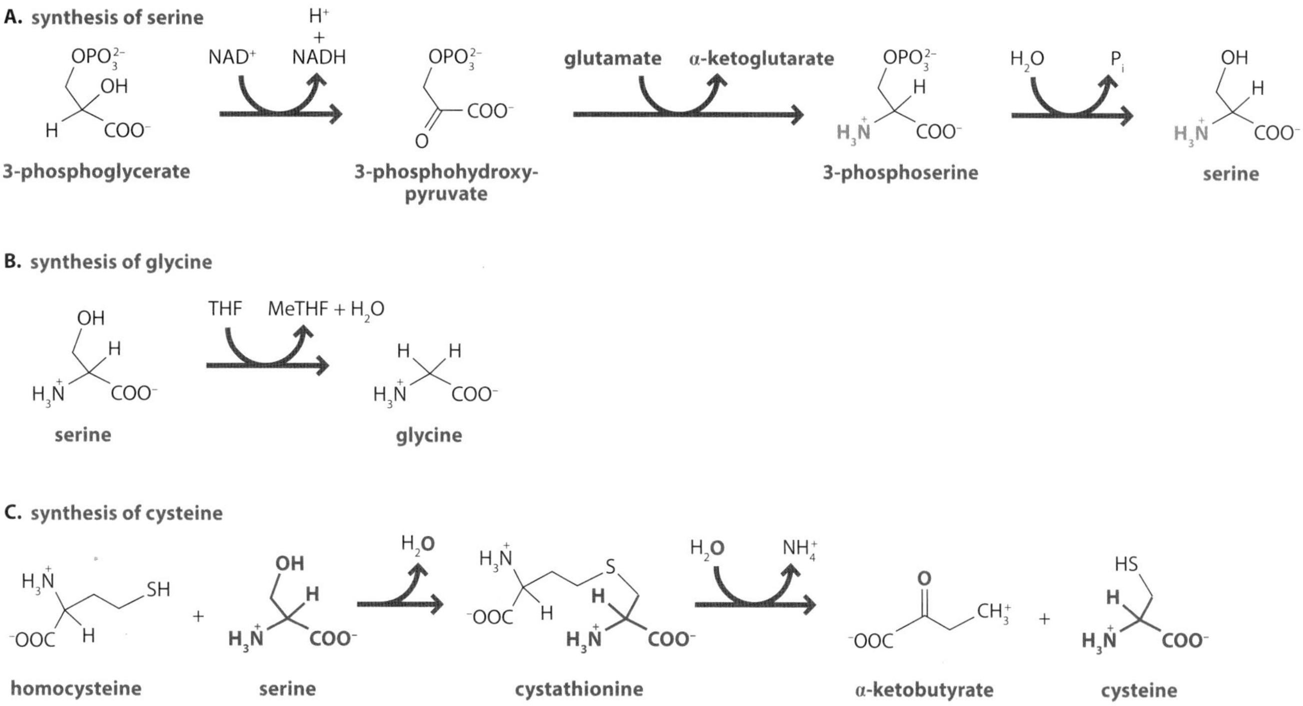

A. synthesis of serine

3-phosphoglycerate → 3-phosphohydroxy-pyruvate → 3-phosphoserine → serine

B. synthesis of glycine

serine → glycine

C. synthesis of cysteine

homocysteine + serine → cystathionine → α-ketobutyrate + cysteine

Figure 13.7 **Synthesis of (A) serine, (B) glycine, and (C) cysteine.**
Abbreviations: MeTHF, methylenetetrahydrofolate; THF, tetrahydrofolate. THF is a cofactor that acts as an acceptor or donor of carbon groups in a variety of biochemical reactions. During the conversion of serine into glycine, THF accepts the CH_2 group from serine, with release of two H^+ ions. This reaction converts THF into MeTHF.

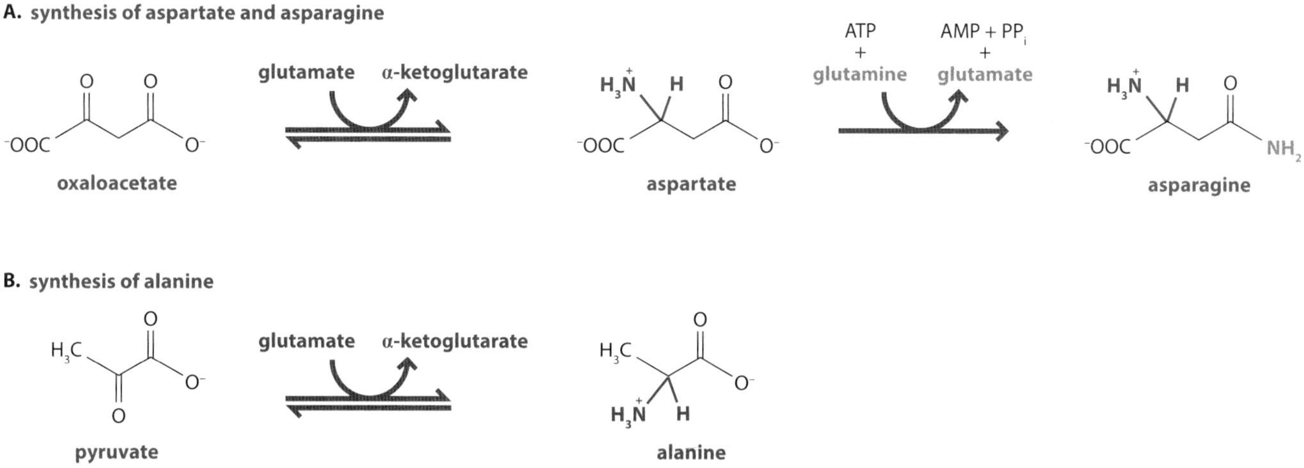

A. synthesis of aspartate and asparagine

oxaloacetate → aspartate → asparagine

B. synthesis of alanine

pyruvate → alanine

Figure 13.8 **Synthesis of (A) aspartate and asparagine, and (B) alanine.**

This just leaves us with tyrosine, which humans obtain by hydroxylation of phenylalanine (*Fig. 13.9*). Humans cannot make phenylalanine so those molecules that are converted into tyrosine are obtained from our diet. Tyrosine could therefore be looked on as an *essential* amino acid, because if the phenylalanine content of the diet is low then the individual also suffers a deficiency in tyrosine.

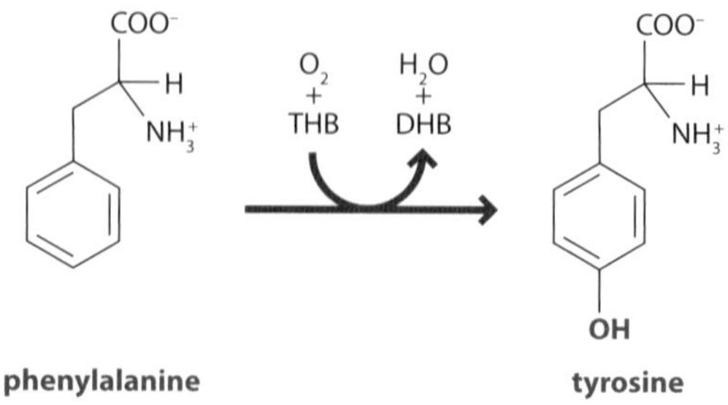

Figure 13.9 Synthesis of tyrosine from phenylalanine.
Abbreviations: DHB, dihydrobiopterin; THB, tetrahydrobiopterin. THB is a cofactor that acts as the H⁺ donor in this reaction.

phenylalanine   tyrosine

Those amino acids that humans cannot synthesize require longer pathways

For humans, there are nine essential amino acids that we must obtain from our diet (see *Table 13.1*). Plants make all 20 amino acids, in varying proportions, and some of the essential ones can also be obtained from eggs, fish or dairy products. Each of the nine amino acids that humans do not synthesize requires quite a lengthy biosynthetic pathway, and those pathways have variations in different species. We will examine the versions that operate in *Escherichia coli*, bacteria being the organisms used in the original research that unraveled these branching and interconnected series of reactions.

Four of the essential amino acids (isoleucine, lysine, methionine and threonine) can be synthesized from aspartate, which bacteria, like humans, make by transamination of oxaloacetate. Two steps give aspartate β-semialdehyde, at which point the pathway branches, with one route leading via seven intermediates to lysine, and the other to homoserine, where the pathway branches again to give methionine in one direction and threonine in the other (*Fig. 13.10A*). Deamination of threonine can then give α-ketobutyrate, which combines with pyruvate at the start of a five-step pathway leading to isoleucine. The five enzymes that catalyze these steps also catalyze a parallel pathway that begins with the linking of two pyruvates to give α-acetolactate, and then continues to yield valine (*Fig. 13.10B*). An intermediate in this set of reactions is α-ketoisovalerate, which is the starting point for a side branch of four additional reactions leading to leucine.

A similar, but less tortuous, branched pathway, results in synthesis of the three aromatic amino acids, phenylalanine, tryptophan and tyrosine. The starting point is condensation of phosphoenolpyruvate, an intermediate in glycolysis, with erythrose 4-phosphate, from the pentose phosphate pathway. Six further steps, including cyclization of the linear compound formed at the start of the pathway, gives chorismate (*Fig. 13.11*). At this point, the pathway branches. In one branch, isomerization of chorismate gives prephenate, which can be converted into either phenylalanine or tyrosine. The second branch from chorismate leads through five intermediates to tryptophan. This amino acid has a two-ring side-chain, the second of these rings constructed from **phosphoribosyl pyrophosphate** (**PRPP**), a ribose that has a diphosphate group attached to carbon number 1, and a monophosphate to carbon 5 (*see Fig. 13.12*).

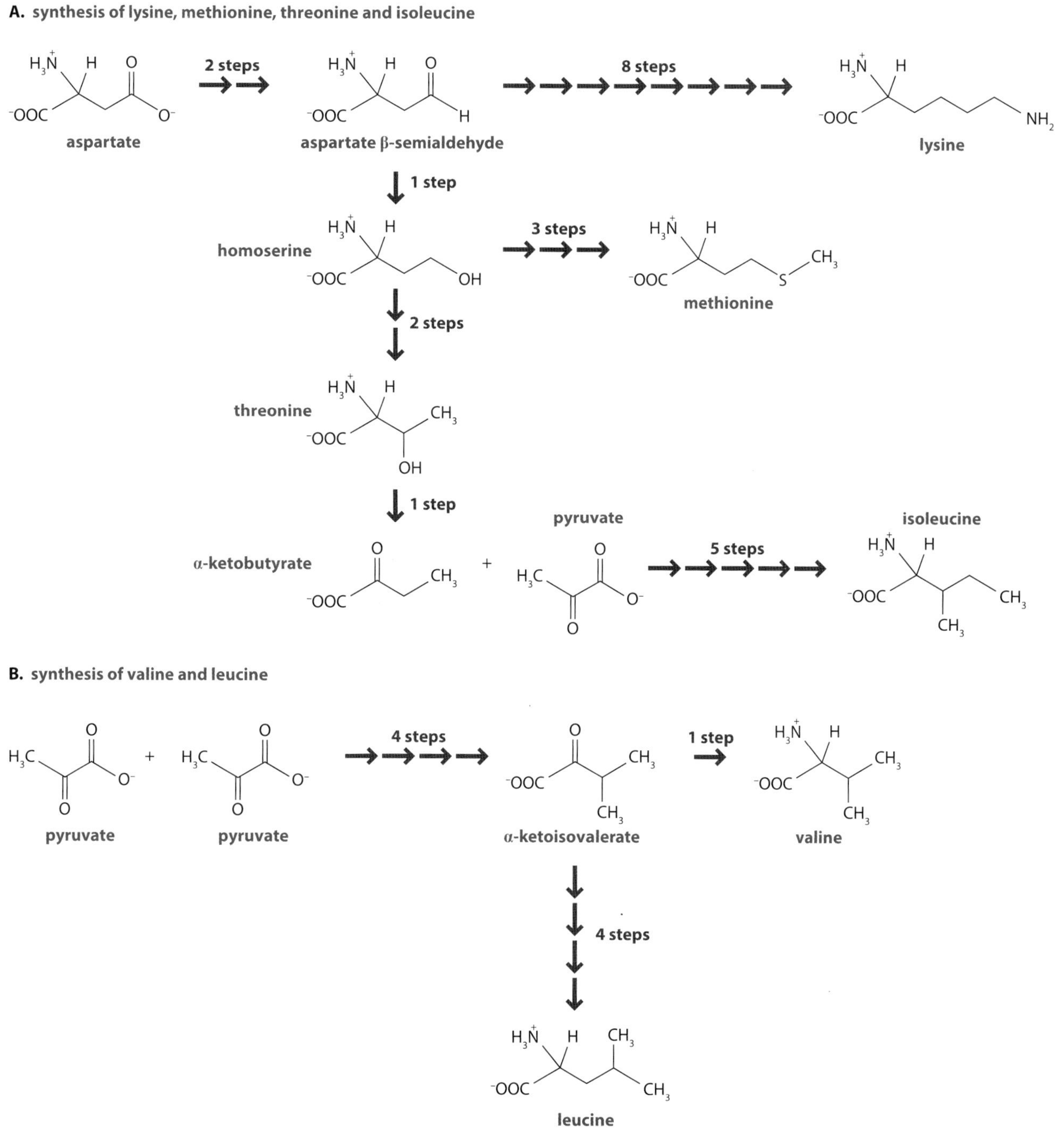

A. synthesis of lysine, methionine, threonine and isoleucine

B. synthesis of valine and leucine

Figure 13.10 **Synthesis of (A) lysine, methionine, threonine and isoleucine, and (B) valine and leucine.**

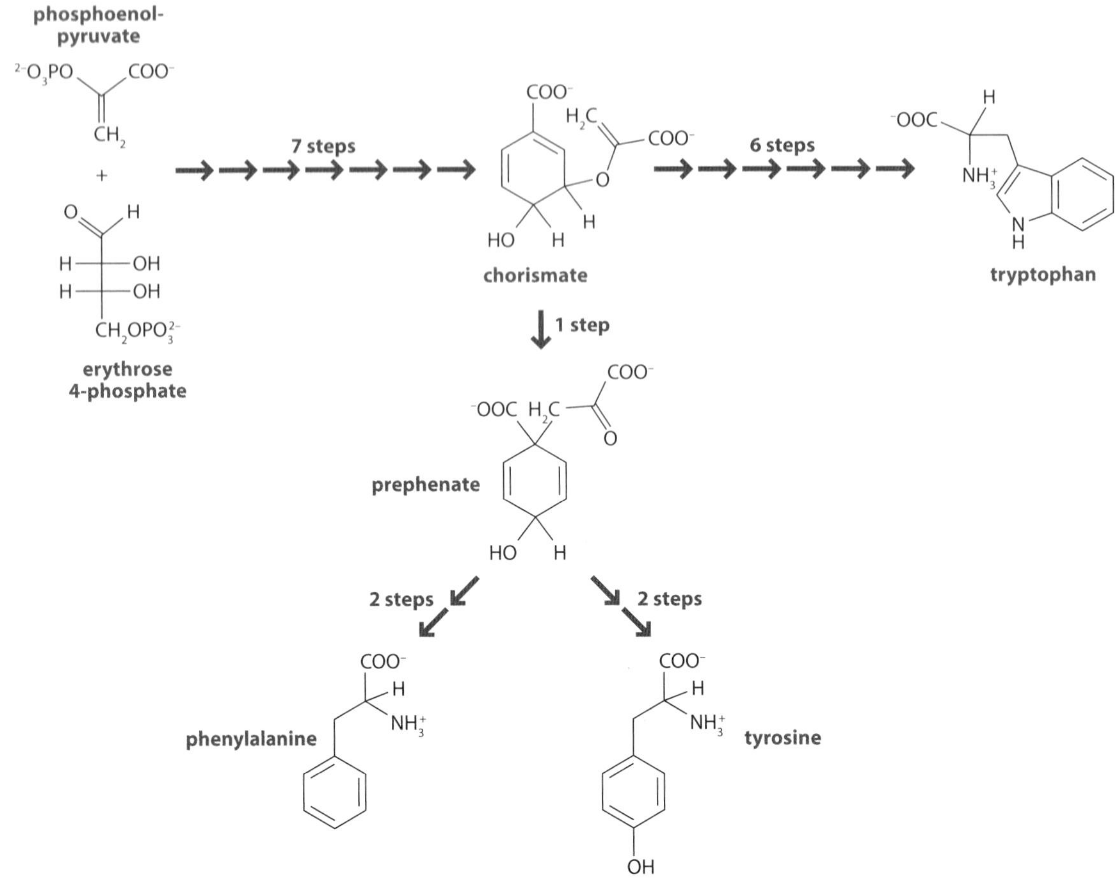

Figure 13.11 **Synthesis of phenylalanine, tyrosine and tryptophan.**

One amino acid, histidine, is still unaccounted for. PRPP is also a substrate in this pathway, but ribose 5-phosphate is usually viewed as the initial precursor, this intermediate in the pentose phosphate pathway being converted into PRPP by the enzyme ribose-phosphate diphosphokinase. Nine additional steps, each catalyzed by a different enzyme, and involving inputs from ATP and glutamine, eventually result in histidine (*Fig. 13.12*).

Branched pathways require careful regulation

The purpose of amino acid biosynthesis is to ensure that the organism has an adequate supply of these substrates for protein synthesis. The supply must also be balanced, so that each amino acid is available in the appropriate amount, with individual surpluses

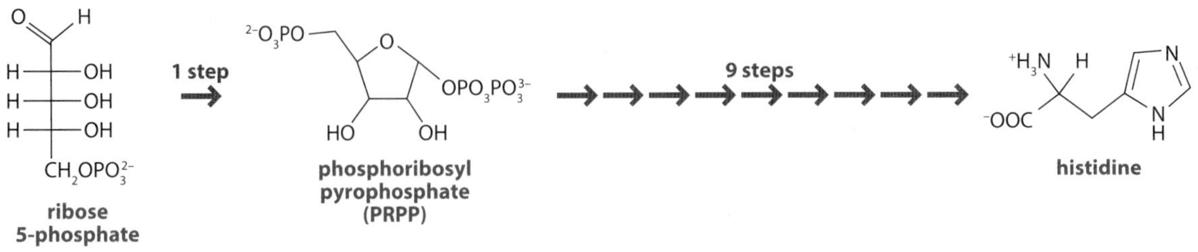

Figure 13.12 **Synthesis of histidine.**

Box 13.3 Genetically modified crops that are resistant to a herbicide that disrupts aromatic amino acid synthesis

Herbicides are extensively used by farmers and horticulturists to control weeds and protect their crops and ornamental plants. One of the most widely used herbicides is glyphosate, which is looked on as environmentally friendly, because it is non-toxic to insects and to animals and has a short residence time in soils, breaking down over a period of a few days into harmless products. Glyphosate is a competitive inhibitor of enolpyruvylshikimate 3-phosphate synthase (EPSPS), the enzyme that catalyzes the last but one step in the pathway that leads from phosphoenolpyruvate and erythrose 4-phosphate to chorismate. Treatment with glyphosate therefore prevents the plant from making chorismate, which in turn means that no phenylalanine, tyrosine or tryptophan is made. Without these amino acids, the plant dies.

phosphoenol-
pyruvate
+
erythrose
4-phosphate
→ → → → → shikimate 5-phosphate

phosphoenol-
pyruvate   P$_i$

EPSPS

enolpyruvylshikimate
3-phosphate → chorismate

inhibits

glyphosate

Although harmless to insects and animals, glyphosate kills all plants and not just weeds, and so has to be applied carefully to avoid harming crops. For farmers, this adds significantly to the cost of producing their harvest. Plant biotechnologists have therefore explored ways of using genetic engineering to produce versions of crop plants that are resistant to the toxic effects of glyphosate.

Initially, genetic engineering was used to generate plants that made greater than normal amounts of the enzyme EPSPS, in the expectation that these plants would be able to withstand higher doses of glyphosate than non-engineered plants. However, this approach was unsuccessful. Although engineered plants that made up to 80 times the normal amount of EPSPS were obtained, the resulting increase in glyphosate tolerance was not sufficient to protect these plants from herbicide application in the field.

A search was therefore carried out for an organism whose EPSPS enzyme is resistant to glyphosate inhibition and whose EPSPS gene might therefore be used to confer resistance on a crop plant. After testing the enzymes from various bacteria, as well as mutant forms of *Petunia* that display glyphosate resistance, the EPSPS enzyme from *Agrobacterium* strain CP4 was chosen. This enzyme has both a high catalytic activity and high resistance to the herbicide. The gene for the *Agrobacterium* EPSPS enzyme was cloned, using techniques that we will study in *Section 19.3.1*, and then transferred into soybean plants. These plants, called "Roundup Ready" after the trade name of the herbicide, have a threefold higher resistance to glyphosate compared to unmodified soybean. Roundup Ready versions of soybean and maize are now being grown routinely in the USA and other parts of the world.

Roundup Ready plants are glyphosate resistant, but they do not actually detoxify glyphosate, which means that the herbicide can accumulate in the plant tissues. Glyphosate is not poisonous to humans or other animals, so the use of such plants as food or forage is not a concern, but accumulation of the herbicide can interfere with reproduction of the plant. Other ways of increasing glyphosate resistance by genetic engineering are therefore being sought. One approach involves an enzyme called glyphosate *N*-acetyltransferase (GAT), from the bacterium *Bacillus licheniformis*. This enzyme detoxifies glyphosate by attaching an acetyl group to the herbicide molecule.

glyphosate

acetylglyphosate

Different strains of the bacterium synthesize different types of GAT, but none of these enzymes detoxifies glyphosate at a high enough rate to be of value if transferred to a genetically modified crop. A technique called **DNA shuffling** has therefore been used to create an artificial gene coding for a highly active GAT enzyme. DNA shuffling involves taking parts of genes from different bacterial strains and reassembling these parts to create new gene variants. The new variants are then tested to identify any that specify a more active GAT enzyme. The genes for these active enzymes are then used in a second round of shuffling. After 11 rounds, a gene specifying a GAT with 10 000 times the activity of the enzymes present in the original *B. licheniformis* strains was obtained.

original GAT genes

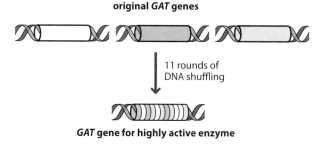

11 rounds of
DNA shuffling

GAT gene for highly active enzyme

The new GAT gene was introduced into maize. The resulting plants were found to tolerate levels of glyphosate six times higher than the amount normally used by farmers to control weeds, without any reduction in the productivity of the plant. This new way of engineering glyphosate resistance is now being exploited in the development of herbicide-resistant soybean and canola (a cultivar of oilseed rape).

The regulation of gene expression, which controls the amount of an individual enzyme that is present in a cell, is described in *Chapter 17*.

and deficiencies avoided. Most organisms achieve this via a sophisticated set of regulatory mechanisms involving control over the synthesis of key enzymes in the biosynthetic pathway, and feedback inhibition of the activities of those enzymes. We will study the control of enzyme synthesis in a later chapter. Here we will look at the feedback regulation.

As we have seen with other metabolic pathways, regulation is usually exerted on the enzyme at the first 'committed' step in the pathway. This is the first step that leads to an intermediate that is used for no purpose other than synthesis of the product. The synthesis of a molecule of this intermediate therefore represents a commitment to make a molecule of the product. If we take the synthesis of isoleucine from threonine as an example, we see from *Figure 13.10A* that the first committed step on this pathway is the deamination of threonine to give α-ketobutyrate, which is catalyzed by **threonine dehydratase**. This enzyme is inhibited by isoleucine via an allosteric regulatory mechanism (*Fig. 13.13A*). The more isoleucine that is present, the greater is the degree to which the activity of threonine dehydratase is inhibited. Isoleucine therefore exerts feedback control over its own synthesis.

With branched pathways, an additional level of sophistication is needed. In the branched pathway for synthesis of the aromatic amino acids we see, as we would anticipate, feedback control by each of the three products – phenylalanine, tryptophan and tyrosine – over the committed steps leading to their individual syntheses. So, for example, tryptophan inhibits anthranilate synthase, the first of the enzymes on the branch leading from chorismate exclusively to tryptophan (*Fig. 13.13B*). But these three feedback systems each operate downstream of chorismate. This means that we need a mechanism for ensuring energy and substrates are not wasted in synthesizing chorismate when there are surpluses of phenylalanine, tryptophan and tyrosine. So, each of these amino acids also regulates the activity of the enzyme 2-keto-3-deoxy-D-arabinoheptulosonate 7-phosphate synthase, which we can abbreviate to **DAHP synthase**. This enzyme catalyzes the condensation of phosphoenolpyruvate

A. control of isoleucine synthesis

B. control of aromatic amino acid synthesis

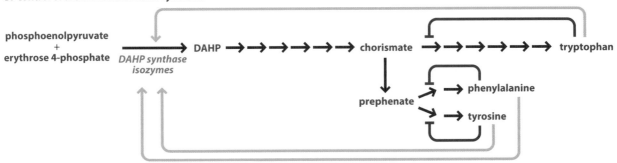

Figure 13.13 Feedback regulation of amino acid synthesis.
(A) Control of threonine synthesis. (B) Control of aromatic amino acid synthesis. In (B) the red lines indicate the feedback inhibition that each of these three amino acids exerts on the committed step in its synthesis pathway, and the green lines indicate feedback control of the DAHP synthase isozymes.

and erythrose 4-phosphate, yielding DAHP, at the committed step in the part of the pathway leading to chorismate.

Regulation of DAHP synthase achieves the necessary feedback inhibition of chorismate synthesis, but what if one of the amino acids is in excess whereas the other two are deficient? Now we would have the undesirable situation where the presence of excess phenylalanine, for example, caused a shut-down of chorismate synthesis, so that deficiencies in tryptophan and tyrosine could not be corrected. This situation is avoided because there are three versions of DAHP synthase. These **isozymes** have similar amino acid sequences and three-dimensional structures, and each catalyzes the same reaction, the conversion of phosphoenolpyruvate and erythrose 4-phosphate to DAHP. Each isozyme is subject to feedback inhibition by a different amino acid, with phenylalanine controlling the activity of one of the DAHP isozymes, tryptophan controlling the second, and tyrosine the third. In this way, the three amino acids that are the end products of this branched system can each individually ensure that the common part of the pathway operates at the appropriate rate.

13.2.2 Synthesis of nucleotides

Nucleotides are the monomeric components of DNA and RNA, and ribonucleotides (those nucleotides found in RNA molecules) include ATP, which we have encountered repeatedly as a carrier of energy. The purine or pyrimidine component of a nucleotide is a single- or double-ring structure, partly made of nitrogen atoms. Nucleotides are therefore nitrogen-containing compounds and we must explore how they are synthesized from the supplies of organic nitrogen present in the cell.

There are two distinct pathways for nucleotide synthesis. The first is the **salvage pathway**, in which purines and pyrimidines released from nucleotides that are being degraded are salvaged and re-used to make new nucleotides. The sugar–phosphate component of the new nucleotides is provided by the phosphorylated ribose called PRPP. Replacement of the diphosphate at carbon 1 of PRPP with a purine or pyrimidine base results in a nucleoside monophosphate (*Fig. 13.14*). If this is AMP then it can be converted to ADP by the enzyme **adenylate kinase**, using up a molecule of ATP in the process. The two resulting ADP molecules are then phosphorylated during glycolysis or via the electron transport chain. ATP is also used to convert other nucleoside monophosphates to their diphosphate forms (e.g. GMP to GDP), although a different enzyme, **nucleoside monophosphate kinase**, is involved. The resulting nucleoside diphosphates are phosphorylated to their triphosphates by a third enzyme, **nucleoside diphosphate kinase**. This enzyme can use any nucleoside triphosphate as the phosphate donor, but ATP is most frequently used simply because it is present in higher concentrations in the cell. All of these reactions, since they begin with PRPP, result in the synthesis of ribonucleotides, the versions present in RNA and also the ones that act as energy carriers. Deoxyribonucleotides, the components of DNA, are obtained from the ribonucleotides by reduction of the 2'-carbon of the ribose sugar, by **ribonucleotide reductase**.

The base components of ribonucleotides can also be obtained by *de novo* **synthesis**. These pathways do not use pre-existing purine and pyrimidine bases but instead construct them from smaller precursors. The single-ring pyrimidines, cytosine and uracil, are formed from aspartate and **carbamoyl phosphate**, the latter built up from bicarbonate, an amino group from glutamine, and a phosphate from ATP (*see Section 13.3.2*). The enzyme **aspartate transcarbamoylase** joins aspartate and carbamoyl phosphate into a linear intermediate that gives rise to orotate (*Fig. 13.15*). Orotate is then attached to PRPP and its carboxyl group replaced with a hydrogen to give UMP. Kinase enzymes convert this to UTP, and some of the UTP is further metabolized to CTP by addition of an amino group from glutamine, catalyzed by **cytidylate synthetase**.

Refer to *Figure 4.3* for the structures of the five nucleotide bases.

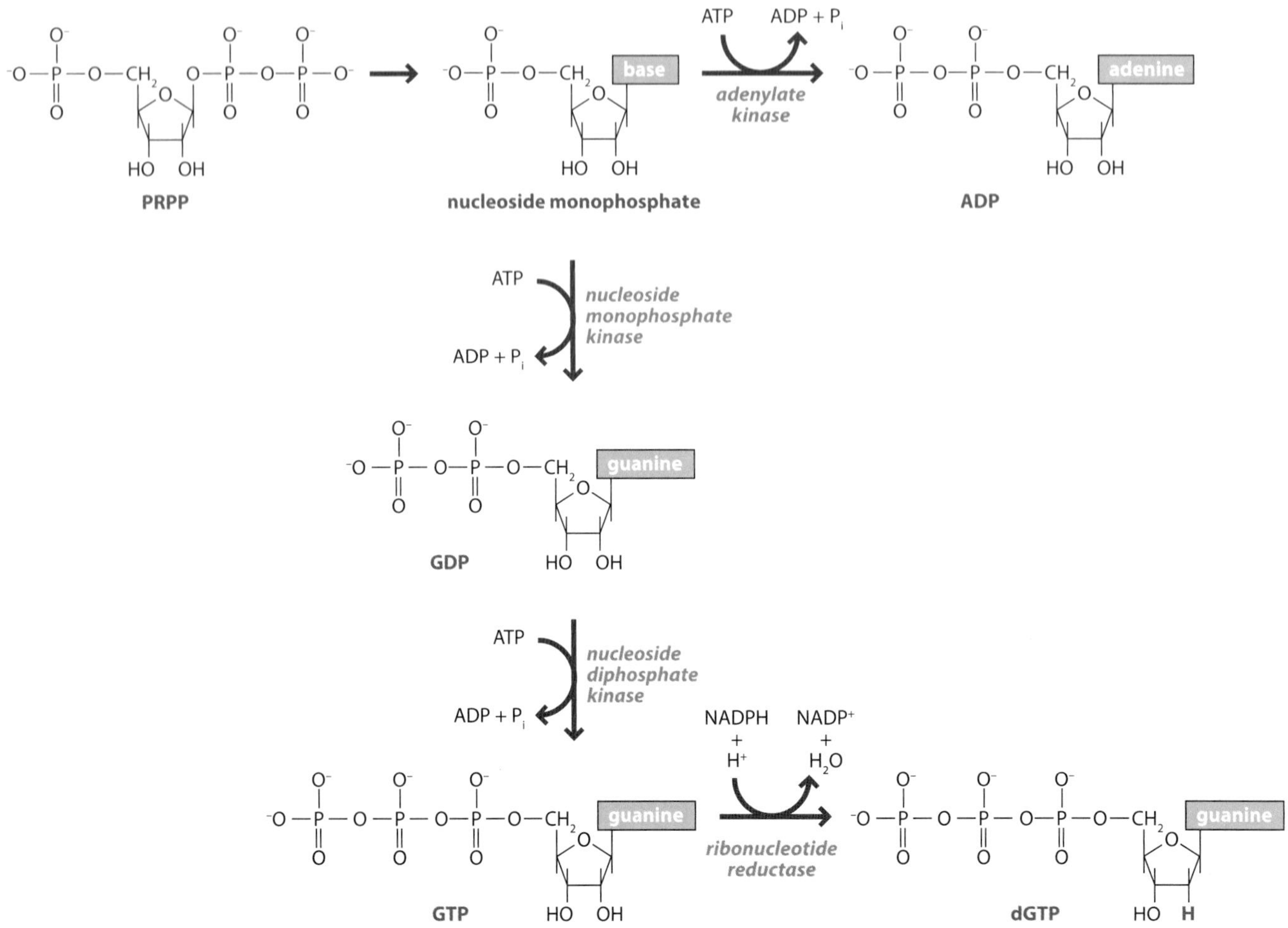

Figure 13.14 The salvage pathway for nucleotide synthesis.
The conversion of GMP to GDP, GTP and then dGTP is shown in detail. The cytosine and thymine nucleotides can be synthesized in the same way, as can ATP and dATP from the ADP produced by adenylate kinase.

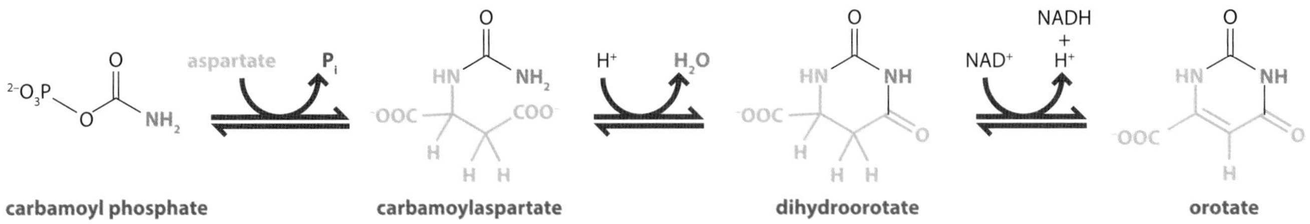

Figure 13.15 Synthesis of orotate, part of the *de novo* pathway for production of pyrimidine nucleotides.

De novo synthesis of the two-ring purines makes use of carbon and nitrogen atoms from aspartate, glycine, formate, glutamine and carbonate ions. The pathway is lengthy and, unlike *de novo* pyrimidine synthesis, does not involve formation of a complete ring structure that is then attached to PRPP. Instead, the purine rings are built up step by step on the PRPP molecule.

Following completion, some of the ribonucleotides obtained by *de novo* synthesis can be converted to their deoxyribonucleotide versions by ribonucleotide reductase, reducing the 2' carbon of the ribose sugar, as described for the salvage pathway. One final step is then needed to produce the thymine-containing deoxynucleotides that have no direct counterparts among the ribonucleotides synthesized in this way. This final step is achieved by **thymidylate synthase**, which adds a methyl group to uracil, converting this pyrimidine into thymine.

13.2.3 Tetrapyrrole synthesis

The tetrapyrrole compounds include heme, which is a cofactor in various proteins including hemoglobin and the cytochrome family of enzymes, and chlorophyll, which is of course the central component of the photosynthetic pathway. A tetrapyrrole is any compound with four pyrrole units, each pyrrole comprising a ring of one nitrogen and four carbon atoms. In the tetrapyrroles that we have just mentioned, the four pyrroles themselves form a ring structure, with a central metal ion held in place by interactions with the four nitrogens. In heme the metal is iron, and in chlorophyll it is magnesium.

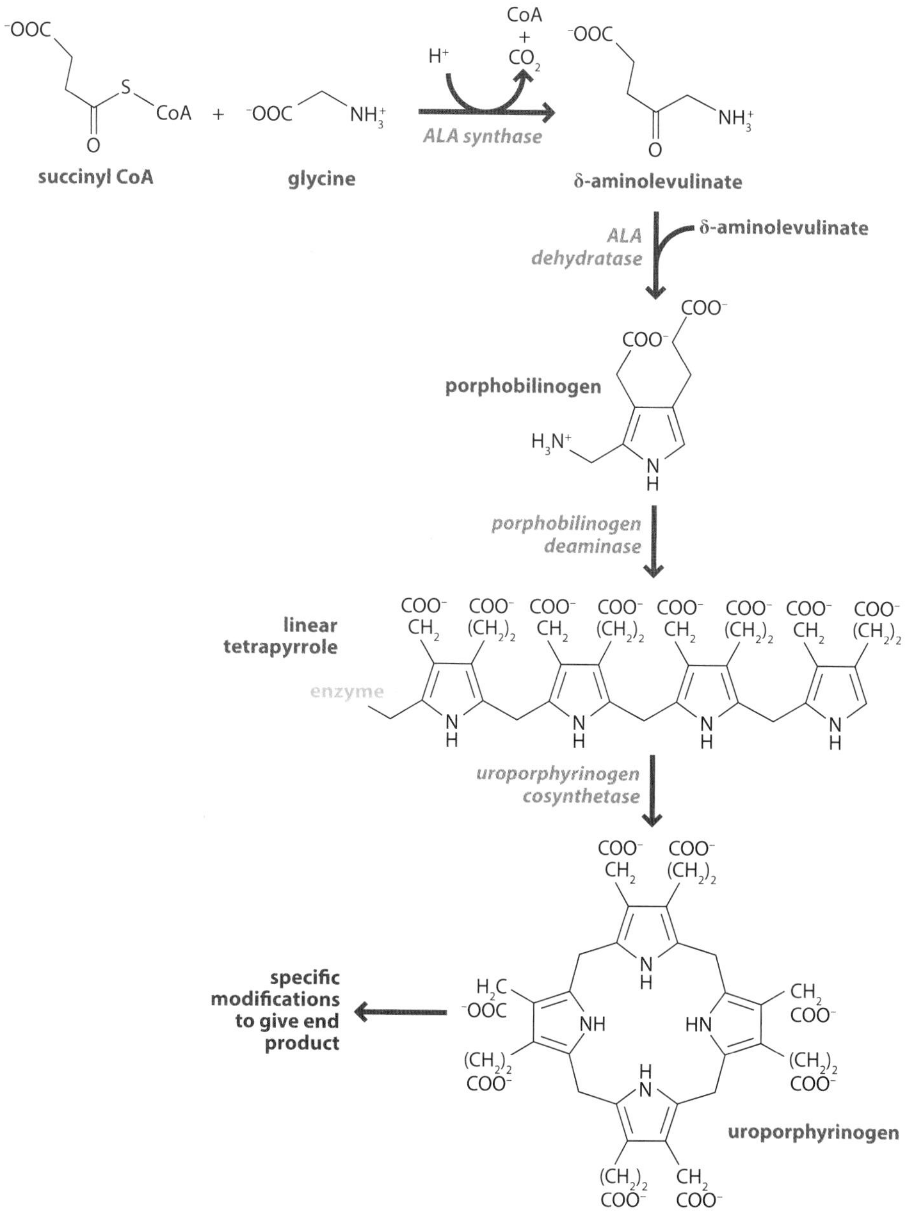

See *Figure 3.25* for the structure of heme and *Figure 10.5* for chlorophyll.

Figure 13.16 Synthesis of tetrapyrrole compounds. Although the linear tetrapyrrole remains attached to the porphobilinogen deaminase enzyme until it circularizes, circularization also requires the activity of the cosynthetase enzyme.

Box 13.4 **Nucleotide synthesis is a target for cancer chemotherapy**

One of the chemotherapy drugs most commonly used in the treatment of cancer is 5-fluorouracil. This compound is an analog of uracil.

5-fluorouracil   **uracil**

After administration, 5-fluorouracil is metabolized into a deoxynucleotide monophosphate. As an analog of dUMP, this 5-fluorouracil nucleotide acts as an irreversible inhibitor of thymidylate synthase, forming a stable complex with the enzyme and its 5,10-methylenetetrahydrofolate cofactor. Treatment with 5-fluorouracil therefore results in a deficiency in dTTP, so cells are unable to replicate or repair their DNA. This leads to what is called 'thymineless death'. 5-fluorouracil is usually administered systemically, but healthy tissues are relatively unaffected because most of their cells are not actively dividing. The drug therefore targets the rapidly dividing cells in cancerous tissues.

The tetrapyrrole synthesis pathway comprises two stages. The first stage results in synthesis of a single pyrrole, four of which are combined together during stage 2 to give the tetrapyrrole. In animals, the substrates for pyrrole synthesis are glycine (which contributes the nitrogen atom) and succinyl CoA. These are combined together by **ALA synthase** to give δ-aminolevulinate (ALA) (*Fig. 13.16*). In plants, this step is slightly different and uses glutamate instead of glycine as the substrate. Two ALA molecules are then joined by **ALA dehydratase**, this reaction forming porphobilinogen, which is a type of pyrrole.

The second stage commences with a series of condensation reactions, catalyzed by **porphobilinogen deaminase**, which results in a linear tetrapyrrole that is immediately cyclized into uroporphyrinogen. Modifications to the groups attached to the pyrrole units, and insertion of the iron or magnesium atom, complete the process by converting uroporphyrinogen into heme or chlorophyll.

13.3 Degradation of nitrogen-containing compounds

So far we have focused only on the synthesis of nitrogen-containing compounds and not looked at any of the pathways by which these compounds are broken down. The pathways for nucleotides and tetrapyrrole do not need to distract us unduly. Many of the purine and pyrimidine bases released during nucleotide degradation are reused via the salvage pathway. Adenines and guanines that are excess to requirements are converted to a different type of purine, **uric acid** (*Fig. 13.17*), which is excreted. The pyrimidines, cytosine, thymine and uracil, are broken down more comprehensively, with the nitrogen being converted to ammonium ions. Excess tetrapyrroles are converted to linear compounds called **bile pigments**. In plants, these are utilized as photosensors, such as **phytochrome**, which coordinate the plant's physiological and biochemical responses to light. Animals are far less adaptable and the bile pigments they synthesize in the liver and spleen are further metabolized and then excreted. Excretion is a strong theme in the degradation of nitrogen-containing compounds, because most organisms lack the ability to store excess nitrogen, and so must simply get rid of it.

Two aspects of nitrogen catabolism are more important and we must study these in more detail. First, the degradation of amino acids gives rise to carbohydrates that can be used in energy generation. Animals acquire 10–15% of their energy supplies in this way, either through breakdown of amino acids derived from dietary protein or from the degradation of cellular proteins. Most of the nitrogen in amino acids is released as ammonia, supplementing the much smaller amount of ammonia

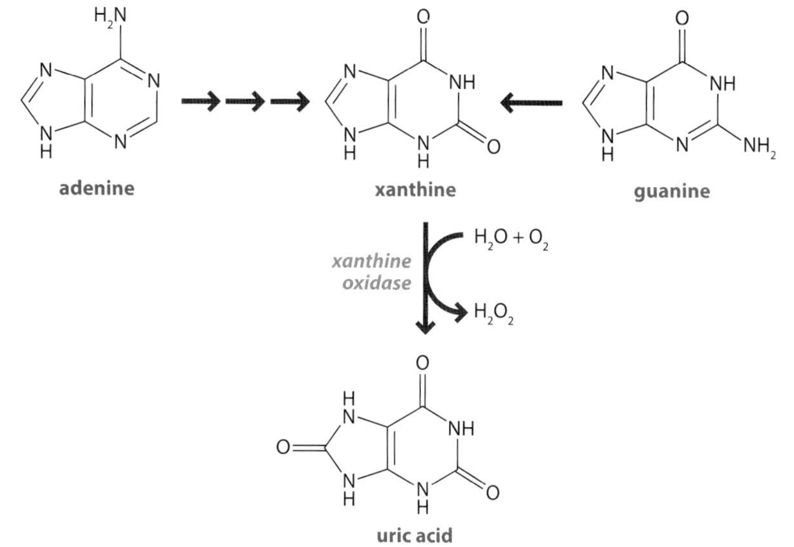

Figure 13.17 The conversion of adenine and guanine into uric acid.

adenine   xanthine   guanine

xanthine oxidase $H_2O + O_2$

H_2O_2

uric acid

Figure 13.17 The conversion of adenine and guanine into uric acid.

generated by pyrimidine degradation. Ammonia is toxic to most organisms and so must be excreted. The **urea cycle**, which results in the detoxification and excretion of ammonia, is therefore the second aspect of nitrogen catabolism that we must study.

13.3.1 Degradation of amino acids

Most amino acid degradation occurs in the liver, with only those three amino acids with branched side-chains – isoleucine, leucine and valine – being broken down in muscles and other energy-requiring tissues. The 20 amino acids follow a variety of degradation pathways, the diversity mirroring the complexity of the pathways by which the amino acids were initially synthesized. We will focus on two key issues, the removal and fate of the nitrogen component of each amino acid, and the ways in which the resulting carbon skeletons are utilized.

The nitrogen in amino acids is released as ammonia

The nitrogen contained in amino acids is ultimately converted to ammonia and then excreted. This refers to both the amino group that all amino acids possess, and the nitrogen-containing groups that a few carry within their side-chains. The side-chain nitrogens are dealt with by specific enzymes: for example, **asparaginase** removes the amide group from asparagine, giving aspartic acid and ammonia (*Fig. 13.18*). The amino groups are removed by a more uniform process involving transamination, which takes place mainly in the liver in mammals. For each amino acid, this transamination moves the amino group to α-ketoglutarate, generating glutamate.

There are a family of **transaminase** enzymes that carry out these amino transfers. Each of these enzymes has a **pyridoxal phosphate** cofactor, derived from vitamin B6. The cofactor is initially bound to the amino group of a lysine within the transaminase polypeptide. Entry of the amino acid to be degraded into the active site displaces the pyridoxal phosphate to lysine linkage, the amino acid now being bound, through its amino group, to the cofactor (*Fig. 13.19*). The bound amino acid is then hydrolyzed, cleaving the amino group and releasing the remainder of the amino acid as an α-keto acid.

The amino group is now transferred to α-ketoglutarate, and the resulting glutamate molecule detaches, freeing the pyridoxal phosphate to re-form its linkage with the

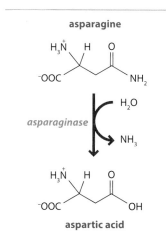

asparagine

asparaginase H_2O

NH_3

aspartic acid

Figure 13.18 Deamidation of asparagine by asparaginase.

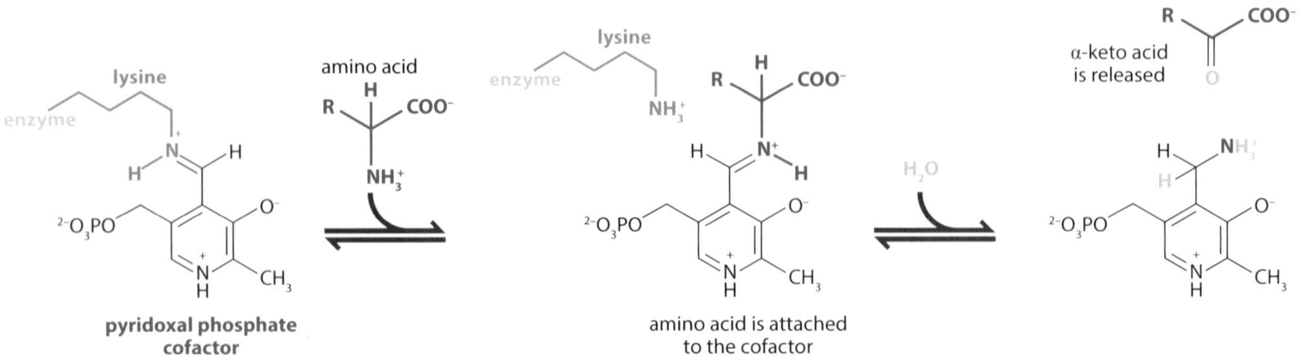

Figure 13.19 **The role of pyridoxal phosphate in deamination of an amino acid.**

original lysine. The glutamate is then oxidized back to α-ketoglutarate with release of the amino group as ammonia. This is the reversal of the first step of nitrogen assimilation (*see Section 13.2.1*), and uses the same enzyme, **glutamate dehydrogenase**. The only difference is that when working in the degradation direction, the enzyme generates a molecule of NADH, whereas during glutamate synthesis it used a molecule of NADPH. This is an important control point in amino acid degradation: glutamate dehydrogenase is inhibited by ATP and GTP, and stimulated by ADP and GDP. This means that when energy resources are low, and ADP and GDP predominate, glutamate dehydrogenase activity increases. More amino acids are therefore oxidized, releasing their carbon skeletons for use as energy supplies.

The carbon skeletons of amino acids are degraded to products that can enter the TCA cycle

Although individual amino acids follow diverse degradation pathways, these all converge into just six major end products, all of which can enter the TCA cycle. The products are pyruvate, oxaloacetate, α-ketoglutarate, succinyl CoA, fumarate and acetyl CoA (*Fig. 13.20*). Because of the complexity of the breakdown pathways, a single amino acid can contribute more than one of these products.

The products of amino acid breakdown can therefore enter the TCA cycle and directly contribute to energy generation. Alternatively they can be used to increase the body's energy reserves. This can be done in either of two ways, depending on the identity of the end product. Pyruvate, oxaloacetate, α-ketoglutarate, succinyl CoA and fumarate can each be channeled into the gluconeogenesis pathway and hence ultimately converted into glucose. The amino acids that give rise to these end products are therefore called **glucogenic**.

Those amino acids that give rise to acetyl CoA can also contribute to **ketone body** synthesis and are therefore called **ketogenic**. They make an important contribution to ketone body synthesis, although fatty acid breakdown is the main source of the acetyl CoA that is used for this purpose. Ketone bodies are produced from excess acetyl CoA by the process called **ketogenesis**. The ketogenesis pathway initially parallels the synthesis of cholesterol, two molecules of acetyl CoA combining to form acetoacetyl CoA, which then links with a third acetyl CoA to give HMG CoA. The next step in cholesterol synthesis is reduction of HMG CoA, but in ketogenesis this compound loses an acetyl CoA group to yield acetoacetate. This might appear to be a convoluted process, given that one of the initial substrates was acetoacetyl CoA, but this is the only biochemical means of removing CoA from acetoacetyl CoA. Some of the acetoacetate undergoes enzymatic conversion to D-3-hydroxybutyrate, and some spontaneously decarboxylates (without any enzyme involvement) to give acetone (*Fig. 13.21*). A

Figure 13.20 Entry of amino acid breakdown products into the TCA cycle.
Amino acids can be classified as glucogenic or ketogenic (or both) depending on whether their breakdown products can contribute to gluconeogenesis or ketone body formation, respectively. Those giving rise to pyruvate can contribute to both pathways depending on how the pyruvate is used.

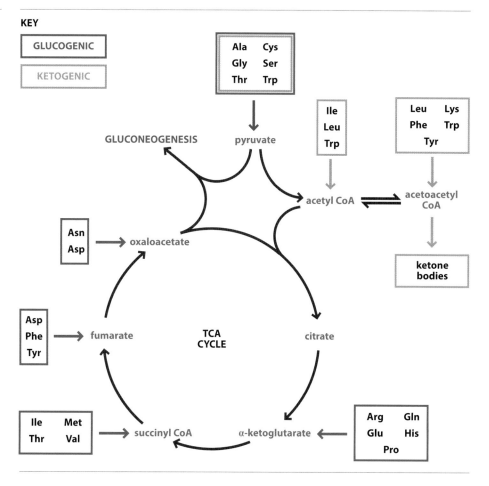

ketone body comprises a mixture of acetoacetate, D-3-hydroxybutyrate and acetone. After synthesis in the liver, ketone bodies are transported to the heart and brain, where the acetoacetate and D-3-hydroxybutyrate are reconverted to acetyl CoA, which enters the TCA cycle and hence contributes to energy generation in these tissues. Indeed, the heart has a preference for ketone bodies rather than glucose as its energy source.

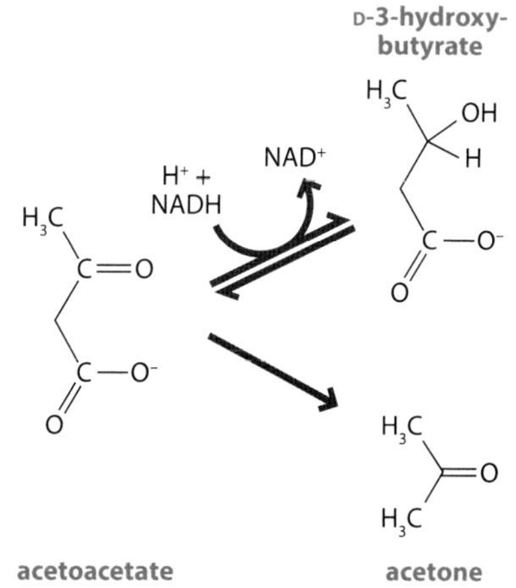

Figure 13.21 The components of ketone bodies.

13.3.2 The urea cycle

Ammonia is an important product of nitrogen catabolism. Some is re-used in the biosynthesis of new nitrogen-containing compounds, but the remainder must be excreted. Different organisms use one of three ways to get rid of this excess ammonia:

- **Ammonotelic** species include many aquatic invertebrates and these organisms simply excrete the ammonia into the water in which they live.

- **Uricotelic** organisms excrete nitrogen in the form of uric acid. These species include birds, snakes, land-dwelling reptiles, and arthropods such as insects.

- **Ureotelic** organisms include mammals, amphibians and some fish, which convert ammonia into urea, which is excreted in the urine.

The pathway by which humans and other ureotelic organisms convert ammonia to urea is called the urea cycle. We will follow through the steps in the urea cycle and then examine how the cycle is controlled.

The urea cycle enables excess nitrogen to be excreted as urea

The urea cycle is carried out in the liver, with the resulting urea passed via the bloodstream to the kidneys from where it is excreted in the form of urine. The pathway is shown in outline in *Figure 13.22*. The individual steps are as follows:

Step 1. Ammonia enters the urea cycle after conversion to carbamoyl phosphate, which requires addition of a bicarbonate ion and a phosphate group from ATP. A second ATP is required to provide energy. The enzyme that catalyzes this reaction is **carbamoyl phosphate synthetase**.

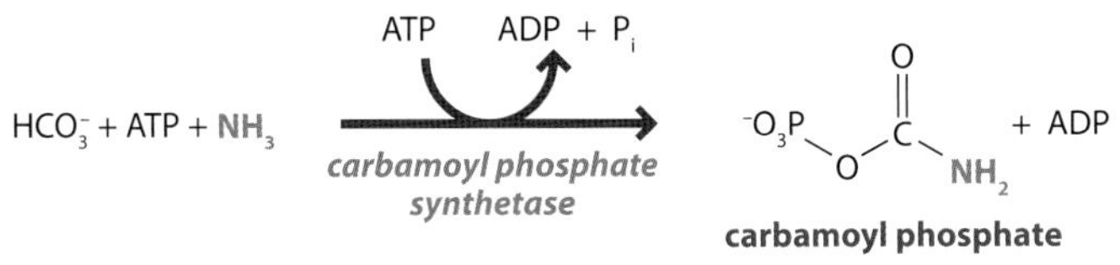

Step 2. The carbamoyl group is now transferred to an ornithine molecule by **ornithine transcarbamoylase**, to give **citrulline**.

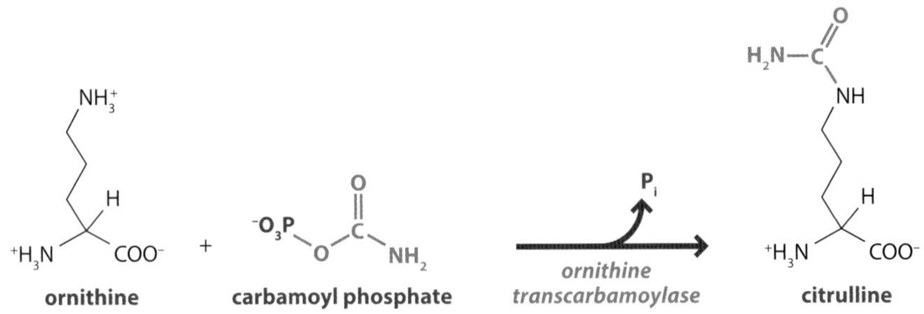

Both ornithine and citrulline are amino acids, but not ones that are used in protein synthesis. *Steps 1* and *2* take place in the mitochondrial matrix, but the remainder of the urea cycle occurs in the cytoplasm. The citrulline is therefore transported out of the mitochondrion.

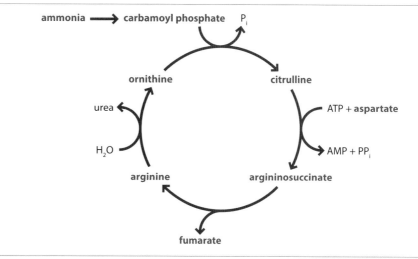

Figure 13.22 Outline of the urea cycle.

Step 3. A condensation reaction between citrulline and aspartate gives argininosuccinate. An ATP is converted to AMP in this reaction catalyzed by **argininosuccinate synthetase**.

Step 4. The carbon skeleton of aspartate is removed from argininosuccinate by **argininosuccinase**. The products are arginine and fumarate.

The outcome of *Steps 3* and *4* is to transfer the amino group of the aspartate to citrulline. The aspartate is converted to fumarate, a non-nitrogenous carbohydrate, and citrulline is converted to arginine.

Step 5. **Arginase** cleaves arginine, producing a molecule of urea and regenerating the ornithine used up at the start of the cycle.

The ornithine shuttles back into the mitochondrion so that the cycle can recommence.

Note that during the urea cycle three ATPs are consumed, but the energy expenditure is the equivalent of four ATPs, because one of the ATPs is converted to AMP.

Regulation of the urea cycle involves a metabolic dead end

The rate at which the urea cycle needs to operate depends on several factors, including the composition of the diet. A protein-rich diet requires rapid operation of the urea cycle to deal with the excess ammonia that is generated as the amino acids are broken down so that their carbon skeletons can be used for energy.

Long-term alterations in the activity of the urea cycle, such as might be demanded by a change in diet, are brought about largely by regulating the synthesis of the enzymes involved in the cycle. More immediate short-term control is exerted by *N*-acetylglutamate, which activates carbamoyl phosphate synthetase, the enzyme that catalyzes step 1 of the cycle. Why is *N*-acetylglutamate the regulatory molecule when this compound is not an intermediate in the urea cycle? The answer is that arginine,

---

Box 13.5 **Diseases associated with defects in nitrogen metabolism**

Several human diseases arise because of a breakdown in one or other of the nitrogen metabolism pathways.

- **Phenylketonuria**, which affects approximately one in 10 000 Caucasians, is caused by a defect in the gene for phenylalanine hydroxylase. This is the enzyme that converts phenylalanine to tyrosine (see *Fig. 13.9*). The resulting tyrosine deficiency can be countered by food supplements, and the build-up of excessive phenylalanine is avoided by dietary control and medication. If untreated in childhood, phenylketonuria can lead to mental retardation, but these symptoms are rarely seen in developed countries because of effective screening for the disease in newborn children. By testing for elevated phenylalanine levels compared to those of tyrosine, infants suffering from phenylketonuria can be identified and immediately placed on the appropriate treatment regime.

- **Gunther disease** is much less common than phenylketonuria, with a prevalence of only one in 1 000 000. It is caused by a mutation in the gene for uroporphyrinogen cosynthetase, the enzyme that converts the linear tetrapyrrole into uroporphyrinogen during synthesis of heme (see *Fig. 13.16*). The resulting heme deficiency leads to anemia, but patients also develop skin disorders due to accumulation of unusual tetrapyrrole compounds that are synthesized instead of uroporphyrinogen.

- **Gout** is caused by excessive uric acid in the blood. Deposition of uric acid crystals in the joints and tendons, especially in the big toe, leads to the excruciating pain associated with the disease. Gout was traditionally associated with over-consumption of protein-rich foods and alcohol, but its dietary causes are now looked on as complex. Several genetic disorders also give a predisposition to gout, such as Lesch–Nyhan syndrome, in which the gene for hypoxanthine-guanine phosphoribosyltransferase (HGPRT) is defective. This enzyme is involved in the purine salvage pathway. If HGPRT is inactive, adenine and guanine released from nucleotides that are being degraded cannot be reused to make new nucleotides, and instead are converted into uric acid. The excessive amounts of uric acid that arise in this way lead to the symptoms of gout.

- **Hyperammonemia** occurs when there is an excess of ammonia in the blood. It is usually caused by a failure of the urea cycle to convert ammonia into urea, either because of a genetic defect in one of the enzymes of the urea cycle, or as a side-effect of liver disease such as hepatitis infection. Hyperammonemia is a serious condition that causes brain damage and likely death.

Figure 13.23 The role of N-acetylglutamate in regulation of the urea cycle.
(A) In plants and bacteria, N-acetylglutamate is the substrate for a pathway leading to synthesis of arginine, the latter regulating N-acetylglutamate synthase by feedback inhibition. (B) In mammals and other vertebrates, N-acetylglutamate is not metabolized to arginine. This amino acid still regulates N-acetylglutamate synthase, but by activation not inhibition. N-acetylglutamate then activates carbamoyl phosphate synthase. The events that led to arginine becoming an activator rather than inhibitor of N-acetylglutamate synthase are not understood.

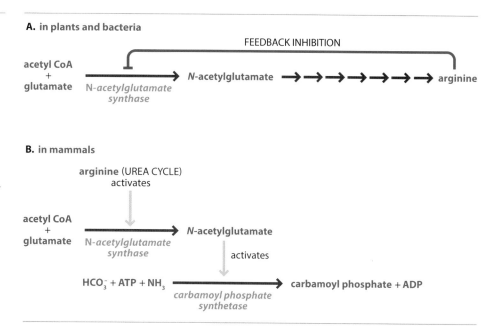

which is a urea cycle intermediate, regulates the activity of **N-acetylglutamate synthase**, the enzyme that synthesizes N-acetylglutamate from acetyl CoA and glutamate. In mammals, this is a relic of an earlier evolutionary period, when arginine could be synthesized from glutamate by a pathway that began with synthesis of N-acetylglutamate. In plants and bacteria this pathway still operates, but in mammals it is a dead end that terminates with N-acetylglutamate (*Fig. 13.23*). This compound now only serves as the regulator of carbamoyl phosphate synthetase, activating this enzyme when its own synthesis is activated by arginine, an excess of arginine signaling that the urea cycle needs to be up-regulated.

A link with the TCA cycle enables some of the energy used in the urea cycle to be recovered

One of the intermediates in the urea cycle is fumarate, which is also formed during the TCA cycle. Fumarate therefore provides a link, referred to as the **aspartate–argininosuccinate shunt**, that enables the two cycles to be interconnected. One outcome of this link is that some of the ATP molecules used up during the urea cycle can be recovered.

The urea and TCA cycles are physically separate, because the TCA cycle occurs in the mitochondria whereas the urea cycle, at least the part that generates fumarate, takes place in the cytoplasm. The fumarate from the urea cycle can be converted to malate by a cytoplasmic isozyme of fumarase, and the malate then transported into the mitochondrion, crossing the inner mitochondrial membrane through the malate–α-ketoglutarate carrier protein, part of the malate–aspartate shuttle system (*Fig. 13.24*).

We studied the malate–aspartate shuttle in *Section 9.2.5*.

Once inside the mitochondrion, malate is oxidized by malate dehydrogenase in step 8 of the TCA cycle, generating oxaloacetate and a molecule of NADH. This NADH can then be used to generate three ATPs via the electron transport chain. We noted above that a single round of the urea cycle consumed four equivalents of ATP, so the aspartate–argininosuccinate shunt enables three-quarters of the energy drain of the urea cycle to be recovered.

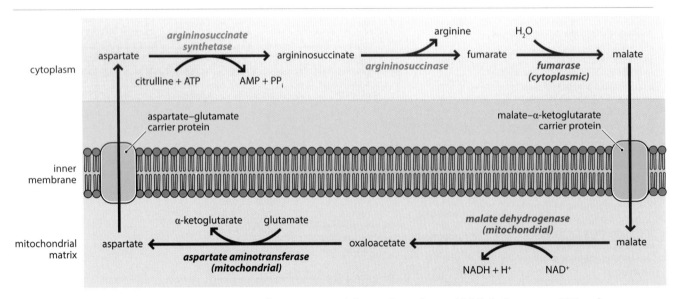

Figure 13.24 The aspartate–argininosuccinate shunt, which links the urea and TCA cycles.
Enzymes for steps in the urea cycle are shown in blue, and those for steps in the TCA cycle are in red.

To complete the shunt, the fumarate removed from the cytoplasm must be replenished. This is achieved by transamination of the oxaloacetate into aspartate by aspartate aminotransferase, and passage of this molecule through the aspartate–glutamate carrier protein and out of the mitochondrion. Once in the cytoplasm, the aspartate can participate in steps 3 and 4 of the urea cycle, regenerating fumarate.

Further reading

Castle LA, Siehl DL and Gorton R (2004) Discovery and directed evolution of a glyphosate tolerance gene. *Science* **304**, 1151–4. *Engineering the glyphosate N-acetyltransferase gene.*

Cheng Q (2008) Perspectives in biological nitrogen fixation research. *Journal of Integrative Plant Biology* **50**, 786–98.

Eliot AC and Kirsch JF (2004) Pyridoxal phosphate enzymes: mechanistic, structural, and evolutionary considerations. *Annual Review of Biochemistry* **73**, 383–415.

Holden HM, Thoden JB and Raushel FM (1999) Carbamoyl phosphate synthetase: an amazing biochemical odyssey from substrate to product. *Cellular and Molecular Life Sciences* **56**, 507–22.

Huang M and Graves LM (2003) *De novo* synthesis of pyrimidine nucleotides: emerging interfaces with signal transduction pathways. *Cellular and Molecular Life Sciences* **60**, 321–36.

Jackson MJ (1986) Mammalian urea cycle enzymes. *Annual Review of Genetics* **20**, 431–64.

Kornberg H (2000) Krebs and his trinity of cycles. *Nature Reviews Molecular Cell Biology* **1**, 225–8. *The discovery of the urea cycle.*

Li M, Li C, Allen A, Stanley CA and Smith TJ (2012) The structure and allosteric regulation of mammalian glutamate dehydrogenase. *Archives of Biochemistry and Biophysics* **519**, 69–80.

Longley DB, Harkin DP and Johnston PG (2003) 5-Fluorouracil: mechanisms of action and clinical strategies. *Nature Reviews Cancer* **3**, 330–8.

Maeda H and Dudareva N (2012) The shikimate pathway and aromatic amino acid biosynthesis in plants. *Annual Review of Plant Biology* **63**, 73–105.

Morris SM (2002) Regulation of enzymes of the urea cycle and arginine metabolism. *Annual Review of Nutrition* **22**, 87–105.

Oldroyd GED, Murray JD, Poole PS and Downie JA (2011) The rules of engagement in the legume–rhizobial symbiosis. *Annual Review of Genetics* **45**, 119–44.

Pollegioni L, Schonbrunn E and Siehl D (2011) Molecular basis of glyphosate resistance – different approaches through protein engineering. *FEBS Journal*, **278**, 2753–66.

Seefeldt LC, Hoffman BM and Dean DR (2009) Mechanism of Mo-dependent nitrogenase. *Annual Review of Biochemistry* **78**, 701–22.

Tanaka R and Tanaka A (2007) Tetrapyrrole biosynthesis in higher plants. *Annual Review of Plant Biology* **58**, 321–46.

Umbarger HE (1978) Amino acid biosynthesis and its regulation. *Annual Review of Biochemistry* **47**, 533–606.

van Spronsen FJ (2010) Phenylketonuria: a 21st century perspective. *Nature Reviews Endocrinology* **6**, 509–14.

Xu Y-F, Létisse F, Absalan F, *et al*. (2013) Nucleotide degradation and ribose salvage in yeast. *Molecular Systems Biology* **9**, 665.

Self-assessment questions

Multiple choice questions

Only one answer is correct for each question.
Answers can be found on the website:
www.scionpublishing.com/biochemistry

1. What is the name given to species able to fix nitrogen?
   (a) Legumes
   (b) Diazotrophs
   (c) Nitrifiers
   (d) Symbionts

2. Which one of the following groups of organisms do **not** include nitrogen fixers?
   (a) Mycobacteria
   (b) Rhizobia
   (c) Frankia
   (d) Cyanobacteria

3. Which of the following statements is **incorrect** with regard to nitrogen fixation?
   (a) Nitrogen-fixing bacteria detect the presence of flavonoids secreted by the roots of a suitable host plant
   (b) The bacteria enter the plant roots via a modified root hair
   (c) Infection induces plant cell division, forming a specialized structure called a root nodule
   (d) Within a nodule the bacteria die, releasing nitrogen-fixing enzymes

4. Which of these statements describes the nitrogenase enzyme?
   (a) Four subunits, two of type α and two β, with a single MoFe center
   (b) Four subunits, two of type α and two β, with a single FeS cluster
   (c) Two identical subunits with a single MoFe center
   (d) Two identical subunits with a single FeS cluster

5. The nitrogenase complex is protected from oxygen by what?
   (a) Leghemoglobin
   (b) Myoglobin
   (c) A heme prosthetic group contained in the complex
   (d) Catalase

6. What is the prosthetic group of nitrite reductase called?
   (a) Leghemoglobin
   (b) Heme
   (c) Phosphopantetheine
   (d) Siroheme

7. Which of the following is **not** a non-essential amino acid?
   (a) Alanine
   (b) Arginine
   (c) Histidine
   (d) Proline

8. Which of the following is an essential amino acid?
   (a) Cysteine
   (b) Glutamate
   (c) Glycine
   (d) Threonine

9. What is the initial product formed by reaction of ammonia with α-ketoglutarate?
   (a) Glutamate
   (b) Glutamine
   (c) Arginine
   (d) Ornithine

10. Which three amino acids are derived from 3-phosphoglycerate?
    (a) Aspartate, asparagine and alanine
    (b) Alanine, phenylalanine and tyrosine
    (c) Leucine, isoleucine and valine
    (d) Serine, glycine and cysteine

11. Which amino acid is obtained by transamination of pyruvate?
    (a) Alanine
    (b) Arginine
    (c) Valine
    (d) Threonine

12. Which of the following statements is **incorrect** regarding synthesis of phenylalanine, tryptophan and tyrosine in bacteria?
    (a) The starting point is condensation of phosphoenolpyruvate
    (b) The pathway branches at chorismate
    (c) Tyrosine is obtained by oxidation of phenylalanine
    (d) The side-chain of tryptophan is constructed from phosphoribosyl pyrophosphate

13. What are the different versions of DAHP synthase, which respond to allosteric control by different amino acids, called?
    (a) Multimeric proteins
    (b) Regulons
    (c) Isozymes
    (d) Isologs

14. Which of the following statements is **incorrect** regarding the salvage pathway for nucleotide synthesis?
    (a) The sugar–phosphate component of the new nucleotides is provided by phosphoribosyl pyrophosphate
    (b) AMP is converted to ADP by adenylate kinase
    (c) Deoxyribonucleotides are obtained from ribonucleotides by reduction by ribonucleotide reductase
    (d) GTP cannot be made by this pathway

15. In *de novo* nucleotide synthesis, which compound combines with aspartate to form cytosine and uracil?
    (a) Carbamoyl phosphate
    (b) Orotate
    (c) Glutamine
    (d) Phosphoribosyl pyrophosphate

16. Which of these enzymes is **not** involved in tetrapyrrole synthesis?
    (a) ALA synthase
    (b) ALA dehydratase
    (c) Porphobilinogen deaminase
    (d) Argininosuccinase

17. Adenines and guanines that are excess to requirements are converted to which compound?
    (a) Urea
    (b) Uric acid
    (c) Bile pigments
    (d) Phytochrome

18. Excess tetrapyrroles are converted into which compound?
    (a) Urea
    (b) Uric acid
    (c) Bile pigments
    (d) Phytochrome

19. What cofactor is possessed by the transaminase enzymes involved in amino acid degradation?
    (a) Siroheme
    (b) Pyridoxal phosphate
    (c) Phosphopantetheine
    (d) Heme

20. Which of the following is a glucogenic amino acid?
    (a) Leucine
    (b) Lysine
    (c) Methionine
    (d) All three are glucogenic

21. Which of the following is a ketogenic amino acid?
    (a) Arginine
    (b) Glutamine
    (c) Histidine
    (d) Leucine

22. What name is given to species, including many aquatic invertebrates, which excrete excess ammonia into the water in which they live?
    (a) Ammonotelic
    (b) Diazotrophic
    (c) Ureotelic
    (d) Uricotelic

23. Which of the following is **not** an intermediate in the urea cycle?
    (a) Citrulline
    (b) Arginine
    (c) Oxaloacetate
    (d) Ornithine

24. The link between the urea and TCA cycles is provided by what?
    (a) The aspartate–argininosuccinate shunt
    (b) The *N*-acetylglutamate shunt
    (c) Involvement of oxaloacetate in both cycles
    (d) Utilization of urea as an energy source

Short answer questions

These questions do not require additional reading.

1. Explain what is meant by 'nitrogen fixation' and describe the various organisms that are able to fix nitrogen.

2. Describe the mode of action of the bacterial nitrogenase complex.

3. Draw the reactions involved in conversion of ammonia into glutamine.

4. Explain the difference between a non-essential and essential amino acid, and describe the non-essential amino acids that are synthesized in humans.

5. Outline the pathways for synthesis of the essential amino acids in bacteria, and highlight the key features of the regulation of these pathways.

6. Distinguish between the salvage and *de novo* pathways for nucleotide synthesis.

7. Outline how heme is synthesized.

8. Describe what happens to the amino groups during breakdown of amino acids.

9. Distinguish between a glucogenic and ketogenic amino acid, and explain the basis to these two terms.

10. Write an essay on the urea cycle.

Self-study questions

These questions will require calculation, additional reading and/or internet research.

1. Hemoglobin and related compounds such as leghemoglobin are present in animals, plants and bacteria. The only group of organisms that do not possess hemoglobin are the archaea. Based on this information, what can you infer about the early evolution of hemoglobin and how can the origin of hemoglobin be related to changes in the atmospheric content of the planet?

2. Lesch–Nyhan syndrome is a rare but serious inherited disorder associated with mental impairment and behavioral disturbances. Lesch–Nyhan patients have elevated amounts of uric acid and phosphoribosyl pyrophosphate in their blood and urine. What is the most likely biochemical basis to the syndrome?

3. Explain why the Atkins and other low-carbohydrate diets are referred to as 'ketogenic'.

4. The link between the TCA and urea cycles is sometimes referred to as the 'Krebs bicycle'. This is because the urea cycle was first described in 1932 by Hans Krebs and Kurt Henseleit. Five years later, Krebs discovered the TCA cycle, which is frequently called the 'Krebs cycle'. Draw a diagram illustrating the Krebs bicycle and explain why, in reality, it has three wheels rather than two.

5. Annotate these three biochemical pathways to show the points at which you would expect the various products to exert feedback regulation over their synthesis. Indicate any steps that you predict will be catalyzed by a group of isozymes.

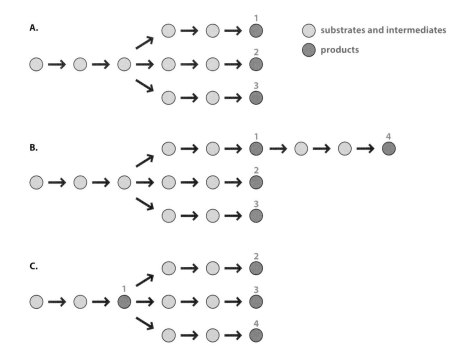

CHAPTER 14

DNA replication and repair

STUDY GOALS

After reading this chapter you will:

- understand the central importance of complementary base pairing in replication of a double helix into two exact copies

- be able to describe how the prepriming complex is assembled at the *E. coli* origin of replication

- know the distinctive features of replication origins in yeast and humans

- understand the key features of replication processes that do not involve replication forks

- be able to describe the role of DNA topoisomerases in DNA replication

- be able to distinguish between different types of DNA polymerase and describe the mode of action of a DNA-dependent DNA polymerase

- understand the implications that 5′→3′ DNA synthesis has for replication of the lagging strand of a DNA molecule

- appreciate that DNA polymerases require a primer and know how this priming problem is solved in *E. coli* and humans

- be able to describe the events occurring at the replication fork in *E. coli* and humans, and in particular know how the Okazaki fragments are joined together

- know why the two replisomes replicating an *E. coli* genome meet within a defined region of this DNA molecule

- be able to describe how telomerase prevents the shortening that can occur when linear DNA molecules are replicated

- understand why DNA repair is important

- be able to describe how the mismatch repair system corrects replication errors, and in particular know how the daughter strand is distinguished

- know the main features of the base and nucleotide excision repair pathways

- be able to describe how single- and double-stranded DNA breaks are repaired

- be able to give an example of a post-replicative repair process

The preceding chapters have revealed the crucial roles that enzymes play in catalyzing the metabolic reactions occurring in a cell. The identities and activities of those enzymes determine which metabolic reactions can take place and how those reactions respond to regulatory signals. The crucial feature underlying the catalytic properties of an enzyme and its response to regulatory signals is its amino acid sequence. The amino acid sequence specifies the three-dimensional structure of the enzyme, including the precise arrangement of chemical groups at its active site and at binding sites for regulatory molecules. The catalytic activity and regulatory responses of an enzyme are therefore determined by its amino acid sequence. The same general principle is also true for proteins that are not enzymes. Their biological activity, whether this is a structural, motor, transport, storage, protective or regulatory function, is specified by the amino acid sequence of the protein.

The mechanism by which the cell makes proteins with specific amino acid sequences is therefore of central importance in biochemistry. This is a question that involves not just proteins but also nucleic acids, because the information needed to synthesize individual proteins is carried by the cell's DNA molecules. Utilization of this information involves a two-stage process called **gene expression** (*Fig. 14.1*). During the first stage of gene expression, the nucleotide sequence of a gene is copied into an RNA molecule, and in the second stage the nucleotide sequence of this RNA is used to direct the order in which amino acids are joined together to make a protein.

We will study the synthesis of RNA and of proteins in *Chapters 15* and *16*, respectively. Before doing so we must focus our attention on the DNA molecules that contain, within their nucleotide sequences, the information for making these RNAs and proteins. We must examine how DNA molecules are replicated, so a cell can pass copies to its daughter cells when it divides, and how a DNA molecule is repaired when it becomes chemically damaged.

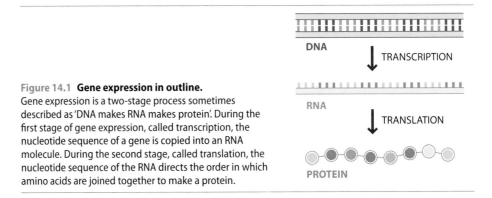

Figure 14.1 Gene expression in outline.
Gene expression is a two-stage process sometimes described as 'DNA makes RNA makes protein'. During the first stage of gene expression, called transcription, the nucleotide sequence of a gene is copied into an RNA molecule. During the second stage, called translation, the nucleotide sequence of the RNA directs the order in which amino acids are joined together to make a protein.

14.1 DNA replication

One of the reasons why the discovery of the double helix structure of DNA is looked upon as one of the major breakthroughs in biology is because this structure immediately indicates how a DNA molecule can be replicated into two identical copies. In their paper describing the double helix, published in *Nature* on 25 April 1953, Watson and Crick wrote:

> "It has not escaped our notice that the specific pairing we have postulated immediately suggests a possible copying mechanism for the genetic material."

This statement is looked on as one of the most understated comments in the biological literature, bearing in mind that the way in which genes form copies of themselves was, at that time, one of the great mysteries of life.

The key to the replication of DNA is the pairing that occurs between bases that are adjacent to one another in the two polynucleotides of the helix. The rules that guide this base pairing are that adenine always pairs with thymine, and cytosine always pairs with guanine. Because of the base pairing, the sequences of the two polynucleotides in a double helix are complementary, with the sequence of one polynucleotide reflecting the sequence of the other. This in turn means that separation of the two polynucleotides, followed by DNA synthesis using the separated polynucleotides as templates, results in two exact copies of the original parent double helix (*Fig. 14.2*).

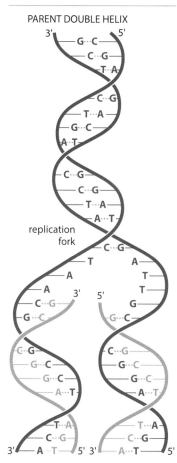

PARENT DOUBLE HELIX

replication
fork

TWO DAUGHTER DOUBLE HELICES

Figure 14.2 DNA replication, as predicted by Watson and Crick.
The polynucleotides of the parent double helix, shown in dark red, act as templates for synthesis of new strands of DNA, shown in pink. The sequences of these new strands are set by the base pairing rules. The result is two exact copies of the original parent double helix. Note that in both the parent and two daughter double helices the polynucleotides are antiparallel, meaning that they run in different directions, one being orientated in the 5′→3′ direction, and the other in the 3′→5′ direction (see *Section 4.1.2*).

For the structures of the base pairs, and other important features of the DNA double helix, see *Section 4.1.2.*

DNA replication is therefore an elegantly straightforward process, at least in outline. The detailed mechanism is, of course, more complex. We will deal with the topic in a logical fashion, dividing the process into three stages:

- The **initiation phase**, during which the machinery for replication is assembled at those points within a DNA molecule at which replication will begin.

- The **elongation phase**, during which the new polynucleotides are synthesized.

- The **termination phase**, which completes the process.

14.1.1 Initiation of DNA replication

When a DNA molecule is being replicated, only a limited region is ever in a non-base-paired form. The breakage in base pairing starts at a distinct position, called the **origin of replication**. Replication origins are best understood in bacteria, so we will begin with *Escherichia coli*.

The E. coli *DNA molecule has a single origin of replication*

The *E. coli* genome comprises a circular DNA molecule of 4.64 Mb. This DNA molecule has a single origin of replication, spanning approximately 245 bp. The DNA in this region includes four copies of a short nine-nucleotide sequence, each of which acts as the binding site for a **DnaA protein**. Once all four copies of DnaA are bound, they recruit other DnaA proteins to the origin, forming a barrel of about 30 proteins in total (*Fig. 14.3*). The DNA is wound around the barrel in such a way that torsional stress is introduced into this region of the double helix. The result is that the hydrogen bonding between the two polynucleotides is disrupted and a short non-base-paired segment opens up. The breakage begins in a part of the replication origin that is **AT rich**, meaning that there is a high proportion of adenine–thymine base pairs. Remember that each A–T base pair is held together by just two hydrogen bonds, whereas a G–C pair has three hydrogen bonds. AT-rich regions of a double helix are therefore less stable than GC-rich ones, or ones with equal amounts of the two types of base pair, and are therefore easier to disrupt by torsional stress, as applied by the barrel of DnaA proteins.

Opening up of the helix initiates a series of events that result in the construction of a pair of **replication forks** within the origin of replication. A replication fork is a point at which the double helix separates into two unpaired strands, and is the position where synthesis of new polynucleotides will take place (*Fig. 14.4*). The first step in formation of the replication forks is the attachment of a **prepriming complex** at each of these two positions. Each prepriming complex initially comprises six copies of the DnaB protein and six copies of DnaC, but the latter has a transitory role and is released as soon as the complex is formed. Its function is probably simply to aid the attachment of DnaB.

DnaB is a **helicase**, an enzyme that can break base pairs. DnaB can therefore increase the size of the open region by breaking base pairs so that the replication forks move further apart. The non-base-paired polynucleotides become coated with **single strand binding proteins** (**SSBs**), which appear to play a dual role. They prevent the polynucleotides from immediately base pairing with one another again, and they protect the polynucleotides from attack by nucleases present in *E. coli* cells which naturally degrade single-stranded DNA.

Now, the enzymes involved in the elongation phase of DNA replication are able to attach. The replication forks begin to move away from the origin and DNA copying begins.

Figure 14.3 Opening up of the DNA helix at the *E. coli* origin of replication.

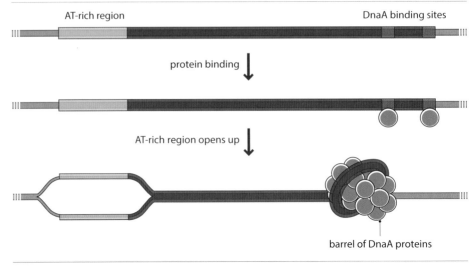

barrel of DnaA proteins

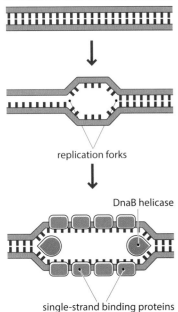

replication forks

DnaB helicase

single-strand binding proteins

Figure 14.4 Completion of the initiation phase of DNA replication in *E. coli*.
Opening of the helix generates two replication forks, the positions at which new DNA synthesis will take place. Single-strand binding proteins attach to the non-base-paired polynucleotides, and the DnaB helicase begins to extend the open region by breaking base pairs so that the replication forks move further apart.

The mitochondrial genome was described in *Section 2.1.3*.

Eukaryotic DNA has multiple replication origins

Most eukaryotic DNA molecules are linear rather than circular. These linear molecules are not replicated by a simple process beginning at one end and proceeding to the other. Instead they have multiple internal origins of replication. The frequency with which these occur depends on the species. The yeast *Saccharomyces cerevisiae*, for example, has about 400 origins, which means that each fork replicates approximately 15 kb of DNA. Human DNA has about 20 000 origins, each copying about 80 kb.

Each of the 400 yeast origins comprises a similar 200 bp DNA sequence that includes the binding sites for a set of six proteins that together make up the **origin recognition complex**. These proteins are not directly equivalent to the bacterial DnaA versions, because they are permanently bound to the DNA, remaining attached to yeast origins even when replication is not being initiated. They are thought to mediate the attachment of additional proteins which are needed to construct the replication forks that emerge from each origin.

The replication origins within the human genome are less easy to identify, but we believe that there are about 20 000 of these altogether. In humans and other mammals, these origins are called **initiation regions**, the name indicating that they are not particularly well defined. In fact, some researchers believe that replication is initiated by protein structures that have specific positions in a mammalian nucleus, with the initiation regions simply being those DNA segments located close to these protein structures in the three-dimensional organization of the nucleus.

Some circular molecules are replicated without replication forks

DNA copying via a pair of replication forks is the predominant system in living cells, being the mechanism by which both the nuclear DNA molecules of eukaryotes and the nucleoid DNAs of prokaryotes are replicated. There are, however, two alternative processes, used by important types of circular DNA molecule.

The first of these unusual modes of replication is called **strand displacement replication**. This mechanism is thought to be used by some types of **plasmid** (small circular DNA molecules often present in bacterial cells) and for many years was also believed to be the mode of replication for the human mitochondrial genome. In these DNA molecules, the point at which replication begins is marked by a **D loop**, a short region where the double helix is disrupted by an RNA molecule that is base-paired to one of the DNA strands. When replication begins, this RNA molecule is extended by adding DNA nucleotides to its 3′ end, the process continuing until one strand of the

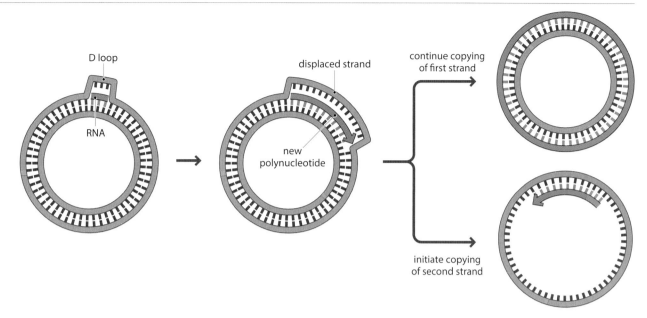

Figure 14.5 Strand displacement replication.
The D loop contains a short RNA molecule (shown in blue) that acts as the starting point for synthesis of the first new polynucleotide (shown in orange). This polynucleotide is extended until one strand of the circular molecule has been completely copied. The result is the double-stranded molecule at the top right. The second strand therefore becomes displaced, and is copied by attachment of a second RNA molecule which acts as the starting point for synthesis of the second new polynucleotide (bottom right).

circular molecule has been completely copied (*Fig. 14.5*). The other strand therefore becomes displaced, and is copied by attachment of a second RNA molecule which acts as the starting point for synthesis of the second new polynucleotide.

It is not obvious what advantage displacement replication has compared with the standard replication fork system. In contrast, the second type of unusual replication process, called **rolling circle replication**, has benefits when the objective is to rapidly synthesize multiple copies of a circular DNA molecule. Rolling circle replication initiates at a nick which is made in one of the parent polynucleotides. The free 3′ end that results is extended, displacing the 5′ end of the polynucleotide. Continued DNA synthesis 'rolls off' a complete copy of the molecule (*Fig. 14.6*), and further synthesis eventually results in a series of identical molecules linked head to tail. These molecules are single stranded and linear, but are converted to double-stranded circular molecules by complementary strand synthesis, followed by cleavage at the junction points and circularization of the resulting segments. Rolling circle replication is used by various types of virus, the best example being the bacterial virus or **bacteriophage** called **lambda**. The rapid generation of new DNA molecules that is possible by this method helps the virus to make multiple copies of itself very rapidly after infecting a host cell.

14.1.2 The elongation phase of DNA replication

Now we will examine the events that occur at the replication forks and which result in the synthesis of new polynucleotides. The first issue that we must address is the apparent inability of the replication forks to progress more than a short distance along a DNA molecule – this is known as the **topological problem**.

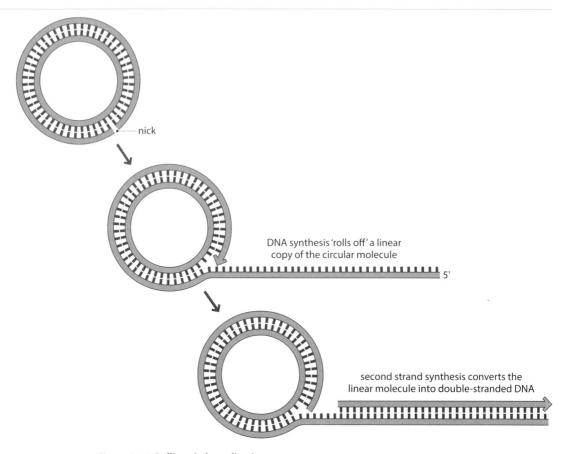

nick

DNA synthesis 'rolls off' a linear
copy of the circular molecule

5'

second strand synthesis converts the
linear molecule into double-stranded DNA

Figure 14.6 Rolling circle replication.
Synthesis of a single linear copy of the circular molecule is shown. In reality, a series of linear copies, linked
head to tail, can be rolled off the circular molecule.

The topological problem is solved by DNA topoisomerases

The topological problem in DNA replication arises because the double helix is a
helix, which means that the two polynucleotides are wound around one another and
they cannot be separated simply by pulling them apart. If, however, the helix has to
be unwound, then a considerable amount of rotation of the molecule is required.
There is one turn of the B-form of the double helix for every 10 bp, which means that
complete replication of the DNA molecule in human chromosome 1, which contains
250 million base pairs, would require 25 million rotations. The unwinding of a linear
DNA molecule is not physically impossible, but a circular double-stranded molecule,
having no free ends, cannot rotate at all. However, circular molecules are able to
replicate. So how is this topological problem solved?

It is solved by a group of enzymes called **DNA topoisomerases**. These enzymes
enable the two polynucleotides in a double helix to be separated without rotating the
helix. There are two types of DNA topoisomerase, which achieve this feat in slightly
different ways:

- **Type I DNA topoisomerases** introduce a break in one polynucleotide and pass the
  second polynucleotide through the gap that is formed. The two ends of the broken
  strand are then rejoined (*Fig. 14.7*).

- **Type II DNA topoisomerases** break both strands of the double helix, creating
  a gap through which a second segment of the helix is passed. Once the
  second segment has passed through, the gap is sealed by rejoining the two
  polynucleotides (*Fig. 14.8*).

Figure 14.7 **The mode of action of a Type I topoisomerase.**

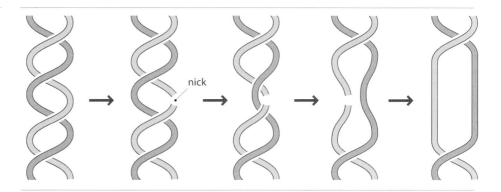

Despite their different mechanisms, both Type I and Type II topoisomerases achieve the same end result. They enable the two polynucleotides to be separated without the helix having to rotate (*Fig. 14.9*). But the repeated breakage and rejoining reactions needed to completely separate the polynucleotides in a long DNA molecule raise their own problems. How do these enzymes ensure that the two members of a pair of cut ends remain close to one another so that they can be rejoined? Part of the answer is that one end of each cut polynucleotide becomes covalently attached to a tyrosine amino acid at the active site of the enzyme, ensuring that this end of the polynucleotide is held tightly in place while the free ends are being manipulated.

Figure 14.8 **The mode of action of a Type II topoisomerase.**

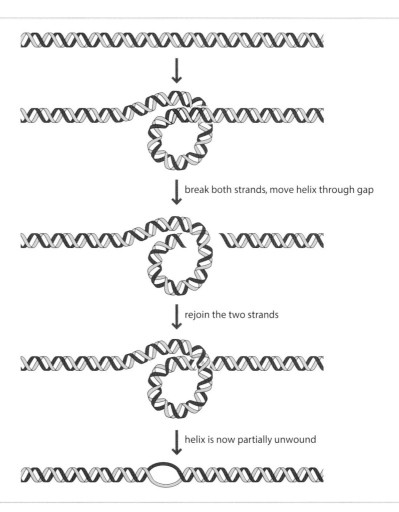

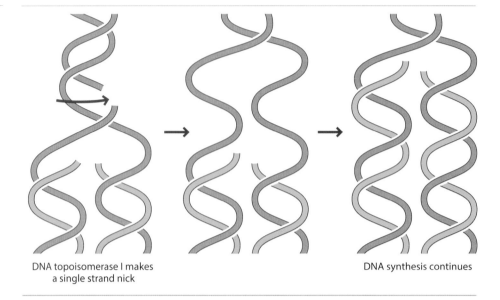

DNA topoisomerase I makes a single strand nick

DNA synthesis continues

Type I and II topoisomerases are subdivided according to the precise chemical structure of the polynucleotide–tyrosine linkage. With IA and IIA enzymes, the link involves a phosphate group attached to the free 5′ end of the cut polynucleotide, and with IB and IIB enzymes the linkage is via a phosphate group at the 3′ end. Both types are present in eukaryotes, but IB and IIB enzymes are very uncommon in bacteria. Most topoisomerases are only able to unwind DNA, but prokaryotic Type IIA enzymes, such as the bacterial DNA gyrase and the archaeal reverse gyrase, can carry out the reverse reaction and introduce extra turns into DNA molecules.

DNA polymerases carry out template-dependent DNA synthesis

The central role in DNA replication is performed by the enzymes which synthesize new polynucleotides. An enzyme that makes a new DNA polynucleotide using an

---

Box 14.1 **Supercoiled DNA**

As well as their role in DNA replication, topoisomerases are also responsible for generating **supercoiled DNA**. Supercoiling occurs when additional turns are introduced into a circular DNA molecule (positive supercoiling) or if turns are removed (negative supercoiling). The torsional stress that results from over- or under-winding the double helix causes a circular molecule to wind around itself to form a more compact supercoiled structure.

Supercoiling enables a circular DNA molecule to be packaged into a small space. For example, the circular *E. coli* DNA molecule has a circumference of approximately 1.6 mm, but after being supercoiled can be packaged into a cell that has dimensions of just 1 μm × 2 μm (see *Box 4.5*).

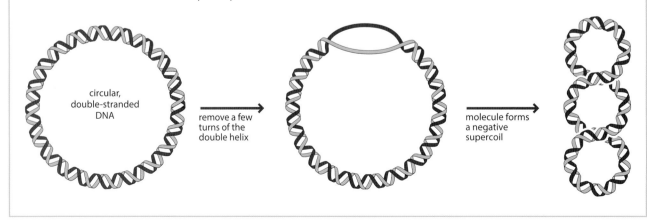

circular, double-stranded DNA

remove a few turns of the double helix

molecule forms a negative supercoil

existing DNA strand as a template (template-dependent DNA synthesis) is called a **DNA-dependent DNA polymerase**. We can abbreviate the name simply to 'DNA polymerase', so long as we recognize that there are also **RNA-dependent DNA polymerases**, which make DNA copies of RNA molecules, and which are particularly important during the replication of certain types of virus.

The reaction catalyzed by a DNA polymerase is shown in detail in *Figure 14.10*. During each nucleotide addition, the α- and β-phosphates are removed from the incoming nucleotide, and the hydroxyl group is removed from the 3'-carbon of the nucleotide at the 3' end of the growing polynucleotide. This results in the loss of a

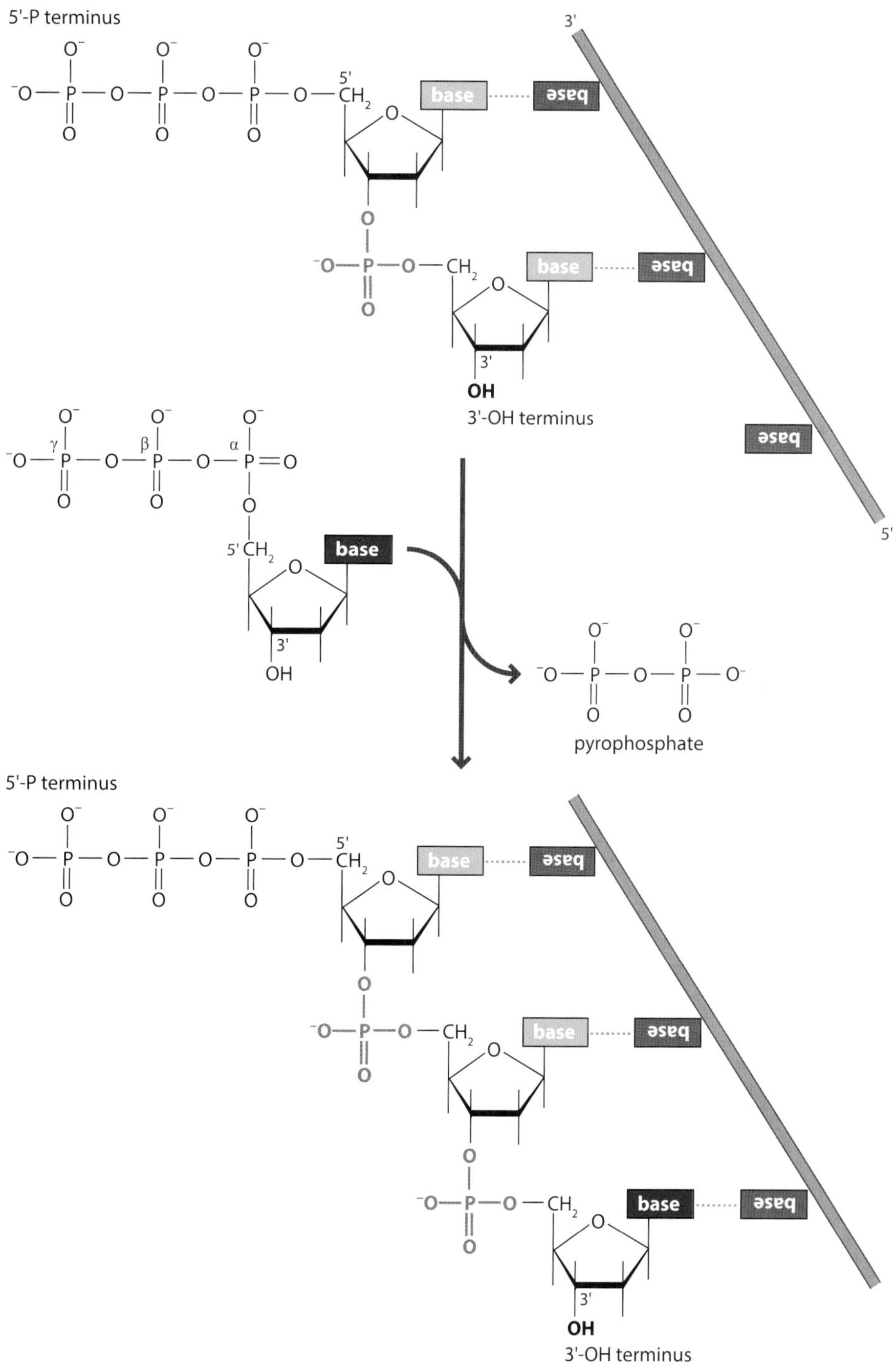

Figure 14.10 The reaction catalyzed by a DNA-dependent DNA polymerase.

pyrophosphate molecule for each phosphodiester bond that is formed. The chemical reaction is guided by the presence of the DNA template, which directs the order in which the individual nucleotides are polymerized, A base pairing with T, and G with C. The new polynucleotide is therefore built up, step-by-step, in the 5'→3' direction, new nucleotides being added to the free 3' end of the growing strand. Remember that in order to base pair, complementary polynucleotides must be antiparallel. This means that the template strand must be read in the 3'→5' direction.

Bacteria such as *E. coli* have five DNA polymerases, called I, II, III, IV and V. **DNA polymerase III** is the enzyme responsible for most of the template-dependent polynucleotide synthesis during DNA replication, with **DNA polymerase I** performing a less extensive, but still vital, function, as we will see below. The other three DNA polymerases are used to repair damaged DNA. DNA polymerase I is just a single polypeptide but DNA polymerase III is multisubunit, with a molecular mass of approximately 900 kDa. The subunit called α is the one that is responsible for synthesizing the new polynucleotide, with the other subunits playing ancillary roles in the replication process. For example, the ε subunit specifies a 3'→5' exonuclease activity. This means that DNA polymerase III, as well as synthesizing DNA in the 5'→3' direction, can degrade DNA in the opposite direction. This is called a **proofreading** function, because it enables the polymerase to correct errors by removing nucleotides that have been inserted incorrectly (*Fig. 14.11*). Another important subunit in DNA polymerase III is β. This one acts as a 'sliding clamp' holding the polymerase complex tightly to the template strand, but at the same time allowing it to move along that strand as it makes the new polynucleotide.

Eukaryotes have at least 15 DNA polymerases, which in mammals are given Greek letters. The main replicating enzymes are **DNA polymerase δ** and **DNA polymerase ε**, which work in conjunction with an accessory protein called the **proliferating cell**

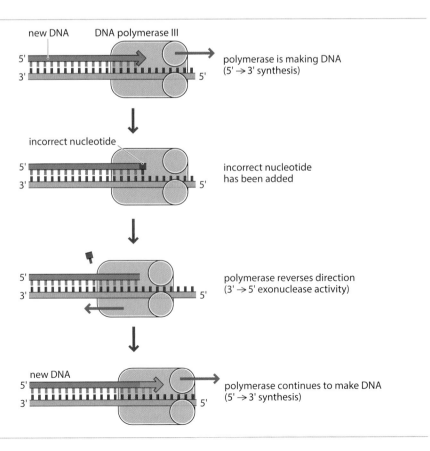

Figure 14.11 Proofreading during DNA replication.
The 3'→5' exonuclease activity enables DNA polymerase III to reverse its direction and remove an incorrect nucleotide that it has added during 5'→3' DNA synthesis.

new DNA   DNA polymerase III

polymerase is making DNA (5'→3' synthesis)

incorrect nucleotide

incorrect nucleotide has been added

polymerase reverses direction (3'→5' exonuclease activity)

new DNA

polymerase continues to make DNA (5'→3' synthesis)

DNA polymerases

A DNA polymerase is an enzyme that can polymerize nucleotides to make a DNA molecule. Most DNA polymerases are DNA-dependent, which means that they use an existing DNA polynucleotide to direct the order in which the nucleotides are polymerized. A few are RNA-dependent and so make DNA copies of RNA molecules, and one type of DNA polymerase, called **terminal deoxynucleotidyl transferase**, is template-independent, adding nucleotides randomly to the ends of existing DNA molecules.

DNA polymerases form a diverse group of enzymes. They can be divided into seven families, the members of a family having structural similarities suggesting a common evolutionary origin. The families are as follows:

- **Family A** includes DNA polymerase I of bacteria and the eukaryotic DNA polymerases γ, which replicates the mitochondrial DNA, and θ, which helps repair double-strand breaks in chromosomal DNA molecules.
- **Family B** includes DNA polymerases α, δ and ε, which are the main enzymes involved in DNA replication in eukaryotes. Other family members are bacterial DNA polymerase II and eukaryotic DNA polymerase ζ whose main roles are in replication of damaged DNA.
- **Family C** has a single member, DNA polymerase III, the main replicating enzyme in bacteria.
- **Family D** polymerases are responsible for DNA replication in some archaea species.
- **Family X** includes DNA polymerases β, σ, λ and μ, which are eukaryotic DNA repair enzymes. DNA polymerase β is involved in base excision repair (see *Fig. 14.26*) and λ and μ repair double-strand DNA breaks. The function of DNA polymerase σ is not understood. Family X also includes terminal deoxynucleotidyl transferase, the template-independent DNA polymerase.

- **Family Y** comprises low-fidelity DNA polymerases capable of replicating damaged segments of DNA, though in an error-prone manner. The members are bacterial DNA polymerases IV and V, and eukaryotic polymerases η, ι and κ. Error-prone replication results in mutations, because the strand being synthesized might not be a direct complement of the original sequence of the template. However, a few mutations can be tolerated if this is the last–best chance of replicating a highly damaged DNA molecule, and the only alternative is the death of the cell.

- **Family RT** are the **reverse transcriptases**, the RNA-dependent DNA polymerases. Many of these enzymes are involved in replication of **retroviruses** such as the human immunodeficiency viruses. A retrovirus contains an RNA version of the viral genome, which is copied to DNA by reverse transcriptase after the virus has infected its host cell. The DNA copy of the genome integrates into one of the host cell chromosomes where it can remain for several cell divisions, being passed to the daughter cells when the chromosomal DNA is replicated. The retrovirus completes its infection cycle by excision of its DNA genome from the chromosome, followed by copying into RNA which is packaged into virus particles.

nuclear antigen (**PCNA**). PCNA is the functional equivalent of the β subunit of *E. coli* DNA polymerase III and holds the enzyme tightly to the DNA that is being copied. **DNA polymerase** α also has an important function in DNA replication, and **DNA polymerase** γ is responsible for replicating the DNA molecules that are present in mitochondria. As with the prokaryotic enzymes, most of the other eukaryotic DNA polymerases are involved in the repair of damaged DNA.

DNA polymerases have limitations that complicate DNA replication

Although DNA polymerases have evolved to replicate DNA molecules, these enzymes have two limitations that complicate the way in which replication occurs in the cell. The first is that all DNA-dependent DNA polymerases synthesize DNA exclusively in the 5'→3' direction. Enzymes able to make DNA in the opposite direction, 3'→5', have never been discovered, and probably do not exist. This means that during polynucleotide synthesis, the template strand is read in the 3'→5' direction, with the DNA polymerase synthesizing its complementary strand in the 5'→3' direction. A problem therefore arises because the two polynucleotides in a double helix are antiparallel, one running in one direction and one in the other. One of the strands of the parent double helix can therefore be synthesized continuously as the replication fork progresses along the molecule (*Fig. 14.12*). This is called the **leading strand**. In contrast, the second strand of the parent molecule, which we call the **lagging strand**, cannot be copied in a continuous fashion. To do so would require polynucleotide synthesis in the 3'→5' direction. To get around this problem it is necessary for the lagging strand to be replicated in sections. Each section covers just that part of the lagging strand that is exposed at the replication fork. When the replication fork progresses further into the parent double helix, a second section of the lagging strand

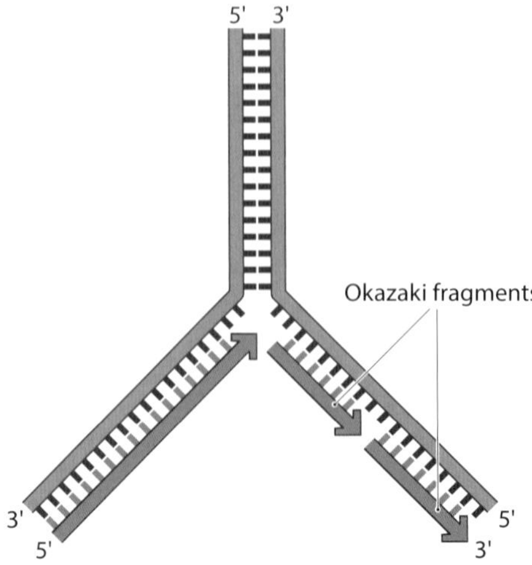

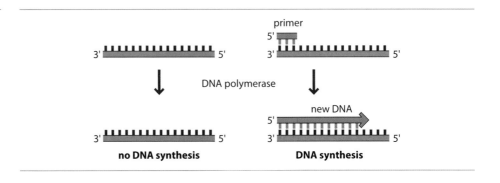

Figure 14.12 The distinction between the leading and lagging strands.

is exposed and replicated, and so on. The result is that, initially at least, the copy that is made of the lagging strand exists as a series of disconnected segments, called **Okazaki fragments**, which are named after their discoverers, Reiji and Tsuneko Okazaki. We will see later how these are joined together to make a continuous polynucleotide.

The second difficulty that arises during replication is that a DNA polymerase cannot begin to synthesize a polynucleotide unless there is already a short double-stranded region to act as a **primer** (*Fig. 14.13*). How is this primer made? The answer is that it is synthesized by an **RNA polymerase**. This is an enzyme that carries out template-dependent RNA synthesis and, unlike a DNA polymerase, it can begin to make its RNA copy on a 'naked' template, rather than only being able to extend an existing polynucleotide. In bacteria, this RNA polymerase is called **primase**, and it makes a primer that is between 4 and 15 nucleotides in length (*Fig. 14.14A*). Once the primer has been completed, polynucleotide synthesis is continued by DNA polymerase III. In eukaryotes the primase is a part of the DNA polymerase α enzyme. The primase makes an RNA primer of 8–12 nucleotides, and DNA polymerase α then extends the primer by the addition of about 20 nucleotides of DNA. This DNA stretch often has a few ribonucleotides mixed in. After completion of the RNA–DNA primer, DNA synthesis is continued by DNA polymerase ε on the leading strand, and DNA polymerase δ on the lagging strand (*Fig. 14.14B*).

Priming occurs just once on the leading strand because, once primed, the leading-strand copy can be synthesized continuously until replication of the parent molecule

Figure 14.13 A primer is needed to initiate DNA synthesis by a DNA polymerase.

Figure 14.14 Priming of DNA synthesis in (A) bacteria, and (B) eukaryotes.

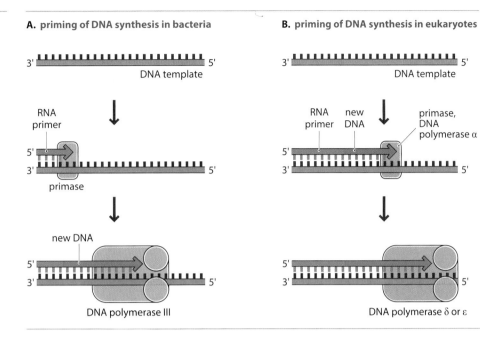

A. priming of DNA synthesis in bacteria

DNA template

RNA primer

primase

new DNA

DNA polymerase III

B. priming of DNA synthesis in eukaryotes

DNA template

RNA primer   new DNA   primase, DNA polymerase α

DNA polymerase δ or ε

is completed, or until a replication fork traveling in the other direction, from a different origin, is met. In contrast, on the lagging strand, priming must occur every time a new Okazaki fragment is initiated. The Okazaki fragments in *E. coli* are 1000–2000 nucleotides in length, so approximately 4000 priming events are needed every time the cell's DNA is replicated. In eukaryotes the Okazaki fragments are much shorter, perhaps less than 200 nucleotides in length, and priming has to occur even more frequently.

The events occurring at the replication fork

Now that we have examined what DNA polymerases do, and what they cannot do, we can turn our attention to the events occurring at the replication fork. These are the events that result in the actual replication of a double-stranded DNA molecule. We begin by studying these events in bacteria, and then we will review the special features of eukaryotic replication.

Box 14.3   **Why does a DNA polymerase require a primer?**

The need for a primer complicates the replication process because it means that a DNA polymerase cannot initiate DNA synthesis on an entirely single-stranded template. RNA polymerases are able to place the first nucleotide of the new strand onto a 'naked' template, so it is not mechanistically impossible for an enzyme to perform this function. Why therefore have DNA polymerases not evolved this ability?

One suggestion is that the requirement for the primer relates to the proofreading function conferred by the 3'→5' exonuclease activity of a DNA polymerase. Proofreading increases the accuracy of replication by enabling a nucleotide that has been inserted incorrectly at the 3' end of a growing DNA strand to be removed before the strand is extended any further. DNA synthesis can therefore be looked on as a step-by-step process,

with the polymerase able to perform either of two reactions after each nucleotide addition:
- If the 3' nucleotide is not base-paired to the template (i.e. an error has been made), then the polymerase uses its 3'→5' exonuclease activity to remove that nucleotide.
- If the 3' nucleotide is base-paired to the template (i.e. it is the correct nucleotide), then the enzyme uses its polymerase activity to add the next nucleotide.

This means that, because of the proofreading function, a DNA polymerase can only carry out strand synthesis if the 3' nucleotide is base-paired to the template. If the template is entirely single stranded then, by definition, there is no base-paired 3' nucleotide. This might be the reason why a 'naked' template cannot be copied by a DNA polymerase.

The prepriming complex that is assembled at the *E. coli* origin of replication is converted to a **primosome** by addition of the primase, which immediately synthesizes the primer for the leading strand. DNA polymerase III then begins to make the leading strand copy by extending that primer by 1000–2000 nucleotides. The primase now makes an RNA primer on the lagging strand, adjacent to the replication fork, and a second copy of DNA polymerase III synthesizes the first Okazaki fragment. This means that there are two DNA polymerase III molecules attached to the parent DNA molecule, one copying its leading strand and one copying the lagging strand.

The combination of these two DNA polymerases, along with the primase, is called the **replisome**. The two DNA polymerases probably face the same direction, which means that the lagging strand has to form a loop in order for DNA synthesis on the two strands to proceed in parallel as the replisome moves along the DNA molecule (*Fig. 14.15*). As the replisome moves along the molecule, it is preceded by DNA topoisomerases that separate the strands of the parent DNA, and DnaB helicase enzymes that break the base pairs. After the replisome has progressed another 1000–2000 base pairs, a second Okazaki fragment is primed. And so the process continues until the entire *E. coli* DNA molecule has been copied.

We are left with one last challenge. After the replisome has passed, adjacent Okazaki fragments must be joined to one another. This is not straightforward because one member of each pair of adjacent Okazaki fragments still has its RNA primer attached at the point where the joining should take place. The primer is at the 5′ end of each Okazaki fragment, and so could be removed by an enzyme possessing a 5′→3′ exonuclease activity. Some DNA polymerases possess such an activity, but DNA polymerase III is not one of them. DNA polymerase III therefore continues making DNA until it reaches the 5′ end of the next Okazaki fragment in the chain (*Fig. 14.16A*). DNA polymerase III then detaches from the lagging strand, its place being taken by DNA polymerase I, which does have a 5′→3′ exonuclease activity. DNA polymerase I uses this activity to remove the primer of the Okazaki fragment that has just been reached, extending the 3′ end of the adjacent Okazaki fragment into the region of the lagging strand that is thus exposed. The two Okazaki fragments now abut, with the terminal regions of both composed entirely of DNA. All that remains is for the missing phosphodiester bond to be put in place by a **DNA ligase**, linking the two fragments and completing replication of this region of the lagging strand.

In eukaryotes, the enzyme replicating the lagging strand is DNA polymerase δ and, like DNA polymerase III, this enzyme lacks a 5′→3′ exonuclease activity. Unfortunately, there is no eukaryotic DNA polymerase that has this activity, so a different way of

Figure 14.15 Parallel replication of the leading and lagging strands. There are two copies of DNA polymerase III, one for each strand. It is thought that the lagging strand loops through its copy of the polymerase, so the leading and lagging strands are replicated in parallel as the two polymerase enzymes move in the same direction along the parent DNA molecule.

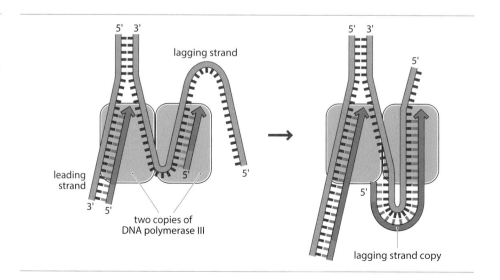

Figure 14.16 Joining of Okazaki fragments in (A) bacteria, and (B) eukaryotes.

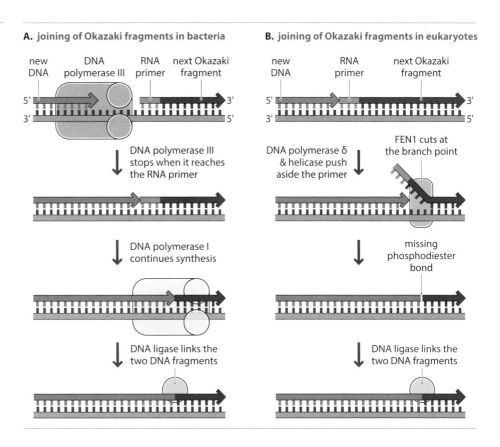

removing the RNA primers of the Okazaki fragments is needed. The enzyme that does this job is the 'flap endonuclease', **FEN1**, which associates with DNA polymerase δ as it approaches the RNA primer of an Okazaki fragment. The base pairs holding the primer to the lagging strand are broken by a helicase enzyme, enabling the primer to be pushed aside by DNA polymerase δ as it extends the adjacent Okazaki fragment into the exposed region (*Fig. 14.16B*). The resulting flap is then cut off by FEN1, whose endonuclease activity enables it to cut the phosphodiester bond at the branch point at the base of the flap. Once again, a DNA ligase joins the adjacent Okazaki fragments.

14.1.3 Termination of replication

Replication terminates when the two strands of the parent DNA molecule have been completely copied. The DNA polymerases and other proteins involved in the replication process detach and the daughter DNA molecules are released. Termination is therefore an uncomplicated event, but there are still two aspects of this stage of replication that we should explore. The first of these is the elegant mechanism used by a bacterium to ensure that the two replisomes traveling in different directions around its circular DNA molecule meet at the appropriate place.

Tus proteins capture replication forks

The circular DNA molecules present in bacteria are copied bidirectionally from a single origin of replication (*Fig. 14.17A*). If the two replisomes travel around the molecule at the same speed then they should meet at a position diametrically opposite the origin of replication. The progress of one or both replisomes might, however, be impeded by other activities that are taking place on the DNA molecule, such as the presence of RNA polymerase enzymes that are copying genes into RNA at the start of the gene expression process. DNA synthesis occurs at approximately five times the

Figure 14.17 **Termination of DNA replication in bacteria.**
(A) A bacterial DNA molecule is copied bidirectionally from a single origin of replication. (B) The terminator sequences ensure that both replication forks meet in a small region of the DNA molecule. The arrows indicate the direction in which each terminator sequence can be passed by a replication fork.

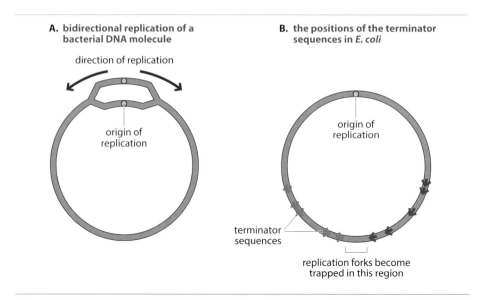

rate of RNA synthesis, so a replisome can easily overtake an RNA polymerase, but this probably does not happen. Instead, it is thought that the replisome pauses behind the RNA polymerase, proceeding only after the RNA molecule has been completed and the RNA polymerase has detached.

If one replisome is delayed because it encounters multiple positions where RNA synthesis is occurring, then it might be possible for the other replisome to overshoot the halfway point and continue replication on the other side of the DNA molecule. It is not obvious why this should be undesirable, but it does not happen. Instead, the replisomes become trapped within a region bounded by **terminator sequences**. There are ten of these in the *E. coli* DNA molecule (*Fig. 14.17B*).

Each terminator sequence acts as the binding site for a **Tus (terminator utilization substance) protein**. When approaching from one direction, the replisome is able to bypass the Tus protein and continue its progress along the DNA (*Fig. 14.18*). Coming

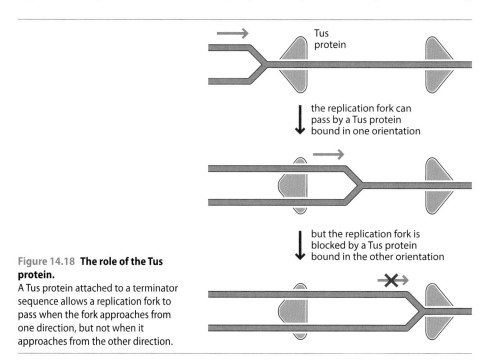

Figure 14.18 **The role of the Tus protein.**
A Tus protein attached to a terminator sequence allows a replication fork to pass when the fork approaches from one direction, but not when it approaches from the other direction.

from the other direction, the progress of the replisome is blocked by the Tus protein. The orientations of the bound Tus proteins on the *E. coli* DNA molecule are such that both replisomes become trapped within a relatively short region. Termination of replication therefore always occurs within this region.

Okazaki fragments cause problems at the ends of linear DNA molecules

The second aspect of the termination of replication that we will consider is the way in which eukaryotic cells ensure that their linear DNA molecules do not decrease in length every time they are replicated. Shortening occurs if the extreme 3′ end of the lagging strand is not copied because the final Okazaki fragment cannot be primed. This

---

Box 14.4 The interaction between Tus proteins and the replisome

When bound to a terminator sequence, a Tus protein allows a replication fork to pass if the replisome is moving in one direction, but blocks progress if the fork approaches from the opposite direction. The directionality is set by the orientation of the Tus protein on the double helix, but exactly how does the protein prevent progression of the replisome? One hypothesis is that the Tus protein interacts with the DnaB helicase, which is breaking the base pairs ahead of the replisome and hence is directly responsible for progress of the replisome along the DNA molecule. According to this model, the DnaB helicase is able to bypass a Tus protein when it approaches from the 'permissive' direction, but its progress is impeded if it approaches from the non-permissive direction.

Recent research has provided support for the alternative possibility that the key interactions are between Tus and the polynucleotides at the replication fork, rather than between Tus and the DnaB helicase or any other protein of the replisome. The progress of individual replication forks has been studied in an experimental system in which the replisome proteins are absent. Without these proteins, a fork will not move naturally along a DNA double helix. Instead, a magnetic bead is attached to the end of one of the polynucleotides, while the end of the second polynucleotide is immobilized by attachment to a solid support. The two polynucleotides are then pulled apart by manipulating the magnetic bead with a **magnetic tweezer**. This is a device

comprising a set of magnets whose positions and field strengths can be varied in such a way that the magnetic bead, and its attached polynucleotide, can be moved about in a controlled manner. By moving the magnetic bead away from the solid support, the helix becomes opened up, producing a fork that can be moved along the helix simply by pulling the ends of the two polynucleotides further apart.

What happens if the DNA molecule being manipulated in this way contains a terminator sequence to which a Tus protein is bound? If the orientation of the terminator sequence is such that the fork approaches the Tus protein from the permissive direction, then movement of the fork is not impeded. On the other hand, when the fork approaches from the non-permissive direction, its progress is prevented by the Tus protein.

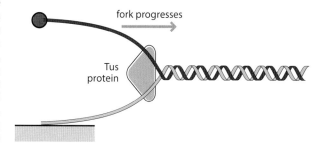

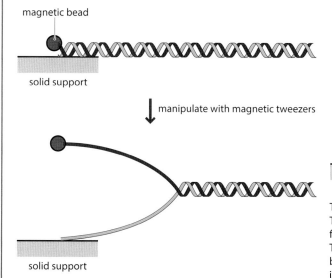

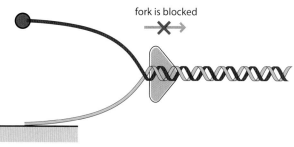

These innovative experiments suggest that interactions between Tus and the DNA being replicated are at least partly responsible for the ability of Tus to block progression of the replication fork. The results do not exclude a role for a protein–protein interaction between Tus and DnaB, but they do show that such an interaction is unlikely to be the full explanation of the mode of action of Tus.

Figure 14.19 A problem arising during replication of the end of a linear DNA molecule.
The parent DNA molecule is replicated in the normal way, but the lagging strand copy is incomplete because the last Okazaki fragment is not made. The resulting daughter molecule has a 3′ overhang and, when replicated, gives rise to a granddaughter molecule that is shorter than the original parent.

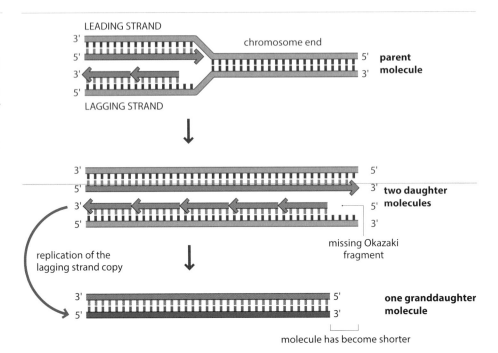

will happen if the natural position for the priming site is beyond the end of the lagging strand (*Fig. 14.19*). During replication of the lagging strand of a human chromosome, primers for Okazaki fragments are synthesized at positions approximately 200 bp apart. If an Okazaki fragment begins at a position less than 200 bp from the 3′ end of the lagging strand, then there will not be room for another fragment next to it, and the equivalent segment of the lagging strand will not be copied. The absence of this

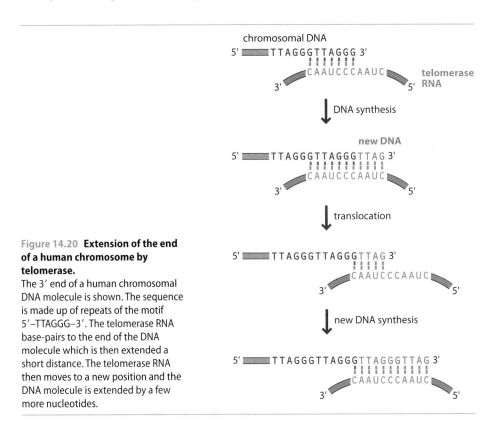

Figure 14.20 Extension of the end of a human chromosome by telomerase.
The 3′ end of a human chromosomal DNA molecule is shown. The sequence is made up of repeats of the motif 5′–TTAGGG–3′. The telomerase RNA base-pairs to the end of the DNA molecule which is then extended a short distance. The telomerase RNA then moves to a new position and the DNA molecule is extended by a few more nucleotides.

RESEARCH HIGHLIGHT

Box 14.5 Telomerase and cancer

In healthy adult humans, telomerase is only active in **stem cells**. These are progenitor cells that divide continually throughout the lifetime of an organism, producing new cells to maintain organs and tissues. In other cell lines, successive cycles of DNA replication and cell division are accompanied by a gradual shortening of the ends of each DNA molecule. Eventually, this shortening becomes so severe that the cell becomes **senescent**, remaining alive but unable to undergo any additional divisions.

Senescence is thought to be a protective mechanism that counters the tendency of a cell lineage to accumulate defects such as broken or rearranged chromosomes. Eventually, these defects could lead to the cells becoming malfunctional. The senescence process prevents this by ensuring that the lineage is terminated before the danger point is reached.

A typical feature of a cancerous cell line is that the cells do not become senescent and instead divide continuously. With several types of cancer, this absence of senescence is associated with activation of telomerase. Cancer biologists have therefore wondered whether telomerase could be a target for drugs designed to combat cancer. The idea is that inactivation of telomerase would induce senescence of the cancerous cells and hence prevent their proliferation. To test this possibility, the protein component of telomerase has been used to prepare vaccines that contain antibodies that bind to and inactivate any telomerase enzymes that they encounter. Clinical trials have shown that these anti-telomerase vaccines are able to reduce the number of cancerous cells circulating in the bloodstream of a patient, decreasing the chances of the cancer spreading to other parts of the body. The next question is whether these vaccines can also have an inhibitory influence on other aspects of cancer progression, such as the growth of tumors.

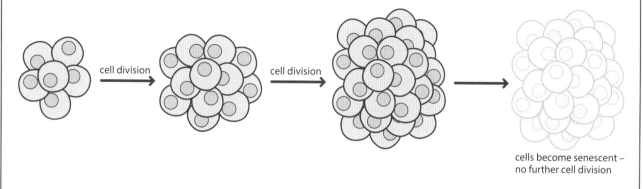

cell division

cell division

cells become senescent – no further cell division

Okazaki fragment means that the lagging strand copy is shorter than it should be. The daughter double-stranded DNA molecule will therefore have a short overhang – one strand is longer than the other. When the shorter strand is copied in the next round of replication, it will give rise to a granddaughter molecule that lacks a segment that was present in the original parent double helix.

This problem is avoided by the action of an unusual enzyme called **telomerase**. This enzyme is unusual because it has two subunits, one of which is made of RNA. In the human enzyme the RNA component is 450 nucleotides in length and contains, near its 5′ end, the sequence 5′–CUAACCCUAAC–3′. The central part of this sequence is complementary to the DNA sequence 5′–TTAGGG–3′, which is repeated multiple times at the ends of each of the human chromosomal DNA molecules. The function of telomerase is to correct the shortening of the DNA ends that occurs because of missing Okazaki fragments. The telomerase RNA binds to the 3′ overhang on the daughter molecule, and acts as a template that enables the enzyme to extend the overhang by a few nucleotides (*Fig. 14.20*). The enzyme then moves its RNA to a new position slightly further along the DNA, so another short stretch of DNA can be added. The process is repeated until the overhang has been made long enough for the final Okazaki fragment to be synthesized.

14.2 DNA repair

During the gene expression pathway, the nucleotide sequence of a gene is first copied into an RNA molecule, and the sequence of this RNA then specifies the sequence of amino acids in a protein (see *Fig. 14.1*). We will look at the details of this process, and the way in which the **genetic code** dictates the translation of the nucleotide sequence to amino acid sequence, in the next two chapters. We do not, however, need to understand all the details of gene expression in order to appreciate how important it is that DNA molecules are replicated accurately. If the replication process introduces any changes into the sequence of one of the daughter polynucleotides, then it is possible that the amino acid sequence of a protein will become altered. Such an alteration is called a **mutation**, and it could cause a change in the activity of the protein, possibly even a complete loss of function. The result could conceivably be lethal for the cell.

A similar problem would arise if, after it has been replicated, the nucleotide sequence of a gene is altered by exposure to a chemical or physical agent that changes the structure of one or more nucleotides. Such agents are called **mutagens**, and they include a range of chemicals that are present in the environment, some man-made and some natural in origin, as well as many types of radiation, including the UV radiation present in sunlight. The more damaging of these mutagens, such as ionizing radiation, introduce chemical changes that prevent a DNA molecule from being replicated, or may even break a DNA molecule into two or more fragments.

To avoid the undesirable consequences of mistakes in DNA replication or damage resulting from the action of mutagens, all organisms have **DNA repair** mechanisms that correct the vast majority of the replication errors and chemical changes that occur in their DNA molecules. These repair mechanisms are the subject of the remainder of this chapter.

14.2.1 Correcting errors in DNA replication

During DNA replication, the sequence of the new strand of DNA that is being made is determined by complementary base pairing with the template polynucleotide. The DNA polymerases that carry out replication use various approaches to ensure that the correct nucleotide is inserted at each position in the growing polynucleotide. The identity of the nucleotide is checked when it is first bound to DNA polymerase and checked again when the nucleotide is moved to the active site of the enzyme. At either of these stages, the enzyme is able to reject the nucleotide if it recognizes that it is not the right one. Should an incorrect nucleotide evade this surveillance system and become attached to the 3' end of the polynucleotide, then it can still be removed by the proofreading function provided by the 3'→5' exonuclease possessed by most DNA polymerases (see *Fig. 14.11*). Despite these precautions, some errors do creep through. Even then all is not lost, because most cells possess a repair system that is able to detect and correct such replication errors after the replisome has moved on.

To correct a replication error, the parent and daughter strands must be distinguished

A replication error leads to a **mismatch** in a daughter DNA molecule, a position where there is no base pairing because the adjacent nucleotides are not complementary (*Fig. 14.21*). To correct the error, the nucleotide in the daughter strand (the new polynucleotide that was synthesized during replication) must be excised and replaced with the correct nucleotide. This means that the parent and daughter polynucleotides must be distinguished. But one DNA strand looks very much like any other, so how can the daughter one be recognized?

In bacteria the answer is that some of the nucleotides in the DNA molecule are modified by attachment of methyl groups. In *E. coli* there are two enzymes that add these methyl groups to DNA (*Fig. 14.22*):

- The **DNA adenine methylase (Dam)** converts adenine to 6-methyladenine in the sequence 5'–GATC–3'.

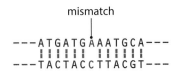

mismatch

```
---ATGATGAAATGCA---
   ||||||| ||||||
---TACTACCTTACGT---
```

Figure 14.21 A mismatch position in a double-stranded DNA molecule.

Box 14.6 **Base tautomerism can result in replication errors**

DNA replication is a highly accurate process, in part because of the proofreading activity possessed by many DNA polymerases. It is, however, possible for errors to occur even when the polymerase is applying the base pairing rules correctly. This is because of **tautomerism**, which is the ability of each nucleotide base to take up alternative isomeric structures in which the constituent atoms are bonded together in slightly different ways. For example, thymine has two tautomers, called the *keto* and *enol* forms, which differ in the positioning of atoms around nitrogen number 3 and carbon number 4.

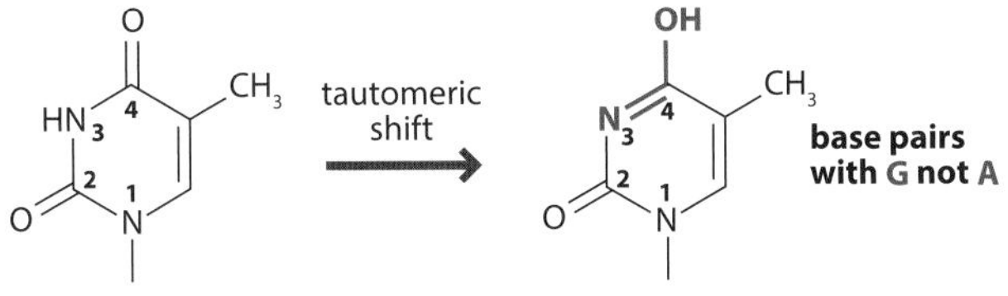

therefore have a replication error, with a G present at a position which should be an A. The same problem can occur with adenine, the rare *imino* tautomer of this base forming a pair with cytosine, and with guanine, the *enol* version of which pairs with thymine.

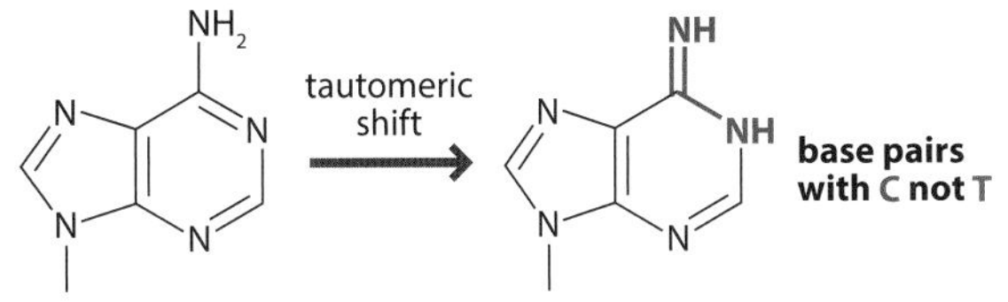

amino-adenine   *imino*-adenine

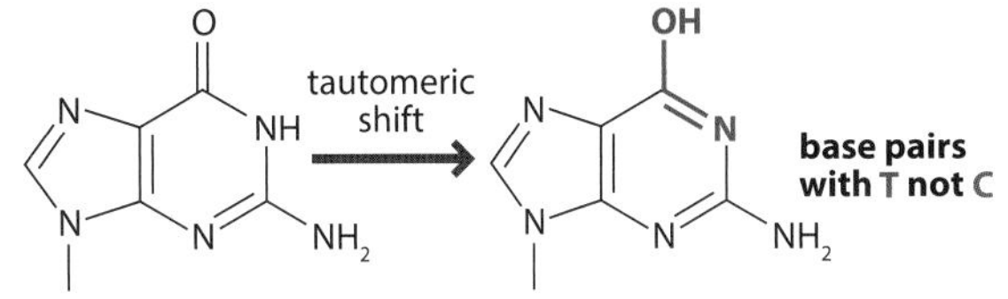

keto-thymine   *enol*-thymine

The two tautomers of thymine are able to interconvert. The equilibrium is biased very much toward the *keto* form but every now and then the *enol* version of thymine occurs in the template DNA at the precise time that the replication fork is moving past. This causes a problem, because *enol*-thymine base pairs with guanine rather than adenine. The daughter polynucleotide will

keto-guanine   *enol*-guanine

Cytosine also has *amino* and *imino* tautomers, but cytosine tautomerism does not result in a replication error because both versions pair with guanine.

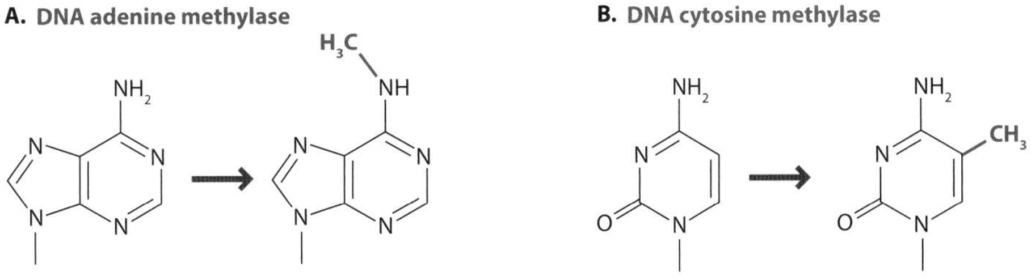

Figure 14.22 **The reactions catalyzed by the methylases, (A) Dam and (B) Dcm.**

A. DNA adenine methylase

adenine   6-methyladenine

B. DNA cytosine methylase

cytosine   5-methylcytosine

- The **DNA cytosine methylase (Dcm)** converts cytosines to 5-methylcytosines in 5'–CCAGG–3' and 5'–CCTGG–3'.

The conversion of adenine to 6-methyladenine or cytosine to 5-methylcytosine does not result in mutations because these changes do not affect the base pairing properties of the nucleotides. A 6-methyladenine nucleotide in one strand still base pairs with a thymine in the other strand, and 5-methylcytosine still base pairs with guanine. The added methyl groups do, however, act as markers for the parent strand in a newly replicated double helix. This is because, immediately after its synthesis, the daughter strand is unmethylated, with the methyl groups not being attached until after the replisome has passed. This provides a brief window of opportunity during which the repair enzymes can scan the DNA for mismatches, distinguishing the parent and daughter strands because the former is methylated whereas the latter is not (*Fig. 14.23*).

Methylation might be used in the same way to direct mismatch repair in eukaryotes, but this is not certain. The problem is that in some eukaryotes, such as yeast and fruit flies, the DNA is not extensively methylated and the frequency of markers might be

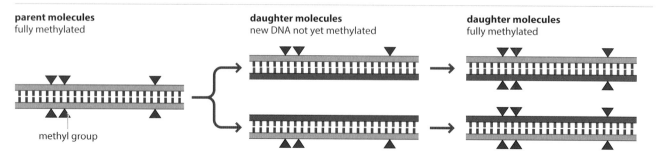

Figure 14.23 Methylation of bacterial DNA during replication.
Methylation of newly synthesized DNA does not occur immediately after replication, providing a window of opportunity for the mismatch repair proteins to recognize the daughter strands and correct replication errors.

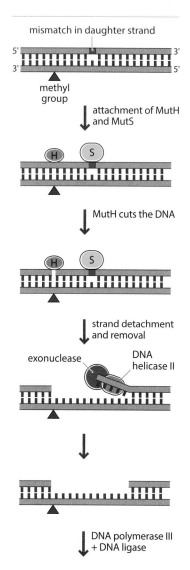

Figure 14.24 Long patch mismatch repair in *E. coli*.

too low to enable the parent and daughter polynucleotides to be distinguished in all parts of a DNA molecule. It is possible that in these organisms the repair enzymes are more closely associated with the replisome, so that repair is coupled with DNA synthesis in such a way that the repair enzymes automatically act only on the daughter polynucleotides.

Mismatch repair involves excision of the incorrect part of the daughter polynucleotide

Mismatch repair is an example of an **excision repair** process. Excision repair involves removal of a segment of the polynucleotide containing the offending nucleotide, followed by resynthesis of the correct nucleotide sequence by a DNA polymerase.

E. coli has at least three mismatch repair systems, called 'long patch', 'short patch' and 'very short patch', the names indicating the lengths of the excised and resynthesized segments. In the long patch system, up to 2000 nucleotides are excised and resynthesized. The key players in the process are two Mut proteins, whose roles are as follows (*Fig. 14.24*):

• **MutS** recognizes and binds to the mismatch position.

• **MutH** distinguishes the daughter polynucleotide by binding to the unmethylated 5'–GATC–3' sequence to one side of the mismatch. It then cuts one of the phosphodiester bonds next to the guanine nucleotide in this sequence.

Once the cut has been made, DNA helicase II begins to break the base pairs holding the error-containing part of the daughter polynucleotide to the parent strand. An exonuclease follows behind the helicase, so that as the strand is detached it is degraded nucleotide by nucleotide from its free end.

The excision results in a gap in the daughter strand. This is filled in by DNA polymerase III, which extends the polynucleotide whose 3' end forms one side of the gap, using base pairing to the parent strand to ensure that this time the correct daughter sequence is synthesized. After the gap has been filled, the final phosphodiester bond is put in place by a DNA ligase.

14.2.2 Repair of damaged nucleotides

Nucleotides which have become damaged by reaction with chemicals or exposure to physical mutagens can also be repaired by excision processes. These include a mechanism for removing and replacing just a single damaged nucleotide base, and others that remove longer pieces of damaged DNA.

Base excision is used to repair individual damaged nucleotides

Base excision is used to repair nucleotides whose bases have suffered relatively minor damage, for example, by deamination or alkylation. Deaminating agents present in the environment include nitrous acid, which is generated in the atmosphere from nitrogen and can accumulate in closed spaces such as poorly ventilated rooms. Nitrous acid deaminates adenine to hypoxanthine, cytosine to uracil, and guanine to xanthine (*Fig. 14.25*). Thymine has no amino group and so cannot be deaminated. Hypoxanthine and uracil result in changes in the base pairing properties of the damaged nucleotide, and xanthine prevents passage of the replisome and hence blocks replication. Alkylating agents include methyl halide compounds, some of which have been used as pesticides. Although the addition of methyl groups at some positions on a nucleotide are harmless, other types of methylation result in crosslinks between the two strands of a DNA molecule, which will of course prevent progress of the replisome.

The base excision process is initiated by a **DNA glycosylase** enzyme, which cleaves the β-*N*-glycosidic bond between a damaged base and the sugar component of the nucleotide (*Fig. 14.26A*). There are various types of DNA glycosylase, each with a limited specificity. The specificities of the glycosylases possessed by a cell determine the range of damaged nucleotides that can be repaired in this way. Most organisms are able to deal with deaminated bases such as uracil and hypoxanthine, oxidation products such as 5-hydroxycytosine and thymine glycol, and methylated bases such as 3-methyladenine, 7-methylguanine and 2-methylcytosine.

Removal of a damaged base by the DNA glycosylase creates an **AP** (apurinic/apyrimidinic) or **baseless site**. The AP site is then converted to a single nucleotide gap by removal of the ribose unit (*Fig. 14.26B*). In most cases this is done by an **AP endonuclease**, which cuts the phosphodiester bond on the 5′ side of the AP site. The AP endonuclease might also cut the 3′-phosphodiester bond, completely excising the sugar, or this step might be carried out by a separate **phosphodiesterase** (an enzyme that breaks phosphodiester bonds) or, in eukaryotes, by the lyase activity possessed by DNA polymerase β.

A lyase is an enzyme that breaks chemical bonds by a process other than oxidation or hydrolysis (see *Section 7.1.3*).

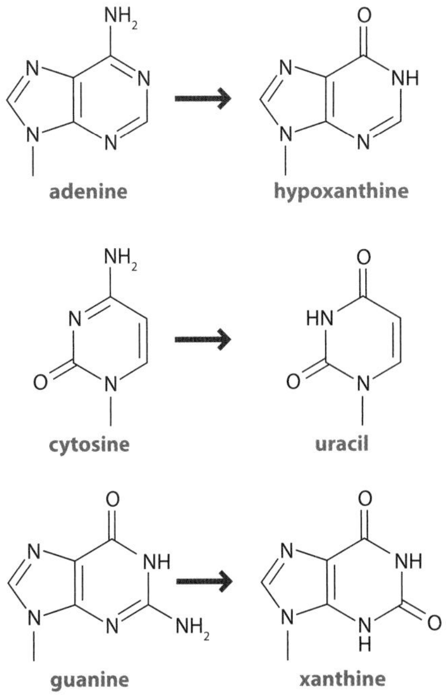

Figure 14.25 **Deamination products of adenine, cytosine and guanine.**

Figure 14.26 Base excision repair.
(A) The role of DNA glycosylase.
(B) Outline of the base excision repair pathway.

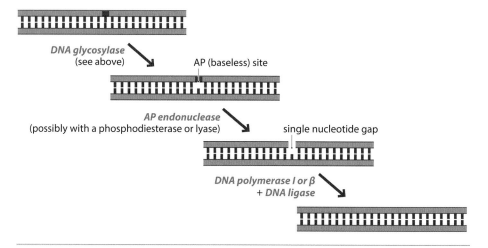

A. the role of DNA glycosylase

B. outline of base excision repair

Box 14.7 **Photoreactivation repair of cyclobutyl dimers**

Cyclobutyl dimers resulting from UV damage can also be repaired directly, by a light-dependent system called **photoreactivation**. In *E. coli*, the process involves the enzyme called **DNA photolyase**. This enzyme binds to cyclobutyl dimers and, when stimulated by light with a wavelength between 300 and 500 nm, converts the dimers back to the original nucleotides. There are two types of photolyase, one type containing a folate cofactor and the other containing a flavin compound. Both types of cofactor capture light energy which is used to reduce FADH to FADH$_2$. An electron is then transferred from FADH$_2$ to the cyclobutyl dimer, causing the latter to revert back to a pair of thymines.

Photoreactivation is a widespread but not universal type of repair. It is known in many but not all bacteria and in a few eukaryotes, including some vertebrates, but is absent in humans and other placental mammals. A similar type of photoreactivation involves the **(6–4) photoproduct photolyase** and results in repair of (6–4) lesions. Neither *E. coli* nor humans have this enzyme but it is possessed by a variety of other organisms.

Figure 14.27 **Photoproducts that result from exposure of DNA to UV irradiation.**

A segment of a polynucleotide containing two adjacent thymine bases is shown. A thymine dimer contains two UV-induced covalent bonds, one linking the carbons at position 6 and the other linking the carbons at position 5. The (6–4) lesion involves formation of a covalent bond between carbons 4 and 6 of the adjacent nucleotides.

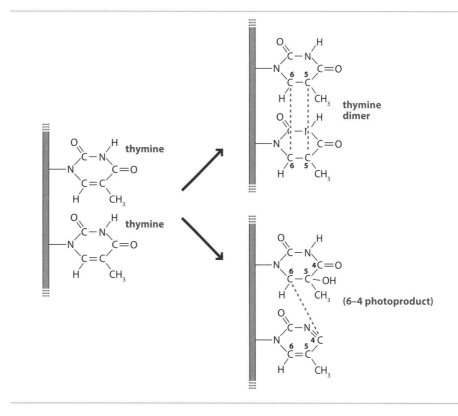

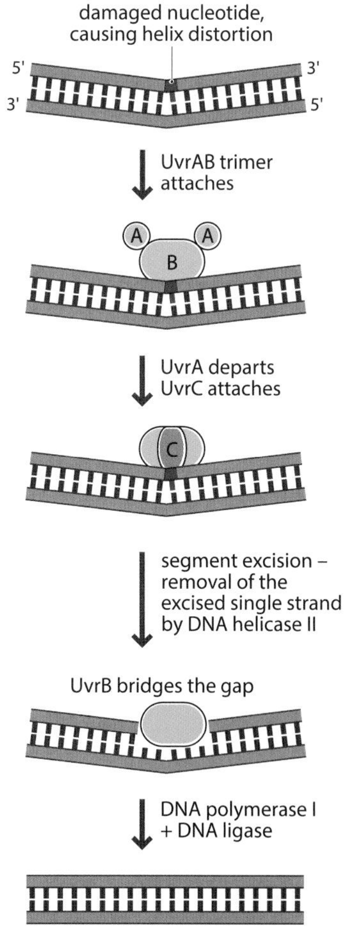

Figure 14.28 **Short patch nucleotide excision repair in *E. coli*.**

Alternatively, some DNA glycosylases dispense with the need for an endonuclease by making the cut themselves, although this will be at the 3′ side of the AP site. Again the sugar is removed by a phosphodiesterase or by DNA polymerase β. The single nucleotide gap is then refilled by DNA polymerase I in bacteria and by DNA polymerase β in eukaryotes, and the final phosphodiester bond added by a DNA ligase.

More extensive types of damage are repaired by nucleotide excision

Nucleotide excision is the main repair system for more extreme forms of DNA damage, including the **photoproducts** that result from exposure to UV radiation. The most common type of photoproduct is a **cyclobutyl dimer**, which results from the dimerization of adjacent pyrimidine bases, especially if these are both thymines (*Fig. 14.27A*). Another type of UV-induced photoproduct is the **(6–4) lesion** in which carbons 4 and 6 of adjacent pyrimidines become covalently linked (*Fig. 14.27B*).

Nucleotide excision repair is similar in many respects to mismatch repair, involving recognition of the damaged part of the DNA by special enzymes, followed by removal and resynthesis of a segment of one of the polynucleotides. An example is the **short patch** process of *E. coli*, so called because the region of polynucleotide that is excised and subsequently 'patched' is relatively short, usually just 12 nucleotides in length. Short patch repair occurs when a trimer comprising two UvrA proteins and one copy of UvrB attaches to the DNA at the damaged site (*Fig. 14.28*). The Uvr proteins are not thought to distinguish individual types of damage, but simply search for regions where the double helix has become distorted, as will happen if a pair of bases form a dimer. Once the damage has been located, the UvrA proteins disassociate and UvrC attaches, forming a UvrBC dimer. UvrC, possibly in conjunction with UvrB, then cuts the polynucleotide either side of the damaged site. The first cut is made at the fifth phosphodiester bond to one side of the damaged nucleotide, and the second cut at

---

Box 14.8 Defects in DNA repair underlie a number of important human diseases

DNA repair plays an essential role in maintaining the integrity of chromosomal DNA molecules. It is therefore no surprise that a number of severe human diseases are caused by defects in one or more of the repair processes. An example is xeroderma pigmentosum, which is caused by mutations that affect several of the proteins involved in nucleotide excision repair. One of the symptoms of this disease is hypersensitivity to UV radiation, because nucleotide excision is the only way in which human cells can repair UV-induced damage such as cyclobutyl dimers and (6–4) lesions. This means that patients suffering from xeroderma pigmentosum often also develop skin cancer.

Hereditary breast–ovarian cancer is a second disease that results from a breakdown in DNA repair. This type of cancer is associated with two proteins, BRCA1 and BRCA2, both of which are involved in repairing breaks in DNA molecules. BRCA1 also plays a role in the mismatch repair process.

Other diseases associated with defects in DNA repair include:
- Hereditary nonpolyposis colorectal cancer (HNPCC), which is caused by dysfunction of the mismatch repair process. HNPCC is a syndrome associated with an increased risk of colon, endometrial and various other types of cancer.
- Ataxia telangiectasia, which results from defects in the process by which damaged sites in DNA molecules are detected. The symptoms of ataxia telangiectasia include sensitivity to ionizing radiation.
- Bloom syndrome and Werner syndrome, which are caused by inactivation of different members of a family of DNA helicases that may have a role in nonhomologous end-joining.
- Spinocerebellar ataxia, which results from defects in the pathway used to repair single-stranded breaks.

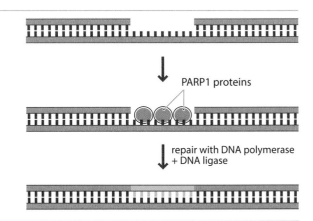

Figure 14.29 **Repair of a single-stranded break in human DNA.**

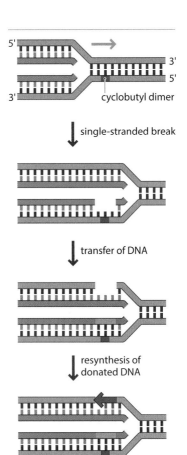

Figure 14.30 **Post-replicative repair of a single-stranded break.**
The break is caused by the presence of a cyclobutyl dimer in the DNA being replicated.

the eighth phosphodiester bond in the other direction, resulting in the 12 nucleotide excision. From this point on, the process is very similar to mismatch repair, the excised segment being detached by DNA helicase II, and the gap filled by DNA polymerase I followed by DNA ligase.

E. coli also has a **long patch** nucleotide excision repair system that involves Uvr proteins but differs in that the piece of DNA that is excised can be up to 2 kb in length. Eukaryotes have just one type of nucleotide excision pathway, resulting in replacement of 24–29 nucleotides of DNA, and unrelated to either of the excision pathways in bacteria.

14.2.3 Repair of DNA breaks

Finally we must consider what happens if one or both polynucleotides in a DNA molecule become broken. A single-stranded break, resulting in loss of part of one of the polynucleotides, can result from some types of oxidative damage. Repair is straightforward: the exposed part of the intact strand is coated with protective single strand binding proteins, which in humans are called **PARP1** proteins, and the break filled in by a DNA polymerase and ligase (*Fig. 14.29*).

Double-stranded breaks are generated by exposure to ionizing radiation and some chemical mutagens, and breakages can also occur during DNA replication. A double-stranded break is more serious than a single-stranded one, because the break converts the molecule into two separate fragments which have to be brought back together again in order for the break to be repaired. One way of repairing a double-stranded break is by **nonhomologous end-joining** (**NHEJ**). This involves a pair of proteins, called Ku, which bind to the DNA ends either side of the break. The individual Ku proteins have an affinity for one another, which means that the two broken ends of the DNA molecule are brought into proximity and can be joined back together by a DNA ligase.

A break can also arise if a damaged part of a DNA molecule evades the repair system and is encountered by a replisome. An example is when the replisome attempts to copy a piece of DNA that contains a cyclobutyl dimer. When a cyclobutyl dimer is encountered, the parent strand cannot be copied and the DNA polymerase simply jumps ahead to the nearest undamaged region where it restarts the replication process. The result is that a single-stranded break appears in one of the newly synthesized polynucleotides (*Fig. 14.30*). This break cannot be filled in by a polymerase because the parent strand is still damaged and cannot be copied. Instead, it is possible to fill in the break by replacing it with the equivalent segment of DNA from the parent polynucleotide present in the second daughter double helix. The gap that is now present in the second helix is refilled by a DNA polymerase, using the undamaged daughter polynucleotide within this helix as the template. This is a type of **post-replicative repair** process, and transfer of the segment of DNA from one polynucleotide to another is an example of **recombination**.

Further reading

Berghuis BA, Dulin D, Xu Z–Q, *et al.* (2015) Strand separation establishes a sustained lock at the Tus–*Ter* replication fork barrier. *Nature Chemical Biology* **11**, 579–85. *Using molecular tweezers to study the interaction between Tus and the replisome.*

Burgers PMJ (2009) Polymerase dynamics at the eukaryotic DNA replication fork. *Journal of Biological Chemistry* **284**, 4041–5.

Cech TR (2004) Beginning to understand the end of the chromosome. *Cell* **116**, 273–9. *Reviews all aspects of telomerase.*

David SS, O'Shea VL and Kundu S (2007) Base-excision repair of oxidative DNA damage. *Nature* **447**, 941–50.

Drake JW, Glickman BW and Ripley LS (1983) Updating the theory of mutation. *American Scientist* **71**, 621–30. *A general review of mutation.*

Hearst JE (1995) The structure of photolyase: using photon energy for DNA repair. *Science* **268**, 1858–9.

Hübscher U, Nasheuer H-P and Syväoja JE (2000) Eukaryotic DNA polymerases: a growing family. *Trends in Biochemical Science* **25**, 143–7.

Johnson A and O'Donnell M (2005) Cellular DNA replicases: components and dynamics at the replication fork. *Annual Review of Biochemistry* **74**, 283–315.

Kornberg A (1960) Biologic synthesis of deoxyribonucleic acid. *Science* **131**, 1503–8. *A description of DNA polymerase I.*

Lehmann AR (1995) Nucleotide excision repair and the link with transcription. *Trends in Biochemical Science* **20**, 402–5.

Li G-M (2008) Mechanisms and functions of DNA mismatch repair. *Cell Research* **18**, 85–98.

Lieber MR (2010) The mechanism of double-strand break repair by the nonhomologous DNA end-joining pathway. *Annual Review of Biochemistry* **79**, 181–211.

Mott ML and Berger JM (2007) DNA replication initiation: mechanisms and regulation in bacteria. *Nature Reviews Microbiology* **5**, 343–54.

O'Driscoll M (2012) Diseases associated with defective responses to DNA damage. *Cold Spring Harbor Perspectives in Biology* **4**, 411–35.

Okazaki T and Okazaki R (1969) Mechanisms of DNA chain growth. *Proceedings of the National Academy of Sciences USA* **64**, 1242–8. *The discovery of Okazaki fragments.*

Pomerantz RT and O'Donnell M (2007) Replisome mechanics: insights into a twin polymerase machine. *Trends in Microbiology* **15**, 156–64.

Ruiz-Masó J, Machón C, Bordanaba-Ruiseco L, Espinosa M, Coll M and Del Solar G (2015) Plasmid rolling-circle replication. *Microbiology Spectrum* **3**(1):PLAS-0035-2014.

Shay JW and Wright WE (2006) Telomerase therapeutics for cancer: challenges and new directions. *Nature Reviews Drug Discovery* **5**, 577–84. *Methods for inhibiting telomerase in order to treat cancer.*

Wang JC (2002) Cellular roles of DNA topoisomerases: a molecular perspective. *Nature Reviews Molecular Cell Biology* **3**, 430–40.

Self-assessment questions

Multiple choice questions

Only one answer is correct for each question. Answers can be found on the website: www.scionpublishing.com/biochemistry

1. The *E. coli* origin of replication spans approximately how far?
   (a) 25 bp
   (b) 75 bp
   (c) 150 bp
   (d) 245 bp

2. What are the proteins that form a barrel structure at the *E. coli* origin of replication called?
   (a) DnaA
   (b) DnaB
   (c) SSBs
   (d) Helicase

3. The prepriming complex in *E. coli* is initially made up of which proteins?
   (a) DnaA and DnaB
   (b) DnaB and DnaC
   (c) DnaB and SSBs
   (d) DnaC and SSBs

4. What is the function of a helicase?
   (a) To unwind the double helix
   (b) To prevent the two single strands from forming base pairs
   (c) To break base pairs
   (d) To prime DNA synthesis

5. Human DNA has approximately how many origins of replication?
   (a) 200
   (b) 2000
   (c) 20 000
   (d) 200 000

6. Which of the following statements is **incorrect** with regard to rolling circle replication?

   (a) It is the process used to replicate the human mitochondrial genome
   (b) It initiates at a nick in one of the parent polynucleotides
   (c) It results in a series of identical molecules linked head to tail
   (d) It is used by various types of virus

7. What is the function of a DNA topoisomerase?
   (a) To unwind the double helix
   (b) To prevent the two single strands from forming base pairs
   (c) To break base pairs
   (d) To prime DNA synthesis

8. Which enzyme is responsible for most of the template-dependent polynucleotide synthesis during DNA replication in *E. coli*?
   (a) DNA polymerase I
   (b) DNA polymerase II
   (c) DNA polymerase III
   (d) DNA polymerase IV

9. The proofreading function of a DNA polymerase uses which activity?
   (a) $3' \rightarrow 5'$ exonuclease
   (b) $5' \rightarrow 3'$ exonuclease
   (c) $3' \rightarrow 5'$ polymerase
   (d) $5' \rightarrow 3'$ polymerase

10. Which of the following statements is **incorrect** with regard to eukaryotic DNA polymerases?
    (a) The main replicating enzymes are DNA polymerase δ and DNA polymerase ε
    (b) DNA polymerase α synthesizes the primers for DNA synthesis
    (c) DNA polymerase γ replicates the mitochondrial DNA
    (d) DNA polymerase β replicates the lagging strand

11. The enzyme that synthesizes the primers for DNA replication in *E. coli* is called what?
    (a) DNA polymerase I
    (b) DNA polymerase III
    (c) Primase
    (d) RNase

12. How long are the Okazaki fragments in *E. coli*?
    (a) Less than 200 nucleotides
    (b) 200–300 nucleotides
    (c) 500–1000 nucleotides
    (d) 1000–2000 nucleotides

13. How long are the Okazaki fragments in humans?
    (a) Less than 200 nucleotides
    (b) 200–300 nucleotides
    (c) 500–1000 nucleotides
    (d) 1000–2000 nucleotides

14. The enzyme that removes the Okazaki fragments in *E. coli* is called what?
    (a) DNA polymerase I
    (b) DNA polymerase II
    (c) DNA polymerase III
    (d) DNA ligase

15. Which of the following statements is **incorrect** with regard to termination of replication in *E. coli*?
    (a) There are ten terminator sequences
    (b) Each terminator sequence is the binding site for a Tus protein
    (c) A Tus protein contains an RNA subunit
    (d) When approaching from one direction, the replisome is able to bypass the Tus protein

16. Which of the following is **not** a feature of telomerase enzymes?
    (a) It has two subunits
    (b) It contains an RNA subunit of 550 nucleotides
    (c) In humans, it synthesizes repeats of the sequence 5′–TTAGGG–3′
    (d) It is looked on as a possible target for anti-cancer drugs

17. What type of DNA modification enables the daughter strand to be recognized during mismatch repair?
    (a) Acetylation
    (b) Methylation
    (c) Phosphorylation
    (d) Deamination

18. MutS and MutH are involved in which repair process in *E. coli*?
    (a) Base excision repair
    (b) Mismatch repair
    (c) Nucleotide excision repair
    (d) Repair of double-strand breaks

19. Deamination of adenine results in what?
    (a) Xanthine
    (b) Uracil
    (c) Thymine
    (d) Hypoxanthine

20. In base excision repair, the enzyme that cleaves the β-*N*-glycosidic bond between a damaged base and the sugar component of the nucleotide is called a what?
    (a) DNA glycolyase
    (b) DNA glycosylase
    (c) AP endonuclease
    (d) Phosphodiesterase

21. Which of the following is **not** a type of photoproduct?
    (a) DNA photolyase
    (b) Thymine dimer
    (c) Cyclobutyl dimer
    (d) (6–4) lesion

22. UvrA and UvrB are involved in which type of repair process in *E. coli*?
    (a) Base excision repair
    (b) Mismatch repair
    (c) Nucleotide excision repair
    (d) Repair of double-strand breaks

23. What are the proteins that protect single-stranded regions prior to repair in humans called?
    (a) Ku proteins
    (b) NHEJ proteins
    (c) Mut proteins
    (d) PARP1 proteins

24. The repair process for single-stranded breaks which involves recombination is called what?
    (a) Post-replicative repair
    (b) Replicative repair
    (c) Nonhomologous end-joining
    (d) Site-specific recombination

Short answer questions

These questions do not require additional reading.

1. Explain why Watson and Crick believed that the double helix structure made obvious the mode of DNA replication.

2. Distinguish between the roles of replication origins in *E. coli*, yeast and humans.

3. What is the role of a DNA topoisomerase in DNA replication? How do the modes of action of Type I and Type II topoisomerases differ?

4. Why does the lagging strand of a DNA molecule have to be replicated in sections, and how are these sections joined together in *E. coli* and humans?

5. Describe how the priming problem is solved in *E. coli* and humans.

6. Explain why the ends of a linear DNA molecule do not gradually become shortened every time the molecule is replicated.

7. Explain why all organisms require DNA repair processes.

8. Describe how errors of replication are corrected and how this repair process ensures that only the daughter strand is repaired.

9. Distinguish between the base and nucleotide excision repair pathways.

10. Describe two pathways for repair of a single-stranded break in a DNA molecule.

Self-study questions

These questions will require calculation, additional reading and/or internet research.

1. DNA replication is described as semi-conservative because each daughter molecule contains one polynucleotide derived from the original molecule and one newly synthesized strand. Two other possible modes of replication are conservative, in which one daughter molecule contains both parent polynucleotides and the other daughter contains both newly synthesized strands, and dispersive, in which each strand of each daughter molecule is composed partly of the original polynucleotide and partly of newly synthesized polynucleotide. Devise an experiment which would confirm that in living cells DNA replication does indeed follow the semi-conservative process, rather than being conservative or dispersive.

2. Would it be possible to replicate the DNA molecules present in living cells if DNA topoisomerases did not exist?

3. Construct a hypothesis to explain why all DNA polymerases require a primer in order to initiate synthesis of a new polynucleotide. Can your hypothesis be tested?

4. In some eukaryotes the mismatch repair process is able to recognize the daughter strand of a double helix even though the two strands lack distinctive methylation patterns. Propose a mechanism by which the daughter strand can be recognized in the absence of methylation. How would you test your hypothesis?

5. Why do defects in DNA repair often lead to cancer?

CHAPTER 15

RNA synthesis

After reading this chapter you will:

- be able to describe the different types of RNA and their roles in the cell

- understand the role of the promoter sequence in initiation of transcription

- know the differences between the structures of bacterial and eukaryotic promoters

- be able to describe how transcription is initiated in *Escherichia coli*

- be able to describe the roles of transcription factors and other associated proteins during initiation of transcription in eukaryotes

- know how RNA is synthesized by the *E. coli* RNA polymerase

- understand the role of elongation factors during RNA synthesis in eukaryotes

- know the structure and mode of synthesis of the cap structure present in eukaryotic mRNAs

- understand the role of stem–loop structures in termination of transcription in *E. coli*

- be able to explain the difference between intrinsic and Rho-dependent termination of transcription in *E. coli*

- understand the role of polyadenylation in termination of eukaryotic mRNA transcription

- be able to describe how rRNAs and tRNAs are processed by cutting and trimming events

- know the key features of discontinuous genes and be able to describe the splicing pathway for GU–AG introns

- know how autocatalytic Group I introns are spliced

- be able to describe how tRNAs and rRNAs are chemically modified in bacteria and eukaryotes

In the next two chapters we will learn how the nucleotide sequence of a DNA molecule is used to direct the synthesis of RNA and protein molecules. This process, which we call gene expression, is conventionally divided into two stages (see *Fig. 14.1*). The first stage, which results in synthesis of an RNA molecule, is called **transcription**. Transcription is a copying reaction, with the nucleotide sequence of the RNA molecule being determined, according to the base pairing rules, by the DNA sequence of the gene that is being copied. For some genes the RNA transcript is the end product of gene expression. For others the transcript is a message that directs the second stage of gene expression, called **translation**. During translation, this RNA molecule (called a **messenger** or **mRNA**) directs the synthesis of a protein whose amino acid sequence is determined by the nucleotide sequence of the mRNA.

In this chapter we will explore the first stage of gene expression, resulting in synthesis of RNAs.

15.1 Transcription of DNA into RNA

As with DNA replication, transcription is most easily dealt with in a logical fashion, dividing the process into three stages:

- **Initiation of transcription**, during which the machinery for transcription is assembled at the start of a gene.

- **RNA synthesis**, during which the transcript is made.

- **Termination of transcription**, which completes the process, releasing the RNA transcript.

With transcription we also have to consider what happens to the RNA molecule after it has been made. This is because many RNA molecules undergo processing events before they are able to carry out their function in the cell. These events include the removal of some segments and/or the attachment of additional chemical groups. Before we do any of these things, we must distinguish between the different types of RNA molecule that are made by transcription.

15.1.1 Coding and noncoding RNAs

The mRNAs that are subsequently translated into protein are called **coding RNAs**, and the genes from which they are transcribed are called **protein-coding genes**. We might therefore look on mRNA as the most important type of RNA, but in fact it is only a small fraction of a cell's RNA, usually no more than 4% of the total. The remainder consists of **noncoding RNA**. These RNAs are not translated into protein, but they still have important functional roles in the cell.

The most abundant of the noncoding RNAs is **ribosomal** or **rRNA**. There are only four different rRNA molecules in humans, but each is present in many copies and together they make up over 80% of the total RNA in an actively dividing cell. Ribosomal RNAs are components of **ribosomes**, the structures on which protein synthesis takes place. The second most abundant type of noncoding RNA is **transfer** or **tRNA**. These tRNAs are small molecules that are also involved in protein synthesis, carrying amino acids to the ribosome and ensuring these are linked together in the order specified by the nucleotide sequence of the mRNA that is being translated. Most organisms synthesize between 30 and 50 different tRNAs.

Ribosomal and transfer RNAs are present in all organisms. Other types of noncoding RNA are only present in eukaryotes. The most important of these are:

- **Small nuclear RNA** (**snRNA**) which, as the name implies, is found in the nucleus. There are 15–20 different types of snRNA, most of which are involved in RNA processing, in particular the removal of segments called **introns** from mRNAs.

- **Small nucleolar RNAs** (**snoRNAs**) are found in the nucleoli, the parts of the nucleus in which rRNA is transcribed. These RNAs are also involved in RNA processing, specifically the addition of extra chemical groups to rRNA molecules.

- **Micro RNAs** (**miRNAs**) and **small interfering RNAs** (**siRNAs**) are involved in the control of gene expression.

15.1.2 Initiation of transcription

Within a DNA molecule, it is only the genes that are transcribed. Most of the **intergenic DNA** – the regions between the genes – is never copied into RNA (*Fig. 15.1*). The *E. coli* DNA molecule has about 4400 genes and there are over 45 000 genes in the human genome. This means that on every DNA molecule there are many positions at which transcription must be initiated, these positions lying at or just in front ('upstream') of

We will study the functions of noncoding RNAs later in the book: see *Section 16.2.1* for rRNA, *Section 16.1.2* for tRNA, *Section 15.2.2* for snRNA, *Section 15.2.3* for snoRNA, and *Section 17.2.1* for miRNA.

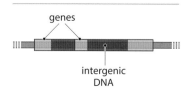

Figure 15.1 Intergenic DNA is the non-transcribed regions between genes.

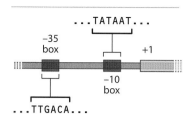

Figure 15.2 The two components of an *E. coli* promoter.

the beginning of genes. The first question that we must tackle is how the enzymes and other proteins involved in transcription identify these start points.

The start points for transcription are marked by promoter sequences

The positions where transcription should begin can be recognized because they contain specific nucleotide sequences that do not occur anywhere else in the DNA. These sequences are called **promoters**.

Promoters were first identified in *E. coli* by comparing the nucleotide sequences immediately upstream of over 1000 genes. This analysis revealed that the *E. coli* promoter has two separate components, called the −35 box and the −10 box (*Fig. 15.2*). The names indicate the positions of the sequences relative to the point at which RNA synthesis begins, the nucleotide at this transcription start point being numbered +1. The −35 box therefore lies approximately 35 nucleotides upstream of the position where transcription begins.

The sequence of the −35 box is 5′–TTGACA–3′, and that of the −10 box is 5′–TATAAT–3′. These are **consensus sequences**, which means that they describe the 'average' of all promoter sequences in *E. coli*, the actual sequence upstream of any particular gene having its own slight variations (*Table 15.1*). These variations affect the efficiency of the promoter, where efficiency is defined as the number of productive initiations that are promoted per second, a productive initiation being one that results in the synthesis of a full-length transcript. The most efficient promoters (called **strong promoters**) direct 1000 times as many productive initiations as the weakest promoters. We refer to these as differences in the **basal rate** of transcript initiation.

Table 15.1. Sequences of *E. coli* promoters

Gene	Protein product	Promoter sequence	
		−35 box	−10 box
Consensus	–	5′–TTGACA–3′	5′–TATAAT–3′
argF	Ornithine transcarbamoylase	5′–TTGTGA–3′	5′–AATAAT–3′
can	Carbonic anhydrase	5′–TTTAAA–3′	5′–TATATT–3′
dnaB	DnaB helicase	5′–TCGTCA–3′	5′–TAAAGT–3′
gcd	Glucose dehydrogenase	5′–ATGACG–3′	5′–TATAAT–3′
gltA	Citrate synthase	5′–TTGACA–3′	5′–TACAAA–3′
ligB	DNA ligase	5′–GTCACA–3′	5′–TAAAAG–3′

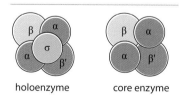

holoenzyme   core enzyme

Figure 15.3 The holoenzyme and core versions of the *E. coli* RNA polymerase.

The spacing between the −35 and −10 boxes is important because it places the two motifs on the same face of the double helix, facilitating their interaction with the **DNA-dependent RNA polymerase** that binds to the DNA in order to carry out transcription. Each *E. coli* cell contains about 7000 RNA polymerase molecules, between 2000 and 5000 of which are actively involved in transcription at any one time. The enzyme has a multisubunit structure described as $\alpha_2\beta\beta'\sigma$, meaning that each molecule is made up of two α subunits plus one each of β, the related β', and σ (*Fig. 15.3*). The σ subunit is responsible for recognizing the promoter sequence, and disassociates from the enzyme soon after it has attached to the DNA. Departure of the σ subunit converts the **holoenzyme** version of RNA polymerase ($\alpha_2\beta\beta'\sigma$) to the **core enzyme** ($\alpha_2\beta\beta'$), which performs the actual RNA synthesis.

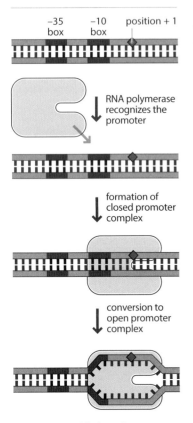

Figure 15.4 Initiation of transcription in *E. coli*.
The RNA polymerase is not drawn to scale. In reality it covers approximately 80 base pairs of the DNA molecule.

See *Section 17.1.2* for the role of the upstream elements in transcription initiation.

Attachment of the RNA polymerase involves interactions between the σ subunit and the –35 and –10 boxes of the promoter, and results in the **closed promoter complex**, with the RNA polymerase covering a region of some 80 base pairs, from upstream of the –35 box to downstream of the –10 box (*Fig. 15.4*). The closed complex is then converted to an **open promoter complex**, opening up the double-stranded DNA by breakage of approximately 13 base pairs spanning the region from the –10 box to just beyond the transcription start site. Note that the consensus sequence for the –10 box is made up entirely of adenine and thymine nucleotides, which form relatively weak base pairs comprising just two hydrogen bonds each. This region will therefore be easier to open up than other parts of the DNA where GC base pairs are present.

After the open promoter complex has been formed, the RNA polymerase moves away from the promoter by transcribing the DNA downstream. It appears, however, that some attempts by the polymerase to leave the promoter region are unsuccessful and lead to truncated transcripts that are degraded soon after they are synthesized. True completion of the initiation stage of transcription therefore occurs when the RNA polymerase has made a stable attachment to the DNA and has begun to synthesize a full-length transcript.

Promoters in eukaryotes have more complex structures

Eukaryotes have three different types of RNA polymerase, each responsible for transcribing a different set of genes (*Table 15.2*). Protein-coding genes are transcribed by **RNA polymerase II**. The promoters for these genes are more complex than those in *E. coli*. As well as the **core promoter**, which is the site at which an RNA polymerase II enzyme attaches, they include additional short sequences found at positions further upstream, sometimes several thousand nucleotides away from the transcription start site (*Fig. 15.5*). Transcription can be initiated in the absence of the upstream elements, but only in an inefficient way.

The RNA polymerase II core promoter consists of two main segments:

- The –25 or **TATA box**, which has the consensus sequence 5'–TATAWAAR–3'. In this sequence 'W' indicates either A or T, which are equally likely to occur at this position, and 'R' indicates a purine, A or G.

- The **initiator (Inr) sequence**, which is located around nucleotide +1. In mammals the consensus of the Inr sequence is 5'–YCANTYY–3', where 'Y' is a pyrimidine, C or T, and 'N' is any of the four nucleotides.

Table 15.2. Functions of the three eukaryotic RNA polymerases

Polymerase	Genes transcribed
RNA polymerase I	28S, 5.8S and 18S rRNA genes
RNA polymerase II	Protein-coding genes, most snRNA genes, miRNA genes
RNA polymerase III	Genes for 5S rRNA, tRNAs, and various small RNAs

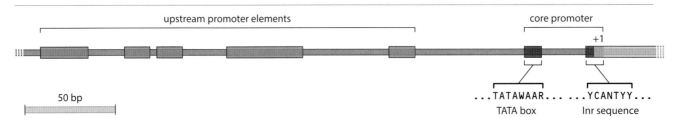

Figure 15.5 A typical promoter for RNA polymerase II.
The core promoter is made up of two segments: the TATA box and Inr sequence. The consensus sequences of these two segments are shown. Abbreviations: 'N' indicates any of the four nucleotides, 'R' indicates an A or G, 'W' indicates an A or T, and 'Y' indicates a C or T.

Box 15.1 Promoters for RNA polymerase I and RNA polymerase III

The promoters for the three eukaryotic RNA polymerases have their own distinctive features, enabling each polymerase to recognize the specific set of genes that it transcribes. We have already considered the structure of the RNA polymerase II promoter. Here we will look at the promoters for the other two polymerases. In vertebrates the details are as follows:

- RNA polymerase I promoters have two components: a core promoter spanning the transcription start point, between nucleotides −45 and +20, and an upstream control element (UCE) about 100 bp further upstream. Both sequences are initially recognized by a small protein called the upstream binding factor (UBF). Once attached to the DNA, the UBF acts as the binding site for RNA polymerase I and the other components of the transcription complex for this enzyme.

- RNA polymerase III promoters are variable, falling into at least three categories. Two of these categories are unusual in that the important sequences are located within the genes whose transcription they promote. These sequences span 50–100 bp and comprise one or two conserved boxes. The third category of RNA polymerase III promoter is similar to those for RNA polymerase II, having a TATA box, a proximal sequence element (PSE) located between positions −45 and −60, and a range of upstream promoter elements.

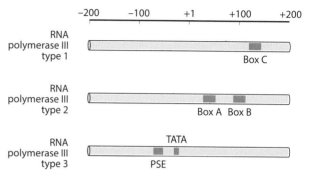

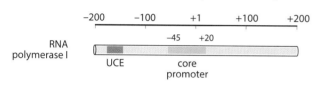

Some genes transcribed by RNA polymerase II have only one of these two components of the core promoter, and some, surprisingly, have neither. The latter are called 'null' genes. They are still transcribed, although the start position for transcription is more variable than for a gene with a TATA and/or Inr sequence.

With some protein-coding genes, an extra level of complexity arises because there may be two or more **alternative promoters**. This means that transcription of the gene can begin at two or more different sites, giving rise to mRNAs of different lengths. The human gene for the dystrophin protein is an example. This protein has been extensively studied because changes in its structure, which arise due to mutations in the dystrophin gene, can result in the disease called Duchenne muscular dystrophy. The dystrophin gene is very long, one of the largest in the human genome, spanning over 2.4 Mb of DNA. It has at least seven alternative promoters, which direct synthesis of mRNAs of different lengths, which in turn specify polypeptides with different numbers of amino acids (*Fig. 15.6*). The alternative promoters are active in different parts of the body, such as the brain, muscle and the retina, enabling different versions of the dystrophin protein to be made in these various tissues. The biochemical properties of these variants are matched to the needs of the cells in which they are synthesized.

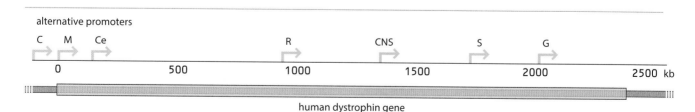

Figure 15.6 Alternative promoters for the human dystrophin gene.
Abbreviations indicate the tissue within which each promoter is active: C, cortical tissue; M, muscle; Ce, cerebellum; R, retinal tissue (and also brain and cardiac tissue); CNS, central nervous system (and also kidney); S, Schwann cells; G, general (most other tissues).

With some genes, alternative promoters are used to generate related versions of a protein at different stages in development, or even to enable a single gene to direct synthesis of two or more proteins at the same time in a single tissue. For example, over 10 500 promoters are active in human fibroblast cells, but these promoters direct transcription of fewer than 8000 genes. A substantial number of genes in these cells are therefore being expressed from two or more promoters simultaneously.

RNA polymerase II does not directly recognize its promoter

In outline, the events involved in initiation of transcription are the same in bacteria and eukaryotes. As in bacteria, attachment of RNA polymerase II to the core promoter results in a closed promoter complex that is converted to an open complex by breakage of base pairs around the start point for transcription. There are, however, important differences when we examine the process in eukaryotes in more detail. The most important of these differences is that RNA polymerase II does not itself recognize the core promoter. Instead, the initial attachment is made by the **TATA-binding protein** or **TBP**, which, as its name indicates, binds to the TATA box. The TBP is a component of a larger protein called **transcription factor IID** or **TFIID**, which has at least 12 additional subunits called **TBP-associated factors** or **TAFs**.

Structural studies have shown that TBP has a saddle-like shape that wraps partially around the DNA molecule, forming a platform onto which RNA polymerase II attaches (*Fig. 15.7*). Correct positioning of RNA polymerase onto this platform is aided by two more transcription factors, TFIIB and TFIIF. Once RNA polymerase II is in place, the open promoter complex forms. This step requires TFIIE and TFIIH, the latter of which is a helicase which breaks the base pairs around the transcription initiation site, opening up the DNA molecule at this point. RNA synthesis can now begin, though only after RNA polymerase II has been activated. This involves addition of phosphate groups to the largest subunit of the enzyme, specifically to a series of amino acids within the

Figure 15.7 The attachment of TBP to a DNA molecule.
The TBP protein is shown in purple and the two strands of the DNA molecule in blue and red. TBP forms a platform onto which RNA polymerase II attaches.
Reproduced from *Biochemistry* 8th edition by Berg *et al.* (© 2015, WH Freeman and Company) and used with permission of the publisher.

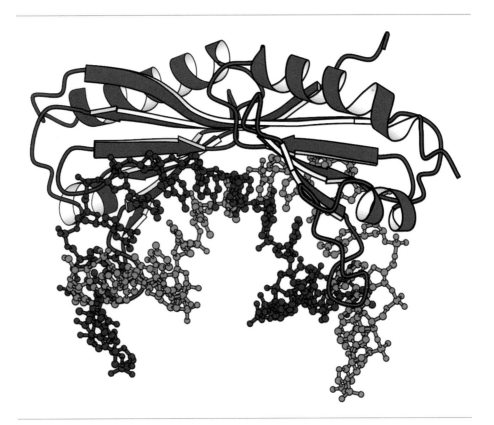

Figure 15.8 Activation of RNA polymerase II by phosphorylation of the C-terminal domain.

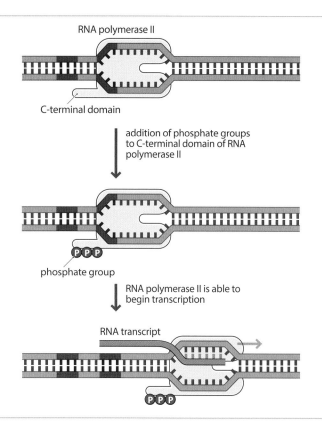

The added complexity of the initiation events in eukaryotes enables the transcription rate of individual genes to respond to many different types of regulatory regime. We will explore these regulatory processes in *Section 17.1.2*.

part of the protein referred to as the **C-terminal domain** (**CTD**). Once these phosphates have been attached, the polymerase is able to leave the initiation complex and begin synthesizing RNA (*Fig. 15.8*).

After the polymerase has departed, some of the transcription factors remain attached to the promoter. A second RNA polymerase II enzyme can therefore attach without the need to construct an entirely new initiation complex. This means that once a gene is switched on, transcripts can be initiated from its promoter with relative ease until a new set of signals switches the gene off.

15.1.3 The RNA synthesis phase of transcription

Once successful initiation has been achieved, the RNA polymerase begins to synthesize the transcript. The reaction is equivalent to the synthesis of DNA during replication, with the RNA being assembled by addition of nucleotides to the 3′ end of the growing molecule and with each addition releasing a pyrophosphate. The order of nucleotide additions is dictated by the base pairing rules using the DNA polynucleotide as the template (*Fig. 15.9*).

As the RNA is synthesized, a transcription bubble moves along the DNA

During RNA synthesis, the *E. coli* RNA polymerase covers about 30–40 bp of the template DNA. This region includes a **transcription bubble** of 12–14 bp where the DNA base pairs have temporarily been broken by helicase enzymes. Within the transcription bubble the growing transcript is held to the template strand of the DNA by approximately eight RNA–DNA base pairs (*Fig. 15.10*). Structural studies have shown that the DNA molecule lies between the β and β′ subunits of the RNA polymerase, within a trough on the enclosed surface of β′. The active site for RNA synthesis also

Figure 15.9 The reaction catalyzed by a DNA-dependent RNA polymerase.
Compare with the equivalent reaction for a DNA-dependent DNA polymerase (see *Fig. 14.10*).

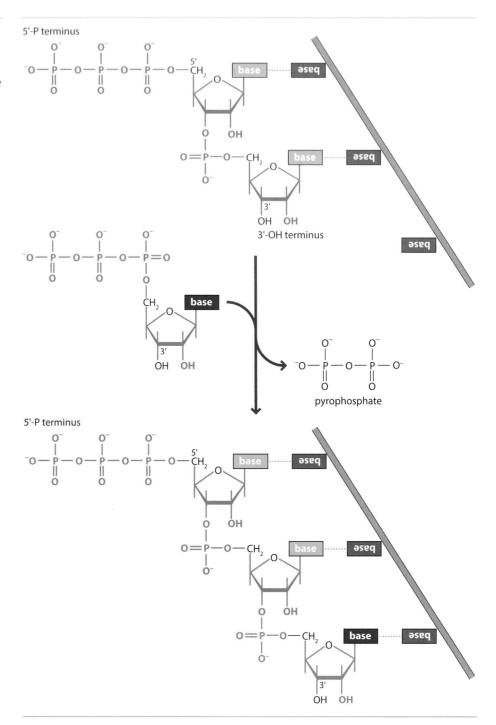

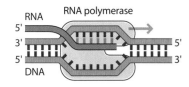

Figure 15.10 RNA synthesis in *E. coli*.
The arrow indicates the direction in which the polymerase moves along the DNA.

lies between these two subunits. The RNA transcript extrudes from the polymerase via a channel formed partly by the β and partly by the β′ subunit (*Fig. 15.11*).

The bacterial RNA polymerase can synthesize RNA at a rate of several hundred nucleotides per minute. The average *E. coli* gene, which is just a few thousand nucleotides in length, can therefore be transcribed in a few minutes. RNA polymerase II has a more rapid synthesis rate, up to 2000 nucleotides per minute, but can take hours to synthesize a single RNA because many eukaryotic genes are much longer than bacterial ones. For example, the 2400 kb transcript of the human dystrophin gene takes about 20 hours to synthesize.

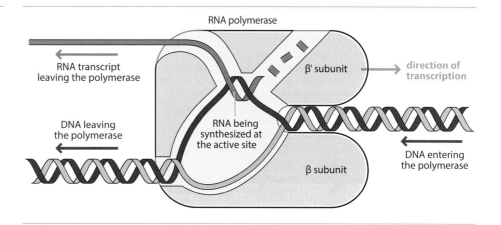

Figure 15.11 **Synthesis of RNA by the *E. coli* RNA polymerase.**
The β and β′ subunits of the RNA polymerase are depicted in pale blue, the double helix is colored green and black, and the RNA transcript is blue.

Transcription does not proceed at a constant rate. Instead, periods of rapid elongation are interspersed by brief pauses lasting a few milliseconds. During a pause the active site of the polymerase undergoes a slight structural rearrangement. The polymerase might also move back along the template by a few nucleotides, an event referred to as **backtracking**. Pauses and backtracking occur randomly, rather than being caused by any particular feature of the template DNA, and might play an equivalent role to proofreading during DNA replication, ensuring that errors made during RNA synthesis are corrected. However, the error-correcting process is not particularly efficient, because RNA polymerase makes an uncorrected mistake once every 10^4–10^5

Box 15.2 **Rifamycins are important antibiotics that block bacterial RNA synthesis**

The bacterial RNA polymerase is the target for the rifamycin family of antibiotics, which include rifampicin and rifabutin. These compounds bind to the trough between the β and β′ subunits of the enzyme, adjacent to the active site where RNA synthesis takes place. They do not interfere directly with phosphodiester bond formation, but instead form a physical blockage in the channel through which the RNA transcript emerges from the polymerase. As a result of this 'steric occlusion' the length of the RNA that can be synthesized is limited to 2–3 nucleotides.

Rifamycins are synthesized by *Amycolatopsis rifamycinica*, a member of the Actinomycetales, the group of soil bacteria that are the source of many antibiotics. Some rifamycins are also made by chemical synthesis. They are particularly effective against mycobacteria, which include the causative agents of tuberculosis and leprosy, and can also be used to treat Legionnaires' disease and some types of meningitis.

As with all antibiotics, the use of rifamycins has to be carefully controlled to avoid the build-up of resistant strains of bacteria. These resistant varieties arise because of mutations in the *rpoB* gene, which codes for the β subunit of RNA polymerase. The resulting amino acid changes in the β subunit do not affect the catalytic activity of RNA polymerase, but change the structure of the β subunit in such a way that the antibiotic is no longer able to bind.

rifampicin blocks RNA elongation

RNA being synthesized at the active site

See *Section 17.2.1* for details of the processes that result in degradation of defective RNAs.

nucleotides, much higher than the rate for most DNA polymerases (typically one error per 10^9 nucleotides). Some transcripts are therefore defective and are degraded soon after they have been synthesized. This is not a problem because many transcripts are made from each gene, so there are always plenty that are error-free.

The great length of some eukaryotic genes means that the transcription complex must be very stable, otherwise it might break apart before the transcript has been completed. When RNA polymerase II carries out RNA synthesis *in vitro* (i.e. under laboratory conditions in a test tube), it does not display the necessary stability. Its polymerization rate is reduced to less than 300 nucleotides per minute and the transcripts that it makes are relatively short. In the nucleus, RNA synthesis is more rapid and the resulting transcripts are much longer than those achievable in the test tube. This is because, in the nucleus, the attachment of RNA polymerase II to the template DNA is stabilized by **elongation factors**. These proteins, of which there are at least 13 different types in mammalian cells, associate with the polymerase after it has moved away from the promoter and remain attached until RNA synthesis is complete.

An RNA polymerase II transcript has a cap structure at its 5′ end

In one respect, the transcript elongation phases in bacteria and eukaryotes are radically different. RNA molecules made by the *E. coli* RNA polymerase have a triphosphate at their 5′ ends, this being the 5′ triphosphate of the very first nucleotide to be base-paired to the transcription initiation site (see *Fig. 15.9*). The 5′ end of the RNA can therefore be denoted as *pppNpN…*, where 'N' is the sugar–base component of the nucleotide and '*p*' represents a phosphate group. In contrast, the RNA molecules made in eukaryotes by RNA polymerase II have a more complicated chemical structure at their 5′ ends. This structure is called the **cap**, and is described as 7–Me*GpppNpN…*, where '7–MeG' is a nucleotide carrying the modified base 7-methylguanine.

The cap structure is attached to the 5′ end of the RNA molecule after the polymerase has left the promoter region, but before the RNA has reached 30 nucleotides in length. To begin the capping process, a GTP is attached to the extreme 5′ end of the RNA, by the enzyme **guanylyl transferase**. The reaction is between the 5′ triphosphate of the terminal nucleotide and the triphosphate of the incoming GTP. The γ phosphate of the terminal nucleotide is removed, as are the β and γ phosphates of the GTP, resulting in a 5′–5′ bond (*Fig. 15.12A*). The new terminal guanosine is then converted into 7-methylguanosine by the attachment of a methyl group to nitrogen number 7 of the purine ring. This modification is catalyzed by **guanine methyltransferase**. The two capping enzymes, guanylyl transferase and guanine methyltransferase, make attachments with the RNA polymerase, and it is possible that they are intrinsic components of the transcription complex during the early stages of RNA synthesis.

The structure described as 7–Me*Gppp*N is called a **type 0 cap**. In unicellular eukaryotes such as yeast, the capping reaction stops when the type 0 structure has been made. In higher eukaryotes, including humans, at least one of two additional steps usually takes place. In the first of these the second nucleotide in the RNA is modified by replacement of the 2′-OH group of the ribose with a methyl group (*Fig. 15.12B*). The resulting structure is called the **type 1 cap**. If this second nucleotide is an adenosine, then at this stage a methyl group might also be added to nitrogen number 6 of the purine. The type 1 cap can therefore involve both ribose and base methylation. In addition, a second 2′-OH methylation of the ribose might occur at the third nucleotide position. This reaction results in a **type 2 cap**.

The role of the cap structure in protein synthesis is described in *Section 16.2.2*.

What is the function of the cap structures added to the transcripts made by RNA polymerase II? Most of these transcripts are mRNAs, copies of protein-coding genes, and the cap structure plays a role when these transcripts are used to direct the synthesis of proteins.

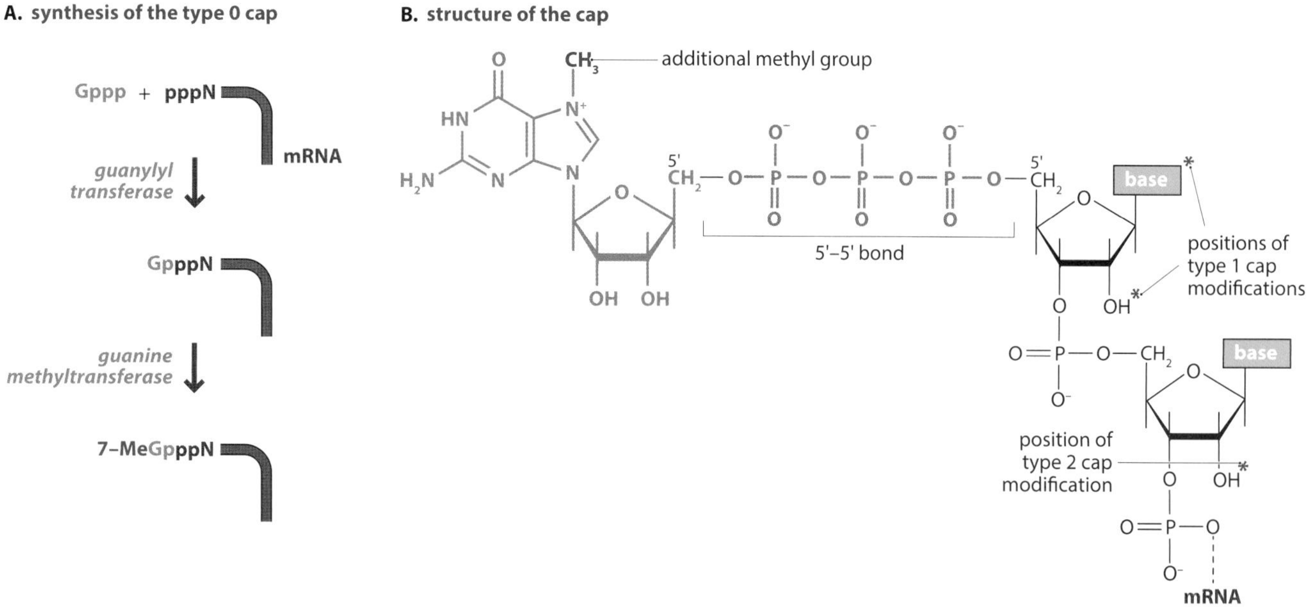

A. synthesis of the type 0 cap

B. structure of the cap

Figure 15.12 **Capping of a eukaryotic mRNA.**
(A) Synthesis of the type 0 cap. (B) The detailed structure of the type 0 cap, showing the positions at which additional modifications are made to give the type 1 and type 2 structures.

15.1.4 Termination of transcription

We have seen that during the elongation phase of transcription, the RNA polymerase pauses frequently, possibly backtracking slightly as it does so. We believe that at each pause the polymerase makes a choice, in a molecular sense, about whether to continue making an RNA molecule or to terminate the transcription process. Which choice is taken depends on whether continued synthesis or termination is more favorable in thermodynamic terms. This means that termination occurs when the polymerase reaches a position on the DNA molecule where its detachment is more thermodynamically favorable than continued RNA synthesis.

Stem–loop structures favor detachment over RNA synthesis

One of the factors that changes the thermodynamics of the transcription process is the structure of the RNA molecule that is being made. In particular, if the RNA can form a stem–loop structure (see *Fig. 4.12*), then these intramolecular base pairs might form in preference to the base pairs between the transcript and the template strand of the DNA. When this happens, detachment of the RNA from the template, which results in termination of transcription, might be more favorable than continued RNA synthesis.

Most positions in the *E. coli* DNA molecule where termination of transcription occurs contain an inverted repeat sequence which, when transcribed, can be folded into a stable stem–loop structure in the RNA. At about half of these positions, the inverted repeat sequence is followed immediately by a run of adenine nucleotides. These positions are called **intrinsic terminators** (*Fig. 15.13A*). The run of adenines is thought to reduce further the stability of the transcription complex at the termination position, because the base pairing between RNA and DNA after the stem–loop has formed will be predominantly of the relatively weak A–U type (*Fig. 15.13B*). Some studies suggest that the RNA stem–loop makes contact with a flap structure on the outer surface of the RNA polymerase β subunit. This flap is adjacent to the exit point of the channel through which the RNA emerges from the polymerase (*Fig. 15.14*).

Figure 15.13 **Termination of transcription at an intrinsic terminator in *E. coli*.**
(A) The structure of an intrinsic terminator, showing how a pair of inverted repeats in the DNA sequence can give rise to a stem–loop in the RNA that is transcribed. (B) Base pairing at an intrinsic terminator. Formation of the RNA stem–loop means that the transcript is held to the DNA by relatively weak A–U base pairs.

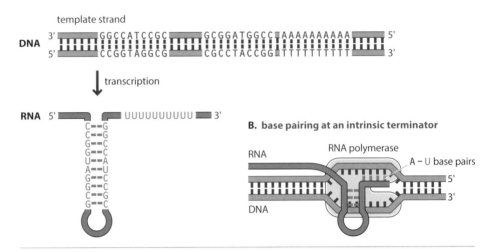

Movement of the flap is thought to affect the positioning of amino acids within the active site, promoting the breakage of the DNA–RNA base pairs and the termination of transcription.

At other termination positions on the *E. coli* DNA, the inverted repeat sequence specifies an RNA stem–loop that is less stable than those involved in intrinsic termination. These terminators are called **Rho dependent**, and work in a very different manner to the intrinsic versions. They require the activity of a protein called **Rho**, which attaches to the RNA while it is being synthesized. The Rho protein then moves along the transcript towards the polymerase (*Fig. 15.15*). If the polymerase can keep ahead of the pursuing Rho, then the transcript continues to be synthesized, but once Rho catches up, transcription stops. This is because Rho is a helicase and so breaks the base pairs holding the RNA to the DNA template. Rho is only able to catch the polymerase when the latter reaches the termination site, probably because formation of the stem–loop structure delays the polymerase briefly. Once caught, the polymerase detaches and the transcript is released.

Termination of eukaryotic mRNA synthesis is combined with polyadenylation

In eukaryotes, termination of transcription by RNA polymerase II does not involve an RNA stem–loop structure, but instead is accompanied by the addition of a **poly(A) tail**

Figure 15.14 **Possible role of the RNA polymerase flap structure in termination of transcription in *E. coli*.**

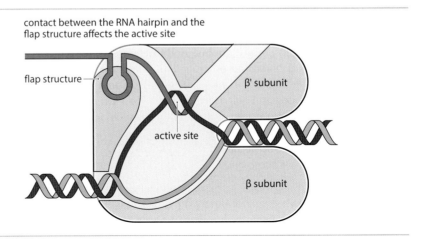

Figure 15.15 **Termination of transcription at a Rho-dependent terminator in E. coli.**

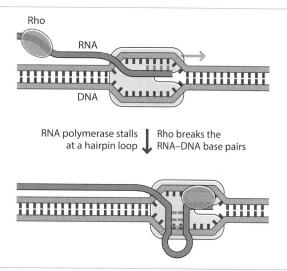

RNA polymerase stalls at a hairpin loop | Rho breaks the RNA–DNA base pairs

to the transcript. The poly(A) tail is a series of up to 250 adenine nucleotides that are placed at the 3' end of the RNA. These adenines are not specified by the DNA and are added to the transcript by a template-independent RNA polymerase called **poly(A) polymerase**.

Poly(A) polymerase does not act at the extreme 3' end of the transcript, but at an internal site which is cut to create a new 3' end to which the poly(A) tail is added. In mammals, the position at which this cut is made is located 10–30 nucleotides downstream of a signal sequence, which is almost always 5'–AAUAAA–3' (*Fig. 15.16*). This sequence acts as the binding site for a multisubunit protein called the **cleavage and polyadenylation specificity factor** (**CPSF**). A second protein complex, the **cleavage stimulation factor** (**CstF**), attaches just downstream of the signal sequence. Poly(A) polymerase then associates with bound CPSF and CstF and synthesizes the poly(A) tail.

Polyadenylation was once looked on as a 'post-transcriptional' event but it is now known to be an inherent part of the mechanism for termination of transcription by RNA polymerase II. CPSF attaches to the polymerase during the initiation stage and rides along the template with it. As soon as the poly(A) signal sequence is transcribed, CPSF leaves the polymerase and binds to the RNA, initiating the polyadenylation reaction. Both CPSF and CstF form contacts with the C-terminal domain of RNA polymerase II, with the nature of these contacts changing when the poly(A) signal sequence is

Figure 15.16 **Polyadenylation of a eukaryotic mRNA.**

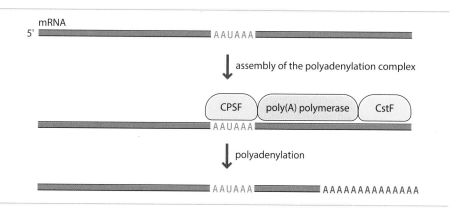

reached. These changes are thought to affect RNA polymerase II in some way, so that termination becomes favored over continued RNA synthesis. As a result, transcription stops soon after the poly(A) signal sequence has been transcribed.

15.2 RNA processing

Now that we understand how RNA molecules are synthesized we can move on to study how the initial products of transcription, called the **primary transcripts**, are processed into functional RNAs. These processing events are important because bacterial mRNA is the only type of RNA that is functional as soon as it has been synthesized. All other types of RNA must undergo processing events before taking up their roles in the cell. There are three types of these processing events:

- Some noncoding RNAs are cut at specific positions to release the functional molecules, and these and other noncoding RNAs might also be processed by removal of short segments from the ends of the molecules ('end-trimming').

- RNA transcripts of eukaryotic protein-coding genes, as well as various noncoding RNAs, are processed by removal of internal segments called introns.

- Ribosomal and transfer RNAs are modified by the addition of new chemical groups at specific positions.

15.2.1 Processing of noncoding RNA by cutting and end-trimming

First, we will consider how the primary transcripts of noncoding RNA are converted to the mature molecules by a combination of internal cutting and end-trimming. As we study these events we must keep one very important factor in mind. All of these processing reactions must be carried out with precision. The cuts must be made at exactly the right positions in the precursor molecules, because if they are not then the RNAs that result will have extra sequences, or may lack segments. If this happens, then they might not be able to carry out their required functions in the cell.

Ribosomal RNAs are transcribed as long precursor molecules

Bacteria synthesize three rRNA molecules, with lengths of 2904, 1541 and 120 nucleotides. Traditionally, these rRNAs are referred to by their **sedimentation coefficients**, which are measures of the rate at which they migrate through a dense solution during **density gradient centrifugation**. According to this notation, the bacterial rRNAs are 23S, 16S and 5S.

The bacterial ribosome contains one copy of each of the 23S, 16S and 5S rRNAs. This means that the bacterium needs to make equal amounts of each rRNA. This is achieved by having the three rRNAs linked together in a single transcription unit (*Fig. 15.17A*), seven copies of which are distributed around the *E. coli* DNA molecule. The product of transcription, the primary transcript, is therefore a long RNA precursor, the **pre-rRNA**, containing each rRNA separated by short spacers. This pre-rRNA has a sedimentation coefficient of 30S, containing the rRNAs in the order 16S–23S–5S. Note that sedimentation coefficients are not additive because they depend on shape as well as mass, hence it is possible for the intact pre-RNA to have an S value that is different from the sum of its three components.

A series of cutting and trimming events produce the mature rRNAs. The cuts are made by a variety of endoribonucleases, most of which cut RNA molecules specifically at double-stranded regions. They therefore cut the pre-rRNA by digesting short segments of double-stranded RNA formed by base pairing between different parts of the precursor (*Fig. 15.17B*). The base pairing, which of course is determined

Box 15.3 **Density gradient centrifugation**

Density gradient centrifugation is one of a series of techniques that were devised for studying cell components in the 1920s, when high-speed centrifuges were first developed. To begin this procedure, a sucrose solution is layered into a centrifuge tube so that a concentration gradient is formed, with the solution being more concentrated and hence more dense towards the bottom of the tube. A cell extract is placed on the top of the solution

cell extract
added to top
of gradient

centrifuge at
500 000 × g
for several
hours

increasing
sucrose
concentration

cell components
form bands
depending on their
sedimentation
coefficients

and the tube centrifuged at 500 000 × g or more for several hours. Under these conditions, the rate of migration of a cell component through the gradient depends on its **sedimentation coefficient**, which in turn depends on its molecular mass and shape. The sedimentation coefficient is expressed as a Svedberg (S) value, after the Swedish chemist The Svedberg, who built the first ultracentrifuge in the early 1920s.

In a second type of density gradient centrifugation (known as isopycnic centrifugation), a solution such as 8 M cesium chloride is used which is substantially denser than the sucrose solution used to measure S values. The starting solution is uniform and the gradient is established during the centrifugation. Cellular components migrate down through the centrifuge tube but molecules such as DNA and proteins do not reach the bottom. Instead these come to rest at a position where their **buoyant density** matches the density of the cesium chloride solution. This technique is able to separate DNA fragments of different base compositions as well as DNA molecules with different conformations, such as the supercoiled and non-supercoiled versions of a circular molecule.

by the sequence of the pre-rRNA, therefore ensures that these cuts are made at the correct positions. Synthesis of the mature rRNAs is then completed by exonuclease enzymes, which trim the ends left by the endoribonucleases.

In eukaryotes there are four rRNAs, of 28S (4718 nucleotides), 18S (1874 nucleotides), 5.8S (160 nucleotides) and 5S (120 nucleotides). As in bacteria, each ribosome contains one copy of each rRNA, but in eukaryotes only three of these are synthesized as a single unit, the odd one out being the 5S rRNA. The pre-rRNA

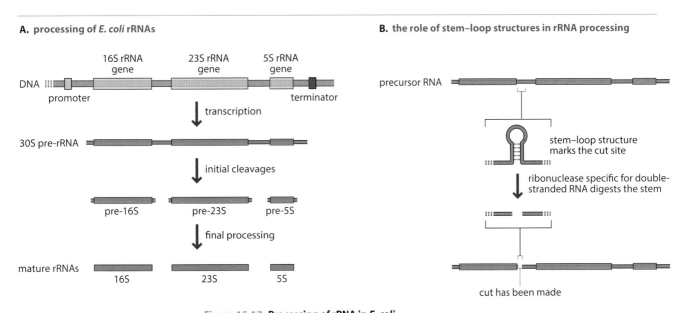

A. processing of E. coli rRNAs

B. the role of stem–loop structures in rRNA processing

Figure 15.17 **Processing of rRNA in E. coli.**
(A) The primary rRNA transcript is processed by a series of cleavages followed by trimming of the ends of the resulting molecules. (B) The positions at which cuts are made are marked by stem–loop structures.

containing the 23S, 18S and 5.8S molecules is transcribed by RNA polymerase I, and then processed by cutting and trimming events that are very similar to those that we have just studied for the bacterial pre-rRNA. The 5S rRNA genes, which are not a part of the main transcription unit, are transcribed by RNA polymerase III.

Transfer RNAs are also processed by cutting and trimming

Transfer RNAs are also transcribed initially as precursor molecules which are subsequently cut and trimmed to release the mature molecules. In *E. coli* there are several separate tRNA transcription units, some containing just one tRNA gene and some with as many as seven different tRNA genes in a cluster. There are also either one or two tRNAs located between the 16S and 23S genes in each of the seven rRNA transcription units.

See *Section 4.1.2* for the tRNA cloverleaf structure.

Each of the **pre-tRNAs** is processed in a similar way (*Fig. 15.18*). Before processing begins, the tRNA adopts its base-paired cloverleaf structure. Two additional stem–loop structures also form, one on either side of the tRNA. Processing begins with a cut by ribonuclease E or F forming a new 3′ end just upstream of one of the hairpins. Ribonuclease D, which is an exonuclease, trims seven nucleotides from this new 3′ end and then pauses while ribonuclease P makes a cut at the start of the cloverleaf, forming the 5′ end of the mature tRNA. Ribonuclease D then removes two more nucleotides, creating the 3′ end of the mature molecule.

Ribonuclease P is an example of a ribozyme, an enzyme made of RNA (see *Section 7.1.1*).

All tRNAs end with the trinucleotide 5′–CCA–3′. With some tRNAs this terminal sequence is present in the pre-RNA and is not removed by ribonuclease D, but with others this sequence is absent, or is removed by the processing ribonucleases. If absent or removed during processing, then the sequence has to be added by one or more template-independent RNA polymerases such as **tRNA nucleotidyltransferase**.

Figure 15.18 Processing of a pre-tRNA in *E. coli*.

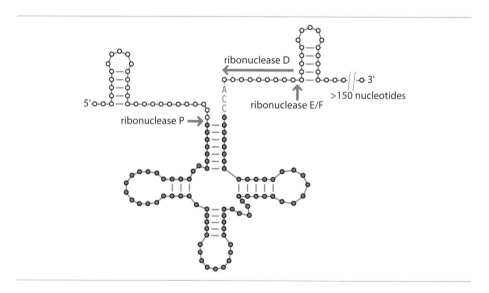

15.2.2 Removal of introns from eukaryotic pre-mRNA

Next we turn our attention to what is usually looked on as the most important category of RNA processing. This is the removal of introns from the primary transcripts of protein-coding genes in eukaryotes.

Many eukaryotic genes are discontinuous

The classical view is that a gene is a single continuous segment of DNA, with a **colinear** relationship between the sequence of nucleotides in the gene and the order

Figure 15.19 **Relationships between a gene and its polypeptide.**
(A) A colinear relationship between a gene and its polypeptide, with the series of nucleotides in the gene, read in the 5′→3′ direction, having a direct relationship with the amino sequence of the polypeptide, as read from the amino- to carboxyl-terminus. (B) The non-colinear relationship that is seen with a discontinuous gene.

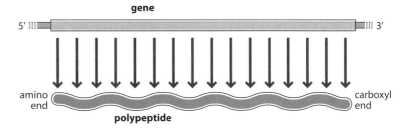

A. colinear relationship between gene and polypeptide

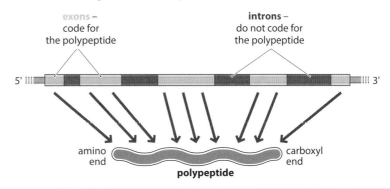

B. discontinuous gene – relationship is not colinear

of amino acids in the polypeptide specified by the gene (*Fig. 15.19A*). During the 1950s and 1960s, geneticists strove to design experiments to prove that genes and proteins were colinear. When successful experiments were eventually carried out with *E. coli*, the results were presented as confirming an assumption that was always looked on as correct. It was therefore a surprise when in the late 1970s it was realized that many genes in eukaryotes are not colinear with their proteins.

The particular genes for which colinearity does not hold are called **discontinuous genes**. In a discontinuous gene (also called a split or mosaic gene) the DNA that specifies the amino acid sequence of the polypeptide is divided into segments, called **expressed sequences** or **exons** (*Fig. 15.19B*). The DNA between a pair of exons is called an **intervening sequence** or **intron**. The intron is still made of As, Cs, Gs and Ts, but the DNA sequence within an intron does not contribute to the amino acid sequence of the protein encoded by the gene. Discontinuous genes are common in eukaryotes. More than 95% of all human genes have at least one intron, and the average number is nine.

Few rules can be established for the distribution of introns in protein-coding genes, beyond the fact that introns are less common in lower eukaryotes such as yeast. The 6000 genes in the yeast genome contain only 239 introns in total, whereas many individual mammalian genes contain 50 or more introns. In many discontinuous genes, the introns are much longer than the exons. In the longest genes, the introns added together make up over 90% of the length of the gene.

During transcription, all of the segments of the gene, both the exons and introns, are transcribed into the pre-mRNA (*Fig. 15.20*). Before this RNA molecule can be used to direct synthesis of a polypeptide, the introns must be removed and the exons linked together. This process, called **splicing**, must be performed with absolute precision. If a cut is made just one nucleotide away from the actual exon–intron boundary position, then the mRNA that is made will be nonfunctional.

Splicing must be accurate because loss of a nucleotide will disrupt the genetic code used when an mRNA is translated into protein. This will become clear when we study the genetic code in *Section 16.1.1*.

Figure 15.20 The pre-mRNA transcribed from a discontinuous gene must be spliced to give the functional mRNA.

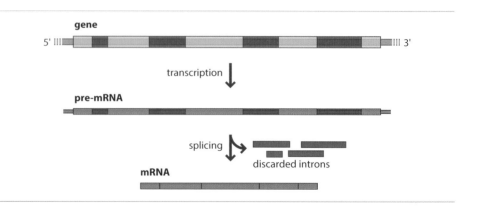

Precision during splicing is ensured by the positioning of special sequences

The high degree of precision that is needed when introns are cut out of a pre-mRNA is ensured by the presence of special sequences at the exon–intron boundary positions. With most pre-mRNA introns, the first two nucleotides of the intron sequence are 5'–GU–3' and the last two are 5'–AG–3' (*Fig. 15.21*). They are therefore called **GU–AG introns**.

Figure 15.21 The positions of the 5'–GU–3' and 5'–AG–3' sequences in a pre-mRNA intron.

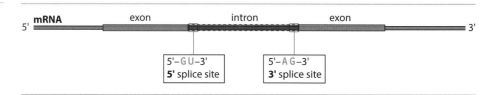

The conserved GU and AG motifs were recognized soon after the first introns were studied. As intron sequences began to accumulate in the databases it was realized that the GU and AG motifs are actually the internal nucleotides of longer conserved sequences that are present at the upstream and downstream boundaries of every intron. As with the promoter sequences of *E. coli*, these sequences are not exactly the same for every intron, and so are represented as consensus sequences. For the upstream splice site, which is also called the **5' splice site** or the **donor site**, the consensus is 5'–AG↓GUAAGU–3', the arrow indicating the position where the cut is made at this end of the intron during splicing. At the downstream position, called the **3' splice site** or **acceptor site**, the consensus is rather less well defined, being 5'–PyPyPyPyPyPyNCAG↓–3'. In this notation, 'Py' is either of the two pyrimidine nucleotides found in RNA (C or U), and 'N' is any of the four nucleotides.

In vertebrates, the motifs at the 5' and 3' splice sites are the only conserved sequences that are present in an intron. There is, however, a **polypyrimidine tract**, made up of a high proportion of cytosine and uracil nucleotides, located just upstream of the 3' end of the intron sequence. In yeast introns, there is also a 5'–UACUAAC–3' sequence, exactly the same in every one of the introns in this genome, located between 18 and 140 nucleotides upstream of the 3' splice site.

Splicing is complicated by the topological issues

When considered as a biochemical reaction, intron splicing is a straightforward two-step process (*Fig. 15.22*).

- During the first step, the 5' splice site is cut by a **transesterification** reaction promoted by the 2'-OH group of an adenosine nucleotide located within the intron sequence. In yeast this nucleotide is the last A in the 5'-UACUAAC-3' sequence. Cleavage of the phosphodiester bond at the 5' splice site is accompanied by

Figure 15.22 Intron splicing in outline.

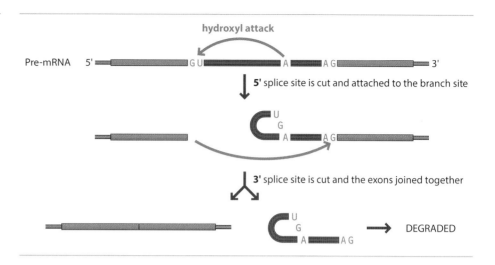

hydroxyl attack

Pre-mRNA 5' — G U — A — A G — 3'

5' splice site is cut and attached to the branch site

3' splice site is cut and the exons joined together

DEGRADED

formation of a new 5′–2′ phosphodiester bond linking the first nucleotide of the intron (the G of the 5′–GU–3′ motif) with the internal adenosine. The intron becomes looped back on itself to form a lariat structure.

- In the second step, the 3′ splice site is cleaved by a second transesterification, this one promoted by the 3′-OH group attached to the end of the upstream exon. This group attacks the phosphodiester bond at the 3′ splice site, cleaving it and releasing the intron as the lariat structure. The 3′ end of the upstream exon joins to the newly formed 5′ end of the downstream exon, completing the splicing process.

The complications in splicing result from topological issues. Some introns are thousands of nucleotides in length, which means that the two splice sites might be 100 nm or more apart if the mRNA is in the form of a linear chain. A means is therefore needed of bringing the splice sites into proximity. This is achieved by the combined action of five structures called **small nuclear ribonucleoproteins (snRNPs)**. Each snRNP contains several proteins and one of the noncoding snRNAs. There are a number of different snRNAs in vertebrate nuclei, the ones involved in splicing being the U1, U2, U4, U5 and U6 snRNAs. These are short, uracil-rich molecules, 106–185 nucleotides in length. Together with other accessory proteins, the snRNPs attach to specific positions in the mRNA and form a series of complexes, the most important of which is the **spliceosome**, the structure within which the actual splicing reactions occur.

Splicing initiates with formation of **complex A** (*Fig. 15.23*). This complex comprises U1-snRNP, which binds to the 5′ splice site, partly by RNA–RNA base pairing, and U2-snRNP, which attaches to the branch site, probably not by base pairing but instead by an interaction between the branch site and one of the proteins associated with the

---

Box 15.4 Transesterification **PRINCIPLES OF CHEMISTRY**

Transesterification is the reaction between an ester and an alcohol. During a transesterification the two reactants exchange R groups.

ROH + [ester] → ROH + [ester]

alcohol   ester   alcohol   ester

During the first step of intron splicing, the alcohol group is provided by the hydroxyl attached to the 2′-carbon of the adenosine at the branch point, and the ester group is a component of the phosphodiester bond at the 5′ splice site. During the second step, the alcohol is the 3′–hydroxyl at the end of the upstream exon and the ester is the phosphodiester bond at the 3′ splice site.

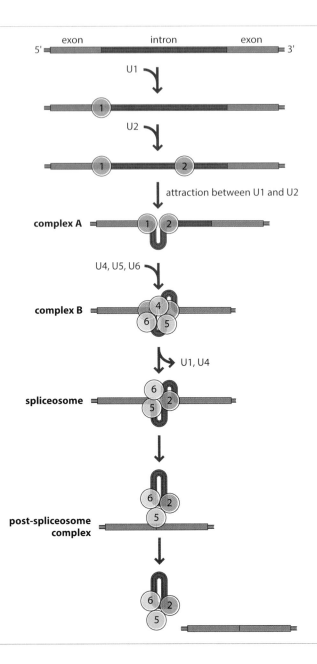

U2-snRNP. The U1- and U2-snRNPs have an affinity for each other, and this draws the 5′ splice site towards the branch point. **Complex B** is then formed when the U4-, U5- and U6-snRNPs attach to the intron. Their arrival results in additional interactions that bring the 3′ splice site close to the 5′ site and the branch point. U1- and U4-snRNPs then leave the complex, giving rise to the spliceosome. All three key positions in the intron are now in proximity and the cutting and joining reactions can take place, catalyzed by the U2- and U6-snRNPs. The initial product of the splicing reaction is a **post-spliceosome complex**, which dissociates into the spliced mRNA and the intron lariat, the latter still attached to the U2- U5- and U6-snRNPs.

Some noncoding RNAs contain a different type of intron, which is autocatalytic

There are several different types of intron, some related to GU–AG introns but others with quite different characteristic features. One family, called **Group I**, is particularly interesting because introns of this type are able to catalyze their own splicing reaction.

Box 15.5 **The 'minor spliceosome'**

A small number of pre-mRNA introns in eukaryotes do not fall into the GU–AG class. Based on their boundary sequences, these introns were originally called the AU–AC type, but now that more examples have been characterized it has become clear that the boundary sequences are variable, and not all contain the AU and AC motifs. The underlying biochemistry of the splicing reaction is identical in both GU–AG and AU–AC introns, the latter being spliced by two transesterifications, the first initiated by the 2'-OH on the final adenosine of the sequence 5'–UCCUUAAC–3', which is almost invariant in this type of intron.

Although the biochemistry is the same, the special sequence features of AU–AC introns mean that a different set of snRNPs is needed. Only the U5-snRNP is involved in the splicing mechanisms of both types of intron. For AU–AC introns, the roles of the U1- and U2-snRNPs are taken by the U11- and U12-snRNPs, respectively, and the U4- and U6-snRNPs are replaced by the U4atac- and U6atac-snRNPs.

Because they are so uncommon, the splicing complex for AU–AC introns is called the **minor spliceosome**.

Group I introns are most common in the DNA molecules present in mitochondria and chloroplasts, but the first one to be discovered was located in the nuclear rRNA of *Tetrahymena*, a ciliated protozoan. The biochemistry underlying the splicing reaction for Group I introns is similar to that of pre-mRNA introns in that two transesterifications are involved. The first of these, resulting in cleavage of the 5' splice site, is induced not by a nucleotide within the intron, as is the case with GU–AG introns, but by a free nucleoside or nucleotide, any one of guanosine, GMP, GDP or GTP (*Fig. 15.24*). The 3'-OH of this cofactor attacks the phosphodiester bond at the 5' splice site, cleaving it, with transfer of the guanosine to the 5' end of the intron. The second transesterification involves the 3'-OH at the free end of the upstream exon, which attacks the phosphodiester bond at the 3' splice site. This results in the 3' site being cut, enabling the exons to be joined. The intron is released as a linear structure, which may undergo additional transesterifications, leading to circular products, as part of its degradation process.

The remarkable feature of the Group I splicing reaction is that it proceeds in the absence of proteins and hence is autocatalytic, with the RNA itself possessing enzymatic activity. This was the first example of an RNA enzyme or ribozyme to be discovered, back in the early 1980s. The **self-splicing** activity of Group I introns depends on the base-paired structure taken up by the RNA. This structure comprises nine major base-paired regions, two of which form a pair of domains at the active site of the ribozyme. The two splice sites are brought into proximity by interactions between other parts of the secondary structure. Although this RNA structure is sufficient for splicing, with some Group I introns the stability of the ribozyme is thought to be enhanced by non-catalytic proteins that bind to it.

Figure 15.24 The splicing reaction for an autocatalytic Group I intron. The 'G' that initiates the splicing reaction can be either guanosine, GMP, GDP or GTP.

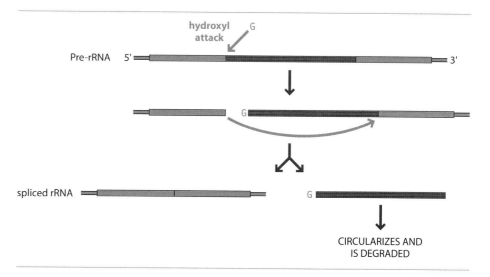

Box 15.6 **Alternative splicing**

When introns were first discovered it was thought that a discontinuous gene would have just one **splicing pathway**, in which all of the exons were joined together to give a single mRNA. We now know that some discontinuous genes have two or more **alternative splicing** pathways, which means that the pre-mRNA can be processed in a variety of ways to give a series of mRNAs made up of different combinations of exons. These mRNAs will direct synthesis of related but different proteins.

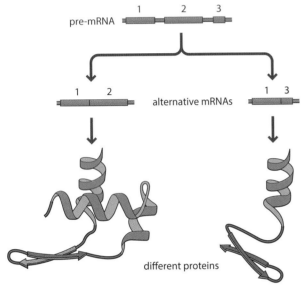

pre-mRNA

alternative mRNAs

different proteins

Alternative splicing is rare in some eukaryotes, but is relatively common in vertebrates, with approximately 75% of all the human protein-coding genes being able to follow two or more splicing pathways. An example is provided by the human calcitonin/CGRP gene, which has two splicing pathways that give rise to quite different proteins. The first of these is calcitonin, a short peptide hormone that is made in the thyroid and, in conjunction with the parathyroid hormone, regulates the calcium ion concentration in the bloodstream. The second product of the gene is the calcitonin gene-related peptide (CGRP), which is a neurotransmitter active in sensory neurons and involved in the pain response. The calcitonin/CGRP gene has six exons:

- Exon 1 covers the part of the mRNA between the transcription start site and the actual start of the gene.
- Exons 2 and 3 specify a **signal peptide**. This is a short sequence of amino acids that enables the protein to enter the endoplasmic reticulum. Once inside the endoplasmic reticulum, the signal peptide is cut off and the protein moves to the Golgi apparatus prior to secretion from the cell. We will study this and other aspects of protein targeting in *Section 16.4.*
- Exon 4 codes for the remainder of the calcitonin protein.
- Exons 5 and 6 code for CGRP.

The pre-mRNA of the calcitonin/CGRP gene therefore follows two tissue-specific splicing pathways. In the thyroid, exons 1–2–3–4 are spliced together to give the precursor form of calcitonin, containing the signal peptide and the peptide hormone. In nervous tissue, exons 1–2–3–5–6 are joined to give the mRNA for the CGRP neurotransmitter.

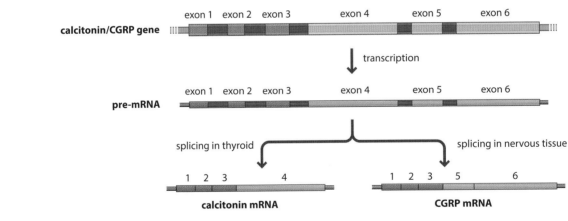

15.2.3 Chemical modification of noncoding RNA

> We studied these chemical modifications in *Section 4.1.3.*

As well as being processed by cutting and joining reactions, some noncoding RNAs are also chemically modified. We have already looked at the nature of these chemical modifications, and their effect on tRNA molecules. Now we will explore the events that occur during pre-rRNA processing.

Transfer RNAs display a range of chemical modifications, but with rRNAs the majority of these changes are either the conversion of uracil to pseudouracil (see *Fig. 4.15*), or 2′-O-methylation, in which the hydrogen of the –OH group attached to

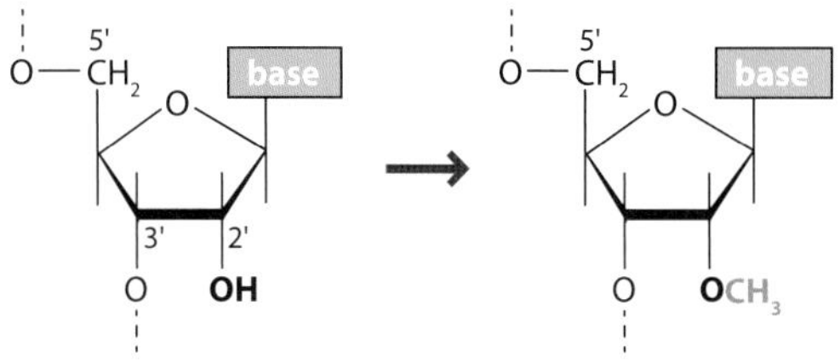

Figure 15.25 **2′-O-methylation of a nucleotide.**

the 2′-carbon of the ribose is replaced by a methyl group (*Fig. 15.25*). The four human rRNAs undergo pseudouridinylation at 95 positions and methylation at another 106, in other words about one modification for every 35 nucleotides. Each of these alterations is at a specific position, always the same in every copy of the rRNA.

In bacteria, rRNAs are modified by enzymes that directly recognize the sequence and/or base-paired structure of the regions of RNA that contain the nucleotides to be modified. In eukaryotes the modification positions do not display any sequence or structural similarity, and a more sophisticated process is needed to ensure positional specificity. This process involves the small nucleolar RNAs or snoRNAs.

Small nucleolar RNAs base pair to the regions on the rRNA where modifications must be made. The first snoRNAs to be discovered were those that direct the methylation reactions. Each of these snoRNAs forms a few base pairs with the rRNA, this pairing always located immediately upstream of a sequence called the D box, which is present in all snoRNAs (*Fig. 15.26*). The nucleotide that will be modified forms the base pair that is five positions away from the D box. The conversion of uracil to pseudouracil also involves snoRNAs. These do not have D boxes but still have conserved motifs that can be recognized by the modifying enzyme, targeting the modification at the correct nucleotide in the rRNA.

There is a different snoRNA for each modified position in a pre-rRNA, except for a few sites that are close enough to be dealt with together. This means that humans make approximately 200 different snoRNAs. A few of these are transcribed from snoRNA genes by RNA polymerase III, but most are specified by sequences within the introns of protein-coding genes, and hence are transcribed by RNA polymerase II, and are released by cutting up these introns after splicing.

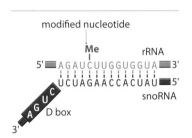

Figure 15.26 **The role of a snoRNA in rRNA methylation.**
Note that the interaction between the rRNA and the snoRNA involves a G–U base pair, which is permissible between RNA polynucleotides.

Further reading

Banerjee S, Chalissery J, Bandey I and Sen R (2006) Rho-dependent transcription termination: more questions than answers. *Journal of Microbiology* **44**, 11–22.

Bujard H (1980) The interaction of *E. coli* RNA polymerase with promoters. *Trends in Biochemical Sciences* **5**, 274–8.

Buratowski S (2009) Progression through the RNA polymerase II CTD cycle. *Molecular Cell* **36**, 541–6. *Describes the role of the C-terminal domain of RNA polymerase II during transcription.*

Cech TR (1990) Self-splicing of group I introns. *Annual Review of Biochemistry* **59**, 543–68. *Review of self-splicing introns.*

Cougot N, van Dijk E, Babajko S and Séraphin B (2004) 'Cap-tabolism'. *Trends in Biochemical Science* **29**, 436–44. *mRNA capping.*

Green MR (2000) TBP-associated factors (TAF$_{II}$s): multiple, selective transcriptional mediators in common complexes. *Trends in Biochemical Science* **25**, 59–63.

Kandah E, Trowitzsch S, Gupta K, Haffke M and Berger I (2014) More pieces to the puzzle: recent structural insights into class II transcription initiation. *Current Opinions in Structural Biology* **24**, 91–7.

Klug A (2001) A marvellous machine for making messages. *Science* **292**, 1844–6. *Description of the bacterial RNA polymerase.*

Manley JL and Takagaki Y (1996) The end of the message – another link between yeast and mammals. *Science* **274**, 1481–2. *Polyadenylation.*

Matera AG and Wang Z (2014) A day in the life of the spliceosome. *Nature Reviews Molecular Cell Biology* **15,** 108–21.

Padgett RA, Grabowski PJ, Konarska MM and Sharp PA (1985) Splicing messenger RNA precursors: branch sites and lariat RNAs. *Trends in Biochemical Sciences,* **10**, 154–7. *A good summary of the basic details of intron splicing.*

Saecker RM, Record MY and deHaseth PL (2011) Mechanism of bacterial transcription initiation: RNA polymerase–promoter binding, isomerization to initiation-competent open complexes, and initiation of RNA synthesis. *Journal of Molecular Biology* **412**, 754–71.

Tollervey D (1996) Small nucleolar RNAs guide ribosomal RNA methylation. *Science* **273**, 1056–7.

Tora L and Timmers HT (2010) The TATA box regulates TATA-binding protein (TBP) dynamics *in vivo*. *Trends in Biochemical Science* **35**, 309–14. *Details of the recognition of the TATA box by TBP.*

Toulokhonov I, Artsimovitch I and Landick R (2001) Allosteric control of RNA polymerase by a site that contacts nascent RNA hairpins. *Science* **292,** 730–3. *A model for termination of transcription in bacteria involving the flap structure on the outer surface of the RNA polymerase.*

Travers AA and Burgess RR (1969) Cyclic re-use of the RNA polymerase sigma factor. *Nature* **222**, 537–40. *The first demonstration of the role of the σ subunit.*

Venema J and Tollervey D (1999) Ribosome synthesis in *Saccharomyces cerevisiae*. *Annual Review of Genetics* **33**, 261–311. *Extensive details on rRNA processing.*

Wahl MC, Will CL and Lührmann R (2009) The spliceosome: design principles of a dynamic RNP machine. *Cell* **136**, 701–18. *Review of intron splicing with focus on the role of the spliceosome.*

Self-assessment questions

Multiple choice questions

Only one answer is correct for each question. Answers can be found on the website: www.scionpublishing.com/biochemistry

1. Which of the following is **not** a type of noncoding RNA?
- **(a)** rRNA
- **(b)** tRNA
- **(c)** miRNA
- **(d)** mRNA

2. Which of the following is the consensus sequence for the −35 box of an *E. coli* promoter?
- (a) 5′–TTGACA–3′
- (b) 5′–TGGACA–3′
- (c) 5′–TCGACA–3′
- (d) 5′–TAGACA–3′

3. Which of the following is the consensus sequence for the −10 box of an *E. coli* promoter?
- **(a)** 5′–TATTAT–3′
- **(b)** 5′–TAAAAT–3′
- **(c)** 5′–TATAAT–3′
- **(d)** 5′–TTTAAT–3′

4. How is the structure of the *E. coli* RNA polymerase described?
- **(a)** $\alpha_2\beta\beta'\sigma$
- **(b)** $\alpha\beta_2\beta'\sigma$
- **(c)** $\alpha_2\beta_2\sigma$
- **(d)** $\alpha\beta\beta'\sigma$

5. Which of the following statements is **incorrect** with regard to transcription initiation in *E. coli*?
- **(a)** Attachment of the RNA polymerase involves interactions between the σ subunit and the −35 and −10 boxes of the promoter
- **(b)** The RNA polymerase covers a region of some 20 base pairs
- **(c)** The closed complex forms first and is then converted to an open promoter complex
- **(d)** The σ subunit is responsible for recognizing the promoter sequence, and disassociates from the enzyme soon after it has attached to the DNA

6. Which types of genes are transcribed by RNA polymerase II?
   (a) Protein-coding genes, most snRNA genes, miRNA genes
   (b) 28S, 5.8S and 18S rRNA genes
   (c) 5S rRNA, tRNAs, and various small RNA genes
   (d) All the above genes

7. What is the name of the sequence located around nucleotide +1 in the promoter for a eukaryotic protein-coding gene?
   (a) Initiation codon
   (b) TATA box
   (c) Alternative promoter
   (d) Initiator sequence

8. What is the name of the protein that recognizes and binds to the TATA box of the promoter for a eukaryotic protein-coding gene?
   (a) TBP
   (b) TAF
   (c) CTD
   (d) RNA polymerase II

9. How is RNA polymerase II activated?
   (a) Addition of phosphate groups to the C-terminal domain of the smallest subunit
   (b) Dissociation of the σ subunit
   (c) Addition of phosphate groups to the C-terminal domain of the largest subunit
   (d) Dissociation of the mediator protein

10. How large is the transcription bubble in *E. coli*?
    (a) 8–10 bp
    (b) 12–14 bp
    (c) 20–24 bp
    (d) 80 bp

11. How rapidly does RNA polymerase II synthesize RNA?
    (a) Up to 1000 nucleotides per minute
    (b) Up to 2000 nucleotides per minute
    (c) Over 10 000 nucleotides per minute
    (d) Over 50 000 nucleotides per minute

12. Which of the following statements is **correct** regarding the cap structure?
    (a) A GTP is attached to the extreme 5′ end of the RNA by the enzyme guanine methyltransferase
    (b) Capping occurs after the RNA has reached 30 nucleotides in length
    (c) The initial reaction is between the 5′ triphosphate of the terminal nucleotide and the triphosphate of the incoming GTP
    (d) The cap includes 5-methylguanosine

13. An intrinsic terminator in *E. coli* typically contains which features?
    (a) An inverted repeat sequence
    (b) A run of adenine nucleotides
    (c) A sequence that can form a stem–loop structure in the RNA transcript
    (d) All of the above

14. What is Rho in a Rho-dependent terminator?
    (a) A helicase protein
    (b) A subunit of the RNA polymerase
    (c) A stem–loop structure
    (d) A topoisomerase

15. Which of the following statements is **incorrect** regarding polyadenylation of a eukaryotic mRNA?
    (a) The poly(A) tail is a series of up to 250 adenine nucleotides that are placed at the 5′ end of the RNA
    (b) These adenines are added by poly(A) polymerase
    (c) The mRNA is cut prior to polyadenylation
    (d) The CPSF and CstF proteins are involved in polyadenylation

16. What is the sedimentation coefficient of the *E. coli* pre-rRNA?
    (a) 5S
    (b) 18S
    (c) 23S
    (d) 30S

17. During tRNA processing in *E. coli*, which enzyme makes the cut that forms the 5′ end of the mature tRNA?
    (a) Ribonuclease D
    (b) Ribonuclease E
    (c) Ribonuclease F
    (d) Ribonuclease P

18. In a discontinuous gene, the segments that do not contribute to the amino acid sequence of the protein are called what?
    (a) Exons
    (b) Extrons
    (c) Intons
    (d) Introns

19. What is the consensus sequence of the 5′ splice site of a GU–AG intron?
    (a) 5′–AGG↓UAAGU–3′
    (b) 5′–AG↓GUAAGU–3′
    (c) 5′–AGGUAAGU↓–3′
    (d) 5′–↓AGGUAAGU–3′

20. What are the biochemical reactions occurring during splicing of a GU–AG intron called?
    (a) Transformations
    (b) Oxidations
    (c) Transesterifications
    (d) Transoxidations

21. What is the structure within which the splicing reactions for a GU–AG intron occur called?
    (a) Spliceosome
    (b) Intron
    (c) snRNP
    (d) Ribosome

22. Which of the following statements is **incorrect** with regard to splicing of a Group I intron?

    (a) Two transesterifications are involved

    (b) Cleavage of the 5' splice site is induced by a nucleotide within the intron

    (c) The RNA of the intron possesses catalytic activity

    (d) The intron is released as a linear structure

23. Which two types of chemical modification are most common in rRNAs?

    (a) Conversion of uracil to 4-thiouracil and 2'-*O*-methylation

    (b) Conversion of uracil to 4-thiouracil and 3'-*O*-methylation

    (c) Conversion of uracil to pseudouracil and 2'-*O*-methylation

    (d) Conversion of uracil to pseudouracil and 3'-*O*-methylation

24. How are most of the human snoRNAs synthesized?

    (a) Transcribed by RNA polymerase I

    (b) Transcribed by RNA polymerase II

    (c) Transcribed by RNA polymerase III

    (d) By cutting up spliced introns

Short answer questions

These questions do not require additional reading.

1. Outline the functions of the various types of noncoding RNA found in eukaryotic cells.

2. Describe the structure and role of the *E. coli* promoter.

3. Draw a series of diagrams illustrating the initiation of transcription in *E. coli*. Carefully distinguish the role of the σ subunit of RNA polymerase in this process.

4. How does the structure of the promoter for a protein-coding gene in a eukaryote differ from that of *E. coli*?

5. Describe the roles of the β and β' subunits during elongation of an RNA transcript by the *E. coli* RNA polymerase. What is the flap structure and what role is this structure thought to play during transcription?

6. Using diagrams, describe how the cap and poly(A) tail are added to a eukaryotic mRNA.

7. Distinguish between the intrinsic and Rho-dependent methods for transcript termination in *E. coli*.

8. Outline the cutting and trimming events that occur during processing of rRNA and tRNA transcripts.

9. Give a detailed description of the splicing process for a GU–AG intron.

10. Describe the role of snoRNAs in processing of noncoding RNA.

Self-study questions

These questions will require calculation, additional reading and/or internet research.

1. Construct a hypothesis to explain why eukaryotes have three RNA polymerases. Can your hypothesis be tested?

2. Current thinking views transcription as a discontinuous process, with the polymerase pausing regularly and making a 'choice' between continuing elongation by adding more nucleotides to the transcript, or terminating by dissociating from the template. Which choice is selected depends on which alternative is more favorable in thermodynamic terms. Evaluate this view of transcription.

3. Discontinuous genes are common in higher organisms but virtually absent in bacteria. Discuss the possible reasons for this.

4. To what extent has the study of AU–AC introns provided insights into the details of GU–AG intron splicing?

5. Discuss the reasons why tRNA and rRNA molecules are chemically modified.

CHAPTER 16

Protein synthesis

After reading this chapter you will:

- understand the key features of the genetic code including the code variations that exist in some species

- know how an amino acid is attached to a tRNA and how errors are avoided during this process

- be able to describe unusual types of aminoacylation, where the amino acid initially attached to a tRNA is not the one specified by that tRNA

- appreciate the importance of codon–anticodon interactions in deciphering the genetic code, and understand the role of wobble in this process

- be able to describe the composition of the bacterial and eukaryotic ribosome

- know how the three-dimensional structure of the ribosome is related to the role of the ribosome in protein synthesis

- be able to describe the events occurring during initiation of translation in *Escherichia coli* and eukaryotes, and in particular know the roles of the various initiation factors

- be able to describe the events occurring during the elongation phase of translation in *E. coli* and eukaryotes

- understand how translation is terminated in *E. coli* and eukaryotes

- be able to give examples of proteins that are processed by proteolytic cleavage

- be able to give examples of chemical modifications that are made to various proteins

- understand the role of chemical modification of proteins in signal transduction pathways and in control of gene expression by histones

- understand the role of sorting sequences in protein targeting

- be able to describe the exocytosis pathway for secreted proteins, and know the diversions from this pathway for proteins destined for various cellular membranes and for lysosomes

In the second stage in the gene expression pathway, an mRNA molecule directs the synthesis of a protein. This process is called **translation**, because the sequence of nucleotides in the mRNA is *translated* into the sequence of amino acids in the resulting protein molecule. There are four different aspects of protein synthesis that we must study:

- First, we must understand the **genetic code**, which is the set of rules that specifies how a nucleotide sequence is converted into an amino acid sequence. We must examine both the features of the genetic code and the way in which the rules laid down by the code are enforced during translation of an individual mRNA.

- Secondly, we must examine the mechanics of the protein synthesis process in order to understand how amino acids are assembled into a polypeptide chain.

- Thirdly, we must study the **post-translational processing** events that are needed to synthesize the functional versions of some proteins. These events include the removal of segments from one or both ends of a polypeptide, cleavage of a large protein into smaller segments, and chemical modification of certain amino acids.

- Finally, we must understand how **protein targeting** results in transport of a protein from its assembly site to the place in the cell where it will perform its function.

16.1　The genetic code

The genetic code is the set of rules that governs the way in which the nucleotide sequence of an mRNA specifies the amino acid sequence of a protein. First, we will study the features of the genetic code, and then we will examine how the rules of the code are enforced during translation of an mRNA.

16.1.1　The features of the genetic code

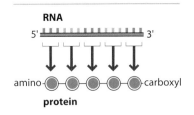

Figure 16.1　The relationship between an mRNA and its protein. Each set of three adjacent nucleotides is a codon that specifies a single amino acid in the protein.

Deciphering the genetic code was one of the major preoccupations of biochemists during the early 1960s. This work established that there is a linear relationship between a gene and the protein it specifies, meaning that the order of nucleotides in the gene correlates with the order of amino acids in the corresponding protein (see *Fig. 15.19*). Other experiments showed that three adjacent nucleotides in an mRNA form a code word, or **codon**, with each codon specifying the identity of a single amino acid in the resulting protein (*Fig. 16.1*). Gradually, the meaning of each of the triplet codons was worked out, the culmination of this whole research programme coming in 1966 when the genetic code operating in the bacterium *Escherichia coli* was completely solved.

The genetic code is degenerate and includes punctuation codons

A triplet code based on four letters has $4^3 = 64$ combinations (AAA, AAT, TAT, GCA, etc.). As there are just 20 amino acids that are specified by the genetic code, we can predict that the code must be **degenerate,** which in this context means that some amino acids will be specified by more than one codon. This is in fact the case. Eighteen of the 20 amino acids have more than one codon, the two exceptions being methionine, which is coded just by AUG, and tryptophan, whose only codon is UGG (*Fig. 16.2*). Most synonymous codons are grouped into families. For example, GGA, GGU, GGG and GGC all code for glycine, so in effect the code word for glycine is GGN, where 'N' is any of the four nucleotides. This similarity between synonymous codons is relevant to the way the code is deciphered during protein synthesis, as we will see when we examine the role of tRNA later in this chapter.

Four of the 64 triplets act as **punctuation codons**. These codons indicate the start and end of the nucleotide sequence that must be translated into protein. They are needed because the **open reading frame**, the part of the mRNA that codes for the amino acid sequence, does not begin with the very first nucleotide of the mRNA, nor does it end with the last nucleotide (*Fig. 16.3*). It is preceded by a noncoding **leader segment**, and followed by a similarly noncoding **trailer** segment, also called the **5'–** and **3'–untranslated regions** (**UTR**), respectively. **Initiation** and **termination codons** are therefore needed to mark the start and end of the open reading frame.

The triplet AUG is the initiation codon for most mRNAs. This triplet codes for methionine, so most newly synthesized polypeptides have this amino acid at the amino terminus, though the methionine may subsequently be removed after the protein has been made. There is only one codon for methionine, so methionines that

Figure 16.2 The genetic code.

UUU UUC	phe	UCU UCC	ser	UAU UAC	tyr	UGU UGC	cys
UUA UUG	leu	UCA UCG		UAA	stop	UGA	stop
				UAG		UGG	trp
CUU CUC CUA CUG	leu	CCU CCC CCA CCG	pro	CAU CAC	his	CGU CGC CGA CGG	arg
				CAA CAG	gln		
AUU AUC	ile	ACU ACC	thr	AAU AAC	asn	AGU AGC	ser
AUA		ACA ACG		AAA AAG	lys	AGA AGG	arg
AUG	met						
GUU GUC GUA GUG	val	GCU GCC GCA GCG	ala	GAU GAC	asp	GGU GGC GGA GGG	gly
				GAA GAG	glu		

are located within a polypeptide are also specified by AUG codons. We will see later in this chapter how the initiation codon is distinguished from these internal AUG codons when the mRNA is translated.

Three triplets, UAA, UAG and UGA, act as termination codons, with one of these always occurring at the end of the open reading frame, at the point where translation must stop. In *E. coli*, these are the only three triplets that do not specify an amino acid.

There are variations to the genetic code

When the genetic code for *E. coli* was worked out in the 1960s, it was assumed that the code would be the same in all organisms. It was difficult to imagine how the code could ever change, because giving a new meaning to any single codon would result in widespread disruption of the amino acid sequences of an organism's proteins. So it was thought that the genetic code must have become established at a very early stage in evolution, and then become 'frozen' so that it would be the same in all modern day species.

This assumption has turned out to be incorrect. The code shown in *Figure 16.2* is correct for the vast majority of genes in the vast majority of organisms, but is by no means universal. Deviations were first discovered when the short DNA molecules present in mitochondria were first studied. It was shown that several of the genes present in the human mitochondrial DNA have internal UGA codons which do not specify the end of the open reading frame. In these genes, UGA codes for tryptophan. Three other deviations from the standard genetic code were also found when the nucleotide sequences of human mitochondrial genes were compared with the amino acid sequences of the corresponding proteins. Two codons AGA and AGG, which usually specify arginine, are termination codons in these genes, and AUA codes for methionine rather than isoleucine (*Table 16.1*). Similar code deviations are known in the mitochondrial genes of other species, and some have been discovered in the nuclear genomes of unicellular eukaryotes such as protozoa and yeast. Deviations

Figure 16.3 The structure of an mRNA, showing the positions of the punctuation codons.

Table 16.1. Deviations from the standard genetic code seen with human mitochondrial genes

Codon	Should code for	Actually codes for
UGA	Stop	Tryptophan
AGA, AGG	Arginine	Stop
AUA	Isoleucine	Methionine

are less common among prokaryotes, but examples are known in *Micrococcus* and *Mycoplasma* species.

A second type of code variation is **context-dependent codon reassignment**, which occurs when the protein contains selenocysteine or pyrrolysine (see *Fig. 3.10*). These are two unusual amino acids that are not looked on as members of the standard set of 20, but which are found in a few proteins. Proteins containing pyrrolysine are rare, and are probably only present in some archaea and a very small number of bacteria, but proteins containing selenocysteine are widespread in many organisms (*Table 16.2*). An example is the enzyme **glutathione peroxidase**, which helps protect mammalian cells against oxidative damage. Selenocysteine is coded by UGA, which therefore has a dual meaning because it is still used as a termination codon in the organisms concerned. A UGA codon that specifies selenocysteine is distinguished by the presence of a stem–loop structure in the mRNA, positioned just downstream of the codon in prokaryotes and in the trailer region in eukaryotes (*Fig. 16.4*). Recognition of the selenocysteine codon requires interaction between the stem–loop and a special protein that is involved in translation of these mRNAs. A similar system probably operates with pyrrolysine, which is specified by another of the termination codons, UAG.

Table 16.2. Examples of proteins that contain selenocysteine

Organism	Protein	Protein function
Mammals	Glutathione peroxidase	Conversion of H_2O_2 to H_2O (see *Box 11.6*)
	Thioredoxin reductase	Regeneration of thioredoxin after its reduction, e.g. during regulation of the Calvin cycle (see *Fig. 10.17*)
	Iodothyronine deiodinase	Activation and deactivation of thyroid hormones
Bacteria	Formate dehydrogenase	Oxidation of formate to CO_2
	Glycine reductase	Reductive deamination of glycine linked to substrate-level phosphorylation
	Proline reductase	Reduction of proline linked to substrate-level phosphorylation
Archaea	Formylmethanofuran dehydrogenase	Reduction of CO_2 as part of the methanogenesis pathway resulting in conversion of CO_2 to methane

Figure 16.4 Context-dependent reassignment of a UGA codon.
A UGA codon that specifies selenocysteine is distinguished by the stem–loop structure, which is positioned in the mRNA just downstream of the codon in prokaryotes, as shown here, or in the trailer region of a eukaryotic gene.

16.1.2 How the genetic code is enforced during protein synthesis

Now that we understand the features of the genetic code, we can move on to address the crucial question of how the rules embodied in the code are enforced when an mRNA is translated into a polypeptide. The key players in this enforcement are tRNAs, which form a physical link between the mRNA that is being read and the protein that is being made (*Fig. 16.5*).

Each triplet codon is recognized by a tRNA, which delivers the appropriate amino acid to the growing end of the polypeptide chain. To understand how tRNAs play this role we must study two topics:

- **Aminoacylation**, which is the process by which the correct amino acid is attached to a tRNA.

- **Codon–anticodon recognition**, which is the interaction between the tRNA and the codon on an mRNA.

The specificity of aminoacylation is ensured by aminoacyl-tRNA synthetases

Aminoacylation is catalyzed by an **aminoacyl-tRNA synthetase** and results in the amino acid becoming linked to the nucleotide at the 3' end of the tRNA, within the part of the cloverleaf structure called the acceptor arm (*Fig. 16.6*). This 3'–terminal nucleotide is always an adenine. Aminoacylation is a two-step process:

Step 1. An activated aminoacyl-AMP intermediate (*Fig. 16.7*) is formed by reaction between the amino acid and ATP.

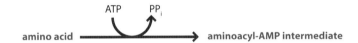

We examined tRNA structure in *Section 4.1.2.*

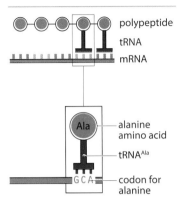

Figure 16.5 The role of tRNAs in protein synthesis.
See *Box 16.1* for the notation used when naming tRNAs.

Figure 16.6 The attachment of an amino acid to a tRNA.
The cloverleaf structure of the tRNA is shown, and the names given to the different parts of the base-paired structure are indicated. In this example, the amino acid is attached to the 2'-OH of the terminal nucleotide . This is the linkage produced by a class I aminoacyl-tRNA synthetase. A class II aminoacyl-tRNA synthetase attaches the amino acid to the 3'-OH group.

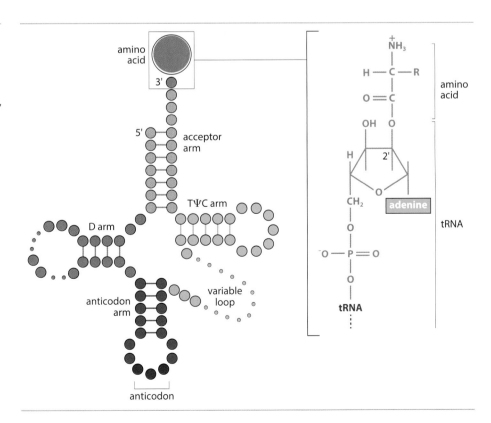

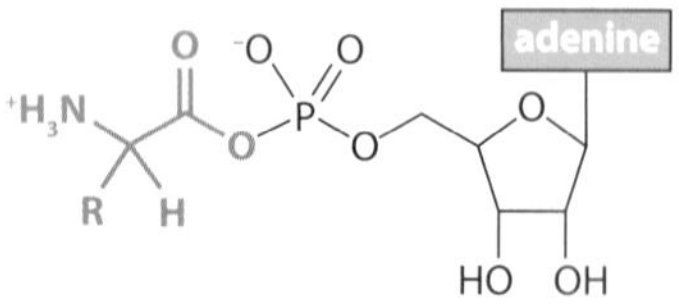

Figure 16.7 The aminoacyl-AMP compound formed as an intermediate during the aminoacylation reaction. The amino acid is shown in blue.

Step 2. The amino acid is then transferred to the 3′ end of the tRNA, releasing AMP.

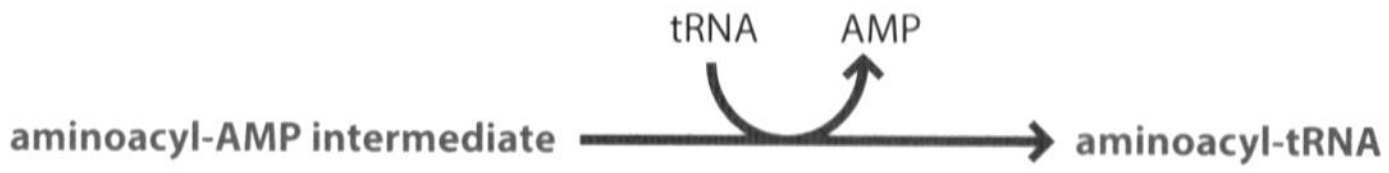

Bacteria contain 30–45 different tRNAs and eukaryotes have up to 50. As there are only 20 amino acids, some groups of tRNAs must be specific for the same amino acid. These groups are called **isoaccepting tRNAs**. However, most organisms have only 20 aminoacyl-tRNA synthetases, one for each amino acid, which means that groups of isoaccepting tRNAs are aminoacylated by a single enzyme.

Surprisingly, the 20 aminoacyl-tRNA synthetases do not make up a single family of enzymes. Instead, there are two distinct types of aminoacyl-tRNA synthetase, with different biochemical properties, the most apparent of which is the nature of the linkage that is made between the amino acid and its tRNA. Class I enzymes attach the amino acid to the 2′-OH group of the terminal nucleotide of the tRNA (as shown in *Fig. 16.6*), whereas Class II enzymes attach the amino acid to the 3′-OH group.

In order for the rules of the genetic code to be enforced, the correct amino acid must be attached to the correct tRNA. The required accuracy is achieved by an extensive interaction between the aminoacyl-tRNA synthetase and the acceptor arm and anticodon loop of the tRNA, as well as individual nucleotides in the D and TψC arms. The interaction between the enzyme and the amino acid is less extensive, simply because an amino acid is much smaller than a tRNA. This means that there are particular problems distinguishing between those amino acids that are structurally similar, such as isoleucine and valine. Errors do therefore occur, at a very low rate for most amino acids, but possibly as frequently as one aminoacylation in 80 for structurally similar pairs. Most errors are corrected by the aminoacyl-tRNA synthetase itself, by an editing process that is distinct from aminoacylation, involving different contacts with the tRNA.

Unusual types of aminoacylation

A few unusual types of aminoacylation are known, these involving variations of the two-step process described above. The commonest example is when the aminoacyl-tRNA synthetase initially attaches an incorrect amino acid to a tRNA, and then changes it to the correct one by a second, separate chemical reaction. This is the way in which *Bacillus megaterium* aminoacylates its tRNA^Gln molecules. This bacterium does not have an aminoacyl-tRNA synthetase that is able to work with glutamine. Instead, the tRNA is aminoacylated with glutamic acid, which is then converted to glutamine by a transamination reaction that is catalyzed by an aminotransferase enzyme (*Fig. 16.8A*). The same process is used by various other bacteria (although not *Escherichia coli*) and by the archaea.

A similar series of events results in synthesis of tRNAs aminoacylated with selenocysteine, required for the context-dependent decoding of UGA codons when

Remember that the cloverleaf is a two-dimensional representation of tRNA structure. In reality, each tRNA adopts the more compact three-dimensional configuration shown in *Figure 4.14*.

Box 16.1 The notation for distinguishing between different tRNAs

The standard format for distinguishing different tRNAs is as follows:
- The amino acid specificity of a tRNA is indicated with a superscript suffix. For example, tRNA^Gly is a tRNA for glycine, and tRNA^Ala is a tRNA for alanine.

- Numbers are used to distinguish different isoaccepting tRNAs. For example, two isoaccepting tRNAs specific for glycine would be written as tRNA^Gly1 and tRNA^Gly2.

A. synthesis of tRNA^{Gln} in some prokaryotes

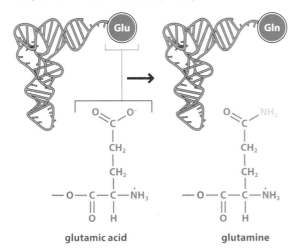

glutamic acid   glutamine

B. synthesis of tRNA aminoacylated with selenocysteine

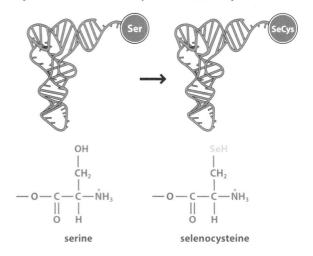

serine   selenocysteine

C. synthesis of tRNA aminoacylated with *N*-formylmethionine

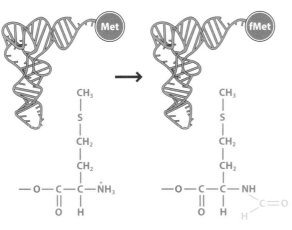

methionine   *N*-formylmethionine

Figure 16.8 Unusual types of aminoacylation.
(A) Synthesis of tRNA^{Gln} (a tRNA aminoacylated with glutamine) in some prokaryotes. (B) Synthesis of tRNA aminoacylated with selenocysteine (SeCys). (C) Synthesis of tRNA aminoacylated with *N*-formylmethionine (fMet). In these drawings, the three-dimensional configuration of the aminoacyl-tRNAs are shown.

selenoproteins such as glutathione peroxidase are being made. These codons are recognized by a special tRNA that is specific for selenocysteine, but there is no aminoacyl-tRNA synthetase which is able to attach selenocysteine to this tRNA. Instead, the tRNA is aminoacylated with serine by the seryl-tRNA synthetase. The serine is then modified by replacement of the –OH group of the serine with an –SeH, to give selenocysteine (*Fig. 16.8B*).

Pyrrolysine is the second unusual amino acid that is incorporated into proteins by context-dependent codon reassignment, but this does not require an unusual aminoacylation process because the organisms that use pyrrolysine possess a specific aminoacyl-tRNA synthetase that directly attaches the amino acid to its tRNA. There is, however, one other unusual aminoacylation that occurs in most bacteria. This involves the tRNA that recognizes AUG initiation codons, and which is aminoacylated with methionine, the amino acid specified by AUG. After attachment to its tRNA, the methionine is modified by addition of a formyl group (–CHO) to produce *N*-formylmethionine (*Fig. 16.8C*). We will return to *N*-formylmethionine when we examine its role in the initiation of translation, later in this chapter.

Codons are read by base pairing between tRNA and mRNA

Attachment of the correct amino acid to a tRNA is the first stage of the process by which the rules of the genetic code are enforced. To complete the process, the tRNA must now recognize and attach to a codon that specifies the amino acid that it carries.

Codon recognition involves a triplet of nucleotides called the **anticodon**, which is located within the anticodon loop of the tRNA (see *Fig. 16.6*). The anticodon is complementary to the codon, and so attaches to it by base pairing (*Fig. 16.9*). The rules of the genetic code are therefore enforced because the anticodon present on a particular tRNA is one that is complementary to a codon for the amino acid that the tRNA is carrying.

There are 61 codons specifying amino acids, but no more than 50 different tRNAs in eukaryotes, and fewer in bacteria. Therefore, some tRNAs must be able to recognize more than one codon. How can this be achieved while retaining the specificity that is essential if the rules of the code are to be followed? The answer lies with the process called **wobble**. Because the anticodon is contained within a loop of the RNA, the triplet of nucleotides is slightly curved. This means that the anticodon cannot make an entirely uniform alignment with the codon. As a result, a nonstandard base pair can form at the 'wobble position', between the third nucleotide of the codon and the first nucleotide of the anticodon (position 34 in *Fig. 16.9*). However, the base pairing rules do not become totally flexible at the wobble position, because only a few types of unusual base pairs are allowed. The two commonest examples are as follows (*Fig. 16.10*):

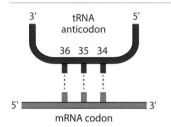

Figure 16.9 The interaction between a codon and its anticodon. The numbers refer to the positions of the nucleotides within the tRNA sequence, position 1 being the nucleotide at the extreme 5′ end.

- **G–U base pairs.** By allowing G to pair with U as well as C, an anticodon with the sequence ♦♦G can base pair with both ♦♦C and ♦♦U. Similarly, the anticodon ♦♦U can base pair with both ♦♦A and ♦♦G. As a result, all four members of a codon family (e.g. GCN, all coding for alanine) can be decoded by just two tRNAs.

- **Anticodons containing inosine**, one of the modified nucleotides present in tRNA. Inosine can base pair with A, C and U. The anticodon UAI (where 'I' is inosine) can therefore pair with each of AUA, AUC and AUU, allowing all three codons for isoleucine to be decoded by a single tRNA.

Wobble reduces the number of tRNAs needed in a cell by enabling one tRNA to read two or possibly three codons. This means that bacteria can decode their mRNAs with as few as 30 tRNAs. Eukaryotes also make use of wobble but in a more restricted way. The human genome, for example, encodes 48 tRNAs, 16 of which use wobble to decode two codons each, with the remaining 32 tRNAs specific for

Figure 16.10 Examples of wobble. (A) Wobble involving G–U base pairs. In these two examples, wobble enables the four-codon family for alanine to be decoded by just two tRNAs. (B) Wobble involving inosine enables the three codons for isoleucine to be read by a single tRNA. In each drawing, the nucleotides at the wobble position are highlighted in red.

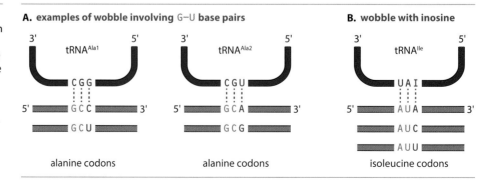

Wobble and alternative initiation codons

The examples of wobble shown in *Figure 16.10* illustrate the flexibility in the base pairing rules that can occur when a polypeptide chain is being extended. As we will see in *Section 16.2.1*, during polypeptide synthesis each consecutive aminoacyl-tRNA enters the part of the ribosome referred to as the A site, and it is within the A site that wobble between the anticodon of the tRNA and the codon on the mRNA can occur.

Each ribosome also has a second tRNA-binding pocket, called the P site, but the only tRNA that can enter directly into this site is the initiator-tRNA that carries methionine in eukaryotes and *N*-formylmethionine in bacteria. The geometries of the A and P sites are different, and in the P site the nature of the contacts between the anticodon of the tRNA and the initiation codon on the mRNA are such that more extreme forms of wobble are possible. Experiments have shown that the initiator-tRNA of eukaryotes can recognize any version of the initiation codon in which two of the three nucleotides correspond to the 'correct'

AUG sequence. This means that in addition to AUG, nine other triplets (CUG, GUG, UUG, AAG, ACG, AGG, AUA, AUC, AUU) can be used as initiation codons. Of these, CUG is the variant that is most efficient at translation initiation, with AAG and AGG being the least effective.

Although the nine alternative triplets can be used as initiation codons with differing degrees of efficiency in experimental systems, alternative initiation codons appear to be used very rarely in eukaryotes. In mammals, just over 20 genes are known to have non-AUG initiation codons. Most of these genes also have a genuine AUG initiation codon a few nucleotides downstream, and the non-standard codon is probably used infrequently. In bacteria, non-AUG initiation codons are more common, with approximately 20% of all bacterial genes using either GUG or UUG for this purpose, the non-standard triplet often being the only initiation codon for its gene.

just a single triplet. Other genomes use more extreme forms of wobble. Translation of mammalian mitochondrial mRNAs requires only 22 tRNAs. With some of these tRNAs the nucleotide in the wobble position of the anticodon is virtually redundant because it can base pair with any nucleotide, enabling all four codons of a family to be recognized by the same tRNA. This phenomenon has been called **superwobble**.

16.2 The mechanics of protein synthesis

Now we move on to the actual process by which an mRNA is translated into a polypeptide. Protein synthesis occurs within the structures called **ribosomes**. Therefore, to understand how proteins are made we must understand what ribosomes are and what they do.

16.2.1 Ribosomes

Ribosomes were originally looked on as passive partners in protein synthesis, simply acting as the structures within which mRNAs are translated into polypeptides. This view has changed over the years and we now realize that ribosomes play two important active roles in protein synthesis:

- Ribosomes *coordinate* protein synthesis by placing the mRNA, aminoacyl-tRNAs and associated protein factors in their correct positions relative to one another.

- Components of ribosomes *catalyze* at least some of the chemical reactions occurring during protein synthesis.

Both of these roles are critically dependent on the structure of the ribosome. Our examination of how proteins are synthesized must therefore begin with a study of ribosome structure.

Ribosomes are made of rRNA and protein

An *E. coli* cell contains approximately 20 000 ribosomes distributed throughout its cytoplasm. The average human cell contains more than a million ribosomes, some free in the cytoplasm and some attached to the outer surface of the endoplasmic reticulum. Ribosomes were first observed in the early decades of the twentieth century as tiny particles almost too small to be visible by light microscopy. In the 1940s and 1950s, when electron microscopy was developed, ribosomes were revealed more clearly as oval shaped structures, with dimensions of 29 nm × 21 nm in bacteria, and a little larger, on average 32 nm × 22 nm, in eukaryotes.

Much of the early progress in understanding the detailed structure of the ribosome was made by analyzing the particles by density gradient centrifugation. These studies showed that eukaryotic ribosomes have a sedimentation coefficient of 80S, and bacterial ones, reflecting their smaller size, are 70S. Each type of ribosome can be broken down into smaller components (*Fig. 16.11*):

- A ribosome comprises two subunits. In eukaryotes these subunits are 60S and 40S, and in bacteria they are 50S and 30S. Remember that sedimentation coefficients are not additive, so the sum of the two subunits can be greater than the S value for the intact ribosome.

- The large subunit contains three rRNAs in eukaryotes (the 28S, 5.8S and 5S rRNAs) but only two in bacteria (23S and 5S rRNAs).

- The small subunit contains a single rRNA in both types of organism. This is an 18S rRNA in eukaryotes and a 16S rRNA in bacteria.

- Both subunits contain a variety of **ribosomal proteins**. There are 50 in the eukaryotic large subunit and 33 in the small one, and in bacteria, 34 in the large subunit and 21 in the small subunit. The ribosomal proteins of the small subunit are called S1, S2, etc., and those of the large subunit are L1, L2, etc. There is just one of each protein per ribosome, except for L7 and L12, which are present as dimers.

The three-dimensional structure of the ribosome

Identifying the components of the ribosome takes us only a little way towards understanding the structural features that underlie the role of the ribosome in protein synthesis. We need to know how those components are assembled together to form a three-dimensional structure.

Density gradient centrifugation and the measurement of sedimentation coefficients were described in *Box 15.3*.

Figure 16.11 The composition of eukaryotic and bacterial ribosomes. There are some variations in the number of ribosomal proteins in different species. The details given are for human ribosomes and those in *E. coli.*

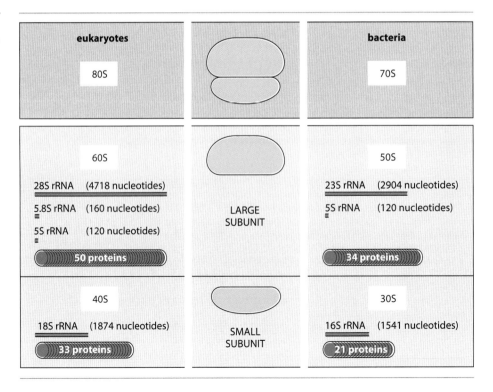

The first steps towards understanding how the rRNA and protein components of the ribosome are assembled were made by examining the nucleotide sequences of rRNA molecules and predicting how individual molecules might fold up through intramolecular base pairing. This work showed that each type of rRNA could adopt a compact structure made up of various stem–loops, with only small regions of the rRNA not participating in internal base pairing. The next step was to identify the binding positions for the ribosomal proteins on the base-paired rRNA structures. These experiments revealed that most ribosomal proteins make contacts with segments of an rRNA that appear to be some distance apart in the two-dimensional representation of the rRNA structure. Clearly the base-paired rRNA is folded up into a more complex three-dimensional structure, with ribosomal proteins making contacts at specific positions within that three-dimensional structure.

For many years, progress in moving from two- to three-dimensional representations of ribosomes was slow. Ribosomes are so small that they are close to the resolution limit of the electron microscope, and in the early days of this technique the best that could be achieved were approximate three-dimensional reconstructions built up by analyzing frustratingly fuzzy images. As electron microscopy gradually became more sophisticated, the overall structure of the ribosome was resolved in greater detail, but the main breakthroughs came when **X-ray diffraction analysis** was used to study purified ribosomes. As a result of these analyses, detailed structures are now known for entire ribosomes, including ones attached to mRNA and tRNAs (*Fig. 16.12A*).

What has all of this structural work told us about the mechanics of protein synthesis? We know that the attachment between the two ribosome subunits is temporary and, when not actively participating in protein synthesis, ribosomes dissociate into their subunits which remain in the cytoplasm waiting to be used for a new round of translation. In bacteria, attachment of the two subunits to one another results in the formation of two sites at which aminoacyl-tRNAs can bind. These are called the **P** or **peptidyl site** and the **A** or **aminoacyl site**. The P site is occupied by the aminoacyl-tRNA whose amino acid has just been attached to the end of the growing polypeptide, and the A site is entered by the aminoacyl-tRNA carrying the next amino acid that will be used. There is also a third site, the **E** or **exit site**, through which the tRNA departs after its amino acid has been attached to the polypeptide (*Fig. 16.12B*). The structures revealed by X-ray diffraction analysis show that these sites are located in the cavity between the large and small subunits of the ribosome, the codon–anticodon interaction being associated with the small subunit and the aminoacyl end of the

For a detailed description of X-ray diffraction analysis, see *Section 18.1.3*.

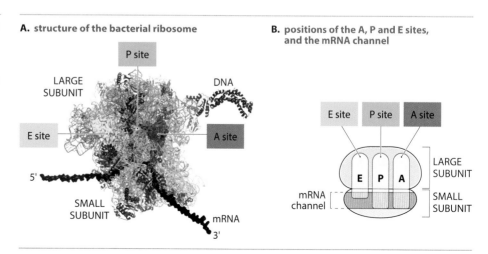

Figure 16.12 Detailed structure of a bacterial ribosome.
(A) Structure of a ribosome in the process of translating an mRNA. The tRNAs positioned in the A, P and E sites are indicated in red, green and yellow, respectively. (B) A diagram showing the relative positions of the A, P and E sites and the channel through which the mRNA is translocated.
(A) reproduced with permission from Macmillan Publishers Ltd from *Nature*, **461**: 1234–42, TM Schmeing and V Ramakrishnan, What recent ribosome structures have revealed about the mechanism of translation; © 2009.

tRNA with the large subunit. The mRNA threads through a channel formed mainly by the small subunit.

The structural work has also revealed some of the dynamics of protein synthesis. As an mRNA is translated, the ribosome must move along the polynucleotide, three nucleotides at a time, so each codon is placed, one after the other, at the correct position between the two subunits. This is called **translocation**. Electron microscopy of ribosomes at intermediate stages in translocation shows that, in order to move along the mRNA, the ribosome adopts a less compact structure, with the two subunits rotating slightly in opposite directions. This opens up the space between them and enables the ribosome to slide along the mRNA.

16.2.2 Translation of an mRNA into a polypeptide

We will now follow through the series of events involved in translation of an mRNA into a polypeptide. These events are similar in bacteria and eukaryotes, though the details are different, most strikingly during the initiation phase.

In bacteria, the ribosome assembles directly onto the initiation codon

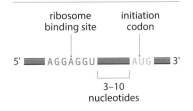

Figure 16.13 The bacterial ribosome binding site.

In bacteria, translation of an mRNA begins with attachment of the small subunit to the **ribosome binding site**, also called the **Shine–Dalgarno sequence**. This is a short sequence within the mRNA, located 3–10 nucleotides upstream of the initiation codon (*Fig. 16.13*). The consensus sequence for the ribosome binding site in *E. coli* is 5'–AGGAGGU–3', although variations occur in the actual sequences that are found upstream of individual genes (*Table 16.3*). The ribosome binding site is complementary to a region at the 3' end of the 16S rRNA, the rRNA that is present in the small subunit. It is therefore thought that base pairing between the mRNA and rRNA is involved in the initial attachment of the small subunit to the ribosome binding site. Attachment of the small subunit is aided by **initiation factor** IF-3. Initiation factors are ancillary proteins that are not permanent components of the ribosome, but which attach at the appropriate times in order to perform their functions (*Table 16.4*). We will meet the other two bacterial initiation factors, IF-1 and IF-2, in a moment.

The ribosome is quite large compared with an mRNA, and so after attachment to the binding site the small subunit covers several tens of nucleotides (*Figure 16.14*). This region includes the initiation codon which, as we already know, is usually AUG and so codes for methionine. The initiation codon is recognized by a special initiator tRNA which carries a methionine that has been modified by attachment of a formyl group (see *Fig. 16.8C*). The initiator tRNA is brought to the small subunit of the ribosome by a second initiation factor, IF-2, along with a molecule of GTP. The attachment of IF-2 is thought to be facilitated by IF-1, although the precise role of IF-1 is unclear, and its main function might be to prevent premature attachment of the large subunit.

Once the initiator tRNA is in place, the initiation phase of translation is completed by attachment of the large subunit of the ribosome. This step requires energy, which is provided by hydrolysis of the molecule of GTP that was previously brought into the complex by IF-2. Attachment of the large subunit is also accompanied by release of the three initiation factors.

Table 16.3. Examples of ribosome binding site sequences in *E. coli*

Gene	Codes for	Ribosome binding sequence	Nucleotides to initiation codon
E. coli consensus	–	5'–AGGAGGU–3'	3–10
Lactose operon	Lactose utilization enzymes	5'–AGGA–3'	7
galE	Hexose 1-phosphate uridyltransferase	5'–GGAG–3'	6
rplJ	Ribosomal protein L10	5'–AGGAG–3'	8

Table 16.4. Functions of the initiation factors in bacteria and eukaryotes

Factor	Function
Bacteria	
IF-1	Unclear; might assist the entry of IF-2 into the initiation complex and prevent premature attachment of the large subunit
IF-2	Directs the initiator tRNA^{Met} to its correct position in the initiation complex, and hydrolyzes GTP to release energy needed in order for the large subunit to attach
IF-3	Mediates reassociation of the large and small subunits of the ribosome
Eukaryotes	
eIF-1	Component of the pre-initiation complex; major role in recognition of the initiation codon
eIF-1A	Component of the pre-initiation complex; aids scanning and initiation codon recognition
eIF-2	Binds to the initiator tRNA^{Met} within the ternary complex component of the pre-initiation complex
eIF-2B	Regenerates the eIF-2–GTP complex
eIF-3	Component of the pre-initiation complex; makes direct contact with eIF-4G and so forms the link with the cap binding complex
eIF-4A	Component of the cap binding complex; a helicase that aids scanning by breaking intramolecular base pairs in the mRNA
eIF-4B	Aids scanning, possibly by acting as a helicase that breaks intramolecular base pairs in the mRNA
eIF-4E	Component of the cap binding complex, possibly the component that makes direct contact with the cap structure at the 5′ end of the mRNA
eIF-4F	The cap binding complex, comprising eIF-4A, eIF-4E and eIF-4G, which makes the primary contact with the cap structure at the 5′ end of the mRNA
eIF-4G	Component of the cap binding complex; forms a bridge between the cap binding complex and eIF-3 in the pre-initiation complex; in at least some organisms, eIF-4G also forms an association with the poly(A) tail, via the polyadenylate-binding protein
eIF-4H	In mammals, aids scanning in a manner similar to eIF-4B
eIF-5B	Brings a molecule of GTP to the initiation complex, and aids release of the other initiation factors at the completion of initiation
eIF-6	Associated with the large subunit of the ribosome; prevents large subunits from attaching to small subunits in the cytoplasm

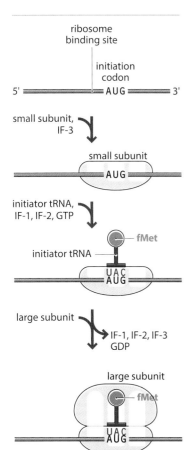

Figure 16.14 Initiation of translation in *E. coli*.

In eukaryotes, the small subunit scans along the mRNA to find the initiation codon

In eukaryotes, the small subunit of the ribosome uses a radically different method in order to locate the initiation codon within an mRNA. Rather than attaching directly to the initiation codon, the small subunit attaches to the 5′ end of the mRNA and then **scans** along the molecule until it locates the initiation codon. In doing so, the small subunit often bypasses AUG triplets that are not the initiation codon. It is able to recognize the correct AUG triplet because it is embedded in a short sequence, consensus 5′–ACCAUGG–3′, that distinguishes the initiation codon from the spurious AUG triplets. This sequence is called the **Kozak consensus**.

Although straightforward in outline, the initiation process in eukaryotes is complex, probably because it is a major control point that determines how rapidly polypeptides are translated from an individual mRNA. The first step involves assembly of the **pre-initiation complex**, which comprises (*Fig. 16.15A*):

- The small subunit of the ribosome.

- A 'ternary complex' made up of the initiation factor eIF-2 (see *Table 16.4*) bound to the initiator tRNA and a molecule of GTP. As in bacteria, the initiator tRNA is distinct from the normal tRNA^{Met} that recognizes internal AUG codons but, unlike bacteria, it carries an ordinary methionine, not the *N*-formylated version.

- Three additional initiation factors, eIF-1, eIF-1A, and eIF-3.

Figure 16.15 Initiation of translation in eukaryotes.
(A) Attachment of the pre-initiation complex to the mRNA. (B) Scanning of the initiation complex in search of the initiation codon. For clarity, some of the initiation factors, as well as the GTP molecule that is part of the pre-initiation complex but is not hydrolyzed until the complex reaches the initiation codon, are not shown.

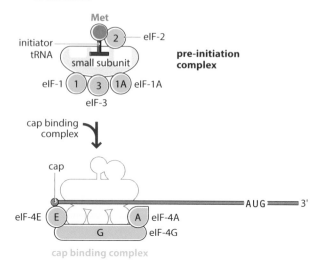

A. attachment of the pre-initiation complex to the mRNA

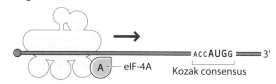

B. scanning

See *Section 15.1.3* for a description of the cap structure.

After assembly, the pre-initiation complex attaches to the 5′ end of the mRNA. Remember that in eukaryotes, mRNAs are made by RNA polymerase II, which means that each one has a cap structure at its 5′ end. This cap appears to be the recognition signal that enables the pre-initiation complex to distinguish mRNAs from other types of RNA in the cytoplasm, and to identify the correct end of the mRNA.

Attachment of the pre-initiation complex to the cap structure requires the **cap binding complex** (sometimes called eIF-4F), which is made up of the initiation factors eIF-4A, eIF-4E and eIF-4G, and is influenced by the poly(A) tail, at the distant 3′ end of the mRNA. This interaction is thought to be mediated by the **polyadenylate-binding protein (PADP)**, which is attached to the poly(A) tail. In yeast and plants it has been shown that PADP can form an association with eIF-4G, this association requiring the mRNA to bend back on itself. With artificially uncapped mRNAs, the PADP interaction is sufficient to attach the pre-initiation complex onto the 5′ end of the mRNA, but under normal circumstances the cap structure and poly(A) tail probably work together. The poly(A) tail could have an important regulatory role, because the length of the tail appears to be correlated with the number of times a particular mRNA is translated per second.

After attachment to the mRNA, the **initiation complex**, as it is now called, begins to scan along the molecule in search of the initiation codon (*Fig. 16.15B*). The leader regions of eukaryotic mRNAs can be several tens, or even hundreds, of nucleotides in length and often contain segments that form stem–loops and other base-paired structures. These segments are probably opened up by eIF-4A, which is a helicase and so is able to break intramolecular base pairs in the mRNA, enabling the initiation complex to pass unhindered. Energy for scanning is provided by ATP hydrolysis; the

---

Box 16.3 Internal ribosome entry sites – initiation of eukaryotic translation without scanning

For many years it was thought that the scanning system was the only process by which translation of a eukaryotic mRNA could be initiated. This assumption was shown to be incorrect in the late 1980s, when picornaviruses were found to use a novel type of translation initiation mechanism. These 'pico-RNA-viruses' are small viruses with an RNA genome. After infection of a cell, the RNA is replicated, with some of the products acting as mRNAs. These mRNAs are not capped but instead have an **internal ribosome entry site** (**IRES**) which is similar in function to the ribosome binding site of bacteria, with the small subunit of the host cell's ribosome attaching directly to the IRES. Attachment to the IRES requires very few if any of the standard eukaryotic initiation factors, but is aided by **IRES trans-acting factors** (**ITAFs**). These are RNA binding proteins with various functions in uninfected cells, which the virus recruits for its own purposes after infection.

The virus's own proteins include three with proteinase activity, which cleave and hence inactivate cellular proteins in order to disrupt the host cell biochemistry. The targets for proteinase attack include the initiation factor eIF-4G, which is central to the scanning process. The presence of IRESs on their mRNAs therefore means that picornaviruses can block protein synthesis in the host cell without affecting translation of their own mRNAs.

Some mammalian mRNAs also have IRESs and so can be translated by a process independent of scanning. It has proved difficult to be certain how many genes can be expressed in this way, because IRESs have variable sequences, which means that they are difficult to recognize simply by examining the nucleotide sequence of an mRNA. Many of those mRNAs known to have IRESs are transcripts of genes involved in the cellular response to stresses such as nutrient limitation or the accumulation of misfolded proteins. As we will see in *Section 17.1.3*, one component of the stress response is a global down-regulation of protein synthesis, which is achieved by phosphorylation of eIF-2, preventing this initiation factor from forming a functional ternary complex. Cap-dependent protein synthesis is therefore inhibited, enabling cellular resources to be directed towards synthesis of the stress response proteins via IRES-mediated translation of their mRNAs.

---

amount of ATP that is needed is dependent on the number of intramolecular base pairs that have to be broken before the complex reaches the initiation codon.

Recognition of the initiation codon is mediated by eIF-1, and results in the initiation complex switching to a 'closed' conformation, which forms a tighter attachment to the mRNA, with the anticodon of the initiator tRNA becoming base-paired to the initiation codon. The change from the scanning to the closed version of the complex requires energy, which is provided by hydrolysis of the GTP molecule that was present in the original ternary complex.

Once the closed version of the initiation complex has been formed, we can move to the final stage of initiation which, as in prokaryotes, involves attachment of the large ribosomal subunit and release of the various initiation factors. This requires hydrolysis of a second molecule of GTP, which is brought into the complex by eIF-5B.

Synthesis of the polypeptide

The initiation phase of translation is completed when the large subunit of the ribosome attaches to the small subunit, forming a complete ribosome positioned over the initiation codon. The events that follow are very similar in both bacteria and eukaryotes. The key to understanding these events lies with the roles of the peptidyl and aminoacyl sites, located between the two subunits of the ribosome.

To begin with, the P site is occupied by the initiator tRNA carrying methionine in eukaryotes and *N*-formylmethionine in bacteria, and base-paired to the initiation codon (*Fig. 16.16A*). At this stage, the A site is empty, but covers the second codon in the open reading frame. The A site becomes filled with the appropriate aminoacyl-tRNA, which in *E. coli* is brought into position by the **elongation factor** EF-1A, which carries a molecule of GTP (*Table 16.5*). One of the roles of EF-1A is to make sure that the tRNA which enters the A site is one that is carrying the correct amino acid. If the aminoacylation process has gone wrong, and the tRNA carries an incorrect amino acid, then EF-1A is able to reject the tRNA. At the other end of the tRNA, the specificity of the codon–anticodon interaction is ensured by contacts that form between the tRNA, mRNA and the small subunit rRNA. These contacts are able to discriminate between a codon–anticodon interaction in which all three base pairs have formed,

Figure 16.16 The elongation step of translation in *E. coli*.
In this example, the second amino acid in the polypeptide is threonine, coded by ACA.

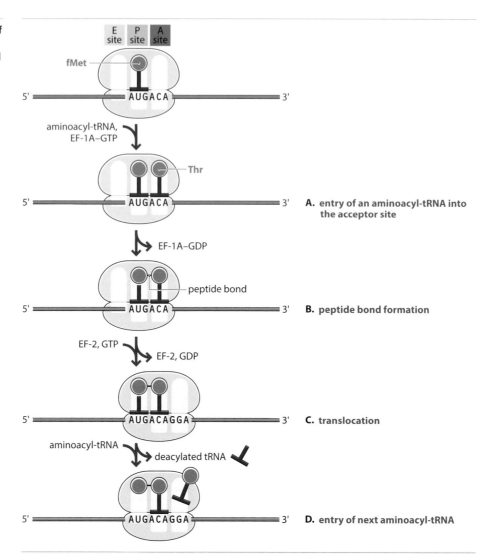

A. entry of an aminoacyl-tRNA into the acceptor site

B. peptide bond formation

C. translocation

D. entry of next aminoacyl-tRNA

Table 16.5. Functions of the elongation factors in bacteria and eukaryotes

Factor	Function
Bacteria	
EF-1A	Directs the next aminoacyl-tRNA to the A site in the ribosome
EF-1B	Acts as a **nucleotide exchange factor**; after the GTP attached to EF-1A has been hydrolyzed, EF-1B replaces the resulting GDP molecule with a new GTP
EF-2	Mediates translocation
Eukaryotes	
eEF-1	Directs the next aminoacyl-tRNA to the A site in the ribosome; a subunit of eEF-1 provides the nucleotide exchange function needed to regenerate the GTP that is hydrolyzed prior to eEF-1 departure from the ribosome
eEF-2	Mediates translocation
The bacterial elongation factors have recently been renamed. The older designations were EF-Tu, EF-Ts and EF-G for EF-1A, EF-1B and EF-2, respectively.	

and one in which there are one or more mispairs, the latter signaling that the wrong tRNA is present. Once the correct aminoacyl-tRNA is base-paired to its codon, the GTP attached to EF-1A is hydrolyzed, causing the elongation factor, now carrying a GDP, to undergo a conformational change and leave the ribosome. In the cytoplasm, EF-1B converts the GDP back to GTP, regenerating the active EF-1A molecule.

Departure of EF-1A signals to the ribosome that the correct aminoacyl-tRNA has entered the A site. The first peptide bond can now be formed (*Fig. 16.16B*). This step is catalyzed by a **peptidyl transferase** enzyme, which releases the amino acid from the initiator tRNA and forms a peptide bond between this amino acid and the one attached to the tRNA in the A site. In both bacteria and eukaryotes, peptidyl transferase is a ribozyme – an RNA enzyme. The catalytic activity for peptide bond formation is therefore provided not by a protein but by an RNA, in this case the largest of the rRNAs present in the large subunit of the ribosome.

The outcome of the events described so far is a dipeptide, whose sequence corresponds to the first two codons in the open reading frame, and which is attached to the tRNA in the A site. The next step is translocation (*Fig. 16.16C*). The ribosome moves three nucleotides along the mRNA. As it does so, three things happen at once:

- The dipeptide-tRNA moves from the A to the P site.

- The deacylated initiator tRNA that was occupying the P site moves to the E site.

- The A site becomes positioned over the next codon in the open reading frame.

Translocation requires hydrolysis of a molecule of GTP and is mediated by EF-2. It results in the A site becoming vacant, allowing a new aminoacyl-tRNA to enter and base pair to the next codon. As it does so, the deacylated tRNA in the E site is ejected from the ribosome (*Figure 16.16D*). The elongation cycle is now repeated, and continues until the end of the open reading frame is reached.

After several cycles of elongation the ribosome has moved away from the initiation codon, and a second ribosome can attach and begin to synthesize another copy of the protein. This resulting structure is called a **polyribosome** or **polysome**, an mRNA that is being translated by several ribosomes at once.

Termination of polypeptide synthesis

The elongation cycle continues until the ribosome reaches the termination codon at the end of the open reading frame. There are no tRNA molecules with anticodons able to base pair with any of the termination codons. Instead, a protein **release factor** enters the A site (*Fig. 16.17*). Bacteria have three release factors:

- RF-1, which recognizes the termination codons UAA and UAG.

- RF-2, which recognizes UAA and UGA.

- RF-3, which stimulates release of RF1 and RF2 from the ribosome after termination, in a reaction requiring energy from the hydrolysis of GTP.

Eukaryotes have just two release factors. The first of these, eRF-1 recognizes all three termination codons, and the second, eRF-3, plays the same role as the bacterial RF-3. Although eRF-1 is a protein, its shape is very similar to that of a tRNA, which led to suggestions that a release factor is able to enter the A site by mimicking a tRNA. This is an attractive model, but other studies suggest that the release factor adopts a different conformation when associated with a ribosome, one that is less similar to the shape of a tRNA.

Entry of a release factor into the A site terminates synthesis of the polypeptide, which is released from the ribosome. At this stage the ribosome is still attached to the mRNA. In bacteria, the completion of translation requires an additional protein called **ribosome recycling factor** (**RRF**). Like eRF-1, RRF has a tRNA-like structure, but whether it enters either the P or A site is unknown. Whatever its mode of action, the result is disassociation of the ribosome into its subunits, with release of the mRNA and the final deacylated tRNA. Disassociation requires energy, provided by GTP hydrolysis catalyzed by elongation factor EF-2, and also requires initiation factor IF-3 to prevent the subunits from attaching together again. A eukaryotic equivalent of RRF has not

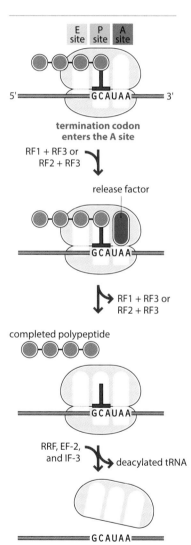

Figure 16.17 Termination of translation in *E. coli*.
In the example shown, the final amino acid of the polypeptide is alanine, coded by GCA, and the termination codon is UAA.

Box 16.4 Antibiotics that target the bacterial ribosome

Many of the most important antibiotics available to us today exert their effect by binding to the bacterial ribosome and inhibiting one or more stages of protein synthesis. The ability to make proteins is, of course, a fundamental requirement for any living cell, so antibiotics that target the ribosome are effective at inhibiting bacterial growth and possibly killing the bacterium. The differences between ribosome structure in bacteria and eukaryotes are such that many of these antibiotics are unable to attach to the equivalent parts of the eukaryotic ribosome, and therefore have only limited toxic effects in humans. This is particularly true of those antibiotics that affect the elongation phase of polypeptide synthesis. These antibiotics include:

- Streptomycin and tetracycline, which prevent entry of an aminoacylated tRNA into the A site.
- Chloramphenicol and puromycin, which inhibit peptide bond formation.
- The aminoglycoside antibiotics (for example, neomycin and hygromycin B), as well as erythromycin, which interfere with translocation.

There are also antibiotics, such as avilamycin, edeine and evemimicin, that target translation initiation, but these are less useful because they tend to have toxic side-effects for the patient. Several antibiotics have more general effects, such as fusidic acid, which disrupts both the elongation and termination stages of translation, and blasticidin S, which inhibits both peptide bond formation and ribosome recycling.

Antibiotics have played an important role in research into protein synthesis. Back in the 1960s, biochemists learned how to prepare cell extracts capable of carrying out protein synthesis *in vitro*. Usually these **cell-free translation** systems are prepared from germinating wheat seeds or from rabbit reticulocyte cells, both of which are exceptionally active in protein synthesis. The extracts contain ribosomes, tRNAs, and all the other molecules needed for protein synthesis. Addition of mRNA molecules, which can themselves be synthesized from nucleotide subunits *in vitro*, activates the cell-free system so polypeptides are made. Addition of an antibiotic has the opposite effect, halting translation of the mRNA by inhibiting one or other stage of protein synthesis. Examination of ribosomes that have become halted at a particular point can then reveal details about the translation process. For example, fusidic acid has been used to obtain ribosomes that have stalled midway through the translocation step. X-ray diffraction analysis of these ribosomes has revealed the precise positioning of the elongation factor EF-2 at this critical point in protein synthesis.

Although antibiotics are important in research, their greatest value is of course their ability to halt bacterial infections. The late twentieth century has been described as the 'golden age' of antibiotics, when many of the diseases that killed millions during the early decades of the century were brought under control through the use of antibiotics. Clinicians are now worried that this period is coming to an end because many bacteria have developed resistance to antibiotics, and infectious diseases are becoming more common again. Resistance often occurs because of mutations in the bacterial genes for the rRNAs or ribosomal proteins, with these mutations not affecting the ability of the ribosome to carry out protein synthesis, but preventing binding of the antibiotic. Other routes to resistance include mutations that prevent an antibiotic being taken up by the bacterium or increase the effectiveness of enzymes capable of degrading the antibiotic either before or after it has entered the cell.

Attempts to counter bacterial resistance include the design of synthetic antibiotics and an increasingly intensive search for new natural ones. The latter strategy can still be effective even though more and more of the soil bacteria that synthesize antibiotics have been tested and their products examined. Recently, a new antibiotic, orthoformimycin, was isolated after screening extracts prepared from 4400 bacterial species and 450 fungi. Orthoformimycin inhibits translocation, apparently in a novel way not seen with other known compounds, and hence might be a genuinely new antibiotic to which disease-causing bacteria are not yet resistant.

An alternative to the search for new natural antibiotics is to make novel compounds by chemical means. This can be achieved by introducing chemical modifications into natural antibiotics to give **semi-synthetic** derivatives with increased effectiveness. An example is solithromycin, a semi-synthetic antibiotic obtained by chemical modification of erythromycin, which was one of the first natural antibiotics to be discovered, back in 1949, and has been important in treating respiratory tract infections and some sexually transmitted diseases. It interferes with translocation by binding to the outer surface of the ribosome close to the exit tunnel through which the polypeptide emerges. The new side-chain present in solithromycin, attached to carbons 11 and 12, is thought to form an additional contact with the ribosome that increases the stability of the ribosome–antibiotic complex and defeats some of the anti-erythromycin mutations that have evolved among bacteria.

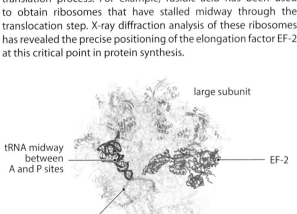

large subunit

tRNA midway between A and P sites

EF-2

mRNA

small subunit

Reproduced from *Science*, 2013; **340**: 1236086; J. Zhou *et al*. Crystal structures of EF-G–ribosome complexes trapped in intermediate states of translocation, with permission from AAAS.

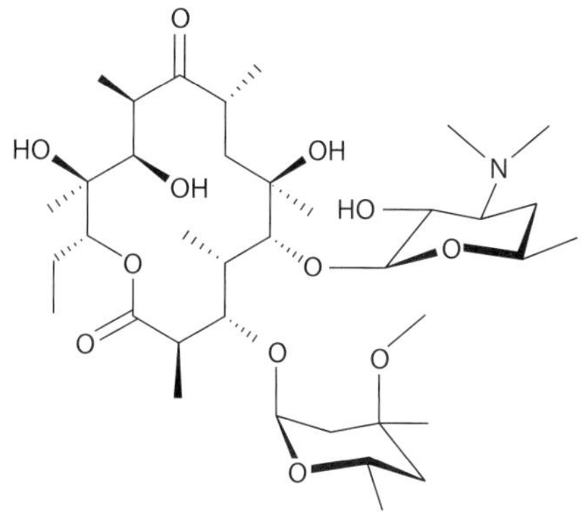

erythromycin

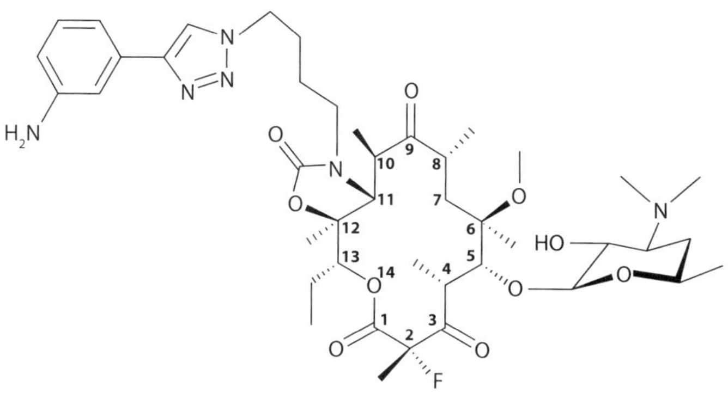

solithromycin

As well as semi-synthetic approaches based on existing antibiotics, attempts are also being made to use our knowledge of ribosome structure and the detailed mechanism of protein synthesis to design entirely new antibiotics with properties not so far discovered in nature. There is particular interest in designing bacterial equivalents of cycloheximide, which inhibits eukaryotic protein synthesis by interfering with tRNA release through the E site of the ribosome. No natural antibiotics capable of inhibiting this step in bacterial protein synthesis have been discovered, so a synthetic compound with the necessary properties might evade the existing resistance mechanisms.

been identified, and in eukaryotes disassociation is thought to require the combined activities of the release factors and other ancillary proteins.

16.3 Post-translational processing of proteins

The initial product of translation is a linear, unfolded polypeptide. To become active, the protein must adopt its correct three-dimensional structure. In some cases, the protein must also undergo additional processing events. These might include one or both of the following:

- **Proteolytic cleavage**, which might result in removal of segments from one or both ends of the polypeptide, or might cut the polypeptide into a number of different segments, all or some of which are active.

- **Chemical modification** of individual amino acids in the polypeptide.

A processing event might occur as the polypeptide is being folded, in which case the event might be necessary for the protein to take up its correct tertiary structure. Alternatively, a fully folded protein might undergo processing, possibly to activate an inactive form of the protein, or to modify its function in some way.

16.3.1 Processing by proteolytic cleavage

Proteolytic cleavage is a common post-translational processing event in eukaryotes, but is less frequently encountered in bacteria. Among its functions are formation of an active protein from an inactive precursor, and cleavage of **polyproteins** into segments, all or some of which are active proteins.

Protein activation by proteolytic cleavage

Processing by cleavage is common with secreted polypeptides whose biochemical activities might be deleterious to the cell producing the protein. These proteins are synthesized in an inactive form and then activated after secretion, so the cell is not

Phospholipase A2 catalyzes the third step in the triacylglycerol breakdown pathway shown in *Figure 12.11*.

harmed. An example is provided by melittin, the most abundant protein in bee venom. Melittin is a small protein, just 26 amino acids in length, with a simple tertiary structure comprising a pair of short α-helices (*Fig. 16.18A*). It inhibits a number of cellular proteins, but its venomous activity is due mainly to stimulation of phospholipase A2, which is the enzyme that removes the fatty acid from carbon number 2 of glycerol during triacylglycerol breakdown. Phospholipase overactivity results in membrane disruption and cell lysis, leading to the typical symptoms of a bee sting.

The bee, quite naturally, wishes to avoid these deleterious effects occurring within those of its cells that produce melittin. The protein is therefore synthesized as an inactive precursor, called promelittin. This precursor has an additional 22 amino acids at its N terminus, the presence of this pre-sequence preventing the protein from adopting its active structure. The pre-sequence is removed by an extracellular protease present in the bee's venom gland. This protease specifically removes amino-terminal dipeptides with the sequence X–Y, where X is alanine, aspartic acid or glutamic acid, and Y is alanine or proline. The pre-sequence is made up of 11 of these dipeptides in series, and so is removed, dipeptide by dipeptide, until the mature, active melittin component is reached (*Fig. 16.18B*). The melittin sequence lacks the dipeptide motifs, and so is untouched by the protease.

Proteolytic processing is also used to convert the inactive precursor forms of many hormones into the active protein. As an example, we will look at the synthesis of insulin, the protein made in the islets of Langerhans in the vertebrate pancreas and responsible for controlling blood sugar levels. Insulin is synthesized as preproinsulin, which is 105 amino acids in length (*Fig. 16.19*). The 24 amino acids at the amino terminus of preproinsulin are a **signal peptide**, a highly hydrophobic segment that directs the protein to the rough endoplasmic reticulum. As the protein crosses the membrane and moves into the lumen of the endoplasmic reticulum, the signal peptide is cleaved. We will look at signal peptides more closely later in this chapter when we study protein targeting.

Once inside the endoplasmic reticulum, the proinsulin molecule that results from signal peptide cleavage takes up a tertiary structure resembling that of active insulin, this process including the formation of three disulfide bonds. The folded prohormone is then transported to the Golgi apparatus where it encounters two endopeptidases called **prohormone convertases**. These excise a central segment called the C protein, leaving the two active parts of the protein, the A and B chains, linked together by two of the three disulfide bonds. In the very last processing step, two additional amino acids are removed from the carboxyl termini of the A and B chains, by the carboxypeptidase E protein.

A third important example of enzyme activation by proteolytic cleavage involves the digestive enzymes **trypsin** and **chymotrypsin**. These two enzymes are proteases

---

Figure 16.18 Post-translational processing of melittin.
(A) The tertiary structure of melittin. (B) Processing of the precursor promelittin. In part (B), the amino acids of the pre-sequence are shown in red and those of melittin in blue.

A. the structure of melittin

B. processing sites in promelittin

cut sites

A P E P E P A P E P E A E A D A E A D P E A G I G A V L K V L T T G L P A L I S W I K R K R Q Q G

Figure 16.19 **Processing of preproinsulin.**

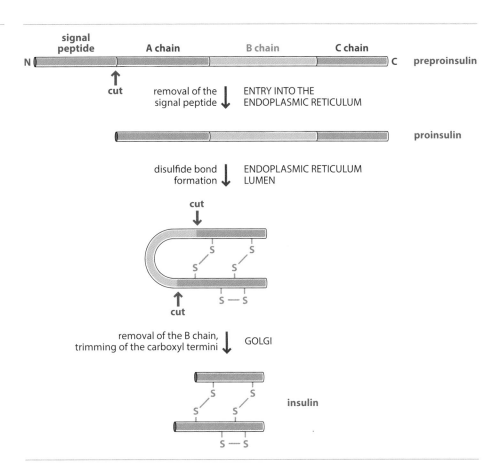

that help break down dietary proteins. Both are synthesized in the pancreas as inactive precursors, called trypsinogen and chymotrypsinogen, and then secreted into the duodenum. **Enteropeptidase**, which is secreted by the mucosal cells of the duodenum, removes the first 15 amino acids of trypsinogen, converting this protein into active trypsin. Trypsin then makes a cut at the same position in chymotrypsinogen, yielding active chymotrypsin. Processing occurs in the duodenum because the presence of the active enzymes in the pancreas would result in damage to the pancreatic cells, leading to **pancreatic self-digestion**.

Proteolytic processing of polyproteins

A **polyprotein** is a long polypeptide that contains a series of mature proteins linked together in head-to-tail fashion. Cleavage of the polyprotein releases the individual proteins, which may have very different functions from one another.

Polyproteins are not uncommon in eukaryotes. Several types of virus that infect eukaryotic cells use polyproteins as a way of reducing the sizes of their genomes, a single polyprotein gene with one promoter and one termination sequence taking up less space than a series of individual genes. The human immunodeficiency virus HIV-1 is an example. During its replication cycle, HIV-1 synthesizes the Gag polyprotein, which has a molecular mass of 55 kDa. The polyprotein is cleaved into four proteins, the largest of these being 231 amino acids in length, as well as two short spacer peptides (*Fig. 16.20*). Three of the four proteins form structural components of the HIV capsid and the fourth, called p6, is involved in the process by which virus particles are released from the cell. HIV-1 also synthesizes a larger polyprotein of 160 kDa, called Gag–Pol. As the name indicates, Gag–Pol is an extended version of Gag, the additional segment being processed to give two enzymes involved in replication

Figure 16.20 **The Gag polyprotein of HIV-1.**
The structures of the four proteins contained in the polyprotein are shown. SP1 and SP2 are spacer peptides that do not appear to have any function after the polyprotein has been cleaved.
Reproduced with permission from Macmillan Publishers Ltd: *Nature Reviews Microbiology*, **13**: 484–96, EO Freed, HIV-1 assembly, release and maturation; © 2015.

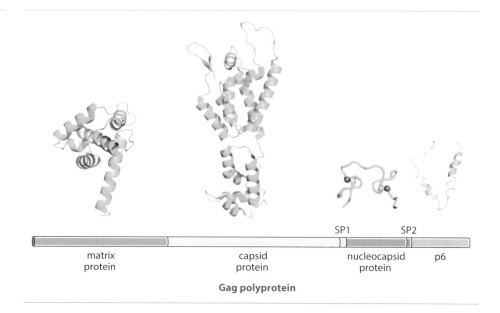

Gag polyprotein

of the HIV genome, as well as the protease that makes the cuts in the Gag and Gag–Pol polyproteins. A few molecules of this protease are stored in each of the new virus particles that are produced, and hence are available to cut up the polyproteins synthesized during the next round of virus replication.

Polyproteins are also involved in the synthesis of peptide hormones in vertebrates. An example is proopiomelanocortin. This protein is initially made as a precursor of 267 amino acids, 26 of which form a signal peptide that is removed when the protein is transferred from the cytoplasm to secretory vesicles inside the cell. Proopiomelanocortin has a variety of internal cleavage sites recognized by prohormone convertases, but not all of these sites are cleaved in every tissue, the combinations and hence the products formed depending on the identities of the convertases that are present (*Figure 16.21*). For example, in the corticotropic cells of the anterior pituitary gland, adrenocorticotropic hormone and the lipotropins are produced. In the melanotropic

---

Box 16.5 **Synthesis of the Gag and Gag–Pol polyproteins**

The Gag and Gag–Pol polyproteins are produced at a ratio of approximately 20 to 1, reflecting the relative requirements for the mature proteins derived from Gag and Pol for HIV replication. More copies of the capsid proteins contained in the Gag polyprotein are needed than the replicative enzymes contained in Pol. In order for the Pol extension to be translated from the Gag–Pol mRNA, the ribosome making the Gag polyprotein must undergo a **frameshift** at the end of the Gag sequence. This involves the ribosome moving back one nucleotide as it approaches the termination codon of the Gag polyprotein. As a consequence, a new set of codons are read, leading into the Pol sequence.

Spontaneous frameshifting can occur during translation of any mRNA, and usually is deleterious because the part of the polypeptide that is synthesized after the frameshift will have the incorrect amino acid sequence. The frequency of spontaneous frameshifting is very low, and the aberrant proteins made in this way are simply degraded by the cell. **Programmed frameshifting**, as occurs with the Gag–Pol mRNA, is induced by the presence of a stem–loop or other base-paired structure that forms in the mRNA immediately downstream of the frameshift site. The base-paired structure might increase the frequency of frameshifting by stalling progress of the ribosome, or might act as the binding site for a protein that regulates frameshifting. Programmed frameshifting occurs during synthesis of proteins of several different types of virus, and examples are also known in bacteria and eukaryotes.

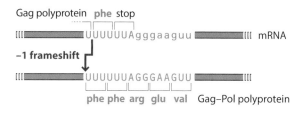

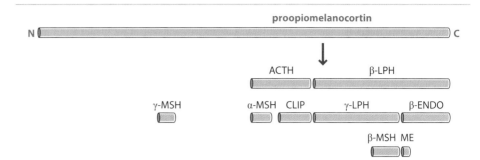

Figure 16.21 Processing of the proopiomelanocortin polyprotein.
Abbreviations: ACTH, adrenocorticotropic hormone; CLIP, corticotropin-like intermediary peptide; ENDO, endorphin; LPH, lipotropin; ME, met-enkephalin; MSH, melanotropin. Two additional peptides are not shown: one of these is an intermediate in the processing events leading to γ-MSH, and the function of the other is unknown. Note that although met-enkephalin can theoretically be obtained by processing of proopiomelanocortin, as shown here, most met-enkephalin made by humans is probably obtained from a different peptide hormone precursor called proenkephalin.

cells of the intermediate lobe of the pituitary, a different set of cleavage sites are used, generating the melanotropins. Altogether, 11 different peptides can be obtained by different patterns of proteolytic cleavage of proopiomelanocortin.

16.3.2 Chemical modification of proteins

As we saw earlier, 20 amino acids are specified by the genetic code, and two others – selenocysteine and pyrrolysine – can be incorporated into a polypeptide during translation, by context-dependent reassignment of UGA and UAG, respectively, which usually act as termination codons. But these 22 amino acids are by no means the only ones found in proteins. This is because of post-translational chemical modification, in which one or more of the amino acids in the original polypeptide chain are converted to more complex structures by addition of new chemical groups. The simpler types of modification occur in all organisms, but the more complex ones are rare in bacteria.

Chemical modifications often have regulatory roles

The simplest types of chemical modification involve addition of a small chemical group (e.g. an acetyl, methyl or phosphate group) to an amino acid side-chain, or to the amino or carboxyl groups of the terminal amino acids in a polypeptide. Over 150 different modified amino acids have been documented in different proteins (*Table 16.6*).

In some cases, the amino acid that is modified is located within the active site of an enzyme, the modification providing the active site with a novel functionality that is utilized during the biochemical reaction catalyzed by the enzyme. We encountered an example of a modified amino acid at an active site when we studied the fixation of carbon dioxide during the dark reactions of photosynthesis. Carbon dioxide fixation is catalyzed by ribulose bisphosphate carboxylase (Rubisco). The active site of Rubisco includes a lysine that has been modified by addition of a carboxyl group to give the carbamoyl derivative. The modified lysine binds a magnesium ion, which plays the central role in placing the reactants in the correct relative positions for carbon fixation to occur. This particular amino acid modification is therefore essential for the activity of Rubisco. Additionally, it provides a means of regulating the activity of the enzyme, with the nonproductive binding of ribulose 1,5-bisphosphate to the unmodified lysine preventing Rubisco activity in low light conditions.

We studied the role of the carbamoyl-lysine in Rubisco activity in *Section 10.3.1*.

Table 16.6. Examples of post-translational chemical modifications

Modification	Amino acids that are modified	Examples
Addition of small chemical groups		
Acetylation	Lysine	Histones
Methylation	Lysine	Histones
Phosphorylation	Serine, threonine, tyrosine	Some proteins involved in signal transduction
Hydroxylation	Proline, lysine	Collagen
N-formylation	*N*-terminal glycine	Melittin
Addition of sugar side-chains (see *Section 6.1.3*)		
O-linked glycosylation	Serine, threonine	Many membrane proteins and secreted proteins
N-linked glycosylation	Asparagine	Many membrane proteins and secreted proteins
Addition of lipid side-chains		
Acylation	Serine, threonine, cysteine	Many membrane proteins
N-myristoylation	*N*-terminal glycine	Some protein kinases involved in signal transduction
Addition of biotin		
Biotinylation	Lysine	Various carboxylase enzymes

Chemical modifications also play an important regulatory role in signal transduction pathways. Often, transduction of the regulatory signal involves a cascade of enzyme modifications, usually the attachment of phosphate groups to target proteins by phosphorylating enzymes such as protein kinase A. The signal transduction pathways by which epinephrine and insulin influence glycogen synthesis and breakdown are good examples (see *Figs 11.7* and *11.8*), as is the MAP kinase pathway (see *Fig. 5.28*). Phosphorylation sometimes also plays a role in induction of the signaling pathway by the transmembrane receptor protein. An example is provided by the epidermal growth factor receptor (EGFR). Attachment of epidermal growth factor to the external

Figure 16.22 Dimerization leads to autophosphorylation of the EGFR. Dimerization is induced by attachment of an epidermal growth factor molecule to each of a pair of receptor monomers.

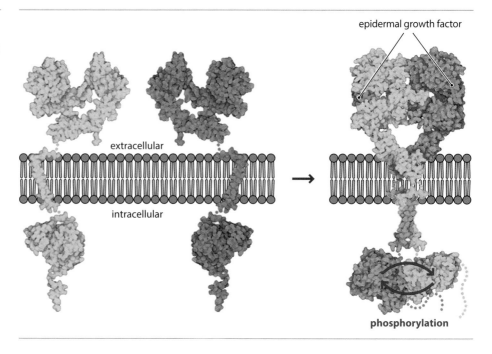

surface of the receptor causes two receptor monomers to come together to form a dimer (*Fig. 16.22*). Formation of this dimer induces each monomer to phosphorylate its partner, resulting in phosphates being added to several tyrosine amino acids on the internal side of the receptor. The presence of these phosphates is recognized by intracellular proteins which attach to the receptor, initiating a signal transduction cascade which, in the case of EGFR, results in cell growth and proliferation. Because of its autophosphorylation activity, EGFR is called a **receptor tyrosine kinase**.

Chemical modification of histones influences gene expression

Histone proteins provide a particularly sophisticated example of the regulatory effects of chemical modification. Histones are components of nucleosomes, the structures that associate with DNA to form the lowest level of DNA packaging in eukaryotic nuclei. The next level of packaging (the 30 nm chromatin fiber) involves interactions between individual nucleosomes. These nucleosomal interactions depend on the pattern of chemical modifications displayed by amino acids in the N-terminal regions of the histone proteins, these regions projecting outside of the nucleosome (see *Fig. 4.19*).

The best studied of these modifications is lysine acetylation, which reduces the affinity between individual nucleosomes. The histones in highly packaged DNA are generally unacetylated, whereas those in less packaged regions are acetylated. Whether or not a histone is acetylated depends on the balance between the activities of two types of enzyme, the **histone acetyltransferases** (**HATs**), which add acetyl groups to histones, and the **histone deacetylases** (**HDACs**), which remove these groups. Conversion of a region of DNA into a packaged conformation is one of the ways in which groups of genes can be switched off, and histone deacetylation is thought to play an important role in this process.

Other modifications to histones include methylation of lysines and arginines, phosphorylation of serines, and addition of the small, common ('ubiquitous') protein called **ubiquitin** to lysines in the C-terminal regions. Altogether, 29 sites in the N- and C-terminal regions of the four core histones (H2A, H2B, H3 and H4) are known to be subject to chemical modification of one type or another (*Fig. 16.23A*). These modifications interact with one another to determine the degree of packaging taken up by a particular stretch of DNA. For example, methylation of lysine-9 (the lysine nine amino acids from the N terminus) of histone H3 forms a binding site for the HP1

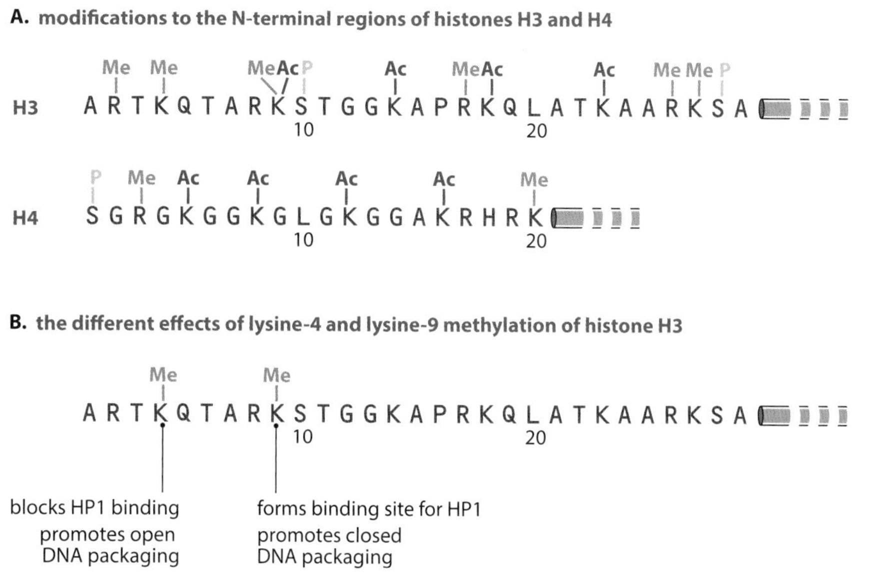

See *Section 4.2.1* for the structures of nucleosomes and the 30 nm chromatin fiber.

Figure 16.23 Histone modification. (A) Modifications known to occur in the N-terminal regions of mammalian histones H3 and H4. Abbreviations: Ac, acetylation; Me, methylation; P, phosphorylation. (B) The differential effects of methylation of lysines 4 and 9 of histone H3.

A. modifications to the N-terminal regions of histones H3 and H4

B. the different effects of lysine-4 and lysine-9 methylation of histone H3

protein which induces DNA packaging, but this event is blocked by the presence of two or three methyl groups attached to lysine-4 (*Fig. 16.23B*). Methylation of lysine-4 therefore promotes a more open degree of packaging, enabling the genes in that stretch of DNA to be expressed. The variety of possible histone modifications, and different interactions that can occur, has led to the suggestion that there is a **histone code**, by which the pattern of chemical modifications specifies which sets of genes are expressed at a particular time.

16.4 Protein targeting

After it has been synthesized, a protein must find its way to the place in the cell where it will carry out its function. In eukaryotes, proteins are synthesized by ribosomes that either float freely in the cytoplasmic matrix or are attached to the outer surface of the rough endoplasmic reticulum. Some proteins remain in the cytoplasm, but others must be transported into mitochondria or chloroplasts, some all the way into the matrix within the organelle, and others to an insertion site in one of the membranes surrounding these organelles. Other proteins must be transported to the insides of organelles such as the nucleus, lysosomes or peroxisomes, and still others will be inserted into the nuclear membrane or the plasma membrane surrounding the cell. Finally, secreted proteins must be transported outside of the cell. Similar issues also apply to prokaryotes, though being less complex these cells have fewer destinations to which proteins need to be transported.

The term **protein targeting** is used to describe the various processes by which proteins reach their correct destinations. These processes make up the final step in the lengthy pathway leading from gene to functional protein.

16.4.1 The role of sorting sequences in protein targeting

The protein targeting mechanism of a eukaryotic cell is made up of a series of transport pathways that lead to the various cellular and extracellular locations to which different proteins must be taken. Which pathway an individual protein follows depends on the identity of one or more **sorting sequences** present within the protein, these being amino acid sequences that specify which transport route(s) must be followed. Most of these sorting sequences are contiguous stretches of amino acids, but a few comprise motifs that are separated in the polypeptide chain but are brought together when the protein folds. The only proteins that lack sorting sequences are those that remain in the cytoplasmic matrix. These are made by cytoplasmic ribosomes and carry out their roles close to their synthesis site. All other proteins follow one or other of the targeting pathways.

Nuclear, mitochondrial and chloroplast proteins are synthesized in the cytoplasmic matrix

As well as those proteins that will remain in the cytoplasm, the ribosomes present in the cytoplasmic matrix also synthesize proteins that are eventually destined for the nucleus, mitochondria and chloroplasts. Those that must be transported into the nucleus contain a **nuclear localization signal**, which is 6–20 amino acids in length, with a high proportion of the positively charged amino acids lysine and arginine. In the shorter versions of the nuclear localization signal, these positively charged amino acids are distributed throughout the signal sequence, but in the longer examples the signal sequence might have an internal, uncharged or partially charged region (*Fig. 16.24*). The nuclear localization signal is recognized by an **importin** protein, which aids transfer of proteins through a nuclear pore complex and into the nucleoplasm.

Figure 16.24 The nuclear localization signals of (A) SV40 T antigen, and (B) nucleoplasmin. The SV40 T antigen is imported into the nucleus after a cell has been infected with SV40 virus. Nucleoplasmin is a chaperone protein with various roles in the nucleus, including nucleosome and ribosome assembly. Positively charged amino acids are shown in yellow.

A. SV40 T antigen

B. nucleoplasmin

Cytoplasmic ribosomes also make proteins that are transported into the mitochondria. The main access points for proteins into a mitochondrion are the **TOM** (**translocator outer membrane**) and **TIM** (**translocator inner membrane**) **complexes**. However, the mitochondrion cannot be looked on as a single destination because different targeting pathways are needed for proteins located in the outer membrane, the intermembrane space, the inner membrane, and the mitochondrial matrix. Further complexity is added by the presence of alternative routes to the same destination: outer membrane proteins, for example, can be inserted directly from the cytoplasm or after transport into the intermembrane space. The pathway followed by a mitochondrial protein depends on the identity of the sorting sequences that it carries. Proteins destined for the outer membrane or intermembrane space have internal sorting sequences of various types, but those proteins that insert into the inner mitochondrial membrane or pass through to the mitochondrial matrix have an N-terminal sorting signal called a **mitochondrial targeting sequence**. This sequence is usually 10–70 amino acids in length, and is made up of a mixture of nonpolar and positively charged amino acids that form an **amphipathic helix**, a type of α-helix that lies on the surface of a protein. The nonpolar amino acids are on the side of the helix that makes contact with the protein, and the charged amino acids are on the other side, exposed to the aqueous environment (*Fig. 16.25*). Transport to the inner membrane or matrix occurs before the protein becomes fully folded, attachment of Hsp70 chaperone proteins preventing folding while the protein is still in the cytoplasm, and is accompanied by cleavage of the targeting sequence.

Figure 16.25 The matrix targeting sequence of cytochrome *c* oxidase. (A) The amino acids present in the matrix targeting sequence. (B) The α-helix formed by the matrix targeting sequence. Charged amino acids are shown in red and nonpolar ones in yellow. Other amino acids are shown in blue. The charged amino acids are mainly on one side of the helix, and the nonpolar amino acids on the opposite side.

A. the cytochrome *c* oxidase matrix targeting sequence

B. the amphipathic helix formed by the targeting sequence

Depending on their combination of sorting sequences, mitochondrial proteins follow one of several routes to their final location (*Fig. 16.26*):

• Proteins destined for the outer mitochondrial membrane are either halted as they attempt to enter the TOM complex and diverted directly to the outer membrane, or pass through TOM into the intermembrane space before being inserted into the outer membrane.

• Some intermembrane space proteins move directly to their location after passing through TOM. Others cross into the matrix and are inserted into the inner membrane in such a way that a segment protrudes into the intermembrane space. This segment is cleaved off to give the functional protein.

Figure 16.26 **Targeting of mitochondrial proteins.**

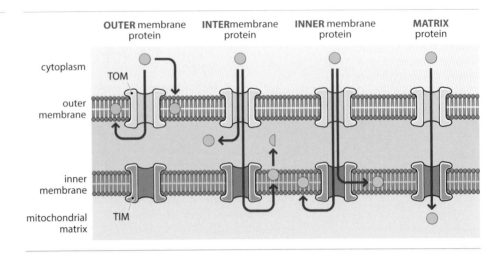

- Proteins for the inner mitochondrial membrane are either integrated into the membrane by one of the TIM complexes, or enter the mitochondrial matrix and are then inserted back into the inner membrane.
- Matrix proteins pass through both the TOM and TIM complexes.

Chloroplast proteins are targeted to their destinations in a similar way. The sorting sequences are called **transit peptides**, the particular combination carried by a protein specifying routes to the stroma, inner and outer membranes, and intermembrane space. Additionally, there is a **luminal targeting sequence**, which directs a protein to the thylakoids.

Secreted proteins are made on the rough endoplasmic reticulum

Proteins that are to be secreted by the cell are synthesized by ribosomes located on the outer surface of the rough endoplasmic reticulum. The transport mechanism for these proteins, which is called **exocytosis**, takes the following route (*Fig. 16.27*):

Figure 16.27 **The exocytosis pathway.**

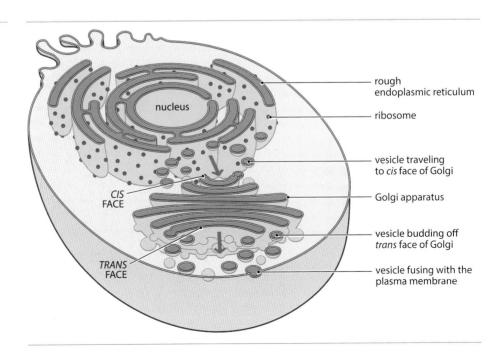

- As it is synthesized the protein passes through the endoplasmic reticulum membrane and into the lumen.

- A vesicle then buds off from the endoplasmic reticulum, carrying the protein to the *cis* face of the Golgi apparatus.

- Next the protein is transferred from cisterna to cisterna and hence to the *trans* face of the Golgi apparatus. Many secretory proteins become glycosylated during this stage.

- Another vesicle buds off the *trans* face of the Golgi, carrying the protein to the plasma membrane.

- Fusion of the vesicle with the plasma membrane transfers the protein outside of the cell.

To enter the exocytosis pathway, a protein must carry a sorting sequence called a **signal peptide**. This is an N-terminal sequence, usually of 5–30 amino acids, with a central region rich in hydrophobic amino acids and able to form an α-helix. Usually this helical region is preceded by a series of positively charged amino acids. The signal peptide directs the protein through the membrane of the endoplasmic reticulum, passage through this membrane occurring as the polypeptide is being synthesized.

Initially, the ribosome is free in the cytoplasm, but it becomes directed to the endoplasmic reticulum as soon as the N terminus of the polypeptide, containing the signal peptide, has been translated (*Fig. 16.28*). Transfer to the endoplasmic reticulum

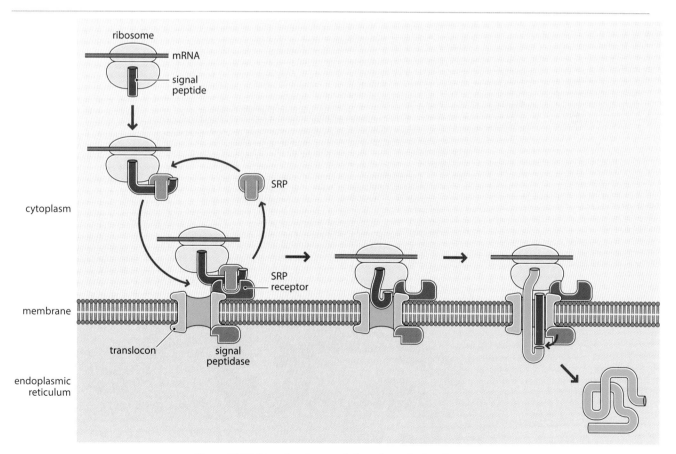

Figure 16.28 Transfer of a protein into the endoplasmic reticulum.
The signal peptide component of the protein that is being synthesized is shown in red, and the remainder of this protein in green.

is mediated by the **signal recognition particle** (**SRP**), which is made up of a short noncoding RNA molecule and six proteins. Attachment of the SRP to the signal peptide causes translation to pause, and directs the ribosome to an **SRP receptor** located on the surface of the endoplasmic reticulum. The signal peptide then enters a pore in the membrane called a **translocon**. Translation resumes, the polypeptide threading through the translocon and into the lumen of the endoplasmic reticulum. The signal peptide is then cleaved off by a **signal peptidase** on the inner surface of the membrane, and the polypeptide begins to fold.

Diversions from the exocytosis pathway lead proteins to other destinations

If the only sorting sequence carried by a secreted protein is the signal peptide, then the protein continues through the exocytosis pathway to the outside of the cell. If other sorting sequences are present then the protein can become diverted from this pathway to an alternative location. For example, the presence of a **retention signal** at the C terminus of the polypeptide results in the protein being located in the endoplasmic reticulum. If this retention signal is the sequence lysine–aspartic acid–glutamic acid–leucine, then the protein is located in the lumen of the endoplasmic reticulum. This is called the **KDEL sequence**, KDEL being the one-letter abbreviations for this series of amino acids. A modified version of the KDEL sequence directs a protein to the membrane of the endoplasmic reticulum. Although called a *retention* signal, these proteins are not, strictly speaking, retained in the endoplasmic reticulum. They are passed to the *cis* face of the Golgi apparatus along with other proteins that carry an N-terminal signal peptide, but are then sorted out by proteins that recognize the KDEL sequence and passed back to the endoplasmic reticulum.

Many proteins, especially cell-surface receptors and transport proteins, are located in the plasma membrane. These proteins become inserted into the membrane of

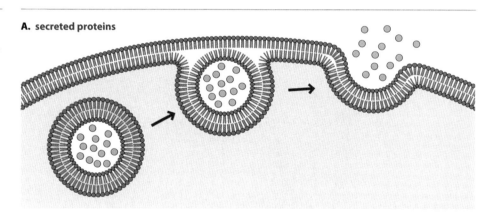

Figure 16.29 The final stage of exocytosis for (A) secreted proteins, and (B) plasma membrane proteins.

A. secreted proteins

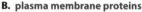

B. plasma membrane proteins

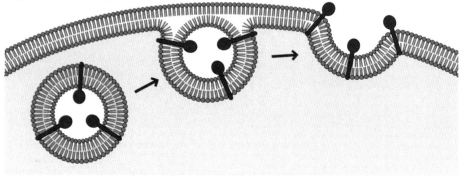

Figure 16.30 Targeting of proteins to the membrane of the endoplasmic reticulum.
The location of the signal peptide and stop-transfer sequence(s) are shown for Type I, Type II and Type III membrane proteins. After following the exocytosis pathway, these proteins become part of the plasma membrane.

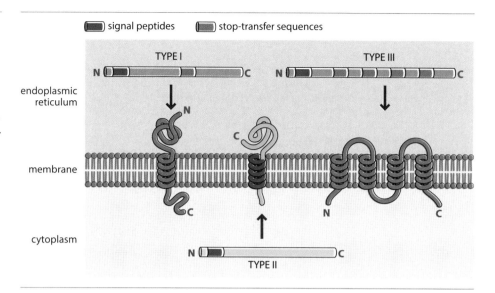

the endoplasmic reticulum as they are being synthesized and passed through the translocon. They then follow the exocytosis pathway but, being attached to the membrane, they are not secreted when the vesicle carrying them from the Golgi apparatus fuses with the plasma membrane. Instead they become part of the plasma membrane (*Fig. 16.29*). These proteins possess a variety of sorting sequences, each specific for a different type of positioning within the plasma membrane (*Figure 16.30*):

- The presence of an internal hydrophobic region called a **stop-transfer sequence** halts transfer of the polypeptide across the membrane of the endoplasmic reticulum, and then acts as an anchor, holding that protein in position. These are called **Type I membrane proteins**. They are integral membrane proteins that span the membrane once.

- Other proteins have multiple stop-transfer sequences, which result in them crossing the membrane more than once. These are called **Type III membrane proteins**.

- A special type of signal peptide at the N terminus, which is not recognized by signal peptidase and hence is not cleaved from the protein, acts as an anchor to hold a **Type II membrane protein** to the inner surface of the endoplasmic reticulum.

Figure 16.31 Targeting of proteins to the lysosomes.

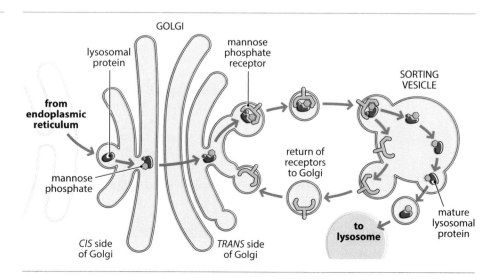

Box 16.6 **Protein targeting in bacteria**

Although most bacteria lack internal membrane-bound organelles, they synthesize many proteins that are secreted from the cell or inserted into the plasma membrane, and targeting pathways are needed to move these proteins to their destinations. These pathways are similar to the eukaryotic process for transfer of a protein into the endoplasmic reticulum. The bacterial protein carries a signal peptide that interacts with an SRP which directs the ribosome to the plasma membrane. The

protein is then transferred across the membrane via a translocon or, if it contains the appropriate sorting signals, it is inserted in to the plasma membrane.

Some proteins are targeted to the bacterial cell wall, which lies immediately outside of the plasma membrane. These proteins have a C-terminal sorting sequence that is cleaved after transfer across the membrane.

Finally, we will look at targeting of proteins to the lysosomes, organelles involved in the breakdown of various waste compounds. The degradative enzymes contained in lysosomes carry signal peptides that result in their initial transfer into the endoplasmic reticulum. They then move to the Golgi where they are tagged by attachment of a mannose 6-phosphate unit to one or more asparagine amino acids. The tagged proteins are then recognized by a **mannose 6-phosphate receptor** protein located on the inner surface of the cisternae at the *trans* side of the Golgi apparatus (*Fig. 16.31*). The vesicles that bud from this region of the Golgi then fuse with **sorting vesicles**, whose contents are acidic. The low pH stimulates the release of the proteins from the mannose 6-phosphate receptors and a phosphatase removes the phosphate groups from the mannose 6-phosphate tags. A further round of budding generates one set of vesicles that return the receptors to the Golgi, and a second set that take the lysosomal proteins to the lysosomes. So this is a slightly different type of targeting system in that the sorting signal is not an amino acid sequence contained in the protein, but the mannose 6-phosphate tag that is attached to the protein by a type of post-translational chemical modification.

Further reading

Agris PF, Vendeix FAP and Graham WD (2007) tRNA's wobble decoding of the genome: 40 years of modification. *Journal of Molecular Biology* **366**, 1–13. *Review of the development of the wobble hypothesis.*

Akopian D, Shen K, Zhang X and Shan S (2013) Signal recognition particle: an essential protein targeting machine. *Annual Review of Biochemistry* **82**, 693–721.

Caskey CT (1980) Peptide chain termination. *Trends in Biochemical Sciences* **5**, 234–7.

Clark B (1980) The elongation step of protein biosynthesis. *Trends in Biochemical Sciences* **5**, 207–10.

Hall BD (1979) Mitochondria spring surprises. *Nature* **282**, 129–30. *Reviews the first reports of unusual genetic codes in mitochondrial genes.*

Hellen CUT and Sarnow P (2001) Internal ribosome entry sites in eukaryotic mRNA molecules. *Genes and Development* **15**, 1593–612.

Hunt T (1980) The initiation of protein synthesis. *Trends in Biochemical Sciences* **5**, 178–81.

Jacks T, Power MD, Masiarz FR, Luciw FR, Barr PJ and Varmus HE (1988) Characterization of ribosomal frameshifting in HIV-1 *gag-pol* expression. *Nature* **331**, 280–3.

Jackson RJ, Hellen CUT and Pestova TV (2010) The mechanism of eukaryotic translation initiation and principles of its regulation. *Nature Reviews Molecular Cell Biology* **11**, 113–27.

Jenuwein T and Allis CD (2001) Translating the histone code. *Science* **293**, 1074–80.

Kapp LD and Lorsch JR (2004) The molecular mechanics of eukaryotic translation. *Annual Review of Biochemistry* **73**, 657–704.

Ling J, Reynolds N and Ibba M (2009) Aminoacyl-tRNA synthesis and translational quality control. *Annual Review of Microbiology* **63**, 61–76. *The role of aminoacyl-tRNA synthetases with focus on how accuracy of aminoacylation is ensured.*

RajBhandary UL (1997) Once there were twenty. *Proceedings of the National Academy of Sciences USA* **94, 11761–3.** *Summarizes unusual types of aminoacylation.*

Schmeing TM and Ramakrishnan V (2009) What recent ribosome structures have revealed about the mechanism of translation. *Nature* **461**, 1234–42.

Smith AI and Funder JW (1988) Proopiomelanocortin processing in the pituitary, central nervous system, and peripheral tissues. *Endocrine Reviews* **9**, 159–79.

Stojanovski D, Bohnert M, Pfanner N and van der Laan M (2015) Mechanisms of protein sorting in mitochondria. *Cold Spring Harbor Perspectives in Biology* **4**:a011320.

Wilson DN (2014) Ribosome-targeting antibiotics and mechanisms of bacterial resistance. *Nature Reviews Microbiology* **12**, 35–48.

Wilson DN and Cate JHD (2015) The structure and function of the eukaryotic ribosome. *Cold Spring Harbor Perspectives in Biology* **4**:a011536.

Zhou J, Lancaster L, Donohue JP and Noller HF (2013) Crystal structures of EF-G–ribosome complexes trapped in intermediate states of translocation. *Science* **340**(6140):1236086.

Self-assessment questions

Multiple choice questions

Only one answer is correct for each question. Answers can be found on the website: www.scionpublishing.com/biochemistry

1. What does the term 'degenerate' mean with regard to the genetic code?
   (a) Some codons have two or more meanings
   (b) The code is universal
   (c) Some amino acids are specified by more than one codon
   (d) The code contains punctuation codons

2. Which of these combinations are the termination codons in the standard genetic code?
   (a) UAA, UAG, UGA
   (b) UAA, UAG, UAC
   (c) UAG, UGA, UGG
   (d) UGA, UGC, UGG

3. Which of the following is not a nonstandard codon meaning in human mitochondria?
   (a) UGA codes for tryptophan
   (b) AGA codes for stop
   (c) CCA codes for stop
   (d) AUA codes for methionine

4. A class I aminoacyl-tRNA synthetase links the amino acid to which carbon on the terminal nucleotide of the tRNA?
   (a) 2′
   (b) 3′
   (c) 4′
   (d) 5′

5. In *Bacillus megaterium*, tRNA^{Gln} molecules are initially aminoacylated with what?
   (a) Serine
   (b) Methionine
   (c) Glutamic acid
   (d) Glutamine

6. Due to wobble, the anticodon UAI, where I is inosine, can base pair with which codons?
   (a) AUA, AUC and AUG
   (b) AUC, AUG and AUU
   (c) AUA, AUG and AUU
   (d) AUA, AUC and AUU

7. Which of the following describes the composition of the large subunit of the *E. coli* ribosome?
   (a) 3 rRNAs and 50 proteins
   (b) 1 rRNA and 21 proteins
   (c) 2 rRNAs and 34 proteins
   (d) 1 rRNA and 33 proteins

8. What is the function of the E site in a bacterial ribosome?
   (a) It is occupied by the aminoacyl-tRNA whose amino acid has just been attached to the end of the growing polypeptide
   (b) It is entered by the aminoacyl-tRNA carrying the next amino acid that will be used
   (c) It is the site through which the tRNA departs after its amino acid has been attached to the polypeptide
   (d) It is the site of peptide bond formation

9. What is the consensus sequence for the *E. coli* ribosome binding site?
   (a) AGGGGGU
   (b) AGGAGGU
   (c) TATAAT
   (d) AGCGCGCA

10. What is the *E. coli* initiation factor which mediates association of the large and small subunits of the ribosome called?
    (a) IF-i
    (b) IF-2
    (c) IF-3
    (d) IF-4

11. Which of the following statements is **incorrect** regarding initiation of translation in eukaryotes?
    (a) The pre-initiation complex binds to the cap structure on the mRNA
    (b) The cap binding complex is made up of eIF-4A, eIF-4E and eIF-4G
    (c) During scanning, stem–loop structures are opened up by eIF-4A
    (d) Recognition of the initiation codon is mediated by eIF-2

12. Which of the following is the role of elongation factor EF-1A during polypeptide synthesis in *E. coli*?
    (a) Acts as a nucleotide exchange factor
    (b) Mediates translocation
    (c) Directs the next aminoacyl-tRNA to the A site in the ribosome
    (d) Synthesizes the peptide bonds

13. What is the role of the *E. coli* ribosome recycling factor?
    (a) Disassociation of the ribosome into subunits after completion of translation
    (b) Hydrolysis of GTP needed for ribosome disassociation
    (c) Synthesis of the final peptide bond in the polypeptide that is being made
    (d) Transfer of the ribosome to the start of a new mRNA

14. Processing of melittin involves proteolytic cleavage yielding peptides containing how many amino acids?
    (a) 2
    (b) 3
    (c) 5
    (d) 9

15. What are endopeptidases that process prohormones called?
    (a) Restriction endopeptidases
    (b) Polyproteins
    (c) Prohormone convertases
    (d) Enteropeptidases

16. What process is responsible for synthesis of the Gag–Pol rather than Gag polyprotein of HIV-1?
    (a) The ribosome makes a frameshift at the end of the Gag sequence
    (b) The HIV-1 mRNA is spliced

(c) A stem–loop forms in the mRNA
(d) The Gag–Pol protein is cleaved by an endopeptidase

17. Which of the following is **not** obtained by processing of proopiomelanocortin?
    (a) Lipotropin
    (b) Endorphin
    (c) Adrenocorticotropic hormone
    (d) Thyroid-stimulating hormone

18. The epidermal growth factor receptor is an example of a what?
    (a) Prohormone convertase
    (b) Receptor tyrosine kinase
    (c) Histone acetyltransferase
    (d) Histone deacylase

19. Ubiquitin attaches to which amino acids in histone proteins?
    (a) C-terminal lysines
    (b) N-terminal lysines
    (c) N-terminal lysines and serines
    (d) N-terminal lysines and arginines

20. Which of the following statements is **incorrect** with regard to the nuclear localization signal present on proteins that are transported to the nucleus?
    (a) 6–20 amino acids in length
    (b) High proportion of the positively charged amino acids lysine and arginine
    (c) Cleaved off as the protein crosses the nuclear membrane
    (d) Recognized by an importin protein

21. Where are translocator outer membrane complexes located?
    (a) Mitochondrial outer membranes
    (b) Plasma membrane
    (c) Endoplasmic reticulum
    (d) Nuclear membrane

22. What is the name of the structure, made up of a short noncoding RNA molecule and six proteins, that aids transfer of proteins to the endoplasmic reticulum?
    (a) Signal recognition particle
    (b) SRP receptor
    (c) Translocon
    (d) Signal peptidase

23. What is the role of the KDEL sequence?
    (a) Directs a protein to the plasma membrane
    (b) Recognition sequence for the SRP receptor
    (c) Directs a protein to the membrane of the endoplasmic reticulum
    (d) Directs a protein to inner mitochondrial membrane

24. What is the tag that directs a protein to a lysosome?
    (a) Signal peptide
    (b) Mannose 6-phosphate
    (c) Ubiquitin
    (d) Stop-transfer sequence

Short answer questions

These questions do not require additional reading.

1. Describe the key features of the genetic code, including those variations that occur in certain genetic systems.

2. What is meant by 'aminoacylation' and how is the accuracy of this process ensured?

3. Explain why most species have fewer than 64 different tRNAs.

4. Outline our current knowledge of the structure of the ribosome and describe how this structure relates to the role of the ribosome in protein synthesis.

5. Give a detailed account of the initiation of translation of an mRNA in (A) E. coli, and (B) eukaryotes, paying particular attention to how the initiation codon is located and the roles of the initiation factors.

6. Compare the events occurring during termination of translation in E. coli and in eukaryotes.

7. At which stages in translation of an E. coli mRNA is energy required, and how is this energy provided?

8. Give examples of proteins that are processed by (A) proteolytic cleavage, and (B) chemical modification.

9. Describe the role of sorting sequences in protein targeting.

10. Give a detailed account of the targeting pathway for a lysosomal protein.

Self-study questions

These questions will require calculation, additional reading and/or internet research.

1. Most organisms display a distinct codon bias in their genes. For example, leucine is specified by six codons in the genetic code (TTA, TTG, CTT, CTC, CTA, and CTG), but in human genes leucine is most frequently coded by CTG and is only rarely specified by TTA or CTA. It has been suggested that a gene that contains a relatively high number of unfavored codons might be expressed at a relatively slow rate. Explain the thinking behind this hypothesis and discuss its ramifications.

2. The 20 amino acids specified by the genetic code are not the only ones found in living cells. Devise a hypothesis to explain why these 20 amino acids are the only ones that are specified by the genetic code. Can your hypothesis be tested?

3. Discuss the connection between wobble and the degeneracy of the genetic code.

4. There appears to be no biological reason why a DNA polynucleotide could not be directly translated into protein, without the intermediary role played by mRNA. What advantages do eukaryotic cells gain from the existence of mRNA?

5. Speculate on the reasons why the poly(A) tail of a eukaryotic mRNA is involved in initiation of translation of the mRNA.

CHAPTER 17

Control of gene expression

After reading this chapter you will:

- appreciate the importance of control of gene expression in remodeling of the proteome and in differentiation and development

- understand the role of alternative σ subunits in control of transcription initiation in bacteria

- be able to describe how expression of the lactose operon of *Escherichia coli* is regulated by the lactose repressor and the catabolite activator protein

- be able to explain how regulatory proteins control initiation of transcription in eukaryotes, and in particular understand the role of the mediator protein

- be able to describe how signal transduction pathways and steroid hormones control gene expression in eukaryotes

- understand how expression of the *E. coli* tryptophan operon is regulated by attenuation

- know how global and transcript-specific regulation of translation occurs in bacteria and eukaryotes

- appreciate the importance of mRNA and protein turnover in the regulation of gene pathways

- be able to describe the pathways for nonspecific mRNA turnover in bacteria and eukaryotes

- understand how the eukaryotic silencing complex degrades specific mRNAs

- be able to describe the roles of ubiquitin and the proteasome in degradation of proteins

Only a few of the genes in a cell are active all of the time. These are the so-called **housekeeping genes**, which specify RNA or protein products that the cell always needs. For example, most cells continually synthesize ribosomes and so have a continuous requirement for transcription of the rRNA and ribosomal protein genes. Similarly, genes coding for enzymes such as RNA polymerase or those involved in the basic metabolic pathways, such as glycolysis, are active in virtually all cells all of the time.

The products of other genes have more specialized roles and these genes are expressed only under certain circumstances. When their products are not needed, these genes are switched off. All organisms are therefore able to regulate the expression of their genes, so that only those whose products are needed are active at any particular time. Additionally, the expression level of those genes that are switched on can be modulated so that the rate at which the gene product is synthesized precisely matches the requirements of the cell.

The notion that gene expression can be regulated is a simple concept, but it has broad implications (*Fig. 17.1*):

- Regulation of gene expression enables the **proteome**, the collection of proteins in a cell, to be remodeled in response to changing conditions. Even the simplest unicellular organisms are able to remodel their proteomes to take account of changes in the environment. This means that their biochemical capabilities, as represented by the repertoire of enzymes that they possess, are continually in tune with the available nutrient supply and the prevailing physical and chemical conditions. Cells in multicellular organisms are equally responsive to changes in the extracellular environment, the only difference being that the major stimuli include hormones and growth factors as well as nutrients.

- The inactivation of particular sets of genes leads to cellular **differentiation**, the adoption by the cell of a specialized physiological role. Only those genes required in order for the cell to pursue its specialized role are switched on. We usually associate differentiation with multicellular organisms, in which a variety of specialized cell types (over 250 types in humans) are organized into tissues and organs. Differentiation also occurs in many unicellular organisms, an example being the production of spore cells by bacteria such as *Bacillus*.

- Regulation of gene expression underlies the **development** of an organism. Assembly of complex multicellular structures, and of the organism as a whole, not only requires coordination of gene expression in different cells, but also demands that the gene expression pattern in a single cell, or group of related cells, changes over time.

Figure 17.1 Outcomes of gene regulation.
(A) Bacteria and eukaryotic cells are able to remodel their proteomes in response to a changing environment. In this example, the environmental change has resulted in up-regulation of the gene for the blue protein. (B) Specialized cells express different genes. (C) Different genes are expressed at different stages of an organism's developmental pathway.

A. remodeling of a bacterial proteome

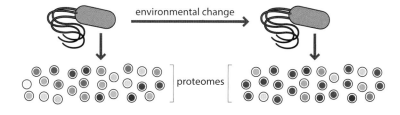

B. specialized cell types express different genes

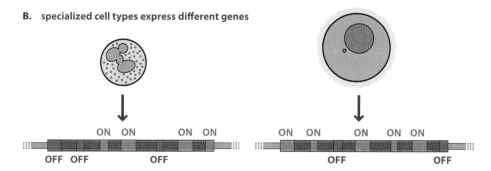

C. different genes are expressed at different stages of development

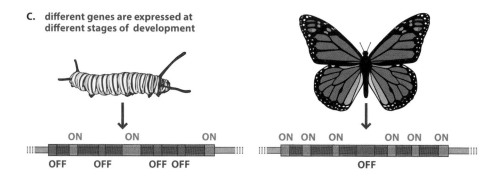

Individual organisms use many different strategies for controlling the expression of their genes, and throughout the living world there are examples of regulation applied to virtually every event in the pathway from gene to processed protein. In this chapter we will examine the most important of these control mechanisms.

17.1 Regulation of the gene expression pathway

Although many different steps in the gene expression pathway are subject to regulation, one step in particular appears to be of greatest importance. This is the step at the very beginning of the process, when an RNA polymerase attaches to the DNA in order to start transcribing a gene. The key events that determine if a gene is switched on or off take place during the initiation of transcription. Later events in the gene expression pathway are able to change the rate of expression of a gene that is switched on, but initiation of transcription is the primary control point, and so this is the step that we must study in greatest detail.

17.1.1 Regulation of the initiation of transcription in bacteria

First we will look at how transcript initiation is regulated in bacteria such as *Escherichia coli*. This will enable us to establish some key principles which will help us when we move on to the more complex regulatory processes that occur in eukaryotic cells.

In bacteria, initiation of transcription is regulated in two distinct ways:

- By changing the subunit composition of the RNA polymerase.

- Through the influence of regulatory proteins that determine whether or not the polymerase can bind to the DNA upstream of a gene.

Alternative σ subunits result in different patterns of gene expression

The bacterial RNA polymerase has a multisubunit structure described as $\alpha_2\beta\beta'\sigma$, with the σ subunit responsible for recognizing the promoter sequence. The promoter sequence is the short series of nucleotides that marks the position upstream of a gene where the enzyme must attach to the DNA in order to begin transcription. One of the critical interactions that result in attachment of the RNA polymerase is between the –35 box of the promoter and a 20 amino acid segment of the σ subunit which forms a secondary structure called a **helix–turn–helix motif**. As the name suggests, this motif comprises two α-helices separated by a β-turn (*Fig. 17.2*). One of the helices, called the **recognition helix**, is positioned on the surface of the σ subunit in an orientation that enables it to fit inside the major groove of the DNA molecule. Within the major groove the helix makes contacts with atoms present in the base components of the nucleotides. Because of the specificity of these contacts, the σ subunit is only able to bind to certain combinations of nucleotides, namely those found in the sequence of the –35 box of the promoter. The σ subunit is therefore a **sequence-specific DNA-binding protein**, its sequence specificity directing the RNA polymerase to the promoter located upstream of a gene.

In *E. coli*, the standard σ subunit, which recognizes a –35 box with the consensus sequence 5'–TTGACA–3', is called σ^{70}, the '70' indicating its molecular mass in kilodaltons. *E. coli* can also make a variety of other σ subunits, each one specific for a different –35 sequence. An example is the σ^{32} subunit, which is synthesized when the bacterium is exposed to a heat shock. This subunit recognizes a –35 sequence which is found upstream of genes coding for special chaperones that protect proteins from heat degradation, as well as DNA repair enzymes that are needed when the bacterium encounters high temperatures (*Fig. 17.3*). The bacterium is therefore able to switch on a whole range of new genes by making one simple alteration to the structure of its

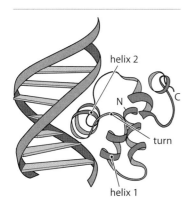

Figure 17.2 The helix–turn–helix motif.
The illustration shows the orientation of the helix–turn–helix motif (in blue) of a DNA-binding protein in the major groove of a DNA double helix.

Figure 17.3 Regulation of gene expression by the σ³² subunit in E. coli.
(A) The sequence of the promoter upstream of genes involved in the heat shock response of E. coli. (B) The heat shock promoter is not recognized by the normal E. coli RNA polymerase containing the σ⁷⁰ subunit, but is recognized by the σ³² subunit.

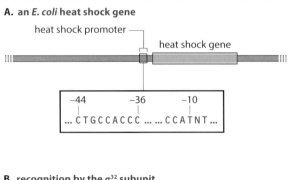

A. an E. coli heat shock gene

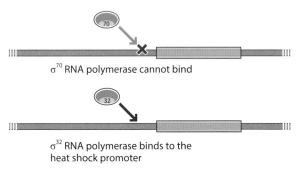

B. recognition by the σ³² subunit

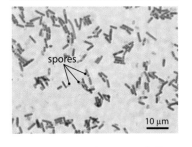

Figure 17.4 Sporulating B. subtilis bacteria.
Image reproduced from Wikimedia.org under a CC BY-SA 3.0 License.

RNA polymerase. Other σ subunits are used during nutrient starvation and nitrogen limitation, again switching on sets of genes whose products are needed under these particular conditions.

This type of gene regulation also underlies a cellular differentiation process displayed by *Bacillus* species. In response to adverse conditions, these bacteria produce spores that are highly resistant to physical and chemical extremes and can survive for years, germinating only when the environmental conditions become favorable (*Fig. 17.4*). The changeover from normal growth to formation of spores is controlled by different σ subunits which switch on the genes needed at each stage of the differentiation pathway. The standard *Bacillus subtilis* σ subunits, the ones used in nonsporulating cells, are called σ^A and σ^H. When sporulation begins, the cell divides into two compartments, one of which will become the spore and the other the mother cell, which dies when the spore is released (*Fig. 17.5*). The σ^A and σ^H subunits are replaced by two new ones, σ^F in the prespore and σ^E in the mother cell. Each of these recognizes its own −35 sequence, which are upstream of genes whose products specify development of the spore or mother cell, respectively. Later in the sporulation process these subunits are replaced by σ^G and σ^K, which switch on the genes needed in the later stages of spore and mother cell formation. The different σ subunits therefore bring about the time-dependent changes in gene expression that underlie differentiation of the bacterium into a spore.

Repressor proteins prevent the polymerase from attaching to the promoter

The different specificities of alternative σ subunits provide a means of switching on new sets of genes in response to changes in the environment, but this regulatory system does not permit any gradation between on and off. Those genes whose promoters are recognized by a particular σ subunit are actively transcribed, and all others are silent. The more detailed regulation needed to ensure that the expression levels of individual genes precisely match the prevailing conditions is provided in a different way, by regulatory proteins that influence the initiation of transcription of those genes.

Figure 17.5 The role of alternative σ subunits during *B. subtilis* sporulation.

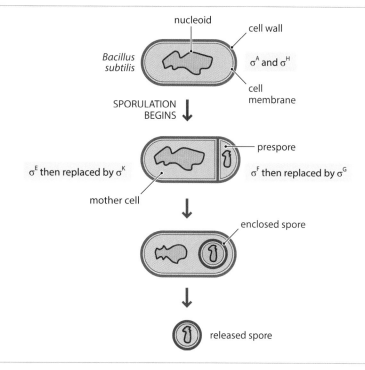

The existence of this type of regulatory protein was first proposed in 1961 by two French geneticists, François Jacob and Jacques Monod, following a lengthy series of genetic experiments. Jacob and Monod studied regulation of the **lactose operon** of *E. coli*. This is a set of three genes that code for the three enzymes needed by the bacterium to utilize lactose as an energy source. These three enzymes are (*Fig. 17.6*):

- **Lactose permease**, which is located in the inner cell membrane and transports lactose into the cell.

- **β-galactosidase**, which catalyzes the splitting of lactose into glucose and galactose. The glucose molecule can directly enter into the glycolysis pathway, and the galactose can enter glycolysis after conversion to glucose.

- **β-galactoside transacetylase**, whose enzymatic function is transfer of an acetyl group from acetyl CoA to a β-galactoside molecule. The β-galactosides are the broad group of compounds to which lactose belongs, many of which can be metabolized by the enzymes of the lactose operon. The exact role of the transacetylase in metabolism of lactose is not understood.

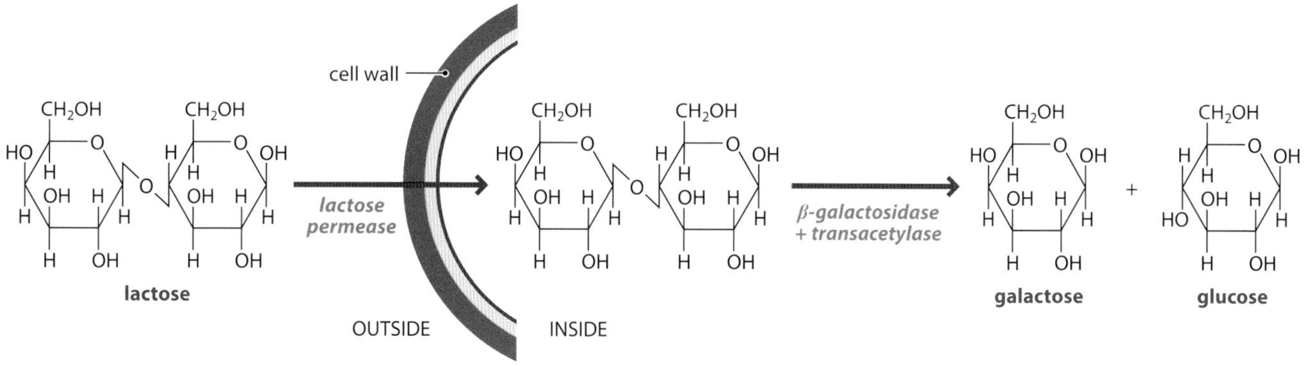

Figure 17.6 Lactose utilization by *E. coli*.

Box 17.1 Transcriptomics – studying changes in gene expression patterns

In addition to understanding the processes that regulate gene expression, it is often important to study the outcomes of those processes. In other words, it is useful to identify the genes that are being expressed in a particular tissue at a particular time, and to explore how that pattern of gene expression changes when the physiological conditions change (e.g. when a tissue responds to stimulation with a hormone) or when a tissue becomes diseased. The key distinction between a gene that is active and one that is not is that the former is transcribed into RNA. Examining the RNA content of a tissue is therefore the most direct way of identifying which genes are being expressed. The RNA content of a tissue is referred to as its **transcriptome** and the study of transcriptomes is called **transcriptomics**. What does transcriptomics involve and what can it tell us?

Transcriptomes are usually studied by **microarray analysis**. A microarray is a small piece of glass onto which a large number of DNA molecules have been applied as spots in an ordered array.

microarray

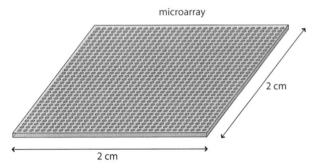

Usually the DNA molecules are single-stranded oligonucleotides of 100–150 nucleotides in length, each of which has been made by chemical synthesis at its appropriate position in the microarray. With the latest technology it is possible to squeeze half a million sites onto a single microarray, each site containing multiple copies of a particular oligonucleotide.

The sequences of the oligonucleotides in the array match the sequences of segments of the genes in the organism whose transcriptome is being studied. This means that each oligonucleotide is able to base pair to the RNA transcript of its target gene.

In practice, RNA is not applied directly to the microarray, because RNAs are easily degraded and difficult to work with in this way. Instead, DNA copies are made by mixing the RNA with the RNA-dependent DNA polymerase called reverse transcriptase and a supply of each of the four deoxynucleotides. One of these nucleotides is chemically modified so that it emits a fluorescent signal. This modification does not affect the ability of the nucleotide to be incorporated into the DNA molecules that are being made, but it means that these molecules are **labeled** – they give off fluorescent signals.

mRNA

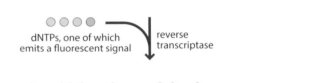

dNTPs, one of which emits a fluorescent signal   reverse transcriptase

fluorescently labeled single-stranded DNA

The labeled DNA is now applied to the microarray. The positions of oligonucleotides that base pair to DNA molecules – and whose target genes are therefore being expressed in the tissue being studied – can now be revealed by scanning the microarray with a fluorescence detector. The relative amounts of the different RNAs in the transcriptome can also be worked out, because abundant RNAs will give rise to more intense fluorescent signals.

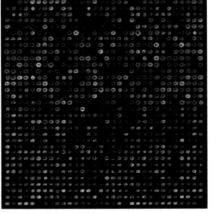

signal intensity

HIGH

LOW

Microarray analysis and other methods for studying transcriptomes have been particularly valuable in research into cancer. Any type of cancer, for example colon or breast, is made up of a variety of different subtypes, and knowing which subtype is presented by a patient is critical to identification of the appropriate treatment regime. Transcriptome studies have enabled each of these subtypes to be associated with a different pattern of gene expression. In some cases, it has been possible to identify particular gene combinations whose expression pattern is diagnostic for a single subtype and hence can be used as a biomarker for that cancer. Biomarkers are particularly valuable as tools in early diagnosis when the chances of arresting the cancer are higher. The type of breast cancer called 'triple-negative' is a good example. The name refers to the absence of expression of the genes for three cell surface receptors that are active in other types of breast cancer. Triple-negative is a particularly aggressive type of breast cancer with a high mortality rate, so its early and accurate diagnosis through transcriptome analysis is important in ensuring that treatment begins as soon as possible.

Figure 17.7 The three genes of the lactose operon are transcribed into a single mRNA.
The gene names are: *lacZ*, β-galactosidase; *lacY*, lactose permease; *lacA*, β-galactoside transacetylase.

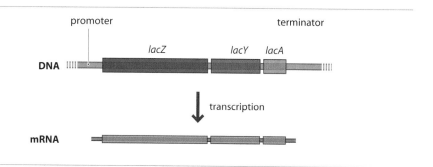

When lactose is absent, only five or so copies of each enzyme are present in the cell, but when the bacterium encounters lactose this number rapidly increases to over 5000. Induction of the three enzymes is coordinate, meaning that each is induced at the same time and to the same extent. This is because the genes for the three enzymes are located together in a single transcription unit, under the control of a single promoter. The three genes are therefore transcribed into a single mRNA (*Fig. 17.7*). The expression of all three genes can therefore be controlled by regulating the events occurring at that single promoter.

Adjacent to the promoter of the lactose operon is a second sequence, called the **operator**, which mediates the regulation of expression of the operon (*Fig. 17.8A*). The operator is the binding site for a regulatory protein, called the **lactose repressor**. When attached to the operator, the repressor prevents the RNA polymerase from binding to the promoter, simply by blocking its access to the relevant segment of DNA.

A. the lactose operator

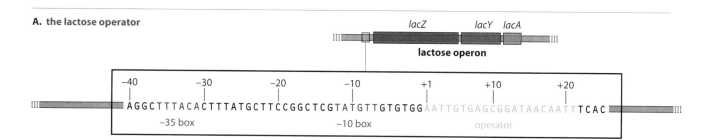

B. the role of the repressor and inducer

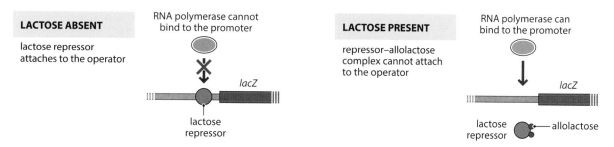

Figure 17.8 Regulation of the lactose operon of *E. coli*.
(A) The operator sequence lies immediately downstream of the promoter for the lactose operon. (B) The role of the lactose repressor protein and the allolactose inducer in controlling access of RNA polymerase to the promoter of the lactose operon.

When lactose is absent, the repressor is bound to the operator and the three lactose genes are switched off (*Fig. 17.8B*). When the bacterium encounters a source of lactose, the repressor detaches and the genes are switched on. How does the repressor respond to the presence of lactose? Initially, a small amount of lactose is transported into the cell and metabolized by the few enzyme molecules that are always present. Alongside the conversion of lactose to glucose and galactose, β-galactosidase also synthesizes **allolactose**, an isomer of lactose. Allolactose is an **inducer** of the lactose operon. When present it binds to the lactose repressor, causing a slight structural change which prevents the repressor from recognizing the operator as a DNA-binding site. The allolactose–repressor complex therefore cannot bind to the operator, enabling the RNA polymerase to gain access to the promoter.

When the supply of lactose is used up and there is no allolactose left to bind to the repressor, the repressor re-attaches to the operator and prevents transcription. Lactose therefore indirectly regulates expression of the genes needed for its metabolism.

Glucose acts as a positive regulator of the lactose operon

From our study of metabolism we are familiar with the notion that a pathway can be regulated by both its substrates and products, the substrates stimulating flow of metabolites along the pathway, and the products inhibiting that flow. Regulation of the lactose operon follows the same bio-logic. As we have just seen, lactose, being the substrate of the metabolic pathway specified by the lactose operon, stimulates

---

Box 17.2 **The allolactose paradox**

The reason why allolactose and not lactose is the inducer of the lactose operon has been debated by biochemists ever since the 1960s, when the details of lactose utilization in *E. coli* were first worked out. Allolactose, like lactose, is a disaccharide comprising D-galactose and D-glucose, but with a β(1→6) rather than β(1→4) linkage.

allolactose

Structural studies of the β-galactosidase reaction mechanism have shown that cleavage of lactose into galactose and glucose results in a transient covalent bond forming between a glutamic acid, at position 537 in the enzyme's polypeptide chain, and carbon number 1 of the galactose molecule. Usually this covalent bond is broken by hydrolysis to release the galactose, but during a few reaction cycles the galactose is transferred to carbon 6 of one of the glucose units that have just been cleaved from the lactose substrate. The result of this minor reaction is allolactose.

The structural studies show how allolactose is synthesized by β-galactosidase, but do not tell us why allolactose and not lactose is the inducer of the operon. Several suggestions have been put forward, including the hypothesis that although the function of the lactose operon is conventionally looked on as

lactose utilization, this may not be its primary role. The argument is that in its natural environment (the mammalian gut) *E. coli* rarely encounters lactose. This is because milk is the only source of lactose and all non-human mammals and many human ones only consume milk during weaning. A β-galactoside is any molecule made up of galactose linked to a second compound by a β-glycosidic bond. A range of β-galactosides are used as substrates by β-galactosidase, and most of these are direct inducers of the operon. An example is β-galactosyl glycerol, a component of plant membranes and hence a significant part of the mammalian diet. So it is possible that the reason why lactose is not an inducer of the operon is because the operon has evolved to enable *E. coli* to utilize β-galactosides other than lactose. If this hypothesis is correct, then it is only by chance that the allolactose by-product of lactose cleavage switches the operon on, enabling lactose to be metabolized.

This hypothesis has become quite popular but recent research has questioned its validity. Not all species of bacteria have β-galactosidase enzymes capable of synthesizing allolactose and even fewer species have the lactose repressor. A phylogenetic study of 1087 species and strains revealed that 53 of these had β-galactosidase enzymes with the structural features necessary for allolactose synthesis, and that among the same 1087 there were 33 species or strains that possessed the lactose repressor. All but one of the 33 bacteria with a lactose repressor had the ability to make allolactose. In other words, there has been a close co-evolution between the lactose repressor and the ability to synthesize allolactose. This finding suggests that allolactose synthesis is not a fortuitous by-product of β-galactosidase activity, but instead has evolved alongside the repressor as an integral part of the regulatory system. The debate regarding the allolactose paradox is not yet over.

LACTOSE PRESENT

repressor–allolactose complex
cannot attach to the operator

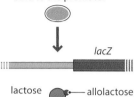

RNA polymerase can
bind to the promoter

lacZ

lactose repressor —— allolactose

but . . . GLUCOSE ALSO PRESENT

expression of the lactose operon
inhibited and no transcription
occurs

Figure 17.9 **When glucose is
present the lactose operon is
switched off, even if there is also a
supply of lactose.**

expression of these three genes. What we have not yet seen, but are about to explore, is the complementary role of glucose, one of the products of the pathway, as an inhibitor of expression of the lactose operon. This complementary effect means that if a bacterium has sufficient glucose to meet its energy needs, then it does not switch on expression of the lactose operon, even if lactose is also available in the environment (*Fig. 17.9*).

Glucose inhibits expression of the lactose operon indirectly via a regulatory protein called the **catabolite activator protein** (**CAP**). Like the lactose repressor, CAP is a DNA-binding protein that attaches to the DNA at a position adjacent to the promoter of the operon. Unlike the repressor, the attachment of CAP does not prevent the polymerase gaining access to the promoter. Instead, CAP interacts with the α subunit of the polymerase and facilitates binding of the RNA polymerase to form a closed promoter complex. When bound to the DNA, CAP therefore stimulates initiation of transcription of the lactose operon.

Glucose has a negative influence on the lactose operon by preventing the attachment of CAP. In the absence of CAP, virtually no productive initiations occur, even if lactose is present and the repressor is also detached. Glucose exerts its influence via a chain of events that begin when this sugar is transported into the cell. Passage of glucose across the inner cell membrane results in dephosphorylation of a membrane-bound protein called **IIA**^{Glc} (*Fig. 17.10*). The dephosphorylated version of IIA^{Glc} inhibits activity of adenylate cyclase, which converts ATP into cAMP. This is important because CAP can only bind to the DNA in the presence of cAMP. By indirectly reducing the levels of cAMP in the cell, the presence of glucose therefore results in detachment of CAP and inactivation of the lactose operon. Only when the glucose levels fall, and the chain of events is reversed, can CAP re-attach, enabling the operon to be expressed, should the bacterium have an available source of lactose.

CAP does not only regulate the lactose operon. Binding sites for the protein are present upstream of other *E. coli* operons that code for enzymes involved in catabolic pathways such as utilization of sugars. At some promoters, in addition to stimulating

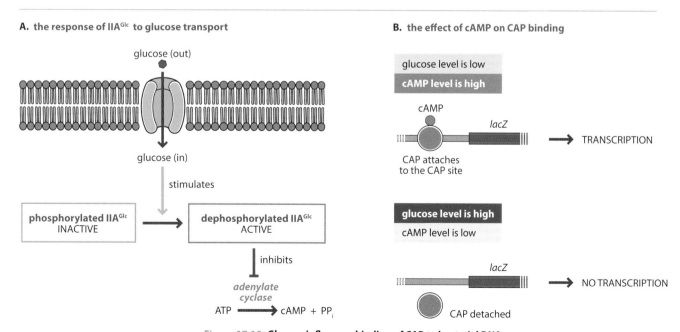

A. the response of IIA^{Glc} **to glucose transport**

glucose (out)

glucose (in)

stimulates

phosphorylated IIA^{Glc}
INACTIVE

dephosphorylated IIA^{Glc}
ACTIVE

inhibits

adenylate cyclase

ATP ——→ cAMP + PP$_i$

B. the effect of cAMP on CAP binding

glucose level is low

cAMP level is high

cAMP

lacZ

TRANSCRIPTION

CAP attaches
to the CAP site

glucose level is high

cAMP level is low

lacZ

NO TRANSCRIPTION

CAP detached

Figure 17.10 **Glucose influences binding of CAP to bacterial DNA.**
(A) Glucose transport results in dephosphorylation of IIA^{Glc} and inhibition of adenylate cyclase activity.
(B) cAMP influences CAP binding.

Box 17.3 **Repressible operons**

The lactose operon is an example of an **inducible operon**, one that is switched on by the regulatory molecule, in this case allolactose. Usually the inducer is a substrate, or an analog of the substrate, for the pathway catalyzed by the enzymes specified by the operon.

Other operons have repressor proteins that respond not to a substrate of the pathway controlled by the operon, but to a product. An example is the tryptophan operon, which contains five genes that specify the set of enzymes needed to convert

chorismate to tryptophan (see *Fig. 17.15*). The regulatory molecule for this operon is tryptophan, which acts as a **co-repressor**. When tryptophan is attached to the tryptophan repressor the latter binds to the operator and prevents RNA polymerase from attaching. When tryptophan levels are low, the co-repressor detaches from the repressor, which in turn detaches from the operator, enabling RNA polymerase to transcribe the operon. The tryptophan operon is therefore switched off in the presence of tryptophan, and switched on when tryptophan is needed. This is an example of a **repressible operon**.

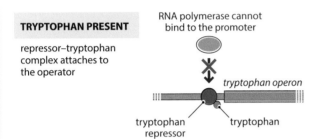

There is one other interesting difference between the lactose and tryptophan repressors. The lactose repressor only regulates expression of the lactose operon, whereas the tryptophan operon has another binding site on the *E. coli* genome. As well as the tryptophan operon, this repressor controls expression of the *aroH* gene, which specifies DAHP synthase, the enzyme that catalyzes the condensation of phosphoenolpyruvate and erythrose 4-phosphate at the commitment step of the pathway leading to synthesis of phenylalanine, tyrosine and tryptophan

(see *Fig. 13.11*). DAHP synthase exists as three isozymes, each of which is subject to feedback inhibition by one of the three amino acids that are products of this pathway (*Fig. 13.13B*). Not surprisingly, *aroH* codes for the isozyme that is controlled by tryptophan. We therefore see interlinking between the feedback control of the commitment step by tryptophan, and regulation of synthesis of the enzyme that catalyzes this step, exerted by the tryptophan repressor.

formation of the closed promoter complex, as occurs at the lactose operon, CAP also makes additional contacts with the RNA polymerase in order to facilitate formation of the open complex and initiation of productive transcription. Whatever the mechanism of CAP action, cAMP is the essential co-activator, enabling the levels of glucose in the cell to influence the extent to which the bacterium utilizes alternative sugar sources.

17.1.2 Regulation of the initiation of transcription in eukaryotes

Our study of the lactose repressor and CAP has introduced us to the key concept that expression of a gene can be controlled by one or more regulatory proteins, which exert their effect by binding to the DNA and influencing the ability of RNA polymerase to recognize the promoter and initiate transcription. This concept applies to eukaryotes as well as bacteria.

Eukaryotic promoters contain binding sites for a variety of regulatory proteins

The main difference between bacteria and eukaryotes, with regard to gene regulation, is that eukaryotic genes respond to a greater diversity of control signals. This means that the promoter regions of most eukaryotic genes contain binding sites for a range of regulatory proteins, which together mediate the response of the gene to the variety of factors that influence its expression. The human insulin gene provides a good example. At least 14 different protein binding sites are present in the 350 bp of DNA adjacent to the TATA box of this gene (*Fig. 17.11*).

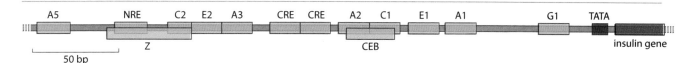

Figure 17.11 Protein binding sites in the promoter of the human insulin gene.
Fourteen binding sites are shown. These comprise basal and cell-specific promoter elements, the latter including the two CRE sites mentioned in the text.

Identifying the functions of the binding sites upstream of a eukaryotic gene is a considerable challenge, but we can divide the sites into two broad groups. The first of these are the **basal promoter elements**. These sites do not respond to any signals from inside or outside of the cell, but instead determine the basal rate of transcription of the gene. This basal rate is the number of productive initiations that occur, per unit time, when expression of the gene is not subject to any other regulatory control. The proteins that bind to the basal promoter elements therefore ensure that when the gene is switched on, but not subject to up- or down-regulation, transcription takes place at an appropriate rate.

As well as the basal elements, many eukaryotic genes also have **cell-specific promoter elements**, which ensure that the gene is expressed in the correct tissues and responds to the appropriate regulatory signals. Two examples in the insulin promoter are the pair of **cAMP response elements (CRE)**, which are binding sites for the **cAMP response element binding (CREB)** protein, which regulates insulin gene expression in response to cellular levels of cAMP. There are also two sites that respond to retinoic acid and thyroid hormone, though these are located further away from the insulin gene, within the 'insulin kilobase upstream' or 'ink' box about 1000 bp from the transcription start site.

Some cell-specific elements are unique to one or just a small number of genes, but others help regulate groups of genes whose products are needed at the same time under particular conditions. An example of the latter in humans is the **heat shock module**, which is recognized by the heat shock protein HSP70. HSP70 is thought to detect cellular damage caused by stresses such as heat shock. When it detects such damage, HSP70 binds to the heat shock modules in the promoters of those genes whose products help repair the damage and protect the cell from further stress. There are also **developmental promoter elements**, which mediate expression of genes that are active at specific developmental stages.

The role of mediator proteins

In order to influence transcription, the attachment of a regulatory protein to a promoter element must have an effect on the activity of the RNA polymerase. The lactose repressor of *E. coli* illustrates a straightforward way in which this can occur, with the repressor simply blocking access of the polymerase to the promoter. The CAP illustrates a second possibility, where a direct contact between regulatory protein and polymerase facilitates initiation of transcription.

Most eukaryotic regulatory proteins contain an **activation domain**, which makes contact with the complex of proteins, including RNA polymerase II, involved in initiation of transcription of a protein-coding gene. Structural studies have shown that although activation domains are variable, most of them fall into one of three categories:

• **Acidic domains**, which are relatively rich in aspartic acid and glutamic acid. This is the commonest category of activation domain.

• **Glutamine-rich domains**.

• **Proline-rich domains**, which are the least common.

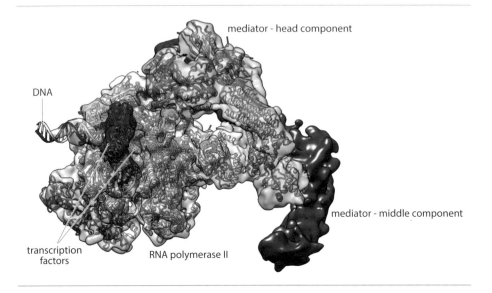

Figure 17.12 The yeast mediator protein attached to the transcription initiation complex. The head and middle components of the mediator protein are shown in light blue and dark blue, respectively. The RNA polymerase is silver and transcription factors are shown in red, green and purple.
Image reproduced by permission from Macmillan Publishers Ltd: *Nature,* **518**: 376; C. Plaschka *et al.* Architecture of the RNA polymerase II–mediator core initiation complex, © 2015.

The contact made between the activation domain and the RNA polymerase complex is not direct. Instead, it is via an intermediary protein called the **mediator**. The mediator protein was first identified in the yeast *Saccharomyces cerevisiae*. In this species, the mediator is made up of 25 subunits, forming a structure with head, middle and tail components. The tail makes contact with the activation domain of the regulatory protein, and the middle and head sections interact with the polymerase complex (*Fig. 17.12*). In humans, the mediator is larger, with over 30 subunits, but its mode of action as a link between regulatory proteins and RNA polymerase is the same.

How does the mediator affect initiation of transcription? Remember that RNA polymerase II must be activated before it can begin to synthesize an RNA transcript. Activation involves addition of phosphate groups to the C-terminal domain (CTD) of the largest subunit of the polymerase. At one time it was thought that the mediator phosphorylates the CTD, but in fact this kinase activity is provided by the Kin28 protein, which is one of the subunits of TFIIH. It is still possible that the mediator influences phosphorylation in an indirect way. In fact, the mediator makes a number of different contacts with the RNA polymerase complex, and its effect on transcript initiation is probably multifaceted. For example, it is present when TBP attaches to the TATA box, and might form part of the platform onto which the remainder of the RNA polymerase complex is constructed.

Regulatory proteins respond to extracellular signals

We have already explored various ways in which extracellular signals, in the form of hormones, growth factors, or other regulatory molecules, can influence biochemical activities within eukaryotic cells. Gene expression is one of the biochemical activities that must respond to these extracellular signals. Many of the signal transduction pathways that we have already studied, as well as activating and deactivating different enzymes, also have an effect on gene expression patterns. The MAP kinase pathway, for example, results in phosphorylation of several types of protein involved in regulation of transcription. Phosphorylation activates these proteins so that they attach to their DNA-binding sites and exert their specific effects on their target genes. Second messengers also have gene regulatory effects, an example being cAMP, which regulates expression of the insulin gene, as well as many others, via the CREB protein.

Important examples of signal transduction pathways that we have studied include the MAP kinase pathway (*Section 5.2.2*) and the pathways by which epinephrine and insulin influence glycogen synthesis and breakdown (*Section 11.1.2*).

Box 17.4 **Zinc fingers**

The nuclear receptor superfamily comprises regulatory proteins that bind directly to DNA at specific sites upstream of the genes that they control. Unlike the σ subunit of the bacterial RNA polymerase, as well as many other DNA-binding proteins, nuclear receptors do not attach to DNA via a helix–turn–helix structure. Instead these proteins contain a different type of DNA-binding motif called a **zinc finger**. These structures are rare in bacterial proteins but common in eukaryotes. Possibly up to 1% of all the proteins made by a mammalian cell contain zinc fingers.

There are several different types of zinc finger. One of the commonest is the **Cys₂His₂** structure, which is made up of 12 or so amino acids, of which two are cysteine and two are histidine. These 12 amino acids form a 'finger', comprising a short two-stranded β-sheet followed by an α-helix, which projects from the surface of the protein. The zinc atom referred to in the name of the structure is held between the sheet and helix, coordinated to the two cysteines and histidines.

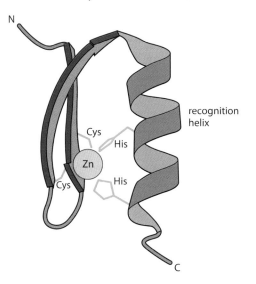

The overall architecture of the zinc finger is such that the α-helix is able to form contacts within the major groove of a DNA molecule, its exact positioning determined by the β-sheet

(which interacts with the sugar–phosphate backbone of the DNA) and the zinc atom (which holds the β-sheet and α-helix in the appropriate positions relative to one another). The α-helix of a Cys₂His₂ finger is therefore a recognition helix, similar to the second helix of the helix–turn–helix structure.

Other versions of the zinc finger have different structures. Those present in the nuclear receptor proteins are a type of **treble clef finger**. They lack a β-sheet component, and instead comprise two α-helices and a series of loops, with two zinc atoms, each held in place by contacts to four cysteines. As with a Cys₂His₂ finger, one of the helices is a recognition helix that can be positioned within the major groove of a DNA molecule.

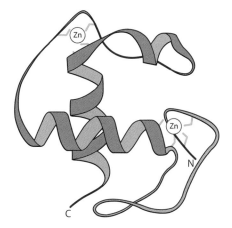

Most DNA-binding proteins are able to recognize and attach to specific sequences in a DNA molecule. This specificity ensures that they bind adjacent to the genes whose expression they influence, and nowhere else in a genome. How the structure of a zinc finger, or any other type of DNA-binding domain, confers sequence specificity is unknown, but the specificity is assumed to derive from the positioning of the amino acids in the recognition helix. The sequence of nucleotides can be identified from the arrangement of atoms within the major groove, and presumably contacts are made between these atoms and partners in the recognition helix of the binding protein.

The signal transduction pathways that we have studied so far have several steps between the extracellular signaling compound and the proteins that regulate gene expression. Some external compounds, on the other hand, have a much more direct link with gene regulation. An example is provided by the **steroid hormones**. These include the sex hormones (estrogens for female sex development, androgens such as testosterone for male sex development), and the glucocorticoid and mineralocorticoid hormones, examples of which are cortisol and aldosterone, respectively (see *Fig. 12.29*). Steroids are hydrophobic and so can pass directly through the cell membrane, rather than transmitting a signal through a cell surface protein. Once inside the cell, each hormone binds to a specific **steroid receptor** protein. The hormone–receptor complex then migrates into the nucleus, where it acts as a regulatory protein, attaching to **hormone response elements** within the promoter regions of target genes.

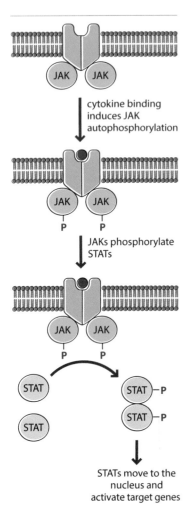

Figure 17.13 The JAK–STAT pathway.

cytokine binding induces JAK autophosphorylation

JAKs phosphorylate STATs

STATs move to the nucleus and activate target genes

We studied the pathway for synthesis of tryptophan in *Section 13.2.1*.

All steroid receptors are structurally similar. Recognition of these similarities has shown that a second set of receptor proteins, the **nuclear receptor superfamily**, belongs to the same general class, although the hormones that they work with are not steroids. As their name suggests, these receptors are located in the nucleus rather than the cytoplasm. They include the receptors for retinoic acid and thyroid hormone, which are involved in regulation of the insulin gene, and vitamin D_3, which is involved in the control of bone development.

Cytokines such as interleukins and interferons, which are extracellular proteins that control cell growth and division, are also able to influence gene expression without their signals being transduced via a lengthy pathway. Cytokines cannot pass through the cell membrane and so influence internal events by binding to a cell surface receptor. These events are mediated by **Janus kinases** (**JAKs**), internal proteins that are associated with cytokine receptors, two JAKs per receptor (*Fig. 17.13*). Cytokine binding induces a conformational change in the receptor, moving its pair of JAKs close enough together to phosphorylate one another. Phosphorylation activates the JAKs which now phosphorylate transcription factors called **STATs** (**signal transducers and activators of transcription**). Phosphorylation causes pairs of STATs to form dimers and then move to the nucleus, where they activate expression of a variety of genes.

17.1.3 Gene regulation after transcript initiation

Although the initiation of transcription appears to be the primary control point for expression of most genes, examples are known where regulation is exerted at virtually every other step in the gene expression pathway. We will now explore two of the most important of these processes.

Bacteria can regulate the termination of transcription

In eukaryotes, transcription occurs in the nucleus and translation in the cytoplasm, and it is not possible to link the two processes together. The RNA transcript must be completely synthesized and transported to the cytoplasm before it can be translated. Bacteria, on the other hand, because they do not have a membrane-bound nucleus, carry out the two stages in gene expression in the same cellular compartment. It is therefore possible for a ribosome to attach to and begin translating an mRNA that is still being transcribed by its RNA polymerase enzyme (*Fig. 17.14*). This linking of transcription and translation is utilized in **attenuation**, a regulatory process that some bacteria use to exert a fine level of control over expression of their operons.

Attenuation is used mainly with operons that specify enzymes involved in amino acid biosynthesis. An example is the tryptophan operon of *E. coli*, which comprises five genes that code for the set of enzymes needed to convert chorismate to tryptophan (*Fig. 17.15*). These five genes are preceded by a short open reading frame (ORF) that

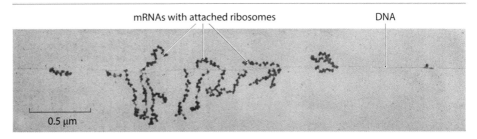

mRNAs with attached ribosomes   DNA

0.5 µm

Figure 17.14 Transcription and translation are linked in bacteria.
This electron micrograph shows several mRNAs being transcribed from *E. coli* DNA. Each of these mRNAs has ribosomes attached to it, which are visible as small dark dots. The mRNAs are therefore being translated even though they have not yet been completely transcribed.
Image from *Science*, 1970; **169**: 392; O.L. Miller *et al.* Visualization of bacterial genes in action, with permission from AAAS.

Figure 17.15 Attenuation control of the *E. coli* tryptophan operon.

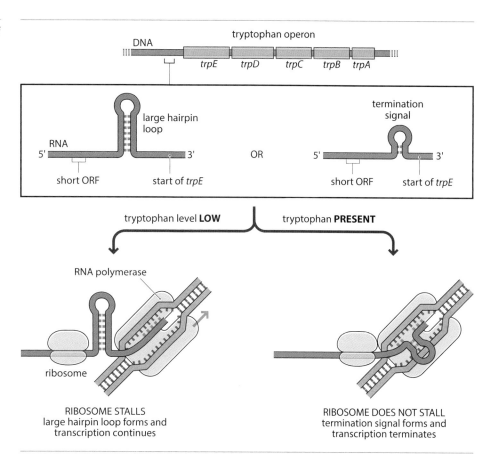

specifies a 14 amino acid peptide that includes two tryptophans. This peptide has no function in the cell, but its synthesis underlies the control system. The ORF is immediately followed by a region that can form two stem–loop structures, but not both at the same time. The smaller of these stem–loops acts as a termination signal, but the larger one, which is closer to the start of the transcript, is more stable because it has more base pairs. Armed with this information, we can examine the steps in the attenuation process:

- Transcription begins and the RNA polymerase progresses to the point where the ORF has been copied into RNA.

- A ribosome attaches to the RNA and begins to translate the ORF into the 14 amino acid peptide.

- To translate the ORF, the ribosome needs two tryptophans. If tryptophan levels in the cell are low, then the ribosome will become held up within the ORF, waiting for two of the available molecules to diffuse, by chance, into the vicinity of the ribosome. If the ribosome stalls in this way, and does not keep up with the polymerase, then the larger stem–loop forms and transcription continues.

- If tryptophan levels are high, then the ribosome translates the ORF without delay, and keeps pace with the RNA polymerase. It is then able to disrupt the larger stem–loop, allowing the termination structure to form. Transcription therefore stops.

So in summary, when tryptophan levels are low, attenuation allows the operon to be transcribed so more tryptophan can be made, but when tryptophan levels are high, attenuation results in premature termination of the transcript, so additional tryptophan is not made.

The *E. coli* tryptophan operon is also controlled by a repressor protein, which switches off the operon when tryptophan is present. It is thought that the repressor is the basic on–off switch, and that attenuation modulates the rate of expression when the operon is switched on. Attenuation therefore ensures that the amount of mRNA that is transcribed is precisely that needed to direct translation of sufficient enzymes to maintain the cellular tryptophan at the appropriate level.

Translation is regulated in both bacteria and eukaryotes

Two different types of regulation can operate during translation. The first of these is **global regulation**, which affects all eukaryotic mRNAs which possess a cap structure. Global regulation is achieved by phosphorylation of the initiation factor eIF-2. The phosphorylated form of eIF-2 cannot bind a GTP molecule, which it must do in order to transport the initiator tRNA to the small subunit of the ribosome. Phosphorylation of eIF-2 occurs during stresses such as heat shock, when translation of the vast majority of mRNAs (all those with cap structures) is down-regulated. The mRNAs for special heat shock proteins, such as HSP70, are unaffected by this type of regulation because they do not have cap structures, and instead are translated via an internal ribosome entry site.

See *Box 16.3* for information on internal ribosome entry sites.

A more directed, **transcript-specific regulation** of translation is also possible with some mRNAs. Examples are known in both bacteria and eukaryotes. In *E. coli*, translation of the mRNAs for several operons coding for ribosomal proteins is regulated by the amount of one or more of those proteins in the cell. The leader region of the mRNA includes a binding site for the ribosomal protein, and when bound this protein blocks attachment of the ribosome and hence prevents translation. The L11–L1 operon is regulated in this way by L1, the second of the two ribosomal proteins coded by the operon (*Fig. 17.16A*). L1 can either attach to its position on the 23S rRNA of the large subunit of the ribosome, or bind to the mRNA and block further translation. The attachment to the rRNA is more stable, and occurs if any of these sites are available. Once they are all filled, L1 binds to its mRNA, blocking translation and hence switching off further synthesis of L1 and L11. Similar events involving other mRNAs ensure that synthesis of each ribosomal protein is coordinated with the amount of free rRNA in the cell.

A mammalian example of transcript-specific regulation of translation involves the mRNA for ferritin, an iron storage protein (*Fig. 17.16B*). When iron levels are low, the regulatory protein IRP-1 attaches to an **iron response element** on the ferritin mRNA, blocking the movement of the ribosome along the transcript. IRP-1 is itself an iron binding protein, the attachment of iron atoms resulting in a conformational change so that the protein no longer recognizes the iron response element. This means that in the

Figure 17.16 Two examples of transcript-specific regulation of translation.
(A) Regulation of ribosomal protein synthesis in bacteria. (B) Regulation of ferritin protein synthesis in mammals.

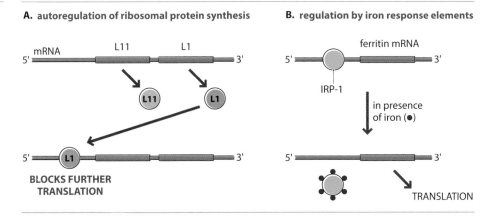

A. autoregulation of ribosomal protein synthesis

B. regulation by iron response elements

presence of iron, IRP-1 detaches from the mRNA, enabling the mRNA to be translated so the amount of ferritin in the cell increases. This regulatory system ensures that at any particular time there are adequate supplies of ferritin to store the amounts of iron that are available to the cell.

17.2   Degradation of mRNA and protein

If a gene is switched off and the mRNA and protein that it specifies are absent from a cell, then switching that gene on will obviously result in an increase in the cellular levels of the mRNA and protein. Up-regulation of gene expression therefore results in up-regulation of the amounts of mRNA and protein. But what happens when the gene is switched off again? Our expectation is that down-regulation of gene expression will result in down-regulation of the mRNA and protein content of the cell. But this is only possible if the existing mRNA and protein are degraded. If not, then the protein activity will still be present.

The amount of an mRNA or a protein in the cell is therefore a balance between its synthesis rate (the number of molecules that are made per unit time) and its degradation rate (how many molecules are broken down per unit time). This balance results in a steady state concentration, and changing either the synthesis or degradation rate will influence that steady state. So far we have studied mechanisms for controlling the rate of synthesis of mRNA and protein. To gain a complete understanding of gene regulation, we must now examine the processes by which mRNAs and proteins are degraded.

17.2.1   RNA degradation

We will examine RNA degradation first, and begin with the processes responsible for nonspecific RNA turnover. These processes act on all mRNAs, and in some cases noncoding RNAs, without discrimination between the transcripts of individual genes.

Several processes are known for nonspecific mRNA turnover

In bacteria, nonspecific mRNA degradation is carried out by the **degradosome**, a multiprotein structure whose components include:

- **Polynucleotide phosphorylase** (**PNPase**), which removes nucleotides sequentially from the 3′ end of an mRNA but, unlike true nucleases, requires inorganic phosphate as a substrate.

- **RNase E**, an endonuclease that makes internal cuts in RNA molecules.

- **RNA helicase B**, which aids degradation by unwinding the double helix structure of the stems of RNA stem–loops.

Enzymes capable of degrading RNA in the 5′→3′ direction have not been identified in bacteria, suggesting that the main degradative process for bacterial RNAs is removal of nucleotides from the 3′ end by enzymes such as PNPase. Most bacterial mRNAs have a stem–loop structure near their 3′ end, the same structure involved in termination of transcription, which will block the progress of PNPase (*Fig. 17.17*). It is therefore assumed that the stem–loop is either disrupted by the RNA helicase prior to arrival of PNPase, or that the region containing the stem–loop is snipped off by RNase E. Either event will enable PNPase to access the remainder of the RNA.

Eukaryotes have a structure equivalent to the degradosome, called the **exosome complex**. An exosome comprises a ring of six proteins, each of which has ribonuclease activity, with three RNA-binding proteins attached to the top of the ring. Other ribonucleases associate with the exosome in a transient manner. It is thought

Figure 17.17 **Possible mechanisms for mRNA degradation by the bacterial degradosome.**

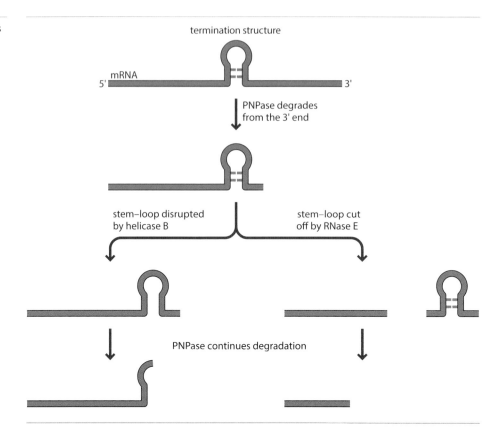

that RNAs to be degraded are initially captured by the binding proteins and then threaded through the channel in the middle of the ring, where they are exposed to the ribonuclease activities of the ring proteins (*Fig. 17.18*).

Exosomes are present in both the cytoplasm and the nucleus. The presence of exosomes in the nucleus indicates that their roles include turnover of RNAs that contain errors due to incorrect transcription or processing. These errors are detected by an **mRNA surveillance** mechanism, which identifies mRNAs that lack a termination codon, which might occur if the DNA has been copied incorrectly, or have a termination codon at an unexpected position, indicating that the exons have been joined together incorrectly during intron splicing. Surveillance involves a complex of proteins which scans mRNAs for these errors and directs aberrant transcripts to the exosome or some other degradation pathway.

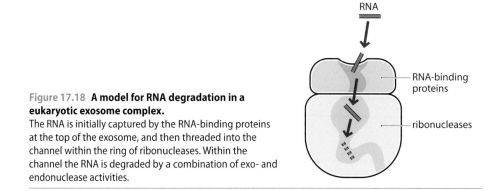

Figure 17.18 **A model for RNA degradation in a eukaryotic exosome complex.**
The RNA is initially captured by the RNA-binding proteins at the top of the exosome, and then threaded into the channel within the ring of ribonucleases. Within the channel the RNA is degraded by a combination of exo- and endonuclease activities.

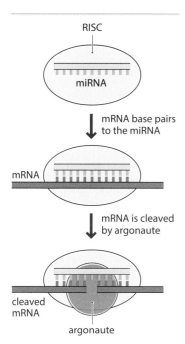

Figure 17.19 **Cleavage of an mRNA in a RISC by the argonaute protein.**

The eukaryotic silencing complex degrades specific mRNAs

Until a few years ago, biochemists had very little information on processes able to degrade individual mRNAs, as would be necessary in order for mRNA turnover to play a significant role in gene regulation. The breakthrough came with discovery of the **RNA-induced silencing complex** (**RISC**). This is a protein–RNA structure, only known in eukaryotes, which cleaves and hence inactivates individual mRNAs. The mRNA becomes bound, by base pairing, to the RNA component of the RISC, which is a 20–25 nucleotide molecule called a **microRNA** (**miRNA**). The mRNA is then cleaved by an endoribonuclease called the **argonaute** protein (*Fig. 17.19*).

The mode of action of a RISC is therefore straightforward. The important question is how the RISC obtains an miRNA complementary to the mRNA that is targeted for degradation. MicroRNAs are transcribed from their genes by RNA polymerase II, initially as precursor molecules several hundred nucleotides in length that can contain up to six mature miRNA molecules. Each miRNA is located in a part of the precursor that forms the stem of a stem–loop structure (*Fig. 17.20*). The stem–loop is cut away from the precursor by **Dicer**, a ribonuclease that cuts double-stranded RNA, and the double-stranded versions of the miRNAs are released by additional cleavages. One strand of each double-stranded miRNA is degraded, and the other incorporated into a RISC. A few miRNAs are obtained by a slightly different method, not by transcription of a gene for a precursor miRNA, but from an intron cut out of the mRNA of a protein-coding gene. Part of the intron RNA folds up to form the stem–loop structure, which is then processed by Dicer as described above.

Human cells are able to make about 1000 miRNAs, but together these can target mRNAs from over 10 000 genes, possibly because mRNAs from different genes share the same miRNA binding sequence, or possibly because a precise match between miRNA and mRNA is not needed in order for the mRNA to be captured by a RISC. Some miRNA genes are located close to the protein-coding genes whose mRNAs are targeted by the miRNA. In these cases it is possible that the same regulatory proteins control both mRNA and miRNA synthesis. This would allow synthesis of the miRNA to be directly coordinated with repression of the protein-coding gene. The mRNA would therefore be degraded immediately after its synthesis is switched off. But in many other cases the miRNA and protein genes are not co-located, and the way in which mRNA synthesis and degradation are coordinated is not clear. This is an area of biochemistry in which important new discoveries are being made every year, and these mysteries are sure to be unraveled before too long.

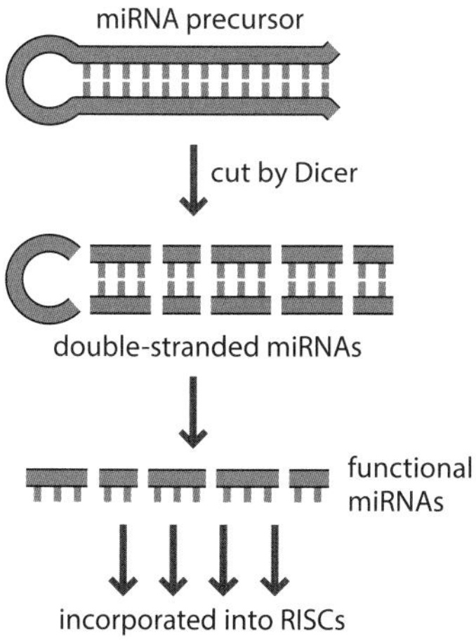

17.2.2 Degradation of proteins

We have a good understanding of the degradation pathways for misfolded proteins and for proteins that have reached the end of their functional life. We are also beginning to recognize how these degradation processes can be directed at individual proteins whose genes have been switched off.

Proteins to be degraded are labeled with ubiquitin molecules

Ubiquitin is an abundant, 'ubiquitous' protein, 76 amino acids in length in humans, that plays a central role in protein degradation by acting as a tag for proteins that must be broken down.

The attachment of ubiquitin to a protein is called **ubiquitination**. This results in a linkage called an **isopeptide bond** being formed between the carboxyl group of the C-terminal amino acid of ubiquitin, which in most species is a glycine, with the amino group present in the side-chain of a lysine located within the protein to be degraded (*Fig. 17.21*). Ubiquitination is a three-step process in which a ubiquitin molecule is

Figure 17.20 **Processing of an miRNA precursor by Dicer.**

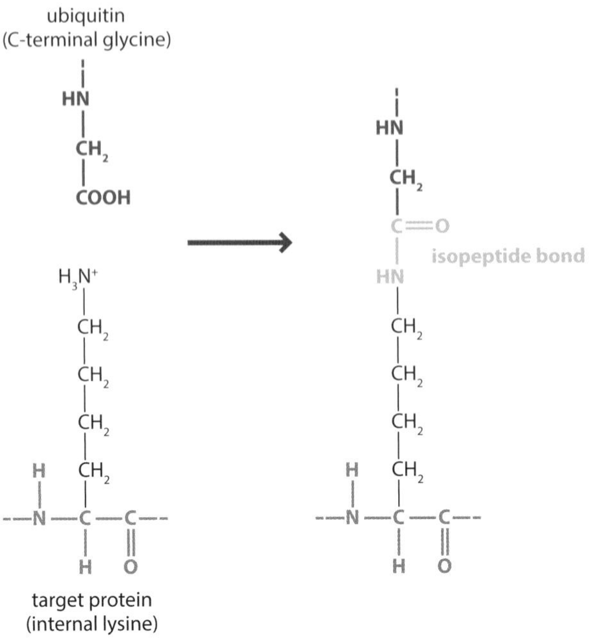

Figure 17.21 Attachment of ubiquitin to an internal lysine in a target protein.

ubiquitin
(C-terminal glycine)

HN

CH₂

COOH

H₃N⁺

CH₂

CH₂

CH₂

H   CH₂

--N—C—C--

H   O

target protein
(internal lysine)

HN

CH₂

C=O   isopeptide bond

HN

CH₂

CH₂

CH₂

H   CH₂

--N—C—C--

H   O

initially attached to an activator protein in an energy-dependent reaction that results in hydrolysis of an ATP. The ubiquitin is then transferred to a conjugating enzyme, and finally a third enzyme, a **ubiquitin ligase**, transfers the ubiquitin to the target protein.

Ubiquitination has a number of functions in addition to labeling proteins for degradation. It can be a signal for movement of a protein to a new location, and is also one of the chemical modifications made to histone proteins as a means of silencing or activating segments of the genome. The different roles of ubiquitination appear to be distinguished by the nature of the structures formed on the target protein by ubiquitin attachment. To act as a label for degradation, chains of linked ubiquitin molecules must be built up on the target protein. Each of these polyubiquitin chains contains a series of ubiquitins, each attached to one of the lysines in the previous unit in the chain (*Fig. 17.22*). As each ubiquitin molecule has seven lysines, a great variety and complexity of polyubiquitin chains can be assembled. In most of the polyubiquitin chains involved in protein degradation, the links involve the second and sixth of the lysines, at positions 11 and 48 in the ubiquitin monomer.

How does the ubiquitination process recognize the correct proteins, those that must be degraded? The answer appears to lie with the specificity of the enzymes that attach ubiquitin to these target proteins. Most species possess just a single activator protein, but have multiple versions of the conjugating enzymes and many types of ubiquitin ligase. In humans, for example, there are 35 conjugating enzymes and several hundred ligases. It is thought that different conjugating enzyme–ligase pairs have specificity for different proteins and for the nature of the chains built up on those proteins. Activation of different enzyme pairs, in response to intra- or extracellular signals, is probably the key to the specific degradation of particular proteins and groups of proteins.

The proteasome is responsible for protein degradation

Ubiquitination labels proteins for degradation but it does not on its own result in the actual breakdown of the protein. To be degraded, a ubiquitinated protein must be moved to a **proteasome**.

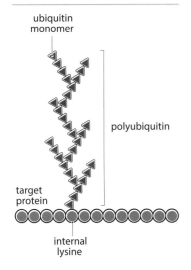

ubiquitin
monomer

polyubiquitin

target
protein

internal
lysine

Figure 17.22 A polyubiquitin chain attached to a protein targeted for degradation.

Box 17.5 Protein and mRNA half-lives

Proteins and mRNAs are continually being turned over in the cell, with existing molecules being degraded and new ones synthesized. The rates of degradation can be expressed as **half-lives** – the period of time required for the amount of an individual type of protein or mRNA to fall to half its initial value, assuming that there is no new synthesis of the molecule.

Half-lives can be measured by **pulse labeling**. The cells being studied are briefly provided with a labeled substrate for protein or mRNA synthesis, such as an amino acid containing the 'heavy' ^{15}N isotope of nitrogen, or radioactive 4-thiouracil, which has a ^{35}S atom rather than an oxygen attached to carbon number 4. Proteins and mRNAs synthesized during the period of pulse labeling will contain the labeled amino acid or nucleotide, but those made before or after will not. The degradation rates of the molecules made during the pulse labeling period can therefore be followed by measuring the amounts of label present in bulk protein or mRNA extracts, or in individual protein or mRNA types, prepared at intervals after the pulse period.

Studies of this type have shown that most bacterial proteins and mRNAs have half-lives of only a few minutes, reflecting the rapid changes in gene expression patterns that can occur in an actively growing bacterium. Eukaryotic molecules last longer (see figure opposite), with median half-lives of 46 h for proteins and 9 h for mRNA in mouse fibroblast cells.

These histograms show that there are striking variations in degradation rates for different proteins and mRNAs. In general, we have little information on the factors that influence the half-life of an individual molecule. Proteins that possess an internal sequence that is rich in proline, glutamic acid, serine and threonine are often ones with short half-lives. This **PEST sequence** (named after the single letter abbreviations for

the four amino acids) might interact with the ubiquitination process in some way to target these proteins for rapid turnover. Eukaryotic mRNA degradation has for some time been thought to be associated with the length of the poly(A) tail, though the mechanism by which the tail length influences degradation has not been described. One possibility is that long-lived mRNAs have an internal uracil-rich sequence to which the poly(A) tail can base pair, forming a stem–loop that prevents degradation by a $3' \rightarrow 5'$ exonuclease.

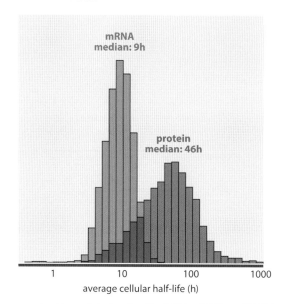

Image reproduced by permission from Macmillan Publishers Ltd: *Nature*, **473**: 337; B. Schwanhäusser *et al.* Global quantification of mammalian gene expression control, © 2011.

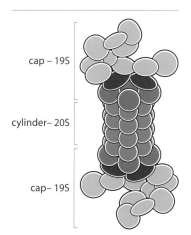

cap – 19S

cylinder– 20S

cap– 19S

Figure 17.23 The eukaryotic proteasome.
The protein components of the two caps are shown in green, orange and red, and those forming the cylinder in blue.

In eukaryotes, the proteasome is a large, multisubunit structure, with a sedimentation coefficient of 26S (*Fig. 17.23*). The main component is a central cylinder with a sedimentation coefficient of 20S, comprising four rings, each made up of seven proteins. The proteins present in the two inner rings are proteases, whose active sites are located on the inner ring surface. This means that a ubiquitinated protein must enter the cylinder in order to be degraded. The proteins that make up the two outer rings do not have protease activities, but instead mediate the entry of ubiquitinated proteins into the proteasome cylinder. They do this in conjunction with a pair of cap structures, one at each end of the cylinder. The commonest cap structures, which have a sedimentation coefficient of 19S and are made up of 19 proteins, mediate the entry of full size proteins into the proteasome. A smaller 11S cap, with just seven proteins, is involved in degradation of shorter peptides.

Ubiquitinated proteins might interact directly with the 19S cap structure of the proteasome, or the interaction might be via a **ubiquitin–receptor protein**. Before entry into the proteasome, the protein to be degraded must be at least partially unfolded and its ubiquitin labels must be removed. These steps require energy, obtained from hydrolysis of ATP, catalyzed by proteins present in the cap. The protein is then moved into the proteasome where it is broken down into peptides, usually of 4–10 amino acids in length. These are released back into the cytoplasm where they are broken down by other proteases into individual amino acids.

The ubiquitin and proteasome system for protein degradation was discovered in eukaryotes but we now know that similar systems exist in at least some bacteria and archaea. The bacterial degradation tag is a 64 amino acid 'ubiquitin-like protein', which is attached to lysine amino acids of the target proteins in a manner similar to eukaryotic ubiquitination. Archaea also have proteasomes of about the same size as the eukaryotic version, but these are less complex, comprising multiple copies of just two proteins. The bacterial proteasome, on the other hand, is smaller and lacks the cap structures present in both the eukaryotic and archaeal versions.

Further reading

Harrison SC and Aggarwal AK (1990) DNA recognition by proteins with the helix–turn–helix motif. *Annual Review of Biochemistry* **59**, 933–69.

Henkin TM (1996) Control of transcription termination in prokaryotes. *Annual Review of Genetics* **30**, 35–57. *A detailed account of attenuation.*

Horvath CM (2000) STAT proteins and transcriptional responses to extracellular signals. *Trends in Biochemical Sciences* **25**, 496–502.

Kim Y-J and Lis JT (2005) Interactions between subunits of *Drosophila* mediator and activator proteins. *Trends in Biochemical Science* **30**, 245–9.

Lopez D, Vlamakis H and Kolter R (2008) Generation of multiple cell types in *Bacillus subtilis*. *FEMS Microbiology Letters* **33**, 152–63. *Describes how different σ subunits are involved in sporulation.*

Losick RL and Sonenshein AL (2001) Turning gene regulation on its head. *Science* **293**, 2018–9. *Describes the attenuation system.*

Mackay JP and Crossley M (1998) Zinc fingers are sticking together. *Trends in Biochemical Science* **23**, 1–4.

Melloui D, Marshak S and Cerasi E (2002) Regulation of insulin gene transcription. *Diabetologia* **45**, 309–26.

Pratt AJ and MacRae IJ (2009) The RNA-induced silencing complex: a versatile gene-silencing machine. *Journal of Biological Chemistry* **264**, 17687–901.

Ptashne M and Gilbert W (1970) Genetic repressors. *Scientific American*, **222**(6), 36–44. *The mode of action of repressors and the methods used in isolation of the proteins.*

Tsai M-J and O'Malley BW (1994) Molecular mechanisms of action of steroid/thyroid receptor superfamily members. *Annual Review of Biochemistry* **63**, 451–86. *Control of gene expression by steroid hormones.*

Vanacova S and Stefl R (2007) The exosome and RNA quality control in the nucleus. *EMBO Reports* **8**, 651–7.

Varshavsky A (1997) The ubiquitin system. *Trends in Biochemical Science* **22**, 383–7.

Voges D, Zwickl P and Baumeister W (1999) The 26S proteasome: a molecular machine designed for controlled proteolysis. *Annual Review of Biochemistry* **68**, 1015–68.

Wek RC, Jiang H-Y and Anthony TG (2006) Coping with stress: eIF2 kinases and translational control. *Biochemical Society Transactions* **34**(1), 7–11. *Global control of translation.*

Wheatley RW, Lo S, Jancewicz LJ, Dugdale ML and Huber RE (2013) Structural explanation for allolactose (*lac* operon inducer) synthesis by *lacZ* β-galactosidase and the evolutionary relationship between allolactose synthesis and the *lac* repressor. *Journal of Biological Chemistry* **288**, 12993–3005.

Zubay G, Schwartz D and Beckwith J (1970) Mechanisms of activation of catabolite-sensitive genes: a positive control system. *Proceedings of the National Academy of Sciences USA*, **66**, 104–10. *An early description of the CAP system.*

Self-assessment questions

Multiple choice questions

Only one answer is correct for each question.
Answers can be found on the website:
www.scionpublishing.com/biochemistry

1. The σ subunit of the *E. coli* RNA polymerase contains which type of DNA-binding structure?
  (a) Helix–helix–turn
  (b) Helix–turn–helix
  (c) Zinc finger
  (d) Heat shock domain

2. What is the standard σ subunit of the *E. coli* RNA polymerase called?
  (a) σ^{32}
  (b) σ^{70}
  (c) σ^{76}
  (d) σ^{90}

3. What is the σ subunit of the *E. coli* RNA polymerase that is used during heat shock called?
  (a) σ^{32}
  (b) σ^{70}
  (c) σ^{76}
  (d) σ^{90}

4. Which bacteria uses alternative σ subunits during its sporulation pathway?
  (a) *Escherichia coli*
  (b) *Bacillus* species
  (c) *Mycobacterium tuberculosis*
  (d) *Sporobacteria*

5. What is the role of the lactose permease of *E. coli*?
  (a) Inducer of the lactose operon
  (b) Mediates the effect of glucose on transcription of the lactose operon
  (c) Splits lactose to galactose and glucose
  (d) Transports lactose into the cell

6. Which statement describes the situation when lactose is present?
  (a) Repressor binds to inducer and prevents transcription of the lactose operon
  (b) Repressor does not bind to inducer and prevents transcription of the lactose operon
  (c) Repressor binds to inducer and allows transcription of the lactose operon
  (d) Repressor does not bind to inducer and allows transcription of the lactose operon

7. What is the repressor binding site called?
  (a) Promoter
  (b) Operator
  (c) CAP binding site
  (d) Operon

8. What is the role of the IIA^{Glc} protein of *E. coli*?
  (a) Inducer of the lactose operon
  (b) Mediates the effect of glucose on transcription of the lactose operon
  (c) Splits lactose to galactose and glucose
  (d) Transports lactose into the cell

9. Which statement describes the situation when glucose and lactose are present?
  (a) CAP and repressor bound, lactose operon is transcribed
  (b) CAP and repressor bound, lactose operon is not transcribed
  (c) Neither CAP nor repressor bound, lactose operon is not transcribed
  (d) CAP but not repressor bound, lactose operon is not transcribed

10. What is a basal promoter element, sometimes found upstream of a eukaryotic protein-coding gene?
  (a) The binding site for the mediator protein
  (b) The binding site for the cAMP response element binding (CREB) protein
  (c) A site that determines the rate of transcription of the gene when not subject to up- or down-regulation
  (d) A part of the heat shock module

11. Which of the following is **not** a type of activation domain?
  (a) Acidic domain
  (b) Basic domain
  (c) Glutamine-rich domain
  (d) Proline-rich domain

12. Which of the following statements is **incorrect** regarding the mediator protein?
  (a) First identified in the yeast *Saccharomyces cerevisiae*
  (b) The tail of the mediator protein makes contact with the activation domain of the regulatory protein
  (c) The middle and head sections interact with the polymerase complex
  (d) Is made up of 25 subunits in humans

13. Which of the following is **not** a steroid hormone?
  (a) Estrogen
  (b) Androgen
  (c) Mineralocorticoid hormone
  (d) Adrenocorticotropic hormone

14. With regard to attenuation control of the *E. coli* tryptophan operon, which statement describes the situation when tryptophan is present?

(a) Ribosome stalls, terminator structure forms, transcription terminates

(b) Ribosome stalls, terminator structure does not form, transcription terminates

(c) Ribosome does not stall, terminator structure forms, transcription terminates

(d) Ribosome does not stall, terminator structure does not form, transcription terminates

15. Global regulation of eukaryotic translation is mediated by phosphorylation of which initiation factor?

(a) eIF-2

(b) eIF-3

(c) eIF-4

(d) eIF-4A

16. Which of the following statements is **correct** regarding transcript-specific control of ferritin synthesis in mammals?

(a) When iron levels are low, the regulatory protein IRP-1 attaches to a basal promoter element on the ferritin mRNA

(b) IRP-1 is an iron binding protein, the attachment of iron atoms resulting in a conformational change so that the protein recognizes the basal promoter element

(c) In the presence of iron, IRP-1 detaches from the mRNA

(d) Detachment of IRP-1 prevents translation of the ferritin mRNA

17. Which of the following is **not** a component of the bacterial degradosome?

(a) Polynucleotide phosphorylase

(b) RNAse E

(c) RNAse P

(d) RNA helicase B

18. What is the eukaryotic equivalent of the degradosome called?

(a) RNA-induced silencing complex

(b) Proteasome

(c) mRNA surveillance particle

(d) Exosome

19. What is the endoribonuclease that cleaves mRNA in the RNA-induced silencing complex called?

(a) RNase E

(b) Argonaute

(c) RNase P

(d) Dicer

20. What is the endoribonuclease that cleaves miRNAs from their precursor molecules called?

(a) RNase E

(b) Argonaute

(c) RNase P

(d) Dicer

21. Human cells make approximately how many miRNAs?

(a) 1000

(b) 5000

(c) 10 000

(d) 50 000

22. What is the median half-life for a eukaryotic protein in mouse fibroblasts?

(a) 45 minutes

(b) 9 hours

(c) 24 hours

(d) 46 hours

23. Which of the following statements is **incorrect** with regard to ubiquitination?

(a) The attachment of ubiquitin to a protein is via an isopeptide bond

(b) The bond involves the N-terminal amino acid of ubiquitin, which in most species is a glycine

(c) Ubiquitin is initially attached to an activator protein

(d) A ubiquitin ligase transfers the ubiquitin to the target protein

24. How many lysines are there in a ubiquitin molecule?

(a) 1

(b) 5

(c) 7

(d) 10

Short answer questions

These questions do not require additional reading.

1. Outline the roles that control of gene expression plays in living organisms.

2. Using examples, explain how alternative σ subunits regulate gene expression in *E. coli*.

3. Give a detailed account of control of the *E. coli* lactose operon by (A) the lactose repressor, and (B) the catabolite activator protein.

4. Summarise the different types of regulatory sequence that can be found in the promoter of a eukaryotic protein-coding gene.

5. What is the role of the mediator protein in gene expression in a eukaryote?

6. Describe how a steroid hormone regulates gene expression in human cells.

7. Give examples of global and transcript-specific regulation of translation in eukaryotes.

8. Outline the pathways for nonspecific mRNA turnover in *E. coli* and in eukaryotes.

9. Give a detailed account of the role of miRNA in mRNA turnover in eukaryotes.

10. Describe the roles of ubiquitin and the proteasome in protein degradation.

Self-study questions

These questions will require calculation, additional reading and/or internet research.

1. With some types of virus, transcription of the host's genes ceases shortly after infection. All of the cell's RNA polymerase enzymes start transcribing the virus genes instead. Suggest events that might underlie this phenomenon.

2. Operons are very convenient systems for achieving coordinated regulation of expression of related genes. Discuss why operons are common in bacteria yet are absent in eukaryotes.

3. The tryptophan operon of *E. coli* is regulated by both a repressor protein and by attenuation. Other operons coding for amino acid biosynthetic enzymes are controlled only by attenuation. Discuss.

4. To what extent is *E. coli* a good model for the regulation of transcription initiation in eukaryotes? Justify your opinion by providing specific examples of how extrapolations from *E. coli* have been helpful or unhelpful in the development of our understanding of equivalent events in eukaryotes.

5. How might the length of the poly(A) tail influence the half-life of a eukaryotic mRNA?

CHAPTER 18

Studying proteins, lipids and carbohydrates

STUDY GOALS

After reading this chapter you will:

- understand the difference between a polyclonal and monoclonal antibody

- know how the antibody–antigen precipitin reaction is exploited in immunoassays, and be able to describe different types of immunoassay

- understand how an enzyme immunoassay is carried out and know the advantages of this technique

- be able to describe the various methods used to separate the proteins in a proteome prior to protein profiling

- understand the role of mass spectrometry in protein profiling

- be able to describe how isotope-coded affinity tags are used to compare the components of two proteomes

- understand how circular dichroism is used to study the secondary structure composition of a protein

- know how a nuclear magnetic spectrum is generated and appreciate the strengths and limitations of NMR in studying protein structure

- understand how an X-ray diffraction pattern can be used to determine the detailed structure of a protein

- be able to describe the principles of gas chromatography and how this method is used to separate lipids

- know how mass spectrometry is used to study lipid structure

- understand how immunological methods, lectins and glycan sequencing are used to study carbohydrate structure

In the final part of this book we will look at the most important of the many different methods that have been developed for studying biomolecules. These methods provide the foundation for the advancement of our understanding of biochemistry, and they are methods that you will use if you decide to pursue a research career in this area of biology.

In this chapter we will explore methods for studying proteins, lipids and carbohydrates, and in the next chapter we will look at methods specifically designed for DNA and RNA. We are able to combine proteins, lipids and carbohydrates into a single chapter because there is a great deal of commonality in the approaches used to study these three types of biomolecule. DNA and RNA, on the other hand, have their own specific technology, centered largely on identification of the nucleotide sequences of these molecules.

18.1 Methods for studying proteins

There are many methods for studying proteins, and in describing these we must make sure we do not get submerged in the diversity of the technology. To avoid this, we will ask three questions:

- How do we determine if a particular protein is present in a cell or tissue?
- How do we identify all the individual members of the entire set of proteins present in a cell or tissue?
- How is the structure of a protein worked out?

By examining how these questions are answered we will become familiar with the methods used to study proteins.

18.1.1 Methods for identifying the presence of an individual protein

First, we will look at the methods used to detect the presence of an individual protein in the mixture of proteins present in a cell or tissue extract. Most of these are **immunological methods**, based on the natural immune response of mammals and other animals.

Immunological methods make use of the reaction between antibody and antigen

The immune response is the physiological process used by animals to provide protection against harmful substances called **antigens**. An antigen is, quite simply, any substance that elicits an immune response. Part of the immune response is the synthesis of **antibodies** by B lymphocytes present in the blood and lymphatic systems. An antibody is a type of a protein, called an **immunoglobulin**, which binds specifically to an antigen, leading to the destruction of the antigen by other components of the immune system (*Fig. 18.1A*). For example, attachment of antibodies to the surface of an invading bacterium activates the **complement system**, a set of enzymes and other proteins that disrupt the bacterial cell membrane, killing the pathogen.

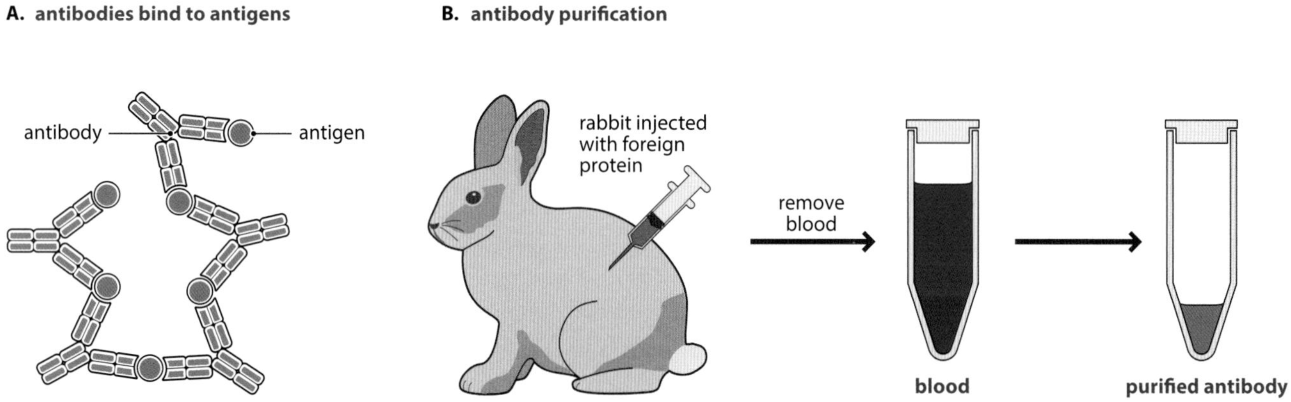

A. antibodies bind to antigens

B. antibody purification

antibody — antigen

rabbit injected with foreign protein

remove blood

blood   purified antibody

Figure 18.1 **Antibodies.**
(A) Antibodies bind to antigens. (B) Purified antibodies can be obtained from a sample of blood taken from a rabbit injected with the foreign protein.

Box 18.1 Immunoglobulins and antibody diversity

Immunoglobulins are synthesized by B lymphocytes and either become attached to the outer surface of the plasma membrane or are secreted into the bloodstream. Each immunoglobulin is a tetramer of four polypeptides, two larger molecules called **heavy chains** and two smaller **light chains**, the latter linked to the heavy chains by disulfide bonds. When joined together, the heavy and light chains form a fork-shaped structure.

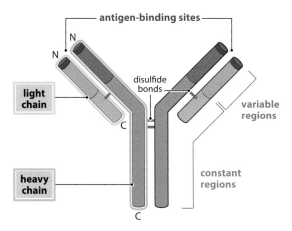

A different immunoglobulin protein is synthesized for every antigen that is encountered. The **variable regions**, which provide an immunoglobulin with its specific antigen-binding properties, are located in the N-terminal regions of the heavy and light chains. The remainder of each chain forms a **constant region** which has a similar amino acid sequence in all immunoglobulins of a particular type.

There are various families and sub-families of heavy chain which distinguish different classes of immunoglobulin. In humans the five main types are:

- **Immunoglobulin M (IgM)** exists as a pentamer in human blood and is the first type of antibody to be synthesized when a new antigen is encountered. This type of immunoglobulin binds tightly to antigenic epitopes on bacteria and other pathogens. It switches on the complement system, and also activates macrophages which engulf and degrade pathogens by the process called **phagocytosis**. IgM molecules have the μ version of the heavy chain.
- **Immunoglobulin G (IgG)** has a γ-type heavy chain. There are several subclasses of the heavy chain, giving rise to IgG_1, IgG_2, IgG_3, etc. IgG is synthesized at a later stage in the immune response and also has a specific role in enabling a mother to provide immunological protection for her fetus and newborn infant. It is the only type of immunoglobulin that is able to pass through the placenta, and it is also secreted into the mother's milk.
- **Immunoglobulin A (IgA)** has the α heavy chain. It is the main type of antibody in tears and saliva.
- **Immunoglobulin E (IgE)** and **immunoglobulin D (IgD)** have less clearly defined roles, though IgE may be important in protecting the body against eukaryotic parasites such as

Plasmodium falciparum, the causative agent of malaria. IgE has an ε heavy chain and IgD has a δ chain.

There are also two variants of the light chain, called κ and λ. Any immunoglobulin molecule can have any combination of light strands – two κ, two λ, or one of each.

Humans are able to make approximately 10^8 different immunoglobulins, each specific for a different antigen epitope. This immense level of variability is possible because of the unusual way in which the mRNAs for the heavy and light chain polypeptides are synthesized. In vertebrate genomes there are no complete genes for the heavy or light chains. Instead, each heavy chain is specified by four gene segments, one segment (C_H) for the constant region and three (V_H, D_H and J_H) for different parts of the variable region. There are multiple copies of each gene segment, each copy specifying a slightly different amino acid sequence. A two-stage process brings these segments together to generate a complete heavy chain mRNA:

- First, an exon coding for the entire variable region is assembled by a DNA recombination event that links together one V_H gene segment with one of D_H and one of J_H.

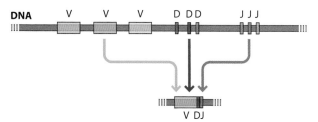

- In the second stage, this V–D–J exon is transcribed and linked to a C_H segment transcript by splicing.

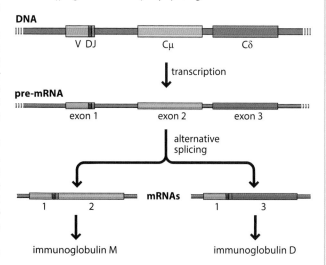

This process creates a complete heavy chain mRNA that is now translated. Light chain mRNAs are produced in a similar way, the only difference being that light chains do not have J segments.

Many proteins act as antigens, especially ones that are foreign to the body and hence are not recognized as being part of the normal set of proteins that the animal synthesizes. This is the aspect of the immune system that it exploited by biochemists in the design of immunological methods for protein detection. If a purified sample of a protein is injected into the bloodstream of a laboratory animal, such as a rabbit, the immune system of the animal responds by synthesizing antibodies that bind specifically to that protein (*Fig. 18.1B*). The amount of the antibody that is present in the rabbit's bloodstream remains high enough over the next few days for substantial quantities to be purified. After purification, the antibody retains its ability to bind to the protein with which the animal was originally challenged.

Most proteins have such complex structures that they stimulate synthesis of not just a single antibody, but several different ones, each one recognizing a different feature, or **epitope**, on the surface of the protein. This collection of reacting immunoglobulins is called a **polyclonal antibody**. To be specific for an individual protein, an immunoglobulin must recognize an epitope that is unique to that protein. Most polyclonal antibodies contain at least some immunoglobulins with this level of specificity, but also others that recognize epitopes that are common surface features shared between different proteins. This means that polyclonal antibodies are rarely entirely specific for the protein against which they are raised. **Monoclonal antibodies**, on the other hand, contain just a single type of immunoglobulin, and will be totally specific for their target antigen, presuming the immunoglobulin recognizes a unique epitope not present on other antigens. Monoclonal antibodies are usually prepared in mice rather than rabbits. After challenging the mouse with the antigen, the spleen, which contains developing B lymphocytes, is removed and the lymphocytes mixed with mouse myeloma cells. Some of the lymphocytes and myeloma cells fuse, creating a **hybridoma**, which possesses both the B-lymphocyte's ability to make immunoglobulin and the myeloma cell's ability to divide indefinitely when placed in a suitable culture medium. Individual hybridomas are therefore cultured in order to provide large quantities of the antibody made by the B-lymphocyte. The antibody is 'monoclonal' because it recognizes a single epitope and is prepared from a clone of hybridoma cells.

A variety of immunological methods have been developed, differing mainly in the way in which the reaction between antibody and antigen is detected. Some of these methods are qualitative, and simply indicate if the target protein is present or not. Others, called **immunoassays**, enable the amount of antigen to be quantified with differing degrees of precision. We will now examine the most important of these immunological methods.

Some immunological methods are based on precipitation of the antibody–antigen complex

Reaction of an antigen with a polyclonal antibody usually results in formation of an insoluble complex, made up of interlinked antibody–antigen networks, which precipitates out of the solution. A number of immunological methods make use of this **precipitin reaction** to detect the presence of the antigen. In the simplest method, the test is carried out in a solution and precipitation is detected by eye, either from the increased cloudiness of the solution, or by the presence of an insoluble pellet after the solution has been centrifuged. The precipitin reaction can be used as a basic immunoassay by measuring the precipitation occurring with increased amounts of antigen when the amount of antibody is kept constant. Increasing the amount of antigen gives larger precipitations until the **zone of equivalence** is reached. This is the point at which the relative amounts of antigen and antibody are optimal for complex formation (*Fig. 18.2*). If the amount of antigen is increased even further, the antibody binding sites become saturated with antigen, so the complexes break down and *less* precipitation is seen.

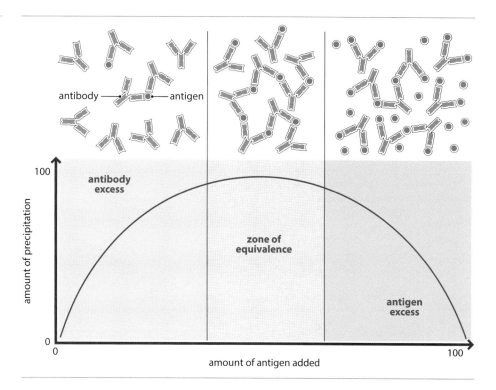

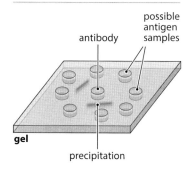

Figure 18.3 **One version of the Ouchterlony technique.**
The antibody is placed in the central well and test solutions in the outer circle of wells. Two precipitin reactions can be seen after staining with Coomassie blue, indicating that the samples in the lower and top left wells contain antigens that cross-react with the antibody. The lack of precipitin reactions with the other samples shows that these are not recognized by the antibody being tested.

Precipitation tests are more commonly carried out in a thin slab of agarose gel, rather than in solution. In the **Ouchterlony technique**, samples of the antibody and antigen are placed in wells in the gel, a centimeter or so apart (*Fig. 18.3*). The two solutions diffuse out of the wells, forming concentration gradients in the gel that eventually overlap. The precipitate forms at the zone of equivalence within the two overlapping concentration gradients. The precipitate can sometimes be seen by eye, or alternatively the gel can be stained with a protein-specific dye such as Coomassie blue, which will reveal the line of precipitation.

The Ouchterlony technique is slow because the antigen and antibody pass through the gel by natural diffusion, which means that it can take hours or days for the concentration gradients to form. **Immunoelectrophoresis** techniques are designed to speed up the process by increasing the rate of movement of the antigen and antibody through the gel. Most proteins have a net negative charge at pH 8.0, and so when placed in an electric field they migrate towards the positive pole, the rate of migration being dependent on their size and charge. Immunoglobulin molecules, unusually, have a neutral charge. Rather than remaining stationary in the gel, they migrate in the opposite direction, towards the negative electrode, by **electroendosmosis**. This process occurs because the agarose molecules in the gel, when exposed to the electric field, become slightly electronegative. Being immobilized these molecules cannot move towards the positive electrode. Instead, the electrical charge is counterbalanced by positively charged water molecules flowing towards the negative electrode. The antibody molecules are carried along in this flow, setting up a concentration gradient of the antibody in the gel. In the **crossover immunoelectrophoresis** (**CIP** or **CIEP**) technique, the antigen and antibody are placed in wells flanked by electrodes (*Fig. 18.4*). When the electric current is switched on, the antigen and antibody molecules migrate towards one another. The antigen protein moves through the gel as a single sharp band, but the antibody forms a concentration gradient. Precipitation occurs at the zone of equivalence within the antibody gradient. Not only is immunoelectrophoresis much quicker than the Ouchterlony method, it is also more sensitive. This is because,

Box 18.2 Electrophoresis

Electrophoresis is the movement of charged molecules in an electric field. Negatively charged molecules migrate towards the positive electrode, and positively charged molecules migrate towards the negative electrode.

In biochemistry, electrophoresis is usually performed in a gel, made either of **agarose** or **polyacrylamide**. Agarose is a polysaccharide of repeating D-galactose and 3,6-anhydro-L-galactopyranose units, which forms a gel after heating in water. The gel consists of a network of pores 100–300 nm in diameter, the size depending on the concentration of agarose. Polyacrylamide is made up of chains of acrylamide monomers (CH_2=CH–CO–NH_2) cross-linked with N,N'-methylenebisacrylamide units (commonly called 'bis'; CH_2=CH–CO–NH–CH_2–NH–CO–CH=CH_2), again forming a gel but with smaller pore sizes, 20–150 nm in diameter. Agarose gels are prepared as a slab on a glass or plastic support or in a capillary tube. Polyacrylamide gels are also prepared in both slab and capillary formats, but the slabs are usually held between two glass plates.

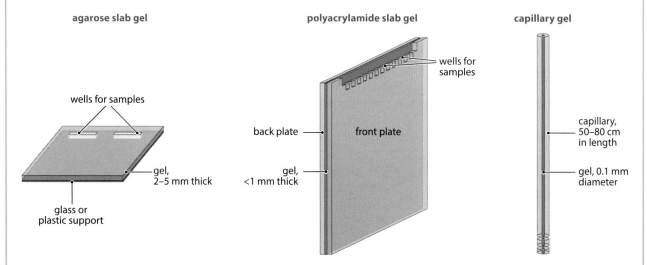

If electrophoresis is carried out in an aqueous solution rather than a gel, then the factors influencing the migration rate are the shape of a molecule and its electric charge. **Gel electrophoresis** enables a greater variety of chemical and physical properties to be used to separate molecules. As well as immunoelectrophoresis, there are three other important types of gel electrophoresis used in studies of proteins and DNA:

- A polyacrylamide gel containing sodium dodecyl sulfate (SDS) is used to separate proteins according to their molecular weights. SDS is an ionic detergent that denatures proteins and coats them with negatively charged detergent molecules. When the electric charge is applied, the proteins move towards the positive electrode at a rate proportional to their size. The smallest proteins move the quickest, because they are able to migrate through the pores in the gel more rapidly than the larger proteins. This technique is called **SDS-polyacrylamide gel electrophoresis** (**SDS-PAGE**). The range of proteins that can be separated depends on the pore size, with gels with smaller pores needed to separate smaller proteins. The pore size is set by the total concentration of monomers (acrylamide + bis) and the ratio of acrylamide to bis, so gels with particular pore sizes can be tailor-made for different requirements.

- DNA molecules can also be separated according to their molecular weights, with polyacrylamide gels being used for molecules up to about 1000 bp and agarose gels for longer ones. It is not necessary to add SDS, because a DNA molecule already carries a negative charge that is proportional to its length, due to the presence of an O^- group within each phosphodiester bond.

- Proteins that have not been treated with SDS can be separated according to their natural charge differences by the technique called **isoelectric focusing**. The gel contains an **immobilized pH gradient**, meaning that the pH gradually changes along the length of the gel. The gradient is established by differential concentrations of weakly acidic or basic compounds that are included in the gel when it is prepared. In this type of gel, a protein migrates to its isoelectric point, the position in the gradient where its net charge is zero.

by moving as a single band, the antigen concentration remains high, rather than becoming diluted out within a concentration gradient.

Enzyme immunoassays are sensitive and quantitative

Although the amount of antigen–antibody complex formed increases with the concentration of the antigen, this quantitative aspect of the precipitin test is only measurable when the reaction is carried out in solution. This means that the

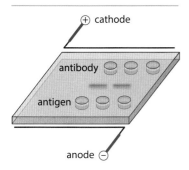

Figure 18.4 Crossover immunoelectrophoresis.
Two precipitin reactions are seen, indicating the two wells which contain antigens that cross-react with the antibody.

Ouchterlony technique and the various immunoelectrophoresis methods are primarily qualitative, enabling the presence of a protein antigen to be detected, but giving only approximate information on the amount of the antigen in a sample. Genuinely quantitative immunoassays are based on different ways of measuring the reaction between antigen and antibody.

The most commonly used immunoassay is the **ELISA** (**enzyme-linked immunosorbent assay**), which is easy to set up and gives rapid results. In this method, the antibody is conjugated to a **reporter enzyme**, an enzyme whose activity can easily be assayed. Horseradish peroxidase (HRP) is an example of a good reporter enzyme, because its activity can be monitored by a simple color test. This enzyme can oxidize various substrates, including artificial ones such as tetramethylbenzidine, which is converted by the enzyme into a blue product (*Fig. 18.5*). The amount of horseradish peroxidase that is present can therefore be measured by assaying the color change.

To carry out an ELISA, the antigen is first adsorbed to the walls of a well in a microtiter plate. In the direct ELISA method, the antibody–HRP conjugate is added to the well and allowed to bind to the antigen (*Fig. 18.6A*). The unbound antibody is washed away, and the amount of the antigen–antibody complex measured by assaying the bound HRP activity.

Alternatively, ELISA can be carried out in an indirect method. Again, the antigen is adsorbed within a microtiter well, but the antibody used to detect the antigen, called the **primary antibody**, is not itself conjugated to HRP. Instead, the amount of antigen–antibody complex that is formed is measured by adding a **secondary antibody**, conjugated to HRP, which recognizes not the antigen but the primary antibody (*Fig. 18.6B*). The secondary antibody is prepared by injecting the primary antibody into an animal, of a different species to the one from which the primary antibody was prepared. So, for example, if the primary antibody was prepared in a rabbit, then the secondary antibody could be obtained by injecting a sample of the primary antibody into a goat. The goat's immune system will look on the primary antibody as a foreign protein antigen, and synthesize the secondary antibody to bind to it.

The indirect ELISA method is more labor intensive, so does it have significant advantages? One benefit is that the secondary antibody, to which the reporter enzyme is conjugated, can be used in many different ELISAs. This secondary antibody is simply 'anti-rabbit immunoglobulin', and so can be used with a range of different primary antibodies, providing all of these have been obtained from rabbits. This means that the lab researcher only has to prepare the primary antibody, and can buy the generic HRP-conjugated secondary antibody from a commercial supplier, avoiding the hassle of having to prepare the conjugate. A second advantage of the indirect method is that it enables greater sensitivity. This is because the secondary antibody, assuming it is polyclonal, will recognize different epitopes on the surface of the primary antibody. As a result, more than one molecule of the secondary antibody will attach to each

Figure 18.5 The chemical basis to the horseradish peroxidase color test.

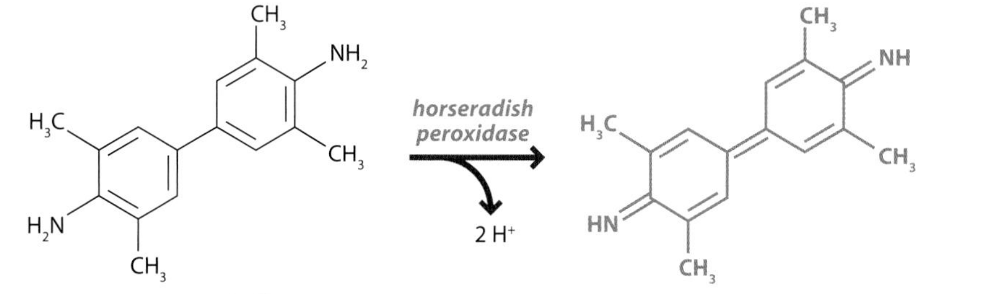

3,3', 5,5'-tetramethylbenzidine → *horseradish peroxidase* (2 H⁺) → 3,3', 5,5'-tetramethylbenzidine diimine

Figure 18.6 (A) Direct, and (B) indirect ELISA methods.

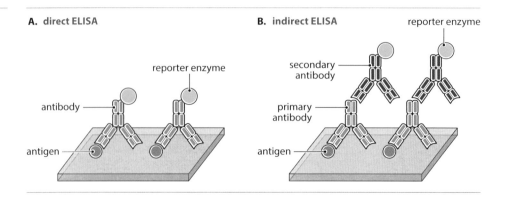

A. direct ELISA

reporter enzyme

antibody

antigen

B. indirect ELISA

reporter enzyme

secondary antibody

primary antibody

antigen

molecule of the primary antibody. As the secondary antibody carries the reporter enzyme, this multiple binding gives rise to a higher amount of signal and hence greater sensitivity.

18.1.2 Studying a proteome

The proteome is the term used to describe the collection of proteins present in a cell or tissue. The composition of the proteome defines the biochemical capability of a cell, so identifying the individual components of a proteome, and the relative amounts of the individual proteins, is a key goal in many research projects.

The methodology used to study proteomes is called **proteomics**. In its broadest context, proteomics includes not just the methods used to identify the proteins in a proteome, but also more advanced techniques aimed at understanding the functions of individual proteins, their localization in the cell, and the interactions between different proteins. The methods that we will study, which are used specifically to identify the components of a proteome, are called **protein profiling** or **expression proteomics**.

The proteins in a proteome must be separated prior to identification

Protein profiling is carried out in two stages:

• In the first stage, the individual proteins in a proteome are separated from one another.

• In the second stage, individual proteins are identified.

The separation of the components of a human proteome can be a challenging task, as some tissues contain as many as 20 000 different proteins. Indeed, with the most complex proteomes, complete separation of all the proteins might not be feasible. The most commonly used separation technique is **two-dimensional electrophoresis** in a polyacrylamide gel. In the first dimension the proteins are separated according to their net charges by isoelectric focusing, and in the second dimension according to their molecular weights by SDS–PAGE (*Fig. 18.7*). After electrophoresis, staining the gel with a protein dye reveals a complex pattern of spots, each one containing a different protein. This two-dimensional approach can separate several thousand proteins in a single gel.

Not all proteomes are as complex as those in typical human cells. In a bacterium, there might be fewer than 1000 proteins being synthesized at any time. Even with eukaryotes, the number of proteins that must be separated can be relatively small, especially if the component of the proteome present in a particular cell fraction (e.g. the mitochondria) is being studied in isolation. In these cases, one-dimensional electrophoresis, in either an SDS or isoelectric focusing gel, might be sufficient to separate the proteins. Alternatively, a type of **column chromatography** can be used.

Figure 18.7 **Two-dimensional polyacrylamide gel electrophoresis.**

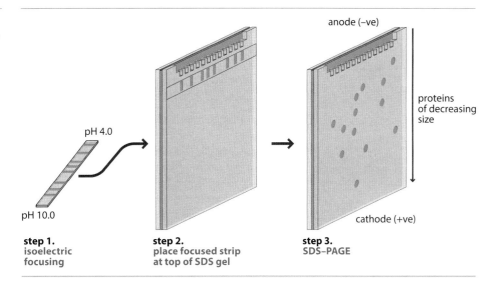

anode (–ve)

pH 4.0

pH 10.0

proteins of decreasing size

cathode (+ve)

step 1.
isoelectric focusing

step 2.
place focused strip at top of SDS gel

step 3.
SDS–PAGE

Column chromatography involves passing the protein mixture through a column packed with a solid matrix of some kind. The proteins in the mixture move through the matrix at different rates, and so become separated into bands. The solution emerging from the column is collected as a series of fractions, with each individual protein present in a different fraction (*Fig. 18.8*).

Several types of column chromatography have been developed, each using a different process for separating the proteins in the starting mixture. The three most important methods are:

• **Gel filtration chromatography**. In this type of chromatography the column is filled with small porous beads, usually made of dextran, polyacrylamide or agarose. As the proteins pass through the column, they enter and exit the pores of the beads. The smaller proteins enter the pores more easily, and are therefore delayed as they

---

Box 18.3 **Chromatography**

Chromatography is a collection of methods for the separation of compounds based on their differential partitioning between **mobile** and **stationary phases**:
• The mobile phase is usually a liquid in which the compounds have been dissolved or a gas in which they have been vaporized.
• The stationary phase is usually a solid matrix of some description, or a liquid that has been adsorbed by a solid matrix, and is often contained in a column or capillary tube through which the mobile phase is passed.

There are, however, alternative forms of chromatography that are, or have been, important in biochemistry, including **paper chromatography**, in which the stationary phase is a strip of filter paper, and **thin layer chromatography**, where the stationary phase is a solid material layered on to a plastic sheet.

In column chromatography, the mobile phase is either pumped through the chromatography column, or moves through the column under the force of gravity. The compounds contained in the mobile phase interact with the stationary phase, the degree of interaction for each compound depending on its physical and/or chemical properties. In the three types of column chromatography described in the text, the differential

partitioning is due to the sizes of the molecules (gel filtration), their charges (ion exchange) or their degrees of hydrophobicity (reverse phase). In gas chromatography, which we will study later in the chapter, the partitioning is between a gaseous mobile and liquid stationary phase, and separation is based on the relative volatility/solubility of the compounds in these two phases.

The degree of interaction of a substance with the stationary phase is called its **partition coefficient**. Although 'partition coefficient' is a central term in chromatography, it relates only to the partitioning of compounds between gaseous and liquid phase, or two immiscible liquid phases (e.g. an aqueous solution and an organic solvent) and so is not strictly appropriate for several of the types of chromatography used in biochemistry.

As the mixture passes through the chromatography column (or along a strip of filter paper or across a thin layer plate) the individual compounds are adsorbed and released by the matrix at rates which depend on their partition coefficients. This means that some compounds pass through the column relatively rapidly and others pass through more slowly. The compounds therefore form bands that elute from the column at different times and can be collected as separate fractions of the mobile phase (see *Fig. 18.8*).

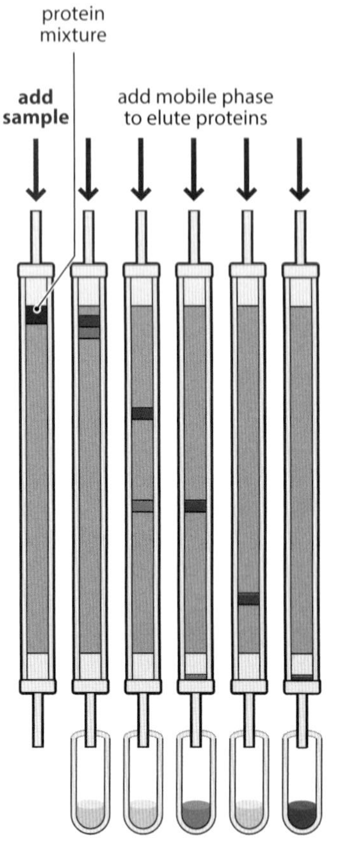

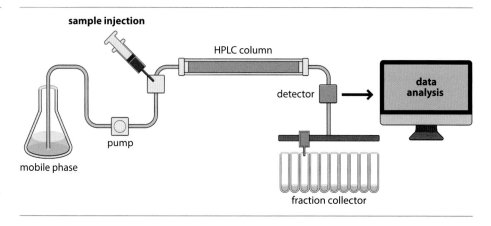

protein
mixture

**add
sample**   add mobile phase
to elute proteins

**Figure 18.8 Column
chromatography.**
The illustration shows a simple
situation in which the mixture contains
just two proteins. In practice tens or
hundreds of fractions might be
collected if the starting sample
contains multiple proteins.

move through the column. In effect, the smaller molecules are able to access a
greater amount of the mobile phase (see *Box 18.3*), which means they spend longer
in the column before they are eluted. The larger proteins, being less able to enter
the beads, pass through the column more quickly. The proteins in the mixture
therefore become separated according to their size, the largest ones eluting from
the column first, and the smallest ones last.

- **Ion exchange chromatography.** This technique separates proteins according to
  their net electric charges. The matrix consists of polystyrene beads that carry either
  positive or negative charges. If the beads are positively charged, then proteins
  with a net negative charge will bind to them, and vice versa. The proteins can be
  eluted with a **salt gradient**, set up by gradually increasing the salt concentration
  of the buffer being passed through the column. The charged salt ions compete
  with the proteins for the binding sites on the beads, so proteins with low charges
  are eluted at low salt concentration, and ones with higher charges at higher salt
  concentrations. The salt gradient therefore separates proteins according to their
  net charges. Alternatively a pH gradient can be used. The net charge of a protein
  depends on the pH, so gradually changing the pH of the mobile phase will result in
  the elution of proteins with different net charges, again achieving their separation.

- **Reverse phase chromatography.** The matrix is silica or other particles whose
  surfaces are covered with nonpolar chemical groups such as hydrocarbons. The
  mobile phase is a mixture of water and an organic solvent such as methanol or
  acetonitrile. Most proteins have hydrophobic areas on their surfaces, which bind
  to the nonpolar matrix, but the stability of this attachment decreases as the organic
  content of the liquid phase increases. Gradually changing the ratio of the organic
  and aqueous components of the mobile phase therefore results in the elution of
  proteins according to their degree of surface hydrophobicity.

Column chromatography can be carried out in a capillary tube with an internal
diameter of less than 1 mm, with the liquid phase being pumped at high pressure. This
is called **high performance liquid chromatography** (**HPLC**), and is designed to achieve
a high resolution between individual proteins (*Fig. 18.9*). Sometimes different types of
column are linked together, each consecutive fraction from one column being fed into
a second column, in which a further round of separation using a different procedure
is carried out. In this way, quite complex mixtures of proteins can be fully separated.

Mass spectrometry is used to identify the separated proteins

The second stage of protein profiling is to identify the individual proteins that have
been separated from the starting mixture. This is achieved by a three-stage process:

- Each protein is treated with a protease that cuts the polypeptide chain at defined
  positions. Trypsin is often used, this protease cutting a polypeptide immediately

**Figure 18.9 High performance
liquid chromatography.**
The diagram shows a typical HPLC
apparatus. The protein mixture is
injected into the system and pumped
through the column along with the
mobile phase solution. Emergence of
proteins from the column is detected,
usually by measuring UV absorbance
at 210–220 nm. Multiple fractions are
collected. These can be of equal
volume, or the data from the detector
can be used to control the
fractionation so that each protein peak
is collected as a single sample of
minimum volume.

after arginine or lysine residues. With most proteins, this results in a mixture of peptides between 5 and 75 amino acids in length.

- The molecular mass of each peptide is determined.

- The molecular masses of the individual peptides are then compared with databases that contain the amino acid sequences of known proteins. Because of the specificity of the protease, the masses of the peptides resulting from cleavage of any protein whose amino acid sequence is known can be predicted. A match between the actual peptides that have been obtained and the predicted peptides listed in the database therefore enables the protein to be identified. The effects on peptide mass of post-translational modifications, such as phosphorylation of individual amino acids, can also be predicted. This means that the identification process is precise enough to distinguish between, for example, the activated and non-activated versions of proteins in a signal transduction pathway.

The first and last stages of this procedure do not present too much of a challenge to any research biochemist. It is easy to cut a protein with a protease and equally easy, thanks to online search tools, to compare the molecular masses of the resulting peptides with those predicted to arise from protease treatment of all known proteins.

The second stage of the process is more demanding. Indeed, it was a very difficult proposition until the mid-2000s when a technique called **peptide mass fingerprinting** was developed. This technique involves a type of **mass spectrometry** called **matrix-assisted laser desorption ionization time of flight** (**MALDI–TOF**). Mass spectrometry is a means of identifying a compound from the **mass-to-charge ratio** (designated m/z) of the ionized form that is produced when molecules of the compound are exposed to a high energy field of some description. In some types of mass spectrometry, the ionization method is relatively harsh and results in fragmentation of the molecules being studied. This is acceptable because a particular compound will fragment in a specific manner, so characterization of the **fragment ions** provides the information needed to identify the starting compound. For peptide mass fingerprinting we do not wish to break the peptides up any further, so matrix-assisted laser desorption, which is a 'soft' ionization method, is used. The mixture of peptides is absorbed into an organic crystalline matrix (often a phenylpropanoid compound called sinapinic acid is used) and excited with a UV laser. The excitation initially ionizes the matrix, with protons then donated to or removed from the peptide molecules, to give the **molecular ions** [M+H]$^+$ and [M–H]$^-$, respectively, where 'M' is the starting compound, one of the peptides in this case (*Fig. 18.10*).

The ionization procedure also results in vaporization of the peptides, which are then accelerated along the tube of the mass spectrometer by an electric field (*Fig. 18.11*).

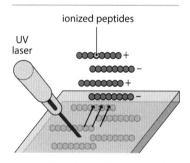

Figure 18.10 Ionization of peptides by matrix-assisted laser desorption.

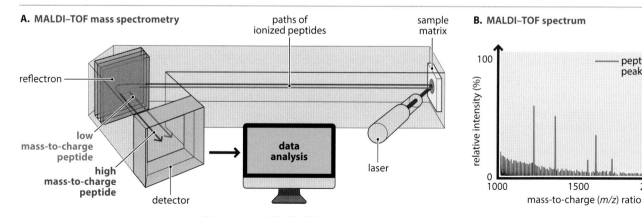

A. MALDI–TOF mass spectrometry

reflectron

low mass-to-charge peptide

high mass-to-charge peptide

detector

data analysis

paths of ionized peptides

sample matrix

laser

B. MALDI–TOF spectrum

peptide peaks

relative intensity (%)

100

0

1000   1500   2000

mass-to-charge (*m/z*) ratio

Figure 18.11 MALDI–TOF mass spectrometry.
(A) The architecture of a typical MALDI–TOF mass spectrometer incorporating a reflectron. (B) A MALDI–TOF spectrum showing peptide peaks whose *m/z* ratios can be read off the x-axis.

The 'time of flight' for an ion (the time taken to reach the detector) depends on its mass-to-charge ratio. As the charge is always +1 or −1, the time of flight can easily be converted to a mass, providing the information that is used to search the databases in order to identify the composition of a particular peptide. The flight path can be direct from the ionization source to the detector, but often the ions are initially directed at a **reflectron**, which reflects the ion beam towards the detector. By enabling a longer flight path to be built into a machine of a defined size, the reflectron also increases the discrimination between peptides of similar masses.

A direct comparison can be made between the compositions of two proteomes

Often the information that a researcher wishes to obtain is not the identity of every protein in a proteome, but the differences between the protein compositions of two different proteomes. This is particularly important if the overall objective is to understand how the biochemistry of a tissue changes in response to a disease such as cancer.

If a particular protein is abundant in one proteome but absent in another, then the difference will be visible simply by looking at the stained gels after two-dimensional electrophoresis. Quite small changes in the relative amounts of different proteins can, however, give rise to significant changes in the biochemical properties of a tissue, and these may not be apparent when the stained gels are examined by eye. To detect these small-scale changes, a more sophisticated way of examining the proteomes must be used. One method makes use of **isotope-coded affinity tags** (**ICATs**). These are chemical groups that can be attached to a protein. In one system, the tags are short hydrocarbon chains that are made in two versions, identical to one another except that one version contains the common ^{12}C isotope of carbon, and the second version contains ^{13}C (*Fig. 18.12*). These tags can be attached to cysteine amino acids in a polypeptide.

How are ICATs used to quantify the differences between two proteomes? The proteins in the proteomes are separated in the normal manner and equivalent proteins from each proteome recovered and treated with protease. One set of peptides is then labeled with ^{12}C tags and the other with the ^{13}C tags. Not all peptides will become labeled because some will lack cysteine amino acids. Those that are labeled are purified and the unlabeled ones discarded. There are various ways in which this can be done, but usually the tags have a terminal biotin group. Biotin is tightly bound by a protein called **avidin** (the chicken egg white protein), so passage of the peptides through a chromatography column that has avidin attached to the matrix will separate the tagged peptides from the untagged ones. The biotin groups on the tagged peptides will bind to the avidin on the matrix, so these peptides are retained in the column, whereas the untagged ones will flow straight through. The avidin–biotin complexes can then be disrupted by increasing the temperature, so the tagged peptides can be collected from the column. Because the ^{12}C and ^{13}C tags have different masses, the m/z ratio of a peptide labeled with a ^{12}C tag will be different from that of an identical

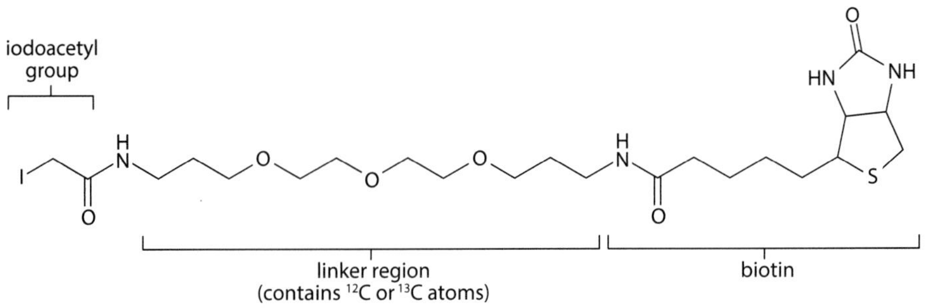

Figure 18.12 A typical isotope-coded affinity tag for proteome studies.
The iodoacetyl group reacts with cysteine and hence forms the attachment to a peptide. The linker region contains either ^{12}C or ^{13}C atoms and so provides the isotope coding function. The terminal biotin group enables tagged peptides to be separated from untagged ones by affinity chromatography.

iodoacetyl group

linker region
(contains ^{12}C or ^{13}C atoms)

biotin

Figure 18.13 Comparing two proteomes using ICATs. In the MALDI–TOF spectrum, peaks resulting from peptides containing ^{12}C atoms are shown in red, and those from peptides containing ^{13}C are shown in blue. The protein under study is approximately 1.5-fold more abundant in the proteome that has been labeled with ^{12}C-ICATs.

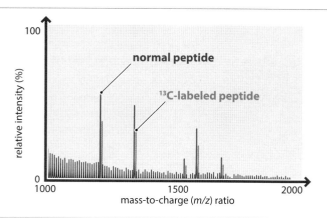

peptide labeled with a ^{13}C tag. The peptides from the two proteomes are therefore run through the mass spectrometer together. A pair of identical peptides (one from each proteome) will occupy slightly different positions on the resulting mass spectrum, because of their distinctive m/z ratios (Fig. 18.13). Comparison of the peak heights allows the relative abundances of each peptide to be estimated.

18.1.3 Studying the structure of a protein

Throughout this book we have seen many examples of the ways in which the biochemical activity of a protein is specified by its three-dimensional structure. We have seen how the fibrous structure of collagen enables this protein to play a structural role in bones and tendons, how the activity of a globular protein such as ribonuclease is lost when the protein is denatured, and regained when the three-dimensional structure reforms, and we have seen how the precise conformation of the active site of an enzyme underlies the ability of that enzyme to catalyze a specific biochemical reaction. It is therefore not in the least surprising that methods for determining the structures of proteins are among the most important of the research tools available to biochemists.

We will study three methods that are used to obtain structural information on proteins. These are:

- **Circular dichroism (CD)**, which can identify the relative amounts of different secondary structures in a protein.

- **Nuclear magnetic resonance (NMR) spectroscopy**, which can provide detailed structures of small proteins.

- **X-ray crystallography**, which can solve the structure of virtually any protein that can be crystallized.

Circular dichroism enables the secondary structure composition of a protein to be estimated

Circular dichroism cannot provide a detailed description of the tertiary structure of a protein, but instead enables the relative amounts of different secondary structural components, such as α-helices, β-sheets and β-turns, to be estimated. This can be useful as the first step in a full structural characterization, but this is not the main application of CD. Instead, CD is most frequently used to assess the structural changes that occur when a protein is exposed to different physical or chemical conditions. An important example is in the study of protein folding, because CD enables the gradual formation of the secondary structural components of a protein to be followed.

Circular dichroism can also be used to identify structural changes that occur during an enzymatic reaction, such as when the enzyme binds its substrate, or when an inhibitor binds to the enzyme. This ability of CD to reveal dynamic changes makes it a valuable technique in biochemical research.

See *Section 3.1.2* for the definition of a chiral carbon.

The data obtained by CD relate specifically to the chiral centers within the protein structure. These include the α-carbons of amino acids, as well as some other structures found in proteins, such as disulfide bonds and some of the aromatic R groups, which can adopt different conformations with chiral properties. Circular dichroism not only identifies the presence of chiral centers, but also gives information on their positions relative to one another. The α-carbons in an α-helix, for example, can be distinguished from those in a β-sheet or β-turn. The relative amounts of these different types of conformation can therefore be estimated.

The key feature of a chiral center is that it displays optical activity. Circular dichroism refers to the effect that the chiral centers within a molecule such as a protein have on circularly polarized light. Depending on its identity and environment, a chiral center will absorb clockwise and/or counterclockwise polarized light of different wavelengths. A CD spectrometer measures this absorption, not for individual centers but for the protein as a whole, by analyzing how a beam of circularly polarized light is affected by passage through a solution of the protein (*Fig. 18.14A*). For α-carbons linked by peptide bonds, the main absorption occurs at wavelengths between 160 and 240 nm, which is in the ultraviolet range of the spectrum. Within this region, the CD spectra resulting from α-helices, β-sheets and random coils are distinctive (*Fig. 18.14B*). Of course, most proteins contain a mixture of these different structures, so the spectrum that is obtained indicates the combined absorbances of a variety of helices, sheets and coils. Interpretation of the spectrum therefore requires 'deconvolution' software that separates the various contributions and reveals the secondary structure composition of the protein.

Box 18.4 Circularly polarized light

Light, like all forms of electromagnetic radiation, is made up of two fields, one electric and the other magnetic, that oscillate at right angles to one another. This is called a **transverse wave**. In natural light, the electric fields of different photons oscillate in different directions, which means that the light is **unpolarized**. Some types of optical filter (including the lenses of certain types of sunglasses) only allow the passage of light waves whose electric fields oscillate along a single vector. This is known as **plane-polarized light**.

In biochemistry, plane-polarized light is used to distinguish the D- and L-isomers of an optically active compound such as an amino acid (see *Fig. 3.5*).

In **circularly polarized light**, the electric field follows a circular vector, with either clockwise or counterclockwise rotation.

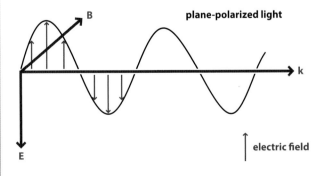

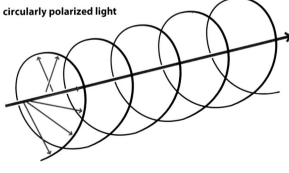

circularly polarized light

Circularly polarized light can be generated by passing plane-polarized light through a device called a quarter-wave plate.

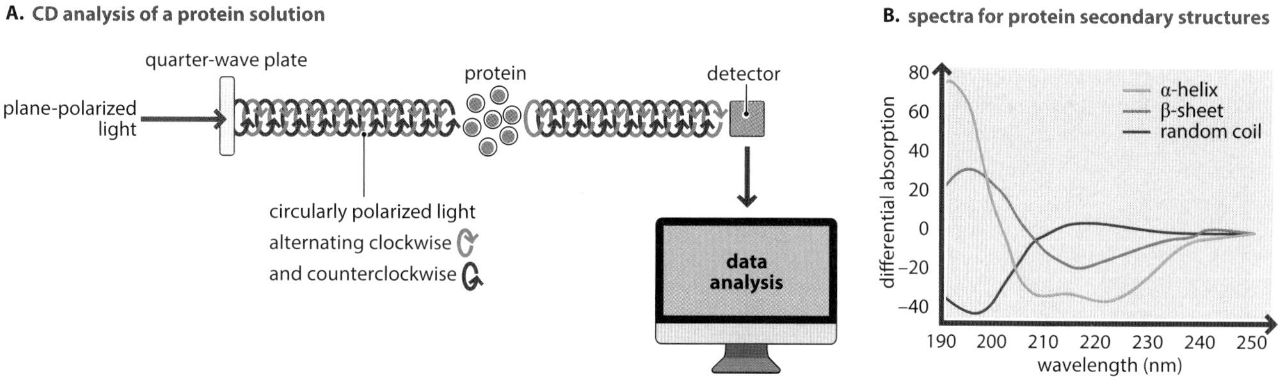

A. CD analysis of a protein solution

quarter-wave plate

plane-polarized light

protein

detector

circularly polarized light
alternating clockwise ↻
and counterclockwise ↺

data analysis

B. spectra for protein secondary structures

— α-helix
— β-sheet
— random coil

differential absorption

190 200 210 220 230 240 250
wavelength (nm)

Figure 18.14 **Circular dichroism.**
(A) Typical apparatus for CD analysis of a protein solution. (B) The differential absorbance spectra for
α-helices, β-sheets and random coils.

NMR spectroscopy is used to study the structures of small proteins

Nuclear magnetic resonance (NMR) spectroscopy is the second important method that is available for structural characterization of proteins. Like CD, the protein under study is in solution, so NMR can be used to study dynamic events such as protein folding. It provides a much greater degree of structural resolution than is possible with CD, enabling precise positioning of chemical groups and a detailed description of tertiary structure. Its main disadvantage, as we will see, is that NMR is only suitable for relatively small proteins.

The principle underlying NMR is that rotation of an atomic nucleus generates a magnetic moment. This magnetic effect means that, when placed in an electromagnetic field, the spinning nucleus can take up either of two orientations (*Fig. 18.15*). These orientations are called α and β, the α spin state being aligned with the magnetic field and hence having a slightly lower energy quotient than the β state, which is aligned against the magnetic field. An NMR spectrometer measures the energy differences between the α and β spin states for individual nuclei, this difference being called the **resonance frequency**. The critical point is that, although each type of nucleus (e.g. ^{1}H, ^{13}C, ^{15}N) has its own specific resonance frequency, the measured frequency is often slightly different from the standard value (typically by less than 10 parts per million). This **chemical shift** occurs because electrons in the vicinity of the rotating nucleus shield it to a certain extent from the magnetic field. The nature of the chemical shift therefore enables the environment of the nucleus to be deduced, yielding the data that are used to build up the structure of the protein. Some types of analysis (called COSY and TOCSY) enable atoms linked by chemical bonds to the spinning nucleus to be identified, whereas others (e.g. NOESY) identify atoms that are close to the spinning nucleus in space but not directly connected to it.

To be suitable for NMR, a chemical nucleus must have an odd number of protons and/or neutrons, otherwise it will not spin when placed in an electromagnetic field. This means that NMR can only be used with nuclei that have an odd number of protons plus neutrons. In protein structural studies, ^{1}H nuclei are initially targeted, the aim being to identify the chemical environments of every hydrogen atom. The resulting data are frequently supplemented by analyses of proteins in which at least some of the carbon and/or nitrogen atoms have been replaced with the rare isotopes ^{13}C and ^{15}N. The combined data from these analyses will be sufficient to enable the protein structure to be worked out, so long as that structure is not overly complex. Problems arise when two or more nuclei, by chance, have very similar chemical shifts. Then it can be difficult to distinguish the environments of those two nuclei, and structural

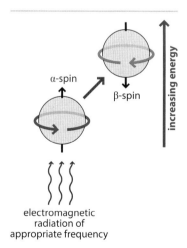

α-spin

β-spin

increasing energy

electromagnetic radiation of appropriate frequency

Figure 18.15 **The basis of NMR.**
A rotating nucleus can take up either of two orientations in an applied electromagnetic field.

Box 18.5 **Interpreting an NMR spectrum**

To illustrate the way in which NMR data are interpreted we will follow through a simple example where a ^1H spectrum has been obtained for a compound whose formula is known to be $C_4H_8O_2$.

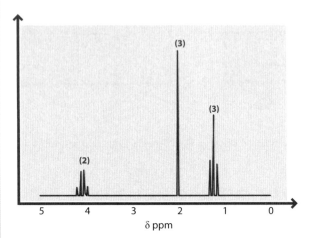

Image redrawn from www.chemguide.co.uk.

As this is a ^1H spectrum, each peak represents one or more hydrogen atoms. The position of a peak on the x-axis indicates the extent of the chemical shift experienced by that hydrogen atom, expressed as δ parts per million (ppm). A chemical shift of 1 ppm would be displayed by a hydrogen atom whose resonance frequency is one part per million less than the standard value.

The spectrum is analyzed step by step as follows:

- There are three clusters of peaks, which means that the hydrogen atoms in this compound exist in three 'environments'. From the chemical formula it seems likely that each environment is a different carbon atom to which one or more hydrogens are attached. Note that this would mean that one of the carbons is not directly attached to any hydrogen atoms.
- If the peak heights in each cluster are added together then the resulting composite peak heights have a ratio of 2:3:3, as indicated by the red numbers on the spectrum. This ratio relates to the number of hydrogens in each environment. As there are eight hydrogens in total, we can conclude that there is one $-CH_2$ group and two $-CH_3$ groups.
- The number of peaks in a cluster is one more than the number of hydrogens in the *adjacent* carbon atom. Therefore:
  ○ The $-CH_2$ at 4.1 ppm has four peaks, and so is adjacent to a carbon with three hydrogens, i.e. a $-CH_3$ group.
  ○ The $-CH_3$ group at 1.3 ppm has three peaks, and so is adjacent to the $-CH_2$. The 1.3 and 4.1 ppm clusters therefore identify an ethyl group, $-CH_2CH_3$.
  ○ The $-CH_3$ group at 2.0 ppm has just one peak, and so is not adjacent to a carbon with attached hydrogens.

We can conclude that the compound is ethyl acetate:

$$CH_3 - C \overset{\displaystyle O}{\underset{\displaystyle O - CH_2 - CH_3}{\Big\langle}}$$

Clearly, ethyl acetate has a much simpler chemical structure than a protein. The spectrum produced by NMR of a typical protein is much more complex, with many more clusters of peaks, and the chemical shifts of individual hydrogens are often affected by more than one adjacent environment. Those environments will also include hydrogen-containing groups other than $-CH_2$ and $-CH_3$, such as hydroxyl and amino groups, adding further complexity to the spectrum. Despite these complications, NMR has developed into an essential tool for studies of proteins and other biomolecules, with structures solved for molecules up to 1000 kDa in size.

---

information is lost. The larger the protein, the greater the probability that pairs and groups of nuclei have similar shifts, and the greater the probability that NMR will fail to provide useful information on the tertiary structure.

X-ray crystallography provides precise structural data for any protein that can be crystallized

X-ray crystallography is the most powerful of the methods available for structural studies of proteins. It gives detailed information on the relative positions of different chemical groups within the protein, enabling the precise conformation of the polypeptide chain to be worked out, along with the positioning of amino acid side-chains, and hence yields a detailed tertiary structure. The one limitation is that the protein must be crystallized before its structure can be studied by this method. For many proteins this is not a problem, as good quality crystals can be obtained from a supersaturated solution. Other proteins, especially membrane proteins which have external hydrophobic regions, are less easy or even impossible to crystallize. Because the protein is crystallized, dynamic studies, which are possible with solution-based techniques such as CD or NMR, are difficult to carry out, although structural changes caused by substrate or inhibitor binding can be assessed by preparing crystals with or without the substrate or inhibitor.

Figure 18.16 X-ray diffraction.
(A) Production of an X-ray diffraction pattern by passing a beam of X-rays through a crystal of the protein being studied. (B) The diffraction pattern obtained with crystals of ribonuclease.

A. production of a diffraction pattern

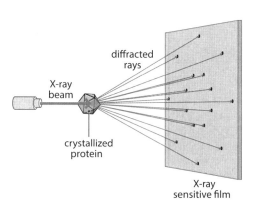

B. X-ray diffraction pattern for ribonuclease

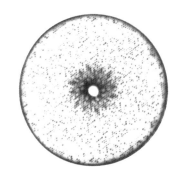

X-ray crystallography is based on **X-ray diffraction**, which is the deflection of X-rays that occurs during their passage through a crystal or other regularly ordered chemical structure. X-rays have very short wavelengths, between 0.01 and 10 nm, similar to the spacings between atoms in chemical structures. When a beam of X-rays is directed onto a crystal, some of the X-rays pass straight through, but others are diffracted and emerge from the crystal at a different angle from that at which they entered (*Fig. 18.16*). The protein molecules within a crystal are positioned in a regular array, which means that different X-rays are diffracted in similar ways. An X-ray sensitive photographic film or electronic detector, placed across the beam after it emerges from the crystal, reveals a series of spots, called an **X-ray diffraction pattern**. The intensities and relative positions of the spots can be used to infer the X-ray deflection angles, and from these data the protein structure can be deduced.

As can be imagined, the challenge of X-ray crystallography lies with the complexity of the diffraction patterns that are obtained with molecules as large as proteins. Even with computational help, the analysis is difficult and time consuming. The aim is to convert the intensities and relative positions of the spots in the X-ray diffraction pattern into an **electron density map** (*Fig. 18.17*). With a protein, the electron density map indicates the conformation of the folded polypeptide. If sufficiently detailed, the electron density map also enables the side-chains of the individual amino acids to be identified and their orientations relative to one another to be established. This in turn allows interactions such as hydrogen bonding to be predicted. In the most successful projects, a resolution of 0.1 nm is possible, which means that structures just 0.1 nm apart in the protein can be distinguished. In proteins, most carbon–carbon bonds are

Figure 18.17 Electron density maps.
(A) Part of the electron density map for ribonuclease. (B) Interpretation of an electron density map at 0.2 nm resolution revealing a tyrosine side-chain.

A. section of the ribonuclease electron density map

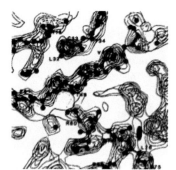

B. interpretation of an electron density map at 0.2 nm resolution revealing a tyrosine side-chain

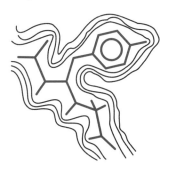

0.1–0.2 nm in length, and carbon–hydrogen bonds are 0.08–0.12 nm. This means that at 0.1 nm resolution, a very detailed three-dimensional model of the protein can be constructed.

18.2 Studying lipids and carbohydrates

Methods for studying lipids and carbohydrates have always been important in biochemistry. In recent years, these methods have assumed additional significance as research has moved away from the study of individual biomolecules and become increasingly focused on large scale studies that attempt to characterize the entire biochemical content of a cell. Proteomics is an example of this field of study, in which the protein content of a cell is examined in order to understand the biochemical capability of that cell, and to investigate how that capability responds to challenges such as disease. To complement the information provided by proteomics, we therefore also need methods for characterizing the lipid and carbohydrate contents. These methods are called **lipidomics** and **glycomics**, respectively. In the remainder of this chapter we will explore the techniques that underlie these two areas of research.

18.2.1 Methods for studying lipids

A number of methods have been developed for studying lipids, but those that are most useful in modern research are ones that enable the individual compounds in a mixture of lipids to be identified and quantified. This is achieved as follows:

• The lipids in the mixture are separated by a chromatography procedure.

• The individual lipids are then identified by mass spectrometry.

The approach is therefore similar to that used in protein profiling, but as we might expect, with lipids being very different to proteins, the details are different.

Box 18.6 Metabolomics

As well as genomics (study of the genome), transcriptomics (study of a cell or tissue's complement of RNA), proteomics (study of the protein content), lipidomics (study of the lipid content) and glycomics (study of the carbohydrate content), biochemists are also interested in **metabolomics**. The **metabolome** is the collection of metabolites within a cell or tissue, with metabolites usually defined as the substrates, intermediates and products of metabolic pathways.

The aim of metabolomics is not just to identify the individual compounds that are present (as is often the case in the other types of 'omics) but also to measure **metabolic flux** – the rate of movement of substrates through individual pathways. Metabolic flux is a valuable concept because it provides a detailed description of the biochemical activity of a cell or tissue. Studies of metabolic flux enable the interconnectedness of different pathways to be explored, and are particularly important in enabling control points to be identified. Changes in metabolic flux occurring during disease provide valuable insights into the

biochemical response to the disease state, and may help in the development of treatments for the disease.

The metabolites in a cell include several different types of compound, so a variety of detection methods must be used in order to identify and quantify each one. The most important of these methods are gas chromatography and HPLC, possibly interfaced with mass spectrometry using both soft and hard ionization techniques.

As well as studies of individual organisms, metabolomics is increasingly being applied to environmental samples. For example, metabolomic studies of the soil around the roots of a plant reveal the biochemical activities of the bacteria and other organisms inhabiting the soil, and can also identify any exudates being secreted by the plant. These biochemical activities can then be linked to parameters such as nutrient cycling and plant productivity.

Gas chromatography is often used to separate the lipids in a mixture

When fatty acids or sterols are being studied, the separation stage is usually carried out by **gas chromatography**. The lipids are volatilized in a carrier gas, usually hydrogen or helium. This carrier gas is the mobile phase, which passes along the chromatography column. The column is very thin (0.1–0.7 mm in diameter) and can be up to 100 m in length. The internal surface of the column is coated with an organic solvent, such as polysiloxane, which is the stationary phase, stationary because it is immobilized within an inert silica matrix (*Figure 18.18*).

The rate at which a lipid passes through the column depends on its partition coefficient, which in turn depends on the relative solubility and volatility of the lipid in the liquid and gas phases, respectively, of the chromatography column. If the lipid is insoluble in the liquid phase then it will pass directly through the column. Other lipids will undergo repeated cycles of adsorption into the liquid phase and release back into the gas phase. The dynamics of this cycling, which is dictated by the partition coefficient, determines how rapidly each compound passes through the column. The individual compounds in a mixture of lipids are therefore separated according to their partition coefficients, and emerge from the end of the column as purified fractions.

The migration rate of individual lipids through a chromatography column depends on temperature. The best resolution of a complex mixture is achieved if the temperature is gradually increased, for example from 40°C to 300°C at a rate of 5°C per minute. At the start of the run, when the temperature is relatively low, those lipids that pass most quickly through the column are retained for sufficient time for the individual compounds to be resolved. As the temperature is increased, the migration rates of the more slowly moving lipids are accelerated, so that these pass through the column quickly enough to be collected within a reasonable time period. A range of lipids with vastly different partition coefficients can therefore be separated in a single run.

The sequential elution of lipids from the chromatography column is depicted as a **chromatogram**, in which the peak heights indicate the relative proportions of the individual compounds in the starting mixture (*Fig. 18.19*). Each lipid has a characteristic retention time, so the chromatogram can also be used to identify the components of the starting mixture, although in practice this is only possible if the mixture is simple and its likely composition already known. In other cases, when there is insufficient information to identify compounds solely from the chromatogram, the individual lipids are analysed further by mass spectrometry.

Identifying lipids by mass spectrometry

Lipids emerging from a gas chromatography column are often ionized by electron bombardment, which results in M^+ molecular ions. Some lipid molecules will also break up in the electron beam, giving a series of fragment ions whose structures are predictable. The *m/z* values of the molecular ion and its daughter fragment ions

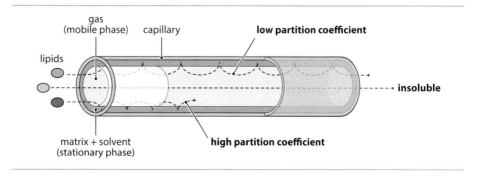

Figure 18.18 Gas chromatography. The passage of three lipids through the capillary column is shown. One lipid (in yellow) is insoluble in the stationary phase and so passes straight through the column. The other two lipids are delayed by repeated adsorption into the stationary phase, their rate of passage depending on their partition coefficients, with the lipid (in green) with lower partition coefficient passing through faster.

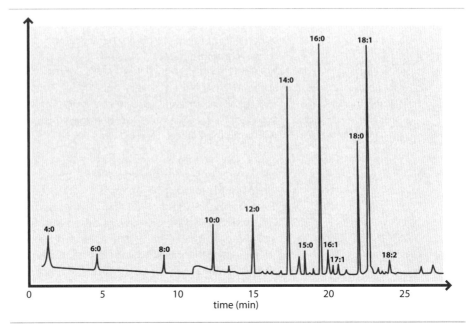

Figure 18.19 Gas chromatogram of milk fatty acids.
The identity of the fatty acid corresponding to each peak is indicated, using the M:N nomenclature (see *Section 5.1.1*).
Image redrawn from AOCS Lipid Library (http://lipidlibrary.aocs.org) with permission.

therefore give a characteristic **mass spectrum** from which the compound can be identified (*Fig. 18.20*). However, identification can be difficult if the lipid breaks into fragments to the extent that the molecular ion is not present, as sometimes happens with electron ionization. A softer chemical ionization process is therefore also used.

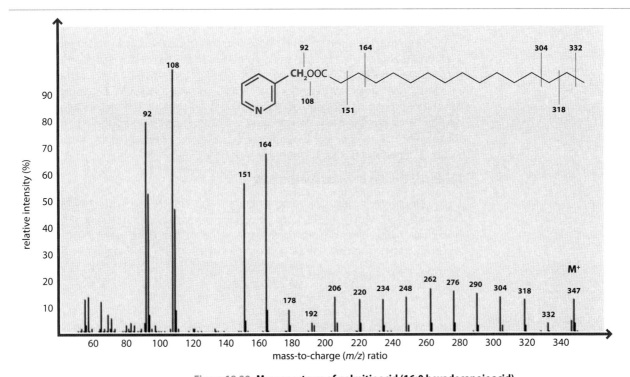

Figure 18.20 Mass spectrum of palmitic acid (16:0 hexadecanoic acid).
Prior to ionization, the fatty acid has been derivatized by attachment of a 3-pyridylcarbinol group (shown in red). The nitrogen in this group is preferentially ionized and so provides a reference point for interpretation of the spectrum. Each peak is labeled with its *m/z* value and the M+ molecular ion is indicated. The breakage points that yield fragment ions are indicated in the molecular structure. The series of ions with *m/z* values from 178 to 290 are fragments with different numbers of –CH$_2$ groups.
Image redrawn from AOCS Lipid Library (http://lipidlibrary.aocs.org) with permission.

A. magnetic sector

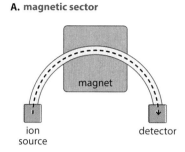

ion source

magnet

detector

B. quadrupole

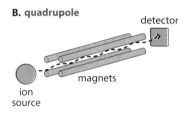

detector

ion source

magnets

Figure 18.21 **Architectures of a (A) magnetic sector, and (B) quadrupole mass analyzer.**

This involves mixing the lipids with an ionized gaseous plasma, usually of methane, ammonia or isobutene, which gives rise to $[M+H]^+$ molecular ions, along with significantly less fragmentation.

There are several types of mass spectrometer, differing in the configuration of the mass analyzer, the part of the instrument that separates the molecules according to their *m/z* values. The two types most frequently used in lipid studies are (*Fig. 18.21*):

• The **magnetic sector mass spectrometer**, in which the mass analyzer is a single or series of magnets through which the ionized molecules are passed. The magnets are arranged so that each ion must follow a curved trajectory in order to avoid hitting the walls of the analyzer. The extent to which the magnetic field deflects an ion depends on the ion's *m/z* value, so most ions hit the walls and only a few pass through the magnet to the detector. The magnetic field can therefore be set up so that only those ions of a particular *m/z* reach the detector, or the field can be gradually changed so that ions of differing *m/z* values can be collected at different times during a single run.

• The **quadrupole** mass analyzer, which has four magnetic rods placed parallel to one another, surrounding a central channel through which the ions must pass. Oscillating electrical fields are applied to the rods, deflecting the ions in a complex way so their trajectories 'wiggle' as they pass through the quadrupole. Again, the field can be set so that only ions of a particular *m/z* value can emerge and be detected, or alternatively the field can be gradually changed so all ions are detected.

The methodology described above is suitable for fatty acids and sterols, but variations are often required when other types of lipid are being analyzed. The first of these variations is the replacement of the gas chromatography stage with HPLC. The underlying basis to HPLC is, as we saw in *Figure 18.9*, the same as gas chromatography except that the mobile phase is a liquid and the stationary phase a solid, often silica. In HPLC, the compounds to be separated are simply dissolved in the liquid phase, and the chromatography is carried out at room temperature. This method is therefore a better choice for triacylglycerols, glycerophospholipids, sphingolipids and eicosanoids, which are too hydrophilic for easy volatilization and/or too unstable for chromatography at high temperature.

For unstable lipids, the ionization method must also be modified, to avoid the molecules breaking down into very small fragment ions. Rather than generating ions by direct exposure to an electron beam or by chemical ionization, a more gentle method called **electrospray ionization** is used. This involves applying a high voltage to the solution emerging from the HPLC, which generates an aerosol of charged droplets which evaporate, transferring their charges to the molecules dissolved within them.

The mass spectrometry stage can also be modified in order to achieve greater discrimination between closely related compounds. In **tandem mass spectrometry**, the mass spectrometer has two or more mass analyzers linked in series. Often, the linked analyzers are of different formats, such as a magnetic sector spectrometer followed by a quadrupole instrument. Prior to each new round of mass spectrometry, the ions undergo additional fragmentation, providing further information on the structure of the starting molecule. Complex molecules, or members of a closely related family of compounds, can therefore be identified.

18.2.2 Studying carbohydrates

Carbohydrates are the most difficult type of biomolecule to study, in part because the structural similarity between different monosaccharides complicates the identification of different compounds. The importance of sugar in the food industry was the initial stimulus for the development of methods for identification and quantification of

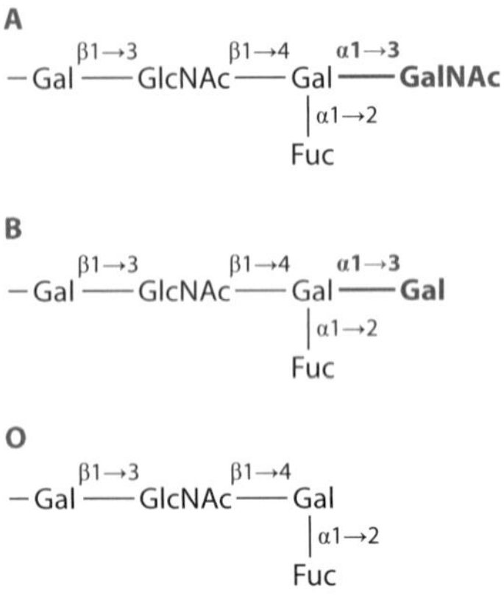

A

$$-\text{Gal} \xrightarrow{\beta1\rightarrow3} \text{GlcNAc} \xrightarrow{\beta1\rightarrow4} \text{Gal} \xrightarrow{\alpha1\rightarrow3} \textbf{GalNAc}$$
$$\Big| \alpha1\rightarrow2$$
$$\text{Fuc}$$

B

$$-\text{Gal} \xrightarrow{\beta1\rightarrow3} \text{GlcNAc} \xrightarrow{\beta1\rightarrow4} \text{Gal} \xrightarrow{\alpha1\rightarrow3} \textbf{Gal}$$
$$\Big| \alpha1\rightarrow2$$
$$\text{Fuc}$$

O

$$-\text{Gal} \xrightarrow{\beta1\rightarrow3} \text{GlcNAc} \xrightarrow{\beta1\rightarrow4} \text{Gal}$$
$$\Big| \alpha1\rightarrow2$$
$$\text{Fuc}$$

Figure 18.22 The A, B and O blood group glycans.
Abbreviations: Gal, galactose; GalNAc, N-acetylgalactosamine; GlcNAc, N-acetylglucosamine; Fuc, fucose. The sugar units distinguishing the three antigens are highlighted.

individual carbohydrates. In recent years, these methods have been supplemented by more sophisticated ones that enable identification of the compositions and structures of glycans attached to proteins. Together, the methodology forms the basis of glycomics, which attempts to describe the entire complement of sugars in a cell or tissue.

Immunological methods can be used to identify carbohydrates, because most carbohydrates display antigenic properties in the same way as proteins. The antigenic properties of carbohydrates have been recognized for many years, and are exploited in the classical blood typing systems. The A, B and O blood groups, for example, are distinguished by the identity of an antigenic glycan attached to a protein present on the surfaces of erythrocytes. In the A type, one of the units of the glycan is *N*-acetylgalactosamine, in the B type it is D-galactose, and in the O group this particular unit is absent (*Fig. 18.22*). The immunological reactivity of carbohydrates means that polyclonal and monoclonal antibodies specific for individual types can be raised, and these antibodies used in precipitin tests and in ELISA systems. Similar approaches can be used with **lectins**, plant or animal proteins with specific monosaccharide-binding properties. An example is **concanavalin A**, from jack bean (*Canavalis ensiformis*), which binds to the terminal α-glucose and α-mannose units in *O*-linked, though not *N*-linked, glycans. Lectins with different specificities are therefore very useful for probing the composition of glycans.

Glycan structure can be studied in greater detail by releasing these oligosaccharides from their proteins. Both *O*- and *N*-linked glycans are released by treatment with hydrazine, and *O*-linked glycans can be specifically removed by borohydride ions. The resulting glycan mixtures are separated by HPLC and individual ones further examined by mass spectrometry or NMR to identify their structures. There are also a variety of glycosidase enzymes that cleave at specific linkages within a glycan. These include exoglycosidases, which remove a terminal sugar, and endoglycosidases, which make internal cuts. Treatment of a glycan with a series of exoglycosidases of different specificities enables the order of sugars in the glycan to be determined. This procedure is called **glycan sequencing**.

Further reading

Alt FW, Blackwell TK and Yancopoulos GD (1987) Development of the primary antibody repertoire. *Science* **238**, 1079–87. *Generation of immunoglobulin diversity.*

Beger RD (2013) A review of the applications of metabolomics in cancer. *Metabolites* **3**, 552–74.

Blanksby SJ and Mitchell TW (2010) Advances in mass spectrometry for lipidomics. *Annual Review of Analytical Chemistry* **3**, 433–65.

Cavanagh J, Fairbrother WJ, Palmer AG and Skelton, NJ (1995) *Protein NMR Spectroscopy: Principles and Practice.* Academic Press, London.

de St Groth SF and Scheidegger D (1980) Production of monoclonal antibodies: strategy and tactics. *Journal of Immunological Methods* **35**, 1–21.

Fenn JB, Mann M, Meng CK, Wong SF and Whitehouse CM (1990) Electrospray ionization – principles and practice. *Mass Spectrometry Reviews* **9**, 37–70.

Garman EF (2014) Developments in X-ray crystallographic structure determination of biological macromolecules. *Science* **343**, 1102–8.

Görg A, Weiss W and Dunn MJ (2004) Current two-dimensional electrophoresis technology for proteomics. *Proteomics* **4**, 3665–85.

Gygi SP, Rist B, Gerber SA, Turecek F, Gelb MH and Aebersold R (1999) Quantitative analysis of complex protein mixtures using isotope-coded affinity tags. *Nature Biotechnology* **17**, 994–9.

Lequin RM (2005) Enzyme immunoassay (EIA) / enzyme-linked immunosorbent assay (ELISA). *Clinical Chemistry* **24**, 15–18.

Murphy RC, Fiedler J and Hevko J (2001) Analysis of non-volatile lipids by mass spectrometry. *Chemical Reviews* **101**, 479–526.

Phizicky E, Bastiaens PIH, Zhu H, Snyder M and Fields S (2003) Protein analysis on a proteomics scale. *Nature* **422**, 208–15. *Reviews all aspects of proteomics.*

Raman R, Raguram S, Venkataraman G, Paulson JC and Sasisekharan R (2005) Glycomics: an integrated systems approach to structure–function relationships of glycans. *Nature Methods* **2**, 817–24.

Ranjbar B and Gill P (2009) Circular dichroism techniques: biomolecular and nanostructural analyses – a review. *Chemical Biology and Drug Design* **74**, 101–20.

Shevchenko A and Simons K (2010) Lipidomics: coming to grips with lipid diversity. *Nature Reviews Molecular Biology* **11**, 593–8.

Walton HF (1976) Ion exchange and liquid column chromatography. *Analytical Chemistry* **48**, 52R–66R.

Self-assessment questions

Multiple choice questions

Only one answer is correct for each question. Answers can be found on the website: www.scionpublishing.com/biochemistry

1. Antibodies are synthesized by which cells?
(a) B lymphocytes
(b) Myeloma cells
(c) Red blood cells
(d) Macrophages

2. Which type of immunoglobulin exists as a pentamer in human blood?
(a) Immunoglobulin A
(b) Immunoglobulin E
(c) Immunoglobulin G
(d) Immunoglobulin M

3. What is the feature on the surface of an antigen that is recognized by an antibody called?
(a) Complement
(b) Epitope
(c) Light chain
(d) Hybridoma

4. The position in the precipitin reaction where the relative amounts of antigen and antibody that are optimal for complex formation is called what?
(a) Zone of equivalence
(b) Complement
(c) Precipitation zone
(d) The Ouchterlony point

5. What is the process that results in movement of immunoglobulin molecules in a gel towards the negative electrode called?
(a) Electrophoresis
(b) Diffusion
(c) Electroendosmosis
(d) Partitioning

6. Which of the following statements is **incorrect** with regard to ELISA?
(a) It is less quantitative than immunoelectrophoresis
(b) One of the antibodies is conjugated to a reporter enzyme
(c) It is more rapid than immunoelectrophoresis
(d) It can be carried out as an indirect process with primary and secondary antibodies

7. Which of the following methods is **not** used in protein profiling?
(a) Two-dimensional gel electrophoresis
(b) Column chromatography
(c) Gas chromatography
(d) Mass spectrometry

8. In which type of chromatography are small porous beads used as the matrix?
(a) Reverse phase
(b) Gel filtration
(c) Gas
(d) Ion exchange

9. In which type of chromatography are polystyrene beads that carry either positive or negative charges used as the matrix?
(a) Reverse phase
(b) Gel filtration
(c) Gas
(d) Ion exchange

10. In which type of chromatography are silica or other particles whose surfaces are covered with nonpolar chemical groups such as hydrocarbons used as the matrix?
(a) Reverse phase
(b) Gel filtration
(c) Gas
(d) Ion exchange

11. Which of the following statements is **incorrect** regarding the use of isotope-coded affinity tags (ICATs)?
   (a) ICATs are chemical groups that can be attached to a protein
   (b) In one system, the pairs of ICATs are distinguished by ^{12}C and ^{13}C labeling
   (c) Usually an ICAT has a terminal biotin group
   (d) A ^{12}C-labeled ICAT will give a peptide a higher m/z ratio than a ^{13}C label

12. Which of the following **cannot** be determined by circular dichroism?
   (a) The relative amounts of different secondary structural components in a protein
   (b) The positions of chiral centers relative to one another
   (c) Structural changes that occur during an enzymatic reaction
   (d) The sequence of amino acids in an α-helix

13. Which of these ions **cannot** be used to generate an NMR spectrum?
   (a) ^1H
   (b) ^{12}C
   (c) ^{13}C
   (d) ^{15}N

14. Which of these is **not** a type of NMR?
   (a) COSY
   (b) NOESY
   (c) NOSEY
   (d) TOCSY

15. Which of the following is **not** achievable by X-ray crystallographic study of a protein?
   (a) Relative positions of chemical groups
   (b) Conformation of the polypeptide chain
   (c) Positioning of amino acid side-chains
   (d) Following protein folding in real time

16. What degree of resolution can be achieved in the most successful X-ray crystallography studies?
   (a) 0.1 nm
   (b) 0.5 nm
   (c) 1.0 nm
   (d) 10 nm

17. What is the rate of movement of substrates through individual pathways called?
   (a) Metabolic flux
   (b) Metabolomics
   (c) Metabolic control
   (d) Metabolic partitioning

18. In gas chromatography, what is the stationary phase?
   (a) Gas
   (b) The solid matrix on the internal surface of the column
   (c) Liquid
   (d) None of the above

19. Chemical ionization gives what type of molecular ion?
   (a) $[M+H]^+$
   (b) $[M+H]^-$
   (c) $[M–H]^+$
   (d) $[M–H]^-$

20. Which of the following is **not** a feature of a quadrupole mass spectrometer?
   (a) Oscillating electrical fields
   (b) Ions 'wiggle' as they pass through the quadrupole
   (c) The field can be gradually changed so ions with different m/z ratios are detected
   (d) The mass analyzer is a single magnet

21. What is the type of gentle ionization procedure used with unstable lipids, in conjunction with HPLC, called?
   (a) Electrospray ionization
   (b) Chemical ionization
   (c) Electron ionization
   (d) Laser-assisted ionization

22. The antigenic glycan that distinguishes the A, B and O blood groups has what features?
   (a) In the A type, one of the units of the glycan is D-galactose, in the B type it is N-acetylgalactosamine, and in the O group this particular unit is absent
   (b) A, N-acetylgalactosamine; B, D-galactose; O, absent
   (c) A, N-acetylgalactosamine; B, D-glucose; O, absent
   (d) A, D-glucose; B, D-galactose; O, absent

23. To which sugars does concanavalin A bind?
   (a) Terminal α-glucose and α-mannose units in O-linked and N-linked glycans
   (b) Terminal α-glucose and α-mannose units in N-linked, though not O-linked, glycans
   (c) Terminal α-glucose and α-mannose units in O-linked, though not N-linked, glycans
   (d) None of the above

24. Treatment with which of the following specifically removes O-linked glycans?
   (a) Hydroxide
   (b) Borohydride ions
   (c) Hydrazine
   (d) Endoglycosidase

Short answer questions

These questions do not require additional reading.

1. Describe the key difference between polyclonal and monoclonal antibodies. How are monoclonal antibodies prepared?

2. Outline the procedures that make use of antibody–antigen precipitation in order to identify the presence of a particular protein in a mixture of proteins.

3. What is 'ELISA' and why is this method more sensitive and accurate than immunoassays that are carried out in a gel?

4. Describe the various methods for protein separation prior to protein profiling.

5. Give a detailed account of how mass spectrometry is used in protein profiling. Include in your answer a summary of a method for comparing the compositions of two proteomes.

6. What is the basis to circular dichroism and what can this method tell you about protein structure?

7. Describe the strengths and weaknesses of (A) NMR, and (B) X-ray crystallography in studies of protein structure.

8. Distinguish between the different methods used to ionize lipids prior to their analysis by mass spectrometry.

9. Using diagrams, illustrate the architectures of typical (A) magnetic sector, and (B) quadrupole mass spectrometers. What is tandem mass spectrometry?

10. Outline the methods used to study the glycome.

Self-study questions

These questions will require calculation, additional reading and/or internet research.

1. What are the relative merits of polyclonal and monoclonal antibodies for each of the types of immunoassay described in *Section 18.1.1*?

2. You have purified a protein from a proteome, digested with trypsin and measured the molecular masses of the resulting six peptides by MALDI–TOF. Five of the peptides give exact matches to a protein in the database, but the mass of the sixth peptide appears to be incorrect. According to the protein sequence this peptide should be SLYSSTIDK, with a mass of 994. The peptide detected by MALDI–TOF has a mass of 1072. What is the likely explanation of the discrepancy between the expected and actual masses of this peptide?

3. The resolution achievable by NMR is directly related to the field strength of the magnet that is used. Explore how this relationship has affected development of NMR over the last 20 years, and speculate about the future potential of the procedure.

4. DNA does not form crystals but X-ray diffraction analysis was very important in the work that led to discovery of the double helix structure. Explain how X-ray diffraction analysis can be used with DNA.

5. A fatty acid has been derivatized by attachment of a 3-pyridylcarbinol group and ionized by electron bombardment. The resulting mass spectrum is shown below. What is the structure of the lipid?

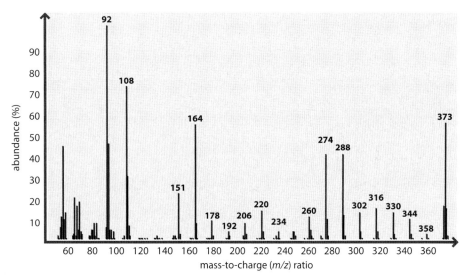

Adapted from *Biological Mass Spectrometry* 1982, **9**, 33 with permission from John Wiley and Sons.

CHAPTER 19
Studying DNA and RNA

STUDY GOALS

After reading this chapter you will:

- be able to describe the different types of nuclease used to manipulate DNA and RNA molecules

- be able to give a detailed description of the key features of restriction endonucleases

- know how DNA ligases are used to join DNA molecules together

- be able to give a detailed description of the polymerase chain reaction, including the quantitative version called real-time PCR

- understand why PCR has become so important in biochemical research

- be able to describe the chain termination and pyrosequencing methods for DNA sequencing

- know the key features of next generation sequencing methodology

- be able to describe the DNA cloning process in outline and be able to give details of how DNA is cloned in the pUC8 cloning vector

- be able to summarize the various ways in which DNA is cloned in different types of eukaryote

- understand the special features of cloning vectors used for synthesis of recombinant protein

- know why bacteria are not always the ideal hosts for recombinant protein synthesis

- understand the strengths and limitations of different eukaryotic cells for recombinant protein production

Biochemists and geneticists have been particularly ingenious in developing methods for studying DNA and RNA molecules. A vast array of techniques are now available for examining and manipulating the expression patterns of individual genes, for transferring genes from one organism to another, and for making directed alterations to the nucleotide sequence of a gene. Most importantly, the technology for working out the order of nucleotides in DNA and RNA molecules, commonly referred to as 'sequencing', has steadily improved since the first practicable methods were devised in the 1970s.

In this chapter we will explore those of the methods for studying DNA and RNA that are of greatest importance to biochemists. First, we will look at the ways in which biochemists use purified enzymes to manipulate DNA and RNA molecules *in vitro*, these manipulations forming the basis of many of the techniques used to study these molecules. Then, we will examine how DNA and RNA are sequenced and how those sequences are interpreted. Finally, we will investigate the methods for **DNA cloning**, which are used to transfer genes from one species to another, and which enable important pharmaceutical proteins such as human insulin to be synthesized by genetically engineered microorganisms.

19.1 Manipulation of DNA and RNA by purified enzymes

Many of the techniques used to study DNA and RNA make use of purified enzymes. Within the cell, these enzymes participate in processes such as DNA replication and repair. After purification, the enzymes continue to carry out their natural reactions when supplied with the appropriate substrates. Although the reactions catalyzed by these enzymes are often straightforward, most are absolutely impossible to perform by standard chemical methods. Purified enzymes are therefore an essential and central component of the methods used to study DNA and RNA. We will begin by looking at the different types of enzyme used in this area of biochemical research. We will then examine in some detail one particular method for manipulating DNA and RNA, the **polymerase chain reaction** (**PCR**). Although simply resulting in the synthesis of multiple copies of a segment of a DNA or RNA molecule, PCR has assumed immense importance in many areas of biological research, including biochemistry.

19.1.1 Types of enzyme used to study DNA and RNA

The three most important types of enzymes used to study DNA and RNA are:

- **Nucleases** – enzymes that cut, shorten or degrade nucleic acid molecules.

- **Ligases** – which join nucleic acid molecules together.

- **Polymerases** – which make copies of molecules.

Nucleases are used to cut DNA and RNA molecules

Nucleases degrade DNA or RNA molecules by breaking the phosphodiester bonds that link one nucleotide to the next in a polynucleotide. There are two different kinds of nuclease (*Fig. 19.1*):

- **Exonucleases** remove nucleotides one at a time from the end of a molecule.

- **Endonucleases** break internal phosphodiester bonds within a molecule.

Some nucleases degrade just one strand of a double-stranded molecule, and others degrade both strands. S1 endonuclease, which is prepared from the fungus *Aspergillus oryzae*, is an example of a single-strand deoxyribonuclease, meaning that it cuts only single-stranded DNA polynucleotides (*Fig. 19.2*). In contrast, deoxyribonuclease I (DNase I), which is prepared from cow pancreas, cuts both single- and double-stranded DNA molecules.

Figure 19.1 The reactions catalyzed by the two different types of nuclease.
(A) An exonuclease, which removes nucleotides from the end of a DNA molecule. (B) An endonuclease, which breaks internal phosphodiester bonds.

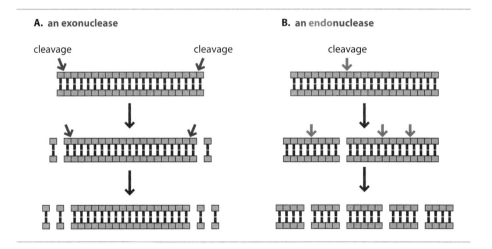

A. an exonuclease

B. an endonuclease

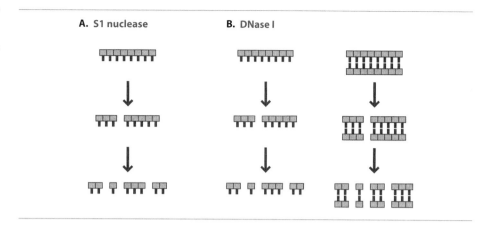

Figure 19.2 The reactions catalyzed by different types of endonuclease. (A) S1 nuclease, which cuts only single-stranded DNA. (B) DNase I, which cuts both single- and double-stranded DNA.

DNase I cuts DNA at any internal phosphodiester bond, so prolonged DNase I treatment gives a mixture of mononucleotides and very short oligonucleotides. However, if a protein is attached to the DNA, then a segment will be undigested. This is because the endonuclease has to gain access to the DNA in order to cut it, and so can attack only the phosphodiester bonds that are not masked by the protein. The nuclease therefore degrades those parts of the DNA molecule that remain exposed after protein attachment, but not the protected segment, which can be recovered intact after the enzyme has been inactivated and the binding proteins removed. We encountered this procedure, called 'nuclease protection', when we studied the way in which DNA associates with nucleosomes in chromatin (see *Fig. 4.16*). Nuclease protection experiments have also been important in identifying the binding sites for proteins that attach to DNA in order to regulate gene expression.

So far we have considered nucleases that work on DNA. A similar set of purified ribonucleases is also available. These include RNase I, from *E. coli*, which is an endonuclease that degrades single-stranded RNA but has no effect on base-paired regions. RNase I can therefore be used identify the double-stranded regions in molecules with one or more stem–loop structures (*Fig. 19.3*). Other endoribonucleases, such as RNase V1, only cut double-stranded RNA.

Restriction enzymes are sequence-specific DNA endonucleases

The most useful type of nuclease would be one that cuts a double-stranded DNA molecule only at specific nucleotide sequences. The positions of these cuts within a DNA molecule could be predicted, assuming that the DNA sequence is known. One or more sequence-specific endonucleases could therefore be used to cut out a defined segment of a DNA molecule, such as an individual gene. **Restriction endonucleases** possess this ability and, not surprisingly, have become the most widely used nucleases in biochemical research.

Strictly speaking, a restriction endonuclease *binds* to a specific nucleotide sequence. With the Type I and III versions of these enzymes, the subsequent cut is made randomly within the region adjacent to the bound enzyme (*Fig. 19.4*). This is a

Figure 19.3 RNase I degrades single-stranded RNA and so can be used to identify double-stranded regions in an RNA molecule.

Box 19.1 What is 'restriction'?

What does the term 'restriction' refer to in 'restriction endonuclease'? During the early 1950s, it was discovered that some types of bacteria are able to withstand infection by bacteriophages. This process was called 'host-controlled restriction' to indicate that the bacterium (the host) could restrict the growth of the bacteriophage. It was subsequently shown that host-controlled restriction occurs when the bacterium synthesizes an enzyme that cuts up the bacteriophage DNA

before it has time to replicate and direct the production of new bacteriophage particles. These cutting enzymes were called 'restriction endonucleases'. Closer study of restriction endonucleases revealed their specificity for particular DNA sequences, the property that makes them so useful today in DNA cloning and other methods that involve manipulation of DNA molecules *in vitro*.

---

useful property, but not as useful as the mode of action of a Type II enzyme. A Type II restriction endonuclease always cuts at the same place, either within the recognition sequence or very close to it. For example, the Type II enzyme called *Eco*RI, which is obtained from *E. coli*, cuts only at the hexanucleotide GAATTC.

Almost 4000 Type II restriction enzymes are known, and several hundred with different recognition sequences can be obtained from commercial suppliers for use in laboratory experiments. Some of these enzymes, like *Eco*RI, have hexanucleotide target sites, but others recognize shorter or longer sequences (*Table 19.1*). A few have degenerate recognition sequences, which means that they cut at any of a family of related sites. An example is *Hin*fI, which recognizes GANTC, where 'N' is any nucleotide, and so cuts at GAATC, GACTC, GAGTC and GATTC.

Restriction enzymes cut double-stranded DNA in two different ways (*Fig. 19.5*):

• Some make a simple double-stranded cut giving a **blunt** or **flush end**.

• Others cut the two DNA strands at different positions, usually two or four nucleotides apart, so that the resulting DNA fragments have short single-stranded overhangs at each end. These are called **sticky** or **cohesive ends**, as base pairing between them can stick the DNA molecule back together again. Some sticky-end cutters give 5′ overhangs (e.g. *Bam*HI, *Sau*3AI, *Hin*fI), others leave 3′ overhangs (e.g. *Pst*I).

Table 19.1. The recognition sequences for some of the most frequently used restriction endonucleases

Enzyme	Organism	Recognition sequence	Blunt or sticky end
*Eco*RI	*Escherichia coli*	GAATTC	Sticky
*Bam*HI	*Bacillus amyloliquefaciens*	GGATCC	Sticky
*Bgl*II	*Bacillus globigii*	AGATCT	Sticky
*Pvu*I	*Proteus vulgaris*	CGATCG	Sticky
*Pvu*II	*Proteus vulgaris*	CAGCTG	Blunt
*Hind*III	*Haemophilus influenzae* R$_d$	AAGCTT	Sticky
*Hin*fI	*Haemophilus influenzae* R$_f$	GANTC	Sticky
*Sau*3A	*Staphylococcus aureus*	GATC	Sticky
*Alu*I	*Arthrobacter luteus*	AGCT	Blunt
*Hae*III	*Haemophilus aegyptius*	GGCC	Blunt
*Not*I	*Nocardia otitidis-caviarum*	GCGGCCGC	Sticky
*Sfi*I	*Streptomyces fimbriatus*	GGCCNNNNNGGCC	Sticky

The recognition sequence is that of one strand, given in the 5′→3′ direction. 'N' indicates any nucleotide. Note that almost all recognition sequences are palindromes: the two strands, when read in opposite directions, have the same nucleotide sequence, for example:

```
            5′–GAATTC–3′
EcoRI         ||||||
            3′–CTTAAG–5′
```

Figure 19.4 Types of cut made by different restriction endonucleases. The position of the recognition sequence is indicated by the orange line.

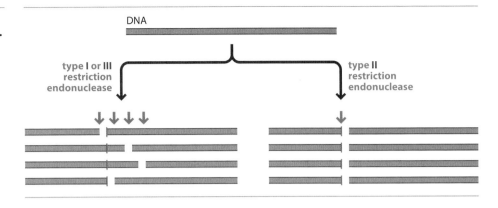

The overhangs left by a sticky-end cutter are often shorter than the recognition sequence. For example, *Bam*HI recognizes GGATCC but leaves just a GATC overhang (see *Fig. 19.5B*). This means that two or more enzymes with different recognition sequences could generate identical sticky ends. *Bgl*II, for example, also gives a GATC overhang, but its recognition sequence, AGATCT, is different to that of *Bam*HI. A third enzyme, *Bcl*I, recognizes TGATCA but leaves the same overhang as *Bam*HI and *Bgl*II, as does *Sau*3AI, whose recognition sequence is just the tetranucleotide GATC (*Figure 19.6*). We will see how the ability to produce fragments with identical sticky ends from enzymes with different recognition sequences is important in research with DNA when we study DNA cloning later in this chapter.

DNA ligases join molecules together

DNA fragments that have been produced by a restriction endonuclease can be joined back together again, or attached to new partners, by a DNA ligase. The reaction requires energy which is provided by adding either ATP or NAD to the reaction mixture, depending on the type of ligase that is being used.

Figure 19.5 Different types of cut made by Type II restriction endonucleases. (A) The difference between blunt and sticky ends. (B) Two types of sticky end: ones with 5′ overhangs and ones with 3′ overhangs.

A. blunt and sticky ends

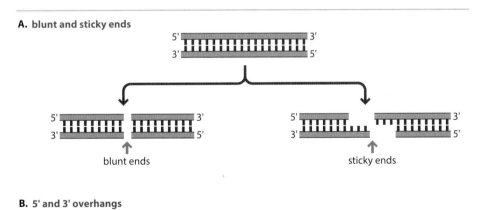

B. 5′ and 3′ overhangs

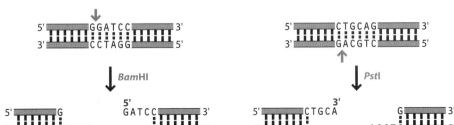

Figure 19.6 **Two restriction enzymes with different recognition sequences can create identical sticky ends.**

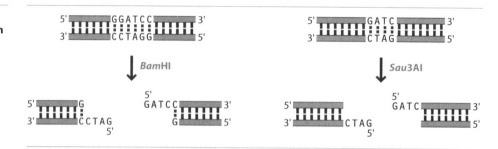

The most widely used DNA ligase is obtained from *E. coli* cells that have been infected with a virus called T4. The natural role of this ligase is to join together Okazaki fragments during replication of the virus DNA. In this natural role, the two molecules to be linked together are base-paired to the template strand of the DNA that is being replicated. In other words, the two ends are held close to one another (*Fig. 19.7*). This is also the case when two restriction fragments with sticky ends are joined together. Although the fragments are initially dispersed in the reaction mixture, and must come together by random diffusion events, once they do so it is likely that transient base pairs will form between the two overhangs. These base pairs persist for sufficient time for a ligase enzyme to attach to the junction and synthesize the pair of phosphodiester bonds that link the two fragments together.

If the molecules are blunt-ended, then ligation is much less efficient. Blunt ends, lacking overhangs, cannot base pair with one another, not even temporarily. Ligation occurs only when the chance diffusion events bring a ligase enzyme into proximity with two ends that just happen to be close to one another in the reaction mixture. To increase the chances of this happening, a high concentration of DNA must be used.

Polymerases are used to make copies of DNA molecules

Polymerases are the third type of enzyme used to manipulate DNA and RNA. DNA synthesis, catalyzed by a polymerase, forms the basis both of PCR and of most sequencing techniques.

> We studied the role of DNA polymerase I, and other DNA polymerases in DNA replication in *Section 14.1.2*.

The most widely used of this type of enzyme is the bacterial DNA polymerase I. As well as being able to synthesize DNA, this enzyme has a 5'→3' exonuclease activity.

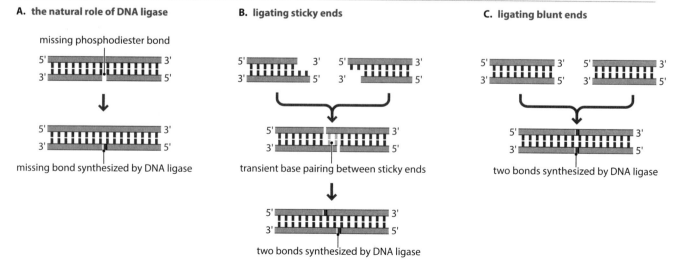

Figure 19.7 **DNA ligase.**
(A) The natural role of DNA ligase, and the use of DNA ligase to join (B) sticky ends, and (C) blunt ends.

Figure 19.8 **Synthesis of a labeled strand of DNA by DNA polymerase I.**

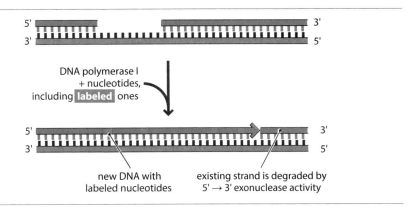

Figure 19.8 **Synthesis of a labeled strand of DNA by DNA polymerase I.**

This means that it can attach to a short single-stranded region in a mainly double-stranded DNA molecule, and then synthesize a completely new strand, degrading the existing strand as it proceeds. This reaction is used to incorporate labeled nucleotides into a DNA molecule, so the DNA becomes labeled and can be tracked through subsequent experiments (*Fig. 19.8*).

DNA polymerase I is often obtained from *E. coli*, but some techniques, including PCR, require a specialized version of this enzyme. This version is obtained from *Thermus aquaticus*, a bacterium that lives in hot springs. Many of this bacterium's enzymes, including its DNA polymerases, are thermostable, meaning that they are resistant to denaturation by heat treatment, and have an optimum working temperature of 70–80°C. The DNA polymerase I of *Thermus aquaticus* is referred to as ***Taq* polymerase** (*Taq* coming from <u>*T*</u>hermus <u>*aq*</u>uaticus).

A different type of polymerase, important in manipulating RNA, is **reverse transcriptase**. This enzyme is involved in the replication of some types of virus whose genomes are made of RNA. During the replication of these virus genomes, the RNA is copied into DNA. This property is used in the laboratory to make DNA copies of RNA molecules (**complementary DNA** or **cDNA**), a procedure known as **cDNA synthesis**.

19.1.2 The polymerase chain reaction

PCR enables any segment of a DNA molecule, up to about 40 kb in length, to be copied repeatedly so that large quantities are obtained. First, we will examine how the procedure is carried out. Then we will ask why so simple a technique has assumed such great importance, not just in biochemistry but in many areas of biology.

PCR results in multiple copies of a targeted region of a DNA molecule

The two key components of a PCR are the thermostable *Taq* polymerase and a pair of short oligonucleotides. The oligonucleotides bind to the target DNA molecule, one to each strand of the double helix. These oligonucleotides, which act as primers for the DNA synthesis reactions, delimit the region that will be amplified. They must therefore be complementary to the target DNA at either side of the segment that is to be copied. The primers are obtained by chemical DNA synthesis.

To begin the reaction, the DNA is mixed with *Taq* polymerase, the two primers, and a supply of nucleotides. The reaction can be carried in a small plastic test tube or in the wells of a microtiter plate, which is placed in a **thermal cycler**, a programmable device that heats and cools the reaction between set temperatures. The reaction is started by heating the mixture to 94°C. At this temperature the hydrogen bonds between the two polynucleotides of the double helix are broken, so the DNA becomes denatured into single-stranded molecules (*Fig. 19.9*). The temperature is then reduced to 50–60°C,

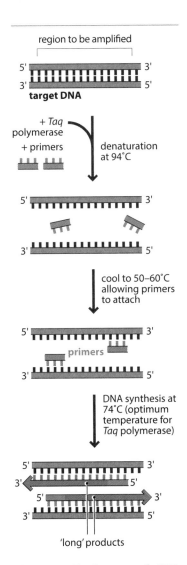

Figure 19.9 **The first stage of a PCR.**

(within figure 19.9:)

region to be amplified

5′ 3′
3′ 5′
target DNA

+ *Taq* polymerase
+ primers

denaturation at 94°C

5′ 3′

3′ 5′

cool to 50–60°C allowing primers to attach

5′ 3′

primers

3′ 5′

DNA synthesis at 74°C (optimum temperature for *Taq* polymerase)

5′ 3′
3′ 5′
5′ 3′
3′ 5′

'long' products

which allows the primers to attach to their binding positions. Next, the temperature is raised to 74°C, within the optimum range for *Taq* polymerase, so that DNA synthesis can begin. In this first stage of the PCR, a set of 'long products' is synthesized from each strand of the target DNA. These long products have identical 5' ends but random 3' ends, the latter set by the positions where DNA synthesis terminates by chance.

The denaturation–annealing–synthesis cycle is now repeated (*Fig. 19.10*). The long products denature and the four resulting strands are copied, giving four double-stranded molecules, two of which are identical to the long products from the first cycle and two of which are made entirely of new DNA. During the third cycle, the latter give rise to 'short products', the 5' and 3' ends of which are both set by the primer annealing positions. In subsequent cycles, the number of short products accumulates exponentially (doubling during each cycle) until one of the components of the reaction becomes depleted. This means that after 30 cycles, there will be over 130 million short products derived from each starting molecule. This equates to several micrograms of PCR product from a few nanograms or less of target DNA.

Figure 19.10 The second and third cycles of a PCR, during which the first short products are synthesized.

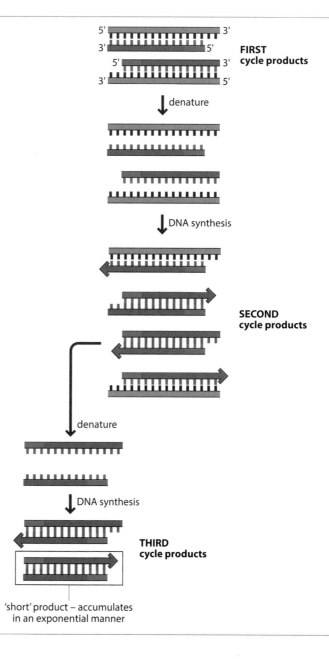

PCR therefore results in the exponential copying of the segment of DNA that is delineated by the primers. It can also be used to amplify a segment of an RNA molecule, the RNA first being converted into cDNA by treatment with reverse transcriptase.

There are just two limitations to PCR. First, the sequence to either side of the DNA or RNA segment that is to be copied must be known. This information is needed in order to synthesize oligonucleotide primers that will attach at the appropriate places in the target molecule. If the sequence is unknown, or cannot be predicted, then PCR cannot be used to study that molecule.

The second limitation relates to the length of DNA that can be amplified. This is determined by the **processivity** of *Taq* polymerase, where processivity refers to the average number of nucleotides that are polymerized before the enzyme detaches from the template DNA. During a PCR, 5 kb of DNA can be copied fairly efficiently, and regions up to 40 kb can be copied using specialized techniques. But many eukaryotic genes are longer than 40 kb, and so would have to be amplified as a series of segments rather than as a single PCR product.

The progression of a PCR can be followed in real time

At the end of a PCR, a sample of the reaction mixture can be examined by agarose gel electrophoresis, sufficient DNA having been produced for the amplified fragment to be visible as a discrete band after staining with a DNA-binding dye.

Alternatively the progression of the reaction can be followed as it is taking place. This format is called **real-time PCR**, because the reaction is followed in real time. There are two ways of carrying out real-time PCR:

- A compound that gives a fluorescent signal when it binds to double-stranded DNA can be included in the PCR mixture. The amount of fluorescent signal will increase during the PCR as more and more DNA is synthesized.

- A short oligonucleotide called a **reporter probe** can be used. The sequence of the reporter probe is designed so that it will base pair with one of the strands of the PCR product. A fluorescent chemical group is attached at one end of the oligonucleotide, and a second group, one that inhibits ('quenches') the fluorescent signal, is attached to the other end. The oligonucleotide is designed in such a way that its two ends base pair to one another, placing the quencher next to the fluorescent group (*Fig. 19.11*). This means that when the oligonucleotide is free in solution, it does not emit any fluorescence. However, base pairing to the PCR product is energetically more favorable, so when the product is present the oligonucleotide opens up and binds to it. Now the quencher is too far away from

Figure 19.11 A reporter probe, as used in one type of real-time PCR.

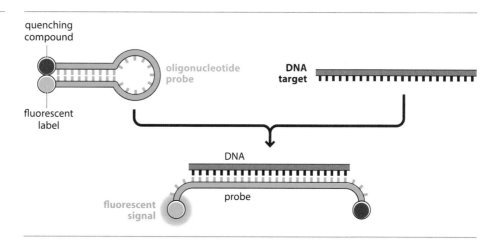

the fluorescent group to inhibit the signal. The amount of fluorescence therefore increases as the PCR proceeds.

Real-time PCR is also commonly called **quantitative PCR** (**qPCR**). This is because the amount of fluorescence at the end of each cycle indicates the amount of PCR product that has been synthesized. This in turn depends on how much template DNA was present at the start of the PCR. The starting amount can therefore be quantified by comparison with control PCRs set up with known amounts of starting DNA. Usually, this comparison is made by identifying the stage in the PCR at which the amount of fluorescent signal reaches a pre-set threshold (*Figure 19.12*). The more rapidly the threshold is reached, the greater the amount of DNA in the starting mixture.

PCR has many applications

To the research biochemist, the ability to amplify segments of a DNA or RNA molecule forms the starting point for many of the more sophisticated methods that are used to study nucleic acids. Later in this chapter we will see how PCR is used in DNA sequencing. PCR also enables the codons in a gene to be altered in a specified way, so that a protein with an altered amino acid sequence and novel biochemical properties can be synthesized.

PCR also underlies many of the important applications of biochemistry in our modern world. In a clinical setting, PCR is used in genetic screening, in which the predisposition of a patient, or even an unborn child, to certain diseases is assessed. PCRs directed at the human globin genes, for example, are used to test for the presence of mutations that might cause the blood disease thalassemia. The primers for these PCRs are easy to design because the sequences of the human globin genes are known, and regions of these that are invariant in the human population have been identified. If the primers are designed so they attach to a pair of these invariant regions, then the PCR will work with any sample of human DNA, even if the sequence between the primers is variable in different individuals. After the PCR, the products are sequenced to determine if any of the thalassemia mutations are present.

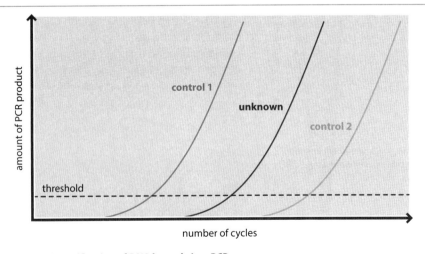

Figure 19.12 Quantification of DNA by real-time PCR.
The graph shows product synthesis during three PCRs, each with a different amount of starting DNA. During a PCR, product accumulates exponentially, the amount present at any particular cycle being proportional to the amount of starting DNA. The blue curve is therefore the PCR with the greatest amount of starting DNA, and the green curve is the one with the least starting DNA. If the amounts of starting DNA in these two PCRs are known, then the amount in a test PCR (red curve) can be quantified by comparison with these controls. The comparison is made by identifying the cycle at which product synthesis moves above a threshold amount, indicated by the horizontal line on the graph.

Box 19.2 Using PCR to alter the codons in a gene

In order to use PCR to change the sequence of a gene, two pairs of primers are needed. One primer of each pair is a perfect match with the gene sequence, and the other contains the nucleotide change that we wish to introduce.

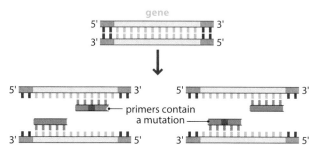

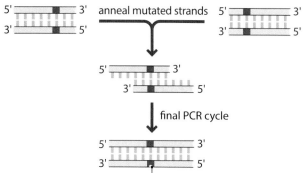

After PCR, the sequence alteration incorporated into the primers will be present in each of the two amplification products as shown below.

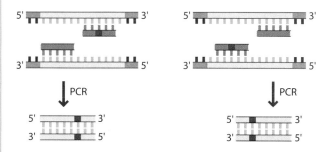

The two PCR products are now mixed together and a final PCR cycle carried out. In this cycle, complementary strands from the two products anneal to one another and are then extended by the polymerase, producing a full-length DNA molecule containing the sequence alteration.

This technique is a type of ***in vitro* mutagenesis**, 'in vitro' because it is carried out in a test tube, and 'mutagenesis' because the outcome is introduction of a mutation into the DNA molecule. The mutation can be at any position, so any codon of a gene can be altered in any desired way. *In vitro* mutagenesis has many applications in biochemistry. For example:

- The altered version of the gene could be placed back in its host organism by cloning, as described later in this chapter (*Section 19.3.1*). The effect of the mutation on the function of the protein coded by the gene could then be studied.
- The altered gene could be cloned in *E. coli* and the product obtained as recombinant protein (*Section 19.3.2*). The protein could then be purified and the effect of the mutation on its structure or activity examined.

In vitro mutagenesis also underlies protein engineering, where the aim is to develop new enzymes for biotechnological purposes. We explored one example of protein engineering in *Box 7.5*, when we looked at how thermostable enzymes are being exploited in the production of biofuels. A second application of protein engineering is in the development of biological washing powders. These detergents contain proteases, such as subtilisin, which digest food residues and other proteins present on the materials that are being cleaned. To improve the performance of biological washing powers, *in vitro* mutagenesis has been used to generate modified types of subtilisin that have greater resistances to the thermal and bleaching (oxidative) stresses encountered in washing machines.

Another clinical application of PCR is in the early detection of virus infections. A positive result indicates that a sample contains the virus and that the person who provided the sample should undergo treatment to prevent onset of the disease. PCR is extremely sensitive, and can give a product even if there is just one copy of the target DNA in the starting mixture. This means that the technique can detect viruses at the earliest stages of an infection, increasing the chances of treatment being successful. If qPCR is used, then information can be obtained on the progress of an infection, and the patient given a robust 'all clear' when the virus is no longer detectable.

PCR has also become hugely important in forensic science. The small amounts of DNA present in hairs and dried bloodstains can be copied, enabling a **genetic profile** to be constructed. Each person's genetic profile is unique, so a match between a suspect's profile and a profile obtained by PCR of samples from a crime scene can lead to the conviction of a criminal. In recent years, researchers have learnt how to increase the sensitivity of PCR so that smaller and smaller amounts of starting DNA can be detected. Evidence from historic crime scenes can therefore yield genetic profiles, leading to convictions in 'cold cases' from the 1990s and earlier.

Similar techniques are used to study **ancient DNA** in archaeological material, such as the bones of extinct humans. PCR has enabled the sequence of the Neanderthal genome to be obtained, providing information on the evolutionary origins of *Homo sapiens*. This work has revealed that some prehistoric members of our species interbred with Neanderthals over 40 000 years ago.

19.2 DNA sequencing

Probably the most important technique used to study nucleic acids is **DNA sequencing** – the method that identifies the precise order of nucleotides in a DNA molecule. Rapid and efficient DNA sequencing methods were first developed in the 1970s. To begin with, these techniques were applied to individual genes, but since the early 1990s an increasing number of entire genome sequences have been obtained. Large scale sequencing was made substantially easier in the mid 2000s by the invention of new automated methodologies, called **next generation sequencing**. We will look first at the conventional methods, which are still used to obtain the sequences of short segments of DNA in the research laboratory. We will then survey the more specialized approaches used in next generation sequencing.

19.2.1 Methodology for DNA sequencing

The conventional approach to DNA sequencing is called the **chain termination** method. This procedure was invented by Frederick Sanger and colleagues in the 1970s and is still widely used today.

Chain termination sequencing makes use of modified nucleotides

Chain termination sequencing can be carried out in several ways, but the most frequently used method involves a reaction very like a PCR. A thermostable DNA polymerase is used, and the steps of the reaction are controlled by cycling between temperatures to allow repeated rounds of strand separation, primer attachment and DNA synthesis. However, compared with PCR there are two critical differences:

- Only one primer is used. This means that the reaction results in multiple copying of just one strand of the target DNA.

- As well as the normal nucleotide substrates for DNA synthesis (dATP, dCTP, dGTP and dTTP), the reaction also contains four modified compounds called 2′,3′-dideoxynucleotides, or simply **dideoxynucleotides** (**ddNTPs**).

The ddNTPs are chain terminating nucleotides. During DNA synthesis, the addition of a nucleotide requires formation of a phosphodiester bond between the 3′-OH group of the last nucleotide in the chain and the 5′-P group of the incoming nucleotide

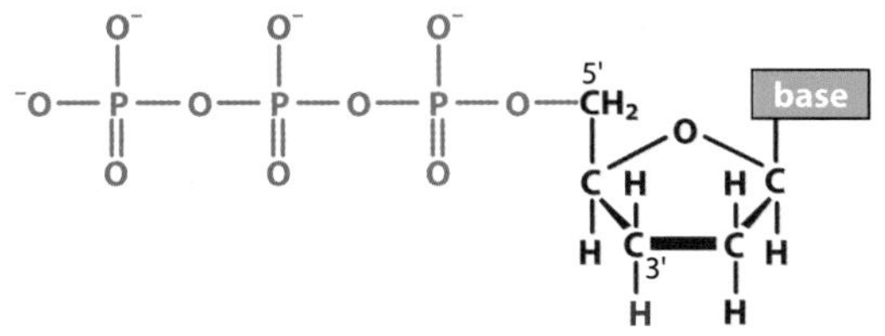

Figure 19.13 The structure of a dideoxynucleotide showing the position where the –OH of a dNTP is replaced by an –H.

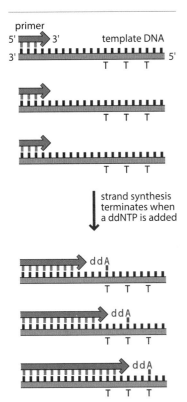

Figure 19.14 **The role of the ddNTPs in a chain termination sequencing experiment.**

(see *Fig. 14.10*). A ddNTP has a normal 5'-P group, and so can be added to the end of a growing polynucleotide. However, it lacks the 3'-OH group needed to form a bond with the next incoming nucleotide (*Fig. 19.13*). This means that, once in place, a ddNTP blocks further strand synthesis. In other words, it causes chain termination.

Although ddNTPs are present in a chain termination sequencing reaction, the normal nucleotides are in excess. This means that DNA synthesis does not stop immediately after it has begun, and instead can continue for anything up to several hundred steps before a ddNTP is incorporated and chain termination occurs. Each cycle of the pseudo-PCR will generate a new set of chain-terminated molecules. At the end of the experiment, there will therefore be a mixture of single-stranded products, of differing lengths, each ending with a ddNTP.

How does this help us work out the sequence of the template DNA? The key point is that the identity of the terminal ddNTP indicates the nucleotide present at that position in the template DNA. If the terminal ddNTP is ddA, for example, then there must be T at that position in the template (*Fig. 19.14*). To work out the sequence of the template we must therefore do two things:

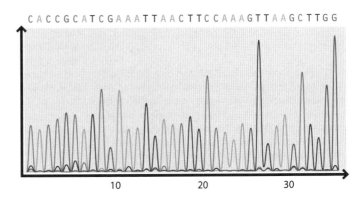

A. identification of the chain-terminated strands

B. the output from the imaging system

Figure 19.15 **Reading the sequence generated by a chain termination experiment.**
(A) Identification of the chain-terminated strands, by virtue of the fluorescent marker attached to each one.
(B) The output from the imaging system. The sequence is represented by a series of peaks, one for each nucleotide position. In this example, a green peak is an A, blue is C, brown is G, and red is T.

- First, separate out the chain-terminated strands according to their lengths, shortest ones first and longest later. This can be achieved by electrophoresis through a thin capillary of polyacrylamide gel. Under appropriate conditions, polynucleotides differing in length by just a single nucleotide can be resolved.

- Secondly, identify which ddNTP is present at the end of each of the chain-terminated strands. This is possible if the ddNTPs used as substrates were labeled with fluorescent markers, a different marker for each of the four ddNTPs.

The sequence is therefore read by a fluorescence detector that identifies the signal emitted by each chain-terminated strand as it passes along the capillary gel (*Fig. 19.15*). In practice, up to 1000 nucleotides can be read in a single experiment.

Pyrosequencing enables direct readout of a DNA sequence

Pyrosequencing is an alternative procedure for sequencing short segments of a DNA molecule. The advantage of this method is that it does not require electrophoresis or any other fragment separation procedure, and so is more rapid than chain termination sequencing.

Pyrosequencing, like the chain termination method, involves synthesis of new strands of DNA from a primer attached at a defined position on a template molecule.

Box 19.3 Did Neanderthals and modern humans meet and interbreed?

RESEARCH HIGHLIGHT

One of the great achievements of next generation sequencing has been the complete sequence of the Neanderthal genome, obtained from ancient DNA preserved in small pieces of a bone from a cave in the Altai mountains of Siberia. Neanderthals are an extinct type of human who lived in Europe and parts of Asia between 200 000 and 30 000 years ago. For most of this period, our ancestors – 'anatomically modern humans' or *Homo sapiens sapiens* – were restricted to Africa, but around 70 000 years ago modern humans ventured out of Africa and began the migrations that would eventually lead to their dispersal across the entire planet. About 45 000 years ago, modern humans arrived in Europe, where they coexisted with Neanderthals for some 15 000 years.

Europe is a big place and the human populations were relatively small at that time, so it is possible that Neanderthals and modern humans rarely encountered one another, especially as Neanderthals were adapted to cold climates and modern humans to warm ones. This has not prevented anthropologists from asking if Neanderthals and modern humans interbred. We believe that Neanderthals were a subspecies of *Homo sapiens*, so interbreeding might have been possible.

Comparisons between our own genome and the genome of Neanderthals suggest that some interbreeding did occur. The genomes of modern Europeans are slightly more similar to the Neanderthal genome than are the genomes of modern Africans. This suggests that some Neanderthal DNA has found its way into the genomes of modern Europeans. If there had been no interbreeding then modern Europeans and Africans should be indistinguishable when compared with Neanderthals.

If interbreeding did occur, then gene variants that evolved in Neanderthals might have been transferred directly to the early European population of modern humans. As Neanderthals were adapted to the relatively harsh climate of Europe, perhaps the

gene variants specified proteins that helped modern humans survive the last Ice Ages and eventually prosper in Europe? Geneticists are beginning to explore this intriguing question by making detailed comparisons of gene variants present in Neanderthals and in modern human populations from different parts of the world. It appears that a particular type of keratin protein possessed by modern Europeans might have been inherited from Neanderthals, possibly changing the hair and skin so that modern Europeans became better able to withstand cold temperatures. However, other aspects of the Neanderthal legacy may have been less advantageous, with some of the gene variants inherited by modern humans being associated with disorders such as Crohn's disease, cirrhosis of the liver, and the autoimmune disease lupus.

It is also becoming clear that a second type of extinct human, the Denisovans, who lived in northern Asia at about the same time as Neanderthals, also contributed to the modern human genome via interbreeding. The most recent estimates are that 1.5–2.1% of the DNA of modern humans from outside of Africa is of Neanderthal origin, and 3.0–6.0% of the genomes of modern inhabitants of Oceania is derived from Denisovans. There is also evidence for interbreeding between Neanderthals and Denisovans, and between Denisovans and an unidentified extinct type of human.

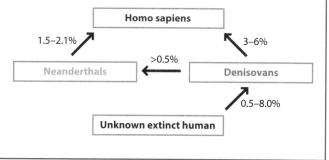

Figure 19.16 **The chemical basis to pyrosequencing.**

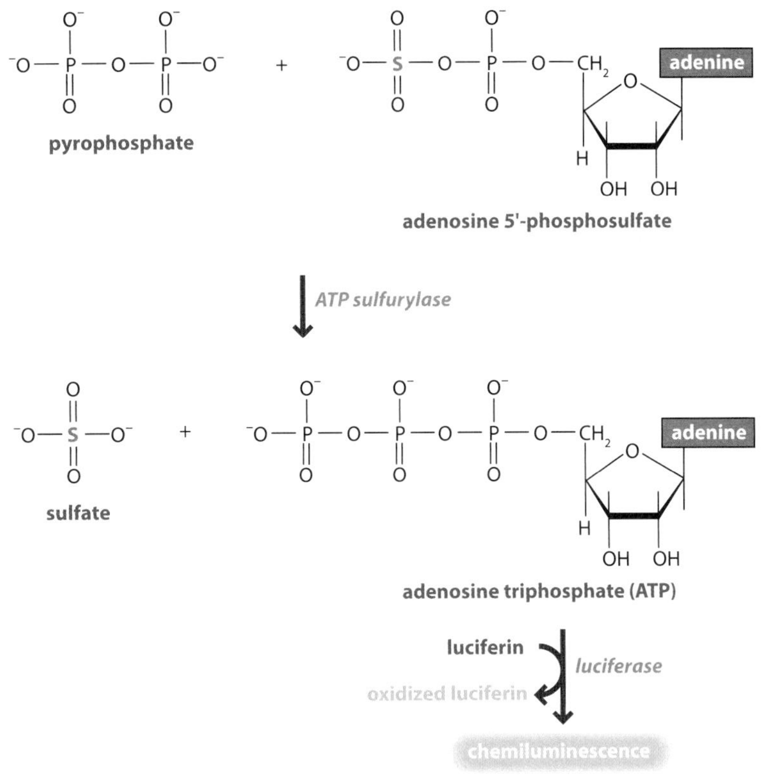

Unlike chain termination sequencing, the template is copied by the DNA polymerase in the normal manner without added ddNTPs. As the new strand is being synthesized, the order in which nucleotides are incorporated is detected. The sequence is therefore read as the reaction proceeds.

This direct sequence readout is achieved in the following way. As we know, the addition of a nucleotide to a growing strand of DNA is accompanied by release of a pyrophosphate molecule. During pyrosequencing, the pyrophosphate is combined with adenosine 5'-phosphosulfate to give ATP in a reaction catalyzed by **ATP sulfurylase** (*Fig. 19.16*). The ATP is then used by a second enzyme, **luciferase**, to oxidize **luciferin**. This is not a single compound, but a family of organic molecules, any one of which emits chemiluminescence when oxidized. As a result of this chain of reactions, every time a nucleotide is added to the growing polynucleotide, there is a flash of chemiluminescence.

How do the repeated chemiluminescent emissions help us work out the nucleotide sequence? The answer is to cycle the addition of nucleotides, A followed by C, then G, and finally T (*Fig. 19.17*). A **nucleotidase** enzyme is also present in the reaction mixture, which means that if a nucleotide is not incorporated into the polynucleotide then it is rapidly degraded before the next one is added. At each step, the particular addition that results in the chemiluminescent signal identifies the nucleotide present at that position in the template DNA. The technique sounds complicated, but it simply requires that a repetitive series of additions be made to the reaction mixture, an operation that is easily automated.

Pyrosequencing is only able to read up to 700 bp of sequence in a single experiment, less than is possible with the chain termination method. Its advantage lies with the ease of automation, which led to it being used in one of the first of the 'next generation'

Figure 19.17 **Pyrosequencing.**
In this example, the cyclical addition of the four nucleotides, with unincorporated ones degraded by nucleotidase, reveals that the sequence is GA.

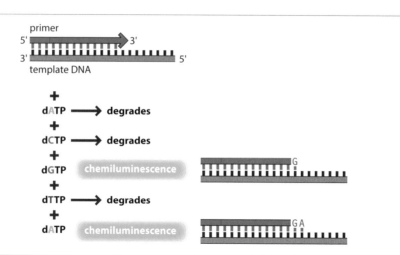

sequencing methods to be devised in the mid 2000s. It is to these new approaches that we now turn our attention.

19.2.2 Next generation sequencing

Until recently, none of the various sequencing methods were able to provide more than about 1000 bp of sequence from a single experiment. This means that over three million experiments would be needed to sequence an entire human genome. Because of this limitation, development of sequencing technology has focused on the design of automated systems in which many sequencing experiments can be carried out at the same time (known as **'massively parallel'** systems). The most successful of these automated systems is able to acquire billions of sequences in a single run, each run taking less than a day to complete.

One of the first of these massively parallel formats made use of pyrosequencing as the underlying sequencing methodology. The DNA to be sequenced is broken into fragments between 300 and 500 bp in length, and these fragments emulsified in an oil–water mixture, so that each droplet of water contains a different fragment (*Fig. 19.18*). The droplets are then placed in the sequencing array and parallel pyrosequencing experiments carried out, with the chemiluminescent signals from each one detected by a similar array of miniaturized detectors.

Another next generation method uses an approach similar to chain termination sequencing. The reaction mixture contains **terminator-dye nucleotides**, which have blocking groups attached to their 3' carbons, and also carry fluorescent labels (*Fig. 19.19*). Unlike chain termination sequencing, only these modified nucleotides are present, so strand synthesis is blocked at the very first nucleotide addition. However, both the blocking group and the fluorescent label are removable, so once the fluorescent signal has been identified, indicating which of the four nucleotides has been incorporated, the block and label are detached. Now the next nucleotide can be added to the strand, and the cycle of identification and deblocking repeated. This system is operated in a massively parallel format with multiple reactions immobilized on a slide.

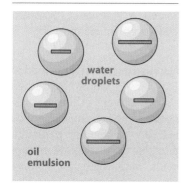

Figure 19.18 **An oil–water emulsion, used in one method of next generation sequencing. Each water droplet contains a single DNA fragment.**

In DNA sequencing jargon, these next generation methods are now being superseded by 'second' and 'third generation' systems. These include one approach that enables 30 000 or more nucleotides to be read as a single continuous sequence, substantially longer than that achievable by any of the other methods. This procedure is called **single molecule real-time sequencing**. A sophisticated optical device called a **zero-mode waveguide** is used to observe the copying of a single DNA template (*Fig. 19.20*).

Figure 19.19 Terminator-dye sequencing.
(A) The structure of a terminator-dye nucleotide. (B) Part of a terminator-dye sequencing experiment.

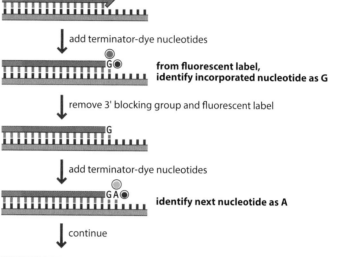

A. a terminator-dye nucleotide

fluorescent label

3' blocking group

B. terminator-dye sequencing

add terminator-dye nucleotides

from fluorescent label, identify incorporated nucleotide as G

remove 3' blocking group and fluorescent label

add terminator-dye nucleotides

identify next nucleotide as A

continue

Figure 19.20 Single molecule real-time DNA sequencing.
Each nucleotide addition is detected with a zero-mode waveguide.

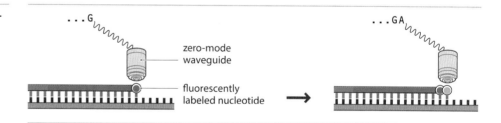

zero-mode waveguide

fluorescently labeled nucleotide

Each nucleotide addition is identified by virtue of its attached fluorescent label. Because the optical system is so precise, there is no need to block the 3' carbon of the nucleotide. The fluorescent signal is detected and the label removed immediately after nucleotide incorporation, so strand synthesis can progress without interruption. This means that the processivity of the polymerase becomes the main factor limiting the length of sequence that can be read.

19.3 DNA cloning

DNA cloning was devised in the 1970s when purified enzymes such as restriction endonucleases and DNA ligases first became available. The ability to cut up DNA molecules in controlled ways, and then join the fragments together in a different order, or to join fragments from completely different species, led to the development of **recombinant DNA technology**, popularly called **genetic engineering**. Among the applications of recombinant DNA technology is the transfer of genes for important pharmaceutical proteins from the human genome into microorganisms such as bacteria or yeast. The resulting **recombinant protein** can then be synthesized in large amounts, providing supplies of insulin, growth factors and other proteins needed to treat ailments such as diabetes and growth disorders.

DNA cloning has evolved into a diverse and complex technology, but the underlying principles remain quite straightforward. We will study those principles and then explore how the methods are used in recombinant protein synthesis.

19.3.1 Methods for DNA cloning

Recombinant DNA technology makes use of restriction endonucleases and a DNA ligase to construct recombinant molecules that are made up of pieces of DNA that are not contiguous in their natural state. In DNA cloning, the recombinant molecule is able to replicate in a host cell, so multiple copies are obtained.

Figure 19.21 An outline of gene cloning.

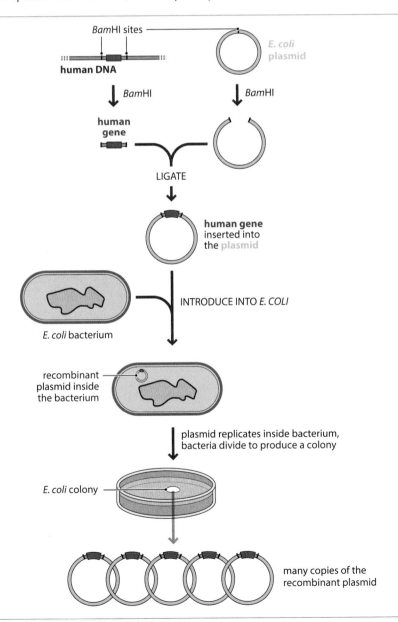

DNA cloning in outline

To illustrate how a DNA cloning experiment is carried out, we will run through a typical series of manipulations that result in construction of a recombinant molecule that is able to replicate inside *E. coli* cells (*Fig. 19.21*):

• First, imagine that a human gene is contained in a single fragment that we have obtained by treatment of human DNA with the restriction enzyme *Bam*HI. The fragment therefore has GATC overhangs.

• Imagine also that we have purified a **plasmid** from *E. coli*. A plasmid is a small circular DNA molecule that is able to replicate inside the bacterium.

• The plasmid contains a single *Bam*HI recognition sequence, and so when we cut it with this restriction enzyme the circle is converted into a linear DNA molecule, again with GATC overhangs.

• Now we mix the human DNA fragment with the linear plasmid and add DNA ligase. This will result in a variety of ligation products, some of which will be **recombinant plasmids**, molecules comprising the circularized plasmid with the human gene inserted into the *Bam*HI restriction site.

• Now we re-introduce the recombinant plasmid into an *E. coli* cell. Once inside the cell, the plasmid replicates until it reaches its natural **copy number**, which for most plasmids is 40–50 copies per cell.

• When the *E. coli* cell divides, its two daughters will each inherit some copies of the recombinant plasmid. Within each daughter cell, the inherited plasmids will replicate until their copy numbers are once again reached.

• More rounds of cell division and plasmid replication will result in a colony of recombinant *E. coli* bacteria, each bacterium containing multiple copies of the plasmid carrying the human gene. The gene has been cloned.

In the experiment that we have just run through, the plasmid acts as a **cloning vector**. Our next task is to understand the properties of cloning vectors in more detail.

Many vectors are based on E. coli plasmids

To understand how cloning vectors are used, we will examine one of the simplest *E. coli* plasmid vectors, called pUC8, which was first designed in the early 1980s and is still widely used today. The vector was constructed by ligating together restriction fragments obtained from three naturally occurring plasmids to give a 2.7 kb circular DNA molecule that carries two *E. coli* genes (*Fig. 19.22*):

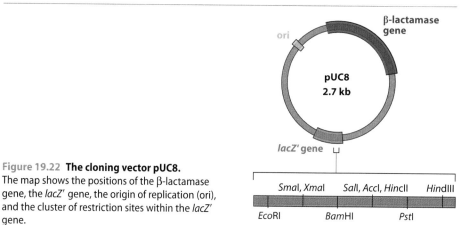

Figure 19.22 The cloning vector pUC8.
The map shows the positions of the β-lactamase gene, the *lacZ'* gene, the origin of replication (ori), and the cluster of restriction sites within the *lacZ'* gene.

- A gene coding for **β-lactamase**, an enzyme that enables *E. coli* to withstand the toxic effect of the antibiotic ampicillin. This gene acts as a **selectable marker** for pUC8: bacteria that contain the plasmid can be 'selected' by inclusion of ampicillin in the growth medium.

- The *lacZ'* gene, which codes for the first 90 amino acids of β-galactosidase, one of the enzymes involved in lactose metabolism. The segment of the enzyme specified by the *lacZ'* gene is called the α-peptide.

pUC8 has a single recognition sequence for *Bam*HI located alongside unique sites for other restriction endonucleases within the *lacZ'* gene. Insertion of new DNA into any of these sites splits the *lacZ'* gene into two segments, which means that the plasmid can no longer direct synthesis of the β-galactosidase α-peptide.

To use pUC8 in cloning we prepare a ligation mixture containing our human DNA fragment and the linear version of the vector. Remember that after ligation there will be a variety of DNA molecules, and we are only interested in one of these – the circular pUC8 vector with inserted human gene. How can we distinguish those *E. coli* cells that take up this particular product? The answer is as follows:

- A bacterium that does not take up any DNA will not acquire the β-lactamase gene, and so will be sensitive to ampicillin.

- Linear ligation products, or circular ones that contain no pUC8 DNA, will either not be taken up by a bacterium, or once inside will not be able to replicate and will be degraded. A bacterium that takes up any of these products will remain ampicillin sensitive.

- A bacterium that takes up a pUC8 molecule that has been circularized by the ligase, but without insertion of the human gene, will have functional β-lactamase and *lacZ'* genes. It will be ampicillin resistant and able to metabolize lactose.

- A bacterium that takes up a circular pUC8 molecule that carries the inserted human gene will have a functional β-lactamase gene but an inactivated *lacZ'* gene. It will be ampicillin resistant but unable to metabolize lactose.

To a microbiologist, distinguishing between these different scenarios is no problem at all. The bacteria are spread onto an agar medium that contains ampicillin and a lactose analog called X-gal (5-bromo-4-chloro-3-indolyl-β-D-galactopyranoside), which β-galactosidase converts into a blue product (*Fig. 19.23*). Only those bacteria

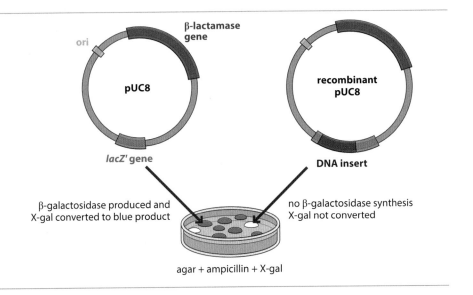

Figure 19.23 Identification of recombinant pUC8 plasmids.

ori

β-lactamase gene

pUC8

lacZ' gene

β-galactosidase produced and X-gal converted to blue product

recombinant pUC8

DNA insert

no β-galactosidase synthesis X-gal not converted

agar + ampicillin + X-gal

that contain pUC8 plasmids, and hence are ampicillin resistant, are able to grow on this medium. Those that contain pUC8 plasmids without the inserted human gene will convert X-gal to its blue product, and so will form blue colonies on the surface of the agar. Those that contain recombinant plasmids, ones which have the cloned human gene, will not be able to synthesize β-galactosidase, and will remain white. Using a sterile wire loop, we remove those white colonies from the agar. We have cloned our human gene.

Genes can also be cloned in eukaryotes

Bacteria are not the only types of organism into which we can introduce new genes by DNA cloning. Vectors have also been developed for the propagation of foreign DNA in most types of eukaryote.

Naturally occurring plasmids are uncommon in eukaryotes, but in those species in which they do occur they have been exploited as the basis to cloning systems. An example is the yeast *Saccharomyces cerevisiae*, some strains of which contain a small plasmid called the **2 μm circle**. Cloning vectors, called **yeast episomal plasmids**, have been developed from the 2 μm circle by adding genes that enable yeast cells carrying the vector to be identified by plating on a selective agar medium, using strategies similar to those that we have followed through for cloning in *E. coli* with pUC8.

The word 'episomal' in the name of this type of vector indicates that they replicate as independent circles of DNA, as is the case for pUC8 and most other bacterial plasmid vectors. This episomal mode of propagation has one disadvantage, especially if the natural copy number of the vector in the cell is quite low. It is possible that when a daughter yeast cell buds off from its parent, then simply by chance it will not contain any copies of the plasmid. This would mean that, over time, the number of cells containing the cloned gene would decline. A second type of vector used with *S. cerevisiae*, called a **yeast integrative plasmid**, is designed to avoid this instability problem by integrating into one of the yeast chromosomes. Once integrated, the cloned DNA becomes a permanent part of the yeast genome, and is very unlikely to be lost, even after many cell divisions.

Integration into the chromosomal DNA is also a feature of the cloning system used with plants. Cloning vectors for plants are derived from the **Ti plasmid**. This is actually a bacterial plasmid, from the soil dwelling *Agrobacterium tumefaciens*, which causes a disease called crown gall when it infects a plant stem (*Fig. 19.24*). During the infection, a part of the Ti plasmid becomes integrated into the plant's chromosomes.

Figure 19.24 Crown gall disease.
The disease is caused by *Agrobacterium tumefaciens*, which enters the plant through a wound near the base of the stem.

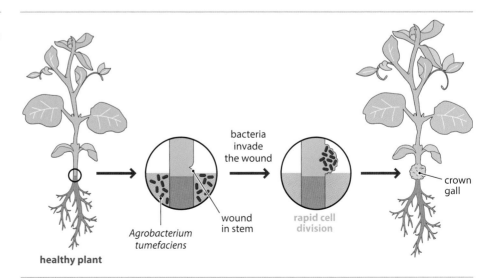

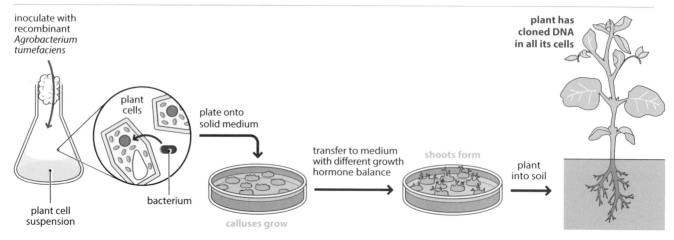

Figure 19.25 Cloning with the Ti plasmid.
A plant cell suspension is inoculated with recombinant *A. tumefaciens*. These are bacteria with Ti plasmids that contain an inserted gene that we wish to clone. The plant cells are plated onto an agar medium where they grow to produce callus – pieces of undifferentiated plant tissue. Re-plating onto medium with a different growth hormone balance induces formation of small shoots, which can then be planted out. Each resulting plant descends from a single cell in the original suspension, so each cell in the plant will contain the Ti plasmid plus inserted gene.

This segment carries a number of genes that are expressed inside the plant's cells and induce various physiological changes that are beneficial for the bacterium. Plant cloning vectors based on the Ti plasmid make use of this natural genetic engineering system. However, the natural infection process is not used because this would result in the cloned DNA only being transferred to the cells around the infection site on the plant stem. Instead, the vector is used with plant cells grown in culture. Those cells that take up the plasmid are then used to regenerate entire plants. In this way, plants that carry the cloned DNA in every cell can be obtained (*Fig. 19.25*).

Plasmids are very rare in animals, so modified virus genomes are employed as cloning vectors. With human cells, **adenoviruses** have been used. These do not integrate into the chromosomes, but they take up semi-permanent residence within the nucleus of an infected cell and are unlikely to be lost during cell division. **Adeno-associated viruses**, which despite their name are quite different from adenoviruses, are also used because these do insert their DNA into a chromosome. It is also possible to clone genes without a vector, and this is the approach that is frequently used with animal cells. DNA can be microinjected directly into a cell nucleus, and some of the DNA will be integrated into the genome (*Fig. 19.26A*). Alternatively, the DNA can be

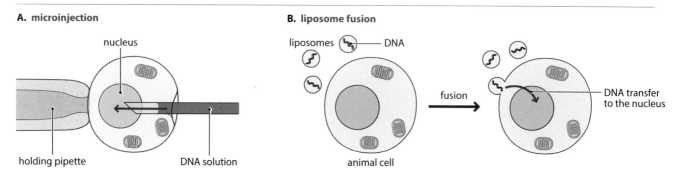

Figure 19.26 Two ways of introducing DNA into an animal cell.
(A) Microinjection of the DNA into the nucleus. (B) Fusion of liposomes containing DNA with the plasma membrane.

Box 19.4 **Integration of a yeast plasmid into a chromosome**

The process that integrates a yeast integrative plasmid into chromosomal DNA is called **homologous recombination**. This is an event that can occur when two DNA molecules have segments in which their nucleotide sequences are identical, or at least closely similar. Breakage of each DNA molecule in the region of similarity, followed by re-union of the cut strands, can lead to DNA being exchanged between the two molecules.

If one molecule is linear and the other circular, then the outcome of homologous recombination is integration of the circular molecule into the linear one. This is what happens when a circular yeast plasmid recombines with a linear DNA molecule from one of the yeast chromosomes.

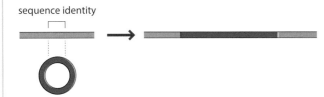

The region of similarity between the yeast integrative plasmid and the chromosomal DNA is provided by a yeast gene that has been inserted into the plasmid. With YIp5, a popular vector of this type, the yeast gene is *URA3*, which codes for orotidine-5′-

phosphate decarboxylase, the enzyme that converts orotidine monophosphate to uridine monophosphate during *de novo* synthesis of pyrimidine nucleotides (see *Section 13.2.2*). The yeast strain that is used has a mutated and hence inactive version of *URA3*, which means that the cells must be provided with uracil in order to survive. After homologous recombination, the chromosomal DNA has two copies of *URA3*, one inactive and one functional.

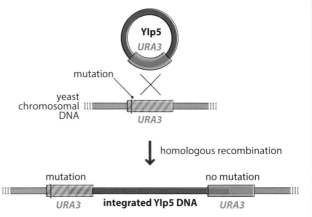

The plasmid copy of *URA3* therefore has a dual role. As well as enabling the recombination to take place, it also acts as a selectable marker for yeast cells that contain the plasmid. Selection is achieved by spreading the yeast cells onto an agar medium that lacks uracil. In the absence of uracil, only those yeast cells that contain the functional *URA3* gene provided by the plasmid are able to make pyrimidine nucleotides and so only these plasmid-containing cells can divide to produce colonies.

enclosed in membrane-bound vesicles called **liposomes**. These are then fused with the plasma membrane of the target cell (*Fig. 19.26B*). Usually, DNA that is transported into a cell via liposome fusion is not stable, though the genes it contains might be retained for a few days or weeks.

19.3.2 Using DNA cloning to obtain recombinant protein

One of the many applications of DNA cloning is in the production of recombinant protein, which is defined as a protein obtained by expression of a cloned gene. In the research laboratory, samples of human and other animal proteins for structural studies are often synthesized from genes cloned in *E. coli* or a eukaryotic microorganism such as yeast. Because these microbes can be grown at high densities in liquid culture vessels, larger amounts of protein can be obtained than is possible by direct purification from human or animal tissue. In industrial settings, huge culture systems, thousands of liters in volume, are used to produce recombinant versions of pharmaceutical proteins such as insulin for commercial purposes.

The production of recombinant protein requires some modifications to the DNA cloning procedure, and presents challenges in ensuring that the resulting protein has the same activity as the natural version. We conclude this chapter by considering these issues.

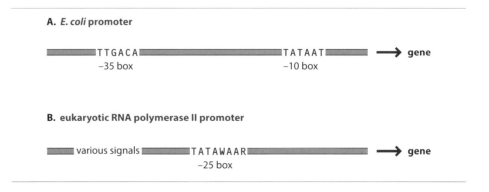

Figure 19.27 Comparison between the promoters for *E. coli* and eukaryotic protein-coding genes. Abbreviations: 'R' is either A or G, and 'W' is either A or T.

Recombinant protein production requires a special type of cloning vector

Simply inserting an animal gene in a cloning vector and transferring it to *E. coli* will not result in recombinant protein synthesis. This is because the promoter sequence upstream of an animal gene, which directs expression of that gene in its natural host, will not be recognized by the *E. coli* RNA polymerase. To illustrate the problem, compare the consensus sequences of the promoter regions of protein-coding genes in *E. coli* and eukaryotes (*Fig. 19.27*). There are similarities, but it is unlikely that an *E. coli* RNA polymerase would be able to attach to a eukaryotic promoter. Most animal genes are therefore inactive in *E. coli*.

To solve this problem, a special type of cloning vector is used. In one of these **expression vectors**, the restriction site into which the animal gene is inserted is located immediately downstream of an *E. coli* promoter sequence (*Fig. 19.28*). The animal gene has to be manipulated carefully, so that its own promoter is cut away, without removing any of the codons of the open reading frame, but this is not too difficult to achieve with the large range of nuclease enzymes available to the genetic engineer. If carried out correctly, the end result is the animal gene inserted into the vector at an appropriate position relative to the *E. coli* promoter.

The *E. coli* promoter has to be chosen with care. Although we wish to achieve as much protein production as possible, there are practical limits arising from the fact that a bacterium is a living organism. The protein might be harmful to the bacteria, in which case we should limit its synthesis so that the toxic levels are not reached. Even if the protein has no harmful effects, a continuous high level of transcription could interfere with plasmid replication, meaning that the cloned gene might not be inherited by all daughter cells. The overall ability of the culture to make the protein would therefore decline.

The ideal promoter is one that can direct a high rate of transcription, but which is also controllable so that the rate of transcription can be set at a lower level if necessary. The promoter of the lactose operon meets these criteria and is often used. It is a strong promoter, and so directs a high rate of transcription. It is also controllable. Remember that in the natural system, the lactose operon is usually switched off, because the repressor protein attaches to the operator sequence, blocking access to the promoter, so the RNA polymerase is unable to bind (see *Fig. 17.8*). If an inducer is added, then the repressor detaches, and transcription can occur. Transfer of the lactose promoter, and the adjacent operator sequence, to an expression vector does not affect this control system. In other words, expression of a cloned animal gene will be subject to exactly the same regulatory regime. There will be no expression until an inducer is added to the culture.

The natural inducer of the lactose promoter is allolactose. This compound is not very stable, and it would be necessary to continually add fresh allolactose in order

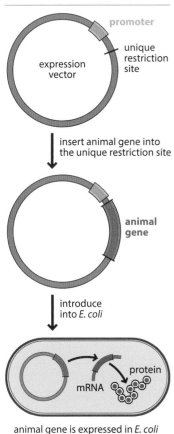

Figure 19.28 Using an expression vector to direct synthesis of an animal protein in *E. coli*.

Box 19.5 Synthesis of recombinant factor VIII protein

The challenges inherent in the production of a recombinant protein are illustrated by the work that has led to the synthesis of recombinant versions of human factor VIII. This protein plays a central role in blood clotting and is defective in the commonest form of hemophilia. Until recently, the only way to treat hemophilia was by injection of purified factor VIII protein, obtained from human blood provided by donors. Purification of factor VIII is a complex procedure and the treatment is expensive. More critically, factor VIII obtained in this way presents its own dangers if the purification method does not rigorously remove virus particles that may be present in the blood. Hepatitis and human immunodeficiency viruses have been passed on to hemophiliacs via factor VIII injections. The two great benefits of a recombinant version of factor VIII, synthesized in a bacterial or eukaryotic host, would therefore be a lower production cost and security from virus contamination.

The factor VIII gene is very large at over 186 kb, and like many human genes is discontinuous, with 26 exons and 25 introns. The presence of introns is a problem because *E. coli* genes are not discontinuous and the bacterium lacks the RNAs and proteins needed for removing introns from pre-mRNA transcripts. This problem can, however, be solved by using reverse transcriptase to prepare a complementary DNA (cDNA) copy of the factor VIII mRNA. The mRNA, after splicing, does not contain introns, so the cDNA is a contiguous series of codons similar in structure to a bacterial gene.

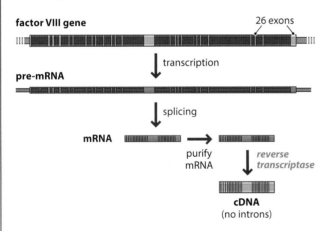

This approach has been used to provide 'bacteria-ready' copies of a number of intron-containing eukaryotic genes and often results in successful synthesis of a recombinant protein. But success is achieved only if the protein does not require extensive processing in order to become active, and in particular does not have to be glycosylated. In humans, the initial factor VIII

translation product is cleaved into two segments to give a dimeric protein, which then has *N*-linked glycans attached at six positions. These processing events do not occur in *E. coli* and so no active protein is produced from the factor VIII cDNA. An alternative host has to be found.

The ideal alternative is a mammalian cell culture, because mammalian cells, even non-human ones, would be expected to process a human protein correctly. Large scale culture systems for animal cells are now available, and although the rates of growth and maximum cell densities are much lower than for bacteria or yeast, limiting the amount of recombinant protein that can be produced, these low yields can be tolerated if mammalian culture is the only way of obtaining the protein in an active form. In the first experiments, hamster cells were used, but only small amounts of factor VIII were synthesized, probably because not all of the initial translation product was processed correctly. In a second attempt, the factor VIII cDNA was divided into two segments, one segment coding for the large subunit polypeptide, and the second for the small subunit. The cDNAs were ligated into expression vectors, downstream of the Ag promoter (a hybrid between the promoters for the chicken β-actin and rabbit β-globin genes) and upstream of a polyadenylation signal from SV40 virus.

After cloning in the hamster cell line, these new constructs directed synthesis of ten times more factor VIII protein than had been obtained in the first experiments. Importantly, the protein was indistinguishable in terms of function from the native form.

More recently, a new approach to factor VIII synthesis has been introduced. This is called **pharming** and involves production of the protein in a farm animal. Pharming is carried out by introducing a gene into a fertilized egg cell, for example by microinjection, which is then implanted into a foster mother. The egg cell divides and the resulting embryo develops into an animal that contains the cloned gene in every cell of its body. To obtain factor VIII by pharming, the complete human cDNA was ligated to the promoter for the whey acidic protein gene of pig. This is a gene that is only active in mammary tissue, the whey acidic protein being a major component of pig milk. Human factor VIII is therefore synthesized by the pig mammary tissue and can be purified from the milk produced by the animal. Factor VIII made in this way appears to be exactly the same as the native human protein and is fully functional in blood clotting assays.

to prevent the cloned gene from being switched off. Instead an artificial inducer such as isopropyl-β-D-thiogalactoside (IPTG) is used (*Fig. 19.29*). IPTG is a β-galactoside, and although its structure is different to that of allolactose it can bind to the repressor. It is much more stable than allolactose and so does not need continual replenishment in order to maintain expression of the animal gene.

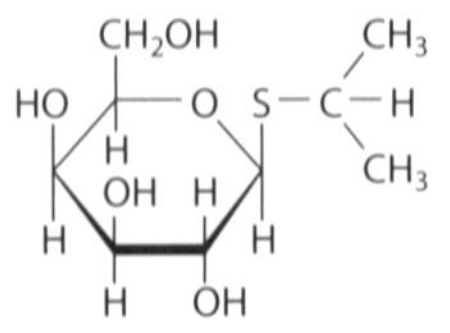

Figure 19.29 Isopropyl-β-D-thiogalactoside.
Compare this structure with that of allolactose, shown in *Box 17.2*.

Bacteria are not always the best hosts for recombinant protein production

The differences between eukaryotic and bacterial promoter sequences are not the only issues that have to be addressed when attempting to use bacteria as the hosts for recombinant protein synthesis. Most animal proteins are larger than bacterial ones, with more sophisticated tertiary structures. Many of these proteins are not folded correctly in *E. coli*, and accumulate as partially folded structures, possibly forming a semi-solid aggregate, called an **inclusion body**, within the bacterium. Inclusion bodies can be recovered from a bacterial extract, and the proteins solubilized, but converting the proteins into their correctly folded forms is usually impossible. A second problem is that bacteria lack the ability to carry out some of the post-translational chemical modifications displayed by animal proteins. In particular, glycosylation is extremely uncommon in bacteria and recombinant proteins synthesized in *E. coli* are never glycosylated correctly. Absence of glycosylation might not impair the function of the protein, but might reduce its stability, and could possibly result in an allergic reaction if the protein is used as a pharmaceutical and injected in a patient's bloodstream.

For these reasons, ways of producing recombinant protein in eukaryotic hosts have been explored. Microbial eukaryotes, such as yeast and filamentous fungi, are attractive alternatives, because they can be grown in culture just like bacteria, although the cell densities that are achievable are lower. Expression vectors are still required because the promoters and other expression signals for animal genes do not in general work efficiently in these lower eukaryotes. *Saccharomyces cerevisiae* is often used, partly because of the well-developed cloning systems based on episomal and integrative plasmids, and also because this yeast is accepted as a safe organism for production of proteins for use in medicines or in foods. The *GAL* promoter, from the gene coding for galactose epimerase, which can be controlled by the level of galactose in the medium, is often used for protein expression.

Although *S. cerevisiae* gives relatively high yields of recombinant animal protein, and usually the proteins are folded correctly, cloning in this yeast does not completely solve the glycosylation problem. A recombinant protein often becomes hyperglycosylated, with the glycans containing more sugar units than are present in the natural animal versions (*Fig. 19.30*). A second species of yeast, *Pichia pastoris*, carries out glycosylation more correctly. The resulting glycans are not identical to those on the natural animal protein, but they are sufficiently similar that the protein does not elicit an allergic reaction. *Pichia pastoris* can synthesize large amounts of recombinant protein, up to 30% of the total cell protein, and most of this protein is secreted into the growth medium. Purification of the protein from the growth medium is much easier, and cheaper, than purification from cell extracts. With this species,

Figure 19.30 *N*-linked glycans in humans and *S. cerevisiae*.
Structures of typical 'high-mannose' glycans are shown. The structure on the right is a hyperglycosylated glycan, sometimes produced by *S. cerevisiae*, which can contain hundreds of mannose units.

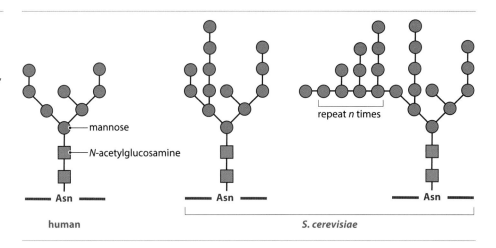

the alcohol oxidase promoter, which is induced by methanol, is often used to drive expression of the cloned gene.

Despite the advances in using eukaryotic microorganisms as hosts for protein production, there are still some proteins, typically ones with complex and essential glycosylation structures, which can only be made in animal cells. Mammalian cell lines derived from humans or hamsters are often used, which means that there are few problems with promoter structure and protein processing. The expression vector is needed only to maximize yields and enable protein synthesis to be regulated. Although this is the most reliable approach for synthesis of active proteins, it is also the most expensive, because the culture systems are more complex than for microorganisms, and protein yields are lower. Quality control measures are also more rigorous, because care must be taken to ensure that cultures do not become contaminated with viruses that might be carried over to the final protein preparation.

Further reading

Broach JR (1982) The yeast 2 μm circle. *Cell* **28**, 203–4.

Çelik E and Çelik P (2012) Production of recombinant proteins by yeast cells. *Biotechnology Advances* **30**, 1108–18.

Chilton MD (1983) A vector for introducing new genes into plants. *Scientific American* 248(June), 50–9. *The Ti plasmid.*

Colosimo A, Goncz KK, Holmes AR, et al. (2000) Transfer and expression of foreign genes in mammalian cells. *Biotechniques* **29**, 314–21.

Crystal RG (2014) Adenovirus: the first effective *in vivo* gene delivery vector. *Human Gene Therapy* **25**, 3–11.

Higuchi R, Dollinger G, Walsh PS and Griffith R (1992) Simultaneous amplification and detection of specific DNA sequences. *Biotechnology* **10**, 413–17. *The first description of real-time PCR.*

Huang C-J, Lin H and Yang X (2012) Industrial production of recombinant therapeutics in *Escherichia coli* and its recent advancements. *Journal of Industrial Microbiology and Biotechnology* **39**, 383–99.

Kaufman RJ, Wasley LC and Dorner AJ (1988) Synthesis, processing, and secretion of recombinant human factor VIII expressed in mammalian cells. *Journal of Biological Chemistry* **263**, 6352–62.

Lee L-Y and Gelvin SB (2008) T-DNA binary vectors and systems. *Plant Physiology* **146**, 325–32.

Päcurar DI, Thordal-Christensen H, Päcurer ML, Pamfil D, Botez C and Bellini C (2011) *Agrobacterium tumefaciens*: from crown gall tumors to genetic transformation. *Physiological and Molecular Plant Pathology* **76**, 76–81.

Paleyanda RK, Velander WH, Lee TK, et al. (1997) Transgenic pigs produce functional human factor VIII in milk. *Nature Biotechnology* **15**, 971–5.

Parent SA, Fenimore CM and Bostian KA (1985) Vector systems for the expression, analysis and cloning of DNA sequences in *S. cerevisiae*. *Yeast* **1**, 83–138.

Pingoud A, Fuxreiter M, Pingoud V and Wende W (2005) Type II restriction endonucleases: structure and mechanism. *Cellular and Molecular Life Sciences* **62**, 685–707.

Ronaghi M, Uhlén M and Nyrén P (1998) A sequencing method based on real-time pyrophosphate. *Science* **281**, 363–5. *Pyrosequencing.*

Saiki RK, Gelfand DH, Stoffel S, et al. (1988) Primer-directed enzymatic amplification of DNA with a thermostable DNA polymerase. *Science* **239**, 487–91. *The first description of PCR with Taq polymerase.*

Sanger F, Nicklen S and Coulson AR (1977) DNA sequencing with chain-terminating inhibitors. *Proceedings of the National Academy of Sciences of the USA* **74**, 5463–7.

Smith HO and Wilcox KW (1970) A restriction enzyme from *Haemophilus influenzae*. *Journal of Molecular Biology* **51**, 379–91. *One of the first full descriptions of a restriction endonuclease.*

van Dijk EL, Auger H, Jaszczyszyn Y and Thermes C (2014) Ten years of next-generation sequencing technology. *Trends in Genetics* **30**, 418–26.

Zhu J (2012) Mammalian cell protein expression for biopharmaceutical production. *Biotechnology Advances* **30**, 1158–70.

Self-assessment questions

Multiple choice questions

Only one answer is correct for each question. Answers can be found on the website: www.scionpublishing.com/biochemistry

1. What is the activity of S1 endonuclease?
   - **(a)** Cuts single- and double-stranded DNA polynucleotides
   - **(b)** Cuts single- and double-stranded DNA and RNA polynucleotides
   - **(c)** Cuts only single-stranded DNA polynucleotides
   - **(d)** Cuts only double-stranded DNA polynucleotides

2. What is the activity of RNase V1?
   - **(a)** Cuts single- and double-stranded DNA polynucleotides
   - **(b)** Cuts single- and double-stranded DNA and RNA polynucleotides
   - **(c)** Cuts only double-stranded RNA polynucleotides
   - **(d)** Cuts only single-stranded RNA polynucleotides

3. Which restriction endonuclease leaves a 3′ overhang?
   - **(a)** *Hin*fI
   - **(b)** *Pvu*II
   - **(c)** *Bam*HI
   - **(d)** *Pst*I

4. Which restriction endonuclease leaves a blunt end?
   - **(a)** *Hin*fI
   - **(b)** *Pvu*II
   - **(c)** *Bam*HI
   - **(d)** *Pst*I

5. Which restriction endonuclease has a degenerate recognition sequence?
   - **(a)** *Hin*fI
   - **(b)** *Pvu*II
   - **(c)** *Bam*HI
   - **(d)** *Pst*I

6. How is the frequency of blunt end ligation by DNA ligase increased?
   - **(a)** By adding more ATP
   - **(b)** By increasing the temperature
   - **(c)** By increasing the DNA concentration
   - **(d)** By adding a precipitant

7. Complementary DNA is synthesized by which DNA polymerase?
   - **(a)** DNA polymerase I
   - **(b)** *Taq* polymerase
   - **(c)** cDNA synthase
   - **(d)** Reverse transcriptase

8. In a PCR, what delimits the region of the target DNA that will be amplified?
   - **(a)** The length of time allowed for DNA synthesis
   - **(b)** The ratio of short and long products
   - **(c)** The identity of the reporter probe
   - **(d)** The annealing positions of the primers

9. Which of the following statements is **incorrect** with regard to a reporter probe used in real-time PCR?
   - **(a)** A fluorescent chemical group is attached at one end of the probe
   - **(b)** The sequence is designed so that the probe will base pair with one of the strands of the PCR product
   - **(c)** When the oligonucleotide is free in solution, it emits fluorescence
   - **(d)** The amount of fluorescence increases as the PCR proceeds

10. What are the chain terminating nucleotides used in DNA sequencing?
    - **(a)** 2′,3′-dideoxynucleotides
    - **(b)** 3′,4′-dideoxynucleotides
    - **(c)** 2′,4′-dideoxynucleotides
    - **(d)** 2′,5′-dideoxynucleotides

11. How many nucleotides can be read in a single chain termination sequencing experiment?
    - **(a)** Up to 100
    - **(b)** Up to 1000
    - **(c)** At least 1000
    - **(d)** Over 5000

12. Which of the following enzymes is **not** used in pyrosequencing?
    - **(a)** ATP sulfurylase
    - **(b)** DNA ligase
    - **(c)** Luciferase
    - **(d)** Nucleotidase

13. Which of the following is **not** used in a next generation sequencing method?
    (a) Terminator-dye nucleotides
    (b) An oil–water emulsion
    (c) A massively parallel array
    (d) Capillary gel electrophoresis

14. The gene for which enzyme, present in the pUC8 plasmid, specifies resistance to ampicillin?
    (a) β-galactosidase
    (b) β-lactamase
    (c) Orotidine-5′-phosphate decarboxylase
    (d) The *lacZ′* gene

15. When pUC8 is used, how are bacteria that contain recombinant plasmids recognized?
    (a) Ampicillin sensitive, able to metabolize lactose
    (b) Ampicillin resistant, able to metabolize lactose
    (c) Ampicillin sensitive, unable to metabolize lactose
    (d) Ampicillin resistant, unable to metabolize lactose

16. What is the process by which a yeast vector integrates into chromosomal DNA called?
    (a) Episome transfer
    (b) Homologous recombination
    (c) Nonhomologous recombination
    (d) Transformation

17. The Ti plasmid is used to clone genes into which type of organism?
    (a) Yeast
    (b) Plants
    (c) Insects
    (d) Animal cells

18. What is a vector used for recombinant protein production called?
    (a) Integrative plasmid
    (b) Ti vector
    (c) Expression vector
    (d) Episome

19. Why is a high level of transcription undesirable when a recombinant protein is being made in *E. coli*?
    (a) The substrates are used up too rapidly

(b) The resulting protein will not be registered as safe for use with humans
(c) The high level of transcription could interfere with plasmid replication
(d) The bacteria will become mutated

20. What is the inducer of the lactose operon, used in recombinant protein production?
    (a) Allolactose
    (b) Lactose
    (c) 5-bromo-4-chloro-3-indolyl-β-D-galactopyranoside
    (d) Isopropyl-β-D-thiogalactoside

21. What are the semi-solid aggregates of partially folded recombinant protein that accumulate in *E. coli* cells called?
    (a) Inclusion bodies
    (b) Speckles
    (c) Protein crystals
    (d) Accumulation spots

22. Why are cDNAs used when animal genes cloned in *E. coli* are being used for recombinant protein production?
    (a) They are shorter so can be transcribed more quickly
    (b) They do not inhibit glycosylation
    (c) Their expression can be controlled by adding an inducer to the medium
    (d) They lack introns

23. When *Pichia pastoris* is used for recombinant protein production, which chemical is used to induce the alcohol oxidase promoter?
    (a) Methanol
    (b) Ethanol
    (c) Butanol
    (d) Acetone

24. The promoter from which gene has been used to drive factor VIII synthesis in the mammary tissue of pigs?
    (a) Chicken β-actin
    (b) SV40 virus
    (c) Galactose epimerase
    (d) Whey acidic protein

Short answer questions

These questions do not require additional reading.

1. Draw a series of diagrams showing the different ways in which nucleases can cut DNA and RNA molecules.

2. Giving examples, describe the important features of restriction endonucleases.

3. Outline the roles of the different types of DNA polymerase that are used in studies of DNA and RNA molecules.

4. Draw a series of diagrams that show the events occurring during PCR, making careful distinction between synthesis of the long and short products. What modifications are made to this procedure in order to carry out real-time PCR?

5. Describe the chain termination method for DNA sequencing.

6. Explain how pyrosequencing and the terminator-dye method are used in next generation sequencing.

7. Draw a diagram that outlines how gene cloning is performed.

8. Give a detailed description of the way in which pUC8 is used to clone an animal gene in *E. coli*.

9. Outline the key features of cloning vectors used with (A) *S. cerevisiae*, (B) plants, and (C) animal cells.

10. Why is *E. coli* not always an ideal host for recombinant protein production and what alternatives are there?

Self-study questions

These questions will require calculation, additional reading and/or internet research.

1. A linear DNA molecule 48.5 kb in length is treated with three restriction endonucleases, singly and in combination. The numbers and lengths of the resulting fragments are as follows:

Enzyme	Number of fragments	Sizes (kb)
*Xba*I	2	24.0, 24.5
*Xho*I	2	15.0, 33.5
*Kpn*I	3	1.5, 17.0, 30.0
*Xba*I + *Xho*I	3	9.0, 15.0, 24.5
*Xba*I + *Kpn*I	4	1.5, 5.0, 17.0, 24.0

From the information provided, work out the positions of the cut sites for these three endonucleases in the DNA molecule. Are there any cut sites that cannot be positioned unambiguously, and if so, what additional information do you need to complete the 'restriction map'?

2. Calculate the numbers of short and long products that would be present after 20, 25 and 30 cycles of a PCR.

3. A 155 bp DNA molecule is randomly broken into overlapping fragments and those fragments sequenced. The resulting sequence 'reads' are as follows:

```
CATGCGCCGATCGAGCGAGC
GCGAGCATCTACTACGTACGTA
CATCGATGCTACTACGTACAGGC
GATGCTACGATGCTGCATGCG
GTACAGGCATGCGCCGATCGAG
CGAGCACTACGATCGATCATCG
TACGTACGTAGCATGCATCGT
CATGCGCCGATCGAGCGAG
ATCGATGCATCGATGCTAC
TACGTAGCATCTACGTACGTAG
```

Is it possible to reconstruct the sequence of the original molecule by searching for overlaps between pairs of reads? If not, then what problem has arisen and how might this problem be solved? [Note that although in this example the starting molecule and the reads are very short, the exercise reproduces a problem that arises when long (>100 kb) segments of eukaryotic DNA are sequenced with reads up to 1000 bp in length.]

4. When DNA is cloned in pUC8, recombinant bacteria (those containing a circular pUC8 molecule that carries an inserted DNA fragment) are identified by plating onto an agar medium containing ampicillin and the lactose analog called X-gal. An older type of cloning vector, called pBR322, also had the gene for ampicillin resistance but did not carry the *lacZ'* gene. Instead, DNA was inserted into a gene for tetracycline resistance present in pBR322. Describe the procedure that would be needed to distinguish bacteria that had taken up a recombinant pBR322 plasmid from those that had taken up a plasmid that had circularized without insertion of new DNA.

5. Discuss the ethical issues raised by the development of pharming as a means of obtaining recombinant proteins.

Glossary

1,2-diacylglycerol (DAG): a component of the second messenger signaling pathway initiated by flow of Ca^+ into a cell.

2 μm circle: a plasmid found in the yeast *Saccharomyces cerevisiae* and used as the basis for a series of cloning vectors.

2,4-dinitrophenol: an uncoupler of the electron transport chain.

3′ or 3′–OH terminus: the end of a polynucleotide that terminates with a hydroxyl group attached to the 3-carbon of the sugar.

3-hydroxy-3-methylglutaryl CoA or HMG CoA: an intermediate in the synthesis of sterol compounds.

3-hydroxyacyl-ACP dehydratase: the enzyme that converts D-3-hydroxybutyryl ACP to crotonyl ACP at step 3 of the fatty acid synthesis pathway.

3′ splice site: the splice site at the 3′ end of an intron.

3′-untranslated region: the untranslated region of an mRNA downstream of the termination codon.

3′,5′-cyclic AMP (cAMP): a modified version of AMP in which an intramolecular phosphodiester bond links the 5′ and 3′ carbons.

5′ or 5′–P terminus: the end of a polynucleotide that terminates with a mono-, di- or triphosphate attached to the 5′-carbon of the sugar.

5′ splice site: the splice site at the 5′ end of an intron.

5′-untranslated region: the untranslated region of an mRNA upstream of the initiation codon.

6-phosphogluconate dehydrogenase: the enzyme that converts 6-phosphogluconate to ribulose 5-phosphate at step 3 of the pentose phosphate pathway.

(6–4) lesion: a dimer between two adjacent pyrimidine bases in a polynucleotide, formed by ultraviolet irradiation.

(6–4) photoproduct photolyase: an enzyme involved in photoreactivation repair.

30 nm fiber: a relatively unpacked form of chromatin consisting of a possibly helical array of nucleosomes in a fiber approximately 30 nm in diameter.

A-form: a structural conformation of the double helix, present but not common in cellular DNA.

Acceptor control: the regulation of the electron transport chain by ADP availability.

Acceptor site: the splice site at the 3′ end of an intron.

Accessory pigment: one of the non-chlorophyll light-harvesting pigments of photosynthetic organisms.

Acetoacetyl CoA: an intermediate in the synthesis of sterol compounds and ketone bodies.

Acetyl CoA: the immediate product of pyruvate breakdown, which acts as the substrate for the TCA cycle.

Acetyl CoA carboxylase: the enzyme that converts acetyl CoA to malonyl CoA at the start of fatty acid synthesis.

Acetyl transferase: the enzyme that transfers acetyl units from acetyl CoA to ACP during fatty acid synthesis.

Acetylcholinesterase: an enzyme present in nerve cells that degrades acetylcholine.

Acid: a compound that releases additional H^+ ions into a water solution and therefore increases the hydronium ion concentration of a solution.

Acidic domain: a type of activation domain.

Aconitase: the enzyme that converts citrate to isocitrate at step 2 of the TCA cycle.

Actinomycetes: a group of filamentous bacteria.

Actinorhizal plants: a group of nitrogen-fixing plants.

Action potential: the wave of depolarization that moves along an axon and results in transmission of a nerve impulse.

Activated carrier molecule: a molecule that acts as a temporary store of free energy.

Activation domain: the part of an activator that makes contact with the initiation complex.

Activation energy or $\Delta G‡$: the difference between the free energy content of the substrates for a reaction and the transition state.

Active site: the position within an enzyme where the substrates bind and the biochemical reaction takes place.

Active transport: movement of a molecule or ion across a membrane by a process that requires energy.

Acyl carrier protein (ACP): the small protein on which fatty acids are synthesized.

Acyl CoA dehydrogenase: the enzyme that converts an acyl CoA to a *trans*-Δ^2-enoyl CoA at step 1 of the fatty acid breakdown pathway, and also converts Δ^4-dienoyl CoA to $\Delta^{2,4}$-dienoyl CoA during breakdown of some unsaturated fatty acids.

Acyl-malonyl-ACP condensing enzyme: the enzyme that converts acetyl ACP and malonyl ACP to acetoacetyl ACP at step 1 of the fatty acid synthesis pathway.

Adenine: one of the purine bases found in DNA and RNA.

Adeno-associated virus: a virus that is unrelated to adenovirus but which is often found in the same infected tissues, because adeno-associated virus makes use of some of the proteins synthesized by adenovirus in order to complete its replication cycle.

Adenosine 5′-triphosphate (ATP): a nucleotide: (1) one of the substrates for synthesis of RNA; (2) an activated carrier molecule.

Adenovirus: an animal virus, derivatives of which have been used to clone genes in mammalian cells.

Adenylate kinase: an enzyme that converts two molecules of ADP to one of ATP and one of AMP.

Adipocyte: fat storage cells found in adipose tissue.

ADP–glucose: an activated form of glucose.

Adrenaline: another name for epinephrine.

Adrenocorticotropic hormone: a hormone synthesized by the anterior pituitary gland that controls various physiological and biochemical activities.

Adrenoleukodystrophy (ALD): a genetic disorder caused by the inability to transport long chain fatty acids into peroxisomes in order to be broken down.

Agarose: a polysaccharide of repeating D-galactose and 3,6-anhydro-L-galactopyranose units, which forms a gel after heating in water.

ALA dehydratase: the enzyme that converts two δ-aminolevulinate molecules to one of porphobilinogen during tetrapyrrole synthesis.

ALA synthase: the enzyme that converts glycine and succinyl CoA to δ-aminolevulinate during tetrapyrrole synthesis.

Alcohol decarboxylase: the enzyme that converts acetaldehyde to ethanol at step 2 of alcoholic fermentation.

Alcoholic fermentation: a biochemical pathway for regenerating NADH under anaerobic conditions, involving the conversion of pyruvate to ethanol.

Aldehyde: an organic compound whose functional group has the structure –CHO.

Aldohexose: an aldose sugar with six carbon atoms.

Aldolase: the enzyme that converts fructose 1,6-bisphosphate to glyceraldehyde 3-phosphate and dihydroxyacetone phosphate at step 4 of the glycolysis pathway.

Aldopentose: an aldose sugar with five carbon atoms.

Aldose: a sugar in which the terminal carbon forms part of a formyl group.

Aldosterone: a steroid hormone involved in control of plasma ion content and blood pressure.

Aldotetrose: an aldose sugar with four carbon atoms.

Aldotriose: an aldose sugar with three carbon atoms.

ALDP: a membrane protein for transport of long chain fatty acids into peroxisomes prior to their breakdown.

Allolactose: an inducer of the lactose operon.

Allosteric enzyme: any enzyme whose activity is influenced by an allosteric effector, which binds to the enzyme at a position separate from the active site.

Allosteric inhibition: non-competitive reversible inhibition.

Allosteric site: the binding site on an enzyme for a non-competitive reversible inhibitor.

α(1→6) glucosidase: the enzyme activity that breaks α(1→6) glycosidic bonds in a branched polysaccharide such as glycogen.

αα motif: a protein motif in which two α-helices lie side by side in antiparallel directions in such a way that their side-chains intermesh.

α-amylase: an enzyme present in saliva which catalyzes the endohydrolysis of α(1→4) bonds in polysaccharides containing three or more glucose units.

α-helix: a common type of protein secondary structure.

α-ketoglutarate dehydrogenase: the enzyme that converts α-ketoglutarate to succinyl CoA at step 4 of the TCA cycle.

Alternative promoter: one of two or more different promoters acting on the same gene.

Alternative splicing: the production of two or more mRNAs from a single pre-mRNA by joining together different combinations of exons.

Amino acid: one of the monomers in a polypeptide.

Amino terminus: the end of a polypeptide that has a free amino group.

Aminoacyl or **A site:** the site in the ribosome occupied by the aminoacyl-tRNA during translation.

Aminoacyl tRNA-synthetase: an enzyme that catalyzes the aminoacylation of one or more tRNAs.

Aminoacylation: attachment of an amino acid to the acceptor arm of a tRNA.

Ammonotelic: species that excrete ammonia into the water in which they live.

AMP-activated protein kinase: an enzyme that phosphorylates and hence inactivates acetyl CoA carboxylase as part of the regulatory process controlling fatty acid synthesis.

Amphipathic helix: a type of α-helix that lies on the surface of a protein.

Amphiphile: a compound with both hydrophilic and hydrophobic properties.

Amphoteric: a compound that acts as both a weak acid and a weak base.

Amylase: the first enzyme activity to be characterized; amylase catalyzes the breakdown of starch to sugar.

Amylose: a component of starch; a linear polymer of D-glucose units linked by α(1→4) glycosidic bonds.

Amylopectin: a component of starch; a branched structure made up of α(1→4) chains and α(1→6) branch points.

Amyloplasts: structures related to chloroplasts that are the sites of stored starch synthesis in plants.

Amylose: a component of starch; a linear polymer of D-glucose units linked by α(1→4) glycosidic bonds.

Anabolism: those biochemical reactions that build up larger molecules from smaller ones.

Ancient DNA: preserved DNA from an archaeological or fossil specimen.

Anomers: two or more compounds with slightly different optical properties but otherwise chemically identical.

Anoxygenic photosynthesis: bacterial photosynthesis that does not use water as the electron donor and hence does not make oxygen.

Antenna complex: the part of a photosystem that captures light energy and channels it to the reaction center.

Antibiotic: a compound that kills or inhibits the growth of bacteria.

Antibody: an immunoglobulin protein that binds to an antigen; as part of the immune response, antibody binding leads to the destruction of the antigen by other components of the immune system.

Anticodon: the triplet of nucleotides, at positions 34–36 in a tRNA molecule, that base-pairs with a codon in an mRNA molecule.

Antigen: any substance that elicits an immune response.

Antiparallel: refers to the arrangement of polynucleotides in the double helix, these running in opposite directions.

Antiparallel β-sheet: a β-sheet in which adjacent strands run in opposite directions.

Antiporter: an active transport protein which couples transport of a molecule or ion against a concentration gradient with the movement, in the opposite direction, of a second ion down a gradient.

AP endonuclease: an enzyme involved in base excision repair.

AP site: a position in a DNA molecule where the base component of the nucleotide is missing.

apoC-II: one of the proteins on the surface of a chylomicron; the activator of lipoprotein lipase.

Apoenzyme: an inactive enzyme, which can be activated by addition of a cofactor.

Apoproteins or **Apolipoproteins:** the protein component of a lipoprotein.

Archaea: one of the two main groups of prokaryotes.

Arginase: the enzyme that converts arginine into ornithine and urea at step 5 of the urea cycle.

Argininosuccinase: the enzyme that converts argininosuccinate into arginine and fumarate at step 4 of the urea cycle.

Argininosuccinate synthetase: the enzyme that converts citrulline and aspartate into argininosuccinate at step 3 of the urea cycle.

Argonaute: an endonuclease that cleaves mRNAs in an RNA-induced silencing complex.

Ascorbic acid: vitamin C, a cofactor for several enzymes including ones involved in collagen synthesis.

Asparaginase: the enzyme that breaks down asparagine by converting it to aspartic acid and ammonia.

Aspartate transaminase: a component of the malate–aspartate shuttle.

Aspartate transcarbamoylase: the enzyme that converts aspartate and carbamoyl phosphate into a linear intermediate that gives rise to orotate during *de novo* synthesis of cytosine and uracil.

Aspartate–argininosuccinate shunt: the link between the TCA and urea cycles.

AT rich: a DNA sequence within which there is a high proportion of adenine–thymine base pairs.

Atherosclerosis: a medical condition commonly known as 'hardening of arteries', thought to be promoted by the deposition of cholesterol and other lipids from low density lipoproteins onto the inner surfaces of blood vessels.

Atomic number: the number of protons in the nucleus of an element.

ATP sulfurylase: an enzyme that adds a pyrophosphate to adenosine 5′-phosphosulfate to give ATP.

ATP-binding cassette (ABC) transporters: a group of P-type pumps which transport a variety of small molecules across membranes.

Attenuation: a process used by some bacteria to regulate expression of an amino acid biosynthetic operon in accordance with the levels of the amino acid in the cell.

Autoinducer: a signaling compound involved in bacterial quorum sensing.

Autotrophs: those organisms able to use light or chemical energy to convert inorganic compounds into energy-containing organic compounds.

Avidin: a protein that has a high affinity for biotin and is used in a detection system for biotinylated probes.

Azolla: a small, aquatic nitrogen-fixing fern.

B-form: the commonest structural conformation of the DNA double helix in living cells.

Bacilli: prokaryotes with rod-shaped cells.

Backtracking: the reversal of an RNA polymerase a short distance along its DNA template strand.

Bacteria: one of the two main groups of prokaryotes.

Bacteriochlorophyll: a porphyrin related to chlorophyll that acts as a light-harvesting pigment in photosynthetic bacteria.

Bacteriophage or **phage:** a virus whose host is a bacterium.

Bacteroid: the differentiated version of a nitrogen-fixing bacterium inside a root nodule.

Basal promoter element: sequence motifs that are present in many eukaryotic promoters and set the basal level of transcription initiation.

Basal rate of transcription: the number of productive initiations of transcription occurring per unit time at a particular promoter.

Base excision repair: a DNA repair process that involves excision and replacement of an abnormal base.

Base pair: the hydrogen-bonded structure formed by two complementary nucleotides. One base pair (bp) is the shortest unit of length for a double-stranded DNA molecule.

Base pairing: the attachment of one polynucleotide to another, or one part of a polynucleotide to another part of the same polynucleotide, by base pairs.

Base stacking: the hydrophobic interactions that occur between adjacent base pairs in a double-stranded DNA molecule.

Baseless site: a position in a DNA molecule where the base component of the nucleotide is missing.

Bases: compounds that decrease the hydronium ion concentration of a solution. Also, the purine or pyrimidine component of a nucleotide.

Beads-on-a-string: an unpacked form of chromatin consisting of nucleosome beads on a string of DNA.

β-adrenergic receptor: the cell-surface receptor protein for epinephrine.

β-carotene: an accessory light-harvesting pigment found in plants.

β-galactosidase: an enzyme that catalyzes the splitting of lactose into glucose and galactose.

β-galactoside transacetylase: an enzyme that catalyzes the transfer of an acetyl group from acetyl CoA to a β-galactoside molecule.

β-ketoacyl-ACP reductase: the enzyme that converts acetoacetyl ACP to D-3-hydroxybutyryl ACP at step 2 of the fatty acid synthesis pathway.

β-ketothiolase: the enzyme that converts a 3-ketoacyl CoA to an acyl CoA plus acetyl CoA at step 4 of the fatty acid breakdown pathway.

β-lactamase: a gene conferring ampicillin resistance, used as a selectable marker for many cloning vectors.

β-N-glycosidic bond: the linkage between the base and sugar of a nucleotide.

β-oxidation: the breakdown pathway for fatty acids.

β-sheet: a common type of protein secondary structure.

βαβ loop: a protein motif made up of two parallel strands of a β-sheet separated by an α-helix.

Bilayer: a double layer of molecules.

Bile acid: a sterol whose side-chain terminates in a carboxyl group; a product of cholesterol breakdown in the liver.

Bile pigments: derivatives of tetrapyrroles that are further metabolized and then excreted.

Binding-charge mechanism: a model for the mechanism of action of the F_0F_1 ATPase in which ATP synthesis is driven by conformational changes in the β subunits of the complex, these conformation changes caused by rotation of the γ subunit.

Binomial nomenclature: the naming system for biological species.

Biochemistry: the study of the chemical processes occurring in living cells and of the compounds involved in those chemical reactions.

Biofilm: a collection of bacteria adhered to one another and to a solid surface, usually embedded in a slimy matrix.

Bioinformatician: a scientist who specializes in bioinformatics: the use of computer methods to study biology.

Biological chemistry: an alternative name for 'biochemistry'.

Biological information: the information contained in the genome of an organism and which directs the development and maintenance of that organism.

Biology: the study of living organisms.

Biotin: vitamin B_7; a prosthetic group for several carboxylase enzymes.

Blunt end: an end of a DNA molecule at which both strands terminate at the same nucleotide position with no single-stranded extension.

Bond energy: the strength of a covalent bond, a measure of the amount of energy needed to break the bond.

Bundle sheath cells: the specialized cells in which the Calvin cycle takes place in C4 plants.

Buoyant density: the density possessed by a molecule or particle when suspended in an aqueous salt or sugar solution.

C-terminal domain (CTD): a component of the largest subunit of RNA polymerase II, important in activation of the polymerase.

C3 plants: those plant species in which Rubisco carries out the initial carbon fixation, producing the three-carbon compound 3-phosphoglycerate.

C4 plants: the group of plants that carry out the Calvin cycle reactions in specialized tissue made up of bundle sheath cells, so called because carbon dioxide is initially fixed as the four-carbon compound oxaloacetate.

Calcitriol: vitamin D; a family of steroid derivatives with various physiological and biochemical roles.

Calmodulin: a protein, activated by Ca^{2+} ions, which regulates a variety of cellular enzymes.

Calvin cycle: the cycle of reactions that result in synthesis of one molecule of glyceraldehyde 3-phosphate from three molecules of carbon dioxide, as part of the dark reactions of photosynthesis.

CAM plants: a group of tropical plant species that fix carbon dioxide as malate at night and then carry out reactions of the Calvin cycle during the day.

cAMP: a modified version of AMP in which an intramolecular phosphodiester bond links the 5′ and 3′ carbons.

cAMP response element (CRE): the binding site, upstream of some eukaryotic protein-coding genes, for the CREB protein.

cAMP response element binding (CREB) protein: a regulatory protein that controls the expression level of target genes in response to cellular cAMP levels.

Cap structure: the chemical modification at the 5′ end of most eukaryotic mRNA molecules.

Cap-binding complex: the complex that makes the initial attachment to the cap structure at the beginning of the scanning phase of eukaryotic translation.

Capsid: the protein coat that surrounds the DNA or RNA genome of a virus.

Carbamoyl: refers to a compound which has an amino group linked to a carbonyl group, as in the carboxylated version of lysine.

Carbamoyl phosphate: a precursor for the *de novo* synthesis of cytosine and uracil.

Carbamoyl phosphate synthetase: the enzyme that converts bicarbonate into carbamoyl phosphate at step 1 of the urea cycle.

Carbohydrate: any compound made up of carbon, hydrogen and oxygen, with the hydrogen and oxygen in a 2:1 ratio; commonly used to refer to saccharide compounds.

Carbon fixation: the conversion of an inorganic form of carbon into an organic form, as occurs during the dark reactions of photosynthesis.

Carboxyl terminus: the end of a polypeptide that has a free carboxyl group.

Carboxysomes: the structures in cyanobacteria within which Rubisco enzymes are located.

Carnitine: a small polar molecule that is attached to long chain and unsaturated fatty acids prior to their transport across the inner mitochondrial membrane.

Carnitine acyltransferase: the enzyme that attaches carnitine to long chain and unsaturated fatty acids prior to their transport across the inner mitochondrial membrane.

Carnitine/acylcarnitine translocase: an integral membrane protein that transports acylcarnitine molecules across the inner mitochondrial membrane.

Carotenoids: a group of lipids that includes several accessory light-harvesting pigments.

Catabolism: the part of metabolism that is devoted to the breakdown of compounds in order to generate energy.

Catabolite activator protein (CAP): a regulatory protein that binds to various sites in a bacterial genome and activates transcription initiation at downstream promoters.

Catalase: an enzyme that converts hydrogen peroxide to water and oxygen, sometimes coupled with detoxification of compounds such as phenols and alcohol.

Catalyst: a compound that increases the rate of a chemical reaction but is not itself transformed by that reaction.

cDNA synthesis: the conversion of RNA into a single- or double-stranded DNA molecule.

Cell envelope: the structure enclosing a bacterial cell, comprising plasma membrane (in all species), cell wall (in most species) and outer membrane (in some species).

Cell membrane: the membrane that encloses a prokaryotic or eukaryotic cell.

Cell wall: an enclosing structure on the external surface of the plasma membrane of the cells of some organisms, usually comprising a rigid layer of polysaccharides.

Cell-free translation system: a cell extract containing all the components required for protein synthesis (i.e. ribosomal subunits, tRNAs, amino acids, enzymes and cofactors) and able to translate added mRNA molecules.

Cell-specific promoter element: sequence motifs present in the promoters of eukaryotic genes that are expressed in just one type of tissue.

Cellular respiration: the biochemical reactions that take place in the mitochondria, using up oxygen and producing carbon dioxide.

Cellulose: a structural homopolysaccharide found in plants and made entirely of D-glucose units.

Cerebroside: a sphingolipid with a simple sugar head group.

cGMP: a modified version of GMP in which an intramolecular phosphodiester bond links the 5′ and 3′ carbons.

Chain termination sequencing: a DNA sequencing method that involves enzymatic synthesis of polynucleotide chains that terminate at specific nucleotide positions.

Chaperonin: a multisubunit protein which forms a structure that aids the folding of other proteins.

Chemical shift: the shift in the resonance frequency of a nucleus resulting from the presence of electrons in the vicinity of that nucleus.

Chemiosmotic theory: the theory that ATP synthesis is driven by the pumping of protons across the inner mitochondrial membrane.

Chief cell: a cell of the stomach lining that secretes pepsin.

Chitin: a structural homopolysaccharide found in some arthropods and made entirely of *N*-acetylglucosamine units.

Chlorophyll: a group of compounds that act as the main light-harvesting pigments of plant chloroplasts.

Chloroplasts: the photosynthetic organelles found in the cells of certain types of eukaryote.

Cholecalciferol: vitamin D_3; a member of the vitamin D group of compounds.

Cholesterol: an animal sterol, a component of some membranes.

Cholic acid: the simplest type of bile acid.

Cholyl CoA: an intermediate in the synthesis of bile acids.

Chorismate: a substrate in the pathway for aromatic amino acid synthesis, at the branch point for those reactions leading to tryptophan and those leading to phenylalanine and tyrosine.

Chromatin: the complex of DNA and histone proteins found in chromosomes.

Chromatogram: a graphical representation of the time-dependent elution of compounds during a chromatography experiment.

Chromatophores: invaginations of the plasma membrane of a purple bacterium, where the light reactions of photosynthesis take place.

Chromosome: one of the DNA–protein structures that contains part of the nuclear genome of a eukaryote. Less accurately, the DNA molecule(s) that contains a prokaryotic genome.

Chromosome territory: the region of a nucleus occupied by a single chromosome.

Chylomicron: the largest type of lipoprotein, which transports dietary triacylglycerols and cholesterol from the intestines to other tissues.

Chylomicron remnant: a cholesterol-rich chylomicron derivative that remains after breakdown of the triacylglycerol content of the original chylomicron.

Chymotrypsin: a protease enzyme secreted by the pancreas and involved in breakdown of proteins in the duodenum.

Circular dichroism: the differential absorption of clockwise and/or counterclockwise polarized light, used as the basis to a method for obtaining information on the structure of a molecule.

Circularly polarized light: refers to light in which the electric field follows a circular vector.

Cisternae: the stacks of membranous plates that make up the Golgi apparatus.

Citrate carrier: a protein that transports citrate across the inner mitochondrial membrane.

Citrate synthase: the enzyme that converts oxaloacetate and acetyl CoA to citrate at step 1 of the TCA cycle.

Citric acid cycle: another name for the TCA cycle.

Citrulline: an intermediate in the urea cycle.

Cleavage and polyadenylation specificity factor (CPSF): a protein that plays an ancillary role during polyadenylation of eukaryotic mRNAs.

Cleavage stimulation factor (CstF): a protein that plays an ancillary role during polyadenylation of eukaryotic mRNAs.

Cloning vector: a DNA molecule that is able to replicate inside a host cell and therefore can be used to clone other fragments of DNA.

Closed promoter complex: the structure formed during the initial step in assembly of the transcription initiation complex. The closed promoter complex consists of the RNA polymerase and/or accessory proteins attached to the promoter, before the DNA has been opened up by breakage of base pairs.

Cloverleaf: a two-dimensional representation of the structure of a tRNA molecule.

Co-repressor: a protein that represses transcription initiation by binding non-specifically to DNA or via protein–protein interactions.

Cocci: prokaryotes with spherical cells.

Coding RNA: an RNA molecule that codes for a protein; an mRNA.

Codon: a triplet of nucleotides coding for a single amino acid.

Codon–anticodon recognition: the interaction between a codon on an mRNA molecule and the corresponding anticodon on a tRNA.

Coenzyme: an organic compound that acts as a cofactor for an enzymatic reaction.

Coenzyme A: a cofactor for several enzymes involved in energy generation or lipid metabolism.

Coenzyme Q (CoQ): the intermediate carrier molecule for electron transfer from Complex I or II to Complex III in the electron transport chain; also called ubiquinone.

Cofactor: an ion or molecule that is required by an enzyme in order for the enzyme to carry out its biochemical reaction.

Cohesive end: an end of a double-stranded DNA molecule where there is a single-stranded extension.

Coiled coil: a protein structure in which two or more α-helices are coiled around one another to form a superhelix.

Colinearity: the relationship between the nucleotide sequence of an exon and the amino acid sequence of the protein segments specified by that exon.

Collagen fingerprinting: a method that uses collagen structure to identify species from fragments of bones.

Column chromatography: chromatography carried out with particles packed in a metal, glass or plastic column.

Commitment step: the first irreversible step in an enzyme-catalyzed pathway that produces an intermediate that is unique to that pathway.

Competitive reversible inhibition: a reversible inhibitor that competes with the substrate for entry into the active site of an enzyme.

Complement system: a set of enzymes and other proteins that disrupt the bacterial cell membrane, leading to the death of the bacterium, as part of the immune response.

Complementary: two polynucleotides that can base pair to form a double-stranded molecule.

Complementary DNA (cDNA): a double-stranded DNA copy of an mRNA molecule.

Complex A: the splicing intermediate comprising the mRNA and U1- and U2-snRNPs.

Complex B: the splicing intermediate comprising the mRNA and U1-, U2-, U4-, U5- and U6-snRNPs.

Complex triacylglycerol: a triacylglycerol in which at least two of the three fatty acid chains are different.

Concanavalin A: a lectin that binds to the terminal α-glucose and α-mannose units in O-linked glycans.

Concerted model: a model for cooperative substrate binding, in which binding of a substrate molecule to one of the enzyme subunits induces the immediate conversion of all the subunits to a new conformation.

Condensation: a chemical reaction that includes expulsion of a molecule of water.

Conjugation: physical contact between two bacteria, usually associated with transfer of DNA from one cell to the other.

Consensus sequence: a nucleotide sequence used to describe a large number of related though non-identical sequences. Each position of the consensus sequence represents the nucleotide most often found at that position in the real sequences.

Context-dependent codon reassignment: refers to the situation whereby the DNA sequence surrounding a codon changes the meaning of that codon.

COOH– or C terminus: the end of a polypeptide that has a free carboxyl group.

Cooperative substrate binding: the situation in which the binding of a substrate molecule to one active site induces a conformational change that facilitates substrate binding at other active sites in an enzyme.

Coordinate bond: a covalent bond in which both members of the shared pair of electrons come from the same atom, as formed in a metalloprotein between the metal ion and an amino acid side-chain.

Coordination center: the metal ion within a coordination sphere.

Coordination sphere: the structure within a metalloprotein comprising the metal ions and the amino acid side-chains to which it is linked.

Copy number: the number of molecules of a plasmid contained in a single cell.

CoQH$_2$–cytochrome c reductase complex: Complex III of the electron transport chain.

Core enzyme: the subunits required in order for an enzyme to carry out its core activity; in particular, the version of the *Escherichia coli* RNA polymerase, subunit composition $\alpha_2\beta\beta'$, that carries out RNA synthesis but is unable to locate promoters efficiently.

Core octamer: the central component of a nucleosome, made up of two subunits each of histones H2A, H2B, H3 and H4, around which DNA is wound.

Core promoter: the position within a eukaryotic promoter where the initiation complex is assembled.

Cori cycle: the combination of glycolysis and lactate production in muscle cells linked to regeneration of pyruvate and glucose in the liver.

Cortisol: a steroid hormone involved in control of blood sugar levels, the immune system and bone growth.

Covalent bond: a bond that forms when two atoms share electrons.

Cristae: infoldings of the inner mitochondrial membrane.

Crossover immunoelectrophoresis (CIP or CIEP): an immunoelectrophoresis technique.

Cyanide-resistant respiration: a version of the electron transport chain in which electrons are passed directly from ubiquinone to oxygen.

Cyanobacteria: a group of photosynthetic bacteria.

Cyclase: an enzyme that synthesizes cAMP or cGMP from ATP or GTP, respectively.

Cyclic photophosphorylation: a modified version of the photosynthetic electron transport chain, in which electrons from plastocyanin are used to return the P700 reaction center to its ground state.

Cyclobutyl dimer: a dimer between two adjacent pyrimidine bases in a polynucleotide, formed by ultraviolet irradiation.

Cys$_2$His$_2$ structure: a type of zinc-finger DNA-binding domain.

Cystic fibrosis transmembrane regulator (CFTR): an ABC transporter, responsible for transport of Cl$^-$ ions out of cells, which when defective gives rise to cystic fibrosis.

Cytidylate synthetase: the enzyme that converts UTP to CTP during *de novo* nucleotide synthesis.

Cytochrome: a protein that contains one or more heme prosthetic groups and acts as an electron carrier.

Cytochrome b_5: part of the enzyme complex that introduces double bonds into the hydrocarbon chain during fatty acid synthesis.

Cytochrome b_6f complex: a complex comprising two iron-containing cytochromes and an iron–sulfur protein; a component of the photosynthetic electron transport chain.

Cytochrome c: the intermediate carrier molecule for electron transfer from Complex III to Complex IV in the electron transport chain.

Cytochrome c oxidase complex: Complex IV of the electron transport chain.

Cytochrome P450: a group of heme-containing enzymes involved in steroid hormone synthesis.

Cytokine: a protein involved in cell signaling.

Cytoplasm: the substance that makes up the internal matrix of a cell.

Cytosine: one of the pyrimidine bases found in DNA and RNA.

D loop: the structure formed when a DNA double helix is invaded by a single-stranded DNA or RNA molecule, which forms a region of base pairing with one of the polynucleotides of the helix.

DAHP synthase: the enzyme that converts phosphoenolpyruvate and erythrose 4-phosphate to DAHP, at the commitment step in the part of the pathway leading to chorismate.

Dalton: the unit of measurement of molecular mass, one Dalton being one-twelfth the mass of a single atom of ^{12}C.

Dark reactions: that part of photosynthesis which uses the energy in ATP and NADPH to synthesize carbohydrates from carbon dioxide and water.

***De novo* nucleotide synthesis:** the synthesis of nucleotide bases from smaller compounds.

***De novo* synthesis:** the synthesis of complex molecules from simple molecules.

Decyclase: an enzyme that synthesizes ATP or GTP from cAMP or cGMP, respectively.

Degenerate: refers to the fact that the genetic code has more than one codon for most amino acids.

Degradosome: a multienzyme complex responsible for degradation of bacterial mRNAs.

$\Delta^{2,4}$-dienoyl CoA reductase: the enzyme that converts $\Delta^{2,4}$-dienoyl CoA to Δ^3-dienoyl CoA during breakdown of some unsaturated fatty acids.

Denaturation: breakdown by chemical or physical means of the non-covalent interactions, such as hydrogen bonding, that maintain the secondary and higher levels of structure of proteins and nucleic acids.

Density gradient centrifugation: a group of techniques in which a cell fraction is centrifuged through a dense solution, in the form of a gradient, so that individual components are separated.

Deoxyribonucleic acid: one of the two forms of nucleic acid in living cells; the genetic material for all cellular life forms and many viruses.

Desaturase: part of the enzyme complex that introduces double bonds into the hydrocarbon chain during fatty acid synthesis.

Detergent: a broad group of compounds with surfactant activities, including fatty acid derivatives with a hydrophobic tail and strongly hydrophilic head group, capable of disrupting a lipid bilayer.

Development: a coordinated series of transient and permanent changes that occurs during the life history of a cell or organism.

Developmental promoter elements: sequences within a eukaryotic promoter which bind proteins that regulate the expression of genes that are active at specific developmental stages.

Diabetes mellitus: a group of diseases characterized by abnormally high blood sugar levels.

Dialysis: the separation of compounds in a liquid based on the differential abilities of those compounds to pass through a membrane.

Diastereomer: a compound with more than one pair of chiral carbons.

Diazotroph: an organism able to carry out nitrogen fixation.

Dicer: a ribonuclease that cuts double-stranded RNA, involved in processing of miRNA precursor molecules.

Dideoxynucleotide (ddNTP): a modified nucleotide that lacks the 3′ hydroxyl group and so prevents further chain elongation when incorporated into a growing polynucleotide.

Differentiation: the adoption by a cell of a specialized biochemical and/or physiological role.

Diglyceride acyltransferase: the enzyme that converts diacylglycerol to triacylglycerol at step 5 of the triacylglycerol synthesis pathway.

Dihydrolipoyl dehydrogenase: a part of the pyruvate dehydrogenase complex.

Dihydrolipoyl transacetylase: a part of the pyruvate dehydrogenase complex.

Dipole: an atom with an uneven electron cloud, resulting in one side of the atom being slightly electropositive and the other side slightly electronegative.

Direct electron transfer: energy transfer in a photosystem, in which a high energy electron is transferred to a neighboring chlorophyll, in return for a low energy electron.

Directed evolution: a set of experimental techniques that are used to obtain novel proteins with improved properties.

Disaccharide: a sugar made up of two linked monosaccharides.

Discontinuous gene: a gene that is split into exons and introns.

Disulfide bond: a covalent bond that forms between two cysteine amino acids.

DNA: deoxyribonucleic acid, one of the two forms of nucleic acid in living cells; the genetic material for all cellular life forms and many viruses.

DNA adenine methylase (Dam): an enzyme involved in methylation of *Escherichia coli* DNA.

DNA cloning: insertion of a fragment of DNA into a cloning vector, and subsequent propagation of the recombinant DNA molecule in a host organism.

DNA cytosine methylase (Dcm): an enzyme involved in methylation of *Escherichia coli* DNA.

DNA glycosylase: an enzyme that cleaves the β-*N*-glycosidic bond between a base and the sugar component of a nucleotide as part of the base excision and mismatch repair processes.

DNA ligase: an enzyme that synthesizes phosphodiester bonds as part of DNA replication, repair and recombination processes.

DNA photolyase: a bacterial enzyme involved in photoreactivation repair.

DNA polymerase: an enzyme that synthesizes DNA.

DNA polymerase I: the bacterial enzyme that completes synthesis of Okazaki fragments during genome replication and is involved in some types of DNA repair.

DNA polymerase III: the main DNA replicating enzyme of bacteria.

DNA polymerase α: the enzyme that primes DNA replication in eukaryotes.

DNA polymerase β: an enzyme involved in some types of DNA repair in eukaryotes.

DNA polymerase γ: the enzyme that replicates mitochondrial DNA molecules.

DNA polymerase δ: the enzyme responsible for replication of the lagging DNA strand in eukaryotes.

DNA polymerase ε: the enzyme responsible for replication of the leading DNA strand in eukaryotes.

DNA repair: the biochemical processes that correct mutations arising from replication errors and the effects of mutagenic agents.

DNA sequencing: a technique for determining the order of nucleotides in a DNA molecule.

DNA shuffling: a PCR-based procedure that results in directed evolution of a DNA sequence.

DNA topoisomerase: an enzyme that introduces or removes turns from the double helix by breakage and reunion of one or both polynucleotides.

DNA-dependent DNA polymerase: an enzyme that makes a DNA copy of a DNA template.

DNA-dependent RNA polymerase: an enzyme that makes an RNA copy of a DNA template.

DnaA protein: a protein that attaches to the bacterial origin of replication and aids in breaking base pairs in this region.

Domain: a distinct segment within the tertiary structure of a protein. Also, one of the three main groups of organism: the bacteria, archaea or eukaryotes.

Donor site: the splice site at the 5′ end of an intron.

Double bond: a covalent bond that results from two atoms sharing two pairs of electrons.

Double helix: the base-paired double-stranded structure that is the natural form of DNA in the cell.

Double membrane: two membranes, one inside the other.

EC number: a four-part number describing the activity of an enzyme in accordance with the nomenclature set by the International Union of Biochemistry and Molecular Biology.

Effector: a small molecule that binds to an enzyme and regulates the activity of that enzyme.

Eicosanoid: a compound derived from arachidonic acid, with hormone-like activity.

Electrochemical gradient: a gradient of electrochemical potential, usually resulting from different concentrations of an ion on either side of a membrane.

Electroendosmosis: the motion of a liquid, such as the buffer in a gel, induced by an electric field.

Electron density map: a plot of the electron density at different positions within a molecule, deduced from an X-ray diffraction pattern.

Electron transport chain: a series of compounds that carry out redox reactions which transfer electrons from donor to acceptor compounds, usually coupled with the movement of protons across a membrane.

Electrophoresis: the movement of charged molecules in an electric field.

Electrospray ionization: a gentle ionization method used during mass spectrometry of unstable lipids.

Electrostatic bond: an interaction between positively and negatively charged chemical groups.

ELISA (enzyme-linked immunosorbent assay): an enzyme-based immunoassay.

Elongation factor: a protein that plays an ancillary role in the elongation step of transcription or translation.

Enantiomers: isomers whose structures are mirror images of one another.

Endergonic: a chemical reaction that requires energy.

Endonuclease: an enzyme that breaks phosphodiester bonds within a nucleic acid molecule.

Endoplasmic reticulum: a network of membranous plates and tubes that pervades the cytoplasm of a eukaryotic cell.

Endosymbiont theory: a theory that states that the mitochondria and chloroplasts of eukaryotic cells are derived from symbiotic prokaryotes.

Energetics: the study of the transformation of energy during a chemical reaction.

Energy coupling: the coupling of an endergonic reaction with a second reaction that generates energy.

Enolase: the enzyme that converts 2-phosphoglycerate to phosphoenolpyruvate at step 9 of the glycolysis pathway.

Enoyl CoA hydratase: the enzyme that converts a *trans-Δ²-* enoyl CoA to a 3-hydroxyacyl CoA at step 2 of the fatty acid breakdown pathway.

Enoyl-ACP reductase: the enzyme that converts crotonyl ACP to butyryl ACP at step 4 of the fatty acid synthesis pathway.

Enteropeptidase: a protease that converts trypsinogen to trypsin and chymotrypsinogen to chymotrypsin.

Entropy: a measure of the degree of disorder of a system.

Enzyme: a protein or, less commonly, RNA that catalyzes a biochemical reaction.

Enzyme kinetics: the study of enzyme-catalyzed reactions, with particular focus on the relationship between the substrate concentration and reaction rate.

Epimers: diastereomers that differ in structure at just one of their chiral carbons.

Epinephrine: a hormone (also called adrenaline) produced by the adrenal glands that controls various cellular activities as part of the 'fight or flight' response.

Epitope: a surface feature on an antigen, recognized by an antibody.

Erythrocyte transporter protein: a uniporter protein that transports glucose into mammalian erythrocytes.

Essential amino acids: those amino acids that cannot be synthesized by a species and hence must be obtained from the diet.

Estradiol: a steroid hormone involved in control of the female reproductive cycle.

Estrogens: a family of steroid hormones involved in control of female sexual development.

Estrone: a steroid hormone involved in control of the female reproductive cycle.

Eukaryote: an organism whose cells contain membrane-bound nuclei.

Excision repair: a DNA repair process that corrects various types of DNA damage by excising and resynthesizing a region of polynucleotide.

Excited: refers to the gain of energy by an electron.

Exciton transfer: energy transfer in a photosystem, in which energy quanta are passed from one chlorophyll molecule to another, the recipient chlorophyll becoming excited and the donor chlorophyll returning to its ground state.

Exergonic: a chemical reaction that releases energy.

Exit or **E site:** a position within a ribosome to which a tRNA moves immediately after deacylation.

Exocytosis: the transport process for secreted proteins.

Exon: a coding region within a discontinuous gene.

Exonuclease: an enzyme that removes nucleotides from one or both ends of a nucleic acid molecule.

Exosome complex: a ring of six proteins, each of which has ribonuclease activity, with three RNA-binding proteins attached to the top of the ring, involved in RNA degradation in eukaryotes.

Expressed sequence: an exon.

Expression proteomics: the methodology used to identify the proteins in a proteome.

Expression vector: a cloning vector designed so that a foreign gene inserted into the vector is expressed in the host organism.

Extracellular space: the material between cells in a tissue.

Extremophile: an organism that is able to live in an environment whose physical and/or chemical conditions are hostile to other organisms.

F_0F_1 ATPase: a multisubunit protein, located in the inner mitochondrial membrane, that utilizes an electrochemical potential across the membrane in order to synthesize ATP.

Facilitated diffusion: a transport process in which a protein moves a molecule from the side of a membrane at which the concentration is higher to that at which the concentration is lower.

Facultative anaerobe: an organism that is able to use oxygen to make ATP, but which can also grow in the absence of oxygen.

Fatty acid: a simple hydrocarbon chain of between four and 36 carbons with their attached hydrogen atoms and a terminal carboxylic group.

Fatty acid synthase: the multifunctional enzyme responsible for fatty acid synthesis in mammals.

Fatty acid thiokinase: an enzyme that converts fatty acids to acyl CoA molecules prior to their breakdown.

Favism: glucose 6-phosphate dehydrogenase deficiency.

Feedback regulation: a system in which an end product controls the rate of its own synthesis by acting as a reversible inhibitor of one of the enzymes that catalyzes an early step in the pathway leading to that end product.

FEN1: the 'flap endonuclease' involved in replication of the lagging strand in eukaryotes.

Ferrodoxin: an FeS protein; a component of the photosynthetic electron transport chain.

Ferrodoxin–thioredoxin reductase: an enzyme that converts thioredoxin to its reduced form.

Fibrous protein: a protein, such as collagen, that is not folded into a tertiary structure.

Filamentous: a bacteriophage or virus capsid in which the protomers are arranged in a helix, producing a rod-shaped structure.

Fimbriae: structures present on the surface of some bacteria, enabling the cells to attach to a solid surface.

Fischer projection: a two-dimensional representation of the tetrahedral arrangement of chemical groups around a carbon atom.

Flagella: structures that provide some cells with motile ability.

Flavin adenine dinucleotide (FAD): a cofactor for several enzymes involved in energy generation.

Flavin mononucleotide (FMN): a cofactor for several enzymes involved in energy generation.

Flavonoids: organic compounds secreted by plant roots that act as attractants for nitrogen-fixing bacteria.

Fluid mosaic model: a model that envisages a membrane as a two-dimensional fluid.

Flush end: an end of a DNA molecule at which both strands terminate at the same nucleotide position with no single-stranded extension.

Folding funnel: a concept used to explain the series of events by which a protein gradually adopts its final structure.

Folding pathway: the series of events, involving partially folded intermediates, that results in an unfolded protein attaining its correct three-dimensional structure.

Food chain: a linear series, starting with a photosynthetic species or other type of autotroph and ending with a top predator, that describes a pathway for energy acquisition by organisms within a food web.

Food web: a network describing how light or chemical energy is directly or indirectly acquired by species within an ecosystem.

Forward reaction: one of the directions of a reversible reaction.

Fragment ion: an ion resulting from fragmentation of a molecule during the ionization phase of mass spectrometry.

Frameshift: the movement of a ribosome from one reading frame to another at an internal position within a gene.

Free energy: the amount of energy possessed by a system that can be converted into work.

Fructokinase: the enzyme that converts fructose to fructose 1-phosphate at step 1 of the fructose 1-phosphate pathway.

Fructose 1-phosphate aldolase: the enzyme that converts fructose 1-phosphate to glyceraldehyde and dihydroxyacetone phosphate at step 2 of the fructose 1-phosphate pathway.

Fructose 1-phosphate pathway: the pathway for entry of fructose into glycolysis used in liver cells.

Fructose 1,6-bisphosphatase: the enzyme that converts fructose 1,6-bisphosphate to fructose 6-phosphate during sucrose synthesis and also during gluconeogenesis.

Fructose bisphosphatase 2: an enzyme that converts fructose 2,6-bisphosphate to fructose 6-phosphate, involved in the substrate-level regulation of glycolysis.

Fucoxanthin: a light-harvesting pigment found in brown algae.

Fumarase: the enzyme that converts fumarate to malate at step 7 of the TCA cycle.

Furanose: the cyclic form of a five-carbon sugar such as ribose.

Futile cycle: a cycle that occurs when there are two metabolic pathways that run in reverse directions, and result in substrate being converted to product and then back to substrate, with a waste of energy.

G protein: a small protein that binds either a molecule of GDP or GTP, the replacement of GDP with GTP activating the protein.

G-protein coupled receptor: a cell-surface receptor protein that responds to the extracellular signal by activating an intracellular G protein.

Galactokinase: the enzyme that converts galactose to galactose 1-phosphate at step 1 of the galactose–glucose interconversion pathway.

Galactose 1-phosphate uridylyl transferase: the enzyme that transfers a uridine group from UDP–glucose to galactose 1-phosphate at step 2 of the galactose–glucose interconversion pathway.

Galactose–glucose interconversion pathway: the pathway for conversion of galactose to glucose by rearrangement of the groups around the chiral carbon.

Ganglioside: a sphingolipid with a complex sugar head group.

Gas chromatography: a chromatographic method in which the mobile phase is gaseous and the stationary phase is a liquid absorbed into a solid matrix.

Gel electrophoresis: electrophoresis performed in a gel so that molecules of similar electrical charge can be separated on the basis of size.

Gel-filtration chromatography: a type of column chromatography using porous beads, which separates compounds according to size.

Gene expression: the series of events by which the biological information carried by a gene is released and made available to the cell.

Gene therapy: a clinical procedure in which a gene or other DNA sequence is used to treat a disease.

Genetic code: the rules that determine which triplet of nucleotides codes for which amino acid during protein synthesis.

Genetic engineering: the use of experimental techniques to produce DNA molecules containing new genes or new combinations of genes.

Genetic profile: the banding pattern revealed after electrophoresis of the products of PCRs directed at a range of microsatellite loci.

Genome: the entire complement of DNA molecules in a cell.

Genus: the taxonomic rank that comprises collections of species.

Gibbs free energy or ***G*:** a measure of the amount of energy possessed by a system that can be converted into work at a constant temperature and volume.

Global regulation: a general down-regulation in protein synthesis that occurs in response to various signals.

Globular protein: a protein whose polypeptide chain is folded into a tertiary structure.

Glucagon: a hormone, synthesized in the pancreas, which raises the concentration of glucose in the bloodstream.

Glucocorticoids: a family of steroid hormones.

Glucogenic: an amino acid whose breakdown products can be used to synthesize glucose.

Glucokinase: the enzyme that converts glucose to glucose 6-phosphate in liver cells.

Gluconeogenesis: a pathway for conversion of pyruvate to glucose.

Glucose: a hexose monosaccharide compound.

Glucose 6-phosphatase: an enzyme that converts glucose 6-phosphate to glucose in the liver.

Glucose 6-phosphate dehydrogenase: the enzyme that converts glucose 6-phosphate to 6-phosphoglucono-δ-lactone at step 1 of the pentose phosphate pathway.

Glucose 6-phosphate dehydrogenase deficiency: an inherited disease also known as favism.

Glutamate dehydrogenase: the enzyme that converts α-ketoglutarate to glutamate at step 1 in the glutamate and glutamine synthesis pathway, and also the reverse reaction during amino acid breakdown.

Glutamine synthase: the enzyme that converts glutamate to glutamine at step 2 in the glutamate and glutamine synthesis pathway.

Glutamine-rich domain: a type of activation domain.

Glutathione: a tripeptide made up of glutamic acid, cysteine and glycine, with an unusual linkage between the first two amino acids.

Glutathione peroxidase: an enzyme that converts reduced glutathione to its oxidized form, coupled with the conversion of hydrogen peroxide to water.

Glutathione reductase: an enzyme that regenerates reduced glutathione from its oxidized form.

Glycan: the oligosaccharide at a single glycosylated position in a glycoprotein.

Glycan sequencing: treatment of a glycan with a series of exoglycosidases of different specificities in order to determine the sequence of sugars in the glycan.

Glyceraldehyde 3-phosphate dehydrogenase: the enzyme that converts glyceraldehyde 3-phosphate to 1,3-bisphosphoglycerate at step 6 of the glycolysis pathway, and which also carries out the reverse reaction at step 3 of the Calvin cycle.

Glycerol 3-phosphate acyltransferase: the enzyme that adds the first and second fatty acid chains to glycerol 3-phosphate during the triacylglycerol synthesis pathway.

Glycerol 3-phosphate dehydrogenase: the enzyme that converts dihydroxyacetone phosphate to glycerol 3-phosphate at step 1 of the triacylglycerol synthesis pathway, and also as part of the glycerol 3-phosphate shuttle.

Glycerol 3-phosphate shuttle: a process that enables cytoplasmic NADH to be used in mitochondrial ATP synthesis.

Glycerol kinase: an enzyme that converts glycerol to glycerol 3-phosphate prior to its conversion to dihydroxyacetone phosphate in the liver.

Glycerophospholipid: a lipid that resembles a triacylglycerol, but with one of the fatty acids replaced by a hydrophilic group attached to the glycerol component by a phosphodiester bond.

Glycocholate: a bile acid; a derivative of cholic acid, which helps to emulsify fats in the diet.

Glycogen: a storage homopolysaccharide found in animals and made entirely of D-glucose units.

Glycogen branching enzyme: an enzyme that synthesizes the $\alpha(1\rightarrow6)$ linkages at the branch points of a glycogen molecule.

Glycogen debranching enzyme: an enzyme that removes glucose units from the branch sites in a glycogen molecule.

Glycogen phosphorylase: an enzyme that removes glucose units one by one from the non-reducing ends of a glycogen molecule.

Glycogen synthase: an enzyme that adds activated glucose units to the non-reducing ends of the growing glycogen molecule.

Glycogenin: the enzyme that primes the synthesis of glycogen.

Glycolipid: a glycosylated lipid.

Glycolysis: the catabolic pathway that generates energy from the breakdown of one molecule of glucose into two molecules of pyruvate.

Glycome: the carbohydrate content of a cell or tissue.

Glycoprotein: a glycosylated protein.

Glycosaminoglycans: a group of extracellular matrix polysaccharides.

Glycosidase: an enzyme that breaks glycosidic bonds.

Glycosylation: the attachment of short chains of sugars to a protein.

Glyoxylate cycle: a cycle of biochemical reactions, occurring in plants and microorganisms, similar to the TCA cycle, but bypassing the reactions from isocitrate to malate.

Glyoxysomes: the plant organelles within which the glyoxylate cycle takes place.

Golgi apparatus: a eukaryotic organelle involved in protein processing.

Gout: a medical condition caused by excessive uric acid in the blood.

Grana: stacks of thylakoids within the stroma of a chloroplast.

Green bacteria: a group of photosynthetic bacteria.

GroEL/GroES complex: a type of chaperonin.

Group I intron: a type of intron found mainly in organelle genes.

GU–AG intron: the commonest type of intron in eukaryotic nuclear genes. The first two nucleotides of the intron are 5′–GU–3′ and the last two are 5′–AG–3′.

Guanine: one of the purine bases found in DNA and RNA.

Guanine methyltransferase: the enzyme that attaches a methyl group to the 5′ end of a eukaryotic mRNA during the capping reaction.

Guanylyl transferase: the enzyme that attaches a GTP to the 5′ end of a eukaryotic mRNA at the start of the capping reaction.

Gunther disease: a defect in uroporphyrinogen cosynthetase.

Haber process: the non-biological process for reduction of nitrogen to ammonia.

Half-life: the time needed for half the atoms or molecules in a sample to decay or be degraded.

Hammerhead: an RNA structure with ribozyme activity that is found in some virusoids and viroids.

Head-and-tail: a bacteriophage capsid made up of an icosahedral head, containing the nucleic acid, and a filamentous tail which facilitates entry of the nucleic acid into the host cell.

Heart attack: a medical condition that might result from deposition of cholesterol onto the inner surfaces of blood vessels.

Heat shock module: a regulatory sequence upstream of genes involved in protection of a cell from heat damage.

Heavy chains: the two larger polypeptides in an immunoglobulin molecule.

Helicase: an enzyme that breaks base pairs in a double-stranded DNA molecule.

Helix–turn–helix motif: a common structural motif for attachment of a protein to a DNA molecule.

Hematopoietic stem cell transplant: replacement of hematopoietic stem cells with donor cells, used to treat some types of genetic disorder.

Heteropolysaccharide: a polysaccharide in which all the monosaccharide units are mixed.

Hexokinase: the enzyme that converts glucose to glucose 6-phosphate at step 1 of the glycolysis pathway.

Hexose monophosphate shunt: another name for the pentose phosphate pathway.

High density lipoproteins (HDLs): lipoproteins that transport cholesterol from the blood to the liver.

High performance liquid chromatography (HPLC): column chromatography carried out in a capillary tube with an internal diameter of less than 1 mm, with the liquid phase being pumped at high pressure.

Histone: one of the basic proteins found in nucleosomes.

Histone acetyltransferase (HAT): an enzyme that attaches acetyl groups to core histones.

Histone code: the hypothesis that the pattern of chemical modification on histone proteins influences various cellular activities by specifying which sets of genes are expressed at a particular time.

Histone deacetylase (HDAC): an enzyme that removes acetyl groups from core histones.

HMG CoA reductase: the enzyme that converts HMG CoA molecules to mevalonate during the cholesterol synthesis pathway.

Holoenzyme: the complex between an enzyme and a cofactor. Also used to denote a complex between a core enzyme and ancillary protein subunits; in particular, the version of the *Escherichia coli* RNA polymerase, subunit composition $\alpha_2\beta\beta'\sigma$, which is able to recognize promoter sequences.

Homologous enzymes: two or more enzymes with identical functions.

Homologous recombination: recombination between two homologous double-stranded DNA molecules, i.e. ones which share extensive nucleotide sequence similarity.

Homopolysaccharide: a polysaccharide in which all the monosaccharide units are identical.

Hormone: a signaling molecule that is secreted into the circulatory system of an animal and which affects biochemical activity in distant tissues.

Hormone response element: a nucleotide sequence within the promoter region of a gene that mediates the regulatory effect of a steroid hormone.

Housekeeping gene: a protein-coding gene that is continually expressed in all or at least most cells of a multicellular organism.

Hsp70 proteins: a family of proteins that bind to hydrophobic regions in other proteins in order to aid their folding.

Hyaluronic acid: a heteropolysaccharide made up of alternating N-acetylglucosamine and D-glucuronic acid units.

Hybridoma: a fusion between a lymphocyte and mouse melanoma cell, which synthesizes a monoclonal antibody.

Hydrocarbon: an organic compound made up entirely of carbon and hydrogen atoms.

Hydrogen bond: an interaction that forms between the slightly electropositive hydrogen atom in a polar group and an electronegative atom.

Hydrolases: enzymes that carry out hydrolysis reactions in which a chemical bond is cleaved by the action of water.

Hydronium ion: H_3O^+, the product of combination of a proton with a water molecule.

Hydrophilic: a chemical group or molecule that is attracted to water and tends to be soluble.

Hydrophobic: a chemical group or molecule that is repelled by water and tends to be insoluble.

Hydroxyacyl CoA dehydrogenase: the enzyme that converts a 3-hydroxyacyl CoA to a 3-ketoacyl CoA at step 3 of the fatty acid breakdown pathway.

Hyperammonemia: a medical condition that occurs when there is excessive ammonia in the blood, possibly because of a defect in the urea cycle.

Hyperglycemia: the situation that arises when blood glucose reaches an abnormally high level.

Hypoglycemia: the situation that arises when blood glucose reaches an abnormally low level.

Icosahedral: a bacteriophage or virus capsid in which the protomers are arranged into a three-dimensional geometric structure that surrounds the nucleic acid.

IIA^{Glc}: a membrane-bound protein in bacteria that is dephosphorylated when glucose is taken up, and which mediates the link between glucose availability and cellular cAMP levels.

Immobilized pH gradient: a pH gradient in an electrophoresis gel, established by differential concentrations of weakly acidic or basic compounds, used in isoelectric focusing.

Immunoassay: a test that makes use of an antibody to quantify the amount of antigen present in a sample.

Immunoelectrophoresis: the procedure used when an immunoassay is carried out in a gel placed in an electric field.

Immunoglobulin: a group of proteins that act as antibodies.

Immunoglobulin A: the main type of immunoglobulin in tears and saliva.

Immunoglobulin D: a type of immunoglobulin with an undefined role in the immune system.

Immunoglobulin E: a type of immunoglobulin that provides protection against parasites.

Immunoglobulin G: a type of immunoglobulin that is synthesized at a later stage in the immune response and also provides immunological protection for the fetus and newborn infant.

Immunoglobulin M: the first type of antibody to be synthesized when a new antigen is encountered; an activator of the complement system and of macrophages.

Immunological methods: experimental methods that make use of purified antibodies.

Importin: a protein that aids transfer of other proteins through a nuclear pore complex and into the nucleoplasm.

***In vitro* mutagenesis:** any one of several techniques used to produce a specified mutation at a predetermined position in a DNA molecule.

Inclusion body: a crystalline or paracrystalline deposit within a cell, often containing substantial quantities of insoluble protein.

Induced fit model: a model for enzyme activity that views the enzyme binding site as a flexible structure, the shape of which changes when the substrate binds.

Inducer: a molecule that induces expression of a gene or operon by binding to a repressor protein and preventing the repressor from attaching to the operator.

Inducible operon: an operon that is switched on by an inducer molecule that prevents the repressor from attaching to its DNA binding site.

Inhibitor: a compound that interferes with the activity of an enzyme, reducing its catalytic rate.

Initial velocity or V_0: the initial, linear rate of an enzyme-catalyzed reaction.

Initiation codon: the codon, usually but not exclusively 5'–AUG–3', found at the start of the coding region of a gene.

Initiation complex: the complex of proteins that initiates transcription. Also the complex that initiates translation.

Initiation factor: a protein that plays an ancillary role during initiation of translation.

Initiation region: a region of eukaryotic chromosomal DNA within which replication initiates at positions that are not clearly defined.

Initiator (Inr) sequence: a component of the RNA polymerase II core promoter.

Inner mitochondrial membrane: the inner of the two membranes of a mitochondrion.

Inosine: a modified version of adenosine, sometimes found at the wobble position of an anticodon.

Inositol-1,4,5-trisphosphate (Ins(1,4,5)P$_3$): a component of the second messenger signaling pathway initiated by flow of Ca^{2+} into a cell.

Insulin: a hormone, synthesized in the pancreas, which lowers the concentration of glucose in the bloodstream.

Insulin-responsive protein kinase: an enzyme in liver cells that activates protein phosphatase in response to extracellular insulin levels.

Integral membrane protein: a protein that forms a tight attachment with a membrane and can only be removed by disrupting the lipid bilayer.

Intergenic DNA: the regions of a genome that do not contain genes; the nucleotide sequence between adjacent genes.

Intermediate density lipoproteins (IDLs): derivatives of very low density lipoproteins, present in the blood.

Intermembrane space: the space between the inner and outer mitochondrial membranes.

Internal ribosome entry site (IRES): a nucleotide sequence that enables the ribosome to assemble at an internal position in some eukaryotic mRNAs.

Interphase: the period between cell divisions.

Intervening sequence: an intron.

Intrinsic terminator: a position in bacterial DNA where termination of transcription occurs without the involvement of Rho.

Intron: a noncoding region within a discontinuous gene.

Ion: a charged atom or molecule.

Ion exchange chromatography: a type of column chromatography using charged beads, which separates compounds according to their net electric charges.

Ionization: the conversion of an uncharged atom or molecule into a form with an electric charge.

Ionophore: a lipid-soluble compound that can carry bound protons through a membrane.

IRES trans-acting factor (ITAF): cellular RNA-binding proteins with various functions, which are used by an infecting virus to aid initiation of protein synthesis at internal ribosome entry sites.

Iron response element: a regulatory sequence upstream of a gene involved in iron uptake or storage.

Iron–sulfur cluster or **FeS cluster:** a cluster of iron atoms coordinated with inorganic sulfur atoms and with the sulfur of a cysteine side-chain.

Iron–sulfur protein or **FeS protein:** a protein containing an iron–sulfur cluster.

Irreversible inhibitor: an inhibitor that has a permanent effect on the activity of an enzyme.

Isoaccepting tRNAs: two or more tRNAs that are aminoacylated with the same amino acid.

Isocitrate dehydrogenase: the enzyme that converts isocitrate to α-ketoglutarate at step 3 of the TCA cycle.

Isocitrate lyase: the enzyme that converts isocitrate to succinate and glyoxylate in the glyoxylate cycle.

Isoelectric focusing: a gel electrophoresis technique that separates proteins according to isoelectric points.

Isoelectric point: the pH at which a molecule has no net electric charge.

Isomerases: enzymes that rearrange the atoms within a molecule.

Isomerization: the rearrangement of atoms within a molecule.

Isomers: two molecules that have identical chemical compositions but different structures.

Isopentenyl pyrophosphate: an intermediate in the synthesis of sterol compounds.

Isopeptide bond: a bond between the carboxyl group of the C-terminal amino acid of one protein and the amino group present in the side-chain of a lysine in a second protein.

Isoprene: a small hydrocarbon, the monomeric unit in a terpene.

Isotope: different versions of an element, with the same numbers of protons but different numbers of neutrons.

Isotope-coded affinity tags (ICATs): markers, containing normal hydrogen and deuterium atoms, used to label individual proteomes prior to analysis by mass spectrometry.

Isozymes: two closely related but distinct enzymes that catalyze the same biochemical reactions.

Janus kinase (JAK): a type of kinase that plays an intermediary role in some types of signal transduction involving STATs.

Joule: the work done by a force of one newton when its point of application moves through a distance of one meter in the direction of the force.

KDEL sequence: a sorting sequence that directs a protein to the lumen of the endoplasmic reticulum.

Ketogenesis: synthesis of ketone bodies.

Ketogenic: an amino acid whose breakdown products can contribute to ketone body synthesis.

Ketohexose: a ketose sugar with six carbon atoms.

Ketone: an organic compound containing a carbonyl group linked to two hydrocarbon groups.

Ketone body: a mixture of acetoacetate, D-3-hydroxybutyrate and acetone generated in the liver from the breakdown products of fatty acids and some amino acids.

Ketopentose: a ketose sugar with five carbon atoms.

Ketose: a sugar in which the terminal carbon forms part of a carbonyl group.

Ketotetrose: a ketose sugar with four carbon atoms.

Ketotriose: a ketose with three carbon atoms.

kiloDalton: 1000 Daltons.

kilojoules per mole or **kJ mol^{-1}:** an SI unit for the amount of energy per amount of material.

K_m or **Michaelis constant:** the substrate concentration at which the rate of enzyme-catalyzed reaction is half of the maximum value; a measure of the affinity of the enzyme for its substrate.

Kozak consensus: the nucleotide sequence surrounding the initiation codon of a eukaryotic mRNA.

Krebs cycle: another name for the TCA cycle.

Labeling: the incorporation of a marker nucleotide into a nucleic acid molecule. The marker is often, but not always, a radioactive or fluorescent label.

Lactate dehydrogenase: the enzyme that converts pyruvate to lactate.

Lactonase: the enzyme that converts 6-phosphoglucono-δ-lactone to 6-phosphogluconate at step 2 of the pentose phosphate pathway.

Lactose operon: the cluster of three genes that code for enzymes involved in utilization of lactose by *Escherichia coli*.

Lactose permease: the protein that transports lactose into a bacterial cell.

Lactose repressor: the regulatory protein that controls transcription of the lactose operon in response to the presence or absence of lactose in the environment.

Lactose tolerance or **lactose persistence:** the continued production of lactase, and hence the ability to digest lactose, after weaning.

Lagging strand: the strand of the double helix which is copied in a discontinuous fashion during genome replication.

Lambda: a bacteriophage that infects *Escherichia coli*, derivatives of which are used as cloning vectors.

Lanosterol: an intermediate in the synthesis of sterol compounds.

Latex: a tree exudate secreted in response to wounding.

Leader segment: the untranslated region of an mRNA upstream of the initiation codon.

Leading strand: the strand of the double helix which is copied in a continuous fashion during genome replication.

Lectins: plant or animal proteins with specific monosaccharide-binding properties.

Leghemoglobin: a protein present in root nodules, with high affinity for oxygen and hence protecting the nitrogenase complex from inhibition by oxygen.

Legume: a group of nitrogen-fixing plants.

Levinthal's paradox: the inability of a protein to find its correct tertiary structure simply by a random search.

Lichens: symbiotic organisms comprising a fungus, a photosynthetic bacterium or alga, and possibly a nitrogen-fixing cyanobacterium.

Life sciences: another name for 'biology'; study of living organisms.

Ligases: enzymes that join molecules together.

Light chains: the two shorter polypeptides in an immunoglobulin molecule.

Light harvesting: the absorption of energy from sunlight by a photosynthetic organism.

Light reactions: that part of photosynthesis that uses energy from sunlight to make ATP and NADPH.

Lineweaver–Burk plot: a graphical depiction of the relationship between the substrate concentration and initial velocity of an enzyme-catalyzed reaction.

Linker DNA: the DNA that links nucleosomes: the 'string' in the 'beads-on-a-string' model for chromatin structure.

Linker histone: a histone, such as H1, that is located outside of the nucleosome core octamer.

Lipases: enzymes that remove the fatty acid chains during triacylglycerol degradation.

Lipid: a member of a broad group of compounds that include fats, oils, waxes, steroids and various resins.

Lipid raft: a relatively stable domain in a membrane where sets of proteins that work together can be co-located.

Lipid-linked protein: a peripheral membrane protein that forms a covalent attachment with a membrane lipid.

Lipidome: the total lipid content of a cell or tissue.

Lipolysis: the process by which triacylglycerols and fatty acids are broken down.

Lipoprotein: a micelle-like particle that consists of a spherical lipid monolayer with various embedded proteins, surrounding a hydrophobic core containing triacylglycerol and cholesterol molecules.

Lipoprotein lipase: an enzyme that breaks down the triacylglycerol content of lipoproteins in muscle and adipose tissue.

Liposome: a small vesicle comprising a lipid bilayer enclosing an internal aqueous compartment; sometimes used to introduce DNA into an animal or plant cell.

Lock and key model: a model for enzyme activity that views the enzyme as possessing a binding pocket on its surface, the shape of which precisely matches the shape of the substrate.

Long patch: a nucleotide excision repair process of *Escherichia coli* that results in excision and resynthesis of up to 2 kb of DNA.

Low density lipoproteins (LDLs): derivatives of intermediate density lipoproteins, lacking apoproteins.

Luciferase: an enzyme which oxidizes luciferin to produce chemiluminescence.

Luciferin: a family of organic compounds from various species, which emit chemiluminescence when oxidized.

Luminal targeting sequence: a sorting sequence that direct a protein to the thylakoids within a chloroplast.

Lyases: enzymes that break chemical bonds by processes other than oxidation and hydrolysis.

Lysosomes: cellular organelles responsible for the degradation of various compounds including the contents of low density lipoproteins.

Macromolecule: a large biological molecule with a mass in excess of 1 kDa.

Magnetic sector mass spectrometer: a mass spectrometer in which the mass analyzer is a single or series of magnets through which the ionized molecules are passed.

Magnetic tweezers: a device comprising a set of magnets whose positions and field strengths can be varied in such a way that a magnetic bead can be moved about in a controlled manner.

Major groove: the larger of the two grooves that spiral around the surface of the B-form of DNA.

Malate dehydrogenase: the enzyme that converts malate to oxaloacetate at step 8 of the TCA cycle, and also carries out the reverse reaction as part of the malate–aspartate shuttle.

Malate synthase: the enzyme that converts glyoxylate and acetyl CoA to malate in the glyoxylate cycle.

Malate–aspartate shuttle: a process that enables cytoplasmic NADH to be used in mitochondrial ATP synthesis.

Malonyl transferase: the enzyme that transfers malonyl units from malonyl CoA to ACP during fatty acid synthesis.

Mannose 6-phosphate receptor: a protein located on the inner surface of the cisternae at the *trans* side of the Golgi apparatus, which recognizes proteins tagged with mannose 6-phosphate.

MAP kinase system: a signal transduction pathway.

Mass number: the total number of protons and neutrons in a nucleus.

Mass spectrometry: an analytical technique in which ions are separated according to their mass-to-charge (*m/z*) ratios.

Mass spectrum: a graphical representation of the *m/z* values of the ions separated by a mass spectrometry experiment.

Mass-to-charge ratio: the basis to separation of ions by mass spectrometry.

Massively parallel system: a high throughput DNA sequencing system in which many individual sequences are generated in parallel.

Matrix-assisted laser desorption ionization time of flight (MALDI-TOF): a type of mass spectrometry used in proteomics.

Mediator: a protein complex that forms a contact between various activators and the C-terminal domain of the largest subunit of RNA polymerase II.

Membrane potential: the electric charge across a membrane.

Mesophyll cells: leaf cells including those cells in which the light reactions of photosynthesis take place.

Mesosome: a small infolding in the plasma membrane of a prokaryotic cell.

Messenger RNA or **mRNA:** the transcript of a protein-coding gene.

Metabolism: the chemical reactions that occur in living organisms.

Metabolome: the complete collection of metabolites present in a cell under a particular set of conditions.

Metalloenzyme: an enzyme that contains a metal ion.

Metalloprotein: a protein that contains a metal ion.

Methylmalonyl CoA mutase: the enzyme that converts methylmalonyl CoA to succinyl CoA during breakdown of odd-numbered fatty acids.

Micelle: a spherical aggregate of amphiphilic molecules, in which the hydrophilic groups are exposed to water and the hydrophobic groups are embedded within the structure.

Michaelis–Menten equation: an equation that indicates the relationship between the substrate concentration, V_{max} and K_m of an enzyme-catalyzed reaction.

Micro RNA or **miRNA:** a class of short RNAs involved in regulation of gene expression in eukaryotes.

Microarray analysis: the use of a microarray – a small piece of glass onto which a large number of DNA molecules have been applied as spots in an ordered array – in transcriptome studies.

Microbiome: the microorganisms that live on or within the human body.

Minor groove: the smaller of the two grooves that spiral around the surface of the B-form of DNA.

Minor spliceosome: the spliceosome for AU–AC introns.

Mismatch: a position in a double-stranded DNA or RNA molecule where base-pairing does not occur because the nucleotides are not complementary; in particular, a non-base-paired position resulting from an error in replication.

Mitochondrial matrix: the central part of a mitochondrion, enclosed by the inner mitochondrial membrane.

Mitochondrial pyruvate carrier: the integral membrane protein that transports pyruvate across the inner mitochondrial membrane.

Mitochondrial shuttle: a transport protein that enables a compound to pass through the inner mitochondrial membrane.

Mitochondrial targeting sequence: a 10–70 amino acid sorting sequence that directs a protein to the mitochondrial matrix.

Mitochondrion: the singular of mitochondria – the energy-generating organelles of eukaryotic cells.

Mobile phase: the movable phase in a chromatography system, usually a liquid in which the compounds have been dissolved or a gas in which they have been vaporized.

Model organism: an organism which is relatively easy to study and hence can be used to obtain information that is relevant to the biology of a second organism that is more difficult to study.

Model-building: an experimental approach in which possible structures of biological molecules are assessed by building scale models of them.

Molecular chaperone: a protein that helps other proteins to fold.

Molecular ion: a type of ion resulting from soft ionization of compounds prior to mass spectrometry.

Molecular mass: the mass of a molecule, calculated as the sum of the masses of the individual atoms making up that molecule.

Molten globule: an intermediate in protein folding, formed by the rapid collapse of a polypeptide into a compact structure, with slightly larger dimensions than the final protein.

Molybdenum–iron or **MoFe center:** an electron-binding cofactor.

Monoclonal antibody: a single type of immunoglobulin, recognizing a single epitope, made by a clone of immune cells.

Monomer: one of the units in a polymeric chain.

Monosaccharide: an individual sugar compound, the monomeric unit in a polysaccharide.

Motif: a combination of secondary structural units in a protein.

Motor protein: a type of protein that can change its shape in a manner that enables an organism to move around.

mRNA surveillance: an RNA degradation process in eukaryotes.

Multicellular: an organism made up of many cells.

Mutagen: a chemical or physical agent that can cause a mutation in a DNA molecule.

Mutarotation: the interconversion between two anomers.

Mutation: an alteration in the nucleotide sequence of a DNA molecule.

MutH: a component of the bacteria mismatch repair system.

MutS: a component of the bacteria mismatch repair system.

Mutualism: a cooperative relationship of mutual benefit to participating species.

Myocardial infarction: heart attack, possibly caused by deposition of cholesterol onto the inner surfaces of blood vessels.

N-acetylglutamate synthase: an enzyme that synthesizes N-acetylglutamate from acetyl CoA and glutamate, involved in control of the urea cycle.

N-linked glycosylation: the attachment of sugar units to an asparagine in a polypeptide.

Na$^+$/Ca^{2+} exchange protein: an antiporter involved in export of Ca^{2+} ions from cells.

Na$^+$/glucose transporter: a symporter involved in uptake of glucose by intestinal cells.

Na$^+$/K$^+$ ATPase: a P-type pump, responsible for maintaining the high potassium and low sodium ion concentration within a mammalian cell.

NADH–CoQ reductase complex: Complex I of the electron transport chain.

NADH–cytochrome b_5 reductase: part of the enzyme complex that introduces double bonds into the hydrocarbon chain during fatty acid synthesis.

NADP reductase: an enzyme that converts NADP$^+$ to NADPH; a component of the photosynthetic electron transport chain.

NADP-linked malate dehydrogenase: an enzyme that converts oxaloacetate to malate in C4 and CAM plants.

NADP-linked malate enzyme: an enzyme that converts malate into pyruvate and carbon dioxide in the bundle sheath cells of C4 plants.

Negative allosteric control: inhibition of enzyme activity by binding of an effector molecule.

Neuron: a nerve cell.

Next generation sequencing: a collection of DNA sequencing methods, each involving a massively parallel strategy.

NH$_2$– or **N terminus:** the end of a polypeptide that has a free amino group.

Niacin: vitamin B$_3$, the precursor of the cofactors NAD$^+$ and NADP$^+$.

Nicotinamide adenine dinucleotide (NAD$^+$): a cofactor for several enzymes involved in energy generation.

Nicotinamide adenine dinucleotide phosphate (NADP$^+$): a cofactor for several enzymes involved in anabolic reactions.

Nitrate reductase: an enzyme that converts nitrate to nitrite.

Nitrate reduction: the biological conversion of soil nitrate into ammonia.

Nitrite reductase: an enzyme that converts nitrite to ammonia.

Nitrogen fixation: the biological conversion of atmospheric nitrogen into ammonia.

Nitrogenase complex: the two-enzyme complex that reduces nitrogen to ammonia.

Nod factor: short oligosaccharides with a fatty acid side-chain secreted by nitrogen-fixing bacteria and detected by suitable host plants.

Noncoding RNA: an RNA molecule that does not code for a protein.

Non-competitive reversible inhibition: a reversible inhibitor that does not compete with the substrate for entry into the active site of an enzyme.

Nonhomologous end-joining (NHEJ): the process for repair of double-strand breaks in DNA molecules.

Non-photochemical energy quenching: transfer of energy from an excited chlorophyll molecule to a quenching compound, resulting in the energy being dissipated as heat.

Norepinephrine: a hormone that controls various physiological and biochemical activities.

Nuclear envelope: the double membrane surrounding the nucleus of a eukaryotic cell.

Nuclear localization signal: a 6–20 amino acid sorting sequence that directs a protein to the nucleus.

Nuclear magnetic resonance (NMR) spectroscopy: a method for studying the structures of molecules based on the magnetic moments generated by rotating nuclei.

Nuclear receptor superfamily: a family of receptor proteins that bind hormones as an intermediate step in modulation of genome activity by these hormones.

Nuclease: an enzyme that degrades a nucleic acid molecule.

Nuclease protection: a technique that uses nuclease digestion to determine the positions of proteins on DNA or RNA molecules.

Nucleic acid: the term first used to describe the acidic chemical compound isolated from the nuclei of eukaryotic cells; now used specifically to describe a polymeric molecule comprising nucleotide monomers such as DNA and RNA.

Nucleoid: the DNA-containing region of a prokaryotic cell.

Nucleolus: the region of the eukaryotic nucleus in which rRNA synthesis occurs.

Nucleoplasm: the equivalent of the cytoplasm, but present within the nucleus of a eukaryotic cell.

Nucleoside: a purine or pyrimidine base attached to a five-carbon sugar.

Nucleoside diphosphate kinase: the enzyme that converts nucleoside diphosphates to their triphosphate forms during the salvage pathway for nucleotide synthesis.

Nucleoside monophosphate kinase: the enzyme that converts nucleoside monophosphates other than AMP to their diphosphate forms during the salvage pathway for nucleotide synthesis.

Nucleosome: the complex of histones and DNA that is the basic structural unit in chromatin.

Nucleotidase: an enzyme that converts a nucleotide into a nucleoside plus phosphate group.

Nucleotide: a purine or pyrimidine base attached to a five-carbon sugar, to which a mono-, di-, or triphosphate is also attached; the monomeric unit of DNA and RNA.

Nucleotide exchange factor: a protein that replaces a nucleotide diphosphate, bound to another protein, with a nucleotide triphosphate.

Nucleotide excision repair: a repair process that corrects various types of DNA damage by excising and resynthesizing a region of a polynucleotide.

Nucleus: the membrane-bound structure of a eukaryotic cell in which the chromosomes are contained.

O-glycosidic bond: the link between the two monosaccharide units in a disaccharide, oligosaccharide or polysaccharide.

O-linked glycosylation: the attachment of sugar units to a serine or threonine in a polypeptide.

Obligate aerobe: an organism that must have oxygen in order to survive.

Obligate anaerobe: an organism that never uses oxygen.

Okazaki fragment: one of the short segments of RNA-primed DNA synthesized during replication of the lagging strand of the double helix.

Oligopeptide: a small polymer of amino acids.

Oligosaccharide: a short polymeric sugar compound.

Omega system: a naming convention for fatty acids.

Open promoter complex: a structure formed during assembly of the transcription initiation complex consisting of the RNA polymerase and/or accessory proteins attached to the promoter, after the DNA has been opened up by breakage of base pairs.

Open reading frame: a series of codons starting with an initiation codon and ending with a termination codon. The part of a protein-coding gene that is translated into protein.

Operator: the nucleotide sequence to which a repressor protein binds to prevent transcription of a gene or operon.

Optical isomers: isomers whose structures are mirror images of one another.

Optical tweezers: a laser device that can be used to manipulate individual molecules.

Orbital: the region of space around an atomic nucleus in which a particular electron is likely to be found.

Organelle: a membrane-bound structure within a eukaryotic cell.

Origin of replication: a site on a DNA molecule where replication initiates.

Origin recognition complex: the set of proteins that binds to an origin of replication in yeast DNA.

Ornithine: an intermediate in the urea cycle, and in synthesis of arginine from glutamate.

Ornithine transcarbamoylase: the enzyme that converts carbamoyl phosphate and ornithine to citrulline at step 2 of the urea cycle.

Osteomalacia: vitamin D deficiency in adults, characterized by a softening or weakening of the bones.

Ouchterlony technique: a gel-based immunoassay.

Outer mitochondrial membrane: the outer of the two membranes of a mitochondrion.

Oxaloacetate: a four-carbon dicarboxylic acid that is one of the substrates and products of the TCA cycle.

Oxidative decarboxylation: the combined oxidization (loss of a pair of electrons) and decarboxylation (loss of CO_2) of a substrate.

Oxidative phosphorylation: generation of ATP from ADP and inorganic phosphate via the electron transport chain.

Oxidoreductases: enzymes that catalyze oxidation or reduction reactions.

P680: the reaction center of photosystem II.

P700: the reaction center of photosystem I.

P-type pump: an ATP-dependent transport protein which forms a transient attachment with the phosphate released by ATP hydrolysis.

Pancreatic self-digestion: the situation that arises if trypsin and chymotrypsin are activated prior to their secretion by the pancreas.

Pantothenic acid: vitamin B_5, the precursor of the cofactor coenzyme A.

Paper chromatography: a chromatography system in which the stationary phase is a strip of filter paper.

Parallel β-sheet: a β-sheet in which all of the strands run in the same direction.

PARP1: protective single-strand binding proteins which aid in repair of single-stranded breaks in DNA molecules.

Partition coefficient: the degree of interaction of a substance with the stationary phase during a chromatographic procedure.

Pentose: a sugar comprising five carbon atoms.

Pentose phosphate pathway: a series of biochemical reactions that generate NADPH.

Peptide: a short polypeptide, less than 50 amino acids in length.

Peptide bond: the chemical link between adjacent amino acids in a polypeptide.

Peptide group: the part of the linkage between two amino acids comprising the two α-carbons and the C, O, N and H atoms in between them.

Peptide mass fingerprinting: identification of a protein by examination of the mass spectrometric properties of peptides generated by treatment with a sequence-specific protease.

Peptidoglycan: the protein–carbohydrate matrix that is the major constituent of the bacterial cell wall.

Peptidyl or **P site:** the site in the ribosome occupied by the tRNA attached to the growing polypeptide during translation.

Peptidyl transferase: the enzyme activity that synthesizes peptide bonds during translation.

Peripheral membrane protein: a protein that forms a relatively loose attachment with a membrane and can be removed without disrupting the lipid bilayer.

PEST sequence: amino acid sequences that influence the degradation of proteins in which they are found.

pH: an inverse measure of the hydronium ion concentration of a solution.

pH optimum: the optimal pH for a chemical reaction.

Phagocytosis: the engulfment and degradation of a bacterium or other pathogen by a macrophage.

Pharming: genetic modification of a farm animal so that the animal synthesizes a recombinant pharmaceutical protein, often in its milk.

Phenylketonuria: a disease caused by a defect in the gene for phenylalanine hydroxylase.

Phosphatase: an enzyme that removes phosphate groups from other enzymes.

Phosphatidate phosphatase: the enzyme that converts phosphatidic acid to diacylglycerol at step 4 of the triacylglycerol synthesis pathway.

Phosphatidic acid: a glycerophospholipid in which the head group is a hydrogen atom.

Phosphatidylglycerol: a glycerophospholipid in which the head group is glycerol.

Phosphatidylinositol-4,5-bisphosphate (PtdIns(4,5)P$_2$): a lipid component of the cell membrane, involved in a second messenger signaling pathway.

Phosphatidylserine: a glycerophospholipid in which the head group is serine.

Phosphodiester bond: the chemical link between adjacent nucleotides in a polynucleotide.

Phosphodiesterase: a type of enzyme that can break phosphodiester bonds.

Phosphoenolpyruvate carboxylase: an enzyme that converts carbon dioxide and phosphoenolpyruvate to oxaloacetate in C4 and CAM plants.

Phosphoenolpyruvate carboxykinase: the enzyme that converts oxaloacetate to phosphoenolpyruvate during gluconeogenesis.

Phosphofructokinase: the enzyme that converts fructose 6-phosphate to fructose 1,6-bisphosphate at step 3 of the glycolysis pathway.

Phosphofructokinase 2: an enzyme that synthesizes fructose 2,6-bisphosphate from fructose 6-phosphate, involved in the substrate-level regulation of glycolysis.

Phosphoglucoisomerase: the enzyme that converts glucose 6-phosphate to fructose 6-phosphate at step 2 of the glycolysis pathway.

Phosphoglucomutase: the enzyme that converts glucose 1-phosphate to glucose 6-phosphate at step 4 of the galactose–glucose interconversion pathway and also during glycogen breakdown, additionally carrying out the reverse reaction during sucrose synthesis.

Phosphogluconate pathway: another name for the pentose phosphate pathway.

Phosphoglycerate kinase: the enzyme that converts 1,3-bisphosphoglycerate to 3-phosphoglycerate at step 7 of the glycolysis pathway, and also the reverse reaction at step 2 of the Calvin cycle.

Phosphoglycerate mutase: the enzyme that converts 3-phosphoglycerate to 2-phosphoglycerate at step 8 of the glycolysis pathway.

Phosphopantetheine: a prosthetic group derived from vitamin B$_5$.

Phosphopentose epimerase: the enzyme that converts ribulose 5-phosphate to xylulose 5-phosphate at step 5 of the pentose phosphate pathway.

Phosphopentose isomerase: the enzyme that converts ribulose 5-phosphate to ribose 5-phosphate at step 4 of the pentose phosphate pathway.

Phosphoribosyl pyrophosphate (PRPP): a ribose that has a diphosphate group attached to carbon number 1 and a monophosphate to carbon 5; an intermediate in synthesis of aromatic amino acids.

Phosphorylase kinase: an enzyme that activates glycogen phosphorylase by adding a phosphate group, converting glycogen phosphorylase *b* into glycogen phosphorylase *a*.

Photophosphorylation: the light-driven synthesis of ATP.

Photoproduct: a modified nucleotide resulting from treatment of DNA with ultraviolet radiation.

Photoprotection: a process that prevents the photosystems from becoming damaged at high light intensities.

Photoreactivation: a DNA repair process in which cyclobutyl dimers and (6–4) photoproducts are corrected by a light-activated enzyme.

Photorespiration: a series of reactions for conversion of 2-phosphoglycolate into 3-phosphoglycerate, used to prevent the accumulation of 2-phosphoglycolate which would result in inhibition of the Calvin cycle.

Photosynthesis: the conversion of sunlight into chemical energy that is stored in carbohydrates such as starch.

Photosystem: multisubunit protein complexes responsible for light harvesting in photosynthetic organisms.

Photosystem I: one of the two photosystems of higher plants.

Photosystem II: one of the two photosystems of higher plants.

Phycobilins: a group of light-harvesting pigments found in many photosynthetic bacteria.

Phytochrome: a plant pigment, equivalent to bile pigments of mammals, which coordinates the plant's physiological and biochemical responses to light.

pI: the isoelectric point, the pH at which a molecule has no net electric charge.

Pili: filamentous structures present on the surface of some bacteria, through which DNA is assumed to pass during conjugation.

pK_a: a measure, on a logarithmic scale, of the acid dissociation constant (K_a) of a compound, which indicates the compound's acidic strength in solution. The pK_a is the pH at which there are an equal number of molecules with the charged and uncharged versions of an ionizable chemical group.

Plane-polarized light: refers to light in which the electric field oscillates along a single vector (plane).

Plasma membrane: the membrane that encloses a prokaryotic or eukaryotic cell.

Plasmid: a usually circular piece of DNA, primarily independent of the host chromosome, often found in bacteria and some other types of cells.

Plastocyanin (PC): a copper-containing protein; a component of the photosynthetic electron transport chain.

Plastoquinone (PQ): a lipid-soluble compound comprising a modified benzene ring, a component of the photosynthetic electron transport chain.

Polarity: the situation that arises if the electrons are not distributed evenly around a chemical group.

Poly(A) polymerase: the enzyme that attaches a poly(A) tail to the 3′ end of a eukaryotic mRNA.

Poly(A) tail: a series of A nucleotides attached to the 3′ end of a eukaryotic mRNA.

Polyacrylamide: a gel made up of chains of acrylamide monomers cross-linked with N,N′-methylenebisacrylamide units.

Polyadenylate-binding protein (PADP): a protein that aids poly(A) polymerase during polyadenylation of eukaryotic mRNAs, and which plays a role in maintenance of the tail after synthesis.

Polyclonal antibody: a mixture of immunoglobulins recognizing multiple epitopes.

Polymer: a compound made up of chains of identical or very similar chemical units.

Polymerase chain reaction (PCR): a technique that enables multiple copies of a DNA molecule to be generated by enzymatic amplification of a target DNA sequence.

Polynucleotide: a single-stranded DNA or RNA molecule.

Polynucleotide phosphorylase (PNPase): a component of the degradosome.

Polypeptide: a polymer of amino acids.

Polyprotein: a translation product consisting of a series of linked proteins which are processed by proteolytic cleavage to release the mature proteins.

Polypyrimidine tract: a pyrimidine-rich region near the 3′ end of a GU–AG intron.

Polyribosome: an mRNA molecule that is being translated by more than one ribosome at the same time.

Polysaccharide: a polymer of monosaccharides.

Polysome: an mRNA molecule that is being translated by more than one ribosome at the same time.

Pore complex: a small channel across the nuclear membrane.

Porin: a transmembrane protein with a barrel-like structure which forms a channel across a membrane.

Porphobilinogen deaminase: the enzyme that converts four porphobilinogen molecules to one of uroporphyrinogen during tetrapyrrole synthesis.

Porphyrin: the class of compounds that includes heme and chlorophyll.

Positive allosteric control: stimulation of enzyme activity by binding of an effector molecule.

Post-replicative repair: a repair process that deals with breaks in daughter DNA molecules that arise as a result of aberrations in the replication process.

Post-spliceosome complex: the initial product of the mRNA splicing reaction.

Post-translational processing: physical and/or chemical modification of a protein that occurs after that protein has been synthesized by translation of an mRNA.

Pre-initiation complex: the structure comprising the small subunit of the ribosome, the initiator tRNA plus ancillary factors that forms the initial association with the mRNA during protein synthesis.

Pre-mRNA: the unspliced version of a eukaryotic mRNA.

Pre-rRNA: the primary transcript of a gene or group of genes specifying rRNA molecules.

Pre-tRNA: the primary transcript of a gene or group of genes specifying tRNA molecules.

Precipitin reaction: an immunoassay carried out in solution.

Pregnenolone: a precursor for steroid hormone synthesis.

Prepriming complex: a complex of proteins formed during initiation of replication in bacteria.

Previtamin D$_3$: an intermediate in the synthesis of vitamin D.

Primary antibody: the antibody that is used to detect the antigen in an indirect ELISA method.

Primary producers: those organisms able to use light or chemical energy to convert inorganic compounds into energy-containing organic compounds.

Primary structure: the sequence of amino acids in a polypeptide.

Primary transcript: the initial product of transcription of a gene or group of genes, subsequently processed to give the mature transcript(s).

Primase: the RNA polymerase enzyme that synthesizes RNA primers during bacterial DNA replication.

Primer: a short series of linked monomers that initiates synthesis of a longer polymer; primers are important in synthesis of DNA and of some polysaccharides.

Primosome: a protein complex involved in genome replication.

Prion: an unusual infectious agent that consists purely of protein.

Processivity: refers to the amount of DNA synthesis that is carried out by a DNA polymerase before dissociation from the template.

Product: a compound produced by a chemical reaction.

Progesterone: a steroid hormone involved in control of pregnancy, menstruation and embryogenesis.

Progestogens: a family of steroid hormones.

Programmed frameshifting: the controlled movement of a ribosome from one reading frame to another at an internal position within a gene.

Prohormone convertases: endopeptidases that cleave prohormones in order to convert these into active hormones.

Prokaryote: an organism whose cells lack a distinct nucleus.

Proline-rich domain: a type of activation domain.

Promoter: the nucleotide sequence, upstream of a gene, that acts as a signal for RNA polymerase binding.

Proofreading: the 3′→5′ exonuclease activity possessed by some DNA polymerases which enables the enzyme to replace a misincorporated nucleotide.

Propionyl CoA carboxylase: the enzyme that converts propionyl CoA to methylmalonyl CoA during breakdown of odd-numbered fatty acids.

Prostaglandin: a type of eicosanoid.

Prosthetic group: an organic or inorganic cofactor that forms a permanent or semi-permanent linkage with an enzyme.

Proteasome: a multisubunit protein structure that is involved in the degradation of other proteins.

Protein: a biomolecule comprising a single or more than one polypeptide.

Protein engineering: various techniques for making directed alterations in protein molecules, often to improve the properties of enzymes used in industrial processes.

Protein kinase A: a family of enzymes that respond to an increase in cellular cAMP levels by phosphorylating a series of target enzymes.

Protein phosphatase: an enzyme that removes phosphate groups from other proteins; for example, from glycogen phosphorylase and glycogen synthase as part of the process that regulates glycogen metabolism.

Protein phosphatase 2A: an enzyme that dephosphorylates and hence activates acetyl CoA carboxylase as part of the regulatory process controlling fatty acid synthesis.

Protein profiling: the methodology used to identify the proteins in a proteome.

Protein targeting: the process that results in transport of a protein from its assembly site to the place in the cell where it will perform its function.

Protein-coding gene: a gene that is transcribed into an mRNA.

Proteome: the collection of proteins synthesized by a living cell.

Proteomics: the collection of techniques used to study the proteome.

Protomer: one of the polypeptide subunits that combine to make the protein coat of a virus.

Pseudomurein: a modified polysaccharide found in the cell walls of archaea.

Pulse labeling: a brief period of labeling carried out at a defined period during the progress of an experiment.

Punctuation codon: a codon that specifies either the start or the end of a gene.

Purine: one of the two types of nitrogenous base found in nucleotides.

Purple bacteria: a group of photosynthetic bacteria.

Pyridoxal phosphate: a derivative of vitamin B_6; a cofactor for transaminase enzymes.

Pyrimidine: one of the two types of nitrogenous base found in nucleotides.

Pyrosequencing: a DNA sequencing method in which addition of a nucleotide to the end of a growing polynucleotide is detected directly by conversion of the released pyrophosphate into a flash of chemiluminescence.

Pyruvate: the three-carbon sugar that is the product of glycolysis.

Pyruvate carboxylase: an enzyme that converts pyruvate to oxaloacetate during gluconeogenesis and also during the shuttling of acetyl CoA from mitochondrion to cytoplasm.

Pyruvate decarboxylase: the enzyme that converts pyruvate to acetaldehyde at step 1 of alcoholic fermentation.

Pyruvate dehydrogenase: an enzyme that binds pyruvate and converts it into acetate, with the release of carbon dioxide; part of the pyruvate dehydrogenase complex.

Pyruvate dehydrogenase complex: a complex of three enzymes that converts pyruvate to acetyl CoA.

Pyruvate dehydrogenase kinase: an enzyme that phosphorylates and hence inactivates the pyruvate dehydrogenase complex.

Pyruvate dehydrogenase phosphatase: an enzyme that dephosphorylates and hence activates the pyruvate dehydrogenase complex.

Pyruvate kinase: the enzyme that converts phosphoenolpyruvate to pyruvate at step 10 of the glycolysis pathway.

Pyruvate-P_i dikinase: an enzyme that converts pyruvate into phosphoenolpyruvate in the mesophyll cells of C4 plants.

Quadrupole mass spectrometer: a mass spectrometer in which the mass analyzer has four magnetic rods placed parallel to one another, surrounding a central channel through which the ions must pass.

Quantitative PCR (qPCR): a method for quantifying the amount of product synthesized during a test PCR by comparison with the amounts synthesized during PCRs with known amounts of starting DNA.

Quaternary structure: the association between different polypeptides to form a multisubunit protein.

Quorum sensing: a process by which bacteria communicate with one another.

R group: the variable group in the structure of an amino acid.

Ramachandran plot: a graphical representation of the possible combinations of the *psi* and *phi* bond angles that can occur within a polypeptide.

Rate constant: a description of the rate associated with an individual step in an enzyme-catalyzed reaction.

Reaction center: the central component of a photosystem, to which energy from sunlight is channeled.

Reactive oxygen species: oxidizing agents that can impair cellular function by damaging membranes and inactivating enzymes.

Real-time PCR: a modification of the standard PCR technique in which synthesis of the product is measured as the PCR proceeds through its series of cycles.

Receptor protein: a protein located in the cell membrane which responds to an external signal by causing a biochemical change within the cell.

Receptor tyrosine kinase: a membrane protein that responds to an extracellular signal by phosphorylating one or more tyrosine amino acids in another protein, this other protein possibly being a second copy of the receptor.

Recognition helix: an α-helix in a DNA-binding protein, one that is responsible for recognition of the target nucleotide sequence.

Recombinant DNA technology: the techniques involved in the construction, study and use of recombinant DNA molecules.

Recombinant plasmid: a plasmid into which a new piece of DNA has been inserted by genetic engineering techniques.

Recombinant protein: a protein synthesized in a recombinant cell as the result of expression of a cloned gene.

Recombination: a large-scale rearrangement of a DNA molecule.

Redox potential: a measure of the affinity of a compound for electrons.

Redox reaction: linked oxidation and reduction reactions, resulting in the loss of oxygen by one compound and the gain of oxygen by a second compound.

Reducing sugar: sugars which, in their linear form, have reducing activity due to the presence of a terminal aldehyde group.

Reflectron: an ion mirror used in some types of mass spectrometer; also used to denote a mass spectrometer that contains an ion mirror.

Regulatory protein: a protein that regulates one or more cellular activities, including the flow of metabolites through a metabolic pathway.

Release factor: a protein that plays an ancillary role during termination of translation.

Replication fork: the region of a double-stranded DNA molecule that is being opened up to enable DNA replication to occur.

Replisome: a complex of proteins involved in genome replication.

Reporter enzyme: an enzyme whose activity can easily be assayed, and which can be therefore be linked to, for example, an antibody and hence used in an immunoassay.

Reporter probe: a short oligonucleotide that gives a fluorescent signal when it hybridizes with a target DNA.

Repressible operon: an operon that is switched off by the repressor working in conjunction with a co-repressor molecule.

Resonance: the redistribution of electrons between adjacent atoms in a molecule.

Resonance energy transfer: energy transfer in a photosystem, in which energy quanta are passed from one chlorophyll molecule to another, the recipient chlorophyll becoming excited and the donor chlorophyll returning to its ground state.

Resonance frequency: the energy difference between the α and β spin states of a nucleus.

Respirasome: an assembly, within the inner mitochondrial membrane, of Complexes I, II and IV of the electron transport chain along with their intermediate carrier molecules.

Respiration: the biochemical reactions that take place in the mitochondria, using up oxygen and producing carbon dioxide.

Respiratory control: the regulation of the electron transport chain by ADP availability.

Restriction endonuclease: an endonuclease that cuts DNA molecules only at a limited number of specific nucleotide sequences.

Retention signal: a sorting sequence that directs a protein to the endoplasmic reticulum.

Retrovirus: a virus with an RNA genome, a DNA copy of which integrates into the genome of its host cell.

Reverse phase chromatography: a type of column chromatography using nonpolar beads, which separates compounds according to their surface hydrophobicity.

Reverse reaction: one of the directions of a reversible reaction.

Reverse transcriptase: an RNA-dependent DNA polymerase, able to synthesize a complementary DNA molecule on a template of single-stranded RNA.

Reversible inhibitor: an inhibitor that has a non-permanent effect on the activity of an enzyme.

Rhizobia: a group of nitrogen-fixing bacteria.

Rho: a protein involved in termination of transcription of some bacterial genes.

Rho-dependent terminator: a position in bacterial DNA where termination of transcription occurs with the involvement of Rho.

Riboflavin: vitamin B_2, the precursor of the cofactors FAD and FMN.

Ribonuclease A: an enzyme that catalyzes the conversion of a polymeric RNA molecule into two shorter molecules, by cutting one of the internal phosphodiester bonds.

Ribonuclease P: an enzyme involved in processing pre-tRNA, whose catalytic activity is a ribozyme.

Ribonucleic acid: one of the two forms of nucleic acid in living cells; the genetic material for some viruses.

Ribonucleotide reductase: the enzyme that converts ribonucleotides to their deoxyribonucleotides during the salvage pathway for nucleotide synthesis.

Ribosomal or **rRNA:** the RNA molecules that are components of ribosomes.

Ribosomal protein: one of the protein components of a ribosome.

Ribosome: one of the protein–RNA complexes on which translation occurs.

Ribosome binding site: the nucleotide sequence that acts as the attachment site for the small subunit of the ribosome during initiation of translation in bacteria.

Ribosome recycling factor (RRF): a protein responsible for disassembly of the ribosome at the end of protein synthesis in bacteria.

Ribozyme: an RNA molecule that has catalytic activity.

Ribulose 5-phosphate kinase: the enzyme that converts ribulose 5-phosphate to ribulose 1,5-bisphosphate at step 5 of the Calvin cycle.

Ribulose bisphosphate carboxylase or **Rubisco:** the enzyme that combines one molecule of carbon dioxide with one of ribulose 1,5-bisphosphate to give two molecules of 3-phosphoglycerate, during the dark reactions of photosynthesis.

Rickets: vitamin D deficiency in children, characterized by a softening or weakening of the bones.

RNA: ribonucleic acid, one of the two forms of nucleic acid in living cells; the genetic material for some viruses.

RNA enzyme: an RNA molecule that has catalytic activity.

RNA helicase B: a component of the degradosome.

RNA polymerase: an enzyme that synthesizes RNA.

RNA polymerase II: the eukaryotic RNA polymerase that transcribes protein-coding genes, most snRNA genes, and miRNA genes.

RNA world: the early period of evolution when all biological reactions were centered on RNA.

RNA-dependent DNA polymerase: an enzyme that makes a DNA copy of an RNA template; a reverse transcriptase.

RNA-induced silencing complex (RISC): a protein–RNA structure in eukaryotes which cleaves and hence inactivates target mRNAs.

RNase E: a component of the degradosome.

Rolling circle replication: a replication process that involves continual synthesis of a polynucleotide which is 'rolled off' a circular template molecule.

Root nodule: a structure within which nitrogen fixation takes place.

Rough endoplasmic reticulum: endoplasmic reticulum that has ribosomes on its outer surface.

Rubisco activase: an enzyme involved in the regulation of ribulose bisphosphate carboxylase activity in response to light intensity.

S-adenosyl methionine (SAM): a cofactor that acts as a donor of a methyl group in several biochemical reactions.

Salt gradient: a gradually increasing concentration of salt that is used to elute compounds according to their net electric charges during ion exchange chromatography.

Salvage pathway: the use of purines and pyrimidines, released from nucleotides that are being degraded, to make new nucleotides.

Saponification: formation of a soap by heating a triacylglycerol with an alkali.

Sarcoplasmic reticulum: a specialized type of smooth endoplasmic reticulum in muscle cells which releases Ca^{2+} ions in response to a nerve impulse.

Satellite RNA: an infectious RNA molecule some 320–400 nucleotides in length which does not encode its own capsid proteins, instead moving from cell to cell within the capsid of a helper virus.

Saturated: a fatty acid that lacks C=C double bonds.

Scanning: a system used during initiation of eukaryotic translation, in which the pre-initiation complex attaches to the 5'-terminal cap structure of the mRNA and then scans along the molecule until it reaches an initiation codon.

SDS-polyacrylamide gel electrophoresis (SDS-PAGE): electrophoresis in a polyacrylamide gel containing sodium dodecyl sulfate (SDS), used to separate polypeptides according to their molecular weights.

Second messenger: an intermediate in a certain type of signal transduction pathway.

Secondary antibody: the antibody that recognizes the primary antibody and is conjugated to the reporter enzyme in an indirect ELISA method.

Secondary structure: a series of conformations, including helices, sheets and turns, that can be adopted by different parts of a polypeptide.

Sedimentation coefficient: the value used to express the rate of migration of a molecule or structure when centrifuged in a dense solution.

Selectable marker: a gene carried by a vector and conferring a recognizable characteristic on a cell containing the vector or a recombinant DNA molecule derived from the vector.

Selective barrier: a barrier, such as a biological membrane, that allows some but not all molecules to pass.

Self-splicing: the ability of Group I introns to splice in the absence of any proteins, indicating that the RNA of the intron has catalytic activity.

Semi-synthetic antibiotic: an antibiotic obtained by making chemical modifications to a natural antibiotic.

Sequence: the order of units in a polymer; e.g. the order of amino acids in a polypeptide.

Sequence-specific DNA-binding protein: a protein that recognizes and binds to a particular sequence on a DNA molecule, often to influence the rate of transcription of an adjacent gene.

Sequential model: a model for cooperative substrate binding, in which binding of a substrate molecule to one of the enzyme subunits induces the conversion of neighboring subunits to a new conformation.

Seven-transmembrane-helix or **7TM protein:** a type of transmembrane protein with seven α-helices forming a barrel-shaped structure, an example being the glucagon receptor protein.

Shine–Dalgarno sequence: the ribosome binding site upstream of an *Escherichia coli* gene.

Short patch repair: a nucleotide excision repair process of *Escherichia coli* that results in excision and resynthesis of about 12 nucleotides of DNA.

Signal peptidase: an enzyme that removes a signal peptide from a protein.

Signal peptide or **signal sequence:** a 5–30 amino acid sorting sequence that directs a protein through the membrane of the endoplasmic reticulum.

Signal recognition particle (SRP): an RNA–protein complex that aids transfer of a protein into the endoplasmic reticulum.

Signal transducers and activators of transcription (STAT): a type of protein that responds to binding of an extracellular signaling compound to a cell surface receptor by activating a transcription factor.

Signal transduction: control of cellular activity via a cell surface receptor that responds to an external signal.

Simple triacylglycerol: a triacylglycerol in which the three fatty acid chains are identical.

Single bond: a covalent bond that results from two atoms sharing one pair of electrons.

Single molecule real-time sequencing: a third generation DNA sequencing method which uses an advanced optical system to observe the addition of individual nucleotides to a growing polynucleotide.

Single strand binding proteins (SSBs): one of the proteins that attach to single-stranded DNA in the region of the replication fork, preventing base pairs forming between the two parent strands before they have been copied.

Singlet oxygen: an excited state of oxygen, denoted by O*, which can give rise to reactive oxygen species such as hydrogen peroxide.

Siroheme: a modified version of heme used as a cofactor by several enzymes involved in the reduction of nitrogen and sulfur-containing compounds.

Small interfering RNA (siRNA): a type of short eukaryotic RNA molecule involved in control of gene expression.

Small nuclear ribonucleoproteins (snRNPs): structures involved in splicing GU–AG and AU–AC introns and in other RNA processing events, comprising one or more snRNA molecules complexed with proteins.

Small nuclear RNA (snRNA): a type of short eukaryotic RNA molecule involved in splicing GU–AG and AU–AC introns and in other RNA processing events.

Small nucleolar RNA (snoRNA): a type of short eukaryotic RNA molecule involved in chemical modification of rRNA.

Smooth endoplasmic reticulum: that part of the endoplasmic reticulum that does not have ribosomes on its outer surface.

Sorting sequence: an amino acid sequence that specifies which transport route a protein must follow.

Sorting vesicle: a vesicle involved in targeting of proteins to lysosomes.

Sphingolipid: an amphiphilic lipid based on sphingosine.

Sphingosine: a long chain hydrocarbon derivative with an internal hydroxyl group.

Spirilla: prokaryotes with cells shaped like coils.

Spliceosome: the protein–RNA complex involved in splicing GU–AG or AU–AC introns.

Splicing: the removal of introns from the primary transcript of a discontinuous gene.

Splicing pathway: the series of events that converts a discontinuous pre-mRNA into a functional mRNA.

Squalene epoxide: an intermediate in the synthesis of sterol compounds.

SRP receptor: a protein located on the surface of the endoplasmic reticulum that aids transfer of a protein into the endoplasmic reticulum.

Standard free energy change or $\Delta G^{0\prime}$: a measure of ΔG for a reaction under standard conditions, at pH 7.0 with each reactant present at equimolar amounts.

Standard redox potential ($E_0\prime$): a measure of the redox potential of a compound under standard conditions, expressed as volts.

Starch: a storage homopolysaccharide found in plants and made entirely of D-glucose units.

Starch branching enzyme: an enzyme that synthesizes the $\alpha(1\rightarrow6)$ links that result in the branched structure of the amylopectin version of starch.

Starch synthase: the enzyme that adds ADP–glucose molecules to the ends of a growing starch molecule.

Stationary phase: the non-movable phase in a chromatography system, usually a solid matrix, or a liquid that has been adsorbed by a solid matrix.

Steady state: the situation that arises when the rate of synthesis of an enzyme–substrate complex equals the rate of its consumption.

Stem cell: a progenitor cell that divides continually throughout the lifetime of an organism, and which can differentiate into one or more types of specialized cell.

Stem–loop: a hairpin structure, consisting of a base-paired stem and a non-base-paired loop, which may form in a polynucleotide.

Stereomers: isomers in which the atoms are joined together in the same sequence, but which differ in the arrangement of atoms around one or more asymmetric centers such as a chiral carbon.

Steric effects: the effects that prevent two atoms from getting too close together and hence limit the possible conformations that any molecule can take up.

Steroid: a sterol derivative in which the hydroxyl attached to the C_3 carbon is replaced with a different chemical group.

Steroid hormone: a steroid with hormonal activity.

Steroid receptor: a protein that binds a steroid hormone after the latter has entered the cell, as an intermediate step in modulation of genome activity.

Sterol: a lipid formed by cyclization of squalene.

Sticky end: an end of a double-stranded DNA molecule where there is a single-stranded extension.

Stigmasterol: a plant sterol.

Stoichiometry: in a chemical reaction, the number of molecules of each reactant that are used compared with the number of molecules of each product that are made.

Stop-transfer sequence: a sorting sequence carried by a Type I membrane protein.

Stored starch synthesis: the synthesis of long-lived starch reserves in amyloplasts.

Strand displacement replication: a mode of replication which involves continuous copying of one strand of the helix, the second strand being displaced and subsequently copied after synthesis of the first daughter strand has been completed.

Stroma: the inner region within a chloroplast.

Strong promoter: an efficient promoter that can direct synthesis of RNA transcripts at a relatively fast rate.

Substrate: a compound that is consumed during a chemical reaction.

Substrate-level phosphorylation: conversion of ADP to ATP (or GDP to GTP) using a phosphate from a phosphorylated intermediate, the latter being one of the substrates of the reaction.

Subviral particle: any one of several types of infectious particle made up of protein and/or nucleic acid, which are considered to be insufficiently complex to be classified as viruses.

Succinate dehydrogenase: the enzyme that converts succinate to fumarate at step 6 of the TCA cycle.

Succinate–CoQ reductase complex: Complex II of the electron transport chain.

Succinyl CoA synthetase: the enzyme that converts succinyl CoA to succinate at step 5 of the TCA cycle.

Sucrose phosphate phosphatase: the enzyme that converts sucrose 6-phosphate to sucrose during sucrose synthesis.

Sucrose phosphate synthase: the enzyme that converts fructose 6-phosphate to sucrose 6-phosphate during sucrose synthesis.

Sugar: a monosaccharide or other short chain carbohydrate.

Sugar pucker: alternative conformations of a sugar ring structure.

Supercoiled DNA or **supercoiling:** a conformational state in which a double helix is overwound or underwound so that superhelical coiling occurs.

Superhelix: a helical structure formed by polymers that are themselves helices.

Superwobble: the extreme form of wobble that occurs in vertebrate mitochondria.

Symporter: an active transport protein which couples transport of a molecule or ion against a concentration gradient with the movement, in the same direction, of a second ion down a gradient.

Synapse: the space at the junction between two adjacent nerve cells.

Tandem mass spectrometry: a type of mass spectrometry that uses two or more mass analyzers linked in series.

***Taq* polymerase:** the thermostable DNA polymerase that is used in PCR.

TATA box: a component of the RNA polymerase II core promoter.

TATA-binding protein or **TBP:** a component of the transcription factor TFIID, the part that recognizes the TATA box of the RNA polymerase II promoter.

Taurocholate: a bile acid; a derivative of cholic acid, which helps to emulsify fats in the diet.

Tautomerism: the spontaneous change of a molecule from one structural isomer to another.

TBP-associated factors or **TAFs:** one of several components of the transcription factor TFIID, playing ancillary roles in recognition of the TATA box.

Telomerase: the enzyme that maintains the ends of eukaryotic chromosomes by synthesizing telomeric repeat sequences.

Temperature optimum: the optimal temperature for a chemical reaction.

Template-dependent DNA synthesis: an enzyme that synthesizes DNA in accordance with the sequence of a template.

Terminal deoxynucleotidyl transferase: an enzyme that adds one or more nucleotides to the 3′ end of a DNA molecule.

Termination codon: one of the three codons that mark the position where translation of an mRNA should stop.

Terminator sequence: one of several sequences on a bacterial genome involved in termination of DNA replication.

Terminator-dye nucleotide: a fluorescently labeled nucleotide that carries a 3′-blocking group, the latter preventing further chain elongation when incorporated into a growing polynucleotide.

Terpene: a variable group of lipid compounds whose structures are based on the small hydrocarbon called isoprene.

Tertiary structure: the overall three-dimensional configuration of a protein.

Testosterone: a steroid hormone involved in regulation of bone and muscle synthesis.

Tetraloop: a stem–loop structure with four base pairs in the stem.

Tetrapyrrole: a compound with four pyrrole units.

Thermal cycler: a programmable device that heats and cools a reaction between pre-set temperatures.

Thermogenin: a proton transport protein present in the inner mitochondrial membrane of brown adipose cells, which acts as an uncoupler of the electron transport chain.

Thermostable: an enzyme that is able to remain active at relatively high temperatures.

Thin layer chromatography: a chromatography system in which the stationary phase is a solid material layered onto a plastic sheet.

Thioesterase: the enzyme that cleaves a completed fatty acid from ACP.

Thiolase: the enzyme that converts two acetyl CoA molecules into one of acetoacetyl CoA during the cholesterol synthesis pathway.

Thiolysis: a chemical reaction in which bond cleavage is driven by a thiol (–SH) group.

Thioredoxin: a small protein involved in several cellular redox reactions, in particular ones resulting in cleavage of disulfide bonds.

Threonine dehydratase: the enzyme that converts threonine to α-ketobutyrate at the commitment step of the pathway for isoleucine synthesis.

Thrombin: a protein involved in blood clotting.

Thromboxane: a type of eicosanoid.

Thylakoid space: the internal region within a thylakoid.

Thylakoids: interconnected membranous structures within the stroma of a chloroplast.

Thymidylate synthase: the enzyme that converts uracil to thymine during *de novo* nucleotide synthesis.

Thymine: one of the pyrimidine bases found in DNA.

Ti plasmid: the large plasmid found in those *Agrobacterium tumefaciens* cells able to direct crown gall formation on certain species of plants.

Topological problem: refers to the need to unwind the double helix in order for DNA replication to occur, and the difficulties that the resulting rotation of the DNA molecule would cause.

Trailer segment: the untranslated region of an mRNA downstream of the termination codon.

Transaldolase: the enzyme that converts glyceraldehyde 3-phosphate and sedoheptulose 7-phosphate to fructose 6-phosphate and erythrose 4-phosphate at step 7 of the pentose phosphate pathway.

Transaminase: an enzyme that catalyzes a transamination reaction.

Transamination: the transfer of an amino group from one compound to another.

Transcript: an RNA copy of a gene.

Transcript-specific regulation: regulatory mechanisms that control protein synthesis by acting on a single transcript or a small group of transcripts coding for related proteins.

Transcription: the synthesis of an RNA copy of a gene.

Transcription bubble: the non-base-paired region of the double helix, maintained by RNA polymerase, within which transcription occurs.

Transcription factor IID or **TFIID:** the protein complex, including the TATA-binding protein, which recognizes the core promoter of a gene transcribed by RNA polymerase II.

Transcriptome: the collection of RNA molecules in a cell or tissue.

Transcriptomics: the various methods used to study a transcriptome.

Transesterification: a type of reaction between an ester and an alcohol.

Transfer or **tRNA:** a small RNA molecule that acts as an adaptor during translation and is responsible for decoding the genetic code.

Transferase: an enzyme that transfers groups from one molecule to another.

Transient or **transitory starch synthesis:** the synthesis of short-lived starch molecules that occurs in chloroplasts.

Transit peptide: sorting sequences that direct proteins to different compartments in a chloroplast.

Transition state: the point in a reaction pathway where the system has the highest free energy content.

Transketolase: the enzyme that converts xylulose 5-phosphate and ribose 5-phosphate (or erythrose 4-phosphate) to glyceraldehyde 3-phosphate and sedoheptulose 7-phosphate (or fructose 6-phosphate) at step 6 (or 8) of the pentose phosphate pathway.

Translation: the synthesis of a polypeptide, the amino acid sequence of which is determined by the nucleotide sequence of an mRNA in accordance with the rules of the genetic code.

Translocation: the movement of a ribosome along an mRNA molecule during translation.

Translocator inner membrane (TIM) complex: an access point for proteins into a mitochondrion.

Translocator outer membrane (TOM) complex: an access point for proteins into a mitochondrion.

Transmembrane protein: an integral membrane protein that spans the entire lipid bilayer.

Transverse wave: a waveform, such as light and other types of electromagnetic radiation, in which the oscillations occur at right angles to the direction of travel and energy transfer.

Treble clef finger: a type of zinc finger that lacks a β-sheet component.

Triacylglycerol or **triacylglyceride:** a lipid comprising three fatty acids attached to a glycerol molecule.

Triacylglycerol synthetase: the enzyme activity resulting from the complex formed between diglyceride acyltransferase and phosphatidate phosphatase, catalyzing steps 4 and 5 of the triacylglycerol synthesis pathway.

Tricarboxylic (TCA) cycle: the cycle of reactions that results in breakdown of the pyruvate molecules resulting from glycolysis.

Triose kinase: the enzyme that converts glyceraldehyde to glyceraldehyde 3-phosphate at step 3 of the fructose 1-phosphate pathway.

Triose phosphate isomerase: the enzyme that converts dihydroxyacetone phosphate to glyceraldehyde 3-phosphate at step 5 of the glycolysis pathway.

Triple helix: a three-stranded superhelix.

tRNA nucleotidyltransferase: the enzyme responsible for the post-transcriptional attachment of the triplet 5′–CCA–3′ to the 3′ end of a tRNA molecule.

Troponin: a protein involved in muscle contraction.

Trypsin: a protease, synthesized by the pancreas, which helps break down dietary protein.

Tryptophan synthase: an enzyme that catalyzes the final two steps in the pathway that results in synthesis of the amino acid tryptophan, converting indole-3-glycerol phosphate to indole and then to tryptophan.

Tus (terminator utilization substance) protein: the protein that binds to a bacterial terminator sequence and mediates termination of DNA replication.

Two-dimensional electrophoresis: a method for separation of proteins used especially in studies of the proteome.

Type 0 cap: the basic cap structure, consisting of 7-methylguanosine attached to the 5′ end of an mRNA.

Type 1 cap: a cap structure comprising the basic 5′-terminal cap plus an additional methylation of the ribose of the second nucleotide.

Type 2 cap: a cap structure comprising the basic 5′-terminal cap plus methylation of the riboses of the second and third nucleotides.

Type I DNA topoisomerase: a DNA topoisomerase that makes a single-stranded break in a double-stranded DNA molecule.

Type I membrane protein: an integral membrane protein that spans the membrane once.

Type II DNA topoisomerase: a DNA topoisomerase that makes a double-stranded break in a double-stranded DNA molecule.

Type II membrane protein: a type of protein with a signal peptide that anchors the protein to the inner surface of the endoplasmic reticulum.

Type III membrane protein: an integral membrane protein that spans the membrane more than once.

Ubiquinol (CoQH₂): the reduced form of ubiquinone.

Ubiquinone: the intermediate carrier molecule for electron transfer from Complex I or II to Complex III in the electron transport chain; also called coenzyme Q (CoQ).

Ubiquitin: a 76-amino-acid protein which, when attached to a second protein, acts as a tag directing that protein for degradation.

Ubiquitin ligase: an enzyme that attaches a ubiquitin molecule to a protein targeted for degradation.

Ubiquitin–receptor protein: a protein associated with the cap structure of a proteasome that aids transfer of ubiquitinated proteins into the proteasome.

Ubiquitination: the attachment of ubiquitin to a protein.

UDP–galactose 4-epimerase: the enzyme that converts UDP–galactose to UDP–glucose at step 3 of the galactose–glucose interconversion pathway.

UDP–glucose pyrophosphorylase: the enzyme that converts glucose 1-phosphate to UDP–glucose during sucrose synthesis.

Uncoupler: a compound that interferes with the electron transport chain by uncoupling the oxidation of NADH and $FADH_2$ from the production of ATP.

Unicellular: an organism comprising just a single cell.

Uniporter: a transport protein that uses facilitated diffusion to move molecules or ions across a membrane.

Unpolarized: refers to natural light; the electric fields of different photons oscillate in different directions.

Unsaturated: a fatty acid that has one or more C=C double bonds.

Uracil: one of the pyrimidine bases found in RNA.

Urea cycle: the pathway used by humans and other ureotelic organisms to convert ammonia to urea.

Urease: the first enzyme to be purified, responsible for the conversion of urea to carbon dioxide and ammonia.

Ureotelic: species that convert ammonia into urea, which is excreted in urine.

Uric acid: a purine, derived from adenine and guanine, which is excreted.

Uricotelic: species that excrete nitrogen in the form of uric acid.

Valency: in simple terms, the number of single bonds that an atom can form. Specifically, the number of atoms of hydrogen that an atom can combine with or displace when forming a compound.

van der Waals forces: weak interactions that involve attractions between two dipolar atoms.

Very low density lipoproteins (VLDLs): lipoproteins synthesized in the liver which transport a variety of lipids to muscle and adipose tissue.

Vesicle: a small membrane-bound sphere.

Viroid: an infectious RNA molecule 240–375 nucleotides in length which contains no genes and never becomes encapsidated, spreading from cell to cell as naked DNA.

Virus: an infective particle, composed of protein and nucleic acid, that must parasitize a host cell in order to replicate.

Virusoid: an infectious RNA molecule some 320–400 nucleotides in length which does not encode its own capsid proteins, instead moving from cell to cell within the capsid of a helper virus.

Vitamin D: a family of steroid derivatives with various physiological and biochemical roles.

Vitamin D$_3$: a member of the vitamin D group of compounds.

V_{max}: the maximum rate of an enzyme-catalyzed reaction.

Voltage-gated ion channel: a transmembrane protein that can change its conformation in response to the electric charge across the membrane.

Vulcanization: a chemical process that results in crosslinks being formed between individual polymers of rubber.

Wobble: the process by which a single tRNA can decode more than one codon.

X-ray crystallography: a method for studying the structures of molecules based on the deflection of X-rays that occurs during their passage through a crystal or other regularly ordered chemical structure.

X-ray diffraction: the deflection of X-rays that occurs during their passage through a crystal or other regularly ordered chemical structure.

X-ray diffraction analysis: the analysis of X-ray diffraction patterns as a means of determining the three-dimensional structure of a large molecule.

X-ray diffraction pattern: the pattern obtained after diffraction of X-rays through a crystal.

Xanthophyll: an accessory light harvesting pigment found in plants.

Xanthophyll cycle: a series of biochemical reactions which result in chemical modification of certain carotenoids to give derivatives with energy quenching properties.

Yeast episomal plasmid (YEp): a yeast vector carrying the 2 μm plasmid origin of replication.

Yeast integrative plasmid (YIp): a yeast vector that relies on integration into the host chromosome for replication.

Z scheme: a graph depicting the changes in redox potential that occur along the photosynthetic electron transport chain.

Z-DNA: a conformation of DNA in which the two polynucleotides are wound into a left-handed helix.

Zero-mode waveguide: a nano-structure that enables individual molecules to be observed.

Zinc finger: a common structural motif for attachment of a protein to a DNA molecule.

Zone of equivalence: the point at which the relative amounts of antigen and antibody are optimal for complex formation.

Zwitterion: a molecule that has no net charge but which has both negative and positive ionized groups.

Index